Siberian Traps and Pt-Cu-Ni Deposits in the Noril'sk Area

Nadezhda A. Krivolutskaya

Siberian Traps and Pt-Cu-Ni Deposits in the Noril'sk Area

With the special cooperation of Alexander Sobolev

Springer

Nadezhda A. Krivolutskaya
Vernadsky Institute of Geochemistry and
Analytical Chemistry
Russian Academy of Sciences
Moscow, Russia

The Work was first published in 2014 by Russian Foundation for Basic Research with the following title:
Evolution of Trap Magmatism and Processes Producing PGE-Cu-Nimineralizationin the Noril'sk Area

ISBN 978-3-319-17204-0 ISBN 978-3-319-17205-7 (eBook)
DOI 10.1007/978-3-319-17205-7

Library of Congress Control Number: 2015937605

Springer Cham Heidelberg New York Dordrecht London

Printed on acid-free paper

Springer International Publishing AG Switzerland is part of Springer Science+Business Media (www.springer.com)

Foreword

This monograph presents a new detailed study of one of the pivotal problems now being discussed by geologists. Siberian traps are unique in several aspects. First, this is the largest magmatic province in the Earth's continental lithosphere. Second, the origin of this province is thought to have caused the greatest environmental catastrophe in the Earth's history at the boundary between the Permian and Triassic. Finally, the province is the location of uniquely large Cu–Ni–Pt sulfide deposits, whose characteristics have no analogs among deposits anywhere else worldwide. As a result, hundreds of studies have been devoted to these traps and to the Permian–Triassic mass extinction, and a few dozen such papers are still published annually. Increadible as it is still possible to conduct and publish an extensive amount of research focused on these problems based on extensive new data and presenting new ideas. The monograph by N. Krivolutskaya is exactly such study.

This study is based on the author's extensive fieldwork in the Noril'sk district and presents extensive new data regarding rock compositions (including their trace element compositions) and the chemistries of their ore and silicate minerals. The analyses were conducted using state-of-the-art analytical techniques: inductively coupled plasma mass spectrometry, laser sampling, and mineral analysis using a special ultraprecise electron microprobe methodic. The monograph presents new data regarding the Pb, Nd, Sr, O, and S isotopic compositions of the rocks and minerals.

Based on the analytical and geological data, N. Krivolutskaya arrives at several conclusions that contradict traditional views regarding the genesis of both the ores and the basalts. This aspect of the monograph is provocative and will stimulate discussion, which is undoubtedly one of its merits.

Below, I list the major disparities. In contrast to the current widespread viewpoint, the author of the monograph denies direct genetic links between the lavas and ore-bearing bodies and argues that the intrusive bodies were produced by individual magmatic pulses. She also believes that the sulfur in the ore bodies was not of sedimentary origin but was rather derived by magmas from a lower crust or from a specific mantle source. The traps are thought not to be mantle products but rather to have been generated via the melting of the lower crust.

Although I do not fully share the viewpoint of the author, the findings and conclusions presented in the monograph and in the latest publications coauthored by her attract serious attention to her work. For example, the new model for the interaction of a hot mantle plume enriched in ancient recycled oceanic crustal with continental lithosphere (Sobolev et al. 2011) allows for the extensive remelting of the lower continental crust and corresponding modification of the parental mantle magmas. Sulfide minerals in this crust are likely to melt, precipitate, and be accumulated in basaltic magmas. Therefore, their unusual sulfur isotopic composition could indeed be inherited from either the ancient recycled crust or the lower continental crust.

I cannot agree with the author's argument that the stable state of the Siberian craton was unfavorable for the emplacement of ancient recycled crustal material during the trap magmatism. The point is that the recycled oceanic crust, whose traces were discerned in the mantle source of the Siberian traps (Sobolev et al. 2007, 2009, 2011), was likely transported upward by the hot mantle plume from the depth between lower mantle and upper core, where this material could have accumulated for a long time independent of the geodynamics of the

Siberian craton. This conclusion is based on the very high temperatures of the parental magmas of the Siberian traps and alkaline rocks in the Maymecha-Kotuy province. Certain other aspects of the study also provoke doubts, but it is well known that the truth ultimately emerges from discussion. I therefore welcome this study and recommend it to those interested in magmatic ore-forming processes and the genesis of magmas and to geologists studying the Noril'sk district.

Alexander Sobolev

Vernadsky Institute of Geochemistry
Russian Academy of Sciences
Moscow, Russia
Institute of Earth Sciences (ISTerre)
J. Fourier University
Grenoble, France
January 20, 2015

Preface

This book contains my new data on the geology, geochemistry, and mineralogy of rocks and ores from the Noril'sk region collected during the last 20 years. Majority of the reported data was obtained personally by me and minority in collaboration with other researchers, i.e., A. V. Sobolev, V. N. Mikhailov, D. V. Kuzmin, and A. A. Ariskin, during our joint fieldworks, sample analysis, and computer modeling. The book also contains several novel unpublished geological large-scale maps that were kindly given by NorilskGeology Ltd.

I tried to examine magmatic rocks of trap formation in Noril'sk area including both lavas and intrusions. The latter are represented by massifs of Talnakh and Noril'sk Troughs since they are the most mineralized in the whole area. By analyzing geological relationships between lavas and intrusions and their geochemistry, I came into conclusion that they should not be considered as members of the same magmatic system. I have demonstrated limited scale of the assimilation of surrounding rocks by magma inside crystallization chambers for the first time. That's why assimilation had no significant impact on the ore formation.

I have not tried to compile all materials accumulated during almost a century of study of the Noril'sk deposits. The understanding of their origin is very ambitious and important task, and it is still open for geologists who specialized in various fields, i.e., geophysics, stratigraphics, mineralogy, etc.

Note that this book comprises about 70 % of the materials from my monograph titled as "Evolution of trap magmatism and processes producing Pt–Cu–Ni mineralization in the Noril'sk area" published in Russian (2014). I preferred to use the nomenclature of rocks common of Noril'sk geologists instead of international one, because it is being used by both domestic and even some international geologists for the last 60 years.

I hope this book will be interesting for all researches of copper–nickel deposits across the globe.

Moscow, Russia
March 20, 2015

Nadezhda A. Krivolutskaya

To Several Generations of Russian Geologists Who Did the Unique Noril'sk Metallogenic Province in Very Severe Conditions

Organization of Geological Works in the Noril'sk Trough
(Noril'sk 1 deposit)
1919–1959

N.N. Urvantsev	N.M. Fedorovsky	A.E. Nelyubin
A.E. Vorontsov	I.A. Korovyakov	M.F. Smirnov
B.N. Rozhkov	Yu.M. Sheiman	A.I. Koreshkov
M.N. Godlevsky	N.S. Zontov	V.S. Domarev
V.K. Kotulsky	G.G. Moor	V.N. Egorov

Discovery of Deposits in the Talnakh Trough
(Talnakh and Oktyabr'skoe deposits)
1960–1965
Laureates of the top USSR decorations (Lenin Prize winners, Heroes of Socialist Labor, Order of Lenin)

Talnakh—1960	*Oktyabr'skoe—1965*
V.F. Kravtsov	G.D. Maslov
V.S. Nesterovsky	V.F. Kravtsov
G.D. Maslov	V.A. Lul'ko
Yu.N. Sedykh	G.I. Saprykin
Yu.D. Kuznetsov	L.L. Vaulin
E.N. Sukhanova	F.A. Starshinov
V.N. Egorov	

1970–2015

V.S. Staroseltsev	V.V. Kurgin	I.A. Tushentsova
V.F. Pzhevsky	O.N. Siminov	L.I. Trofimova
N.E. Kunilov	V.A. Fedorenko	K.K. Shad'ko
Yu.N. Amosov	O.A. Dyuzhikov	N.I. Kokorin
M.I. Savushkin	I.A. Matveev	V.A. Teteryuk
E.V. Sereda	V.N. Mikhailov	V.A. Rad'ko

+ many, many other geologists

Contents

Abbreviations

Minerals

Ol	olivine
Cpx	clinopyroxene
Opx	orthopyroxene
Pl	plagioclase
Sp	spinel
TiMt	titanomagnetite
Po	pyrrhotite
Cp	chalcopyrite
Pn	pentlandite

Basalts, Formations

T_1iv	Ivakinsky
T_1sv	Syverminsky
T_1gd	Gudchikhinsky
T_1hk	Khakanchansky
T_1nd	Nadezhdinsky
T_1mr	Morongovsky
T_1mk	Mokulaevsky
T_1hr	Kharaelakhsky
T_1km	Kumginsky
T_1sm	Samoedsky

Gabbro-dolerites

gd	gabbro-dolerites
Γ	olivine-free gabbro-dolerites
Γп	picritic gabbro-dolerites
Γт	taxitic gabbro-dolerites
Γтп	taxitic–picritic gabbro-dolerites
∇/	gabbro–diorites
Γo	olivine gabbro-dolerites
Γoc	olivine-bearing gabbro-dolerites
T °C	temperature
P	pressure
PGE	elements of platinum group

Institutes

GEOKHI RAS	Vernadsky Institute of Geochemistry and Analytical Chemistry of Russian Academy of Sciences, Moscow, Russia
VSEGEI	A.P. Karpinsky Russian Geological Research Institute, St. Petersburg, Russia
IEM RAS	Institute of Experimental Mineralogy of Russian Academy of Sciences, Moscow, Russia
NorilskGeology Ltd.	Exploration expedition, Noril'sk, Russia
MPI	Max-Planck Institute of Chemistry, Mainz, Germany

Introduction: Formulation of the Problem and Its Urgency

1

In previous years, the genesis of uniquely large mineral deposits became one of the most urgent geological problems because of the key role played by giant deposits: accounting for <5 % of the currently developed mineral deposits, these deposits satisfy 85 % the world's consumption of mineral resources. The development of these deposits should provide a basis for the stable progression of our civilization (Rundkvist and Kravchenko 1996). Another important aspect in the analysis of super-large mineral deposits is their genesis because these deposits are, in fact, giant geochemical anomalies in the Earth's crust (e.g., the PGE concentrations in the Noril'sk ores are six to seven orders of magnitude higher than those in the clarkes).

The problem regarding the mobilization of global concentrations of ore material in the Earth's interior and its further concentration in the form of super-large deposits was actively discussed in recent decades (Ryabchikov 1997; Kogarko 1999; Likhachev 2006) and still remains a matter of heated discussion. The key problem is whether giant deposits were formed by "normal" geological processes or, conversely, resulted from certain unique circumstances (Ovchinnikov 1988; Marakushev et al. 1998; Likhachev 2006). This problem was discussed at meetings of the International Association on the Genesis of Ore Deposits (IAGOD – 1999, 2001, 2004, 2008), Simposium on super-large mineral deposits (Adelaida 2010), Meetings of the International Mineralogical Association (IMA 2000), and other Symposia and Meetings of lower rank (St. Petersburg 1996; Moscow 2000). A project under the International Geological Correlation Program (IGCP 1995–1999) was also devoted to this problem.

In light of the possible solutions to this problem, a very important issue for endogenic ore deposits is the nature of their relationship with magmatism, which remains largely uncertain. It has been convincingly demonstrated that even certain rare-metal deposits were produced by specific magmas, although the overall contribution of magmatism is evaluated at no more than 25 % of all possible geological controls, such as structural, lithological, etc. (Kovalenko et al. 1992, 1993, 2000). It would be logical to suggest that the role of magmatic processes in producing magmatic deposits per se, for which the ores are components of ultrabasite–basite complexes, should be much higher, up to the development of natural ore-bearing magmas, for instance, with the occurrence of high-Fe melts at the El Laco volcano in Chile (Frutos and Oyarzun 1975) and ore-bearing melts at the Juan de Fuca Ridge in the East Pacific (Zhmodik 2002).

Large Pt–Cu–Ni deposits hosted in layered plutons or smaller intrusive bodies (especially in Noril'sk area) have been studied by many researchers (Urvantsev 1927, 1972; Godlevsky 1959; Dodin and Batiev 1971; Dodin 2002; Genkin et al. 1981; Likhachev 1982, 1994; Distler et al. 1988; Dyuzhikov et al. 1988; Gongalsky and Krivolutskaya 1993; Naldrett 1992, 2004, 2005, 2009; Naldrett et al. 1992; Barnes and Maier 2002; Barnes et al. 2006; Gorbunov et al. 1999; Li et al. 2009; Krivolutskaya 2014) and several others. However, many of these publications dealt with metallogenic models of various rank, from the analysis of the global distribution patterns of deposits to the local geological and tectonic controls at certain ore fields (Luznicka 1983; Kutina 2001; Lu Guxian and Yin Jicai 1995), whereas the essence of the relationship between magmatic and ore-forming processes was analyzed much more rarely (Godlevsky and Likhachev 1981, 1997; Campbel et al. 1992; Distler et al. 1988; Naldrett 2005, 2009; Chai and Naldrett 1992; Lightfoot and Keays 1995; Keays and Lightfoot 2007, 2010; Marakushev et al. 1998; Likhachev 1965, 1977,1978, 1982, 1996a, b, 2006; Li et al. 2009).

The magmatic deposits include two major types, which are usually separated from one another in nature: PGE deposits in large layered plutons (Bushveld, Great Dyke, Stillwater, and others) and Cu–Ni sulfide deposits (Sudbury, Voisey's Bay, Jinchuan, Lac des Îles, etc.). Each provides evidence of different relationships between magmatism and ore-forming processes. In only one case both types of ore mineralization are combined within a single intrusion: low-sulfide PGE mineralization and Cu–Ni sulfide ores, which

N.A. Krivolutskaya, *Siberian Traps and Pt-Cu-Ni Deposits in the Noril'sk Area*,
DOI 10.1007/978-3-319-17205-7_1

are also enriched in PGE. These intrusions are in the Noril'sk area in northern central Siberia. These deposits are principally different from other magmatic deposits as they are younger (their age is Early Triassic, in contrast to the Proterozoic age of most analogous deposits) and spatially related to relatively small intrusive bodies (in contrast to the usual association of this type of mineral deposits with large layered plutons).

The Noril'sk area lies in the northern part of the Siberian trap province, which presents a principally important problem regarding the genetic relationships between the deposits and volcanism. The magmatic products include numerous intrusions, whose ore potential broadly varies, from barren massifs to uniquely large reserves of ore mineralization at rare deposits, but the reason why certain massifs are accompanied by ore mineralization and most others are not is still uncertain despite long-term investigations designed to shedding light on this issue. This geological situation makes the Noril'sk area uniquely favorable for elucidating several problems related to the genesis of the ores because this area allows one to compare the various structural, mineralogical, and geochemical features of intrusions with different ore potential. Thus, this area was elected as a first-priority target for studies designed to clarify the possible relationships between magmatism and ore-forming processes. Analogous studies were also conducted at other major Cu–Ni ore provinces in Russia: in northern Transbaikalia, in Karelia and the Kola Peninsula, and in the Eastern Sayan Range in central Siberia.

Although the Noril'sk area was explored for Cu–Ni ores starting in the early 1920s (Sotnikov 1919; Kotulsky 1946; Rogover 1959; Godlevsky 1959; Korovyakov et al. 1963) and the development of the deposits provided the country with strategically important metals during World War II (the very first Ni was obtained at Noril'sk fabric in 1942), it was not an outstanding region until the discovery of the Talnakh and Oktyabr'skoe deposit, which hosted uniquely large reserves of sulfide ores with high PGE grades in the early 1960s, that the Soviet Union became one of the world's leading Pt and Ni producers (Egorov and Sukhanova 1963; Vaulin and Sukhanova 1970; Kravtsov 2003). Now, the Noril'sk area has the world's second largest reserves of Ni and PGE and is the world's second largest producer of these ores, after Bushveld in South Africa and Sudbury in Canada, respectively (Naldrett 2004, 2005). In 2008, the Noril'sk deposits gave 17 % of the world's annual Ni production (no other country has ever produced more than 10 % individually), and the processing of these same ores yielded 30 % of the world's PGE production (Bezhanova and Kyzina 2009; Dodin et al. 2011).

The discovery of a new type of ore has not only changed the situation on the world market of metals but has also advanced the general theory of magmatic ore-forming processes. Thus, the following important problems were formulated: how these deposits are related to volcanism, and are they a regular result of the evolution of any magmatic trap system? Is it reasonable to expect that analogous deposits can be discovered in other trap provinces elsewhere?

The solution of these problems is of paramount applied importance because understanding the genetic links between the Noril'sk deposits and trap volcanism can facilitate the discovery of analogous deposits in other provinces. The great economic importance of the newly discovered Noril'sk deposits stimulated the interest of geologists worldwide in such deposits during the second half of the twentieth century. The evaluation of the ore potential of trap provinces was also performed abroad regarding data on the Noril'sk deposits (Keays and Lightfoot 2007, 2010; Ripley et al. 2010).

A great variety of models were suggested to explain the genesis of the Noril'sk deposits from magmatic to metasomatic and other processes. These hypotheses and models vary broadly from those arguing that the decisive role was played by the introduction and deposition of metals by magmatic melts (Godlevsky 1959; Korovyakov et al. 1963; Likhachev 1965, 1982, 1994; Dyuzhikov et al. 1988; Naldrett 1992; Lightfoot et al. 1993) to those attaching importance to fluid components (Zolotukhin 1997; Zotov 1979, 1989). The concepts of the former group differ in that each investigation assigned principally different roles to the magmatic system, whose products now host the ores. Some researchers believe that this system played a decisive role by providing a portion of parental picrite melt enriched in volatile and ore components (Godlevsky 1959; Zolotukhin et al. 1978; Likhachev 1978, 2006; Genkin et al. 1981; Dyuzhikov et al. 1988; Distler et al. 1988; Zotov 1989; Ryabov et al. 2000, 2014). Other geologists believe that neither the composition of the magma nor its fluid contents played any appreciable role in the origin of the ores because the intrusions were merely one member of the trap system. The primary role in the generation of ore mineralization was thereby played by the duration the melt ascended to the surface and the interaction of this melt with the host rocks (Rad'ko 1991; Naldrett 1992, 2004, 2009; Howkesworth et al. 1995; Arndt et al. 2003; Li et al. 2009). Hence, these hypotheses principally differ in that they suggest that the ores were produced in either a closed or open magmatic system.

The latter hypothesis is now widely accepted: this model for the genesis of Cu–Ni ores (with certain variations depending on the geological specifics of certain magmatic and orebodies) was suggested to explain the origin of the world's largest Cu–Ni deposits, such as Jinchuan in China, Voisey's Bay in Canada, and others. However, despite the attractiveness of this model, which suggests a clear and realistic mechanism that concentrates PGE in sulfides, the model itself has certain disadvantages. In this model there is the absence of evidence of any relationship between the intrusive and volcanic products.

Several aspects of the geology, petrology, and ore-forming processes in the Noril'sk area remain uncertain. These are (1) the relationship between the ore-bearing intrusions and lavas, (2) the composition of the melts that produced the mineralized intrusions, (3) the role of assimilation processes in the origin of the ores, and (4) the sources of the magmas and ore material.

A realistic genetic model for the deposits would be of paramount importance for exploration operations in the Noril'sk area. After the development of these deposits during roughly the past half century, their unique massive ore bodies are now almost completely depleted, which emphasizes the urgently incrementing of the PGE and Ni reserves in Russia. The renewed exploration of high-grade ores in the Noril'sk area requires new methodical and scientific approaches based on modern mineralogical and geochemical techniques. Moreover, gaining insight into the genesis of the Noril'sk deposits should greatly facilitate exploration at other trap provinces elsewhere.

When studying the Noril'sk deposits, we centered on elucidating the general relationships and trends in the evolution of trap magmatism in the area and the location of ore-forming processes during this evolution. Our studies were based on new geological and high-precision geochemical data obtained by using state-of-the-art analytical techniques on the volcanic and intrusive rocks.

Our efforts focused on elucidating the aforementioned major disputable issues related to the genesis of the deposits. Our approaches toward the resolution of these problems were as follows:

1. First, we tried to elucidate the relationship between the ore-bearing intrusions and lavas based on (a) detailed studies of the tuff–lava sequence and its geochemistry in various tectonic structures in the area to reproduce the spatiotemporal evolution of the volcanic processes, (b) distinguishing the major geochemical types of ultrabasite–basite intrusions with different ore potential, and (c) a comparison of the distinctive geochemical features of the lavas and intrusions produced during analogous evolutionary episodes within the magmatic system; we have also compared the geochemistry of the Noril'sk intrusions with that of intrusions in Karelia, the Kola Peninsula, and southern Siberia.
2. The composition of the parental magmas of the Noril'sk intrusions was evaluated "directly," i.e., by studying melt inclusions in olivine and pyroxene, and indirectly via geochemical thermometry techniques (numerical simulations with the COMAGMAT program package).
3. The extent of the assimilation of the host rocks by the basite melts in the magmatic chambers was quantified by a thorough study of sections across contact zones and the petrography and geochemistry of these rocks (including their trace element and radioactive isotopic compositions).
4. The effect of the melt composition on the composition of the ores was studied mostly at the Maslovsky deposit.

The newly obtained data allowed us to develop a qualitative model for the genesis of the Noril'sk deposits.

This publication is based on the results of the author's long-term (1982–2011) studies of the geology, petrography, geochemistry, and mineralogy of the PGE–Cu–Ni deposits in Russia. These studies focused mostly on the Noril'sk area, where the author collected rocks in the course of fieldwork in 1997–2014 (during this work, 13 km of vertical sections through the volcanic sequence and 15 km of vertical sections through intrusive bodies were prepared based on materials collected from exposures and borehole cores). Some samples (namely, samples from the South Kovdor area) for our comparative studies of the ultrabasite–basite complexes were provided for the author (under contractual obligations), courtesy of V. P. Mamontov.

The laboratory analytical studies involved (Annex 1) (1) XRF analyses conducted by I. A. Roshchina and T. V. Romashova at the Vernadsky Institute of Geochemistry and Analytical Chemistry, Russian Academy of Sciences, Moscow, and 208 analyses made by N. S. Baluev at the Institute of Natural Resources, Chita, Russia, and (2) ICP-MS analyses, including (i) ICP-MS whole-rock analyses made at the Institute of Mineralogy, Geochemistry, and Crystal Chemistry of Rare Elements, Moscow (analyst D. Z. Zhuravlev), and at the Institute of Experimental Mineralogy, Russian Academy of Sciences, Chernogolovka (analyst V. K. Karandashev); (ii) LA-ICP-MS analyses of glasses, pyroxenes, and olivine grains at the Max-Planck Institute for Chemistry, Mainz, Germany (analysts D. V. Kuzmin, B. Stoll, and N. A. Krivolutskaya); (iii) electron microprobe analyses on Cameca SX 50 and 100 at the Vernadsky Institute of Geochemistry and Analytical Chemistry, Russian Academy of Sciences, Moscow (analyst N. N. Kononkova), and on JXA 8200 at the Max-Planck Institute for Chemistry in Mainz, Germany (analysts N. A. Krivolutskaya and D. V. Kuzmin); (iv) ion probe analyses on Cameca IMS-4F at the Institute of Microelectronics, Russian Academy of Sciences, Yaroslavl (analysts S. G. Simakin and E. V. Potapov); (v) Raman spectroscopy in Nancy, France (analyst J. Debussy), including analyses of fluid inclusions; (vi) analyses of stable O, H, C, and S isotopes in rocks conducted by B. G. Pokrovsky at the Geological Institute (GIN), Russian Academy of Sciences, Moscow, and by S. G. Kryazhev at the Central Institute of Geological Exploration for Base and Precious Metals in Moscow; (vii) whole-rock analyses for radiogenic Sr, Pb, Sm–Nd, and U–Pb isotopes conducted by B. V. Belyatsky at the All-Russia Research Institute of Geology and Mineral Resources of the World Ocean

(VNIIOkeanologiya) in St. Petersburg, A. A. Plechova at the Vernadsky Institute in Moscow, and Z. Fekiasova at the Max-Planck Institute of Chemistry in Mainz, Germany; and (viii) analyses for PGE and Au conducted by A. A. Yushin at the Institute of Ore-Forming Processes, Mineralogy, and Geochemistry of the Ukrainian National Academy of Sciences, Kiev (fire assay) and I. V. Kurbakova, O. A. Tyutyunnik, and A. D. Chkhetiya at the Vernadsky Institute.

Our experimental studies of the homogenization of melt inclusions were carried out in a muffle chamber at the Institute of the Geology of Ore Deposits, Petrography, Mineralogy, and Geochemistry, Russian Academy of Sciences, Moscow (by author with assistance of A. Babansky and I. Solovova), and in a heating stage (designed by A. V. Sobolev and B. Slutsky, 1984) at the Vernadsky Institute and in a furnace with adjustable oxygen fugacity at the Vernadsky Institute (these experiments were conducted by A. A. Kargaltsev, M. Volovetsky, and N. Svirskaya). The numerical simulations of basaltic magma crystallization were conducted with the COMAGMAT-3.5 (Ariskin and Barmina, 2000) and PETROLOG-2.0 program packages (Danyushevsky et al. 1999).

Acknowledgments Passion for the geology of mineral deposits was imparted to me by Acad. V. I. Smirnov and his colleagues when I was his student and then a PhD student at the Department of Mineral Resources of Moscow State University. Along with Acad. V. I. Smirnov himself were Yu. S. Borodaev, G. F. Yakovlev, N. I. Eremin, V. V. Avdonin, V. I. Starostin, N. E. Sergeeva, T. A. Filitsyna, E. M. Zakharova, and others. The author appreciates help in organizing the field work provided by the following organizations and persons: geologists at Noril'sk Nickel—O. N. Simonov, A. V. Pospelov, and A. A. Shashkov; specialists at Ltd. NorilskGeology—V. V. Kurgin, YU. K. Krakovetsky, L. I. Trofimova, I. N. Tushentsova, E. A. Arshinova, I. A. Matveev, V. P. Strelnikov, V. A. Teteryuk, V. Yu. Wan Chan, K. V. Kradenov, V. A. Rad'ko, K. V. Shishaev, V. A. Pinaev, A. A. Danchenko, S. G. Snisar, K. K. Kovalchuk, O. P. Legezina, G. I. Legezin, S. A. Vilensky, E. V. Sereda, and others; and D. M. Turovtsev, S. F. Sluzhenikin, A. G. Tarasov, S. V. Nistratov, I. V. Khramov, A. V. Rudakova, and K. V. Bychkov. V. N. Mikhailov is thanked for the long-term assistance in the field studying volcanic rocks in the Noril'sk area and for fruitful discussions regarding the genesis of these rocks. Our analytical work could not be successful without the aid of I. A. Roshchina, N. N. Kononkova, D. V. Kuzmin, O. B. Kuzmina, S. G. Simakin, E. V. Potapov, A. A. Plechova, B. V. Belyatsky, V. A. Turkov, S. V. Lutkova, and T. L. Krylova. The author thanks A. A. Ariskin for discussion and help in modeling the crystallization of the magmas of ore-bearing massifs and highly appreciates the friendly and creative atmosphere at the Laboratory for the Geochemistry of Magmatic and Metamorphic Rocks at the Vernadsky Institute. Special thanks are due to the closest colleagues of the author who helped with graphics, I. V. Karlina, T. B. Shlychkova, N.M. Svirskaya, and O. P. Tsameryan. The author is heartily grateful to B. I. Gongalskiy for cooperation in studying basite rocks in ore-bearing complexes in the field and processing and interpreting the results over the course of three decades. Indispensable help in this research was provided by the author by the corresponding member of the Russian Academy of Sciences A. V. Sobolev, who is thanked for continuous keen interest in studying trap magmatism in Siberia.

This study was financially supported by the Russian Foundation for Basic Research (project nos. 00-05-64507-a, 00-05-74508-z, 01-05-74552-z, 03-05-79123-k, 03-05-64578-a, 05-05-74622-z, 07-05-01007-a, 08-05-100092-k, 09-05-01193-a, 09-05-10065-k 10-05-10088-k 10-05-08173-z, 11-05-10076-k, and 12-05-10074-k, 13-05-12110—ofr, 15-05-09250), programs for supporting leading research schools by the President of the Russian Federation (programs NSH-150.2008.5, NSH-3919.2010.5, and NSH-741.2012.5), programs 4, 8, and 9 under the Earth Science Division of the Russian Academy of Sciences, partially program RAS "Deposits of strategic metals in Russia: innovative approaches to forecasting, exploration and exploitation," and the Wolfgang Paul Project (Germany).

References

Ariskin AA, Barmina GS (2000) Simulation of phase Equilibria at basalt magma crystallization. Nauka, Moscow (in Russian)

Arndt NT, Czamanske GK, Walker RJ et al (2003) Geochemistry and origin of the intrusive hosts of the Noril'sk–Talnakh Cu–Ni–PGE sulfide deposits. Econ Geol 98:495–515

Barnes SJ, Maier WD (2002) Platinum-group element distributions in the Rustenberg layered suite of the Bushveld Complex, South Africa. In: Cabri LJ (ed) Geology, geochemistry and Mineral beneficiation of platinum-group elements, special vol 54. Canadian Institute of Mining and Metallurgy, Ottawa, pp 431–458

Barnes S-J, Cox RA, Zientek ML (2006) Platinum-group element, Gold, Silver and Base Metal distribution in compositionally zoned sulfide droplets from the Medvezky Creek Mine, Noril'sk, Russia. Contrib Mineral Petrol 152:187–200

Bezhanova MP, Kyzina LV (2009) The world's resources and mining of the most important mineral deposits. VNIIZarubezhgeologiya, Moscow, 329 p (in Russian)

Campbell IA, Czamanske GK, Fedorenko VA et al (1992) Synchronism of the Siberian traps and the Permian–Triassic boundary. Science 258:1760–176

Chai G, Naldrett AJ (1992) Petrology and geochemistry of the Jinchuan ultramafic intrusion: cumulate of high-Mg basaltic magma. J Petrol 33:1–27

Danyushevsky LV, Della-Pasqua FN, Sokolov S (1999) Re-equilibration of melt inclusions trapped by magnesian olivine phenocrysts from subduction-related magmas: petrological implications. Contrib Mineral Petrol 138(1):68–83

Distler VV, Grokhovskaya TL, Evstigneeva TL, Sluzhenikin SF, Filimonova AA, Dyuzhikov OA (1988) Petrology of magmatic sulfide ore formation. Nauka, Moscow, 232 p (in Russian)

Dodin DA (2002) Metallogeny of the Taimyr-Noril'sk region. Nauka Press, St. Petersburg, 821 p (in Russian)

Dodin DA, Batiev BN (1971) Geology and petrology of the Talnakh differentiated intrusions and their metamorphic aureole. In: Petrology and ore resource potential of the Talnakh and Noril'sk differentiated intrusions. Nedra, Leningrad, pp 31–100 (in Russian)

Dodin DA, Zoloev KK, Koroteev VA, Chernyushev NM (2011) Platinum of Russia: the state and problems. Platin Russ 7:12–51 (in Russian)

Dyuzhikov OA, Distler VV, Strunin BM et al (1988) Geology and ore potential of the Noril'sk ore district. Nauka, Moscow. Dyuzhikov OA, Distler VV, Strunin BM, Mkrtychyan AK, Sherman ML, Sluzhenikin SF, Lurye AM (1992) Geology and metallogeny of sulfide deposits Noril'sk region USSR (trans). Econ Geol Monogr. Ontario, Spec vol 1, 241 p

Egorov VN, Sukhanova EN (1963) The Talnakhskii ore-bearing intrusion in the northwestern Siberian platform. Razved Okhr Nedr, no 1, pp 17–21 (in Russian)

Frutos J, Oyarzun JM (1975) Tectonic and geochemical evidence concerning the genesis of El Laco magnetite lava flow deposit. Econ Geol 70(5):988–990

Genkin AD, Distler VV, Gladyshev GD et al (1981) Sulfide copper–nickel ores of the Noril'sk deposits. Nauka, Moscow (in Russian)

Godlevsky MN (1959) Traps and ore-bearing intrusions of the Noril'sk district. Gosgeoltekhizdat, Moscow, 61p (in Russian)

Godlevsky MN, Likhachev AP (1981) Conditions of formation and evolution of ore-bearing ultrabasic magmas. Zapisky Mineral Soc USSR 6:646–655 (in Russian)

Godlevsky MN, Likhachev AP (1997) Origin and conditions' crystallization of magmas forming Cu-Ni deposits. In: Main parameters of natural processes of endogenic ore-forming processes. Nauka, Novosibirsk, pp 109–118 (in Russian)

Gongalsky BI, Krivolutskaya NA (1993) Chineysky layered pluton. Nauka, Novosibirsk, 187 p (in Russian)

Gorbunov GI, Astaf'ev YA, Goncharov YV, Korchagin AU, Neradovsky YN, Smolkin VF, Sokolov SV, Sharkov EV, Yakovlev YN (1999) Cu-Ni Pechenga deposits. GEOS, Moscow, 236 p (in Russian)

Howkesworth CJ, Lightfoot PC, Fedorenko VA et al (1995) Magma differentiation and mineralization in the Siberian continental flood basalts. Lithos 34:61–81

Keays RR, Lightfoot PC (2007) Siderophile and chalcophile metal variations in tertiary picrites and basalts from West Greenland with implications for the sulphide history of continental flood basalt magmas. Miner Deposita 42:319–336

Keays RR, Lightfoot PC (2010) Crustal sulfur is required to form magmatic Ni–Cu sulfide deposits: evidence from chalcophile element signatures of Siberian and Deccan Trap basalts. Miner Deposita 45:241–257

Kogarko LN (1999) Problems of genesis of giant apatite and rare metals deposits in Kola Peninsula, Russia. Geol Ore Deposit, no 3, pp 387–403

Korovyakov IA, Nelyubin AE, Raikova ZA, Khortova LK (1963) The origin of Noril'sk trap intrusions containing copper–nickel ores. Gosgeotekhizdat, Moscow. Trudy Vses Nauch–Issled Inst Miner Syr'ya, 101p (in Russian)

Kotulsky VK (1946) About the origin of magmatic copper-nickel deposits. Doklady Acad Sci USSR 51:381–384 (in Russian)

Kovalenko BI, Tsareva GM, Khervig RL, Yarmoluk VV (1992) Trace elements and water in melt inclusions (magmas) rare metals alkaline granites. Doklady Acad Sci USSR 326(4):349–353

Kovalenko VI, Bakhteev RK, Yarmoluk VV, Chizhova IA (1993) Quantitative estimation of connection magmatism with endogenic mineralization (by the example of Mongolia). Geol Ore Deposit 35(2):161–176

Kovalenko VI, Salnikova EB, Antipin VS, Yarmoluk VV, Kovach VP, Kotov AB (2000) Unusual association Li-F granites of Soktuy massif (Eastern Transbaikalia): age and magmas' sources. Doklady Acad Sci USSR 327(4):536–540

Kravtsov VF (2003) History discoveries of copper-nickel deposits in the Noril'sk area. Essays on the history of discoveries of mineral deposits of Taimyr. Publishing House of SB RAS, Novosibirsk, pp 21–40 (in Russian)

Krivolutskaya NA (2014) Evolution of trap magmatism and processes producing Pt-Cu-Ni mineralization in the Noril'sk area. KMK, Moscow, 320 p (in Russian)

Kutina J (2001) The role of transregional mantle-rooted structural discontinuities in the concentration of metals, with examples from the United States, Uzbekistan, Burma and other countries. Glob Tect Metall 7(3,4):159–182

Li CS, Ripley EM, Naldrett AJ (2009) A new genetic model for the giant Ni–Cu–PGE sulfide deposits associated with the Siberian flood basalts. Econ Geol 104:291–301

Lightfoot PC, Keays RR (1995) Siderophile and chalcophile metal variations in flood basalts from the Siberian trap, Noril'sk Region: implications for the origin of the Ni-Cu-PGE sulfide ores. Econ Geol, no 100, pp 439–462

Lightfoot PC, Howkesworth CJ, Hergt J, Naldrett AJ, Gorbachev NS, Fedorenko VA, Doherty W (1993) Remobilisation of the continental lithosphere by mantle plumes: major-/trace-element and Sr-, Nd-, and Pb-isotope evidence from picritic and tholeiitic lavas of the Noril'sk district, Siberian trap, Russia. Contrib Mineral Petrol 114:171–188

Likhachev AP (1965) The role of leucocratic gabbro in the origin of Noril'sk differentiated intrusions. Izv Akad Nauk SSSR Ser Geol, no 12, pp 50–66 (in Russian)

Likhachev AP (1977) On the crystallization conditions of trap magmas in the northwestern part of the Siberian platform. Zapiski Vses Mineral O–va, no 5, pp 594–606 (in Russian)

Likhachev AP (1978) On the formation conditions of ore-bearing and non-bearing mafic–ultramafic magmas. Dokl Akad Nauk SSSR 238(2):447–450 (in Russian)

Likhachev AP (1982) Formation conditions of copper–nickel deposits. Sov Geol, no 6, pp 31–46 (in Russian)

Likhachev AP (1994) Ore-bearing intrusions of the Noril'sk region. Proceeding of the Sudbury–Noril'sk symposium, Ontario. Geol Surv Spec 5:185–201

Likhachev AP (1996a) The Kharaelakh intrusion and its PGM–Cu–Ni ores. Rudy et Metals, no 3, pp 48–62 (in Russian)

Likhachev AP (1996b) Emplacement dynamics of the Talnakh ore bearing intrusions and related PGM–Cu–Ni ores. Otechestvennaya Geol, no 8, pp 20–26 (in Russian)

Likhachev AP (2006) Platinum–copper–nickel and platinum deposits. Eslan, Moscow, 496 p (in Russian)

Lu Guxian, Yin Jicai (1995) Effects of tectonic force on hydrostatic pressure during petrogenesis and metallogenesis. Geotectonica et Metallogen 19(2–3):39–47

Luznicka P (1983) Giant ore deposits: a quantitative approach. Glob Tect Metall 2(1,2):41–63

Marakushev AA, Paneyakh NA, Rusiniv VL, Pertsev NN, Zotov IA (1998) Petological models for formation of giant deposits. Geol Ore Deposit 40(3):236–255 (in Russian)

Naldrett AJ (1992) Model for the Ni–Cu–PGE ores of the Noril'sk region and its application to other areas of flood basalts. Econ Geol 87:1945–1962

Naldrett AJ (2004) Magmatic sulphide deposits: geology, geochemistry and exploration. Springer, Berlin/Heidelberg/New York, 727 p

Naldrett AJ (2005) A history of our understanding of magmatic Ni–Cu sulfide deposits. Can Mineral 43:2069–2098

Naldrett AJ (2009) Ore deposits related to flood basalts, Siberia. In: Li C, Ripley EM (eds) New developments in magmatic Cu–Ni and PGE deposits. Geological Publishing House, Beijing, pp 141–179

Naldrett AJ, Lightfoot PC, Fedorenko VA et al (1992) Geology and geochemistry of intrusions and Flood Basalts of the Noril'sk region, USSR, with implication for origin of the Ni–Cu ores. Econ Geol 87:975–1004

Ovchinnikov LN (1988) Origin of ore deposits. Nedra, Moscow, 355 p (in Russian)

Rad'ko VA (1991) Model of dynamic differentiation of intrusive traps at the northwestern Siberian trap. Geol Geophys 32(11):19–27 (in Russian)

Ripley EM, Li C, Craig H et al (2010) Micro-scale S isotope studies of the Kharaelakh intrusion, Noril'sk region, Siberia: constraints on the genesis of coexisting anhydrite and sulfide minerals. Geochim et Cosmochim Acta 74:634–644

Rogover GB (1959) Norilsk deposit 1. Gosgeoltechizdat, Moscow, 168 p (in Russian)

Rundkvist DV, Kravchenko SM (1996) Economic metals' superconcentrations in the lithosphere. Geol Ore Deposit 38(3):298–303

Ryabchikov ID (1997) Global flows of ore metals in deep processes. Geol Ore Deposit 39(5):403–408

Ryabov VV, Shevko AY, Gora MP (2000) Igneous rocks of the Noril'sk district. V.1. Petrology of traps. Nonparel Novosibirsk, vol 1, 2 (in Russian)
Ryabov VV, Shevko AY, Gora MP (2014) Trap magmatism and ore formation in the Siberian Noril'sk region. V 1, 2. Springer
Sobolev AV, Slutsky AB (1984) Composition and crystallization conditions of the parental melt of the Siberian meymechites in connection with the general problem of ultrabasic magmas. Geol Geofiz 12:97–110 (in Russian)
Sobolev AV, Krivolutskaya NA, Kuzmin DV (2009) Petrology of the parental melts and mantle sources of Siberian trap magmatism. Petrology 17:253–286
Sobolev AV, Hofmann AW, Kuzmin DV, Yaxley GM, Arndt NT, Chung S-L, Danyushevsky LV, Elliott T, Frey FA, Garcia MO, Gurenko AA, Kamenetsky VS, Kerr AC, Krivolutskaya NA, Matvienkov VV, Nikogosian IK, Rocholl A, Suschevskaya NM, Teklay M (2007) Estimating the amount of recycled crust in sources of mantle derived melts. Science 316:412–417
Sobolev SV, Sobolev AV, Kuzmin DV, Krivolutskaya NA, Petrunin AG, Arndt NT, Radko VA, Vasilev YR (2011) Linking mantle plumes, large igneous provinces and environmental catastrophes. Nature 477:312–316
Sotnikov AA (1919) About operation of the Noril'sk (Dudinsky) deposits of coal and copper ore in connection with the practical implementation and development of the Northern Sea Route. Tomsk, 98 p (in Russian)
Urvantsev NN (1927) Exploration of Cu-Ni Noril'sk deposit. Izvestiya Geol Committee 44(2) (in Russian)
Urvantsev NN (1972) Some questions on formation of ore-bearing intrusion and ore. In: Cu-Ni ores of Talnakh ore junction. Nedra, Leningrad, pp 100–144 (in Russian)
Vaulin LL, Sukhanova EN (1970) Oktyabrskoe copper-nickel deposit. Prot Prospect 4:48–52 (in Russian)
Zhmodik AS (2002) Composition and thermodynamic conditions of development of magmatic ore mineralization in basalts of the kleft segment of the Juan de Fuca Ridge. Candidate dissertation. Institute of Geology and Geophysics, Novosibirsk, 16 p (in Russian)
Zolotukhin VV (1997) Basaltic pegmatoides of the Noril'sk ore-bearing intrusions and the problem of its origin. Trudy IGG. Novosibirsk, vol 834, 90 p (in Russian)
Zolotukhin VV, Vasil'ev YR, Dyuzhikov OA (1978) Diversity of traps and initial magmas: a case of the Siberian platform. Nauka, Novosibirsk (in Russian)
Zotov IA (1979) The genesis of trap intrusions and metamorphic rocks of Talnakh). Nauka, Moscow (in Russian)
Zotov IA (1989) Transmagmatic fluids in magmatism and ore formation. Nauka, Moscow (in Russian)

2 Overview of the Geology of the Noril'sk Area and the Problems of Ore Genesis

The Pt–Cu–Ni Noril'sk deposits were discovered in the nineteenth century on the SW Siberian Platform. They are located in a paleorift zone and display a structure that differs from that of the Siberian craton basement. Terrigenous and carbonate rocks with anhydrite and coal are exposed on the surface of the Noril'sk area. They are overlapped by volcanic rocks of the trap formation. The stratigraphy of the sediments and tuff–lava deposits is given. The intrusive rocks consist of 10 % plutonic rocks and are represented by ultrabasic to basic intrusions (sills, dykes) of mostly normal alkalinity and rarely of elevated alkalinity. There are two types of sulfide ores: disseminated and massive. These are located near the contact of intrusion with surrounding rocks. There are three main deposits in the Noril'sk area: the Talnakh, Oktyabr'skoe, and Noril'sk deposits. Many other intrusions contain mineralizations in economic quantities.

2.1 Study History of the Noril'sk Mineral Deposits

Extensive studies of the Noril'sk area were launched regarding the exploration of coal for the Northern Sea Route in 1919–1920 and were carried out by the Siberian Geological Committee.

The territory has been known to host ores since the late nineteenth century (Kravtsov 2003), when disseminated sulfide ores were found in shales in the bottom part of Mount Rudnaya, and this event was marked by erecting a witness corner in 1865, in which the characters K & S denoted the discoverers: the gold-mining industrialist A. Kytmanov and merchant K. Sotnikov. The mineral deposit was named Sotnikovskoe (now it corresponds to the outer-contact portion of the Noril'sk 1 deposit), and 200 poods (320 kg) of crude copper had been produced by 1872. K. Sotnikov's grandson, A. Sotnikov then a student at the Tomsk Technological Institute, continued the family business of studying and exploring the deposit. In 1915, he conducted eye-work survey operations at the deposit, drilled a hole, and collected local ores and rocks; in 1919, he published the paper "On the Development of the Noril'sk (Dudinka) Coal and Copper Deposits in Relation to the Development of the Northern Sea Route." The petrography of the ores and rocks was examined by N.N. Urvantsev. N.N. Urvantsev organized the extensive exploration of Cu in the area of Mount Rudnaya in 1919–1926 (Urvantsev 1927). The Noril'sk deposits were thus discovered by the Sotnikovy, or Sotnikov family, and N.N. Urvantsev was the first to organize the systematic exploration of this territory, with these operations continuing under the Soviet administration. In 1923, N.K. Vysotsky identified Pt in ores from the Noril'sk 1 deposit.

The later exploration of the territory was conducted by Noril'skstroy (Chief Geologist A.E. Vorontsov) and other governmental companies and resulted in the discovery of massive ores, an increment in the resources of disseminated ore, and the discovery of the Noril'sk 2, Chernogorsky, and Bolshaya Bar'ernaya deposits. A government order to launch the construction of the Noril'sk Processing Complex was issued in 1935 year.

However, it was no earlier than the 1960s that Noril'sk was considered a unique ore district after the discovery of the Talnakh and Oktyabr'skoe deposits during the systematic exploration operations conducted by the Noril'sk Complex Exploration Expedition. The value of the ores due to their complex multicomponent characteristics and their compactness was several times higher than the value of ores at any other deposit of this type worldwide, including Bushveld and Sudbury (Naldrett 2004, 2005).

2.2 Geological Overview of the Territory

The Noril'sk ore district in the northern Siberian Platform (Fig. 2.1) was exhaustively studied and described in much detail in several publications (Godlevsky 1959; Genkin et al.

N.A. Krivolutskaya, *Siberian Traps and Pt-Cu-Ni Deposits in the Noril'sk Area*,
DOI 10.1007/978-3-319-17205-7_2

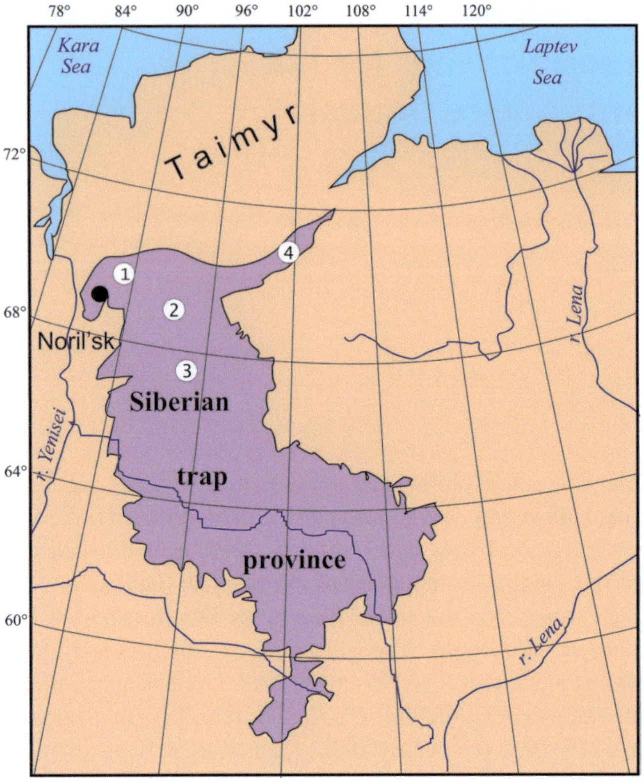

Fig. 2.1 Location map of the Noril'sk area in the Siberian trap province. *1*, Noril'sk area; *2*, Putorana zone; *3*, Tunguska sineclise; *4*, Maimecha–Kotuy province

1981; Dyuzhikov et al. 1988; Distler et al. 1988; Geology and ore… 1994; Sudbury-Noril'sk 1994; Ryabov et al. 2000; Turovtsev 2002; Naldrett 2004; Likhachev 2006; Sluzhenikin et al. 2014). It is bounded by the Yenisei and Yenisei–Khatangsky Troughs to the west and north and by the Tunguska syneclise to the east. It is an individualized tectonic continental crustal block of lower thickness, which consists of a crystalline basement and sedimentary–volcanic cover (Fig. 2.2). The deep structure of the territory is reproduced based mostly on interpretations of the GORIZONT (Vorkuta–Tiksi) and METEORIT (Dyupkun–Dixon) deep seismic traverses.

2.2.1 Deep Geological Structure of the Territory

The western portion of the platform hosts three clearly pronounced major morphostructural provinces: Yenisei–Khatanga, Putorana, and Angara–Tunguska. The structures of the Yenisei–Khatangsky Trough trend sublatitudinally, and the location's crustal thickness varies from 36 to 39 km (Fig. 2.2). The crust at the equant Putorana structure is 45 km thick. The bottom of the crust beneath the Angara–Tunguska province occurs at depths of 39–42 km and occasionally at depths as great as 45 km.

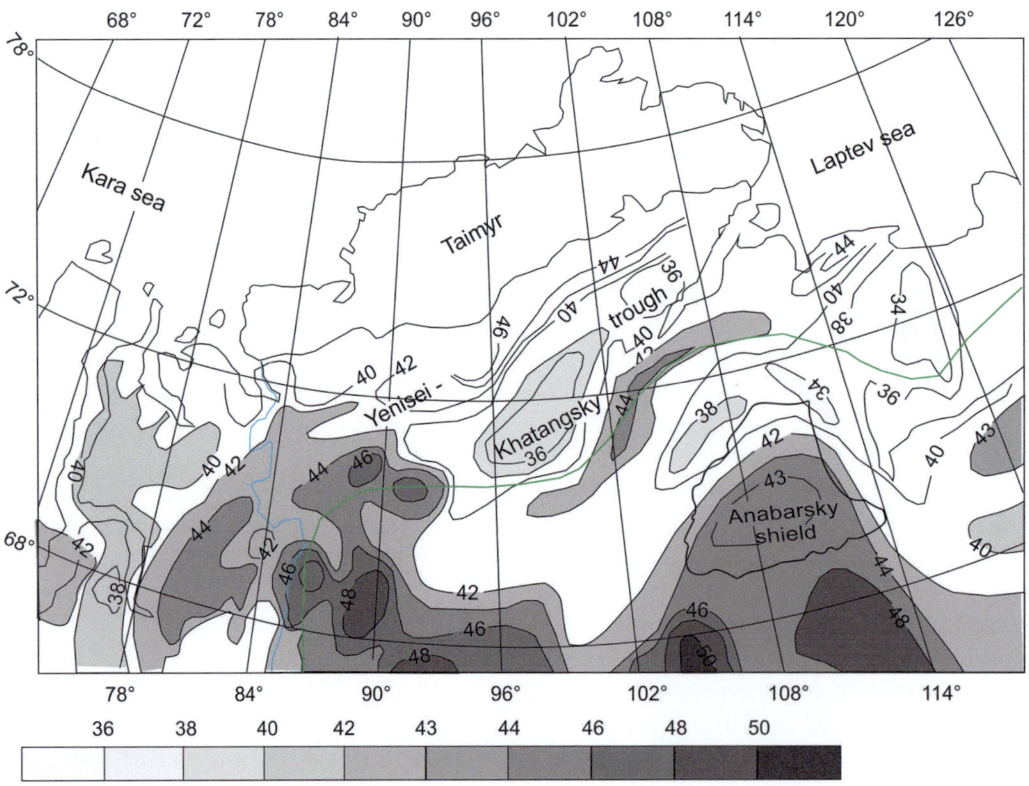

Fig. 2.2 Map of the crust's isopachs for north Siberian Platform Scale – depth up to Moho in km (After Kostyuchenko 2006)

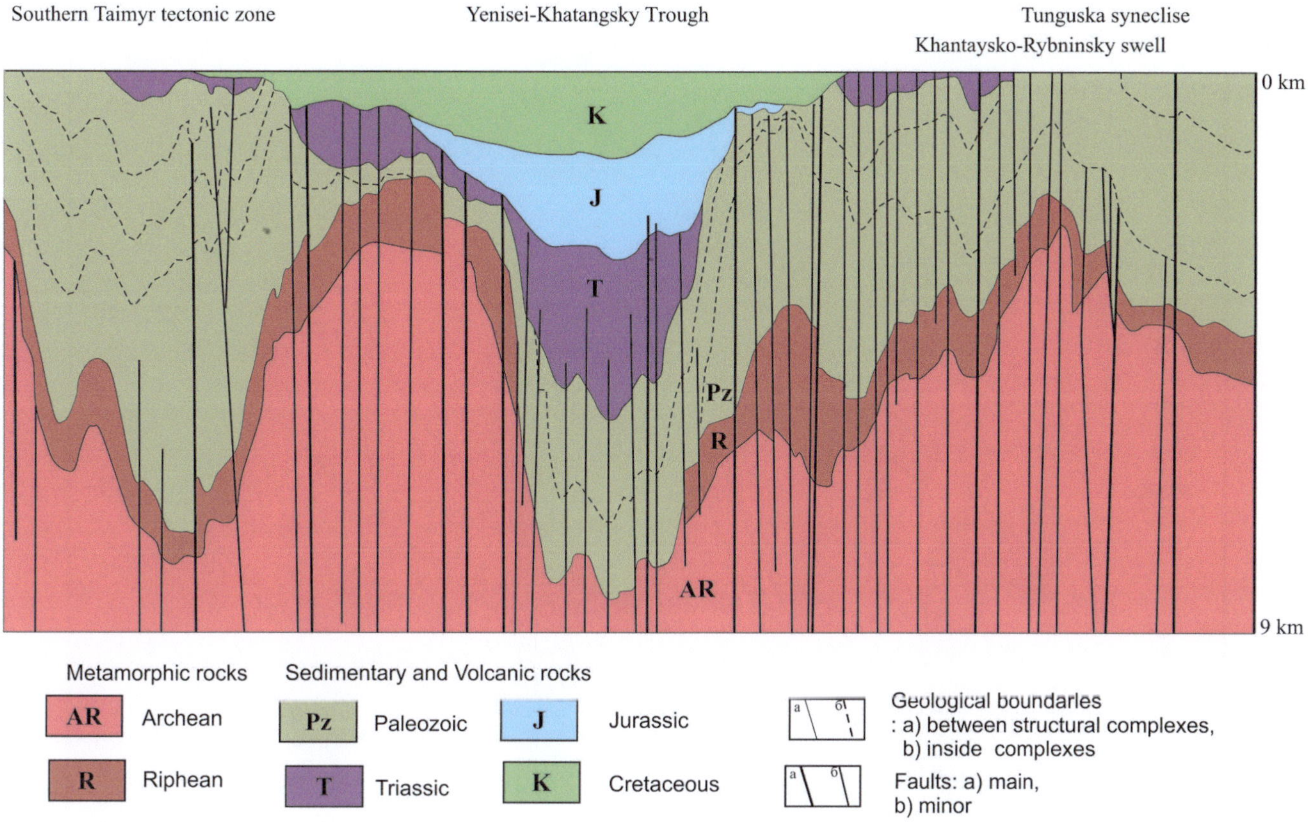

Fig. 2.3 Generalized seismic profile throw the crust along the Dixon-Khantaiskoe Lake line (Based on unpublished data of VNIIOkeangeologiya and others)

Geophysical data indicate that the Yenisei–Khatangsky morphostructural province in the northwestern part of the Siberian Platform (Malitch et al. 1987, 1988) is underlain by a fossil (buried) ancient rift system (ARS), which has a crustal block that is atypical of platforms, remained mobile throughout the whole evolutionary history, and has a deep structure typical of rift systems (Dolgal' 2012). The block is separated by mantle faults from the Tunguska and Taimyr blocks, whose structures are typical platforms. The Igarka–Noril'sk block is a part of a large rift system, as observed in the Dixon–Khilok seismic–geological traverse (Fig. 2.3). The block has high-gradient troughs in its basement, which are filled with thick (up to 15 km) sedimentary–volcanic sequences. The structure of the block is horst–graben, with a significant density of faults, a thicker basaltic layer, significant volumes of mantle volcanic rocks, and a transitional layer between the crust and mantle (with Vp = 7.3 km/s). This layer occurs at a depth of 35 km beneath the Noril'sk district. The eastern boundary of the Igarka–Noril'sk ARS is the Lamsko–Letninsky Fault and its splay Keta–Irbinsky Fault, which separate the rift system from the Archean craton. The northwestern boundary of the rift block is conventionally drawn along the Yenisei Fault, which also defines the eastern flank of the West Siberian Basin. The ARS is traced along the left-hand bank of the Yenisei River and can, perhaps, merge with the modern rifts of the West Siberian Basin. The ARS is characterized by a much thicker cover (than nearby platform structures), with the basement surface occurring at a depth of 12 km. Another noteworthy feature of the ARS is the presence of a low-velocity layer beneath the Moho.

The Igarka–Noril'sk ARS is a continental rift (corresponding to the divergent phase of lithosphere evolution), which comprises blocks and grabens showing evidence of horizontal extension and lithospheric thinning and breakup. It is characterized by a high density of faults and a great volume of erupted mantle material: the total volume of magmatic masses in the Noril'sk–Kharaelakh Trough is evaluated at 50,000–70,000 thousand km^3 (based on its higher density than in other territories), which indicates that the structure is highly permeable (Geology and ore… 1994).

The large folded nappe structures exposed in the modern structure of the Igarka–Noril'sk ARS compose the Riphean Chernorechensky anticline, Khantaysko–Rybninsky uplift, Kulyumbino–Sukharikhinskaya zone of Vendian–Silurian structures, Noril'sk–Kharaelakh Trough, and Dudinka Late Permian–Early Triassic swell (Fig. 2.4). The amplitude of the Khantaysko–Rybninsky swell is 4–5 km, and the over-

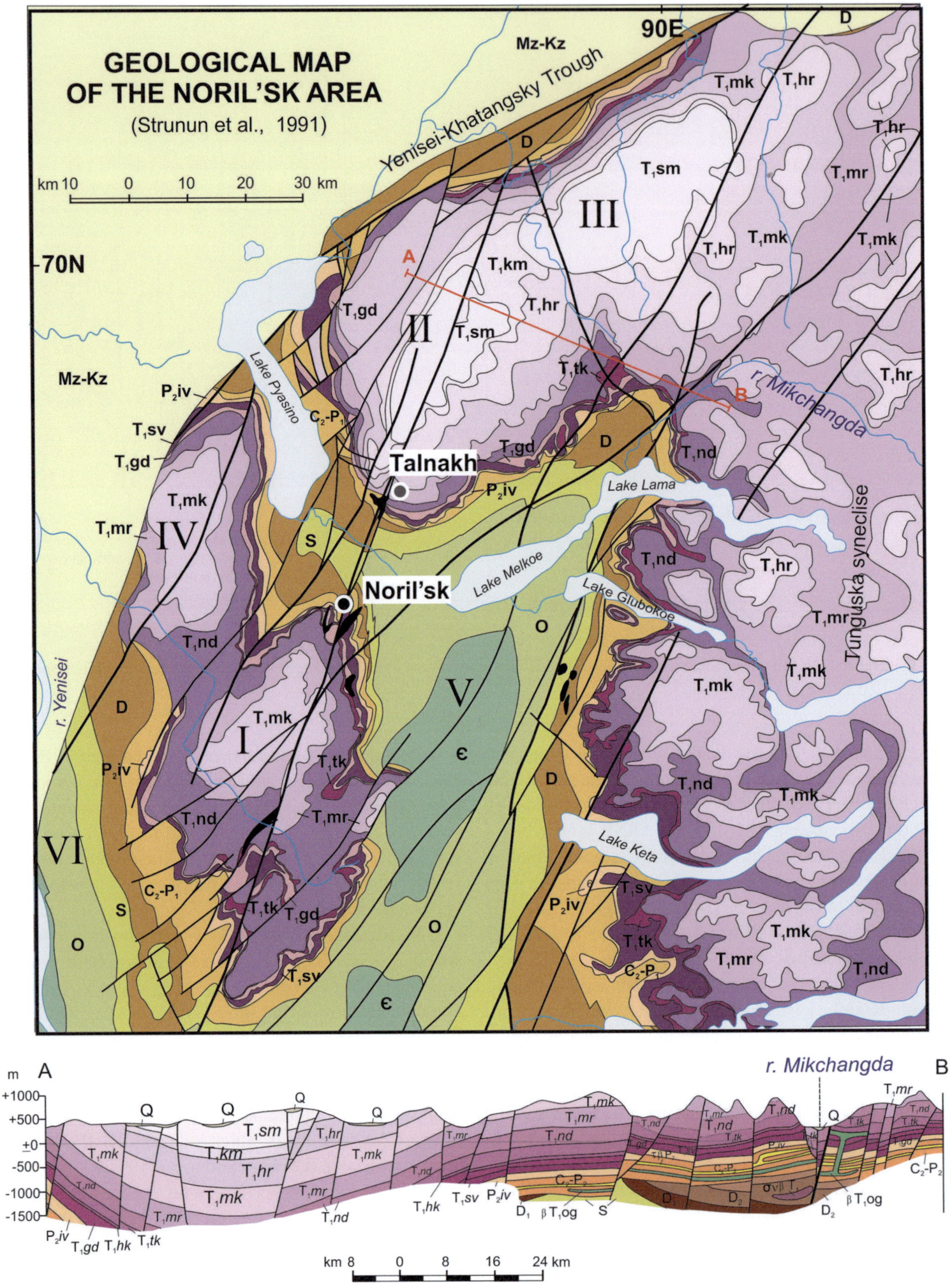

Fig. 2.4 Geological map of the Noril'sk area Modified after Geological map… 1994
Depressions: I—Noril'sk, II—Kharaelakhsky, III—Kumginsky; IV—Yeniseysky; Swells: V—Khantaysko-Rybninsky, VI—Dudinsky; AB—vertical section line; 1–2—faults: 1—Noril'sk-Kharaelakh, 2—Lamsko-Letninsky

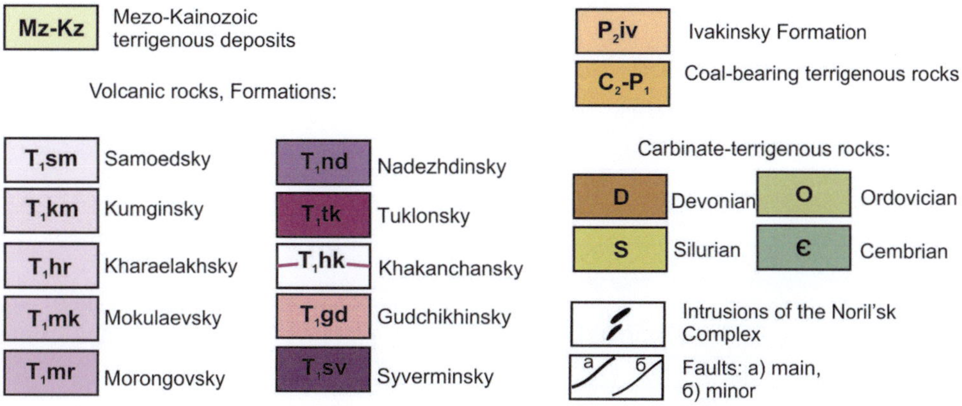

Fig. 2.4 (continued)

printed depressions filled with volcanic rocks are 2–3 km deep. The topography of the top of the Moho is correlated with the types of the surface structures. For example, the crust bottom occurs at a depth of 35 km beneath the central part of the Kharaelakh depression and at a depth of 38 km beneath the Khantaysko–Rybninsky swell.

In the Igarka–Noril'sk block, a very important role is played by deep faults, which predetermine the metallogenic specifics of this block. The Noril'sk–Kharaelakh Fault is one of the most important ones, because it is accompanied by large PGE–Ni–Cu deposits (Noril'sk, Talnakh, and Oktyabr'skoe). The fault trends to the north–northeast for more than 350 km, and the fault zone is 10–20 km wide. The fault is traced in the magnetic field and reaches the mantle according to deep seismic data. The fault is splayed by minor northeast-trending (Mikchangdinsky, Ergalakhsky, and others) and northwest-trending (Talnakh graben zone) faults across throughout its whole length. The main suture of the Noril'sk–Kharaelakh Fault is a mylonite and tectonic breccia zone up to 100 m wide. The vertical displacement along the zone is 400–1,000 m, and the horizontal displacement ranges from 40 to 200 m due to its downdip–strike-slip nature. The southern continuation of the Noril'sk–Kharaelakh Fault is the Igarka–Sukharikhinsky Fault, which is a splay of the large transcontinental Yenisei Fault at the boundary between the Igarka–Noril'sk rift system and the West Siberian Basin.

The another major fault in this territory is the long-lived Lamsko–Letninsky fault. It was reactivated multiple times during rifting in the Riphean, Vendian, Devonian, Late Paleozoic, and Early Mesozoic. The fault is pronounced in the modern structure as a zone (10-1,020 km length) of normal faults, overthrusts, fault-line folds, and grabens, along which changes in the facies and thicknesses of the Phanerozoic rocks can be seen or observed. The fault controls PGE–Ni–Cu deposits (Imangda and others) and the occurrences of magnetite mineralization (Makus).

2.2.2 Stratified Rocks

The platform cover in the Noril'sk region has a two-part structure and is more than 12 km thick. No basement rocks are exposed anywhere in the Noril'sk area (Geology and ore… 1994), and the nature of these rocks can be inferred from fragments of muscovite leucogranites found in the Early Triassic pyroclastic material occurring 20 km south of Noril'sk, with these fragments dated at 1,700–2,200 Ma. Early Proterozoic (?) sedimentary–volcanic and molasse-like rocks from the Riphean folded structural floor, low-grade metamorphosed terrigenous-carbonate and volcanic rocks from the Riphean structural floor, and molasse-like rocks from the Early Vendian floor that were deformed into gently dipping folds were recovered by boreholes in some blocks.

2.2.2.1 Riphean Rocks

At the base are hard, dislocated, weakly metamorphosed sedimentary and volcanic rocks of the Riphean region. The total thickness is approximately 3,500 m (Fig. 2.5). This sequence consists of (from bottom to top) the following rock types: sericite–chlorite schists of the Plakhinsky Formation (Rph) with a thickness more than 700 m; red-colored quartz sandstones of the Gubinsky Formation (Rgb) with a thickness of approximately 1,000 m; dolomites of the Medvezhinsky Formation (Rmd) with a thickness of 1,300 m; and limestone, dolomite, and shale of the Chernorechensky Formation (Rcr, 500 m) (Dyuzhikov et al. 1988). Although these formations are not exposed in the Noril'sk mining district, they are exposed in the explosive volcano near the city of Noril'sk. The absolute age of these rocks is 1,300–1,700 Ma (Nd–Sm dating by the US Geological Survey, Geology and ore… 1994).

2.2.2.2 Vendian Through Lower Cambrian

The Vendian through Lower Cambrian deposits consists of the Rybninsky (Vrb, more than 720 m thick), Gremyakinsky

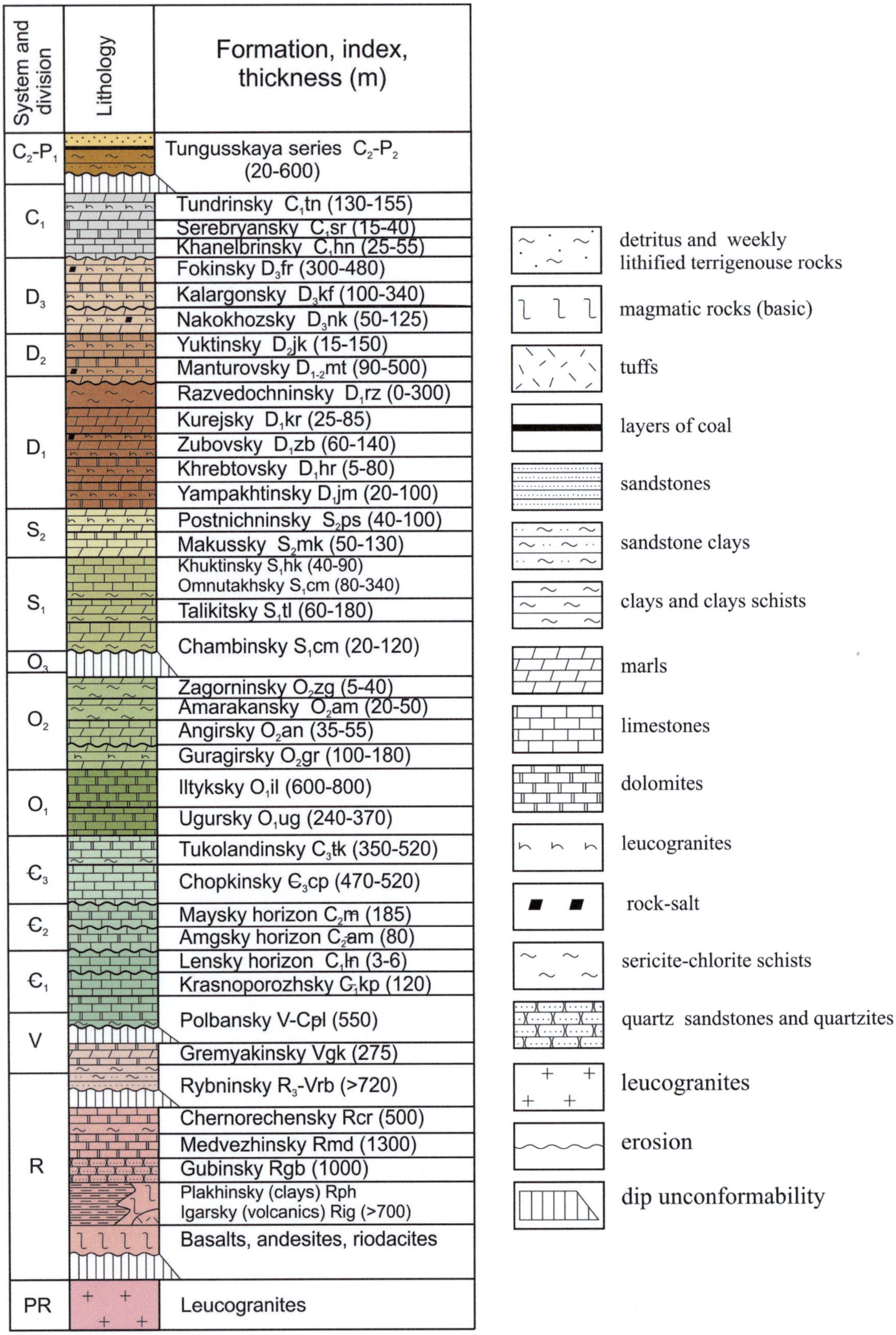

Fig. 2.5 Stratigraphic column for rocks of the Noril'sk area (Geological map… 1994)

(Vgk, thickness of approximately 275 m), and Polbansky (V-Є1pl, thickness of up to 550 m) Formations. The total thickness of this section gradually decreases from the west to the northeast from 1,500 to 420–480 m.

2.2.2.3 Cambrian System, Lower–Upper Division

V. Datsenko, V. Dragunov, V. Savitsky, V. Khomentovsky, and others studied the Cambrian section in this area in detail (Dyuzhikov et al. 1988). Cambrian deposits in the eastern part of this region are characterized by erosion of certain portions and lesser thickness (up to 1,500 m) in comparison with the western part; the most complete section is exposed in the Igarka district. This section consists of the Krasnoporozhsky, Ustbrussky, and Tukolandinsky Formations. The Krasnoporozhsky Formation ($Є_1$kp) overlaps the underlying sediments of the Polbansky suites and clay–carbonate rocks in the form of a wedge of dolomitic limestone containing abundant organic matter. The ustbrussky Formation ($Є_2$ub) unconformably overlies lower the Cambrian formations. The thickness of this formation varies from 150 m (r. Kulyumbe) to 250 m in the southeastern area. Upper Cambrian rocks are represented by two formations: the Chopkinsky ($Є_3$cp) Formation on the bottom and the Tukolandinsky ($Є_3$tk) Formation on the top. Their thicknesses gradually decrease to the northeast from 520 to 470 and 350 m, respectively. Both formations consist of sulfate-rich terrigenous-carbonate rocks.

2.2.2.4 Lower–Middle Ordovician System

Ordovician sediments are present everywhere in the Noril'sk mining district, although they are exposed primarily along the periphery of the Khantaysko–Rybninsky area and the central part of the Dudinsky swell (Fig. 2.4). These deposits are represented by the Uighur, Iltyksky, Guragirsky, Angirsky, Amarakansky, and Zagorninsky Formations (Tesakov 2012). The Uighur Formation (O_1ug) consists of sulfate-rich terrigenous-carbonate rocks with a total thickness of approximately 350 m. The Iltyksky Formation (O_1il), which includes dolomite–limestone and shale–limestone–dolomite deposits, has a thickness of 600 to 800 m. The Guragirsky Formation (O_2gr) is similar in composition to the previously mentioned units and contains sulfate-rich terrigenous-carbonate deposits with sandstones and quartzites. The thickness of this formation ranges from 100 to 180 m. The Amarkansky Formation (O_2am) is characterized by a significantly higher degree of terrigenous input compared with the older (similar in composition) units and has a lesser thickness of up to 50 m. The Zagorninsky Formation (O_2zg) consists of marls, mudstones, and organogenic–detrital limestones. Its thickness ranges from 5 to 40 m depending on the extent of Silurian erosion.

2.2.2.5 Silurian System

The first systematic data regarding the Silurian System of the Noril'sk ore district were gathered by a group of authors in 1982 under the leadership of Yu. Tesakov (Tesakov 2012). N. Baptist studied the reference sections of the district near the Kulyumbe River. The Silurian sediments and Ordovician bedrock are exposed primarily within the Khantaysko–Rybninsky and Dudinsky swells, which bound the Noril'sk Trough on the east and south. Based on their lithological features and paleontological characteristics, the Silurian sediments were subdivided into the Chambinsky, Talikitsky, Omnutahsky, Lower Hyuktinsky, and Upper Makussky Formations. The Chambinsky Formation (S_1čm), which is of middle Landoverskiy age, consists of folded graptolitic mudstone interbedded with organic limestone concretions and pyrite. The thickness of this formation is 100–130 m. It overlies the older (Ordovician) sediments along a major unconformity. The Talikitsky (S_1tl) and Omnutakhsky (S_1om) Formations (middle Late Llandovery time) consist of shale–limestone with a predominance of organic limestone. The thickness of the formation is 205–300 m in the Noril'sk area to 330 m in the Kulyumbe basin (south of the Noril'sk ore district). The Hyuktinsky Formation (S_1hk) consists of massive coral-stromatoporoids limestones with chert nodules throughout and argillaceous beds near the top of the unit. Its thickness ranges from 66 to 90 m near Noril'sk to 130 m in the Kulyumbe basin. Its contacts with the overlying and underlying deposits are comfortable. The Makussky Formation (S_2mk), which is of Ludlow time, is characterized by a complex of limestone with elements of the Rifoidnoy Formation and substantial amounts of dolomite and sulfates at the top of the section. The thickness of this formation is 80–100 m. The Postnichny Formation (S_2ps), of late Ludlow time, consists of the interlayered limestones, dolomites, marls, and anhydrite. In the swells (Dudinsky and Khantaysko–Rybninsky), the total thickness of these rocks does not exceed 75–80 m. The Postnichny Formation conformably overlies the underlying Makussky sediments.

2.2.2.6 Devonian System

The Devonian sediments outcrop as narrow exposures flanking the northern and southern parts of the Noril'sk and Kharaelakh Troughs, west of the Tungusskaya syneclise (Figs. 2.4 and 2.5). The Devonian rocks overlie the Silurian rocks and are most complete in the Lake Lama and Lake Pyasino areas, the Fokina basin, and east of the Imagda-Letninsky deep fault zone (south of Lake Lama and extending to Lake Keta). Based on their lithological features and paleontological characteristics, the Devonian rocks were subdivided into the Yampaktinsky, Khrebtovsky, Zubovsky, Kureysky, and Razvedochninsky Formations (lower part), the Manturovsky and Yuktinsky Formations (middle part), and the Nakohozsky, Kalargonsky and Fokinsky Formations (upper part). The composition of the Devonian sediments is of great importance in understanding the geological structure of the area because nearly all intrusions of the Noril'sk Complex are localized within them. The Yampaktinsky (D_1jm) and Khrebtovsky (D_1hr) Formations consist of similar carbonate rocks interbedded with abundant gypsum and anhydrite, among which are the oldest lenses of celestite in the area. The total thicknesses of these two units are 100 and 80 m, respectively. The Zubovsky Formation (D_1zb) is composed of gray-colored dolomitic marls interbedded with argillaceous dolomites, mudstones, and anhydrite with a total thickness of 100–150 m. The Zubovsky Formation unconformably overlies the Khrebtovsky Formation in the Noril'sk region. The Kureysky Formation (D_1kr) consists of mottled dolomite and calcareous mudstones and marls with rare siltstone and limestone. The thicknesses of all units in this section remain essentially constant and vary within 50–60 m. The contacts with the overlying and underlying formations are conformable.

The Razvedochninsky Formation (D_1rz) is dominated by siltstones, sandstones, and conglomeratic sandstones that comprise a transgressive microcycle. The thickness of these deposits does not exceed 110–150 m, reaches 150–235 m in the troughs, and decreases sharply to the south until fully wedging out. The Manturovsky Formation (D_2mt) overlies the eroded Razvedochninsky Formation and consists of a terrigenous-carbonate section with abundant salt-bearing strata, most of which consist of rock salt. This formation's thickness is 500 m. The Yuktinsky section (D_2jk) is dominated by clastic–carbonate sediments ranging from 12 to 40 m thick; in the troughs, the thickness of sulfate rocks reaches 55 m. The contacts with the underlying and overlying Manturovsky deposits are comfortable. The Nakokhozsky Formation (D_3nk) consists of folded sulfate-rich variegated shale–carbonate rocks with a thickness of 2–60 m that increases in the troughs to 80–130 m. The Kalargonsky Formation (D_3kl) is characterized by a gray-colored terrigenous-carbonate section that includes dolomites, dolomitic marl, dolomite–limestone, and anhydrite dominate in the basins. This formation's thickness is 170–270 m. The Kalargonsky Formation unconformably overlies the Nakokhozsky sediments. The Fokinsky Formation (D_3fk) consists of evaporite sulfate-rich clastic–carbonate sequences, primarily within the troughs, and anhydrite, dolomitic marls interbedded with limestone lenses of rock salt, and clay–carbonate breccias. The thickness of this formation is 220–420 m (approximately 500 m in the western part of the Vologochansky Trough).

2.2.2.7 Lower Carboniferous System

The Lower Carboniferous deposits overlie the Late Devonian formations along an erosional unconformity and are exposed in small areas along the peripheries of the Noril'sk, Vologochansky and Kharaelakh Troughs. These deposits include the Hanelbrinsky, Serebryansky, and Tundrinsky Formations. The Hanelbrinsky (C_1hn) and Serebryansky (C_1sr) Formations consist of folded detrital organic limestone with minor lenses of chert. In the Noril'sk area, these rocks are preserved only in troughs, and their total thickness does not exceed 65–80 m. The Tundrinsky Formation (C_1tn) consists of marls interbedded with gypsum and anhydrite; the degree of terrigenous input increases northward. Its maximum thickness is 150 m west of Noril'sk and reaches 70–114 m in the Vologochansky Trough. It conformably overlies the Serebryansky Formation.

2.2.2.8 Middle Carboniferous System Through Upper Permian System

Rocks of the Middle Carboniferous through Upper Permian consist of the Tunguska Group (C_2–P_2), which overlies the Lower Carboniferous rocks along a prominent angular unconformity. The magnitude of erosion represented by this unconformity varies by 1,000 m within a distance of 5–10 km. The Tunguska group consists of coal-bearing paralic and lacustrine rocks. In composition they vary in degree of granularity: mudstones, siltstones, coarse conglomeratic sandstone, rare thin layers of conglomerates with layers of coals to anthracite. The thickness of the Tunguska Group varies from 20 to 40 m and reaches 600 m in the axial part of the basins. These rocks form large coal basin. Coal is explored in Noril'sk region in Kaerkan coal section (Fig. 2.6a, b). Productive coal horizons underlay basalts of the Ivakinsky volcanic Formation, but there is one coal layer located between two basalt flows, named "Zametnyi," i.e., this is a huge (>1 km length) coal xenolite in basalt, in its upper part (Fig. 2.6c). Main rock types are shown on pictures Fig. 2.6d-f.

The modern stratigraphic subdivision of the Tunguska Group consists of seven suites: the Apsekansky (C_2ap, 0–25 m), Adylkansky (C_{2-3}ad, 0–80 m), Rudninsky (C_3–P_1rd, 0–80 m), Daldykansky (P_1dl, 0–60 m), Shmidtinsky (P_2sm, 8–65 m), Kayerkansky (P_2kr, 15–85 m), and Ambarninsky (P_2am, 0–70 m) suites. All of these suites are separated in one way or another by significant erosional unconformities. Layers of coal in economic quantities are present in the Daldykansky, Shmidtinsky, and Kayerkansky suites. At the top of the Tunguska Group, there is an admixture of pyroclastic material, which is the most prominent feature in the Ambarninsky suite.

Fig. 2.6 Photos of the Tunguska Group rocks
(**a**) coal open pit "KUR 2," (**b**) main coal horizon, (**c**) coal horizon between two basalt flows of the Ivakinsky Formation ("Plast Zametny"), (**d**, **e**) interlayering of conglomerates and sandstones, (**f**) fragments of coal in sandstones

Fig. 2.7 Outcrop of intrusive rocks of the Daldykansky Complex (Mount "Zub")

The *volcanic sequence* is subdivided into 11 formations (from bottom to top): the high-Ti (>2–3 wt % TiO_2 subalkaline and picrite basalts) Ivakinsky, Syverminsky, and Gudchikhinsky Formations and low-Ti (≤1 wt % TiO_2, tholeiites) Khakanchansky, Tuklonsky, Nadezhdinsky, Morongovsky, Mokulaevsky, Kharaelakhsky, Kumginsky, and Samoedsky Formations (Geological map … 1994; Fig. 2.4).

2.2.3 Intrusive Rocks

Most of the ultrabasite–basite intrusions are gently dipping bodies unconformably connected to the host terrigenous-sedimentary and volcanic rocks. It is usually fairly difficult to correlate them with the lavas because contacts between these rocks are rare and the intrusions mostly occur at stratigraphically lower levels than the volcanic rocks, which is largely due to the lithological control of their emplacement (the intrusive rocks are hosted predominantly in the Devonian rocks).

All the intrusive rocks in the Noril'sk area are subdivided into four main complexes. They include reliably identified subalkaline massifs, which are ascribed to the Ergalakhsky Complex and can be correlated with the rocks from the Ivakinsky Formation. The normal alkalinity ultrabasite–basite rocks include rocks from the Daldykansky Complex (Fig. 2.7) which cut the basalts of the upper formations and are close to them in composition because of elevated Ti concentrations. These rocks were likely the youngest in the Mokulaevsky–Samoedsky magmatic episode.

It is hard to attribute all the other numerous and diverse basite rocks to any given complexes or types because of the similarities between their compositions and their uncertain relationships with the lavas. Each rock type was produced within a brief time span (<1 Ma; Kamo et al. 2003; Pavlov et al. 2007), which makes it difficult to date them by isotopic geochemical techniques. The numerous massifs of the Noril'sk area are classified into complexes and subcomplexes (types) based on the structural–textural and chemical features of the rocks and their distribution in the vertical section. The currently adopted classification is underlain by the degree of differentiation of the intrusive bodies (Geology and ore… 1994). For example, numerous (from several dozens to a few hundred) variably differentiated and mineralized massifs with elevated MgO concentrations (8–16 wt % MgO) are subdivided in compliance with this principle into the following types according to their inner structures (lower and upper units of the rocks): Noril'sk (picritic gabbro–dolerites – diorites), lower Talnakh (picritic gabbro–dolerites – olivine gabbro–dolerites), Zubovsky (troctolites – gabbro–dolerites), and Kruglogorsky (olivine gabbro–dolerites – leucogabbro). The weakly differentiated or undifferentiated intrusions are attributed to the Ogonersky Complex. However, this classification of the massifs mostly reflects the crystallization circumstances of their magmas rather than their origin.

The massifs of the Noril'sk ore-bearing complex are typically differentiated (the weighted mean MgO concentration is 10–12 wt %) and variably mineralized. The three chonolith-shaped (up to 12 km long, 2–4 km wide, and from 20–50 to 350 m thick) Noril'sk 1, Talnakh, and Kharaelakh intrusion bodies host uniquely large ore reserves. From bottom to top, intrusions consist of gabbro-dolerites (contactic, taxitic, picritic, olivine and olivine-bearing, olivine-free, upper taxitic, sporadically occurring upper picritic, and upper contact), leucogabbro, and gabbro-diorite. Massive and disseminated ore mineralization is partially constrained to the contacts of the massifs with their host rocks (Godlevsky 1959; Likhachev 1994, 1996, 2006; Genkin et al. 1981; Distler et al. 1988; Ryabov et al. 2014; and others). The Noril'sk deposits are distinguished by the thick metamorphic and metasomatic aureoles around them, with the thicknesses of these aureoles exceeding those of the intrusions themselves (Turovtsev 2002).

2.2.4 Ores of the Noril'sk Deposits

The Noril'sk deposits contain ores of two types: PGE–Cu–Ni sulfide and PGE low sulfide. The latter are hosted in the upper portions of the massifs, in the leucogabbro, the magmatic breccia, and occasionally the picritic gabbro-dolerite units. These ores are typical of the Noril'sk 1 deposit (Sluzhenikin et al. 1994).

The sulfide ores are contained in the contact zone of the intrusion and the host rocks from the lower contact zone (Geology and ore… 1994; Sluzhenikin et al. 2014), although they were also found in the upper inner and outer-contact zones (Oktyabr'skoe deposit). The ores come in two types: disseminated and massive, which occur in both the intrusions and the host rocks. An unusual feature of the mineral deposits in the Noril'sk area is the inconsistency between the thicknesses of the intrusions and related ores: the volume of the ores can consist 15 % of the intrusions' volume; the thickness of massive ores can rich up to 54 m (Likhachev 1996). Given the low sulfur solubility in basaltic melts near the depths corresponding to those of the emplacement and crystallization of the intrusions in the Noril'sk Complex, it is hard to imagine how so much sulfur could be mobilized only from the parental magma that produced the intrusive bodies.

The mineralogy of the ores was documented in much detail by many researchers (Sukhanova 1968; Genkin 1968; Genkin et al. 1981; Distler et al. 1988, 1999; Komarova et al. 2002; Sluzhenikin et al. 2014 et ctr.). The major minerals are pyrrhotite (hexagonal and monoclinic pyrrhotite and troilite), pentlandite, and chalcopyrite-group minerals, which often compose individual ore types and were discovered at the Talnakh deposits by A.A. Filimonova (talnakhite, putoranite, and Ni-bearing putoranite). The Noril'sk ores are distinguished for their great diversity of rare minerals, which have never been found at any other Cu–Ni deposits, and the huge size of PGM. Studies of these ores and the discovery of new minerals in them were contributed by A.D. Genkin (six new minerals), T.L. Evstigneeva (four new minerals), A.A. Filimoniva, V.A. Kovalenker, and other researchers (see Chap. 4).

The three largest currently developed deposits in the Noril'sk district are Oktyabr'skoe, Talnakh, and Noril'sk 1.

References

Distler VV, Grokhovskaya TL, Evstigneeva TL, Sluzhenikin SF, Filimonova AA, Dyuzhikov OA (1988) Petrology of magmatic sulfide ore formation. Nauka, Moscow, 232 p (in Russian)

Distler VV, Sluzhenikin SF, Cabri LJ, Krivolutskaya NA, Turovtsev DM, Golovanova TA et al (1999) Platinum ores of the Noril'sk layered intrusions: the proportion of the magma and fluid concentration. Geol Ore Depos 41:241–265

Dolgal' AS (2012) Implementation of some of the ideas of Academician VN Strakhov in practice interpreting geopotential fields. In: Academician VN Strakhov as a mathematics and Geophysics. Nauka, Moscow, pp 55–78

Dyuzhikov OA, Distler VV, Strunin BM et al (1988) Geology and ore potential of the Noril'sk ore district. Nauka, Moscow. Dyuzhikov OA, Distler VV, Strunin BM, Mkrtychyan AK, ShermanML, Sluzhenikin SF, Lurye AM (1992) Geology and metallogeny of sulfide deposits Noril'sk region USSR (trans). Econ Geol Monogr. Ontario, Spec vol 1, 241 p

Genkin AD (1968) Minerals of platinum metals in ores of the Noril'sk 1 deposit. Nauka, Moscow (in Russian)

Genkin AD, Distler VV, Gladyshev GD et al (1981) Sulfide copper–nickel ores of the Noril'sk deposits. Nauka, Moscow (in Russian)

Geology and ore deposits of the Noril'sk region (1994) In Distler VV, Kunilov VE (eds) Guidebook of the VII international platinum symposium Moscow-Noril'sk. Moskovsky Contact Press, Moscow.

Geological map of the Noril'sk ore district 1:200000 scale. 1994. Ed. B.M. Strunin. St.Peterburg. VSEGEI (in Russian)

Godlevsky MN (1959) Traps and ore-bearing intrusions of the Noril'sk district. Gosgeoltekhizdat, Moscow, 61 p (in Russian)

Kamo SL, Czamanske GK, Amelin Y et al (2003) Rapid eruption of Siberian flood-volcanic rocks and evidence for coincidence with the Permian-Triassic boundary and mass extinction at 251 Ma. Earth Planet Sci Lett 214:75–91

Komarova MZ, Kozyrev SM, Simonov ON, Lyul'ko VA (2002) The PGE mineralization of disseminated sulfide ores of the Noril'sk-Taimyr region. In: Cabri LJ (ed) The geology, geochemistry, mineralogy and mineral beneficiation of platinum-group elements, vol 54. Canadian Institute of Mining, Metallurgy and Petroleum, Westmount, pp 547–567 (in Russian)

Kostyuchenko SL (2006) Structure of the crust and surface mechanisms of formation of near-arctic continental sedimentary basins in Siberia. Reg Geol Metallog 10:125–135 (in Russian)

Kravtsov VF (2003) History discoveries of copper-nickel deposits in the Noril'sk area. Essays on the history of discoveries of mineral deposits of Taimyr. Publishing House of SB RAS, Novosibirsk, pp 21–40 (in Russian)

Likhachev AP (1994) Ore-bearing intrusions of the Noril'sk region. Proceeding of the Sudbury–Noril'sk symposium, Ontario. Geol Surv Spec 5:185–201

Likhachev AP (1996) The Kharaelakh intrusion and its PGM–Cu–Ni ores, Rudy et Metals, no 3, pp 48–62 (in Russian)

Likhachev AP (2006) Platinum–copper–nickel and platinum deposits. Eslan, Moscow, 496 p (in Russian)

Malitch NS, Mironyuk EP, Tuganova EV, Grinson AS, Masaitis VL, Surkov VS et al (1987) Geological structure of the USSR and regularities of distribution of mineral deposits. In: Malitch NS, Masaitis VL, Surkov VS (eds) The Siberian platform 4. Nedra Press, Leningrad (in Russian)

Malitch NS, Grinson AS, Tuganova EV, Chernyshev NM (1988) Rifting of the Siberian platform. In: 28th session of International Geological Congress. Tectonic processes. Nauka Press, Moscow, pp 184–193 (in Russian)

Naldrett AJ (2004) Magmatic sulphide deposits: geology, geochemistry and exploration. Springer, Berlin/Heidelberg/New York, 727 p

Naldrett AJ (2005) A history of our understanding of magmatic Ni-Cu sulfide deposits. Can Mineral 43:2069–2098

Pavlov VE, Courtillot V, Bazhenov ML, Veselovsky RV (2007) Paleomagnetism of the Siberian traps: new data and a new overall 250 Ma pole for Siberia. Tectonophysics 443:72–92

Proceeding of the Sudbury–Noril'sk symposium Ontario (1994) Geol Surv Spec 5:185–201

Ryabov VV, Shevko AYa, Gora MP (2000) Igneous rocks of the Noril'sk district. V.1. Petrology of traps. Nonparel, Novosibirsk, vol 1, 2 (in Russian)

Ryabov VV, Shevko AYa, Gora MP (2014) Trap magmatism and ore formation in the Siberian Noril'sk region,V 1, 2. Springer

Sluzhenikin SF, Distler VV, Dyuzhikov OA, Kravtsov VE, Kunilov VE, Laputina IP, Turovtsev DM (1994) Low sulfide platinum mineralization in the Noril'sk differentiated intrusive bodies. Geol Ore Deposit 36:195–217

Sluzhenikin SF, Krivolutskaya NA, Rad'ko VA, Malitch KN, Distler VV, Fedorenko VA (2014) Ultramafic-mafic intrusions, volcanic rocks and PGE-Cu-Ni sulfide deposits of the Noril'sk Province, Polar Siberia. Field trip guidebook. IGG UB RAS, Yekaterinburg, 87 p

Sotnikov AA (1919) About operation of the Noril'sk (Dudinsky) deposits of coal and copper ore in connection with the practical implementation and development of the Northern Sea Route. Tomsk, 98 p (in Russian)

Sukhanova EN (1968) Zoning of orebodies, intrusions, and tectonomagmatic clusters and its applied implications. In: Geology and mineral resources of the Noril'sk mining district. NTO Tsvetmet, Noril'sk, pp 139–142 (in Russian)

Tesakov YuA (2012) The Silurian basin of the east Siberia. IBGG SB, Novosibirsk vol 1–4 (in Russian)

Turovtsev DM (2002) Contact metamorphism of the Noril'sk intrusions. Nauchny Mir, Moscow, 293 p (in Russian)

Urvantsev NN (1927) Exploration of Cu-Ni Noril'sk deposit. Izvestiya Geol Committee 44(2) (in Russian)

3 Specifics of the Tuff–Lava Sequence: Geological and Geochemical Evidences

The history, subdivision, and characteristics of the volcanic rocks (251 Ma) of the Noril'sk region may be described thusly: the structures of the extrusive rocks differ from one tectonic element to the next (the Kharaelakh and Noril'sk Troughs and the Tunguska syneclise). Many of the basal basalts were studied, particularly with regard to their concentrations of major and rare elements. The lower formations (Ivakinsky, Syverminsky, Gudchikhinsky, and Nadezhdinsky) are located only in the paleorift zone (including Yenisei–Khatangsky depression), whereas the upper formations (Tuklonsky, Morongovsky, Mokulaevsky, Kharaelakhsky, Kumginsky, Samoedsky) are widespread in the Tunguska syneclise. Based on this distribution of volcanic rocks, two main stages of their formation were suggested. Based on their geochemical features, the widespread basalts were deposited in four cycles, and thus, a new schema of their magmatic evolution was constructed.

The Noril'sk ore district sits within the world's largest Siberian trap province, which is one of the largest igneous provinces on the Earth. Large igneous provinces (LIP) are short-lived (usually for no more than a few million years), simultaneous manifestations of magmatic activity that are spread over vast territories on continents or the seafloor (Campbell and Griffiths 1992). Typical LIPs are continental flood basalt (trap) provinces, such as the Siberian, Parana, and Deccan provinces, and ocean plateaus, such as Ontong Java.

The origin of LIPs was explained within the scopes of diverse hypotheses (Sobolev et al. 2009), some of which can be mutually inconsistent. The two currently most widely accepted hypotheses are as follows. According to the model of thermal mantle plumes, LIPs are produced by plume heads via high decompressional melting of mantle peridotite caused by high potential temperatures (White and McKenzie 1995; Campbell and Griffiths 1992; Dobretsov et al. 2008). The proponents of the other hypothesis, which claims to realistically account for the origin of continental LIPs, suggest the delamination and subsidence of the lithosphere and the subsequent filling of the open space with hot convecting mantle material, which had also been affected by high decompressional melting (see, e.g., Elkins-Tanton 2005). The main disadvantage of these hypotheses is that they cannot provide rigorous estimates for the circumstances under which the parental magmas are derived and for the compositions of their mantle sources.

The Siberian trap province is distinguished from LIPs by its grandiose size (it is the world's largest continental basalt LIP of Phanerozoic age; Masaitis 1983; Dobretsov 1997; Yarmolyuk et al. 2000; Reichow et al. 2005) and because its age coincides (within the accuracies of multiple dates) with one of the most significant mass extinction episodes in the Earth's evolutionary history at 251 Ma, i.e., at the boundary between the Paleozoic and Mesozoic eras. Thus, the origin of this province was suggested to be one of the reasons for this catastrophic extinction (Campbell et al. 1992; Kamo et al. 2003; White and Saunders 2005; Ivanov et al. 2013). Finally, the northwestern portion of this province hosts the world's largest PGE-Cu-Ni sulfide deposits (Dodin et al. 1971; Dyuzhikov et al. 1988; Geology and ore… 1994; Naldrett 2004). These unique features of the Siberian traps have yet to be explained within the scope of a single coherent genetic model. The Siberian province encompasses most of the East Siberian Platform, part of the Taimyr Peninsula, and part of the West Siberian Plate (Fig. 3.1). Drilling through the last location resulted in the discovery of deep-sitting (at a depth of 3–7 km) volcanic rocks beneath the sedimentary cover (Masaitis 1983; Al'mukhamedov et al. 2000; Medvedev 2004).

The East Siberian Platform was intensely studied in the 1960s and 1970s, when systematic 1:1,000,000, 1:200,000, and 1:50,000 (at selected areas) survey operations were conducted. Because it is difficult to access the territory and because it was surveyed and studied by various organizations, which did not commonly use state-of-the-art techniques when analyzing rocks and minerals, the volcanic and plutonic rocks in various parts of the territory were subdivided differently

N.A. Krivolutskaya, *Siberian Traps and Pt-Cu-Ni Deposits in the Noril'sk Area*,
DOI 10.1007/978-3-319-17205-7_3

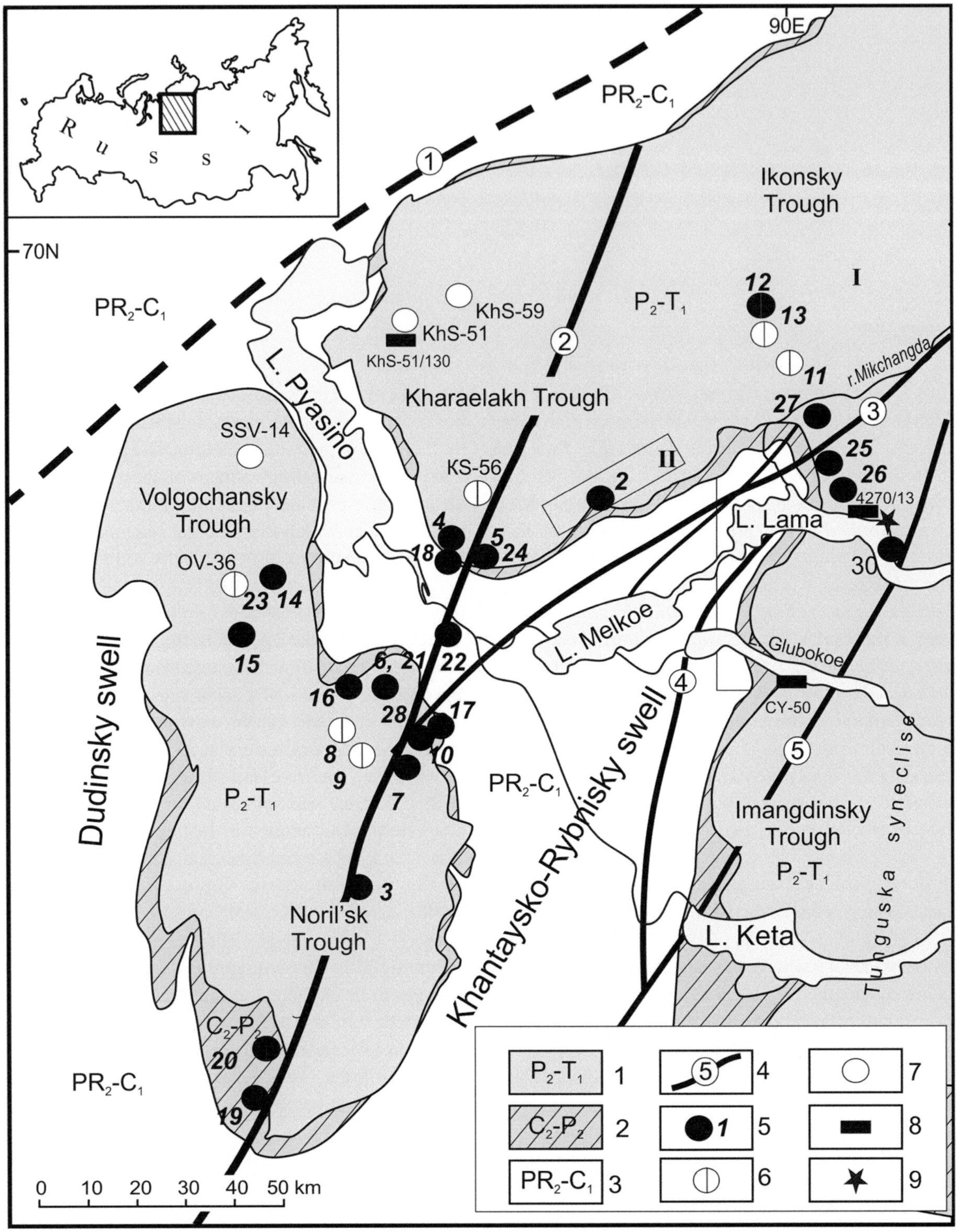

Fig. 3.1 Map of factual material: location of vertical sections in volcanic sequences shown in a schematic tectonic map of the Noril'sk area Modified after Mikhailov (2003, unpublished)
(1) Volcanic rocks of the trap association; (2) Tunguska Group; (3) terrigenous-carbonate rocks; (4) faults (*1*, Yenisei–Khatangsky; *2*, Noril'sk–Kharaelakh; *3*, Mikchangdinsky; *4*, Imangdinsky; *5*, Lamsko–Khatangsky); (5) intrusions and their numbers (see Table 4.1); (6, 7) boreholes in which rocks were studied: *6*, volcanic and intrusive rocks; *7*, volcanic rocks. *Rectangles mark* areas in which volcanic rocks were studied in much detail. (8) Sample sites of the Gudchikhinsky picrites in which melt inclusions were studied and their numbers; (9) location of the geological profile shown in Fig. 3.18. Boreholes SG-9 and SG-32 borrowed from (Lightfoot et al. 1990, 1993, 1994; Fedorenko et al. 1996)

and can hardly be correlated with one another. Only a few papers were published that attempted to synthesize and generalize these diverse materials. Such publications include those by V. Zolotukhin et al. (1978), M. Sharma (1997), and A. Ivanov et al. (2013) and along with the published 1:1,000,000 state map. Most materials related to the geology of this territory are deposited and classified.

Recently obtained high-accuracy and high-precision data on the geochemistry of the rocks in the Noril'sk and Meimecha-Kotuy areas (Fedorenko 1981; Fedorenko et al. 1989, 1996; Lightfoot et al. 1990, 1993, 1994; Arndt et al. 1993, 1998; Fedorenko and Czamanske 1997) led to the principal revision of earlier concepts regarding the structure and evolutionary history of trap magmatism, but the whole situation is still largely obscure because the formations distinguished in various tectonic zones have not yet been correlated based on modern geochemical methods. Previous correlations are not always reliable enough because they were based on the comparison of petrochemically different basalts and tuffs, which were not analyzed for trace elements. Reliable correlations can be drawn for only the subalkaline rocks from early volcanic episodes and some of the tholeiitic porphyritic basalts from the main phase. These materials are obviously insufficient for reproducing the whole evolutionary history of the volcanic processes within the entire province.

Thus, the overall trends and evolutionary relationships of trap magmatism in the area can be evaluated only based on detailed geological and geochemical studies of the volcanic and intrusive rocks in various folded structures. We applied this exact approach to study the volcanic rocks in the Noril'sk area. The results of these studies are presented below.

3.1 History and the Principal Problems in Studying of the Volcanic Rocks in the Noril'sk Area

The Noril'sk area is of paramount importance for understanding the evolutionary history of the territory because of its very thick volcanic rocks (3.7 km), the occurrence of subalkaline and picritic rocks in the vertical section (along with widespread tholeiites), and the presence of unique Pt–Cu–Ni deposits in ultrabasite–basite complexes. These facts facilitate the successful resolution of problems related to the origin of the vast volumes of magmatic rocks in certain crustal blocks and their role in the origin of the superlarge mineral deposits, which have no analogues anywhere else in the world in terms of the reserves and grade of the sulfide ores.

The Noril'sk area has been explored in more detail than the rest of the province because of the former's high ore potential. Its systematic study and exploration were launched in the 1940s–1950s by Yu. A. Speit, V. S. Domogarov, P. S. Fomin, N. A. Timashkov, and others. Their materials later provided a basis for sheets R-46,47 and R-44,45 of the 1:1,000,000 State Geological Map. The "philosophy" and principles for subdividing the rocks in the volcanic sequence were elaborated in the late 1950s and early 1960s during the regional geological survey and exploration. The major formations were distinguished by G. D. Maslov, D. A. Dodin, and Ya. I. Polkin. In 1993, the Editorial Board of Taimyr Geolkom approved *The Reference Legend for 1:50 000 Geological Map, Noril'sk Series*, which is now used as a reference in conducting all types of large-scale exploration operations in the Noril'sk area. The volcanic sequence is subdivided into 11 (progressing upward, Table 3.1) formations, including the high-Ti (TiO_2>2–3 wt %, subalkaline and picritic basalts) Ivakinsky (iv), Syverminsky (sv), and Gudchikhinsky (gd) and low-Ti ($TiO_2 \leq 1$ wt %, tholeiites) Khakanchansky (hk), Tuklonsky (tk), Nadezhdinsky (nd), Morongovsky (mr), Mokulaevsky (mk), Kharaelakhsky (hr), Kumginsky (km), and Samoedsky (sm) (Geological map… 1994). Their principal characteristics (petrography, petrochemistry, geochemistry, and mineralogical composition) are described in a number of publications (Dodin 1967; Dodin et al. 1971; Vasil'ev and Zolotukhin 1975; Zolotukhin et al. 1984; Ryabov et al. 2000, 2014). E. Lind and V. Shchekoturov (1991) determined that the rocks in the Ivakinsky Formation are reversely magnetized and, hence, were produced in the Late Permian, which was shown in the maps.

The tuff–lava sequences typically consist of alternating basaltic flows and beds of pyroclastic and sedimentary–volcanic material. The latter often contain fossil flora, which can be utilized to date the rocks. The tuff–lava sequence is subdivided into two parts: the lower part, comprising the Ivakinsky–Nadezhdinsky Formations and noted for the contrasting composition of its rocks (differentiated lava series of alkaline ultramafic rocks and rocks with normal alkalinity and elevated silicity), and the upper formation (volcanic rocks from the Tuklonsky to Samoedsky Formations with slightly varying chemical composition and corresponding rocks of normal alkalinity). The rocks in the lower unit, which are noted for their contrasting compositional variations, can be reliably subdivided into formations that can be easily identified in the field and mapped. Conversely, the rocks in the upper unit have closely similar petrographic, geochemical, textural, and structural characteristics; hence, it is difficult to further subdivide these rocks. Their subdivision into formations and subformations is not always supported by sound evidence because of the significant variations in the structure and composition of the upper unit along its strike.

Table 3.1 Schematic stratigraphic chart for volcanic rocks in the Noril'sk area

Age, symbol	Formation	Subformation	Thickness, m	Predominant rocks
T_1sm	Samoedsky	T_1sm	0–500	Aphyric, poikilophitic basalts, and tuffs
T_1km	Kumginsky	T_1km	0–200	Aphyric and porphyritic poikilophitic basalts and tuffs
T_1hr	Kharaelakhsky	$T_1kr_2^2$	125	Aphyric and porphyritic basalts and tuffs
		$T_1kr_2^1$	125	Aphyric basalts and tuffs
		$T_1kr_1^2$	80	Plagioglomeroporphyritic basalts
		$T_1kr_1^1$	185	Aphyric, glomeroporphyritic basalts
T_1mk	Mokulaevsky	T_1mk_2	285	Aphyric, glomeroporphyritic basalts and tuffs
		T_1mk_1	300	
T_1mr	Morongovsky	T_1mr_2	445	Aphyric, porphyritic, glomeroporphyritic basalts and tuff breccias
		T_1mr_1	215	
T_1nd	Nadezhdinsky	T_1nd_3	40–72	Glomeroporphyritic basalts
		T_1nd_2	170–250	Porphyritic basalts and tuffs
		T_1nd_1	160–350	Porphyritic and tholeiitic basalts and tuffs
T_1tk	Tuklonsky	T_1tk	0–250	Tholeiitic basalts
T_1hk	Khakanchansky	T_1hk	0.6–22	Tholeiitic basalts, tuffs, and tuffites
T_1gd	Gudchikhinsky	T_1gd_3	0–60	Glomeroporphyritic basalts
		T_1gd_2	0–140	Picrites, tuffs
		T_1gd_1	0–230	Porphyritic basalts and tuffs
T_1sv	Syverminsky	T_1sv	25–130	Tholeiitic basalts
P_2iv	Ivankinsky	P_2iv_2	72 – 100	Andesine basalts
		P_2iv_1	28 – 100	Labrador basalts

Note: Modified after Geological map… (1994)

The absence of reliable marker horizons makes virtually impossible to draw boundaries between individual formations. The only reliable reference key bed is a thick (20–100 m) flow of glomeroporphyritic basalt (Nadayansky flow) in the bottom portion of the Mokulaevsky Formation, which reliably marks the boundary between the Morongovsky and Mokulaevsky Formations in the eastern part of the Noril'sk area and in the Putorana Plateau. In its absence in the western part of the area, the boundary is drawn fairly arbitrarily. It is most difficult to draw boundaries between the intermediate formations: Khakanchansky, Tuklonsky, and Nadezhdinsky, all of which contain abundant tholeiites that occur above one another in the vertical section.

A new phase of studies, during which more precise and accurate data were obtained on the composition and structure of the volcanic rocks in the area, began when it became possible to apply modern geochemical techniques in geology: to study the distributions of trace elements in rocks and to analyze their isotopic characteristics. These studies were first undertaken by I.D. Volkov (1963), Yu.A. Balashov and G.S. Nesterenko (1966), G.S. Nesterenko et al. (1964, 1973, 1991), and others. However, most modern information was obtained over the past two decades (Lightfoot et al. 1990, 1993, 1994; Lightfoot and Keays 1995; Brügmann et al. 1993; Wooden et al. 1993; Howkesworth et al. 1995; Fedorenko et al. 1996; Fedorenko and Czamanske 1997; Kamo et al. 2003; Naldrett 2004; Keays and Lightfoot 2007, 2010). These data demonstrate that the lower and upper tuff–lava units have elevated (>2) and lower (<2) Gd/Yb ratios, respectively. It was established that compositionally similar tholeiites in the Tuklonsky and Nadezhdinsky Formations have principally different multielemental patterns, which more reliably distinguishes them. In addition, the lower portion of the Nadezhdinsky Formation is depleated in base metals than the other formations: for example, the Cu concentrations are 20–30 ppm in these rocks and 80–100 ppm in the rocks from other formations.

Regretfully, these studies were conducted in only the vertical section of the basalts in the Kharaelakh Trough (Boreholes SG-9 and SG-32, sections F15, F16), which was appended with the vertical section at Mount Sunduk (section 1F). At the same time, it was demonstrated (Dodin 1967; Zolotukhin et al. 1986; Staroseltsev 1989; Ryabov et al. 2000) that various structures in the territory consist of different rock sequences; hence, a comprehensive portrait of the magmatic evolution should be based on data from volcanic rocks in various tectonic structures of the territory. The results of such studies can lead researchers to principally different conclusions than those drawn from studies of single separated vertical sections. Examples of such inconsistencies were presented in our earlier publication (Krivolutskaya et al. 2009).

Multiple attempts were undertaken to date various rocks in the area. The very first dates were obtained by the ^{40}Ar–^{39}Ar technique on plagioclase and whole-rock samples from the Ivakinsky, Syverminsky, Gudchikhinsky, and Kharaelakhsky Formations (Renne and Basu 1991; Dalrymple et al. 1995; Campbell et al 1991). All these dates fall within the range of 243.5–245.3 Ma. Surprisingly, the ore-bearing Noril'sk 1 intrusion, which cuts through these formations, was dated at 248.7 ± 2.4 and 249.2 ± 2.4 Ma (Kamo et al. 2003).

The author studied volcanic rocks in the area from primary exposures (some field results were obtained together with geologists from Norilskgeologia during 1:50,000 survey operations conducted under the supervision of V. N. Mikhailov) and from the core material of boreholes (Fig. 3.1). Volcanic rocks were studied mostly in the eastern part of the area, which previously had been only cursorily studied with the application of modern techniques. This territory attracts much attention because it is adjacent to the northern pericline closure of the Khantaysko–Rybninsky swell, which divides the Noril'sk area into two structurally distinct blocks: the Noril'sk–Kharaelakh Trough in the west and the western part of the Tunguska syneclise in the east (Figs. 3.4 and 3.1). As will be demonstrated below, the magmatic evolution was different in these territories; hence, it is particularly interesting to study the structures of the volcanic sequence in the transitional zone from the rift structure to the platform because this territory yielded the most significant data on the volcanic rocks. The major relationships revealed during earlier studies were also confirmed for the eastern part of the Noril'sk area.

Many sections of volcanic rocks have been studied in detail across the Noril'sk region. In the sections each flow and tuff horizon were sampled and analyzed by XRF and ICP-MS. It is not a problem to distinguish individual basaltic flow because their amygdaloidal zones were kept in good conditions. Figure 3.2 demonstrates the layered structure of volcanic pile in the Noril'sk region. We observed these contacts and took samples from the central zone of flows. The examples of boundaries between different flows and tuffs are shown on picture (Fig. 3.3). Basalt's proper feature is columnar jointing especially typical for upper formations; rarely they have fan separateness (Fig. 3.4). Sometimes one can see different types of jointing in one flow, in the lower and the upper its parts (Fig. 3.4b). On the surface of lavas, tracks of flowing and gas channels exist (Fig. 3.5a-c). The surface is represented by lava breccia if the flow was fast and lava crushed the hard crust on the surface (Fig. 3.5d). Volcanoes of the central type are typical only for lower formation, Gudchikhinsky (Fig. 3.6a), Khakanchansky (Fig. 3.6b), and Syverminsky (Fig. 3.6c,d).

The major results can be summarized as follows: (1) the characteristics of the main formations were obtained based on the distribution of trace elements in the rocks; and (2) the vertical section consists of two units, reflecting variations in the composition of the tuff–lava sequence, which will be discussed below.

3.2 Data Obtained on the Volcanic Rocks in the Noril'sk Area

In describing the rocks, I present a detailed characterization of the vertical section, which was made together with A.V. Rudakova (Krivolutskaya and Rudakova 2009, Rudakoba and Krivolutskaya 2009) using the cores from boreholes drilled in the northern part of the Noril'sk depression when the Maslovsky deposit was explored. This vertical section has never been utilized in the studies of any other geochemists.

3.2.1 General Characterization of the Volcanic Rocks in the Area: An Example of the Noril'sk Trough (Lower Formations)

The Noril'sk Trough, whose volcanic rocks were not extensively studied geochemically, is of particular interest because it hosts numerous intrusive bodies of the Noril'sk Complex with disseminated or massive ores (Noril'sk 1, Noril'sk 2, Bolshaya Bar'ernaya, Chernogorsky, and Maslovsky deposits), which are hosted in volcanic rocks located at the highest stratigraphic levels in the Noril'sk area, namely, in the middle portion of the tuff–lava sequence. This geological situation makes it possible to approach the problem regarding the possible comagmatic character of the intrusive and volcanic rocks based on more reliable data because the geologic relationships between the gabbro-dolerites and volcanic rocks can be examined in a single section. In other folded structures, such massifs are hosted at stratigraphically lower levels, such as in the Devonian sedimentary rocks underneath the volcanic sequence (the Kharaelakh and part of the Talnakh intrusions with high-grade PGE-Cu-Ni ores are

Fig. 3.2 Structure of tuff–lava pile
Photos **b-d** of B. Gongalsky

Fig. 3.3 Contacts of the Kharaelaksky tuffs and Tuklonsky basalts (**a-c**) and between two basaltic flows with burnt zone (**d**, *crimson color* of rocks) Gas channels are *inside* bottom of basalts (**c, d**)

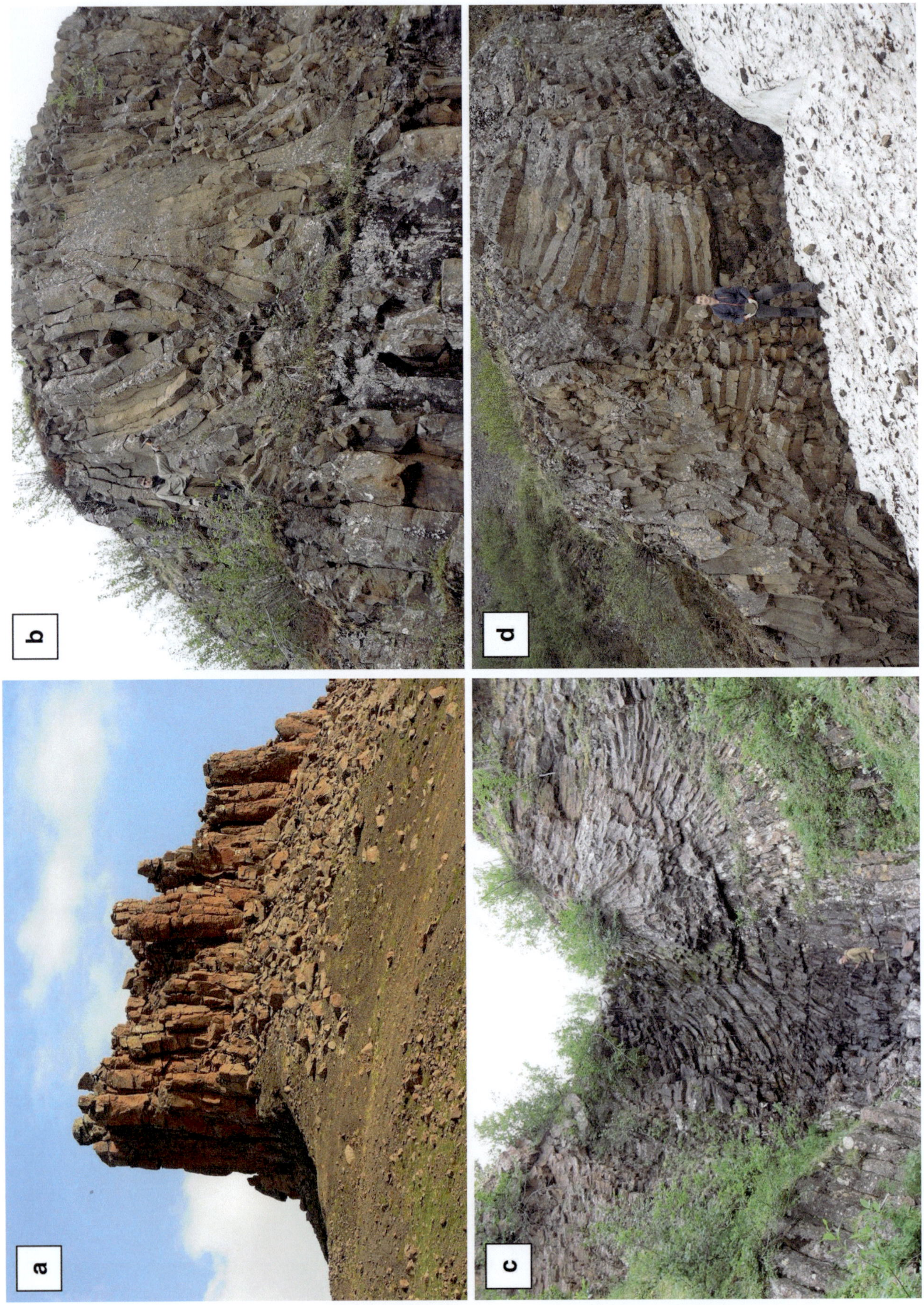

Fig. 3.4 Column (**a**) and fan (**b-d**) jointing in basalts
Formations: **a**—Mokulaevsky, **b**, **d**—Morongovsky, **c**—Ivakinsky

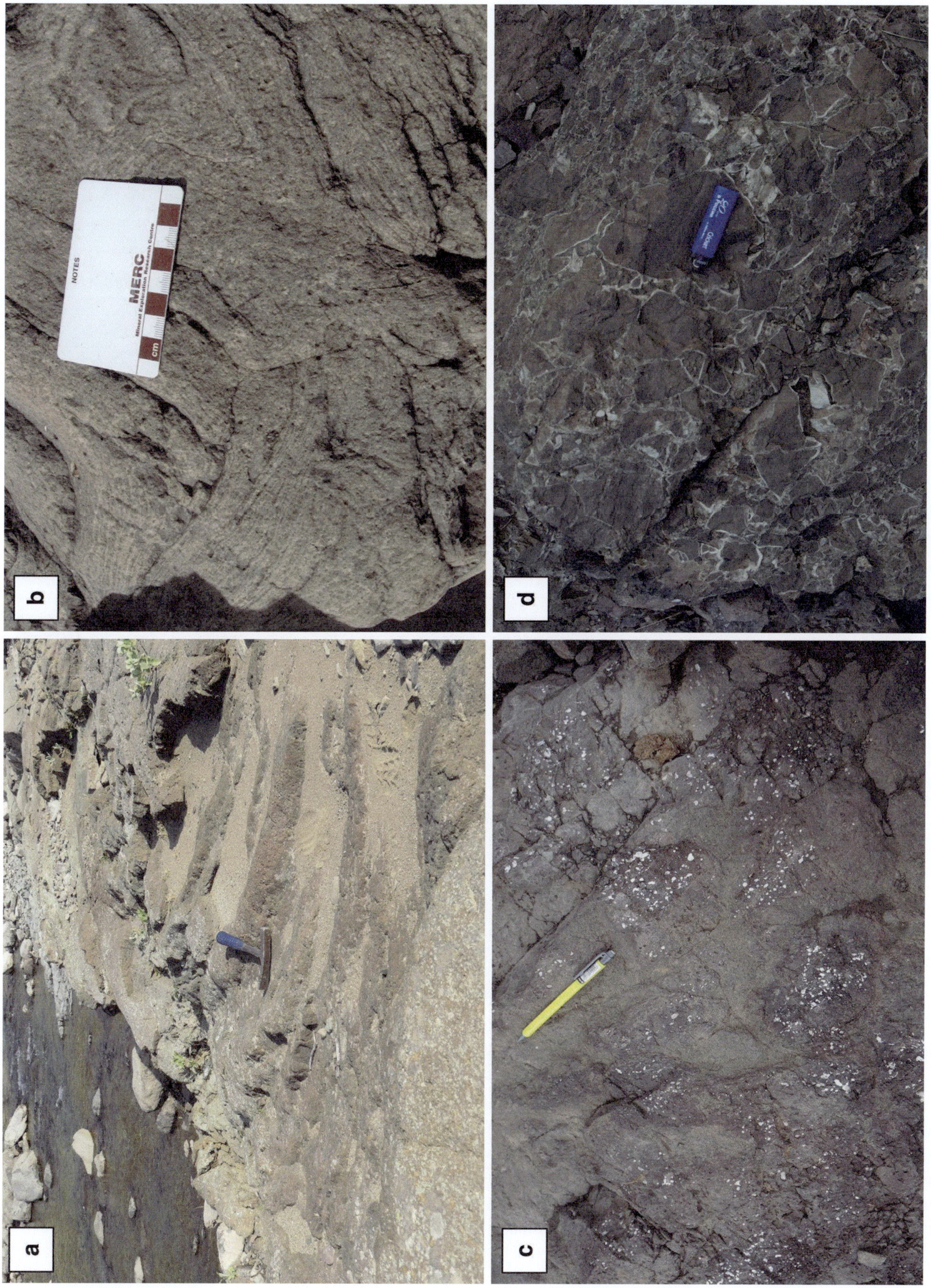

Fig. 3.5 Surface of lava flows
a, **b**—rope lava, **d**—lava breccia, **c**—exits of gas channel on the surface

Fig. 3.6 Volcanoes in the Siberian traps
Formations: (a) Gudchikhinsky, (b) Khakanchansky, (c-d) Syverminsky with bombs of the Ivakinsky basalts

located in the Kharaelakh Trough, and the Vologochansky–Pyasino intrusion with disseminated ore mineralization occurs in the Vologochansky Trough); hence, it is much more difficult to correlate them with the lavas. Moreover, the Noril'sk Trough contains rocks from all the formations that were used in models by various researchers describing the origin of the deposits (Gudchikhinsky, Tuklonsky, Nadezhdinsky, and Morongovsky), which allows one to analyze in detail the problem regarding their comagmatic relationships with intrusions from the Noril'sk Complex.

The volcanic sequences were examined in the core material of boreholes drilled to the north of the Noril'sk Trough and in exposures in the eastern part of the territory. Boreholes OM-6 and OM-25 were selected as references because they penetrated volcanic rocks 1,100 and 700 m thick in the central and eastern parts of the trough, respectively, and were drilled 5 km apart from each other (Fig. 3.1). This allowed us to trace the lateral structural and compositional variations in the rocks in the trough. An important criterion for choosing to study borehole OM-25 in detail was because the volcanic rocks host a thin sill from the Noril'sk Complex (an apophyse of the Maslovsky intrusion) with high-grade disseminated-stringer Cu–Ni mineralization. This offers the possibility of obtaining additional information on the genetic relationships between the intrusive and volcanic rocks in the area.

The detailed investigation of the volcanic rocks in the Noril'sk Trough revealed the presence of 7 of the 11 formations, representing the lower half of the complete basaltoid section of the Noril'sk region (Table 3.1). The thickest section of trap rocks was described in borehole OM-6 in the central part of the structure (Fig. 3.7), and a reduced section was observed in the eastern part of the trough (Fig. 3.8). The assignment of flows to particular formations was based

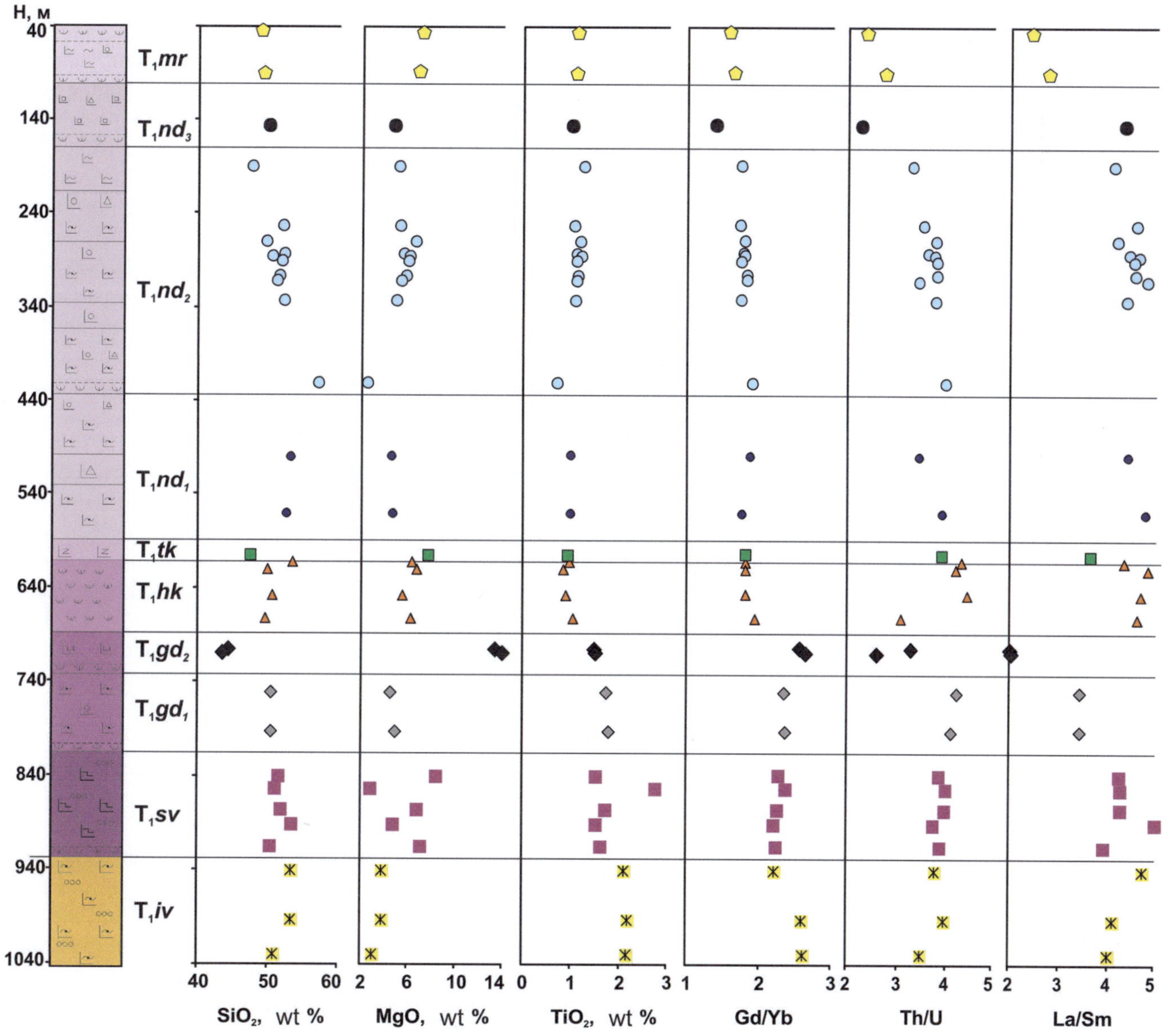

Fig. 3.7 Composite vertical section of the tuff–lava sequence in the Noril'sk Trough (borehole OM-6) After Krivolutskaya and Rudakova (2009)

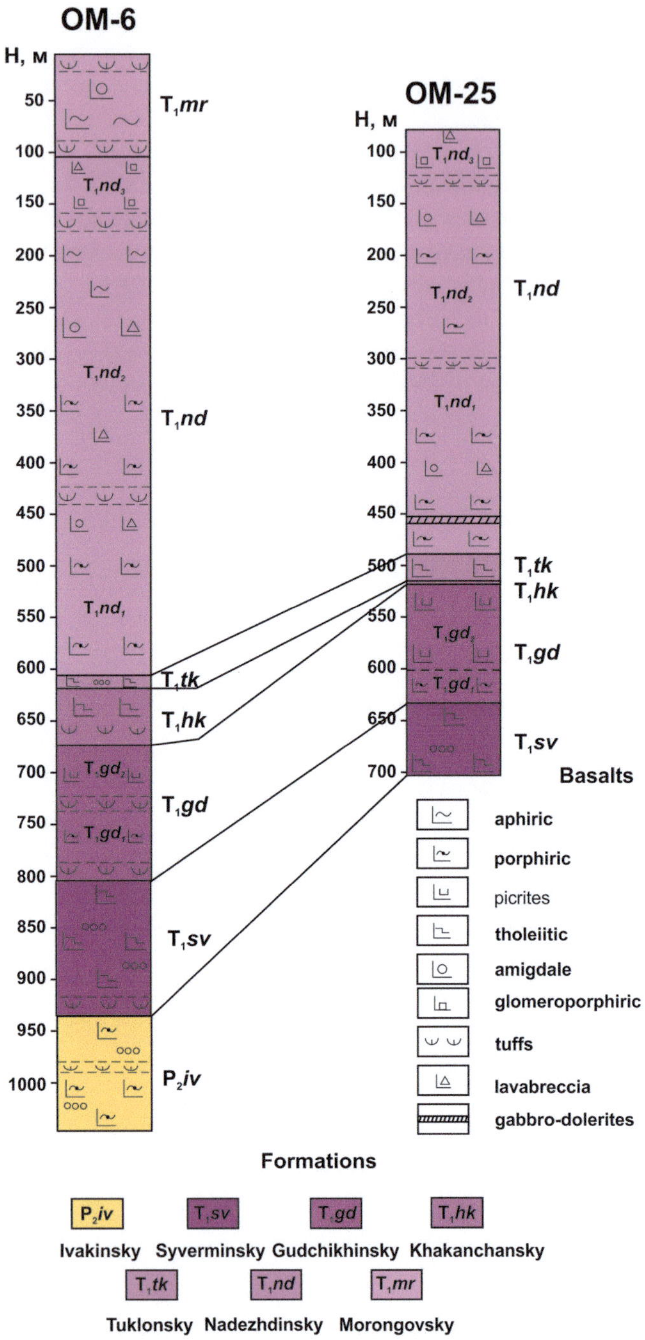

Fig. 3.8 Comparison of tuff–lavas pile structure in different parts of the Noril'sk Trough (boreholes OM-6 and OM-25)
After Krivolutskaya and Rudakova (2009)

on the characteristic morphological, structural, and textural features of their rocks (Dodin et al. 1971), and the boundaries were additionally constrained using geochemical data. It should be noted that almost all the flows have lower and upper (thicker) amygdaloidal zones, which allowed us to distinguish the flows in the section with certainty.

The *Ivakinsky Formation* conformably overlies the terrigenous rocks in the Tunguska Group. It is dominated by fine-grained dark gray to black rocks with a porphyritic texture. In the central part of the trough, the formation consists of two subformations separated by a 2-m-thick tuff layer. The lower subformation is made up of two sheets of porphyritic (labradorite) basalts, which are 19.9 and 7.4 m thick; and the upper subformation includes three flows of more silicic andesine basalts that are 39.1, 18.0, and 15.5 m thick. The base of this unit is represented by a 2 m-thick bed of pelite–psammite tuffs. The total thickness of the formation is up to 105 m in the central part of the trough; there are no such data for its eastern part because borehole OM-25 did not penetrate the Ivakinsky Formation.

The basalts in the *Syverminsky Formation* overlie the porphyritic basalts of the Ivakinsky Formation and are separated from them by a meter-thick bed of pelite–psammite tuffs with an indistinctly laminated structure. They show a tholeiitic texture (Polovinkina 1966) and massive structure. Their distinctive feature is a greenish-gray color related to the development of secondary minerals and thin flows (4–6 m on average); compared with the flows from other formations, they have thicker amygdaloidal zones (1–3 m). The formation includes 18 basaltic flows. Three thin tuff and tuff breccia interbeds (0.4, 0.3, and 1.2 m) were detected between the flows. The total thickness of the formation is 129.5 m in the center and approximately 100 m in the eastern part of the trough (incomplete thickness partly penetrated by borehole OM-25).

The rocks from the *Gudchikhinsky Formation* occur in the stratigraphic section above the Syverminsky basalts and are separated from the latter by an interbed of tuffs of silt–psammite grain size. The thickness of the tuffs is 0.6 m in the central part and 0.3 m southeast of the trough. The rocks in the formation are more melanocratic than other basalt varieties from the underlying and overlying formations, which is related to the presence of strongly altered olivine. The formation is subdivided into two subformations, lower and upper. The lower subformation includes two flows of dark gray polyphyric basalts with a fine-grained texture, which contain glomerophyric intergrowths of plagioclase and olivine phenocrysts that are up to 2 mm in size. The thicknesses of the flows are 34.2 and 9.9 m in the center and 26.2 and 11.7 m in the outer part of the trough. They are separated by a rather thick layer of tuff breccias (13.9 m in borehole OM-6 and 8.7 m in borehole OM-25). The upper subformation consists of medium-grained dark gray picrites with a brown tint and a single thin (0.8 m) tuff interbed. The central section of the trough includes eight basaltic flows, which are 7.8–22.7 m thick. Their number decreases to four in the southeast, and the thickness also decreases substantially (to 2.3–10.0 m). The rocks are strongly altered throughout the section of the formation. In borehole OM-6, the picrites are overlain by a tuff breccia layer (7.2 m). The thickness of the formation ranges from 127 m in the central part to 100.9 m in the periphery of the Noril'sk Trough.

The *Khakanchansky Formation* overlies the Gudchikhinsky rocks and is represented by tuffogenic rocks and basalts with a tholeiitic texture. Its structure is more complex in the center of the trough, which includes three basaltic flows (34 m in total thickness) and a heterogeneous tuff horizon. This horizon is subdivided into three parts differing in the fragment size and bedding character. The lower part (12 m) consists of distinctly bedded tuffs represented by alternating thin (from 1 mm to 3–4 cm) layers of dark gray pelitic rocks and interbeds of greenish-gray psammitic tuffs (10–15 cm). The middle part (3.8 m) consists of greenish-gray tuffs, which also show a laminated structure owing to the presence of rocks of different (compared with the lower part) grain sizes—psephitic (0.1–5 mm) and pelitic (<0.1 mm). The layers of the former are significantly thicker (from 2 m to 15–20 cm) than those of the latter (1–15 mm). The upper part (4 m) consists of dark gray pelitic tuffs with an indistinctly laminated structure. The thickness of the tuffs in the Khakanchansky Formation decreases considerably from the central (19.4 m) to the marginal part of the trough (0.6 m), and the thickness of the basalts decreases to 17 m.

The volcanics of the *Tuklonsky Formation* overlie the Khakanchansky rocks and are represented by light greenish-gray basalts with a fine-grained tholeiitic texture and massive structure. The thickest section of the volcanic sequence observed in borehole OM-6 includes five flows of varying thickness (from bottom to top): 31.3, 7.2, 2.1, and 8.7 m. The strong chloritization of the rocks imparted a uniform greenish color, which distinguishes them from the underlying and overlying rocks. A thin (20 cm) layer of greenish claret psammitic tuffs with a distinctly laminated structure was found in the basaltic sequence. The number of flows decreases to two (16.3 and 25.6 m thick) toward the side of the trough. The lower flow is made up of dark gray sparsely porphyritic basalts with a fine-grained doleritic texture in the groundmass and massive structure, and the upper flow is identical to the tholeiitic basalts in the central part. The flows are separated by a thin (1 m) layer of dark gray psammitic tuffs with weakly manifested bedding. The thickness of the Tuklonsky Formation ranges from 52.3 to 42.9 m.

The *Nadezhdinsky Formation* lies above the Tuklonsky rocks. It is distinguished by the greater thickness of both particular flows and the total section than those of the other seven formations described within the Noril'sk Trough. Most of the rocks in the Nadezhdinsky Formation are characterized by a dark gray to black color and an aphyric texture. Its structure in the central and southwestern parts of the trough is significantly different. In the center, the formation consists of three subformations with thicknesses of (from bottom to top) 171.8, 252.3, and 72.2 m separated by tuff interbeds. The lower subformation includes ten flows (10–15 m thick) of aphyric and sparsely porphyritic (mainly polyphyric) basalts with tuff breccias and tuffs (5–7 m) at the base of the unit. The fragments range in size from pelitic to agglomerate. At the southeastern side of the trough, two flows of tholeiitic basalts (13.2 and 11.4 m) overlain by six flows of aphyric and polyphyric rock varieties appear at the base of the formation. The total thickness of the subformation remains unchanged. The middle subformation begins from a thick (14.3 m) tuff layer, which is overlain by tuff breccias. It includes 22 basalt sheets, 7–14 m thick, which are dominated by aphyric and polyphyric varieties. They are separated by tuff interbeds (1.0–6.7 m thick). The roof of the member is marked by a thick (15.9 m) layer of psephitic and psammitic tuffs and two sheets of aphyric basalts. The structure of the member changes in the east, where it consists of ten sheets (from 4.4 to 35.7 m thick) of dominant polyphyric and glomerophyric basalts (with plagioclase and pyroxene phenocrysts) and occasional basalts with an indistinctly crystalline texture. The upper subformation is significantly different from the lower ones: it is represented by only glomerophyric fine-grained basalts containing up to 25 % intergrown tabular plagioclase grains up to 5 mm in size. They form six flows ranging in thickness from 1.4 to 22.3 m. To the southeast, the thickness of the glomerophyric rock varieties decreases to 40 m, and the thicknesses of the lower and middle subformations are 167.6 and 173.1 m, respectively. The maximum thickness of the Nadezhdinsky Formation (borehole OM-6) is almost 500 m.

The *Morongovsky Formation* crowns the section and rests on the Nadezhdinsky Formation. In addition to basalts, the formation contains considerable amounts of pyroclastic rocks, which distinguishes it from the other formations. It was documented in only the central part of the trough, where it comprises five sheets of aphyric and occasionally porphyritic basalts in the lower subformation and a thick tuff layer in the upper subformation. The dark gray basalts in the lower subformation show a fluidal structure. Porphyritic varieties are the most common, containing minor amounts of plagioclase phenocrysts (up to 5 %) that often form glomerophyric aggregates. Most of the lava flows are underlain by pelitic–psephitic laminated tuffs (3–5 m). The thickness of the lower subformation is 75 m. The upper subformation is composed of laminated green and brownish-red tuffs of varying grain sizes, from pelitic (fragments no larger than 0.5 mm) to agglomerate (tens of centimeters). The fragments are dominated by aphyric and, occasionally, amygdaloidal basalts. The thickness of the subformation is up to 28 m.

A comparison of the sections in the Noril'sk Trough with those described in the eastern slope of the Khantaysko–Rybninsky swell (western part of the Tunguska syneclise) and investigated by us in the lower reaches of the Mikchangda River allowed us to make the following main conclusions. The thicknesses of almost all the formations decrease by

20–25 % from the central part of the trough to its eastern side, toward the western part of the Khantaysko–Rybninsky swell. It is possible that the development of intrusive complexes in different structural units was also characterized by specific features, which could affect the formation of Pt–Cu–Ni mineralization. The analysis of this suggestion requires further detailed investigations focusing on a comparison between the geological structure and the development of the various tectonic units in the region.

3.2.2 General Characterization of the Volcanic Rocks in the Area: An Example of the Mokulaevsky Creek (Upper Formations)

The upper formations were studied in the Kharaelakh Trough because they were eroded in the Noril'sk Trough. Partially these data were published (Sluzhenikin et al. 2014).

The *Morongovsky Formation* (T_1mr). As part of the formation (Fig. 3.9), 13 lava flows and 9 horizons of tuffaceous rocks have been revealed. The total formation thickness is 300 m. It borders the Mokulaevsky Formation along the roof of the thickest aphyric basaltic flow. Dominating in the Morongovsky Formation are basalts with the aphyric structure: the lower part is dominated by thin (10–15 m) flows and the top by thicker flows up to 30–40 m. Two lava flows with glomeroporphyric structure containing large phenocrysts and plagioclase splices have been established. The rock composition is typical for tholeiitic basalts: 30–52 % plagioclase (An_{45-60}), 25–45 % clinopyroxene (Mg# = 68–80), 3–7 % olivine, 3–5 % ore minerals, and 5–22 % glass devitrification products. Tuffs contain angular and rounded fragments with psammitic, gravel and lapilli dimensions presented by variously altered basalts, volcanic glass, plagioclase, pyroxene or quartz grains, etc. Recent basalt fragments are observed in fragmental tuffs. Xenogenic fragments of sedimentary rocks underlying the tuff–lava strata are observed less frequently. Cement is composed of fine-clastic ash material and often entirely chloritized and calcitated. Compared with the Nadezhdinsky Formation, the Morongovsky basalts differ in somewhat elevated titanium levels (TiO_2 > 1 wt %, typical for the Siberian Traps) and low potassium concentrations (0.4 wt % of K_2O on average).

The *Mokulaevsky Formation* (T_1mk). There are 14 lava flows and 2 tuff horizons in the Mokulaevsky Formation. It is characterized by the predominance of effusive rocks over tuffaceous rocks. Total thickness of the Mokulaevsky Formation basalts is 240 m. The upper boundary is on the roof of tuffaceous horizon, above which oligoglomeroporphyritic basalts appear. The latter dominate dramatically in the section. Poikilophitic and aphyric basalt structural varieties are rarely observed. In thick flows the transition is from aphanitic structure in their marginal parts to fine and medium grain in the central part, from micro-grained to coarse-grained poikilophitic structure. The Mokulaevsky Formation basalts are characterized by insignificant variations of their mineral composition without a great variety of microstructures. Among the latter, poikilophitic and doleritic textures are the most abundant, whereas ophitic, tholeiitic, and intersertal textures are less frequent. Prevailing in porphyry phases are plagioclase (An_{65-80}) grains, which often form splices of 2–5 mm in size, with the maximum amount (10–15 %) in the basalts of the upper unit of the low-Mokulaevsky subformation. The bulk of the basalts consists of 35–53 % plagioclase (An_{45-60}), 25–35 % clinopyroxene (Mg# = 64–82), 12.3 % olivine (Fo_{44-67}), ~ 3–7 % ore minerals, and 5–13 % devitrification glass products.

By composition, the Mokulaevsky Formation rocks have elevated (compared to the Morongovsky basalts) titanium contents (above 1.2 wt %), which allows distinguishing them in mapping even in the absence of the Nadayansky horizon allocated at the base of the Mokulaevsky Formation and widely known within limits of the Siberian platform.

The *Kharaelakhsky Formation* (T_1hr). As part of the Kharaelakhsky Formation (Fig. 3.9), 30 lava flows and 4 horizons of alevrite and psammite tuffs have been allocated. The total thickness of the suite is 475 m. Its boundary with the overlying Kumginsky Formation goes along the appearance of fine-grained aphyric basalts in the section. The basalts are characterized by aphyric and glomeroporphyritic structures (in the lower and top parts of the section, respectively) and fluidly texture (Fig. 3.10a). Poikilophitic structure of the bulk, often coarse-grained, is typical for both of them. This helps diagnose formation rocks in the field conditions. The present horizons of tuffaceous rocks occur in the upper section. Phenocrysts in porphyritic basalts are presented by bytownite (An_{75}, up to 3 %) sized 0.4–2 mm and olivine (Fa_{31}, 4 %) sized 0.5–0.8 mm. The groundmass is composed of 50 % plagioclase (An_{65-60}), around 30 % xenomorphic clinopyroxene grains and 3 % olivine (Fa_{44}) grains, 3 % lamellar splices of ilmenite and magnetite, and 6 % volcanic glass.

The *Kumginsky Formation* (T_1km). The Kumginsky Formation has a simple structure: it consists of 6 basalt sheaths/overthrusts with aphyric, porphyry, and glomero- and oligoglomeroporphyritic rock structures and prevailing thin-fine-grained microdolerite and micropoikilophitic textures. A characteristic distinctive feature of the Kumginsky Formation glomeroporphyritic basalts is a large number (10–15 %) of plagioclase agglomerations of 3–5 mm in size. The visible thickness of the described Kumginsky Formation near the Mokulaevsky Creek is 40 m (its upper part is turfed). According to their textural and structural features, the basalts of the tuff–lava strata are very similar to each other. The flow

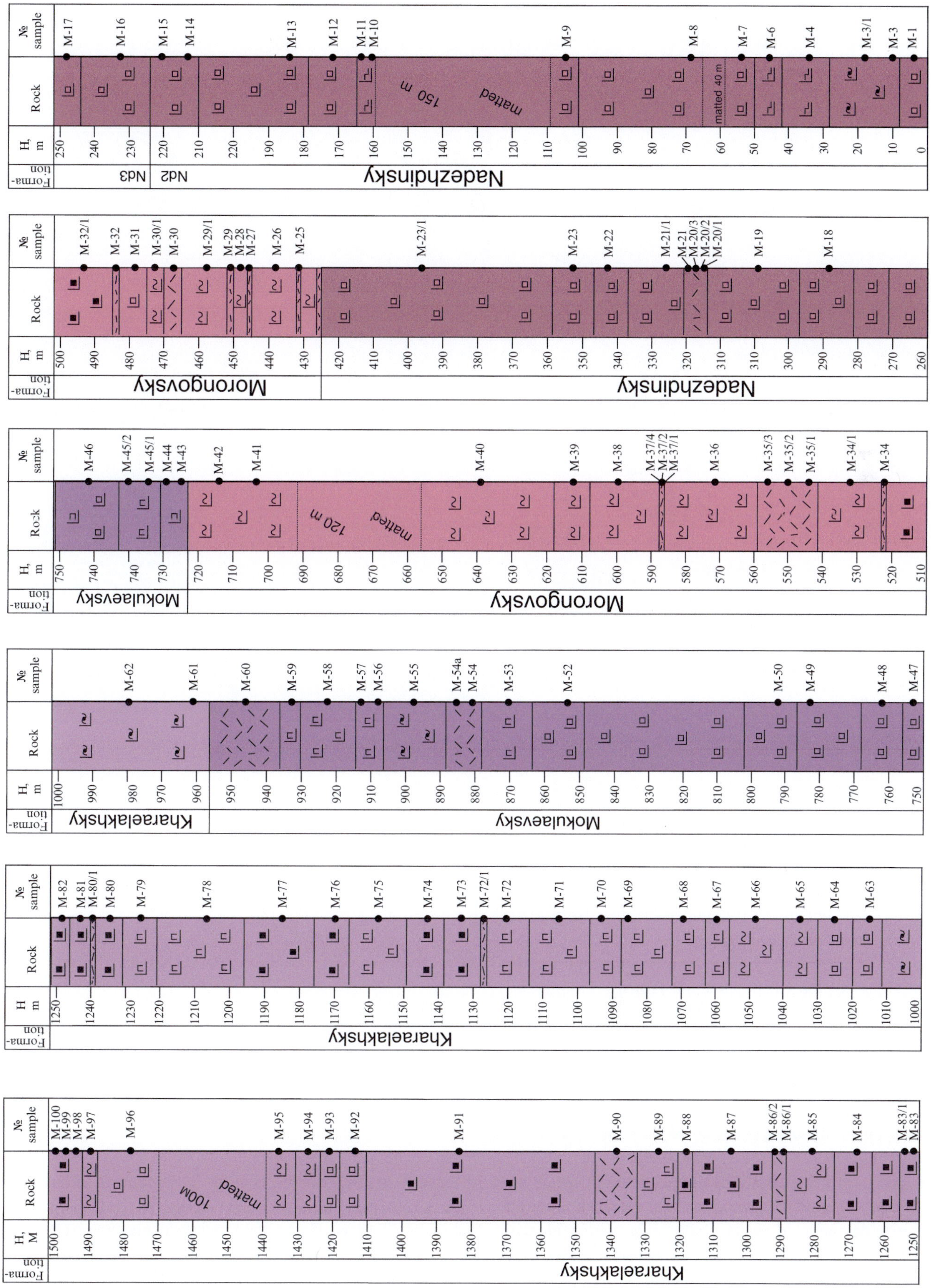

Fig. 3.9 Structure of volcanic rocks in the western part of the Kharaelakh Trough (Mokulay Creek) After Sluzhenikin et al. (2014)

Fig. 3.10 Basalts of the Formations
a—Kharaelakhsky, **b**—Mokulaevsky, upper zone of flow

borders can be established with enough confidence by the presence of the upper amygdaloidal zone, which sometimes has a complex structure (Fig. 3.10b). This can be demonstrated by the example of one of their flows (observation point M-51). It is not always possible to identify the borders of the formations during geological trips.

Rocks of the *Samoedsky Formation* are concentrated in the central part of the Kharaelakh and Ikonsky Troughs, to north from Mokulaevsky Creek. They are represented by aphyric, rarely poikilophitic basalts.

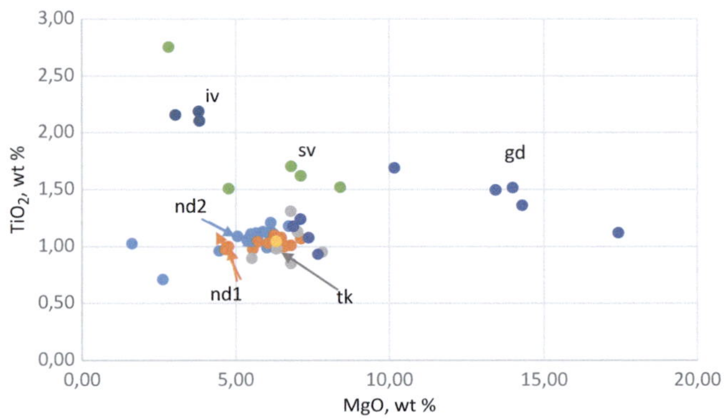

Fig. 3.11 Diagram TiO_2 vs. MgO for volcanic rocks of the Noril'sk Trough

3.2.2.1 Geochemical Characteristics of the Major Elements in the Volcanic Rocks

Specific features of each formation are shown in a series of diagrams (Figs. 3.11, 3.12, 3.13, 3.14, 3.15).

The analysis of the variations in the chemical composition of the volcanic sequence in the Noril'sk Trough reveals a wide range of rocks (Table 3.2), which clearly illustrates the difference between its uniform upper part and the compositionally more contrasting lower part. For instance, a gradual decrease in the TiO_2 content in basalts (Fig. 3.11) from the lower part of the section from the Ivakinsky (2.18 wt %) to Tuklonsky Formations (0.9–1.3 wt %) is changed by the persistently low TiO_2 content (0.9–1.3 wt %) in the upper part. Nonetheless, minor vari-

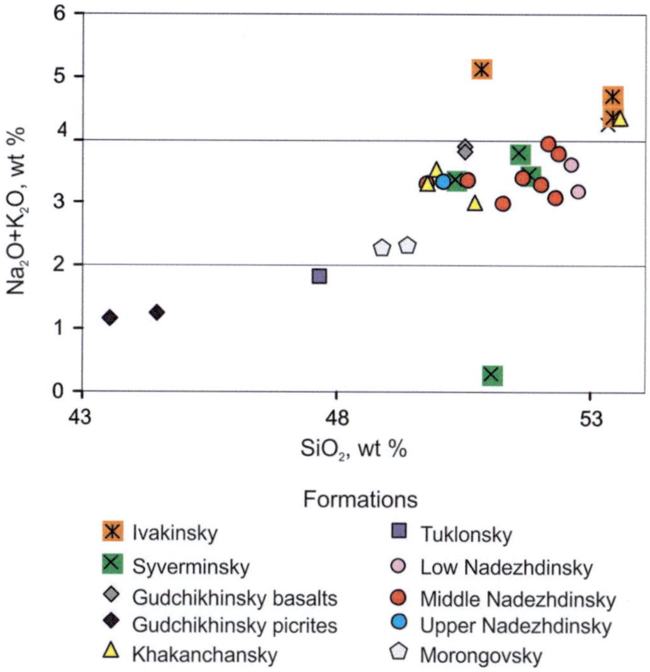

Fig. 3.12 Diagram Na_2O+K_2O vs. SiO_2 for volcanic rocks of the Noril'sk Trough
After Krivolutskaya and Rudakova (2009)

ations in TiO_2 allowed us to distinguish between the youngest rocks in the middle Nadezhdinsky–Morongovsky Formations that are enriched in this component and the older TiO_2-depleted rocks from the lower part of the Nadezhdinsky Formation.

With respect to the SiO_2 content, there is a continuous series from basalt to andesite (46–55 wt % SiO_2). The total alkalis content ranges from 0.4 to 6 wt %, and the MgO content is 3–20 wt %. As observed in the (Na_2O+K_2O)—SiO_2 diagram (Fig. 3.12), the highest alkalinity is characteristic of the lower two formations, Ivakinsky and Syverminsky, which consist of trachybasalts and basaltic trachyandesites. The rocks in the former formation are highly potassic. The basalts in the other formations fall within the field of subalkaline rocks. Considerable variations of the MgO contents were observed for the rocks in the Gudchikhinsky Formation. Significant differences are the characteristic of the Nadezhdinsky Formation, i.e., the compositional points of its constituent basalts are distinct from those of other formations regarding SiO_2, TiO_2, Fe_2O_3, etc. (Fig. 3.13). The rocks of the lower part of Nadezhdinsky Formation are depleted in Cu (Table 3.3). Volcanic rocks of the Morongovsky Formation are also quite clearly distinguished by a number of components from the underlying rocks. The higher basalts of Mokulaevsky, Kharaelakhsky, and Kumginsky Formations are along the section, the more homogeneous they are. They are subdivided effectively only by textural and structural features.

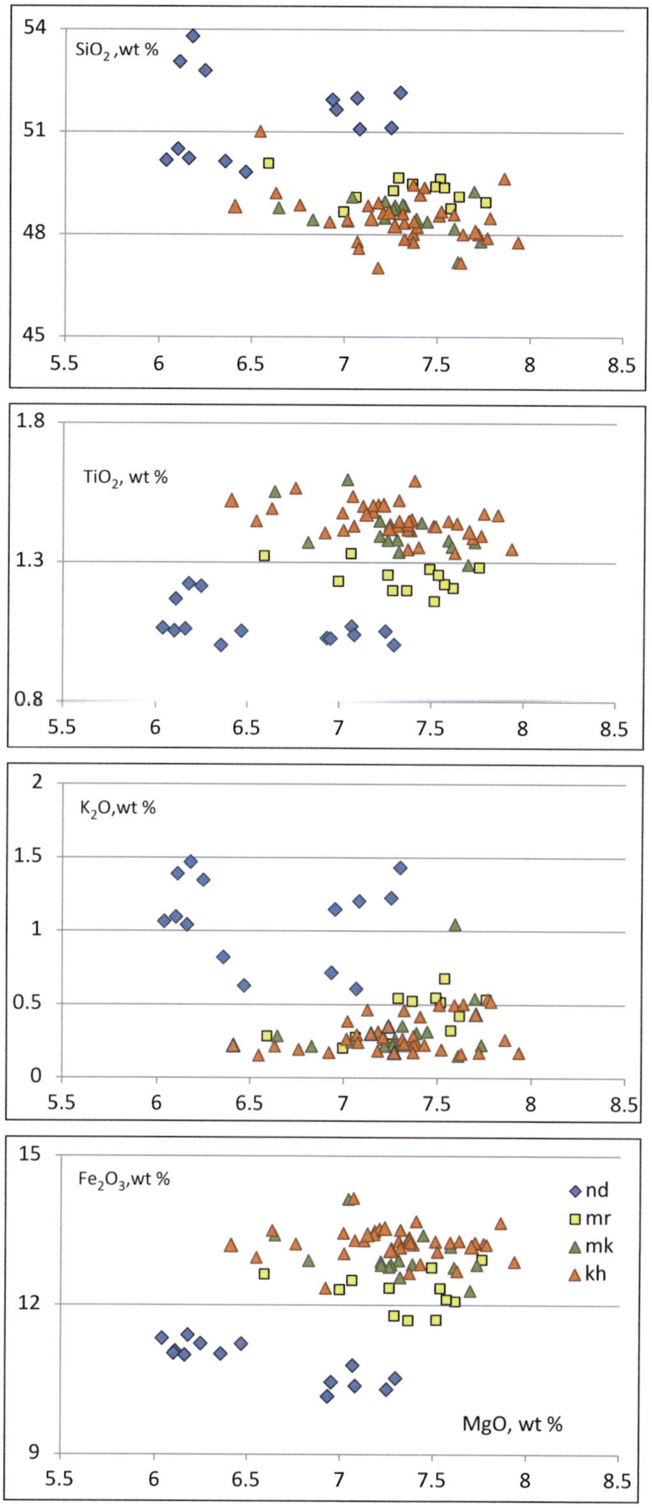

Fig. 3.13 Harker diagrams for volcanic rocks from the Mokulay Creek

3.2.2.2 Trace Elements

The analysis of primitive mantle-normalized trace element patterns in the volcanic rocks from the trap association of the Noril'sk Trough (Fig. 3.14) indicates, on the one hand,

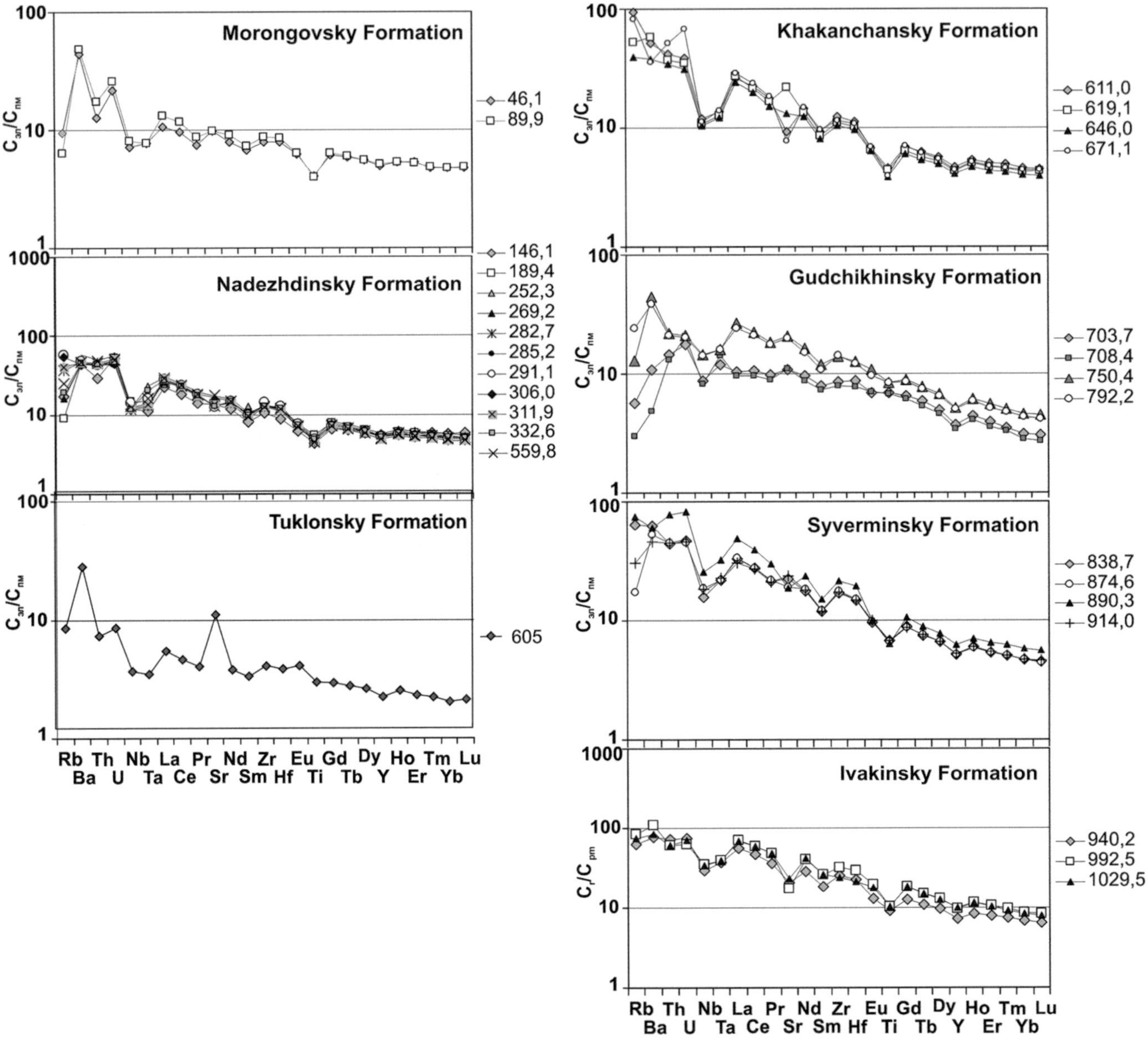

Fig. 3.14 Trace element patterns for volcanic rocks of the Noril'sk Trough (normalized to primitive mantle, Hofmann 1988) After Krivolutskaya and Rudakova (2009)

their similarity (enrichment in incompatible elements and the presence of varying negative anomalies of Ta–Nb, Eu, and Ti) and, on the other hand, the evolution of rock composition with time. The alkali-rich rocks from the lower two formations are generally enriched in all trace elements, especially large-ion lithophile (LIL) elements. The overlying basalts of the Gudchikhinsky Formation do not show such enrichment, but they are similar to the rocks from the lower formations, which have high Gd/Yb ratios. All the overlying rock varieties show flatter heavy REE patterns, which allowed us to distinguish them into a separate group from the complexes of the three lower formations. The differences between the compositions of the rocks from these two parts of the section are illustrated by the MgO, SiO_2, TiO_2, Gd/Yb, and La/Sm variations in the volcanics from borehole OM-6 (Fig. 3.11). The differences are clearly observed in the La/Sm–Gd/Yb diagram (Fig. 3.15a), where the compositions of the three lower formations plot at much higher Gd/Yb values (>2) than the basalts from the other formations. A gradual increase in the magnitude of the Ta–Nb and Ti anomalies from the early to late formations is a general tendency of geochemical variations (Fig. 3.15b). In addition to the aforementioned common features, each formation shows its own distinctive geochemical characteris-

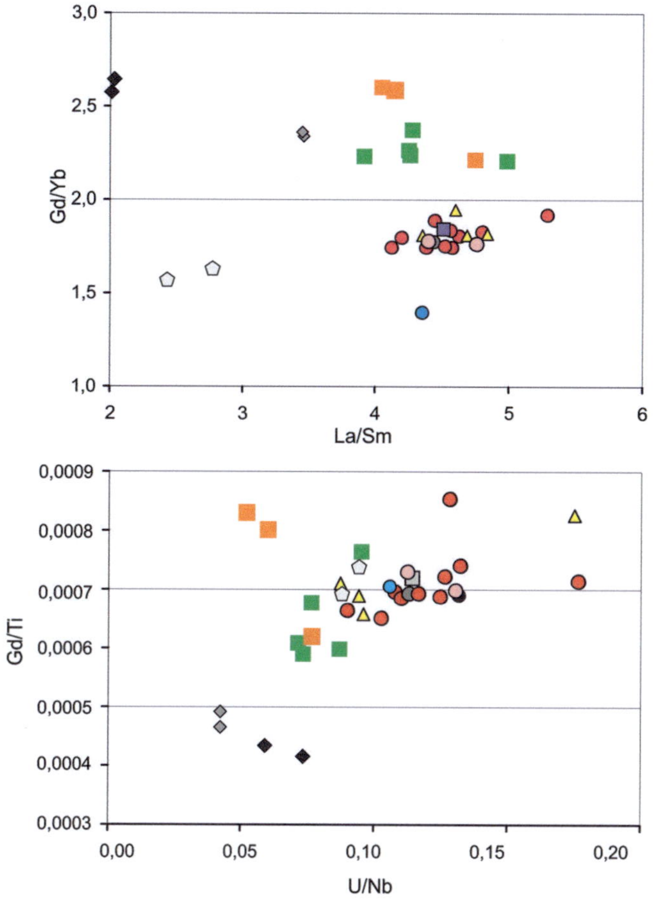

Fig. 3.15 Diagrams La/Sm–Gd/Yb for volcanic rocks of the Noril'sk Trough
Conventions see Fig. 3.12

tics. For instance, the lower two formations show relatively high amounts of all trace elements, whereas the minimum amounts (an order of magnitude lower than in the other formations) are typical of the Gudchikhinsky Formation. The compositions of the volcanic rocks in the Khakanchansky Formation are strongly different from those of the underlying Gudchikhinsky rocks and overlying Tuklonsky rocks, but their distribution patterns are identical to those of the Nadezhdinsky basalts. The compositions of the rocks in the Khakanchansky Formation are in Table 3.2 (analyses 18–21). It should be emphasized that the trace element amounts in the tuffs were determined for the first time for the Noril'sk region. The trace element patterns of the rocks in the Tuklonsky Formation lie somewhat higher than those of the Gudchikhinsky rocks and exhibit a distinct positive Sr anomaly. Their REE patterns have a flatter heavy REE slope. The Nadezhdinsky Formation is distinguished by significant enrichment in incompatible elements (in particular, it shows high La/Sm ratios ranging from 4.6 to 4.9) and the presence of a distinct Ta-Nb anomaly (Fig. 3.16). The rocks in the Morongovsky Formation have less enriched trace element patterns and show identical compositions in the Noril'sk Trough and other structures (Mokulay Greek). The gradient of the distribution spectrum of trace elements and the value of Ta–Nb anomalies are clearly observed.

3.2.3 General Characterization of the Volcanic Rocks in the Area: An Example of the Mikchangdinsky Area

Mikchangdinsky area (Fig. 3.17) is a very interesting territory due to its location, in northern pericline framing of the Khantaysko–Rybninsky swell, which divides the Noril'sk area into two structurally distinct blocks: the Noril'sk–Kharaelakh Trough and the western part of the Tunguska syneclise. Many individual basalt sections have been studied (white lines in Fig. 3.17). Figure 3.18 demonstrates the example of one of them. The best example of basalt sections within this area is in the southern shore of Lake Glubokoe. The bottom portion of the section of volcanic rocks crops out at Mount Sunduk and is described in (Section 1F, Lightfoot et al. 1994; Fedorenko et al. 1996). We prepare a more detailed structural chart of the volcanic rocks (Fig. 3.19) than those published earlier, and we distinguish certain new patterns in the inner structure of the tuff–lava pile. In particular, it was found that the geochemical features of the tuffs in Tuklonsky Formation are analogous to the Nadezhdinsky basalts (Table 3.3). Composite section of volcanic rocks for the eastern part of the Noril'sk area is shown in Fig. 3.20. The lower formations display high Gd/Yb ratios, while for the upper formations, low Gd/Yb is typical.

The analysis of the primitive mantle-normalized trace element patterns of volcanic rocks of the flood basalt association in the Noril'sk area shows, on the one hand, certain similarities between the rocks (richness in incompatible elements and variably pronounced Tb–Nb, Eu, and Ti anomalies) and, on the other hand, an evolution of the composition of the rocks with time (Fig. 3.21a). The rocks of the two lower formations have a subalkaline composition and are generally enriched in trace elements, first of all in LILE. The overlying basalts of the Gudchikhinsky Formation show no such features but are nevertheless similar to the underlying rocks by having elevated Gd/Yb ratios. This is the only rock type in the area that exhibits evidence of derivation from a mantle source (absence of the aforementioned anomalies). All of the overlying rocks have more gently sloping HREE patterns, which led us to distinguish these rocks as a separate group that is different from rocks of the lower three formations. All variety of spectra can be reduced to five types (Fig. 3.21b).

Table 3.2 Chemical composition of the volcanic rocks from the Noril'sk Trough (borehole OM-6)

No	1	2	3	4	5	6	7	8	9
No sample element	46.1	89.9	146	189	252	269	282	285	291
SiO_2	48.9	49.4	50.1	47.8	52.1	49.7	52.4	50.6	52.0
TiO_2	1.12	1.10	1.00	1.26	1.06	1.18	1.12	1.21	1.12
Al_2O_3	16.7	16.2	15.0	14.9	12.9	15.3	13.3	14.4	14.5
Fe_2O_3	12.3	12.4	12.1	13.4	10.9	11.4	12.2	11.8	11.2
MnO	0.19	0.18	0.17	0.17	0.16	0.20	0.15	0.17	0.16
MgO	7.19	6.90	4.86	5.29	5.36	6.72	5.66	6.14	6.04
CaO	11.2	11.2	10.0	10.2	8.98	9.62	8.95	10.6	9.81
Na_2O	1.93	1.95	2.69	2.34	2.94	2.74	2.87	2.18	2.05
K_2O	0.33	0.38	0.63	0.43	0.99	0.56	0.91	1.18	1.22
P_2O_5	0.11	0.11	0.21	0.16	0.14	0.12	0.16	0.13	0.13
LOW	1.45	1.87	2.85	3.18	3.60	2.11	1.48	1.22	1.45
Summa	101	101	99.7	99.2	99.3	99.7	99.3	99.8	99.8
Rb	6.00	4.00	11.00	5.68	31.9	10.2	23.3	26.2	36.3
Ba	304	338	311	318	404	313	328	317	339
Th	1.07	1.48	2.52	3.71	4.18	3.74	3.90	3.65	3.64
U	0.45	0.54	1.10	1.11	1.17	0.98	1.07	0.96	0.95
Nb	5.1	5.70	8.35	10.7	8.82	9.01	8.39	8.67	10.4
Ta	0.32	0.32	0.45	0.64	0.95	0.58	0.56	0.55	0.68
La	7.33	9.07	15.5	18.60	21.0	17.9	18.9	19.6	19.6
Ce	17.1	20.7	32.8	40.6	44.3	41.0	40.8	42.0	41.8
Pr	2.06	2.39	3.94	4.97	5.32	5.06	4.93	5.06	5.00
Sr	207	207	286	273	364	350	275	271	261
Nd	10.6	12.3	16.19	20.09	21.3	19.9	20.0	19.8	19.9
Sm	3.02	3.27	3.57	4.50	4.60	4.27	4.26	4.24	4.33
Zr	89.0	98.0	117	150.8	144	144	144	142	161
Hf	2.43	2.63	2.70	3.81	3.84	3.67	3.80	3.53	3.95
Eu	1.05	1.07	1.05	1.29	1.24	1.18	1.18	1.21	1.28
Ti	5,242	5,148	5,683	7,123	6,504	6,416	6,147	6,302	6,443
Gd	3.63	3.80	3.93	4.62	4.82	4.46	4.43	4.31	4.28
Tb	0.64	0.65	0.68	0.75	0.80	0.74	0.73	0.73	0.73
Dy	4.08	4.08	4.46	4.71	4.90	4.48	4.51	4.41	4.48
Y	22.3	23.1	25.2	24.2	25.6	23.4	23.1	23.4	23.3
Ho	0.88	0.87	0.98	0.98	1.05	0.92	0.96	0.91	0.92
Er	2.54	2.53	2.81	2.76	2.90	2.59	2.66	2.59	2.57
Tm	0.35	0.36	0.43	0.41	0.43	0.38	0.40	0.38	0.38
Yb	2.32	2.33	2.81	2.66	2.77	2.48	2.50	2.39	2.45
Lu	0.35	0.36	0.43	0.37	0.40	0.37	0.38	0.37	0.37
Cu	122	113	68.4	73.2	95.8	54.9	49.3	66.4	41.4
Ni	107	86	65.1	688	48.9	33.2	54.9	47.6	44.6

No	10	11	12	13	14	15	16	17	18
No sample element	306.0	311.9	317.2	332.6	421.0	495.2	559.8	604.0	611.0
SiO_2	51.6	51.2	51.7	52.2	57.2	52.6	52.7	47.6	53.5
TiO_2	1.13	1.11	0.96	1.09	0.71	1.00	1.00	0.95	0.98
Al_2O_3	14.6	14.3	12.9	14.5	13.7	15.2	14.1	15.0	13.4
Fe_2O_3	11.3	10.8	9.19	11.8	7.55	10.9	11.0	10.8	10.6
MnO	0.15	0.18	0.20	0.15	0.11	0.13	0.14	0.17	0.14
MgO	5.88	5.49	4.45	5.05	2.62	4.74	4.70	7.83	6.33
CaO	9.89	10.9	8.21	9.62	3.37	8.86	9.25	11.9	6.40

(continued)

Table 3.2 (continued)

No	10	11	12	13	14	15	16	17	18
No sample element	306.0	311.9	317.2	332.6	421.0	495.2	559.8	604.0	611.0
Na_2O	2.32	2.07	2.66	2.44	0.37	2.64	2.60	1.60	3.25
K_2O	1.06	0.92	3.12	0.66	6.53	1.07	0.53	0.23	1.12
P_2O_5	0.12	0.13	0.11	0.14	0.15	0.14	0.13	0.08	0.13
LOW	1.53	2.50	6.13	1.63	6.77	1.96	3.06	3.43	3.24
Summa	99.8	99.8	99.7	99.4	99.1	99.3	99.3	99.7	99.1
Rb	34.4	25.6	154	12.47	354	21.3	16.2	5.63	60.1
Ba	313	329	535	315	444	306	330	159	357
Th	3.61	3.78	3.21	3.82	6.65	3.77	4.12	0.67	3.54
U	0.94	1.09	1.31	1.00	1.65	1.11	1.04	0.19	0.81
Nb	8.25	8.69	7.41	8.53	12.7	8.46	9.16	3.27	8.56
Ta	0.53	0.56	0.50	0.85	0.91	0.59	0.66	0.18	0.56
La	18.5	19.9	16.6	19.2	28.0	18.0	20.1	4.28	17.8
Ce	39.8	41.8	34.6	41.5	58.8	38.4	42.5	9.74	39.3
Pr	4.82	4.92	4.24	4.99	6.75	4.64	5.10	1.37	4.82
Sr	281	290	251	273	71.2	272	377	230	196
Nd	19.1	19.5	16.8	20.0	26.5	18.9	20.2	6.34	19.6
Sm	4.06	4.14	3.74	4.39	5.29	4.10	4.22	1.86	4.10
Zr	138	145	125	143	200	129	138	55.6	137
Hf	3.44	3.62	3.21	3.71	5.13	3.30	3.61	1.46	3.47
Eu	1.18	1.18	0.91	1.23	1.27	1.13	1.21	0.83	1.13
Ti	5,958	6,233	5,442	6,378	6,012	5,922	5,809	5,040	5,979
Gd	4.13	4.27	3.89	4.41	5.13	4.13	4.23	2.28	4.12
Tb	0.69	0.72	0.61	0.73	0.82	0.68	0.70	0.40	0.67
Dy	4.17	4.37	3.86	4.57	4.98	4.21	4.34	2.60	4.22
Y	22.2	22.2	20.3	23.5	25.8	21.7	22.2	14.1	21.3
Ho	0.88	0.91	0.81	0.97	1.03	0.89	0.92	0.56	0.87
Er	2.40	2.48	2.25	2.67	2.89	2.49	2.55	1.53	2.43
Tm	0.37	0.37	0.33	0.40	0.44	0.37	0.38	0.23	0.36
Yb	2.26	2.35	2.06	2.54	2.68	2.25	2.39	1.42	2.27
Lu	0.35	0.35	0.32	0.37	0.40	0.34	0.35	0.22	0.33
Cu	74.8	69.1	62.0	40.3	32.3	5.06	19.3	63.4	19.2
Ni	<2	44.2	42.1	38.6	166	50.2	69.9	86.7	17.9

No	19	20	21	22	23	24	25	26	27
No sample element	619.1	646.0	671.1	703.7	708.4	750.4	792.2	838.7	852.5
SiO_2	49.9	50.7	49.7	44.4	43.5	50.4	50.4	51.6	51.0
TiO_2	0.85	0.90	1.05	1.50	1.52	1.75	1.79	1.52	2.75
Al_2O_3	14.0	14.7	13.2	9.13	8.65	15.2	15.2	13.7	12.5
Fe_2O_3	11.0	10.6	9.52	14.4	15.3	11.9	11.8	10.9	10.8
MnO	0.15	0.14	0.13	0.12	0.15	0.15	0.14	0.21	0.13
MgO	6.81	5.54	6.32	13.4	13.9	4.57	4.95	8.41	2.81
CaO	8.01	10.3	6.17	8.00	8.11	9.36	8.99	5.41	14.5
Na_2O	2.82	2.29	2.23	1.04	1.06	3.30	3.11	2.82	0.15
K_2O	0.72	0.71	1.02	0.18	0.07	0.60	0.69	0.94	0.14
P_2O_5	0.13	0.10	0.12	0.13	0.12	0.24	0.24	0.20	0.43
LOW	4.63	3.13	9.82	6.64	6.52	1.98	1.84	3.59	3.99
Summa	99.1	99.2	99.3	99.1	99.0	99.5	99.3	99.3	99.4
Rb	33.4	25.1	52.1	3.59	1.89	8.17	15.2	40.7	3.59
Ba	401	265	249	74.9	33.7	306	268	436	26.3

(continued)

Table 3.2 (continued)

No	19	20	21	22	23	24	25	26	27
No sample element	619.1	646.0	671.1	703.7	708.4	750.4	792.2	838.7	852.5
Th	3.13	2.93	4.38	1.23	1.13	1.87	1.78	3.78	7.35
U	0.74	0.65	1.41	0.37	0.43	0.44	0.43	0.97	1.81
Nb	7.68	7.44	8.06	6.24	5.85	10.3	10.0	11.0	23.6
Ta	0.52	0.50	0.57	0.49	0.58	0.65	0.65	0.89	1.64
La	18.4	16.6	19.7	7.18	6.60	18.1	16.6	22.6	41.7
Ce	37.7	34.9	42.1	18.9	17.0	39.3	37.2	48.7	89.5
Pr	4.55	4.20	5.01	2.73	2.46	5.08	4.83	5.94	10.9
Sr	4,614	276	163	229	231	433	419	468	98.0
Nd	18.1	16.7	19.8	13.0	11.8	21.9	20.4	23.9	44.9
Sm	3.80	3.56	4.28	3.55	3.24	5.26	4.84	5.32	9.76
Zr	122	116	128	96.6	87.8	1,566	158	188	341
Hf	3.17	2.95	3.34	2.69	2.41	3.92	3.79	4.60	8.56
Eu	1.09	1.07	1.16	1.17	1.17	1.81	1.63	1.63	2.61
Ti	5,743	5,020	5,105	8,977	8,746	10,860	10,854	8,811	14,055
Gd	3.78	3.57	4.22	3.89	3.63	5.34	5.05	5.26	9.52
Tb	0.62	0.58	0.66	0.63	0.58	0.83	0.81	0.81	1.48
Dy	3.85	3.64	4.04	3.65	3.36	5.00	4.81	4.89	8.77
Y	19.6	18.4	20.2	16.9	15.4	23.5	22.9	23.3	42.0
Ho	0.82	0.76	0.84	0.72	0.67	1.00	0.97	0.99	1.76
Er	2.24	2.09	2.29	1.87	1.70	2.65	2.49	2.59	4.66
Tm	0.34	0.32	0.34	0.26	0.24	0.37	0.35	0.37	0.67
Yb	2.08	1.97	2.17	1.52	1.38	2.28	2.14	2.32	4.02
Lu	0.32	0.29	0.33	0.23	0.20	0.33	0.31	0.33	0.60
Cu	38.1	26.2	117	77.4	81.9	53.6	40.8	22.4	92.4
Ni	41.0	41.7	189	487	680	56.1	40.3	83.9	108

No	28	29	30	31	32	33
No sample element	874.6	890.3	914.0	940.2	992.5	1,029.5
SiO_2	51.7	53.3	50.3	53.4	53.4	50.8
TiO_2	1.70	1.51	1.62	2.10	2.18	2.15
Al_2O_3	12.2	14.4	13.2	13.5	13.2	14.5
Fe_2O_3	11.5	10.3	11.8	12.0	13.4	14.2
MnO	0.14	0.15	0.16	0.13	0.16	0.17
MgO	6.81	4.76	7.12	3.82	3.79	3.04
CaO	8.27	6.99	7.54	7.04	5.34	6.59
Na_2O	2.71	2.67	2.61	2.75	2.96	3.71
K_2O	0.74	1.55	0.78	1.65	1.74	1.41
P_2O_5	0.25	0.19	0.27	0.35	0.80	0.95
LOW	2.89	3.10	3.58	2.15	2.50	1.67
Summa	99.1	99.1	99.0	99.0	99.6	99.3
Rb	10.8	47.8	19.3	39.6	52.7	46.8
Ba	3,649	422	323	531	764	582
Th	3.82	6.58	3.78	6.11	5.21	5.17
U	0.95	1.73	0.97	1.59	1.30	1.47
Nb	13.2	18.1	13.1	20.8	25.3	24.5
Ta	0.90	1.33	0.92	1.51	1.64	1.61
La	23.1	33.2	20.7	38.8	48.7	47.3
Ce	49.2	69.5	47.5	82.1	105	104
Pr	6.08	8.29	5.82	9.93	13.3	13.4

(continued)

Table 3.2 (continued)

No	28	29	30	31	32	33
No sample element	874.6	890.3	914.0	940.2	992.5	1,029.5
Sr	400	395	495	440	365	486
Nd	24.7	32.1	24.1	38.6	54.8	56.1
Sm	5.41	6.67	5.28	8.19	11.76	11.7
Zr	195	2,396	192	281	359	270
Hf	4.67	5.95	4.54	6.94	8.99	6.60
Eu	1.64	1.69	1.68	2.16	3.32	3.00
Ti	8,550	8,224	8,743	12,057	13,313	13,561
Gd	5.20	6.28	5.18	7.50	11.06	10.8
Tb	0.83	0.96	0.83	1.18	1.65	1.61
Dy	4.88	5.76	4.93	7.08	9.61	9.31
Y	23.3	28.4	23.7	33.3	43.7	45.1
Ho	0.97	1.16	0.99	1.39	1.91	1.88
Er	2.59	3.12	2.63	3.77	5.03	4.95
Tm	0.37	0.46	0.38	0.55	0.71	0.68
Yb	2.31	2.84	2.31	3.38	4.26	4.18
Lu	0.32	0.41	0.34	0.48	0.62	0.59
Cu	20.5	22.8	52.5	27.8	29.2	19.0
Ni	124	89.2	56.8	14.5	23.0	65.0

Note: (1) No sample correlates with the depth in borehole OM-6, m (OM-6/46.1); (2) No analyses, formations: 1, 2—T_1*mr*, 3–16 (3—T_1*nd3*, 4–14—T_1*nd2*, 15–16—T_1*nd1*), 17—T_1*tk*, 18–21—T_1*hk*, 22–25—T_1*gd* (22–23—T_1*gd2*, 24–25—T_1*gd1*), 26–30—T_1*sv*, 31–33—P_2*iv* (indexes are given in Table 3.1); (3) oxides are given in wt %, elements—ppm. Analyses were carried out in GEOKHI RAS (major elements by XRF, analyst I. Roshsina; elements by ICP-MS in IMGRE, analyst D. Zhuravlev)
After Krivolutskaya and Rudakova (2009)

3.3 Principal New Data on the Volcanic Rocks

3.3.1 Tholeiitic Basalts of the Central Part of Pile

Rocks of the lowest formations are clearly divided into formations based on their major compositions, whereas such division of volcanic rocks of the upper part of section is often difficult because of the similarities in the petrochemical, structural, and textural characteristics of the lavas and the absence of reliable reference horizons among them. This difficulty is mostly associated with three "transition" formations, namely, the Khakanchansky, Tuklonsky, and Nadezhdinsky Formations (and partially with Ivakinsky), which are typical basalts with tholeiitic textures that are macroscopically indistinguishable and located one above the other (Fig. 3.22). Their recognition is possible only on the basis of correlation of minor element distribution spectra in rocks, which vary significantly from one formation to the next. Thus, the tholeiitic basalt of the Nadezhdinsky Formation is characterized by the high concentrations of LILEs and a $(La/Sm)n$ ratio greater than 3, whereas volcanic rocks of the Tuklonsky Formation are distinguished by a low $(La/Sm)n$ ratio (<1.5). The Khakanchansky Formation, which primarily consists of tuff, also contains rare tholeiitic basalt layers, although their geochemical characterization is not available. The only analysis of tuff is given in (Fedorenko et al. 1996). Due to the importance of these formations, we studied the structure of eight sections of the middle part of the tuff–lava series in various parts of the region: the northern periclinal framework of the Khantaysko–Rybninsky swell (northern side of Lama Lake, downstream along the Mikchangda River), the northwestern part of the Kharaelakh Trough, and the western part of the Tunguska syneclise (Glubokoe Lake). As a result, it was established that tholeiite basalt layers described by previous researchers while studying the composition of the Khakanchansky Formation were abundant only in the central part of the Noril'sk region and were most typical in the area of the Khantaysko–Rybninsky swell. Their composition was that of the tuff alternating with them in the section: based on the concentrations of major and, particularly, minor elements, the basalts are identical to the rocks of the Tuklonsky Formation, whereas the tuff and tuffite are analogous to rocks of the Nadezhdinsky Formation. This finding is supported by findings in one of the sections studied in detail on the north side of Lama Lake (Krivolutskaya 2011). In this case, picritic basalt of the Gudchikhinsky Formation characterized in (Sobolev et al. 2009) is overlain by alternating tuff, tuffite, and tholeiitic basalt (Fig. 3.23), which in the field conditions were presumably related to the Khakanchansky Formation.

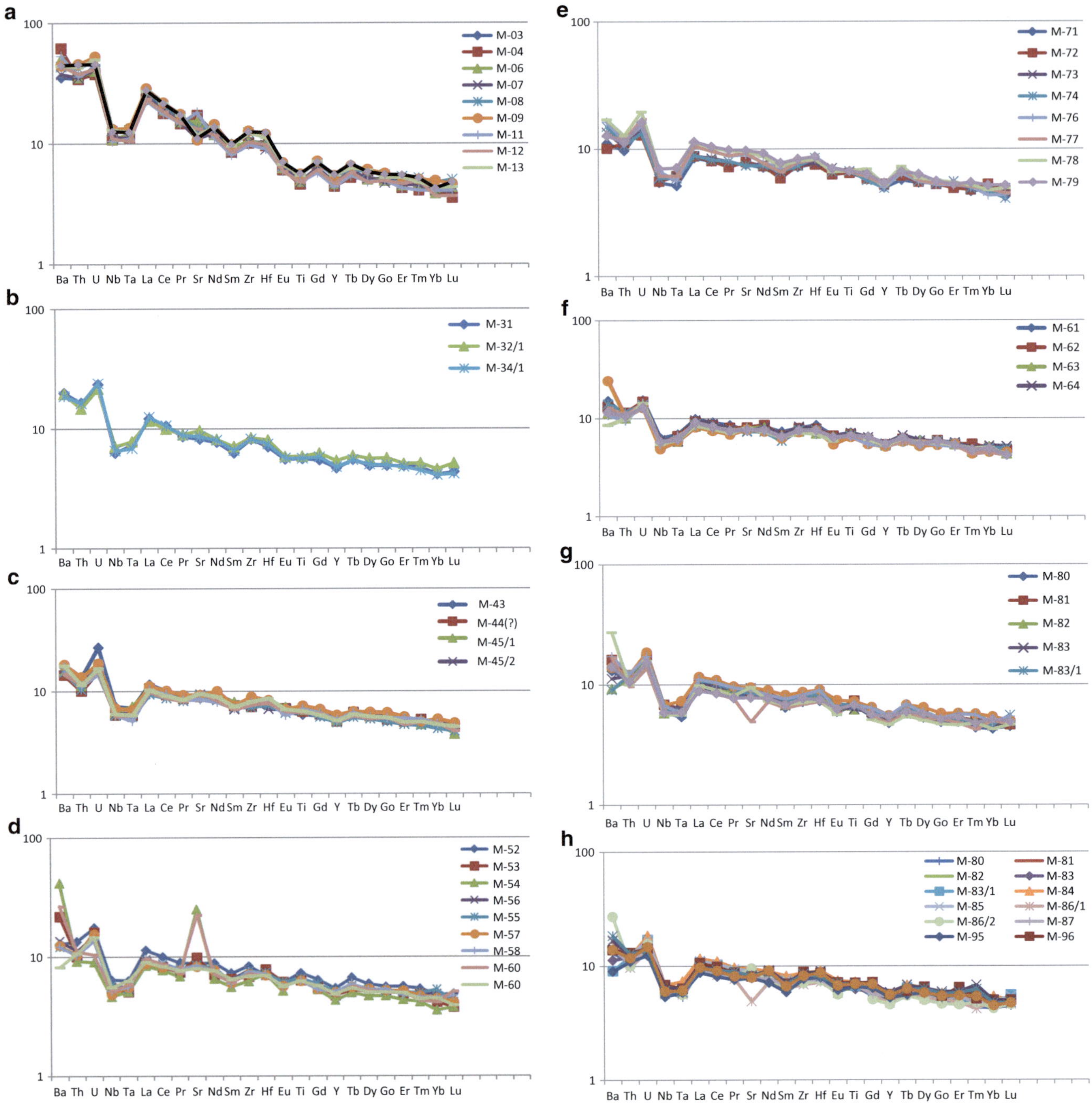

Fig. 3.16 Trace element patterns for volcanic rocks of the Mokulay Creek After Sluzhenikin et al. (2014)

However, their detailed investigation demonstrated that the basalt and tuff had different geochemical characteristics. This is most clearly evident from variations in the (La/Sm)*n* ratio (Fig. 3.23), which significantly differs between rocks of the Nadezhdinsky and Tuklonsky Formations: in the first case, the ratio reaches 3.4, whereas in the second case, it does not exceed 1.4. The spectra strongly enriched in LILEs are typical of rocks of the Nadezhdinsky Formation in comparison with basalt of the Tuklonsky Formation, which is clearly evident from Fig. 3.23. The analogous regularities were established downstream along the Mikchangda River (in the S. Iken River valley), but in this case, the tuff alternates with poikilophitic basalt, which were always macroscopically considered to be products of crystallization of magma of the Tuklonsky Formation. The tuff between them was previously correlated with the Tuklonsky Formation, and the establishment of their characteristics analogous to those of Nadezhdinsky Formation rocks was unexpected.

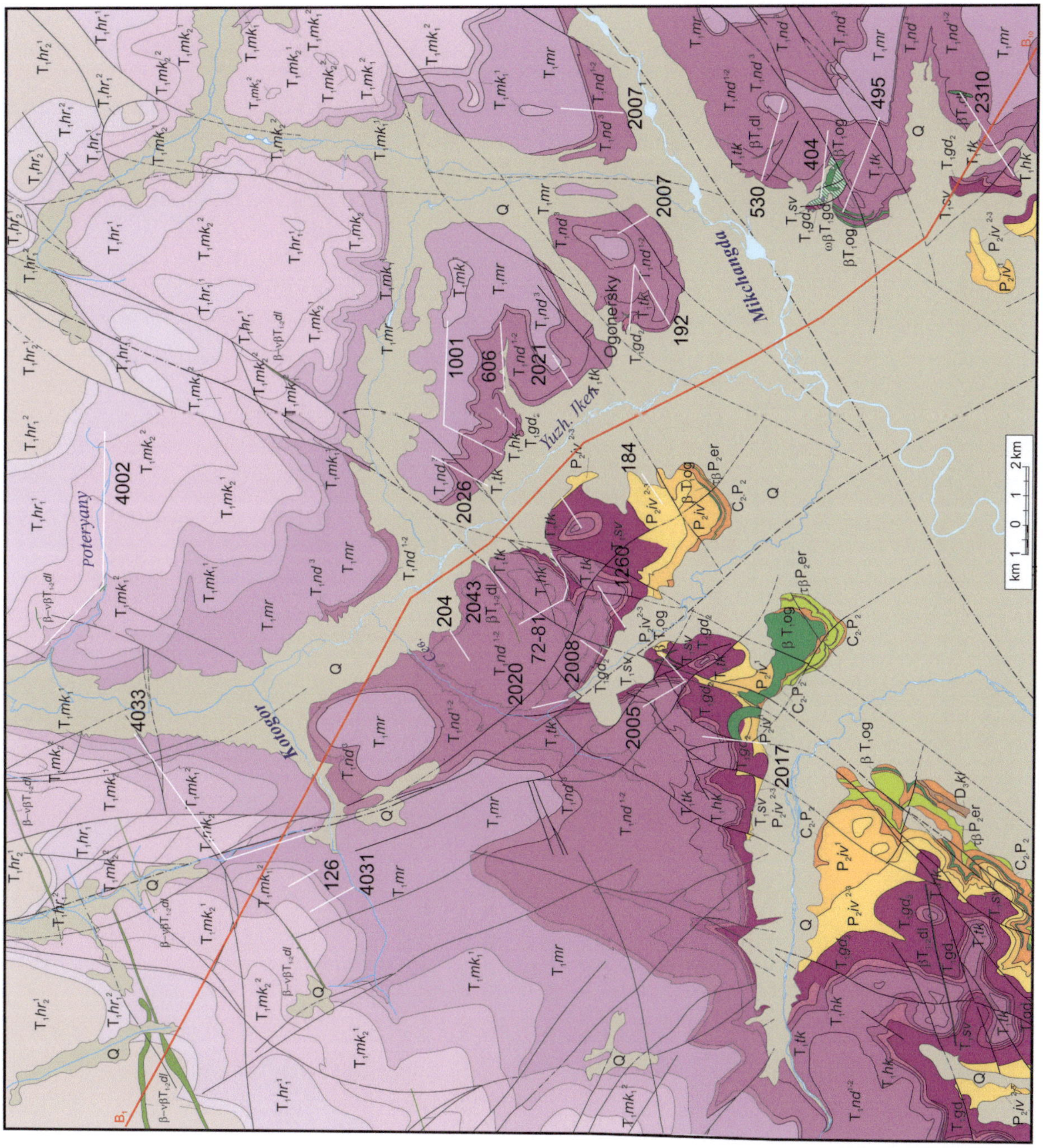

Fig. 3.17 Geological map of the r. Mikchangda lower course (After Mikhailov et al. 2003, unpublished) Conventions see on Fig. 2.4

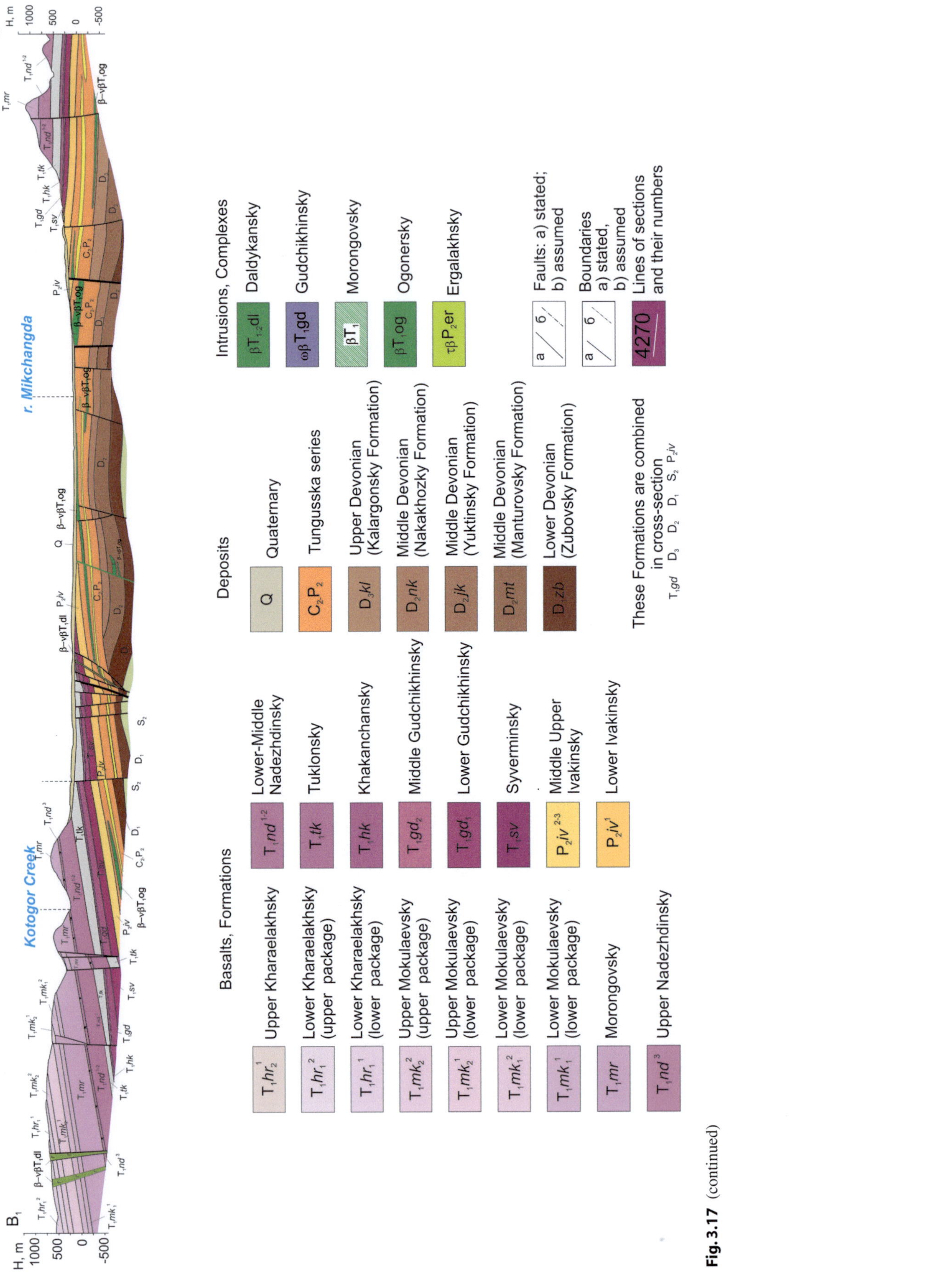

Fig. 3.17 (continued)

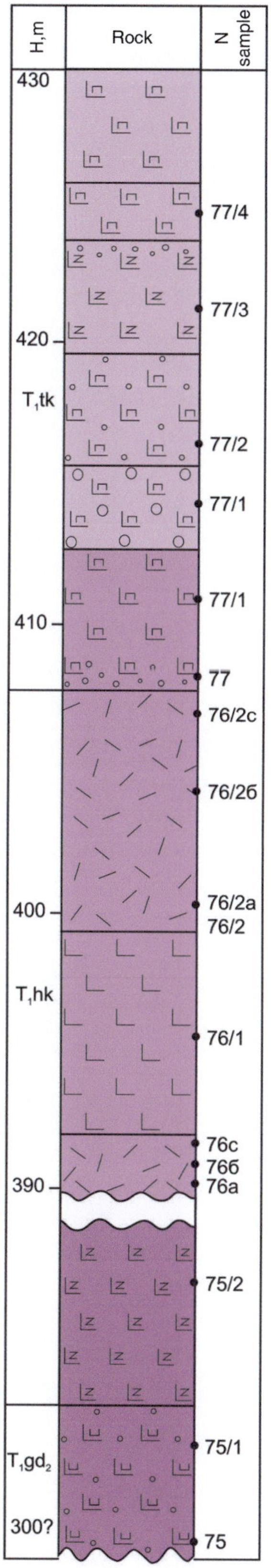

Fig. 3.18 Structure of volcanic rocks in eastern part of the Noril'sk area (nearby borehole MD-38)
After Krivolutskaya (2014)

Thus, during mapping of individual areas of the Noril'sk region, we demonstrated that rocks of the Tuklonsky Formation completely disappeared in its western part, whereas the thickness of volcanic rocks of the Nadezhdinsky Formation increased toward the Yenisei–Khatangsky Trough. The opposite pattern is typical in the eastern part of the region, where the number of flows and the thickness of the Tuklonsky Formation significantly increase, whereas the Nadezhdinsky Formation rapidly thins. It is clearly evident from the isopachs (in Fig. 3.24) that the centers of eruptions of these formations were located in different parts of the region; as this occurred, the area of abundance of volcanic rocks of the Lower Nadezhdinsky Formation slightly exceeded that of the Gudchikhinsky Formation, although their morphologies are essentially identical. This finding provides evidence for inheritance of the development of the Yenisei–Khatangsky rift structure in the northern Noril'sk region.

Based on the data obtained, the history of volcanism in the region was as follows. The first stage of magmatism was controlled by continental riftogenesis and was accompanied by eruption of deep melts (with the high Gd/Yb ratio providing evidence for the presence of garnet in the source). The active magmatic activity (Ivakinsky and Syverminsky Formations) gradually terminated and was localized in the areas adjacent to the Yenisei–Khatangsky Trough, which is most clearly represented by the Gudchikhinsky Formation. The stage of the formation of tuff and basalt of the Nadezhdinsky Formation occurred within the rift structures later. An analogous evolution produced the similar structures of the Western Siberian Platform where subalkaline rocks and basalts with a composition close to that of the Nadezhdinsky Formation were deposited (Al'mukhamedov et al. 1999; Reichow et al. 2005, 2009). The Khakanchansky Formation represents the lower part of the Nadezhdinsky Formation that formed during the explosive period.

Simultaneously with the beginning of the Nadezhdinsky stage, the new magmatic center was formed in the eastern part of the region with a source that differed significantly from that developing at the same time in the west. This new center was characterized by intrusion of tholeiitic and poikilophitic (sometimes picritic) basalt of the Tuklonsky Formation; based on all available parameters, its products of eruption are very close to the primary trappean stage of magmatism (Morongovsky–Samoedsky Formations). Thus, two eruption centers occurred in the post-Gudchikhinsky time; their products were simultaneously deposited in the central part of the region. As a result, we observe the alternating tuff and tuffite of the Khakanchansky (in essence, Nadezhdinsky) Formation and basalt of the Tuklonsky Formation, the tongues of which reached the central part of the region. The early explosive stage was replaced by the effusive stage, during which tholeiitic basalt of the Nadezhdinsky Formation was

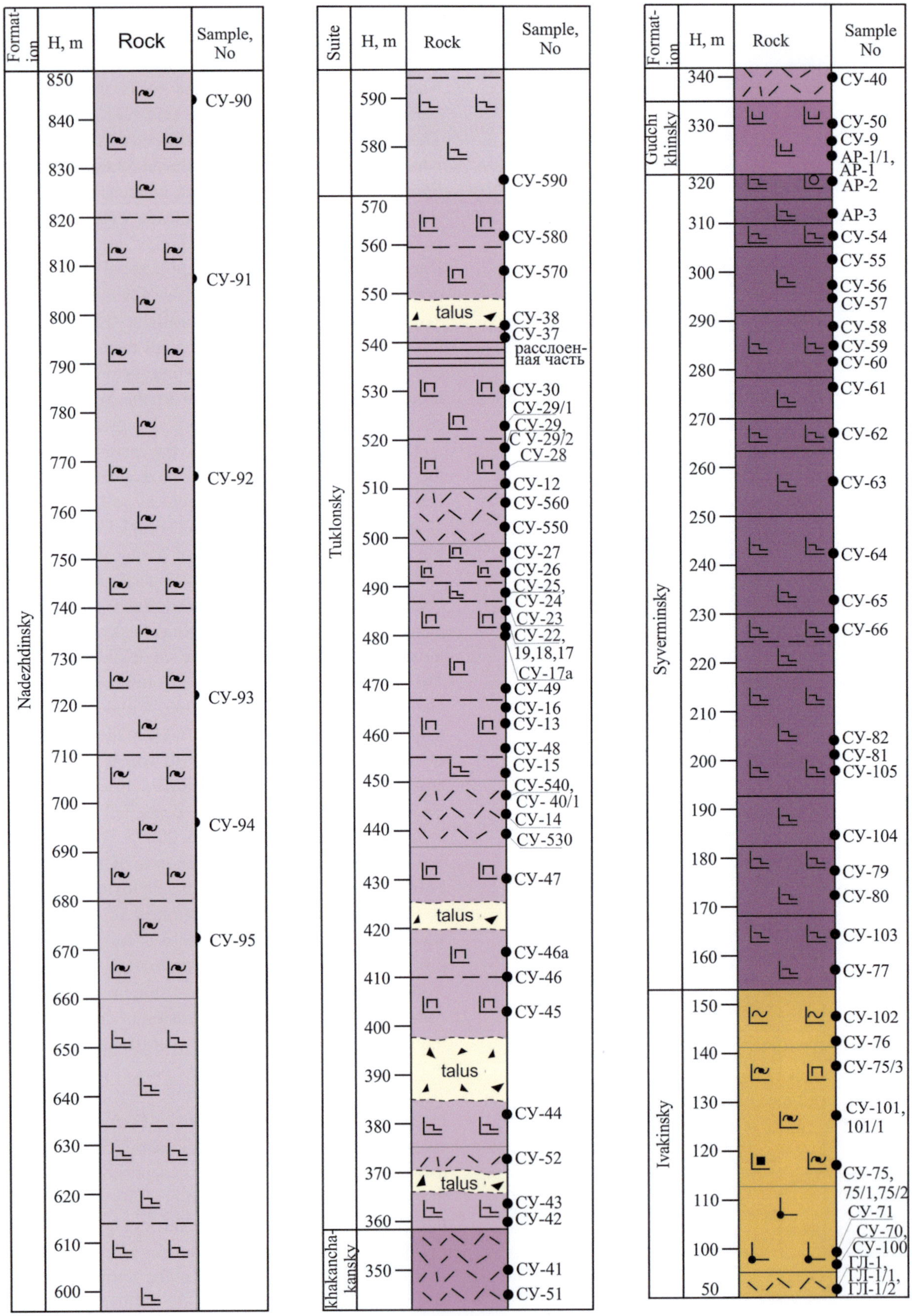

Fig. 3.19 Structure of volcanic rocks in eastern part of the Noril'sk area (Mount Sunduk) СУ-51 – No sample (After Krivolutskaya 2014)

Table 3.3 Composition of the volcanic rocks from the Mokulay Greek

No	1	2	3	4	5	6	7	8	9
No sample	M-3	M-4	M-6	M-7	M-8	M-9	M-11	M-12	M-13
Formation	T_1nd_2	T_1nd_2	T_1nd_2	T_1nd_2	T_1nd_2	T_1nd_2	T_1nd_2	T_1nd_2	T_1nd_2
SiO_2	50.6	51	50.6	51.6	51.88	52.4	50.2	51.4	52.7
Al_2O_3	15.2	14.8	15.3	15.1	14.57	14.7	15.4	14.9	14.6
TiO_2	1.04	0.98	1.04	1.02	1.05	1.19	1.02	1.02	1.16
Fe_2O_3	10.5	10.3	10.2	10.1	10.95	11.1	10.2	10.4	11
MnO	0.17	0.17	0.18	0.16	0.15	0.17	0.19	0.16	0.17
MgO	6.88	7.14	7.18	6.89	6.73	6.02	6.96	6.92	6.07
CaO	11.6	8	8.82	11.1	8.94	9.86	8.79	9.35	10
Na_2O	2.06	3.45	3.15	2.18	2.84	2.35	3.35	2.87	2.31
K_2O	0.59	1.4	1.21	0.71	1.33	1.43	1.18	1.14	1.38
P_2O_5	0.11	0.09	0.11	0.11	0.15	0.13	0.11	0.11	0.12
LOW	1.19	2.68	2.11	1.02	2.21	0.64	2.6	1.69	0.37
Summa	99.94	100.01	99.9	99.99	100.82	99.99	100	99.96	99.88
Ba	310	490	400	250	200	310	410	360	310
Li	8.2	18.5	16.1	10.3	11	8.12	18	13	7.66
Sc	31.8	28.5	30.8	30.5	32.4	30.6	28.7	30.4	30
Co	44.7	43.4	46.4	43.7	45.2	44	42.9	45.9	45.7
Ni	48.7	47	52.2	45.3	37.4	32.1	43	47.2	35.3
Cu	77.4	58	77.7	65.6	51.5	34.9	73.5	73.4	86.7
Zn	89.4	81.4	88.8	87.7	104	95.7	82.9	81.3	91.8
V	214	207	215	201	231	238	204	210	224
Cr	111	112	116	107	42.3	33.4	105	109	39.5
Ga	16.5	15.5	16.8	16.2	18.1	18.3	14.8	16.7	17.7
Rb	15.1	40.6	31.8	8.86	26.8	37	34.5	23.7	29.5
Sr	284	361	322	231	279	226	381	280	248
Y	21.5	19.9	21	20.9	24.1	24.5	19.7	21	24.7
Zr	117	113	118	115	138	142	109	115	136
Nb	8.27	7.71	7.75	7.97	9.1	9.41	7.48	7.81	9.28
Sn	1.16	0.89	1.11	1.08	1.26	1.28	1.04	1.1	1.18
Cs	0.69	0.65	0.8	0.83	0.35	0.69	0.75	0.83	0.43
Ba	246	430	351	262	303	310	370	305	306
La	16.9	16.1	16.3	16.5	18.4	19.6	15.5	16	18.6
Ce	34.1	31.6	33.6	33.4	36.6	38.6	31.2	32.5	38.3
Pr	4.32	4.04	4.2	4.15	4.67	4.83	4.04	4.08	4.74
Nd	17.2	16.5	16.1	16.6	17.9	19.5	15	15.9	18.6
Sm	3.84	3.75	4.03	3.68	4.41	4.34	3.55	3.75	4.21
Eu	1.06	1.01	1.05	1.08	1.1	1.16	0.99	0.98	1.12
Ti	6,291	5,928	6,291	6,170	6,370	7,198	6,170	6,170	7,017
Gd	3.72	3.63	3.68	3.73	4.06	4.23	3.39	3.58	4.21
Tb	0.63	0.56	0.65	0.64	0.66	0.7	0.58	0.62	0.67
Dy	3.77	3.7	3.7	3.75	4.22	4.44	3.53	3.48	4.39
Ho	0.82	0.81	0.8	0.77	0.89	0.92	0.8	0.8	0.85
Er	2.42	2.04	2.15	2.17	2.53	2.48	2.03	2.21	2.48
Tm	0.35	0.3	0.34	0.33	0.38	0.37	0.3	0.36	0.35
Yb	2.03	1.99	1.89	1.99	2.32	2.41	1.98	1.79	2.11
Lu	0.33	0.26	0.32	0.29	0.37	0.34	0.3	0.27	0.32
Hf	3.12	2.96	3.03	2.74	3.45	3.68	2.77	3.01	3.47
Ta	0.45	0.46	0.49	0.46	0.53	0.55	0.45	0.43	0.51
Th	2.91	2.91	3.01	2.98	3.57	3.85	3.05	3.21	3.64

(continued)

Table 3.3 (continued)

No	10	11	12	13	14	15	16	17	18
No sample	M-15	M-17	M-19	M-20/1	M-20/2	M-20/3	M-21	M-22	M-23
Formation	T_1nd_2	T_1nd_3	T_1nd_3	T_1nd_3	T_1nd_3	T_1nd_3	T_1nd_3	T_1nd_3	T_1nd_3
SiO_2	52.2	49.6	49.9	45.34	46.38	47.88	48.81	49.3	49.6
Al_2O_3	14.6	15.6	15.6	11.6	11.14	14.84	13.15	15.4	15.6
TiO_2	1.2	0.99	1.04	1.07	0.91	1.10	1.17	1.04	1.05
Fe_2O_3	11.1	10.9	10.9	10.62	8.82	11.46	12.01	11.1	11.2
MnO	0.17	0.17	0.17	0.16	0.13	0.11	0.16	0.18	0.16
MgO	6.18	6.29	6.03	7.59	6.59	6.87	7.98	6.4	5.97
CaO	10.1	12	11.6	9.46	10.63	5.67	7.66	12.3	11.7
Na_2O	2.33	2.35	2.44	3.24	2.31	3.76	2.67	2.3	2.43
K_2O	1.33	0.81	1.08	2.51	2.46	1.76	1.45	0.62	1.05
P_2O_5	0.13	0.18	0.17	0.12	0.14	0.12	0.10	0.18	0.18
LOW	0.6	0.96	1.01	9.14	11.02	6.95	5.15	1.11	1.06
Summa	99.94	99.85	99.94	100.86	100.55	100.53	100.33	99.93	100
Ba	340	280	260	207	204	270	165	300	330
Li	7.69	5.78	5.26	9.8	11.1	14.9	17	7.36	8.74
Sc	30	32.6	32	29.8	24.6	30.6	30.1	32.9	32.9
Co	45.1	45.2	41.8	48.3	38.2	47.3	50.3	44.7	44.8
Ni	35	83.3	65.1	102	85.9	108	109	102	73.6
Cu	74.4	102	72.3	103	85.6	132	267	128	104
Zn	94.7	94.2	94.6	85.8	80.4	77.1	79.5	119	96.1
V	229	217	209	184	164	200	212	224	219
Cr	36.9	135	80.1	61.8	67	77.6	86	145	85.5
Ga	18.4	15.6	16	13.9	12.3	11.8	15.6	16.3	16.4
Rb	25.4	10.8	22.1	76	92	61.2	62.4	4.67	22.7
Sr	233	283	273	103	87.3	128	209	298	287
Y	25	24.6	26	19.6	18.8	21.7	21.4	26.4	26.5
Zr	140	103	112	94.8	110	132	97.6	114	112
Nb	8.97	7	7.57	5.63	5.53	6.14	6.15	7.48	7.22
Sn	1.2	0.95	1.06	0.91	1.1	1.08	1.07	0.97	0.97
Cs	0.31	0.46	0.45	2.87	2.75	3.58	0.67	0.19	0.53
Ba	310	262	291	285	234	367	247	297	292
La	19	13.5	14.2	9.86	11.2	13.8	11	14	14.3
Ce	38.1	27.8	29	21.3	24.2	26.3	23.4	29.5	29.3
Pr	4.76	3.47	3.63	2.77	2.97	3.17	2.98	3.51	3.72
Nd	18.5	13.5	14	11.7	12.8	13.6	13	14.5	15.2
Sm	4.38	3.37	3.38	2.85	2.79	3.44	3.16	3.28	3.45
Eu	1.17	0.99	1.03	0.89	0.83	0.99	0.99	1.01	0.99
Ti	7,259	5,989	6,291	70,168	67,386	89,767	79,544	6,291	6,351
Gd	4.15	3.52	3.84	3.46	3.23	3.75	3.8	3.96	3.91
Tb	0.73	0.64	0.62	0.59	0.5	0.62	0.67	0.68	0.7
Dy	4.26	3.94	4.18	3.55	3.15	3.62	3.85	4.16	4.3
Ho	0.9	0.9	0.98	0.74	0.68	0.8	0.75	1	0.94
Er	2.62	2.67	2.73	2.17	2.01	2.44	2.33	2.93	2.95
Tm	0.38	0.4	0.43	0.32	0.27	0.35	0.35	0.42	0.45
Yb	2.08	2.38	2.48	1.87	1.68	1.9	2.05	2.55	2.66
Lu	0.35	0.41	0.44	0.29	0.29	0.32	0.32	0.44	0.44
Hf	3.8	2.51	2.66	2.56	2.58	3.26	2.48	2.47	2.72
Ta	0.51	0.37	0.37	0.31	0.33	0.36	0.36	0.38	0.43
Th	3.84	2.04	2.24	1.73	2.84	2.55	3.04	2.14	2.33

(continued)

Table 3.3 (continued)

No	19	20	21	22	23	24	25	26	27
No sample	M-23/1	M-25	M-26	M-27	M-28	M-29	M-29/1	M-30	M-30/1
Formation	T_1nd_3	T_1mr	T_1mr	T_1mr	T_1mr	T_1mr	T_1mr	T_1mr	T_1mr
SiO_2	49.8	47.46	49.1	48.89	49.3	48.45	49.2	47.5	49.6
Al_2O_3	15.6	14.2	15	14.36	14.8	16.4	15.1	13.4	14.3
TiO2	1.05	1.02	1.19	1.32	1.19	1.06	1.15	1.09	1.31
Fe_2O_3	10.9	11.46	11.6	12.42	11.7	11.38	11.6	11.62	12.5
MnO	0.17	0.2	0.21	0.18	0.17	0.16	0.17	0.17	0.19
MgO	6.11	7.71	7.31	6.74	7.24	6.69	7.45	7.69	6.53
CaO	11.6	7.3	12.1	11.3	12.1	11.43	12	8.68	12.2
Na_2O	2.42	3.77	2	2.33	2.05	2.5	1.97	4.51	2.06
K_2O	1.03	1.16	0.52	0.24	0.54	0.5	0.51	0.5	0.28
P_2O_5	0.18	0.10	0.12	0.15	0.12	0.15	0.11	0.15	0.14
LOW	1.04	5.66	0.69	2.91	0.71	2.14	0.76	5.56	0.91
Summa	99.9	100.05	99.84	100.85	99.92	100.87	100.02	100.87	100.02
Ba	290	205	180	190	150	140	120	109	100
Li	9.73	18.6	6.58	11.4	4.47	31.3	6.09	5.76	3.76
Sc	31.4	30.5	36.2	30.8	34.2	31.4	34.1	32.8	35.9
Co	42.6	50	53.2	47.6	50.7	44	52	50.4	47.5
Ni	70.8	104	98.5	101	94.7	96.1	102	99.4	67.3
Cu	89.1	104	113	111	120	143	126	128	133
Zn	92.2	80.7	93.1	71.9	87.2	82.9	87.3	87.1	91
V	213	227	273	240	267	248	252	222	279
Cr	80.3	93	122	97.7	118	111	85	92.8	142
Ga	16.5	13.4	17.3	13.3	16.2	13.3	15.8	14.4	16.1
Rb	21.7	21.9	5.31	82.8	6.24	14.2	6.01	10.2	8.51
Sr	279	131	194	108	196	189	182	36.5	193
Y	26.9	20.9	22.5	21.2	22.8	20.2	20.6	19.5	23.6
Zr	117	94.1	101	101	97.9	89.8	92.7	90.8	108
Nb	8	5.99	6.05	5.61	5.53	5.16	5.2	5.12	6.06
Sn	0.98	0.9	0.96	0.85	0.91	1.03	0.8	0.69	0.95
Cs	0.5	4.73	<0.1	5.63	<0.1	0.26	<0.1	2.35	0.34
Ba	301	369	168	209	162	179	140	44.6	117
La	14.2	10.6	10.6	10	9.86	9.2	8.98	9.14	9.9
Ce	30.1	22.1	23	22.1	21.3	21.2	19.4	19.2	23.1
Pr	3.57	2.81	2.99	2.84	2.74	2.74	2.6	2.59	2.91
Nd	15.6	12	13.2	12	12.4	11.2	11.4	11.3	12.1
Sm	3.47	3.18	3.31	3.13	3.04	2.84	2.98	2.97	3.29
Eu	1	0.97	1.03	1.01	1.03	1.03	1.08	0.9	1.12
Ti	6,351	6,170	7,198	8,021	7,198	6,412	6,956	6,593	7,924
Gd	3.94	3.35	3.94	3.45	3.56	3.46	3.52	3.21	3.85
Tb	0.7	0.56	0.64	0.61	0.62	0.63	0.61	0.55	0.67
Dy	4.36	3.64	3.95	3.57	3.91	3.59	3.45	3.3	4.05
Ho	0.95	0.75	0.84	0.7	0.85	0.7	0.76	0.69	0.88
Er	2.9	2.07	2.54	2.38	2.31	2.32	2.31	2.08	2.7
Tm	0.43	0.37	0.38	0.28	0.37	0.32	0.33	0.32	0.34
Yb	2.72	1.84	2.2	2.05	2.15	1.83	1.96	1.44	2.11
Lu	0.4	0.28	0.36	0.3	0.31	0.3	0.3	0.28	0.38
Hf	2.47	2.4	2.96	2.77	2.44	2.09	2.4	2.15	3.06
Ta	0.42	0.35	0.34	0.34	0.31	0.27	0.28	0.31	0.31
Th	2.42	1.95	1.9	1.84	1.72	1.64	1.6	1.64	1.65

(continued)

Table 3.3 (continued)

No	28	29	30	31	32	33	34	35	36
No sample	M-31	M-32/1	M-34	M-34/1	M-35/1	M-35/2	M-35/3	M-36	M-37/1
Formation	T_1mr	T_1mr	T_1mr	T_1mr	T_1mr	T_1mr	T_1mr	T_1mr	T_1mr
SiO_2	48.8	48.8	47.97	48.3	36.06	38.39	41.54	48.5	47.99
Al_2O_3	15.1	15.2	13.81	15.3	10.22	10.05	11.52	14.9	13.92
TiO2	1.2	1.26	1.16	1.21	0.74	0.88	0.87	1.27	1.12
Fe_2O_3	12	12.6	11.45	12	8.02	8.41	9.48	12.8	11.48
MnO	0.18	0.19	0.16	0.18	0.16	0.17	0.16	0.19	0.19
MgO	7.57	7.4	7.97	7.5	6.6	7.09	7.72	7.69	7.49
CaO	11.6	11.1	4.99	11.9	19.13	17.29	12.77	11	8.59
Na_2O	2.04	2.14	1.81	1.94	2.95	1.2	2.63	2.04	2.55
K_2O	0.42	0.54	3.88	0.32	0.87	2	1.58	0.53	0.88
P_2O_5	0.12	0.13	0.11	0.12	0.11	0.10	0.10	0.13	0.10
LOW	1.03	0.62	6.31	1.18	15.83	15.04	12.34	0.92	6.3
Summa	100.06	99.98	99.64	99.95	100.71	100.63	100.73	99.97	100.62
Ba	130	120	414	120	132	322	156	130	120
Li	5.68	8.25	17.2	5.87	9.7	24.6	12.8	10.9	13.7
Sc	33.6	33.5	30.8	35	20.9	21.1	27.5	33.1	33.3
Co	55.8	51.4	50.2	56.2	36.2	34.8	42.2	52.6	49.9
Ni	122	110	103	128	73.9	71	79	118	117
Cu	109	136	136	106	73.3	89.7	98.9	143	140
Zn	92.4	91.2	84.7	93.2	66.6	60.2	70.8	98	91.2
V	258	296	0.02	254	243	156	176	260	263
Cr	88.1	136	241	94	90.9	121	191	121	153
Ga	16.3	18.4	91.1	16.9	11.1	11.3	14.8	15.9	13.2
Rb	3.89	12.2	1.57	2.66	27.9	59.3	50.8	10.6	22.5
Sr	172	203	183	181	87.8	518	216	156	144
Y	21.2	24.5	331	21.7	17.5	15.8	17.6	22	22.2
Zr	91.1	93.7	22.5	91.5	71.7	65.2	72.4	80.4	83.1
Nb	4.47	5.05	91.2	4.65	4.14	3.92	4.11	4.35	4.35
Sn	0.86	0.78	<0.6	0.87	0.62	0.62	0.73	0.82	0.74
Cs	0.11	0.2	1.02	0.12	4.63	0.79	2.62	0.2	0.24
Ba	138	136	2.23	130	90.3	280	197	124	341
La	8.29	7.96	626	8.54	7.55	7.34	7.43	7.14	6.78
Ce	18.6	17.6	8.95	18.7	15.4	15.8	16.8	15.4	15.3
Pr	2.36	2.46	20.2	2.43	2.01	2.11	2.11	2.16	2.21
Nd	10.4	11	2.63	10.8	8.63	8.89	8.84	9.92	9.8
Sm	2.77	3.14	11.6	2.91	2.06	2.47	2.54	2.67	2.97
Eu	0.93	0.98	3.41	0.95	0.69	0.7	0.76	1.01	0.96
Ti	7,259	7,622	7,261	7,319	4,488	5,335	5,311	7,682	6,805
Gd	3.25	3.71	3.28	3.45	2.58	2.46	2.90	3.13	3.36
Tb	0.59	0.64	3.53	0.59	0.45	0.47	0.48	0.58	0.58
Dy	3.61	4.14	0.63	3.69	2.97	2.74	3.00	3.42	4.01
Ho	0.8	0.93	3.89	0.81	0.62	0.64	0.7	0.84	0.82
Er	2.33	2.44	0.88	2.29	1.88	1.64	1.84	2.36	2.31
Tm	0.35	0.38	2.35	0.33	0.26	0.21	0.28	0.29	0.34
Yb	2.04	2.25	0.34	2.01	1.6	1.43	1.74	2.27	1.94
Lu	0.32	0.38	2.15	0.31	0.25	0.25	0.26	0.31	0.32
Hf	2.17	2.48	0.33	2.29	1.97	1.87	2.11	1.99	2.34
Ta	0.3	0.32	2.31	0.28	0.23	0.21	0.26	0.26	0.26
Th	1.39	1.25	0.25	1.38	1.15	1.25	1.27	1.06	0.92

(continued)

Table 3.3 (continued)

No	37	38	39	40	41	42	43	44	45
No sample	M-37/2	M-37/4	M-38/3	M-38?	M-39	M-41	M-43	M-44	M-45/1
Formation	T_1mr	T_1mr	T_1mr	T_1mr	T_1mr	T_1mr	T_1mk	T_1mk	T_1mk
SiO_2	35.81	46.66	48.7	48.3	48.2	48.8	48.5	47.4	48.2
Al_2O_3	10.08	12.34	15.1	15.4	15.2	15.1	15.1	15.2	15.1
TiO2	0.95	1.28	1.24	1.31	1.22	1.24	1.27	1.36	1.42
Fe_2O_3	9.87	13.18	12.2	12.3	12.2	12.2	12.1	12.7	13
MnO	0.22	0.17	0.2	0.19	0.17	0.18	0.19	0.26	0.18
MgO	7.8	7.71	7.18	6.95	6.93	7.45	7.58	7.67	7.2
CaO	18.81	7.14	12	11.9	12.2	11.2	11.3	11.5	11.5
Na_2O	1.33	2.32	2.1	2.04	1.94	2.08	2.12	2.02	2.17
K_2O	1.52	1.35	0.23	0.27	0.2	0.67	0.53	0.22	0.26
P_2O_5	0.16	0.10	0.13	0.14	0.13	0.13	0.14	0.12	0.13
LOW	14.09	6.23	0.77	1.00	1.39	0.88	1.16	1.51	0.75
Summa	100.64	98.48	99.85	99.8	99.78	99.93	99.99	99.96	99.91
Ba	114	145	130	74	100	160	130	180	160
Li	18.9	21.9	4.83	6.92	4.91	9.05	12.7	9.06	5.98
Sc	25.5	30.4	32.7	33.1	32.4	32.6	32.6	35.4	30.3
Co	45.7	59.1	52.9	51.5	53.1	52.8	52.9	56.6	52.9
Ni	104	135	120	111	123	126	122	123	120
Cu	120	157	130	143	133	132	148	129	129
Zn	72.3	96	90.3	100	97.9	101	99.9	107	102
V	203	246	251	262	253	264	268	265	275
Cr	130	152	116	105	109	124	115	141	150
Ga	14.3	14.7	16.4	15.9	15.8	16.4	16.5	17.6	17.4
Rb	72.2	52.1	6.15	8.71	5.94	13	9.48	<2	<2
Sr	103	110	193	187	184	188	181	187	195
Y	18.8	22.7	21.8	23.1	21.8	22.3	23.8	22.9	23.8
Zr	67.6	80.2	83.7	84.7	81.3	82.9	86.1	79.1	85.4
Nb	3.47	4.32	4.88	5.37	5.02	4.72	5.06	4.2	4.43
Sn	0.89	0.67	0.68	0.73	0.83	0.79	0.83	0.91	0.8
Cs	0.65	0.85	0.32	0.41	0.27	0.23	0.13	<0.1	<0.1
Ba	157	220	104	99.5	62.6	158	127	101	113
La	7.24	7.27	7.61	7.85	7.58	7.43	7.96	7.31	7.26
Ce	14.5	15.3	16.5	17.7	17	16.8	17.8	16.4	17.1
Pr	1.81	2.11	2.23	2.5	2.45	2.37	2.52	2.34	2.39
Nd	8.84	9.65	10.3	11.6	11.2	10.6	11.8	11.5	12.4
Sm	2.15	2.75	2.75	3.1	2.85	2.95	3.26	3.22	3.47
Eu	0.75	0.94	0.88	1.08	1.02	0.99	1.05	1.14	1.17
Ti	5,753	7,743	7,501	7,924	7,380	7,501	7,682	8,227	8,590
Gd	2.76	3.23	3.28	3.47	3.24	3.22	3.43	3.55	3.65
Tb	0.46	0.58	0.61	0.65	0.61	0.59	0.63	0.67	0.65
Dy	2.89	3.91	3.73	4.02	3.85	3.77	4.16	4.16	4.25
Ho	0.74	0.83	0.86	0.85	0.84	0.84	0.84	0.86	0.9
Er	1.8	2.3	2.43	2.49	2.36	2.36	2.52	2.41	2.67
Tm	0.28	0.32	0.33	0.37	0.35	0.32	0.37	0.39	0.35
Yb	1.78	2.35	2.27	2.34	2.19	2.34	2.35	2.35	2.41
Lu	0.29	0.33	0.33	0.3	0.3	0.3	0.35	0.33	0.28
Hf	1.84	2.15	2.16	2.6	2.18	2.01	2.48	2.35	2.47
Ta	0.21	0.23	0.28	0.3	0.29	0.27	0.28	0.24	0.29
Th	0.74	0.94	1.08	1.15	1.14	1.11	1.17	0.86	0.98

(continued)

Table 3.3 (continued)

No	37	38	39	40	41	42	43	44	45
No sample	M-37/2	M-37/4	M-38/3	M-38?	M-39	M-41	M-43	M-44	M-45/1
Formation	T_1mr	T_1mr	T_1mr	T_1mr	T_1mr	T_1mr	T_1mk	T_1mk	T_1mk
SiO_2	35.81	46.66	48.7	48.3	48.2	48.8	48.5	47.4	48.2
Al_2O_3	10.08	12.34	15.1	15.4	15.2	15.1	15.1	15.2	15.1
TiO2	0.95	1.28	1.24	1.31	1.22	1.24	1.27	1.36	1.42
Fe_2O_3	9.87	13.18	12.2	12.3	12.2	12.2	12.1	12.7	13
MnO	0.22	0.17	0.2	0.19	0.17	0.18	0.19	0.26	0.18
MgO	7.8	7.71	7.18	6.95	6.93	7.45	7.58	7.67	7.2
CaO	18.81	7.14	12	11.9	12.2	11.2	11.3	11.5	11.5
Na_2O	1.33	2.32	2.1	2.04	1.94	2.08	2.12	2.02	2.17
K_2O	1.52	1.35	0.23	0.27	0.2	0.67	0.53	0.22	0.26
P_2O_5	0.16	0.10	0.13	0.14	0.13	0.13	0.14	0.12	0.13
LOW	14.09	6.23	0.77	1.00	1.39	0.88	1.16	1.51	0.75
Summa	100.64	98.48	99.85	99.8	99.78	99.93	99.99	99.96	99.91
Ba	114	145	130	74	100	160	130	180	160
Li	18.9	21.9	4.83	6.92	4.91	9.05	12.7	9.06	5.98
Sc	25.5	30.4	32.7	33.1	32.4	32.6	32.6	35.4	30.3
Co	45.7	59.1	52.9	51.5	53.1	52.8	52.9	56.6	52.9
Ni	104	135	120	111	123	126	122	123	120
Cu	120	157	130	143	133	132	148	129	129
Zn	72.3	96	90.3	100	97.9	101	99.9	107	102
V	203	246	251	262	253	264	268	265	275
Cr	130	152	116	105	109	124	115	141	150
Ga	14.3	14.7	16.4	15.9	15.8	16.4	16.5	17.6	17.4
Rb	72.2	52.1	6.15	8.71	5.94	13	9.48	<2	<2
Sr	103	110	193	187	184	188	181	187	195
Y	18.8	22.7	21.8	23.1	21.8	22.3	23.8	22.9	23.8
Zr	67.6	80.2	83.7	84.7	81.3	82.9	86.1	79.1	85.4
Nb	3.47	4.32	4.88	5.37	5.02	4.72	5.06	4.2	4.43
Sn	0.89	0.67	0.68	0.73	0.83	0.79	0.83	0.91	0.8
Cs	0.65	0.85	0.32	0.41	0.27	0.23	0.13	<0.1	<0.1
Ba	157	220	104	99.5	62.6	158	127	101	113
La	7.24	7.27	7.61	7.85	7.58	7.43	7.96	7.31	7.26
Ce	14.5	15.3	16.5	17.7	17	16.8	17.8	16.4	17.1
Pr	1.81	2.11	2.23	2.5	2.45	2.37	2.52	2.34	2.39
Nd	8.84	9.65	10.3	11.6	11.2	10.6	11.8	11.5	12.4
Sm	2.15	2.75	2.75	3.1	2.85	2.95	3.26	3.22	3.47
Eu	0.75	0.94	0.88	1.08	1.02	0.99	1.05	1.14	1.17
Ti	5,753	7,743	7,501	7,924	7,380	7,501	7,682	8,227	8,590
Gd	2.76	3.23	3.28	3.47	3.24	3.22	3.43	3.55	3.65
Tb	0.46	0.58	0.61	0.65	0.61	0.59	0.63	0.67	0.65
Dy	2.89	3.91	3.73	4.02	3.85	3.77	4.16	4.16	4.25
Ho	0.74	0.83	0.86	0.85	0.84	0.84	0.84	0.86	0.9
Er	1.8	2.3	2.43	2.49	2.36	2.36	2.52	2.41	2.67
Tm	0.28	0.32	0.33	0.37	0.35	0.32	0.37	0.39	0.35
Yb	1.78	2.35	2.27	2.34	2.19	2.34	2.35	2.35	2.41
Lu	0.29	0.33	0.33	0.3	0.3	0.3	0.35	0.33	0.28
Hf	1.84	2.15	2.16	2.6	2.18	2.01	2.48	2.35	2.47
Ta	0.21	0.23	0.28	0.3	0.29	0.27	0.28	0.24	0.29
Th	0.74	0.94	1.08	1.15	1.14	1.11	1.17	0.86	0.98

(continued)

Table 3.3 (continued)

No	46	47	48	49	50	51	52	53	54
No sample	M-45/2	M-45/3	M-47	M-48	M-49	M-50	M-52	M-53	M-54
Formation	T_1mk	T_1mk	T_1mk	T_1mk	T_1mk	T_1mk	T_1mk	T_1mk	T_1mk
SiO_2	48.2	48	48.4	48	48.6	48.5	48.3	47.9	48.16
Al_2O_3	15.1	15.7	15.2	15.2	15	15.3	14.7	14.9	12.73
TiO2	1.44	1.4	1.54	1.43	1.33	1.37	1.57	1.37	1.15
Fe_2O_3	12.8	12.7	13.3	13.3	12.5	12.8	13.9	13.1	12.04
MnO	0.18	0.21	0.2	0.19	0.19	0.19	0.2	0.18	0.17
MgO	7.18	7.32	6.6	7.39	7.29	7.26	6.93	7.55	7.72
CaO	11.5	11.5	11.1	11.1	11.9	11.5	11.1	10.3	9.63
Na_2O	2.14	2.17	2.29	2.18	2.08	2.13	2.15	1.92	1.35
K_2O	0.26	0.3	0.28	0.31	0.22	0.35	0.24	1.04	1.97
P_2O_5	0.14	0.13	0.15	0.14	0.13	0.13	0.14	0.13	0.14
LOW	1.01	0.58	0.75	0.62	0.84	0.56	0.86	1.67	4.66
Summa	99.95	100.01	99.81	99.86	100.08	100.09	100.09	100.06	99.73
Ba	130	55	130	130	94	130	125	160	156
Li	4.72	6.22	4.08	7.3	5.86	6.56	6.15	7.24	9.49
Sc	30	31.7	33.1	30.2	31.6	31.1	33.4	30.8	25.9
Co	52.5	57.2	54	54.3	54.2	52.5	57.2	54.9	41.5
Ni	122	138	127	125	125	108	122	121	101
Cu	130	185	219	168	172	196	137	155	121
Zn	99.9	108	110	102	100	97.5	112	101	79.6
V	280	263	283	255	276	273	266	249	163
Cr	160	137	105	109	159	132	97.1	105	147
Ga	18.2	17	18.7	17.9	17.2	17.8	18.4	16.1	14.8
Rb	2.41	3.59	2.00	2.16	7.08	3.3	7.03	18.8	8.4
Sr	197	191	193	175	203	199	186	207	530
Y	24.8	22.3	26.5	23.7	23	23.3	24.8	21.7	19.7
Zr	88.7	78.3	98.1	85.7	82.6	88.3	93.5	76.4	69.9
Nb	4.29	4.17	4.87	4.22	4.12	4.29	4.56	3.87	3.34
Sn	0.95	0.73	0.87	0.77	0.85	0.82	0.89	0.76	0.73
Cs	<0.1	0.17	<0.1	<0.1	0.45	<0.1	0.22	<0.1	1.71
Ba	117	108	127	120	103	124	89.6	154	291
La	7.28	6.42	7.67	7.03	6.77	7.05	7.83	6.25	5.79
Ce	16.7	15.2	17.9	16.4	15.2	15.9	17.8	14.6	13.7
Pr	2.43	2.22	2.53	2.26	2.21	2.27	2.47	2.06	1.89
Nd	11.8	11.1	13.5	10.6	11	11.9	12	9.64	8.81
Sm	2.95	3.07	3.19	3.02	2.95	3.12	3.23	2.9	2.49
Eu	1.12	1.03	1.11	0.98	1.07	1.11	1.02	1.04	0.87
Ti	8,711	8,469	9,315	8,650	8,045	8,287	9,497	8,287	6,962
Gd	3.53	3.33	3.97	3.63	3.55	3.47	3.82	3.29	3.16
Tb	0.67	0.6	0.68	0.66	0.62	0.64	0.72	0.57	0.54
Dy	4.35	3.93	4.57	4.1	3.98	4.15	4.37	3.78	3.46
Ho	0.91	0.82	1.01	0.89	0.9	0.9	0.92	0.83	0.77
Er	2.43	2.26	2.68	2.58	2.25	2.37	2.71	2.39	2.1
Tm	0.34	0.34	0.37	0.39	0.37	0.37	0.4	0.36	0.31
Yb	2.46	2.11	2.62	2.35	2.36	2.3	2.36	2.05	1.75
Lu	0.33	0.3	0.36	0.32	0.3	0.33	0.36	0.28	0.28
Hf	2.07	2.26	2.51	2.47	2.36	2.62	2.16	2.41	2.22
Ta	0.25	0.23	0.27	0.21	0.24	0.24	0.26	0.21	0.21
Th	1.06	0.89	1.17	0.96	0.98	0.98	1.15	0.86	0.78

(continued)

Table 3.3 (continued)

No	55	56	57	58	60	61	62	63	64
No sample	M-55	M-56	M-57	M-58	M-59	M-61	M-62	M-63	M-64
Formation	T_1mk	T_1mk	T_1mk	T_1mk	T_1mk	T_1Kh	T_1Kh	T_1Kh	T_1Kh
SiO_2	48.4	48.1	47.7	47.8	47	48.3	48.6	48.5	47.3
Al_2O_3	15.3	15.6	15.9	15.6	15.4	15.1	15.1	15.1	14.9
TiO2	1.41	1.36	1.35	1.36	1.35	1.52	1.47	1.47	1.52
Fe_2O_3	12.7	12.6	12.7	12.5	12.7	13.5	13.4	13.3	14
MnO	0.2	0.19	0.16	0.18	0.19	0.2	0.21	0.2	0.2
MgO	7.21	7.17	6.73	7.05	7.58	7.32	7.14	6.54	7
CaO	11.7	11.6	11.7	11.6	11.2	10.8	11.3	11.7	11.1
Na_2O	2.16	2.13	2.1	2.07	1.93	2.25	2.27	2.19	2.11
K_2O	0.23	0.22	0.21	0.21	0.15	0.46	0.31	0.21	0.29
P_2O_5	0.13	0.13	0.13	0.13	0.12	0.14	0.13	0.13	0.14
LOW	0.7	0.82	1.32	1.47	2.04	0.5	0.11	0.8	1.28
Summa	100.14	99.92	100	99.97	99.66	100.09	100.04	100.14	99.84
Ba	77	65	72	91	89	170	110	63	130
Li	5.5	4.35	4.59	5.08	7.07	12.2	5.88	3.77	6.26
Sc	34.1	31.6	31.5	32.3	28.1	34.9	31.6	32.2	33.1
Co	55.5	53.7	53.3	52.7	52.9	56.6	52.8	51.5	56.4
Ni	124	122	124	128	124	127	115	86.2	100
Cu	162	158	145	145	146	189	165	185	135
Zn	105	97.8	98.1	94.8	90.4	112	104	106	109
V	252	267	259	256	246	263	277	270	268
Cr	134	134	134	140	123	126	137	113	86.3
Ga	17	17.7	16.9	17.2	16.3	17.4	17	17	18
Rb	2.82	<2	2.23	2.26	<2	9.84	6.08	3.43	3.27
Sr	179	178	176	178	169	159	170	167	154
Y	23.3	23.1	23.2	23.1	22.3	24.3	24.4	24.5	25
Zr	84.7	85.8	80.9	83.9	76.9	88.7	88.2	82.1	91.7
Nb	4.15	3.99	3.48	3.72	3.96	4.47	3.97	3.79	4.31
Sn	0.82	0.8	0.72	0.8	0.72	0.8	0.96	0.89	0.95
Cs	0.18	0.16	<0.1	<0.1	<0.1	0.22	0.28	0.2	<0.1
Ba	84.8	95.7	87.6	85.4	57.5	105	91.8	79.3	87.8
La	6.42	6.54	6.06	6.57	6.22	6.71	6.33	5.72	6.58
Ce	15.6	14.9	14.6	15.1	14.7	16	15.1	14	15.6
Pr	2.18	2.11	2.14	2.17	2.12	2.33	2.2	2.07	2.13
Nd	10.9	10.6	10.5	10.3	10.4	11	11.5	10.9	10.9
Sm	2.75	2.84	2.71	2.64	2.88	3.22	2.95	2.95	3.00
Eu	1.03	0.98	0.96	0.94	0.95	1.11	1.11	1.09	1.06
Ti	8,529	8,227	8,166	8,227	8,166	9,194	8,892	8,892	9,194
Gd	3.53	3.33	3.2	3.17	3.41	3.77	3.56	3.42	3.8
Tb	0.62	0.61	0.6	0.64	0.59	0.67	0.66	0.65	0.72
Dy	3.98	3.88	3.98	3.92	3.7	4.35	4.07	4.32	4.34
Ho	0.84	0.86	0.86	0.88	0.83	0.9	0.97	0.94	0.94
Er	2.42	2.42	2.49	2.3	2.4	2.64	2.6	2.59	2.7
Tm	0.37	0.37	0.35	0.34	0.33	0.39	0.4	0.34	0.36
Yb	2.61	2.4	2.22	2.22	2.22	2.35	2.42	2.51	2.57
Lu	0.31	0.34	0.31	0.35	0.29	0.34	0.35	0.32	0.38
Hf	2.27	2.14	2.16	2.21	2.17	2.61	2.4	2.17	2.41
Ta	0.22	0.23	0.23	0.23	0.26	0.28	0.27	0.24	0.24
Th	0.87	0.95	0.89	0.88	0.91	0.9	0.97	0.86	0.94

(continued)

Table 3.3 (continued)

No	65	66	67	68	69	70	71	72	73
No sample	M-65	M-66	M-67	M-68	M-69	M-70	M-71	M-72	M-73
Formation	T_1Kh	T_1Kh	T_1Kh	T_1Kh	T_1Kh	T_1Kh	T_1Kh	T_1Kh	T_1Kh
SiO_2	47.9	47.4	47.8	48	46.57	47.3	47.9	47.8	48
Al_2O_3	15.2	15.2	15.1	15.2	14.94	15.3	15.4	15	15.4
TiO2	1.41	1.38	1.41	1.42	1.31	1.42	1.4	1.38	1.43
Fe_2O_3	13.1	13.1	13.3	13.2	12.52	13.2	12.9	13.2	13.1
MnO	0.2	0.2	0.19	0.19	0.17	0.17	0.21	0.19	0.2
MgO	7.41	7.69	7.34	7.6	7.53	7.3	6.95	7.69	7.5
CaO	11.1	10.8	11.4	11.4	11.05	11.5	11.6	11.4	11.1
Na_2O	2.06	2.23	2.06	2.17	2.28	2.12	2.13	2.04	2.23
K_2O	0.49	0.53	0.24	0.25	0.16	0.28	0.38	0.17	0.49
P_2O_5	0.12	0.13	0.13	0.13	0.14	0.13	0.13	0.12	0.12
LOW	1.09	1.3	1.04	0.46	3.93	1.32	0.98	0.9	0.64
Summa	100.08	99.96	100.01	100.02	100.61	100.04	99.98	99.89	100.21
Ba	69	180	85	100	100	93	85	74	88
Li	4.59	7.41	3.59	5.37	3.05	3.7	3.16	4.6	7.96
Sc	31.3	32.6	30.9	34	33.5	33.3	30.6	31.7	33.9
Co	52.8	58.4	55.6	57.4	55.8	55.2	52.3	55.6	53.4
Ni	134	145	138	153	141	137	130	134	120
Cu	219	105	480	174	154	125	160	144	160
Zn	101	109	106	113	105	108	103	105	106
V	270	257	258	257	272	277	274	260	274
Cr	171	171	165	178	195	207	196	196	200
Ga	17.1	16.5	17.2	16.7	17.2	18.9	17	17.2	17.4
Rb	7.96	11.1	2.94	<2	2.36	3.39	6.98	2.31	7.47
Sr	156	162	167	163	166	161	158	166	156
Y	23.7	23.3	23.8	23.5	24.1	24.8	22.4	24	23.3
Zr	84.6	81	82.8	81.5	80	84.2	80.3	82.6	83.4
Nb	3.95	3.49	3.9	4.05	3.89	4.11	3.86	3.95	4.14
Sn	0.82	0.91	0.87	0.86	0.8	0.83	0.75	0.82	0.85
Cs	0.13	0.15	0.1	0.11	0.13	<0.1	0.17	0.16	0.3
Ba	97.3	169	75.7	89.3	59.8	80.7	77.7	70.3	94.6
La	5.99	5.59	5.75	6.12	5.8	6.23	5.81	6	6.03
Ce	13.8	13.3	14	14.3	13.9	14.6	14.1	14.2	14.9
Pr	2	1.89	2.03	2.1	1.96	2.06	2.11	1.98	2.19
Nd	9.89	9.95	9.89	10.3	9.88	10.3	9.6	9.87	10.3
Sm	2.6	2.78	2.8	2.81	2.86	2.86	2.7	2.59	2.99
Eu	1.00	0.91	1.07	1.05	0.98	1.07	1.06	1.05	1.18
Ti	8,529	8,348	8,529	8,590	7,960	8,590	8,469	8,348	8,650
Gd	3.52	3.24	3.23	3.64	3.61	3.84	3.51	3.45	3.86
Tb	0.63	0.62	0.61	0.61	0.67	0.68	0.61	0.66	0.7
Dy	3.97	3.81	4.12	4.26	4.06	3.94	3.96	4.02	4.19
Ho	0.95	0.87	0.87	0.92	0.92	0.96	0.88	0.86	0.85
Er	2.62	2.6	2.46	2.52	2.53	2.56	2.5	2.34	2.65
Tm	0.33	0.32	0.34	0.32	0.36	0.35	0.34	0.35	0.35
Yb	2.42	2.22	2.23	2.2	2.46	2.4	2.36	2.59	2.31
Lu	0.36	0.33	0.32	0.31	0.31	0.31	0.31	0.34	0.36
Hf	2.3	2.3	2.11	2.29	2.14	2.33	2.34	2.31	2.37
Ta	0.25	0.24	0.27	0.26	0.25	0.25	0.21	0.25	0.25
Th	0.97	0.88	0.81	0.91	0.84	0.9	0.82	0.95	0.89

(continued)

Table 3.3 (continued)

No	74	75	76	77	78	79	80	81	82
No sample	M-74	M-75	M-76	M-77	M-78	M-79	M-80	M-81	M-82
Formation	T_1Kh	T_1Kh	T_1Kh	T_1Kh	T_1Kh	T_1Kh	T_1Kh	T_1Kh	T_1Kh
SiO_2	47.7	47.5	48.3	48.4	47.8	47.8	47.6	48.5	48.9
Al_2O_3	15	15.6	15.1	15	15.6	15.4	15.5	14.9	14.8
TiO2	1.43	1.34	1.42	1.42	1.44	1.43	1.43	1.57	1.33
Fe_2O_3	13.2	12.8	13.2	13	13.1	13.1	13.3	13.5	12.5
MnO	0.2	0.2	0.19	0.19	0.21	0.21	0.18	0.22	0.19
MgO	7.59	7.89	7.27	7.48	7.33	7.29	7.08	7.31	7.29
CaO	11	11.6	11.4	11.4	11.5	11.5	11.3	11.2	11.4
Na_2O	2.07	2.15	2.16	2.15	2.1	2.05	2.04	2.26	1.96
K_2O	0.5	0.17	0.26	0.19	0.22	0.22	0.24	0.41	0.17
P_2O_5	0.12	0.12	0.13	0.13	0.14	0.14	0.14	0.14	0.12
LOW	1.27	0.55	0.72	0.78	0.56	1.04	1.14	0.1	1.34
Summa	100.08	99.92	100.15	100.14	100	100.18	99.95	100.11	100
Fe_2O_3	5.91	5.26	5.64	5.39	5.11	6	6.45	4.41	6
FeO	6.53	6.79	6.79	6.85	7.18	6.4	6.2	8.16	5.88
Ba	56	100	86	120	54	55	61	150	
Li	6.24	3.34	4.81	5.61	4.71	4.54	4.79	6.02	4.34
Sc	33.5	33.1	32.2	31	30.2	32.7	29.7	31.8	30.9
Co	53.8	54.1	53.5	52	52.7	53.7	53.8	55.1	52
Ni	120	131	117	115	115	121	121	122	121
Cu	154	134	194	151	143	132	123	169	147
Zn	105	102	106	104	104	107	106	113	100
V	265	271	266	265	279	291	235	253	261
Cr	185	183	155	150	161	163	137	135	175
Ga	16.6	16.5	17.3	17.6	18.1	18.8	17	17.1	17.9
Rb	8.16	<2	2.24	<2	4.71	2.89	<2	7.22	4
Sr	155	160	190	195	200	205	174	187	193
Y	22.2	23	22.5	23.6	23.8	24.4	21.5	24	22.6
Zr	81.6	80.4	87.6	87.4	89.6	92.9	80.1	89.3	84.8
Nb	4.2	4.08	4.74	4.36	4.98	4.97	4.2	4.86	4.14
Sn	0.83	0.89	0.85	0.79	0.87	0.85	0.84	0.8	0.77
Cs	0.19	<0.1	0.18	0.37	0.38	0.14	0.12	0.22	0.42
Ba	105	67.5	114	90	119	89.2	91.1	113	64.7
La	6.08	5.95	7.16	7.24	7.73	7.82	6.54	7.19	6.78
Ce	14.5	13.8	17.1	16.9	18.1	18.5	15.6	17.5	16.7
Pr	2.2	2.08	2.41	2.43	2.65	2.71	2.28	2.42	2.35
Nd	9.97	9.32	10.5	10.4	11.7	12.5	10.2	11.3	10.7
Sm	2.93	2.65	3.13	2.91	3.24	3.42	2.85	3.33	3.02
Eu	1.1	0.9	1.14	1.11	1.1	1.17	0.99	1.04	1.06
Ti	8,650	8,106	8,590	8,590	8,711	8,650	8,650	9,497	8,045
Gd	3.35	3.57	3.81	3.57	4.12	3.76	3.22	3.49	3.63
Tb	0.7	0.6	0.69	0.74	0.78	0.71	0.64	0.65	0.62
Dy	4.06	3.88	4.08	4.14	4.31	4.64	3.87	4.23	4.19
Ho	0.87	0.84	0.85	0.87	0.92	0.91	0.79	0.84	0.82
Er	2.64	2.52	2.47	2.52	2.6	2.51	2.32	2.53	2.41
Tm	0.36	0.32	0.37	0.37	0.38	0.4	0.32	0.37	0.37
Yb	2.27	2.22	2.11	2.35	2.32	2.53	2.09	2.42	2.25
Lu	0.3	0.33	0.32	0.33	0.36	0.38	0.33	0.34	0.39
Hf	2.61	2.09	2.58	2.52	2.54	2.67	2.25	2.54	2.37
Ta	0.24	0.21	0.23	0.25	0.29	0.29	0.22	0.26	0.27
Th	0.91	0.92	0.98	0.99	1.08	0.96	0.92	0.97	0.94

(continued)

Table 3.3 (continued)

No	83	84	85	86	87	88	89	90	91
No sample	M-83	M-83/1	M-84	M-85	M-87	M-88	M-89	M-91	M-92
Formation	T_1Kh	T_1Kh	T_1Kh	T_1Kh	T_1Kh	T_1Kh	T_1Kh	T_1Kh	T_1Kh
SiO_2	48.9	50.4	48.4	47.9	47.7	47.3	48.2	48.5	47.8
Al_2O_3	14.7	14.8	15.2	15.5	15.4	15.5	15.8	15.5	15.8
TiO2	1.34	1.43	1.55	1.46	1.45	1.43	1.4	1.49	1.48
Fe_2O_3	12.7	12.8	13.1	13.3	13	13	12.3	13.2	13.3
MnO	0.19	0.19	0.2	0.19	0.18	0.19	0.18	0.2	0.21
MgO	7.36	6.47	6.7	6.94	7.66	7.24	6.9	7.08	7.09
CaO	11.3	10.6	11.1	11.2	10.8	11.3	11.6	10.9	11
Na_2O	2.04	2.06	2.21	2.18	2.08	2.07	2.15	2.21	2.21
K_2O	0.22	0.15	0.19	0.26	0.51	0.23	0.17	0.46	0.27
P_2O_5	0.12	0.14	0.15	0.14	0.13	0.13	0.13	0.13	0.14
LOW	1.1	1.12	1.16	0.82	1.18	1.55	1.21	0.63	0.7
Summa	99.97	100.16	99.96	99.89	100.09	99.94	100.04	100.3	100
Ba	77	67	150	120	100	120	81	89	88
Li	5.13	4.00	4.4	5.71	5.11	4.98	5.23	8.75	3.75
Sc	32.4	30.9	32.5	32.4	32.8	33.5	36	34.2	35.4
Co	53.7	52.6	53	56.4	53.5	55.3	55.6	51.2	51
Ni	123	165	115	135	143	124	130	114	86
Cu	146	198	138	144	93.3	149	154	167	146
Zn	104	110	115	110	103	104	107	99.3	99.1
V	274	247	273	274	274	246	264	294	272
Cr	182	108	97.1	116	173	123	132	141	119
Ga	18	17	18.1	18.8	17.7	17.8	17.9	20	18.5
Rb	7.1	<2	2.28	<2	6.62	<2	<2	10.2	2.36
Sr	197	177	200	190	164	193	182	171	160
Y	23	23.6	25.1	25	24.5	23.6	23.8	26.6	25.2
Zr	87.4	86.9	96.7	91.7	85.7	83.4	86.7	94.1	88.6
Nb	4.49	4.39	4.83	4.78	4.18	4.36	4.13	4.57	4.4
Sn	0.93	0.83	0.94	0.84	0.91	0.7	0.82	0.95	1.1
Cs	0.4	0.27	0.24	0.14	<0.1	<0.1	0.22	0.26	<0.1
Ba	78.6	63.3	97.4	120	98.3	102	76.8	109	97.6
La	7.01	7.45	7.96	7.61	6.07	6.9	6.87	7.15	6.67
Ce	17	18.2	19.4	18.4	15	16.6	16.2	16.8	15.8
Pr	2.53	2.49	2.66	2.6	2.13	2.32	2.33	2.41	2.24
Nd	11.2	11.2	12.1	11.1	10.4	10.8	10.3	10.9	10.9
Sm	2.96	3.28	3.62	3.36	3.01	2.92	3.05	2.97	3.19
Eu	1.08	1.16	1.24	1.05	1.05	1.1	1.07	1.12	1.05
Ti	8,106	8,650	9,376	8,832	8,771	8,650	8,469	9,013	8,953
Gd	3.5	3.61	3.85	3.78	3.35	3.54	3.33	3.7	3.9
Tb	0.61	0.69	0.73	0.74	0.64	0.64	0.63	0.71	0.68
Dy	4.21	4.36	4.73	4.3	4.18	4.2	4.18	4.43	4.33
Ho	0.84	0.85	0.94	0.85	0.84	0.82	0.86	0.93	0.89
Er	2.51	2.48	2.76	2.72	2.57	2.53	2.32	2.65	2.68
Tm	0.35	0.34	0.42	0.41	0.35	0.38	0.39	0.42	0.42
Yb	2.2	2.26	2.65	2.38	2.51	2.32	2.26	2.6	2.27
Lu	0.35	0.41	0.37	0.4	0.36	0.36	0.38	0.43	0.37
Hf	2.68	2.58	2.79	2.76	2.3	2.5	2.44	2.6	2.27
Ta	0.27	0.24	0.3	0.24	0.24	0.22	0.21	0.25	0.24
Th	1.02	1.03	1.03	1.04	0.9	0.95	0.99	1.04	0.93

(continued)

Table 3.3 (continued)

No	92	93	94	95	96	97	98
No sample	M-93	M-94	M-95	M-96	M-98	M-99	M-100
Formation	T_1Kh	T_1km	T_1km	T_1km	T_1km	T_1km	T_1km
SiO_2	46.6	47.8	47.9	48.8	48.5	48.4	48.1
Al_2O_3	16.3	15.3	15.6	15.1	15.1	15.2	15.7
TiO2	1.49	1.4	1.41	1.52	1.5	1.47	1.49
Fe_2O_3	13.3	13.1	13	13.2	13.5	13.4	13.4
MnO	0.18	0.19	0.2	0.19	0.21	0.2	0.2
MgO	7.12	7.66	7.22	6.41	7.22	7.14	7.11
CaO	10.8	11	11.7	11.5	11.2	11.2	11.2
Na_2O	2.18	2.1	2.09	2.19	2.26	2.25	2.25
K_2O	0.18	0.43	0.17	0.22	0.35	0.3	0.3
P_2O_5	0.14	0.12	0.12	0.13	0.13	0.13	0.13
LOW	1.71	1.11	0.79	0.81	0.16	0.39	0.34
Summa	100	100.21	100.2	100.07	100.13	100.08	100.22
Fe_2O_3	8.03	4.82	5.38	6.34	4.4	3.71	3.83
Li	4.05	3.82	5.11	4.49	5.77	5.94	5.71
Sc	36.1	34.4	36.1	38.8	32.7	34.4	33.7
Co	55.1	52.3	53.2	50.3	46.5	49	48.9
Ni	98.1	134	138	96.4	120	128	130
Cu	137	135	166	186	138	145	147
Zn	106	90.9	91.5	99.7	101	106	108
V	290	278	268	291	294	282	272
Cr	101	177	180	121	132	133	127
Ga	19.6	18.7	18.2	19.7	19.2	18.5	18.7
Rb	<2	5.71	3.88	2.91	6	8.82	7.77
Sr	167	164	165	170	173	174	167
Y	26	24.3	23.9	25.6	26.7	25.6	25.3
Zr	89.6	85.1	85.9	98.6	95.3	94.5	91.5
Nb	4.23	4.24	3.83	4.85	4.63	4.32	4.26
Sn	0.92	0.93	0.92	0.93	1.01	0.74	0.88
Cs	<0.1	<0.1	0.2	0.16	0.23	0.29	0.33
Ba	76.6	103	62.6	96.9	117	129	97
La	6.52	6.22	6	7.27	6.79	6.67	6.66
Ce	15.6	15.3	14.3	17.4	16.4	16.3	16
Pr	2.26	2.25	2.08	2.41	2.38	2.45	2.24
Nd	10.2	10.3	9.62	12.2	12.4	11.9	12.3
Sm	3.01	2.77	2.61	3.09	3.17	3.02	2.94
Eu	1.04	0.98	1.03	1.2	1.09	1.17	1.13
Ti	9,013	8,469	8,529	9,194	9,074	8,892	9,013
Gd	3.79	3.79	3.81	4.26	4.07	4.05	4.11
Tb	0.66	0.61	0.6	0.65	0.73	0.67	0.68
Dy	4.66	4.66	4.29	4.79	4.67	4.62	4.27
Ho	0.94	0.84	0.87	0.89	0.97	0.93	0.89
Er	2.74	2.85	2.65	3.09	2.86	2.78	2.63
Tm	0.38	0.4	0.43	0.38	0.5	0.46	0.4
Yb	2.34	2.43	2.4	2.46	2.44	2.32	2.2
Lu	0.42	0.34	0.36	0.37	0.37	0.35	0.35
Hf	2.21	2.43	2.44	2.68	2.77	2.49	2.68
Ta	0.24	0.25	0.24	0.26	0.26	0.24	0.25
Th	0.97	0.97	0.93	1.11	1.04	1.08	1.01

Note: Analyses were carried out in VSEGEI by XRF and ICP-MS methods

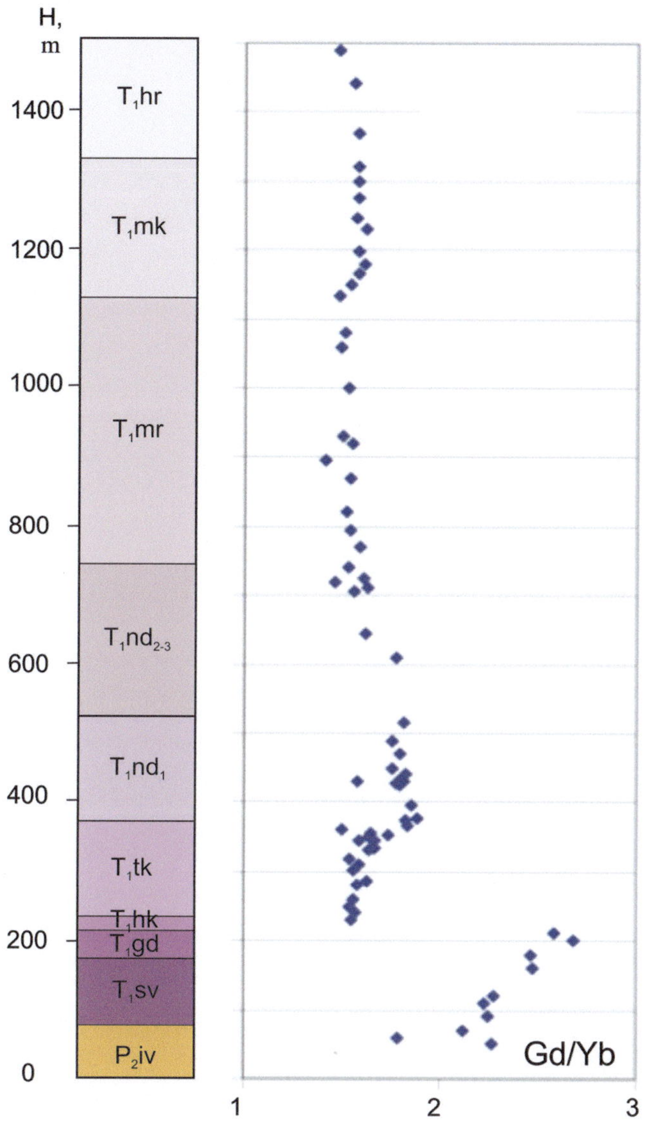

Fig. 3.20 Composite section of the volcanic rocks in eastern part of the Noril'sk area
After Krivolutskaya (2014)

deposited, whereas eruption of Tuklonsky lava practically terminated. In conclusion, we should mention that the Khakanchansky, Tuklonsky, and Nadezhdinsky Formations are not intermediate in time but characterize different tectonic structures and are related to various types of magmatism: riftogenic and proper trappean platform (which occurred only on the Eastern Siberian Platform). Because the Khakanchansky Formation is identical to the Nadezhdinsky Formation in terms of its petrographic and geochemical characteristics, it would be appropriate to consider it as a member of the Nadezhdinsky Formation. The subsequent evolution of magmatism in the Noril'sk region and on the Siberian Platform was controlled by the evolution of magmatism of the Tuklonsky type.

3.3.2 High-Mg Volcanic Rocks

The high-magnesian rocks of the Noril'sk district have attracted considerable attention. They have a primitive composition and may, on the one hand, be considered the parental magmas of the Siberian flood basalts. On the other hand, they host rich Pt–Cu–Ni mineralization in the differentiated ultramafic–mafic massifs (Dyuzhikov et al. 1988). However, partly because of the paucity of data on volcanic varieties, problems associated with the comagmatic origin of picritic basalts and gabbro-dolerites and their role in ore formation currently remained unresolved. Among the 11 previously described formations of the Noril'sk tuff–lava sequence, the picrites and were found only in the Gudchikhinsky and Tuklonsky Formations, which are attributed to the first and second stages, respectively, of volcanic activity in this area. Detailed geochemical investigations suggest an affinity between Noril'sk intrusions and rocks of the second stage of flood volcanism. According to this model, the Noril'sk deposits were formed during the assimilation of the host sedimentary rocks by the ascending Tuklonsky magma (average MgO 8–9 wt %).

It is proposed that this assimilation was responsible for an increase in SiO_2 content and a corresponding decrease in sulfur solubility. This led to the formation of the ore-generating sulfide melt and residual silicate melt depleted in base metals, particularly in Cu, Co, and Ni. This feature is also observed in the lower part of the Nadezhdinsky Formation (hereafter, lower Nadezhdinsky Formation), which is characterized by amounts of these metals that are three times lower than those of other formations (Cu 30–40 ppm) and by increased SiO_2 contents (50.66–52.87 wt %). Therefore, the Nadezhdinsky Formation is considered to be complementary to the ores, whereas the Tuklonsky rocks are considered to be the parental magma of the Nadezhdinsky Formation and ores. Thus, the establishment of primitive magmas for the Tuklonsky and Nadezhdinsky Formations is of great significance in terms of ore formation. It should be noted that the Tuklonsky Formation in the major part of the Noril'sk region primarily consists of tholeiitic basalts with subordinate olivine and picritic rocks that are restricted to the eastern part of the study area. The Tuklonsky section was examined near Lake Glubokoe on Mt. Sunduk by Lightfoot et al. (1994) and Wooden et al. (1993) and by us (this study). This section is characterized by the presence of a layered basaltic flow, which is considered to be a typomorphic feature of the Tuklonsky lavas. Similar rocks were reported near the right tributary of the Mikchangda River along Pyroxenite Creek, where the layered flow was preliminarily assigned to the Tuklonsky or Gudchikhinsky Formation (Geological map… 1994; Ryabov et al. 2000). We evaluated the petrological–geochemical features of the layered flow and underlying and overlying rocks along the lower reaches of the Mikchangda River.

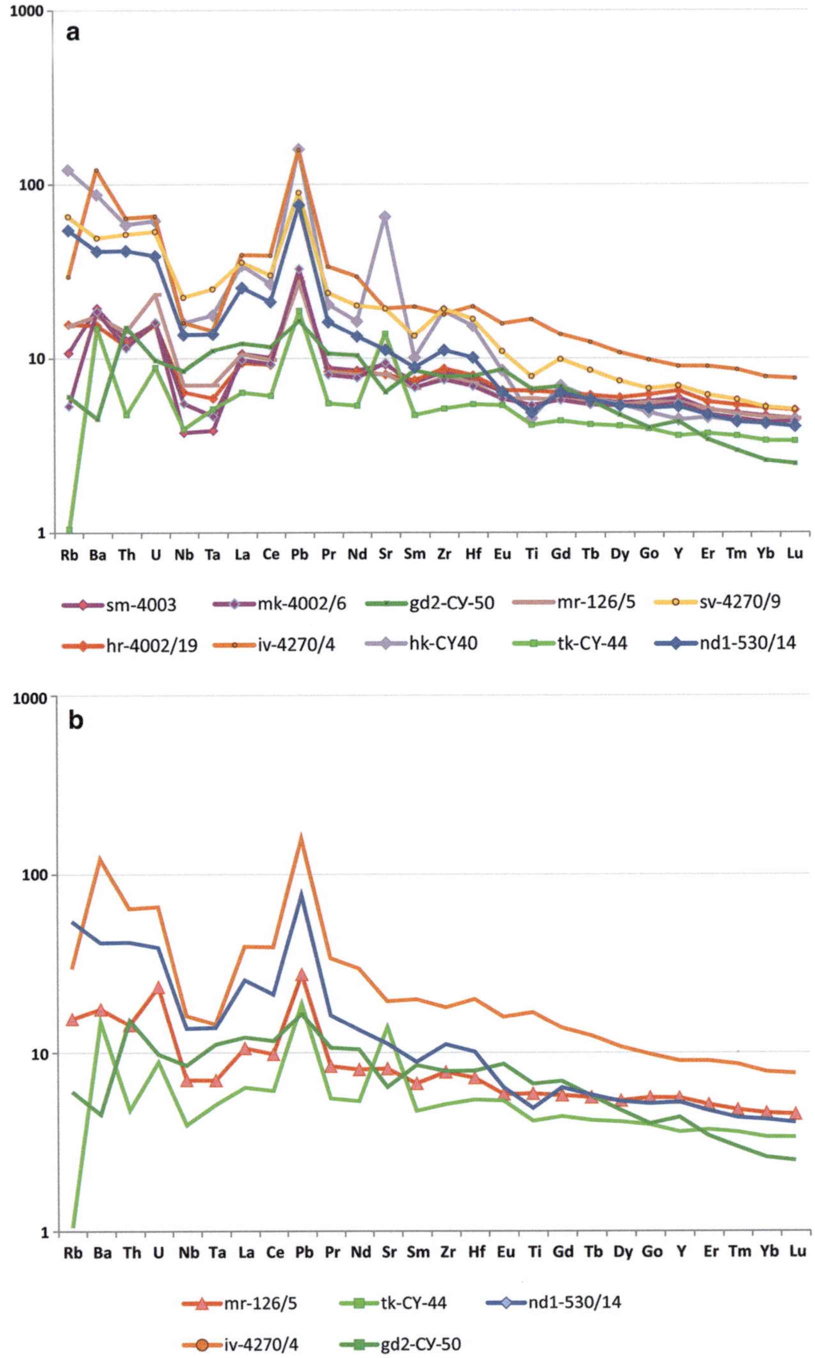

Fig. 3.21 Trace element patterns for volcanic rocks of the Noril'sk area (**a**) all formations, (**b**) main types of spectra. After Krivolutskaya (2014)

Picritic basalts of the Mikchangdinsky flow are assigned to the previously unknown primitive rocks of the Nadezhdinsky Formation. The section along Pyroxenite Creek contains a thick (up to 20 m) aleurolitic to agglomerate tuff sequence at the base followed by two flows of poikilophitic basalts (Fig. 3.25). Based on geochemical data, these rocks are assigned to the Tuklonsky Formation. They, in turn, are overlain by tholeiitic basaltic flows, which grade into a high-magnesium layered flow (Fig. 3.26). The thickness of the flow is approximately 70 m. The lower part is represented by the amygdaloidal zone (up to 0.5 m), whose massive basalts contain numerous hollow gas channels (up to 20 vol.%). This observation confirms the volcanic origin of these rocks. In upward succession, these rocks grade into massive olivine basalts (MgO 12 wt %). They contain individual beds (up to 5 m thick) containing various amounts of spherulitic clinopyroxene (from single aggregates to 80–90 vol.%, size up to 3–5 cm) or dendritic pyroxenes among plagioclase (Fig. 3.27), and thus the rocks are classified as pyroxene-phyric basalts or mikchangdites (Ryabov et al. 2000). The thickest (50 cm)

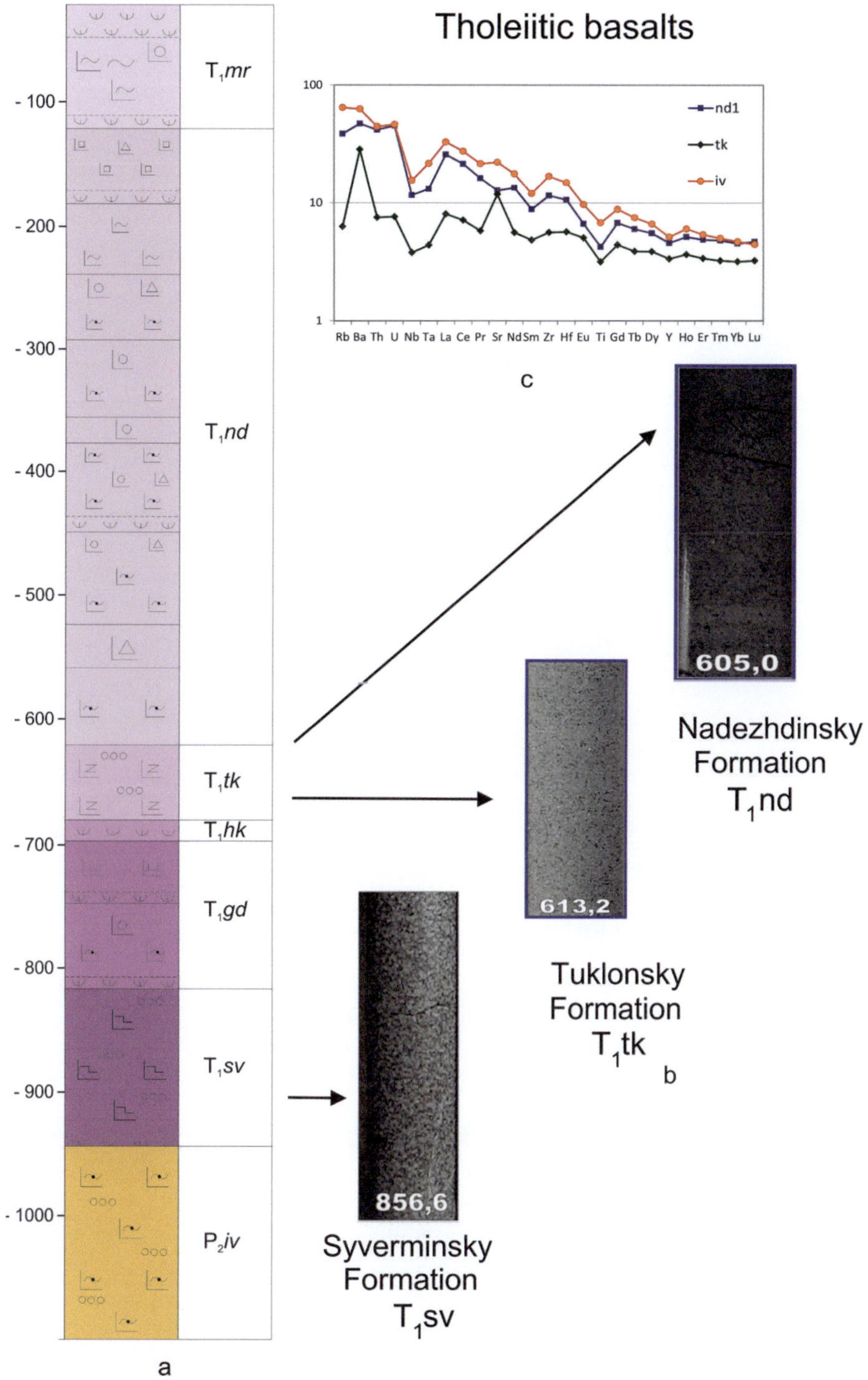

Fig. 3.22 Structure of volcanic rocks
(**a**) in borehole OM-6 (Noril'sk Trough), texture of tholeiitic basalts from different formations (**b**) and their REE patterns (**c**)
After Krivolutskaya (2014)

mikchangdite bed (Fig. 3.27d) is underlain and overlain by picritic basalts (0.3 and 1.5 m thick).

The picritic basalts (sample 530/12, Table 3.4) are massive porphyritic rocks consisting of large (3–4 mm) olivine phenocrysts (up to 50 vol.%) among a fine-grained (0.8–1.5 mm) groundmass (Fig. 3.28). The olivine contains Cr–spinel inclusions. Olivine is cut by fine fissures filled with serpentine and bowlingite. Approximately 20–30 vol.% of the olivine grains remain unaltered. The olivine varies from $Fo_{71.82}$ to $Fo_{77.9}$ and contains 0.09–0.14 wt % NiO and 0.16–0.29 wt % CaO. The groundmass (pyroxene–plagioclase aggregate with biotite and ilmenite) displays a microdoleritic (occasionally poikilophitic) texture. Sulfide segregations are absent from both the olivine grains and groundmass.

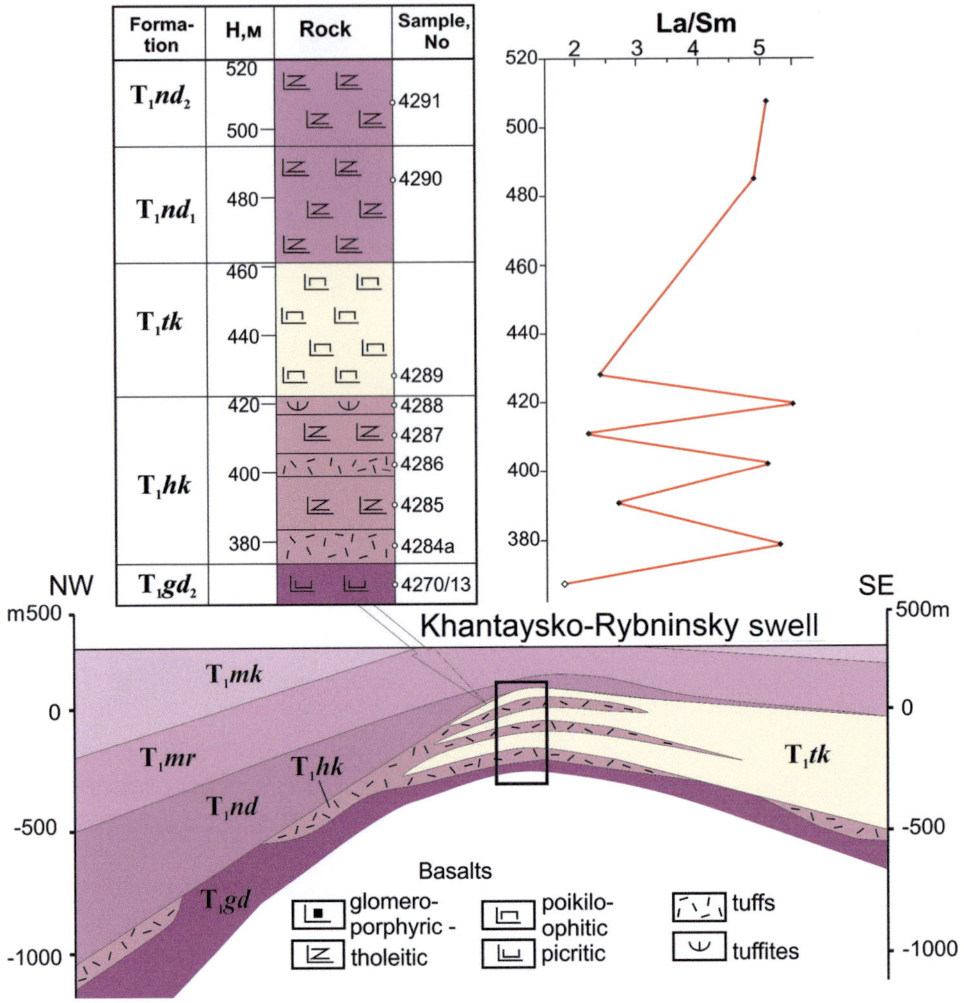

Fig. 3.23 Structure of volcanic rocks (Lake Lama) and variations of their La/Sm ratio
After Krivolutskaya (2014)

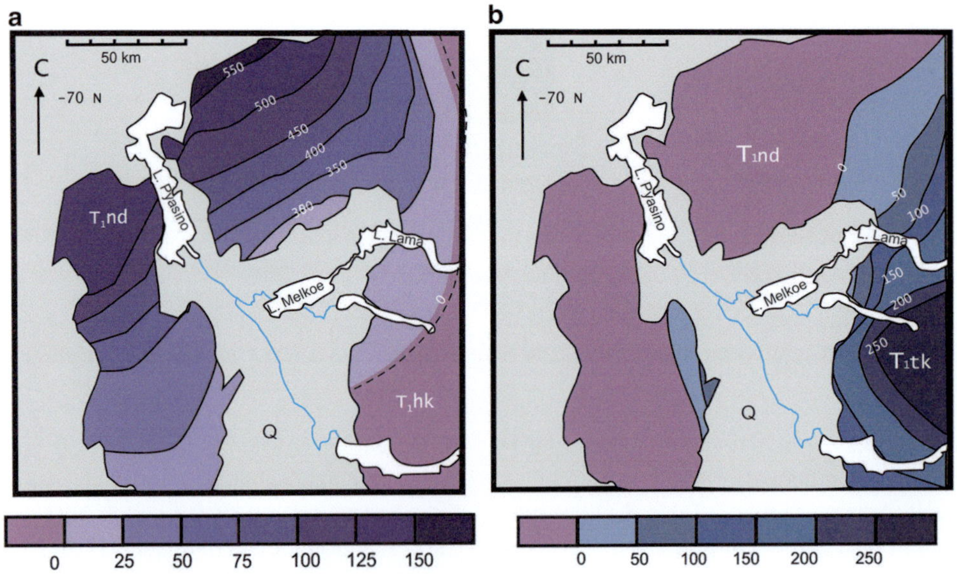

Fig. 3.24 Isopach maps for Nadezhdinsky (a) and Tuklonsky (b) Formations
After Krivolutskaya (2011)

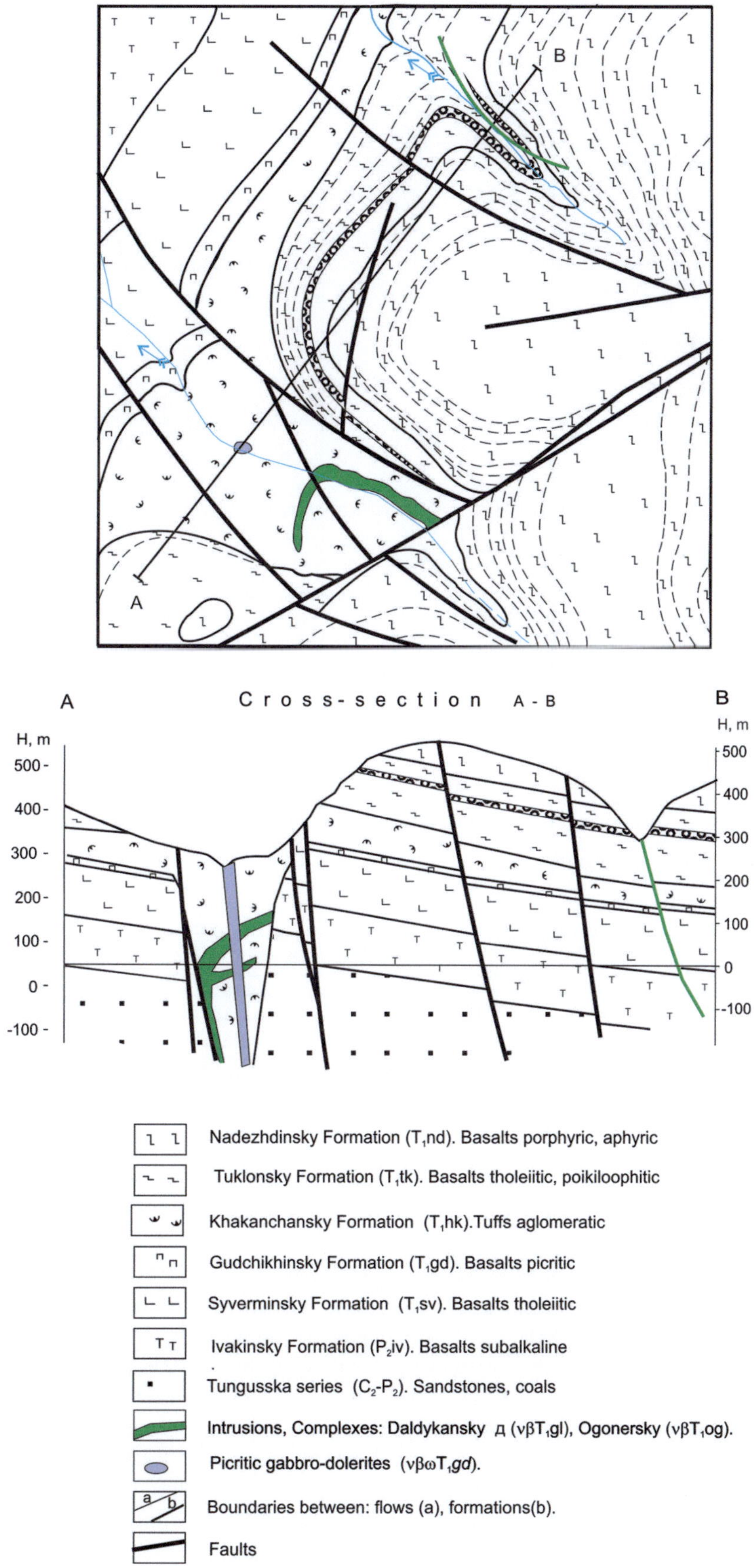

Fig. 3.25 Schematic geological map and cross section of the Mikchangdinsky flow area
Author V.N. Mikhailov

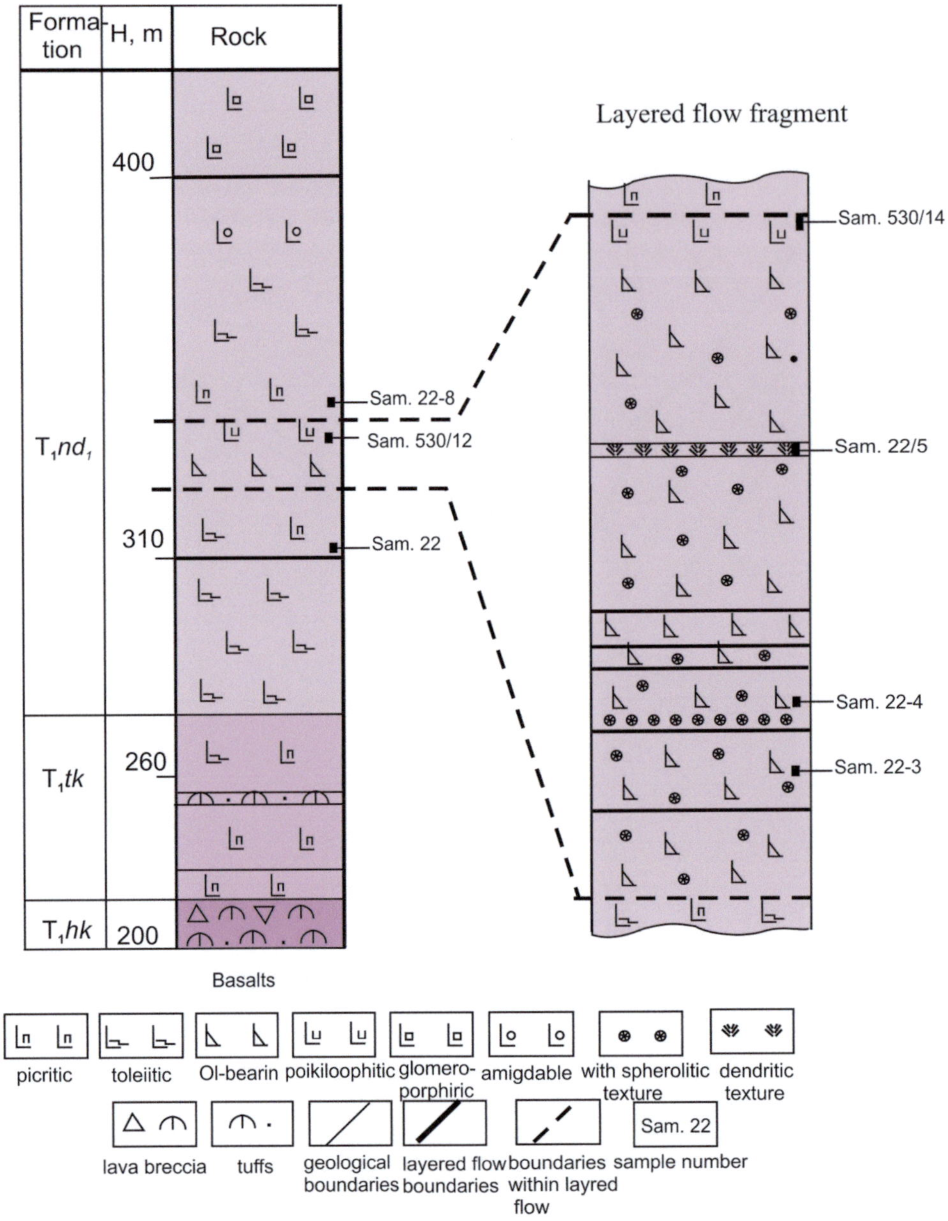

Fig. 3.26 Structure of the Mikchangdinsky layered flow (Pyroxenite Creek) Section was studied with V. Mikhailov. After Krivolutskaya (2014)

We studied the geochemical composition of rocks throughout the section along Pyroxenite Creek, including rocks of the Mikchangdinsky flow (Krivolutskaya et al. 2005, 2012). The major elements in the rocks (27 samples) were analyzed using XRF at the Vernadsky Institute of Geochemistry and Analytical Chemistry (by analysts I. A. Roshchina and T. A. Romashova), and the trace elements were determined by ICP-MS at the Institute of Mineralogy, Geochemistry, and Crystal Chemistry of Rare Elements (by analyst D. Z. Zhuravlev). Representative results from the analyses of rocks of the Mikchangdinsky flow with different Mg indices and underlying and overlying lavas are listed in the table and shown in Fig. 3.29. The compositions of rocks of the Nadezhdinsky Formation (Fedorenko et al. 1996) and of picritic basalts of the Tuklonsky Formation (Sample CY-33) near Lake Glubokoe are shown for comparison. The data obtained make it possible to determine the stratigraphic affiliations of the Mikchangdinsky flow and the underlying and overlying basalts. The rocks correlate well with lavas of the lower Nadezhdinsky Formation, which contain low amounts of base metals, particularly Cu. These rocks are strongly depleted in these elements compared to other formations of the Noril'sk tuff–lavas. The rocks of the Mikchangdinsky flow sharply differ from basalts of the Tuklonsky Formation and overlying Mokulaevsky and Morongovsky Formations. These differences are clearly seen in the La/Sm ratio (Fig. 3.29b), in which rocks of the lower Nadezhdinsky Formation and the Mikchangdinsky flow cluster apart from those of other formations, including the Tuklonsky Formation.

Fig. 3.27 Layering in rocks of the Mikchangdinsky flow
(**a, b**—white layers are pyroxenite basalts, *dark layers* are picritic basalts; **c, d**—layer with dendritic texture) (After Krivolutskaya 2014)

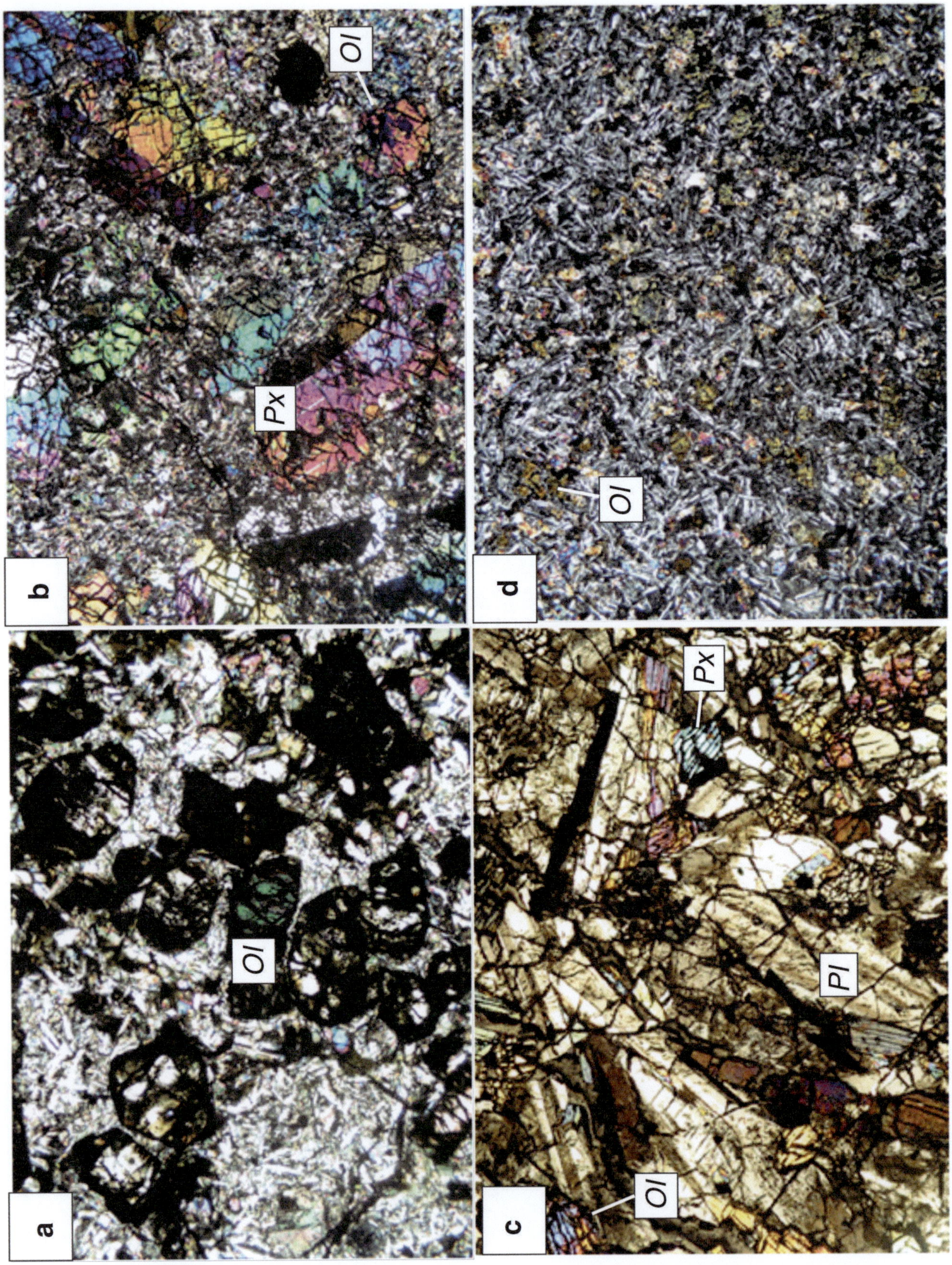

Fig. 3.28 Texture of rocks from the Mikchangdinsky flow (*Ol* olivine, *Px* pyroxene, *Pl* plagioclase). Lower boundary of frame is 2 mm

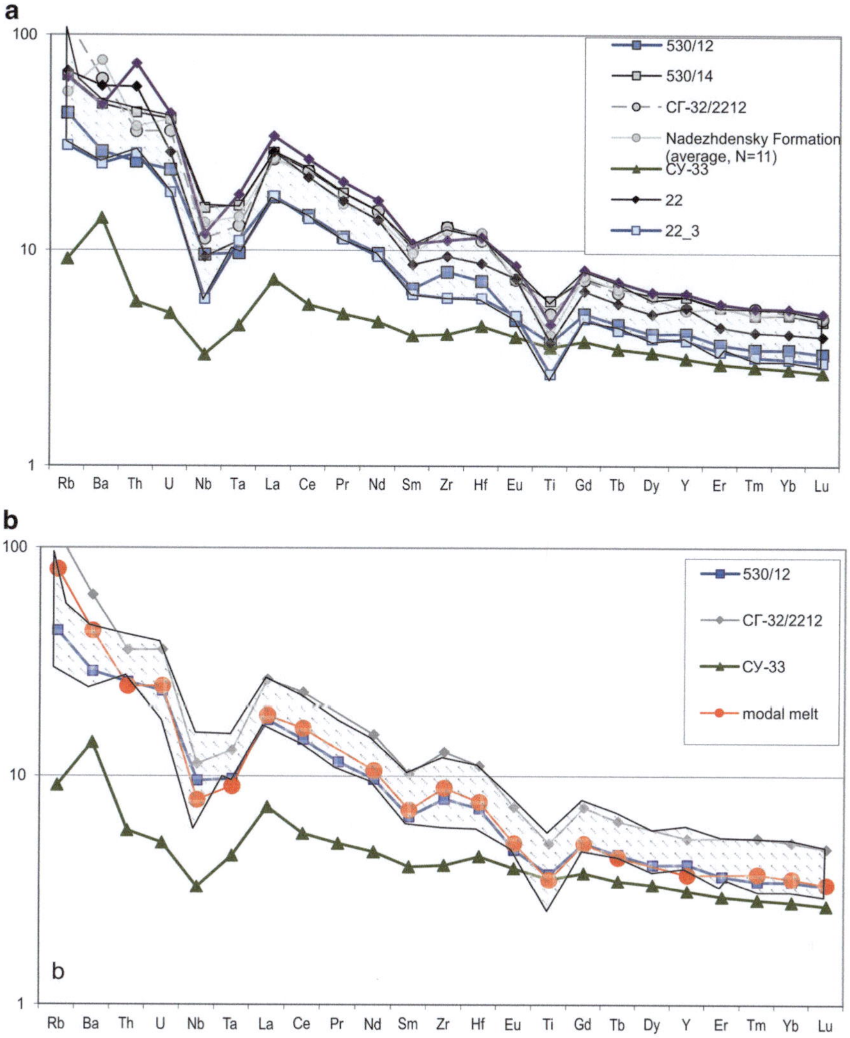

Fig. 3.29 Trace element patterns normalized to primitive mantle (Hofmann 1988) for rocks of the Mikchangdinsky flow
(**a**) Rocks of the Mikchangdinsky flow (*shaded field*): (sample 530/12) picritic basalt, (sample 22–3) olivine basalt, (sample 530/14) tholeiitic basalt; (samples 22 and 22–8) underlying and overlaying basalts, respectively; (Nd_1) lower Nadezhdinsky Formation (average composition based on 11 analyses (Fedorenko et al. 1996); (sample CY-33) picritic basalt from the Tuklonsky Formation. (**b**) *Spidergram* showing similarity of the model basaltic melt (mod) and real picritic basalt from the Mikchangdinsky flow (sample 530/12) (After Krivolutskaya et al. 2005)

The trace element patterns (Fig. 3.29) also indicate the similarity between the Mikchangdinsky flow and host rocks (Table 3.4). These rocks form parallel patterns with levels depending on the Mg index. All of the rocks under investigation are enriched in the most incompatible elements (steep negative slope on the left side of the plot). The Tuklonsky basalts represented by the picritic basalt near Lake Glubokoe, Mt. Sunduk, yield gentler slopes in the trace element plots. The genetic relationship between the high- and low-Mg rocks can be established by least-squares modeling of the olivine fractionation. Because the secondary alteration was intense, calculations were performed only for immobile elements (Nb, Ta, La, Ce, Pr, Nd, Sm, Zr, Hf, Eu, Gd, Dy, Tb, Y, Tm, Yb, and Lu). Primitive mantle-normalized abundances (Hofmann 1988) were used in the model calculations. The concentrations of major and other trace elements were also calculated. Picritic varieties of the Mikchangdinsky flow were obtained by the addition of 32.3 wt % olivine ($Fo_{75.58}$, average of 128 analyses) to the Nadezhdinsky basalts (average composition from 11 analyses (Fedorenko et al. 1996). The composition of the model melt is presented in the Table 3.4 and Fig. 3.29. The most-immobile incompatible elements were reproduced within a 5 % error, indicating the high probability of the applied model. The model was also consistent in terms of K and Rb, indicating their stability during secondary processes. By contrast, the strong difference between the calculated and observed Nb, Ba, Sr, and U concentrations suggests the perfectly mobile behavior of these elements during superimposed processes.

Thus, the picrites of the Mikchangdinsky flow represent the first discovered differentiation products of the most primitive melts within the Nadezhdinsky Formation. This magma was significantly enriched in incompatible elements and was depleted in base metals relative to the Tuklonsky basalt and

Table 3.4 Composition of rocks and model melt for Nadezhdinsky Formation

N п/п	1	2	3	4	5	6	7	8	9	10
N sample	530/12	22–3	530/14	SG-32 2212.5	nd n = 11	CY-33	22	22–8	Mod	Olivine
Formation	nd_1	nd_1	nd_1	nd_1	nd	tk	nd_1	nd_1	melt	Fo = 75.58
SiO_2	48.12	47.62	53.2	53.48	52.2	43.75	50.63	51.2	48.79	38.56
TiO_2	0.65	0.56	0.92	0.92	0.94	0.69	0.92	0.93	0.64	0
$A_{12}O_3$	9.93	11.5	15.23	15.25	16.38	12.12	15.26	14.94	10.59	0
FeO	11.55	9.76	9.41	9.26	9.23	11.12	9.64	9.62	13.27	22.63
MnO	0.17	0.17	0.14	0.16	0.16	0.18	0.15	0.15	0.21	0.33
MgO	17.28	12.93	6.42	6.82	6.32	15.27	7.37	7.05	16.62	39.34
CaO	7.74	10.68	10.18	9.41	10.33	8.51	8.9	10.2	6.61	0.25
Na_2O	0.33	1.27	1.06	2.75	2.35	0.78	2.42	2.35	1.91	0
K_2O	0.65	0.53	1.03	1.82	1	0.23	1.26	0.96	1.26	0
P_2O_5	0.07	0.07	0.13	0.12	0.1	0.06	0.11	0.11	0.08	0.00
Cr_2O_3	0.17	0.12	0.03			0.09	0.04	0.06		0.04
LOW	2.98	3.96	1.56	0.59	2.76	6.37	2.83	1.7		
Summa	99.66	99.19	99.33	102.62	101.77	99.18	99.55	99.28		101.26
Be	0.67	0.45	1.09			0.24	0.70	0.95		
Sc	20.2	27.5	31.6	32.0	30.0	16.9	27.6	32.8	22.2	
Ti	4,049	2,892	6,339	5,516		3,848	4,052	4,930	3,831	
V	194	213	220		212	179	164	198		
Cr	967	1,179	205	268	134	969	309	553		
Mn	1,240	1,117	1,086			1,243	851	1,001		
Co	87.2	56.6	39.6	41.0		79.4	33.5	39.2	28.48	
Ni	349	102	33	76	23	429	44	40	310	
Cu	29.4	23.9	27.1	103.0	32.0	79.4	41.0	29.2	71.5	
Zn	81.8	51.1	75.8	84.0	84.0	73.0	54.5	63.9	58.3	
Ga	11.5	11.2	17.0			11.7	13.9	15.9		
Rb	23.1	16.4	34.5	62.0	29.0	4.9	36.2	33.9	43.1	
Sr	138	156	236	335	283	139	297	230	233	
Y	16.2	15.0	24.0	21.0	20.7	12.4	21.3	24.8	14.6	
Zr	77.2	58.5	124.3	124.0	119.0	39.7	91.1	107.8	86.1	
Nb	5.90	3.68	9.71	7.00	8.3	2.03	5.77	7.34	4.86	
Cs	3.33	0.82	1.14	0.53		4.16	0.61	1.36	0.37	
Ba	174	152	288	376	460	85	349	285	261	
La	10.8	11.0	17.5	16.3	16.4	4.5	17.6	20.8	11.3	
Ce	23.2	22.4	37.6	37.3	35.4	9.0	35.0	42.3	25.9	
Pr	2.79	2.73	4.47		3.97	1.23	4.10	5.03		
Nd	11.5	11.2	18.1	18.1	16.7	5.6	16.4	20.2	12.6	
Sm	2.57	2.42	3.93	3.96	3.75	1.56	3.33	4.16	2.75	
Eu	0.70	0.73	1.07	1.07	1.07	0.58	1.08	1.24	0.74	
Gd	2.60	2.47	3.79	3.74	3.89	1.93	3.32	4.15	2.60	
Tb	0.43	0.39	0.62	0.60	0.62	0.33	0.53	0.67	0.41	
Dy	2.61	2.48	3.93		3.85	2.13	3.23	4.07		
Ho	0.58	0.52	0.85	0.64	0.82	0.44	0.67	0.84	0.44	
Er	1.52	1.41	2.27		2.26	1.23	1.84	2.35		
Tm	0.22	0.20	0.32	0.34	0.32	0.18	0.27	0.34	0.24	
Yb	1.43	1.29	2.08	2.12	2.1	1.16	1.69	2.21	1.47	
Lu	0.21	0.19	0.30	0.31	0.31	0.17	0.25	0.32	0.21	
Hf	1.93	1.61	3.11	2.97	3.22	1.20	2.33	3.07	2.06	
Ta	0.34	0.39	0.56	0.46	0.5	0.16	0.39	0.63	0.32	
W	0.27	0.46	0.33			0.28	0.43	0.66		
Pb	3.41	2.93	5.39	3.83		3.27	4.28	6.83	2.66	
Th	2.09	2.28	3.52	2.89	3.04	0.47	4.64	5.94	2.00	
U	0.47	0.37	0.81	0.71	0.81	0.10	0.57	0.86	0.50	

Note: Formations, *nd* Nadezhdinsky, *tk* Tuklonsky; N sample—samples from Mikchangdinsky flow, Nd = 11- average composition Nadezhdinsky formation and SG-32/2212.5 (Fedorenko et al. 1996). After Krivolutskaya et al. (2005)

Table 3.5 Isotope composition of rocks in the Mikchangdinsky and Sunduksky flows and picrites in Gudchikhinsky Formation

N	N sample	Rb, ppm	Sr, ppm	$^{87}Rb/^{86}Sr$	$^{87}Sr/^{86}Sr$	±2 σ	Sm, ppm	Nd, ppm	$^{147}Sm/^{144}Nd$	$^{143}{}_{Nd}/^{144}Nd$	±2σ	ε_{Nd} (T = 251 Ma)	($^{87}Sr/^{86}Sr$)251 Ma
1	530/11	8.88	293.9	0.0874	0.708610	0.000006	1.45	5.01	0.1751	0.512181	0.000005	−8.2	0.708298
2	530/13	10.58	124.8	0.2454	0.709294	0.000009	1.83	7.75	0.1426	0.512112	0.000007	−8.5	0.708417
3	530/12	22.58	119.2	0.5481	0.710144	0.000013	2.22	10.13	0.1324	0.512089	0.000008	−8.6	0.708188
4	530/5a	26.29	244.6	0.3110	0.709028	0.000010	3.06	13.90	0.1331	0.512125	0.000002	−8.0	0.707918
5	xc51-130	4.22	120.7	0.1012	0.706509	0.000009	1.82	6.57	0.1672	0.512800	0.000006	4.1	0.706147
6	cy-50	2.97	151.5	0.0566	0.703905	0.000004	3.52	12.56	0.1697	0.512865	0.000009	5.3	0.703703
7	4283	4.37	193.5	0.0654	0.705565	0.000011	2.80	10.25	0.1650	0.512783	0.000005	3.9	0.705331
8	4270/13	5.78	229.7	0.0728	0.703528	0.000013	4.53	16.21	0.1691	0.512889	0.000006	5.8	0.703268
9	CY-31	4.02	247.7	0.0470	0.706020	0.000008	0.79	2.74	0.1731	0.512540	0.000013	−1.2	0.705852
10	CY-36	3.06	119.5	0.0741	0.705908	0.000007	0.83	2.77	0.1803	0.512540	0.000009	−1.4	0.705644

Note: Analyses were done in Vernadsky Institute of Geochemistry and Analytical Chemistry, Moscow, Russia. Analyst A.A. Plechova. The relative errors for $^{87}Rb/^{86}Sr$ и $^{147}Sm/^{144}Nd$ are 1 and 0.1 % corresponding; No: 1–4—Nadezhdinsky Formation (Mikchangdinsky flow), 5-8 - Gudchikhinsky Formation, 9–10—Tuklonsky formation (Sunduksky flow) (After Krivolutskaya et al. 2012)

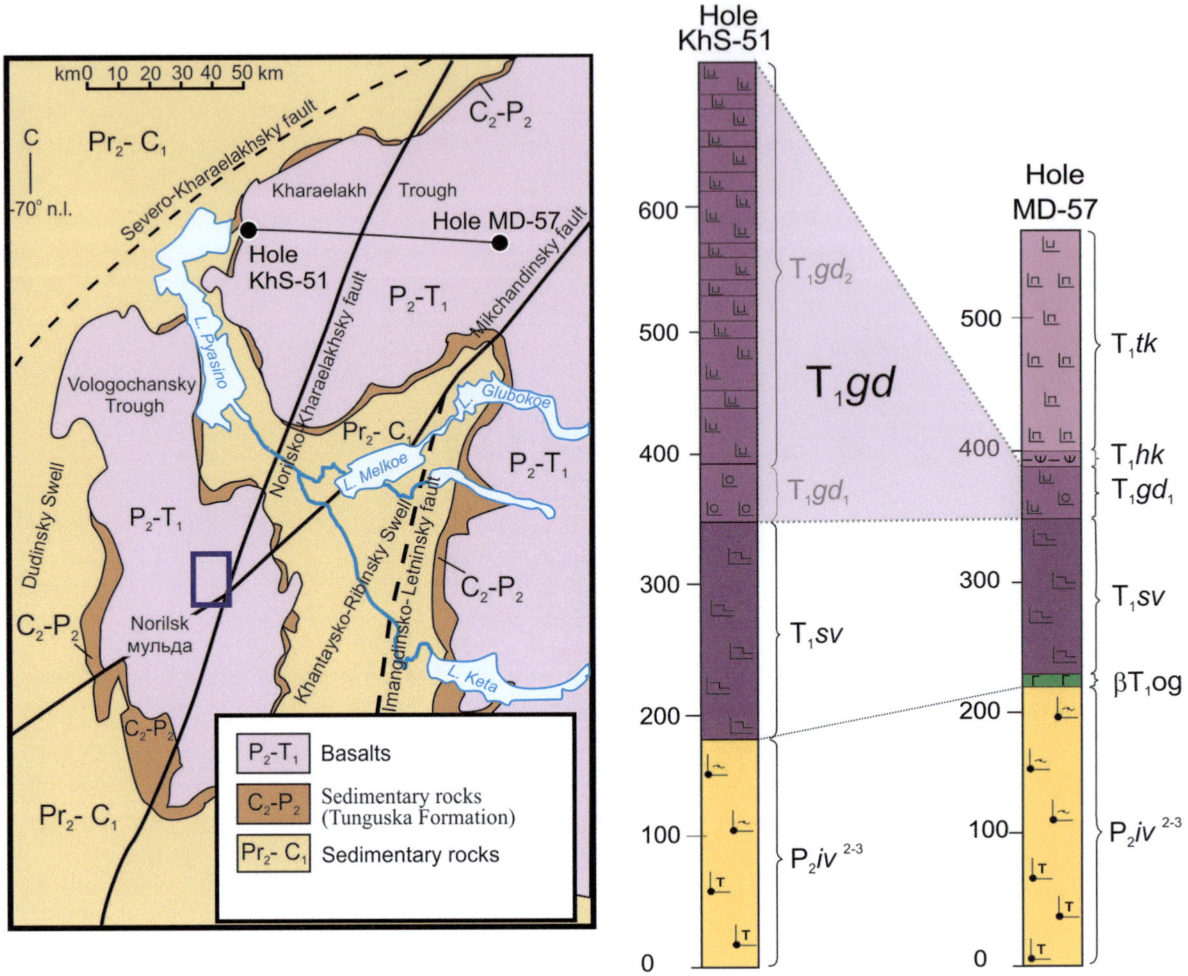

Fig. 3.30 Thickness variation of the Gudchikhinsky Formation
After Krivolutskaya (2014)

had another isotope composition (Table 3.5). Therefore, the results obtained call into question the inferred derivation of the Nadezhdinsky magma from the Tuklonsky basalt and its relationship with the mineralization (Naldrett 1992; Li et al. 2009).

3.3.2.1 Gudchikhinsky Picrites

Key sections were selected in three locations (Fig. 3.1): the western part of the Kharaelakh Trough, the eastern slope of the Khantaysko–Rybninsky swell, and the western flank of the Tunguska syneclise. This investigation allowed us to trace the evolution of the structure and composition of the Gudchikhinsky picrites from west to east. The thickest section of the Gudchikhinsky Formation was investigated in borehole KhS-51 in the western part of the Kharaelakh Trough (Fig. 3.30). The incomplete thickness of the formation (its upper part was eroded away) is 460 m. The lower part of the section is composed of porphyritic and aphyric basalts with an average MgO content of 6 wt %, and the upper part (111 m) is composed of 17 flows of picrites containing up to 24 wt % MgO. The flows are of only moderate thicknesses (4–6 m on average), and their upper portions (0.5–1.3 m) are composed of amygdaloidal varieties, which facilitated the distinguishing of flows in the section. The rocks are extensively altered, and the olivine is almost completely replaced by serpentine or bowlingite. Fresh samples of the picrites were collected from the central part of the thickest and, consequently, least-altered flow, i.e., from the borehole at depths of 118–140 m. One of the samples (KhS-51/130) was studied in detail.

The thickness of the Gudchikhinsky Formation decreases to a few tens of meters on the eastern side of the trough (Fig. 3.31). The structure and composition of the rocks of the formation change dramatically in the eastern slope of the Khantaysko–Rybninsky swell. Their total thickness is only 22 m, and the rocks of the lower part of the formation (basalt porphyries with normal Mg contents) are completely missing from the section. The formation is represented there by two picritic basaltic flows with MgO contents of 12–16 wt %. It should be noted that the character of magmatism changes in this region: numerous volcanoes of central type are present there (Fig. 3.6a); these are a few tens of meters thick and composed of thin (10–15 cm) beds of lava and ash materials. Thus, this rigid anticline structure is

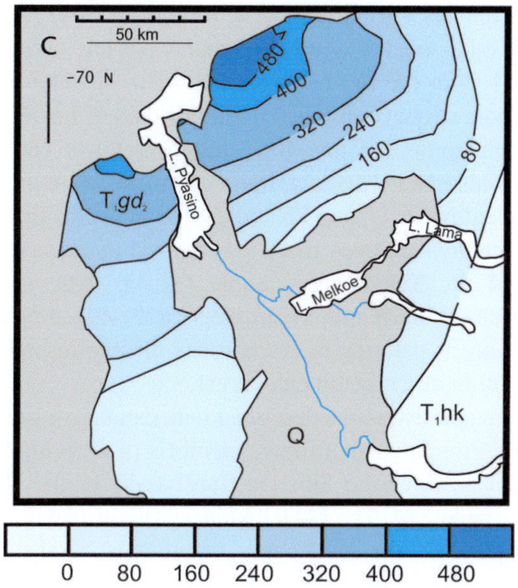

Fig. 3.31 Isopach map for the volcanic rocks of the Gudchikhinsky Formation

dominated by vertical fracture channels, which promoted faster magma ascent to the surface compared with the adjacent trough (which may be responsible for the less extensive crustal contamination of the melts, as in sample 4270/13; see CY-50).

The Tuklonsky and Nadezhdinsky Formations contain very minor amounts of high-magnesium rocks. These rocks were documented only in layered flows (Fig. 3.27). Picritic basalts were found in the Tuklonsky Formation at Sunduk Mountain (southern coast of Lake Glubokoe; Lightfoot et al. 1993) and in the Nadezhdinsky Formation among the rocks of the Mikchangdinsky flow (Krivolutskaya et al. 2005) on the eastern slope of the Khantaysko–Rybninsky swell. These rocks are represented by samples CY-50 (Tk) and 530/12 (Nd_1).

3.3.2.2 Petrography

The picritic basalts of the Gudchikhinsky Formation are macroscopically distinct from other lava varieties, including the high-magnesium rocks of the Tuklonsky and Nadezhdinsky Formations. They are dark gray (sometimes with a violet shade) and medium- to coarse-grained massive rocks consisting of olivine (15–50 vol.%), plagioclase (20–50 %), pyroxene (10–30 %), chrome spinel, and secondary minerals. Their structural, textural, and mineral characteristics display minor variations. The most magnesian picrites from the western part of the Kharaelakh Trough (sample KhS-51/130) have a weak porphyritic structure and a hypidiomorphic texture. Their compositions are dominated by olivine (up to 50 %), whose large euhedral prismatic crystals (2–3 mm) appear as hexagonal or rectangular grains in thin sections. Small (up to 1 mm) oval or rounded grains of this mineral are often present in elongated (up to 3 mm) tabular crystals of plagioclase (25 %) forming a poikilitic texture. Clinopyroxene (13 %) also forms elongated (up to 1 mm) grains, and orthopyroxene occurs in minor amounts (up to 3–4 %). The rock also contains chrome spinel, titanomagnetite, and serpentine group minerals. Picritic basalts from the eastern slope of the Khantaysko–Rybninsky swell (sample 4270/13) have a finer-grained hypidiomorphic texture. Olivine (approximately 20 %) is usually present as small euhedral grains of uniform size (0.5 mm on average) randomly distributed in the rock and is nearly unaffected by secondary alteration. Plagioclase (approximately 50 %) is present as large laths (5–2 mm), and pyroxene (approximately 30 %) forms smaller irregular grains. Typical picritic basalts from the western part of the Tunguska syneclise (sample CY-50) are very similar in mineral composition, texture, and structure to the rock variety described above, but they are better crystallized. The strongly altered subhedral olivine grains average 1–2 mm in size (up to 15 %), and 60–70 % of the grains are replaced by bowlingite or serpentine.

The picritic basalts of the Tuklonsky and Nadezhdinsky Formations are present as 5–80-cm-thick interbeds among the high-magnesium tholeiitic basalts (8–9 wt % MgO) and are similar to each other both in composition and textural features. In particular, they display a weak porphyritic structure and a doleritic texture of the groundmass, which contains radial aggregates of plagioclase and/or pyroxene. The sample of picritic basalt from the Tuklonsky Formation (CY-33) contains nearly completely altered olivine grains (15 %), sheaf-like intergrowths of plagioclase crystals up to 2 mm in size (30 %), and anhedral clinopyroxene grains (55 %). Similar rocks from the Nadezhdinsky Formation (sample 530/12) contain up to 30 vol.% of small (1–2 mm) porphyritic olivine grains. They are uniformly distributed in the groundmass with a doleritic or radial texture composed of small (0.5 mm) plagioclase laths (40 %) and isometric pyroxene grains (30 %) of the same size.

3.3.2.3 Composition of Rocks

The Noril'sk traps are dominated by low-magnesium (MgO < 7 wt %) tholeiitic basalts and contain minor olivine basalts and picrites. Alkaline and subalkaline varieties are rare and are present primarily in the lower formations, i.e., the Ivakinsky and Syverminsky Formations. The composition of the lavas changes systematically from bottom to top, which is consistent with the data from previous studies (Lightfoot et al. 1993; Wooden et al. 1993). The lower formations are significantly depleted in heavy REEs and, correspondingly, display high Gd/Yb ratios, which suggests the presence of garnet in their source (Lightfoot et al. 1993; Wooden et al. 1993; Sharma 1997). This parameter decreases considerably in the rocks directly overlying the Gudchikhinsky Formation, which serves as evidence for the retention of garnet in the source. The Nb/La ratio is significantly variable within the section. As does the Ta/La ratio, this ratio reflects the degree of magma contamination by continental crust (Lightfoot et al. 1993). A Nb/La ratio of less than 1 associated with an undepleted composition implies a negative Nb (Ta) anomaly, which is compelling evidence for the contribution of continental crust (Rudnick 2002). Figure 3.21 demonstrates that most of the Noril'sk trap rocks are contaminated by continental crust. The degree of contamination ranges from minimum values in the rocks of the Gudchikhinsky Formation to maximum values in the Nadezhdinsky basalts (Wooden et al. 1993).

In the Noril'sk region, magnesian rocks are known only in the lower formations, i.e., the Gudchikhinsky, Tuklonsky, and Nadezhdinsky Formations. Their compositions are shown in Fig. 3.32 (Table 3.6). Author has studied these rocks with A. Sobolev and D. Kuzmin (Sobolev et al. 2009). The primary difference between the Gudchikhinsky rocks and the overlying Tuklonsky and Nadezhdinsky basalts is the depletion in heavy REE and, correspondingly, high Gd/Yb ratios. The compositions of two samples from the Gudchikhinsky Formation (4270/13 and CY-50) are similar to each other regarding most elements but are significantly different from the third sample, KhS-51/130. The latter is much more strongly enriched in olivine, and its incompatible element patterns are therefore shifted to lower concentrations. It also bears distinct signs of crustal contamination: enrichment in Th, U, and Pb and depletion in Nb, Ta, and Ti. These features are even more pronounced in the magnesian rocks of the Tuklonsky (sample CY-33) and especially Nadezhdinsky Formations (sample 530/12) (Fig. 3.32b). The trace element patterns of these rocks closely resemble the composition of the continental crust.

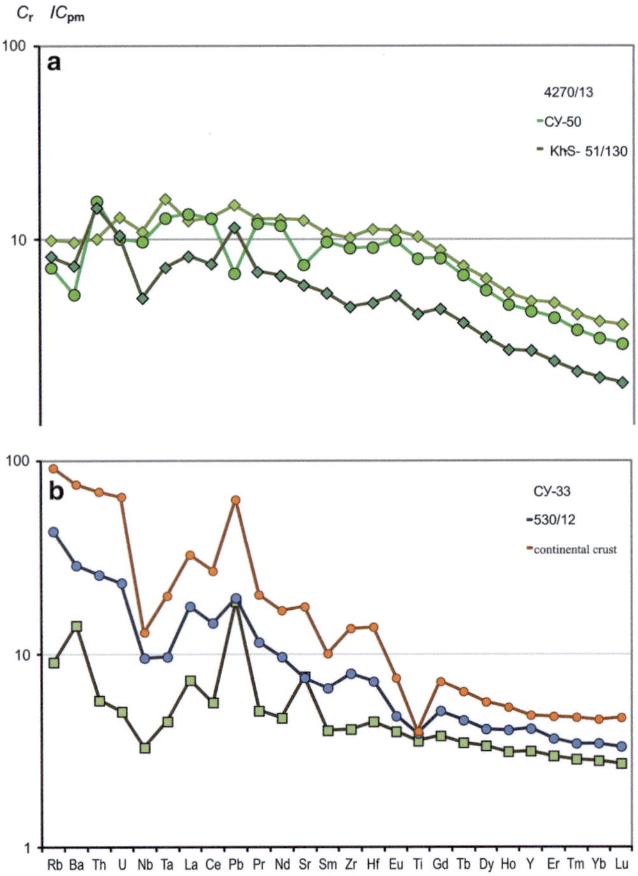

Fig. 3.32 Trace element patterns for volcanic rocks of the Gudchikhinsky Formation (**a**) and for Tuklonsky, Nadezhdinsky Formations, and continental crust (**b**)
Continental crust after Rudnick 2002 (After Sobolev et al. 2009)

The magnesian rocks described here can be considered as the least differentiated primitive members of the primary geochemical types of the Siberian trap basalts of the Noril'sk region (Table 3.6). The Gudchikhinsky picrites are primitive counterparts of the moderately titaniferous basalts, which accounts for ~8 % of the lavas of the section (Fedorenko et al. 1996). With respect to their combined geochemical characteristics, the Tuklonsky picrites correspond to the primary low-titanium type of magmatism, accounting for more than 75 vol.% of the Noril'sk section and differ significantly from other rocks of this suite only in their low Pb isotope ratios (Table 3.7), $^{206}Pb/^{204}Pb$ and $^{208}Pb/^{204}Pb$ (Wooden et al. 1993; Fedorenko et al. 1996). However because crustal contamination provided the primary contribution to the content and isotopic composition of Pb in the Tuklonsky picrites, this difference should be attributed to different compositions of the contaminants rather than the heterogeneity of the primary magmas. The magnesian rocks of the Nadezhdinsky Formation are crystallization products of the most contaminated primitive magmas that produced the rocks of this formation (Krivolutskaya et al. 2005; Reichow et al. 2005).

3.3.2.4 Composition of Spinel

The spinel inclusions in the olivine phenocrysts from the rocks of the Gudchikhinsky Formation (Table 3.8) correspond to a Cr-rich variety ($Cr/(Cr+Al)=0.70\pm0.03$) with high amounts of TiO_2 (1.6 ± 0.4 wt %) and V_2O_3 (0.5 ± 0.1 wt %) and a high Fe^{2+}/Fe^{3+} ratio (5.0 ± 0.6). The composition of the spinel is not correlated with that of the host olivine, which varies within the Fo_{84}–Fo_{79} range. Noteworthy are the high Fe^{2+}/Fe^{3+} ratio and significant amounts of vanadium in the spinel, which indicates unusually low oxygen fugacity in the crystallization environment (Canil 2002).

3.3.2.5 Composition of Olivine

Three groups of olivine phenocryst compositions are clearly distinguished (Table 3.9, Fig. 3.33). The first group consist of high-Ni and low-Mn olivines from the rocks of the Gudchikhinsky Formation, which essentially have no analogues among the olivines of mantle magmas. Forsterite

Table 3.6 Composition of high-Mg rocks in the Noril'sk area

Formation	gd	gd	gd	tk	nd
Sample, No	4270/13	СУ-50	KhS-51/130	СУ-33	530/12
SiO_2	47.33	44.40	44.47	43.75	48.12
TiO_2	2.13	1.58	0.88	0.70	0.65
Al_2O_3	10.14	8.07	6.68	12.12	9.93
Fe_2O_3	13.22	12.94	13.35	12.34	12.82
MnO	0.18	0.19	0.19	0.18	0.18
MgO	13.14	18.85	22.97	15.27	17.28
CaO	9.54	6.58	5.37	8.51	7.74
Na_2O	2.40	0.78	0.62	0.78	0.33
K_2O	0.26	0.10	0.07	0.23	0.65
P_2O_5	0.20	0.17	0.10	0.06	0.08
Cr_2O_3	0.10	0.15	0.18	0.09	0.18
LOW	1.44	5.35	5.40	6.37	2.98
Summa	100.07	99.15	100.28	100.40	100.93
Sc	18.9	17.5	15.9	16.9	29.2
V	260	202	148	179	194
Co	66	83	101	79	87
Ni	211	1,057	1,532	429	349
Cu	66	67	50	79	20
Zn	98	92	82	73	82
Rb	5.3	3.8	4.4	4.9	23.1
Sr	229	135	106	139	138
Y	20	20	12	12	16
Zr	99	88	44	40	77
Nb	6.8	6.0	3.1	2.0	5.9
Cs	2.28	1.59	0.39	4.16	3.33
Ba	59	32	44	85	174
La	7.7	8.3	5.0	4.5	10.8
Ce	21	21	12	9	23
Pr	3.09	2.94	1.65	1.23	2.79
Nd	15	14	8	6	12
Sm	4.2	3.8	2.0	1.6	2.6
Eu	1.63	1.45	0.75	0.58	0.70
Gd	4.54	4.12	2.25	1.93	2.60
Tb	0.69	0.62	0.35	0.33	0.43
Dy	4.00	3.49	2.00	2.13	2.61
Ho	0.76	0.66	0.38	0.44	0.58
Er	1.96	1.64	0.98	1.23	1.52
Tm	0.26	0.22	0.13	0.18	0.22
Yb	1.57	1.28	0.80	1.16	1.43
Lu	0.23	0.18	0.12	0.17	0.21
Hf	3.03	2.44	1.26	1.20	1.93
Ta	0.57	0.45	0.25	0.16	0.34
Pb	2.64	1.17	2.02	3.27	3.41
Th	0.82	1.28	1.19	0.47	2.09
U	0.27	0.21	0.21	0.10	0.47

Note: Formations, *gd* Gudchikhinsky, *tk* Tuklonsky, *nd* Nadezhdinsky. Oxides are given in %, elements—in ppm
After Sobolev et al. (2009)

Table 3.7 Isotope composition of volcanic rocks in the Noril'sk region

No	N sample	Rock	Свита	$^{87}Sr/^{86}Sr$ Measured	$(^{87}Sr/^{86}Sr)t$	$^{143}Nd/^{144}Nd$ Measured	(εNd) t
1	492	Basalt	nd_1	0.711145	0.70989	0.51252	−0.2
2	530/12	Picrobasalt	nd_1	0.709939	0.70820	0.51212	−8.1
3	530/15	Basalt	nd_1	0.708703	0.70823	0.51212	−8.5
4	CY-32	Basalt	tk	0.705716	0.70558	0.51253	−1.3
5	CY-33	Basalt	tk	0.705925	0.70556	0.51262	0.6
6	CY-38	Basalt	tk	0.705846	0.70583	0.51263	0.4
7	CY-50	Picrite	gd_2	0.703539	0.70324	0.51292	6.6
8	KhS-51/131	Picrite	gd_2	0.706141	0.70584	0.51287	5.5
9	KhS-51/129	Picrite	gd_2	0.706361	0.70604	0.51283	4.7
10	KhS-51/99	Picrite	gd_2	n/d	н/о	н/о	н/о
11	KhS-51/108	Picrite	gd_2	0.705326	0.70485	0.51279	4.1
12	KhS-59/1144	Picrobasalt	gd_1	0.705808	0.70527	0.51263	1.0
13	4032	Basalt	mr	0.705303	0.70504	0.51276	3.3
14	4002/14	Basalt	mk	0.70549	0.70525	0.51268	1.8
15	4002/19	Basalt	khr	0.705519	0.70491	0.5128	3.7

No	N sample	U, ppm	Pb, ppm	$^{206}Pb/^{204}Pb$ Measured	$^{207}Pb/^{204}Pb$ Measured	$^{208}Pb/^{204}Pb$ Measured	$^{206}Pb/^{204}Pb$ Measured	$(^{207}Pb/^{204}Pb)t$	$(^{208}Pb/^{204}Pb)t$
1	492	1.03	6.78	18.12	15.53	38.19	17.75	15.51	37.80
2	530/12	0.47	3.41	18.15	15.53	38.36	17.81	15.51	37.87
3	530/15	0.22	1.84	17.89	15.51	38.10	17.59	15.49	37.62
4	CY-32	0.12	1.24	17.45	15.43	37.48	17.20	15.41	37.20
5	CY-33	0.10	3.27	17.43	15.42	37.40	17.35	15.42	37.28
6	CY-38	0.10	1.49	17.39	15.42	37.38	17.23	15.41	37.19
7	CY-50	0.20	1.17	18.75	15.52	38.48	18.34	15.50	37.62
8	KhS-51/131	0.27	2.12	18.84	15.63	38.18	18.52	15.62	37.93
9	KhS-51/129	0.24	2.37	18.79	15.63	38.19	18.54	15.62	37.98
10	KhS-51/99	0.29	1.94	18.35	15.55	38.28	17.97	15.53	37.79
11	KhS-51/108	0.40	1.66	18.47	15.56	38.40	18.21	15.54	38.10
12	KhS-59/1144	0.19	2.58	17.80	15.47	38.28	17.42	15.45	37.65
13	4032	0.45	1.68	18.79	15.55	38.30	18.14	15.51	37.74
14	4002/14	0.36	2.00	18.33	15.52	38.13	17.89	15.49	37.70
15	4002/19	0.33	2.07	18.37	15.52	38.15	17.98	15.50	37.77

Note: Analyses were done in Max-Planck Institute of Chemistry, Mainz, Germany, according method (Fekiacova et al. 2007), analysts—Z. Fekiacova; error—0,005 %; No: 2–3—Mikchangdinsky Flow, 4–6—Sunduksky Flow; *t*—age 251 Ma. Names of Formation are in Table 3.1
After Krivolutskaya et al. (2012)

comprises up to 84 % of the olivines of this group. Magnesian olivines from the rocks of the Nadezhdinsky Formation contain less Ni and more Mn, and the lowest Ni amounts coupled with the highest Mn are characteristic of olivines from the rocks of the Tuklonsky Formation. In the latter two groups, the maximum forsterite component is no higher than 80 %. A characteristic feature of olivine from the rocks of the Nadezhdinsky Formation is the steep decrease in Ni content with decreasing MgO, which is not reflected in the Mn/Fe ratio. *A.* Sobolev (Sobolev et al. 2005, 2007) demonstrated that a Ni excess and Mn deficit in olivine composition compared with the level of equilibrium with mantle peridotites are indicative of the presence of melting products from olivine-free pyroxenite formed by the reaction between recycled crust and peridotite. The Mn/Fe and Ni/(Mg/Fe) ratios of olivine were measured to determine the fraction of melt from a pyroxenite source in the bulk composition of the magma (Sobolev et al. 2011).

This parameter was estimated independently on the basis of the Mn/Fe and Ni/(Mg/Fe) ratios in the olivine phenocrysts from our samples (Fig. 3.33). The olivine phenocrysts from the Gudchikhinsky Formation indicate a nearly pure olivine-free pyroxenite source for these magmas, which was previously noted by Sobolev et al. (2007). The compositions

Table 3.8 Composition of spinel from Gudchikhinsky picrites, wt %

Component	ol5-12-18[a]	ol5-12-20	ol5-12-23	ol5-13-1	ol5-13-11	ol5-13-13	ol5-13-21	ol5-15-1	ol5-15-11
SiO_2	0.36	0.34	0.48	0.34	0.36	0.34	0.38	0.34	0.39
TiO_2	1.70	1.54	1.20	1.36	1.52	1.52	0.93	1.44	2.87
Al_2O_3	14.45	13.40	13.34	13.21	13.50	12.35	10.40	13.22	13.20
Cr_2O_3	45.75	46.53	46.80	46.57	46.70	44.93	49.59	46.56	41.16
V_2O_3	0.33	0.39	0.45	0.49	0.41	0.39	0.44	0.45	0.71
FeO_{com}	27.40	28.51	27.71	29.03	27.40	32.81	28.68	29.38	31.33
MnO	0.26	0.25	0.26	0.30	0.22	0.27	0.30	0.27	0.32
MgO	8.41	7.43	7.08	6.60	8.09	5.98	6.63	6.86	6.98
NiO	0.19	0.16	0.14	0.14	0.17	0.14	0.14	0.16	0.23
ZnO	0.15	0.19	0.22	0.21	0.18	0.21	0.23	0.18	0.24
Summa 1	99.02	98.75	97.74	98.26	98.56	98.94	97.72	98.86	97.58
FeO	22.67	23.76	23.78	24.64	22.75	25.90	23.64	24.55	25.11
Fe_2O_3	5.26	5.28	4.36	4.88	5.16	7.68	5.60	5.36	6.91
Summa 2	99.53	99.27	98.11	98.74	99.05	99.70	98.27	99.38	98.13
$Fe^{2+/3+}$ sp	4.8	5.0	6.1	5.6	4.9	3.7	4.7	5.1	4.0
$Fe^{2+/3+}$ m	21.9	23.1	29.7	26.9	22.5	15.9	21.2	23.7	17.5
Fo	82.17	82.68	81.06	83.47	79.13	82.86	81.42	81.42	81.60

Component	ol5-15-18	ol5-15-19	ol5-16-7	6ol-2-4	6ol-2-6	6ol-2-7	6ol-3-2	6ol-3-7	6ol-3-10
SiO_2	0.42	0.34	0.36	0.37	0.36	0.34	0.34	0.33	0.36
TiO_2	1.51	1.33	2.80	1.56	1.84	1.52	1.59	1.44	1.34
Al_2O_3	13.93	12.52	13.14	13.20	17.81	13.24	13.43	13.39	12.64
Cr_2O_3	46.28	48.34	43.18	45.47	40.82	45.76	44.94	45.62	46.23
V_2O_3	0.45	0.40	0.56	0.48	0.38	0.41	0.45	0.44	0.45
FeO_{com}	28.53	27.94	30.26	30.56	28.90	29.63	30.65	28.92	30.55
MnO	0.25	0.24	0.29	0.27	0.25	0.27	0.26	0.27	0.28
MgO	7.18	7.18	6.43	6.23	8.21	6.86	6.47	7.03	6.41
NiO	0.14	0.13	0.18	0.16	0.21	0.13	0.16	0.15	0.14
ZnO	0.19	0.19	0.23	0.22	0.19	0.19	0.20	0.20	0.23
Summa 1	98.90	98.64	97.53	98.53	98.97	98.34	98.49	97.79	98.63
FeO	24.37	23.81	25.90	25.54	23.59	24.49	25.22	23.95	25.00
Fe_2O_3	4.63	4.59	4.85	5.58	5.90	5.71	6.03	5.52	6.17
Summa 2	99.33	99.08	97.92	99.08	99.55	98.91	99.08	98.33	99.24
$Fe^{2+/3+}$ sp	5.9	5.8	5.9	5.1	4.4	4.8	4.7	4.8	4.5
$Fe^{2+/3+}$ m	28.4	27.9	28.9	23.6	19.8	21.7	21.0	22.0	20.1
Fo, mol %	82.51	82.79	81.67	82.38	80.39	79.55	81.37	80.12	82.23

Note: Calculation FeO, Fe_2O_3 were done according spinel stoichiometry; FeO_{com}-Fe common; Summa 1 и Summa 2 - summa before and after Fe division consequently; $Fe^{+2}/^{+3}$ m- ratio Fe forms in melt, calculated according model (Maurel and Maurel 1982); *Fo* - here and in Table 3.6 - Fo (mol %)—forsterite content in olivine

[a]Number of sample

After Sobolev et al. (2009)

of magnesian olivines from the rocks of the Nadezhdinsky and Tuklonsky Formations indicate a significantly lower contribution from a pyroxenite source. Sulfide melt fractionation in the magmas of the Nadezhdinsky Formation leads to significantly underestimated contributions of the pyroxenite component calculated from the Ni/(Mg/Fe) ratio of olivine but has almost no effect on the estimates based on the Mn/Fe ratio (Fig. 3.33).

3.3.2.6 Experimental Investigations

The melt inclusions were studied using a low-inertia visually controlled heating stage with a purified He atmosphere (Sobolev and Slutsky 1984) following the method of Sobolev and Danyushevsky (1994). The experimental temperature was measured using a $Pt_{90}Rh_{10}$ thermocouple and was checked in every experiment against the melting point of high-purity gold. To minimize H_2O loss from the inclusions, the total

Table 3.9 Composition of olivine from picritic basalts in the Nadezhdinsky and Tuklonsky Formations, wt %

N sample	SiO_2	TiO_2	Al_2O_3	FeO	MnO	MgO	CaO	NiO	CoO	Cr_2O_3	Summa	*Fo*
530-247	39.3		0.03	19.45	0.27	42.0	0.23	0.19	0.041	0.046	101.6	79.4
530-187	39.0	0.01	0.02	19.5	0.27	41.7	0.25	0.15	0.027	0.031	101.1	79.1
530-138	38.9	0.01	0.02	19.6	0.27	41.5	0.25	0.13	0.026	0.038	100.9	79.0
530-53	39.0	0.01	0.03	19.7	0.27	41.6	0.25	0.16	0.026	0.042	101.1	79.0
530-37	39.1	0.01	0.02	19.7	0.27	41.6	0.25	0.12	0.027	0.034	101.3	78.9
530-62	38.9	0.01	0.02	19.8	0.28	41.4	0.25	0.13	0.028	0.046	101.0	78.8
530-41	39.1	0.01	0.02	19.8	0.27	41.5	0.23	0.15	0.026	0.042	101.3	78.8
530-8	39.0	0.01	0.03	19.8	0.27	41.4	0.24	0.16	0.025	0.037	101.1	78.8
530-128	38.9	0.01	0.03	19.9	0.28	41.6	0.25	0.15	0.025	0.044	101.2	78.8
530-30	38.8		0.02	19.7	0.27	41.2	0.23	0.16	0.042	0.041	100.7	78.8
530-159	39.0	0.01	0.02	19.8	0.28	41.4	0.25	0.13	0.025	0.04	101.1	78.8
530-196	38.9	0.01	0.03	19.9	0.27	41.4	0.24	0.14	0.025	0.044	101.1	78.7
530-117	39.0	0.01	0.02	19.9	0.27	41.4	0.25	0.12	0.026	0.035	101.1	78.7
530-85	39.0	0.01	0.03	19.9	0.28	41.5	0.26	0.13	0.026	0.049	101.3	78.7
530-156	38.9	0.01	0.02	20.0	0.28	41.4	0.24	0.11	0.024	0.042	101.2	78.6
530-99	38.9	0.01	0.02	20.0	0.28	41.4	0.25	0.13	0.026	0.034	101.1	78.6
530-100	38.9	0.01	0.04	20.0	0.28	41.4	0.24	0.12	0.027	0.046	101.2	78.6
530-142	39.0	0.01	0.03	20.1	0.27	41.2	0.24	0.13	0.026	0.045	101.0	78.5
530-60	38.8	0.01	0.02	20.1	0.28	41.1	0.25	0.12	0.027	0.038	100.9	78.4
530-126	38.8	0.01	0.15	20.1	0.28	40.8	0.39	0.12	0.026	0.056	100.8	78.3
530-59	38.9	0.01	0.06	20.2	0.28	40.9	0.28	0.15	0.024	0.052	101.0	78.3
530-181	39.0	0.01	0.02	20.3	0.28	41.2	0.25	0.12	0.027	0.036	101.3	78.3
530-191	38.9	0.04	0.02	20.3	0.28	41.1	0.26	0.11	0.027	0.032	101.2	78.3
530-52	38.9	0.01	0.03	20.3	0.28	41.1	0.25	0.12	0.028	0.041	101.1	78.3
530-9	38.8	0.01	0.02	20.3	0.28	41.0	0.27	0.12	0.028	0.045	101.0	78.2
530-146	38.9	0.01	0.02	20.4	0.28	41.1	0.27	0.12	0.027	0.034	101.2	78.1
530-163	39.0	0.01	0.02	20.5	0.29	41.2	0.27	0.12	0.027	0.035	101.6	78.1
530-39	38.8	0.01	0.03	20.6	0.28	40.7	0.25	0.13	0.027	0.041	101.0	77.8
530-14	38.8	0.01	0.02	20.8	0.28	40.6	0.26	0.12	0.026	0.042	101.1	77.6
530-341	38.9		0.01	21.6	0.30	40.1	0.26	0.08	0.045	0.021	101.5	76.7
530-320	38.9		0.02	22.3	0.30	39.8	0.26	0.09	0.048	0.025	101.9	76.0
530-296	38.5		0.01	23.1	0.31	39.2	0.26	0.08	0.047	0.015	101.6	75.1
530-31	38.7		0.02	23.1	0.31	38.8	0.25	0.09	0.048	0.024	101.5	74.9
530-317	38.5		0.02	24.4	0.32	38.1	0.24	0.09	0.048	0.016	101.9	73.5
530-345	38.3		0.00	24.8	0.32	37.8	0.23	0.09	0.049	0.012	101.7	73.0
CY36c-1-4	38.0	0.01	0.02	26.3	0.37	36.2	0.24	0.12	0.031	0.011	101.4	71.0
CY36c-1-2	37.5	0.03	0.01	26.8	0.37	35.5	0.20	0.11	0.032	0.000	100.8	70.2
CY36a-1-9	38.4	0.03	0.01	24.0	0.35	38.0	0.20	0.12	0.030	0.003	101.3	73.8
CY36a-1-8	38.8	0.04	0.02	24.1	0.35	38.2	0.19	0.12	0.030	0.005	101.9	73.8
CY36a-1-7	38.5	0.03	0.01	24.2	0.35	37.9	0.19	0.12	0.030	0.000	101.5	73.5
CY36a-1-5	38.2	0.03	0.01	24.1	0.35	37.5	0.20	0.11	0.029	0.003	100.6	73.4
CY36a-1-4	38.5	0.03	0.01	24.3	0.36	37.3	0.21	0.11	0.031	0.002	101.0	73.2
CY36a-1-2	38.3	0.03	0.01	23.9	0.35	37.9	0.22	0.12	0.028	0.004	101.0	73.8
CY36a-1-15	38.5	0.03	0.01	24.3	0.35	38.2	0.22	0.11	0.029	0.002	101.8	73.6
CY36a-1-14	38.4	0.02	0.02	24.1	0.35	38.2	0.24	0.12	0.030	0.009	101.6	73.8
CY36a-1-13	38.9	0.02	0.02	21.6	0.32	39.8	0.23	0.12	0.028	0.019	101.2	76.6
CY36a-1-10	38.5	0.03	0.01	24.1	0.35	37.9	0.20	0.12	0.030	0.004	101.4	73.7
CY36a-1-1	38.3	0.03	0.01	24.6	0.35	37.3	0.22	0.11	0.032	0.002	101.1	73.0
CY36-1-7	38.1	0.01	0.01	25.9	0.37	36.5	0.25	0.11	0.031	0.002	101.3	71.5
CY36-1-5	38.3	0.03	0.00	26.3	0.38	36.4	0.22	0.11	0.031	0.000	101.9	71.1

(continued)

Table 3.9 (continued)

N sample	SiO_2	TiO_2	Al_2O_3	FeO	MnO	MgO	CaO	NiO	CoO	Cr_2O_3	Summa	*Fo*
CY36-1-1	38.4	0.03	0.01	26.2	0.37	36.4	0.22	0.11	0.031	0.000	101.8	71.3
CY33-3a-9	38.3	0.03	0.01	24.6	0.36	37.3	0.19	0.12	0.030	0.002	100.9	73.0
CY33-3a-8	38.5	0.03	0.00	24.5	0.36	37.2	0.19	0.11	0.029	0.005	101.0	73.1
CY33-3a-7	38.7	0.03	0.01	24.8	0.36	37.5	0.19	0.11	0.032	0.000	101.7	72.9
CY33-3a-6	38.4	0.03	0.02	25.0	0.36	37.3	0.21	0.12	0.033	0.000	101.6	72.7
CY33-3a-5	38.5	0.03	0.00	24.5	0.35	37.2	0.18	0.11	0.031	0.002	101.0	73.1
CY33-3a-4	38.3	0.03	0.01	24.8	0.36	37.3	0.18	0.11	0.031	0.001	101.2	72.8
CY33-3a-3	38.7	0.03	0.01	24.9	0.36	37.1	0.19	0.12	0.029	0.002	101.5	72.6
CY33-3a-26	38.4	0.03	0.01	25.2	0.37	36.7	0.18	0.11	0.029	0.000	101.0	72.2
CY33-3a-25	38.4	0.02	0.00	24.6	0.36	37.1	0.18	0.12	0.027	0.005	100.8	72.9
CY33-3a-23	38.4	0.03	0.01	24.3	0.36	37.3	0.19	0.11	0.032	0.006	100.8	73.2
CY33-3a-22	38.0	0.03	0.01	25.5	0.36	36.7	0.17	0.11	0.030	0.000	100.9	72.0
CY33-3a-21	38.3	0.04	0.01	25.2	0.36	37.3	0.20	0.12	0.031	0.000	101.7	72.5
CY33-3a-20	38.3	0.03	0.01	25.2	0.36	37.3	0.19	0.11	0.030	0.010	101.6	72.6
CY33-3a-19	38.5	0.03	0.00	24.4	0.35	37.2	0.20	0.11	0.029	0.005	100.8	73.1
CY33-3a-17	38.2	0.03	0.01	24.9	0.36	37.4	0.18	0.12	0.031	0.001	101.2	72.8
CY33-3a-16	38.6	0.03	0.01	24.3	0.35	38.2	0.21	0.12	0.032	0.007	101.9	73.7
CY33-3a-15	38.1	0.03	0.00	25.0	0.36	37.3	0.17	0.12	0.030	0.002	101.2	72.7
CY33-3a-14	38.2	0.03	0.00	24.6	0.36	37.1	0.18	0.11	0.029	0.001	100.7	72.9
CY33 3a 13	37.7	0.03	0.00	24.3	0.36	36.9	0.19	0.11	0.030	0.000	99.7	73.0
CY33-3a-12	38.3	0.03	0.01	25.2	0.36	37.2	0.21	0.12	0.030	0.002	101.5	72.5
CY33-3a-11	38.6	0.03	0.01	25.0	0.36	37.4	0.20	0.12	0.030	0.004	101.7	72.7
CY33-3a-10	38.7	0.03	0.01	24.7	0.36	37.5	0.19	0.12	0.030	0.004	101.7	73.0
CY33-3a-1	38.8	0.03	0.00	22.5	0.33	39.0	0.19	0.12	0.029	0.008	101.1	75.6
CY33-3-9	38.4	0.03	0.00	24.1	0.35	37.6	0.20	0.12	0.032	0.009	100.9	73.6
CY33-3-7	38.2	0.03	0.00	25.6	0.37	36.5	0.17	0.11	0.031	0.002	101.0	71.8
CY33-3-5	38.6	0.02	0.01	24.4	0.36	37.4	0.20	0.11	0.031	0.004	101.2	73.2
CY33-3-45	38.1	0.03	0.01	23.8	0.35	37.6	0.18	0.12	0.030	0.005	100.3	73.8
CY33-3-44	38.6	0.02	0.00	23.9	0.35	37.8	0.22	0.11	0.028	0.013	101.1	73.9
CY33-3-43	38.4	0.01	0.02	23.2	0.34	38.3	0.25	0.12	0.030	0.013	100.7	74.7
CY33-3-42	38.5	0.01	0.01	23.3	0.34	38.3	0.24	0.12	0.032	0.009	100.9	74.6
CY33-3-41	38.1	0.03	0.00	25.2	0.36	37.0	0.20	0.11	0.030	0.001	101.1	72.4
CY33-3-4	38.8	0.04	0.02	22.3	0.33	39.0	0.19	0.12	0.028	0.012	101.0	75.7
CY33-3-39	37.8	0.04	0.01	24.7	0.36	36.6	0.21	0.12	0.030	0.002	99.9	72.5
CY33-3-38	38.1	0.03	0.01	24.9	0.36	36.8	0.20	0.12	0.031	0.000	100.6	72.5
CY33-3-37	38.1	0.03	0.00	24.8	0.36	36.8	0.18	0.11	0.032	0.000	100.5	72.5
CY33-3-35	38.2	0.03	0.01	24.5	0.36	37.4	0.21	0.12	0.029	0.002	100.9	73.1
CY33-3-33	38.4	0.03	0.01	24.5	0.36	37.7	0.16	0.12	0.030	0.002	101.3	73.3
CY33-3-32	38.1	0.03	0.01	24.3	0.35	37.3	0.18	0.12	0.030	0.003	100.4	73.2
CY33-3-31	38.0	0.02	0.00	24.3	0.35	37.5	0.19	0.12	0.030	0.005	100.6	73.3
CY33-3-30	38.4	0.02	0.01	24.5	0.35	37.6	0.24	0.11	0.029	0.008	101.2	73.2
CY33-3-3	38.7	0.02	0.01	23.1	0.34	38.4	0.20	0.12	0.028	0.010	100.9	74.8
CY33-3-29	38.4	0.02	0.01	24.3	0.35	37.7	0.22	0.12	0.031	0.006	101.2	73.5
CY33-3-28	38.3	0.03	0.02	24.7	0.35	37.3	0.19	0.12	0.030	0.000	101.0	73.0
CY33-3-27	38.3	0.03	0.00	24.8	0.36	37.2	0.19	0.12	0.028	0.002	101.0	72.8
CY33-3-26	38.4	0.03	0.01	24.8	0.36	37.3	0.21	0.11	0.028	0.001	101.2	72.8
CY33-3-25	38.3	0.03	0.00	24.6	0.36	37.1	0.19	0.12	0.028	0.000	100.7	72.9
CY33-3-24	38.7	0.03	0.01	24.0	0.35	38.1	0.19	0.12	0.031	0.001	101.5	73.9
CY33-3-23	38.5	0.03	0.01	24.2	0.35	37.9	0.18	0.11	0.029	0.000	101.3	73.6
CY33-3-22	38.5	0.03	0.01	24.1	0.35	37.7	0.18	0.12	0.029	0.001	101.1	73.6

(continued)

Table 3.9 (continued)

N sample	SiO_2	TiO_2	Al_2O_3	FeO	MnO	MgO	CaO	NiO	CoO	Cr_2O_3	Summa	*Fo*
CY33-3-21	38.9	0.09	0.02	23.4	0.35	35.8	0.21	0.12	0.020	0.010	98.9	73.2
CY33-3-2	39.1	0.01	0.02	19.2	0.28	41.8	0.25	0.13	0.026	0.035	100.8	79.5
CY33-3-19	38.9	0.03	0.00	22.2	0.33	39.2	0.19	0.12	0.029	0.011	101.0	75.9
CY33-3-18	39.1	0.02	0.01	22.2	0.33	39.5	0.20	0.12	0.029	0.012	101.6	76.1
CY33-3-17	38.8	0.03	0.01	22.2	0.33	38.7	0.22	0.12	0.030	0.013	100.6	75.7
CY33-3-15	39.0	0.02	0.01	20.9	0.31	40.3	0.20	0.13	0.029	0.025	101.0	77.4
CY33-3-14	38.5	0.03	0.00	24.1	0.35	38.0	0.19	0.12	0.030	0.008	101.4	73.8
CY33-3-12	38.7	0.02	0.01	20.6	0.31	40.1	0.26	0.12	0.027	0.037	100.2	77.6
CY33-3-11	38.4	0.03	0.01	24.0	0.36	37.7	0.20	0.12	0.030	0.008	100.9	73.7
CY33-3-10	38.5	0.03	0.01	23.9	0.35	37.9	0.19	0.12	0.030	0.015	101.1	73.9
CY33-1b-8	37.8	0.03	0.00	27.9	0.37	35.5	0.19	0.12	0.033	0.001	101.8	69.4
CY33-1b-3-5	37.9	0.03	0.01	26.7	0.37	35.9	0.20	0.11	0.029	0.017	101.3	70.6
CY33-1b-3-2	38.0	0.03	0.00	27.3	0.38	35.8	0.19	0.11	0.029	0.000	101.8	70.0
CY33-1a-1-8	37.9	0.02	0.00	24.3	0.35	37.5	0.18	0.11	0.029	0.005	100.4	73.4
CY33-1a-1-16	38.0	0.03	0.01	23.7	0.33	37.6	0.16	0.11	0.029	0.020	100.0	73.9
CY33-1a-1-1	37.6	0.03	0.02	25.1	0.35	36.8	0.16	0.11	0.029	0.007	100.3	72.4
CY31-4-1-2	37.4	0.02	0.01	27.7	0.38	35.1	0.17	0.10	0.031	0.001	100.9	69.3
CY31-2a-9	38.5	0.03	0.00	25.3	0.36	37.4	0.18	0.11	0.029	0.004	101.9	72.5
CY31-2a-6	38.2	0.03	0.00	25.4	0.36	37.3	0.18	0.11	0.030	0.004	101.6	72.4
CY31-2a-5	38.4	0.03	0.01	25.4	0.36	37.5	0.17	0.11	0.030	0.001	102.0	72.5
CY31-2a-4	38.3	0.03	0.01	25.4	0.36	37.5	0.16	0.11	0.028	0.004	101.9	72.5
CY31-2a-2	38.2	0.03	0.01	25.1	0.36	37.5	0.18	0.11	0.030	0.005	101.5	72.7
CY1-2a-12	38.1	0.02	0.01	25.8	0.36	36.7	0.17	0.10	0.029	0.000	101.3	71.7
CY31-2a-10	38.3	0.03	0.01	25.2	0.36	37.4	0.17	0.11	0.029	0.001	101.7	72.6
CY31-2a-1	38.4	0.03	0.01	25.1	0.36	37.0	0.20	0.11	0.029	0.002	101.3	72.4
CY31-2-4	38.1	0.02	0.01	26.6	0.37	36.4	0.18	0.10	0.031	0.000	101.9	70.9
CY31-2-3	37.9	0.02	0.01	26.9	0.38	36.1	0.17	0.10	0.030	0.000	101.6	70.5
CY31-2-2-9	37.7	0.02	0.00	25.3	0.36	36.8	0.18	0.10	0.028	0.003	100.5	72.1
CY31-2-2-6	38.0	0.03	0.01	25.7	0.36	37.1	0.16	0.10	0.029	0.002	101.6	72.0
CY31-2-2-5	37.9	0.03	0.01	25.5	0.36	37.0	0.20	0.11	0.027	0.000	101.2	72.1
CY31-2-2-12	37.6	0.02	0.01	25.7	0.37	36.6	0.18	0.10	0.030	0.002	100.6	71.7
CY31-2-2-11	37.8	0.02	0.00	26.2	0.37	36.2	0.17	0.11	0.031	0.001	101.0	71.1
CY31-2-2-10	37.6	0.02	0.00	26.5	0.38	36.2	0.16	0.10	0.031	0.000	101.0	70.9
CY31-2-2-1	38.0	0.03	0.01	25.5	0.36	37.1	0.19	0.10	0.028	0.005	101.4	72.2

Note: Here and in Tables 3.10 and 3.11 empty cells-element was not determined, *Fo – mol.* % forsterite in olivine
After Sobolev et al. (2009)

exposure to temperatures exceeding 1,000 °C was no longer than 15 min. In addition to experiments under visual control, quench experiments were performed using a controlled-atmosphere vertical furnace at the Petrology and Geochemistry Department of J.W. Goethe University of Frankfurt am Maine, Germany. The experiments were carried out with an H_2/CO_2 gas mixture at a temperature of 1,250 °C and oxygen fugacity corresponding to the quartz–fayalite–magnetite (QFM) buffer. Selected olivine crystals were loaded into open platinum capsules and placed in a platinum container directly into the hot zone of the furnace. After a 20-min exposure, the samples were quenched by automatically dropping the platinum container into the cold zone.

Magmatic inclusions were investigated in the olivine from samples KhS-51/130, 4270/13, and SU-50 of the Gudchikhinsky Formation (Sobolev et al. 2009). These inclusions are represented by crystallized melt, spinel, low-density fluid, and occasionally combinations of these phases in varying proportions. Most of the inclusions are not confined to fractures but are randomly distributed within the phenocryst fraction. Such inclusions are interpreted as primary, i.e., trapped during phenocryst growth (Roedder 1984). Most of the melt inclusions have rounded or ellipsoidal shapes. Their typical size ranges from 20 to 80 μm along the long axis. The inclusions are composed of clinopyroxene crystals, interstitial glass, a low-density fluid phase

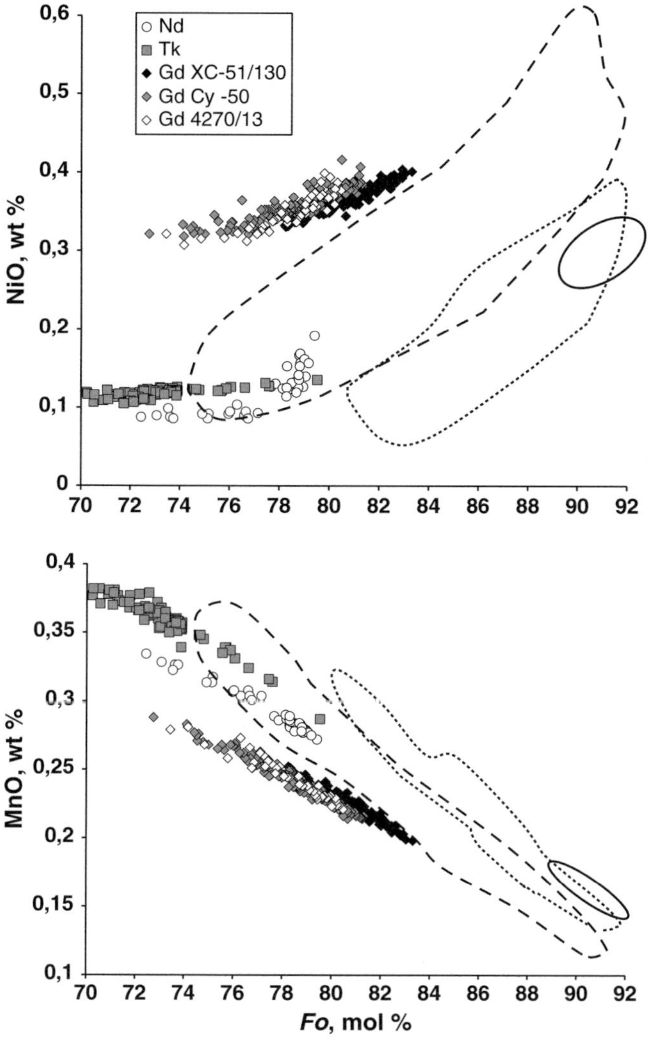

Fig. 3.33 Compositions of olivine phenocrysts from the magnesian Siberian traps of the Gudchikhinsky (*Gd*), Nadezhdinsky (*Nd*), and Tuklonsky (*Tk*)
Formations of the Noril'sk region (After Sobolev et al. 2009). The solid line encloses the compositional field of olivines in equilibrium with peridotite material. The *dotted line* encloses the field of the majority of olivine phenocrysts from mid-ocean ridge basalts. The *dashed line* encloses the field of olivine phenocrysts from intraplate magmas formed beneath the thick lithosphere (more than 70 km). The compositional fields are after Sobolev et al. 2007

(shrinkage voids), spinel crystals, and occasionally tiny droplets of sulfide melt. During heating, the inclusions begin to melt at temperatures of 1,050–1,100 °C, and the last daughter pyroxene crystal disappears at 1,150–1,180 °C. Homogenization (complete dissolution of the fluid phase in the melt) occurs almost simultaneously in all inclusions of a series. Based on the results of 15 experiments, the range of homogenization temperatures is 1,180–1,290 °C. In a few of the homogenized inclusions, a well-shaped spinel crystal was preserved after quenching; such crystals were interpreted as xenocrysts trapped with the melt. To obtain statistically representative data regarding the compositions of the melt inclusions, 100–150 olivine grains from each of samples KhS-51/130, 4270/13, and CY-50 were annealed for 20 min at a temperature of 1,250 °C in the vertical furnace at oxygen fugacity corresponding to the QFM buffer and were quenched. The quenched crystals were sectioned, polished, and examined under a microscope. Homogeneous inclusions or inclusions with a fluid vesicle accounting for less than 0.1 % of the inclusion volume and more than 40 μm in size were selected for analysis.

3.3.2.7 Crystallization Conditions and Melt Composition

The compositions of 37 quenched melt inclusions and their host olivines from the samples of the Gudchikhinsky Formation are given in Table 3.10. Because the quenching temperature (1,250 °C) of inclusions in the vertical furnace generally differed from the temperature of entrapment, the compositions of the inclusions were also recalculated to equilibrium with the host olivine using the model of Ford et al. (1983). There is a significant linear correlation between the amounts of FeO in the inclusions and the host olivine (FeO melt=0.53FeOOl – 0.9, R=0.7), which indicates that the loss of Fe from the inclusions was due to the diffusion redistribution of Fe and Mg between the inclusions and host olivine during cooling (Sobolev and Danyushevsky 1994). This effect can be accounted for by the significant correlation between the amounts of FeO and SiO_2 (FeOtot=30.84—0.369SiO_2, R=0.7) in the rocks of the Gudchikhinsky Formation containing, similar to the melts, 7–15 wt % MgO (GEOROC; http://georoc.mpch-mainz.gwdg.de/georoc/). The Fe^{2+}/Fe^{3+} ratio in the melt was determined from the composition of chrome spinel and the model of spinel–melt Fe^{2+}/Fe^{3+} partitioning (Maurel and Maurel 1982). The corrected values differ only slightly from the compositions of the quenched inclusions, primarily with respect to the Mg and Fe concentrations. Nonetheless, Table 3.11 presents correction coefficients, *K*cor, for olivine-incompatible trace element concentrations in the inclusions. The coefficient was calculated for each inclusion as a ratio of the Al_2O_3 concentrations in the measured and corrected compositions. To obtain a corrected value for the incompatible element content in the trapped melt, values from Table 3.10 should be multiplied by *K*cor from Table 3.11. Hereafter, we will consider only the corrected compositions. The content of Ni in the melt was calculated from the composition of host olivine and the model of olivine–melt equilibrium by Beattie (1993). The temperatures of equilibrium between trapped melts and their host olivine lie within the range of 1,170–1,250 °C, which is near the range of homogenization temperatures for these inclusions (1,180–1,290 °C). The oxygen fugacity is unusually low (2.5–3.0 logarithmic units below the Ni–NiO

Table 3.10 Composition of melt inclusions and host olivines from picrites of the Gudchikhinsky Formation

No	1	2	3	4	5	6	7	8	9	10
Component	Melt inclusions									
SiO_2	50.76	50.92	51.05	51.06	50.25	51.24	51.26	50.52	51.16	50.45
TiO_2	2.29	2.56	2.22	2.24	2.63	2.26	2.28	2.30	2.38	2.35
Al_2O_3	13.01	12.48	13.35	13.09	12.87	13.16	13.33	12.71	13.34	13.27
FeO	9.46	10.05	8.82	8.57	9.02	8.80	8.74	9.66	8.19	8.15
MnO	0.13	0.13	0.11	0.13	0.11	0.12	0.14	0.10	0.13	0.10
MgO	9.85	9.62	9.50	9.71	9.96	9.90	9.88	9.99	9.67	9.74
CaO	10.88	10.22	11.26	11.39	11.18	11.35	11.13	10.91	11.54	11.41
Na_2O	2.33	2.42	2.45	2.23	2.27	2.29	2.40	2.30	2.34	2.33
K_2O	0.37	0.43	0.40	0.50	0.44	0.45	0.39	0.35	0.40	0.39
P_2O_5	0.21	0.25	0.19	0.23	0.25	0.19	0.21	0.21	0.20	0.21
S, wt %	0.04	0.05	0.04	0.06	0.06	0.04	0.05	0.03	0.03	0.04
Cl, wt %	0.04	0.07	0.10	0.20	0.05	0.12	0.07	0.04	0.07	0.05
Summa	99.38	99.20	99.49	99.42	99.09	99.92	99.88	99.12	99.44	98.49
NiO	0.04	0.03	0.04	0.03	0.05	0.04	0.04	0.04	0.04	0.04
Cr_2O_3	0.18	0.10	0.18	0.21	0.17	0.18	0.20	0.18	0.15	0.15
TiO_2, wt %	2.27	2.30	2.05	2.14		2.24	2.23	2.26	2.36	2.21
H_2O, wt %	0.05	0.07	0.03	0.03		0.05	0.03	0.06	0.04	0.03
B	4.0	4.7	5.4	7.1		6.5	4.9	4.0	5.3	4.3
Li	5.8	6.2	4.3	4.9		4.4	4.4	5.6	5.0	4.9
Rb	7.05	9.57	10.85	11.48	10.76	11.71	8.27	7.58	9.31	8.39
Ba	71.7	85.1	83.9	99.7	84.7	84.7	76.1	69.2	77.5	71.6
Th	1.00	1.22	1.26	1.39	1.19	1.24	0.96	1.02	1.15	1.02
U	0.26	0.33	0.35	0.39	0.33	0.33	0.32	0.28	0.31	0.32
Nb	9.85	11.43	10.23	11.71	12.90	10.23	9.48	9.90	10.58	10.10
Ta	0.71	0.75	0.69	0.77	0.83	0.68	0.63	0.66	0.69	0.68
La	10.51	12.21	11.11	12.68	12.49	11.27	10.39	10.27	10.56	10.19
Ce	28.13	31.10	27.96	31.98	32.35	27.31	26.26	26.07	27.97	26.67
Pb	1.35	1.87	1.81	1.90	1.93	1.93	1.58	1.41	1.63	2.46
Pr	4.03	4.75	4.02	4.60	4.86	4.02	3.71	3.94	4.19	3.93
Nd	19.92	23.73	20.80	21.42	23.70	20.11	19.70	19.93	20.19	19.05
Sr	297	368	343	355	387	322	292	299	313	302
Sm	5.56	6.63	6.14	5.41	6.50	5.45	5.64	5.76	5.84	5.53
Zr	146	173	148	153	186	143	132	143	150	147
Hf	3.65	4.63	3.90	4.21	4.62	3.63	3.64	3.74	3.99	3.85
Eu	2.02	2.22	1.96	1.97	2.08	1.82	1.81	1.82	1.98	1.89
Ti	14,108	15,850	13,939	14,608	16,584	14,292	13,765	14,195	15,210	14,787
Gd	6.04	7.29	6.43	6.22	6.85	6.11	5.97	6.25	6.01	6.19
Tb	0.97	1.15	1.01	0.96	1.05	0.93	0.98	0.92	0.92	0.97
Dy	5.67	6.62	5.79	5.82	5.86	5.84	5.29	5.54	5.54	5.56
Ho	1.09	1.26	1.09	1.06	1.08	1.06	0.93	1.06	0.98	0.99
Y	26.83	33.81	29.22	27.43	30.44	27.80	26.43	26.02	26.03	26.40
Er	2.65	3.44	2.94	2.74	2.82	2.59	2.43	2.72	2.56	2.71
Tm	0.35	0.43	0.34	0.33	0.35	0.34	0.32	0.31	0.32	0.32
Yb	1.96	2.75	2.13	2.23	2.04	2.03	1.95	1.94	2.13	2.00
Lu	0.29	0.39	0.31	0.28	0.29	0.27	0.27	0.27	0.26	0.29
	Ol-host									
Fo, mol %	79.62	77.42	80.40	81.24	79.82	80.15	80.43	78.86	81.01	80.45
SiO_2	38.64	38.64	38.90	39.04	39.12	39.21	39.40	39.07	39.23	39.02
TiO_2	0.008	0.01	0.01	0.01	0.01	0.01	0.008	0.008	0.01	0.01

(continued)

Table 3.10 (continued)

No	1	2	3	4	5	6	7	8	9	10
Component	Melt inclusions									
Al_2O_3	0.02	0.02	0.02	0.02	0.02	0.02	0.02	0.02	0.02	0.02
FeO	19.07	21.00	18.47	17.83	19.06	18.77	18.47	19.83	17.95	18.43
MnO	0.23	0.26	0.22	0.21	0.23	0.22	0.22	0.24	0.21	0.22
MgO	41.79	40.40	42.52	43.32	42.31	42.28	42.59	41.51	42.96	42.58
CaO	0.29	0.30	0.29	0.29	0.30	0.30	0.29	0.28	0.29	0.29
NiO	0.36	0.31	0.38	0.38	0.35	0.37	0.38	0.35	0.38	0.38
Cr_2O_3	0.04	0.03	0.04	0.04	0.04	0.04	0.04	0.04	0.04	0.03
Summa	100.47	101.01	100.90	101.20	101.49	101.26	101.46	101.40	101.14	101.03

No	11	12	13	14	15	16	17	18
Component	Melt inclusions							
SiO_2	50.63	50.09	50.39	50.04	50.59	50.53	50.28	50.44
TiO_2	2.29	2.30	2.37	2.42	2.30	2.33	2.36	2.21
Al_2O_3	12.94	12.73	12.84	13.32	12.83	12.80	12.79	13.00
FeO	9.39	9.99	9.69	8.41	9.90	9.71	10.20	9.99
MnO	0.13	0.13	0.14	0.11	0.13	0.13	0.14	0.15
MgO	9.78	9.70	10.05	9.91	9.84	9.77	9.37	9.11
CaO	10.77	11.22	11.21	11.16	10.77	11.17	11.27	10.69
Na_2O	2.46	2.19	2.12	2.26	2.29	2.07	2.07	2.46
K_2O	0.49	0.34	0.34	0.38	0.37	0.36	0.36	0.35
P_2O_5	0.21	0.19	0.22	0.23	0.22	0.20	0.23	0.19
S, wt %	0.05	0.04	0.05	0.06	0.05	0.03	0.06	0.04
Cl, wt %	0.13	0.03	0.03	0.06	0.04	0.04	0.04	0.06
Summa	99.27	98.97	99.48	98.37	99.34	99.14	99.19	98.69
NiO	0.03	0.04	0.04	0.05	0.04	0.04	0.04	0.04
Cr_2O_3	0.12	0.17	0.15	0.15	0.18	0.20	0.18	0.20
TiO_2,,wt %	2.20							
H_2O, wt %	0.04							
B	6.5							
Li	7.4							
Rb	11.32	7.78	7.50	8.90	7.24	8.18	7.27	8.11
Ba	93.7	73.6	71.9	77.0	69.6	76.6	73.6	72.4
Th	1.34	1.00	0.94	1.11	0.97	1.11	1.00	1.03
U	0.36	0.28	0.28	0.32	0.29	0.30	0.30	0.26
Nb	11.27	10.28	10.08	10.55	9.65	9.96	9.91	9.28
Ta	0.73	0.66	0.67	0.72	0.65	0.68	0.70	0.61
La	12.21	10.41	10.43	11.07	10.39	10.36	10.38	10.41
Ce	29.99	26.76	26.76	28.33	26.87	27.11	27.38	26.65
Pb	2.03	1.59	1.46	1.76	1.48	1.57	1.51	1.55
Pr	4.31	3.94	4.01	4.29	4.09	3.98	3.84	4.01
Nd	21.35	20.27	20.30	21.49	18.87	20.46	20.29	18.73
Sr	325	308	299	320	312	313	313	304
Sm	5.79	5.49	5.85	6.01	5.70	5.75	5.69	5.96
Zr	147	147	144	155	149	148	149	141
Hf	4.03	4.09	3.92	4.21	3.84	4.02	3.87	3.75
Eu	1.99	1.97	1.86	1.96	1.87	1.85	1.97	2.03
Ti	14,244	14,927	14,896	15,370	14,218	14,274	14,543	13,566
Gd	6.43	6.20	6.38	6.69	6.13	6.06	5.86	6.32
Tb	1.01	0.95	0.96	0.99	0.99	0.93	0.95	1.03

(continued)

Table 3.10 (continued)

No	11	12	13	14	15	16	17	18
Component	Melt inclusions							
Dy	6.36	5.86	5.74	5.92	5.73	5.71	5.70	5.64
Ho	1.17	1.03	1.04	1.09	1.01	1.06	1.08	1.01
Y	29.75	27.26	26.57	27.20	26.90	26.93	26.95	26.92
Er	3.08	2.65	2.68	2.69	2.55	2.68	2.72	2.60
Tm	0.39	0.33	0.33	0.34	0.33	0.34	0.31	0.34
Yb	2.43	2.05	2.07	2.10	2.15	2.21	2.02	1.99
Lu	0.31	0.28	0.27	0.29	0.27	0.30	0.28	0.30
	Ol-host							
Fo, mol %	78.06	77.07	77.66	81.25	79.85	78.30	79.14	79.36
SiO_2	38.77	38.42	38.94	39.39	39.57	39.11	39.43	38.70
TiO_2	0.008	0.01	0.01	0.01	0.009	0.01	0.009	0.01
Al_2O_3	0.02	0.03	0.02	0.02	0.02	0.02	0.02	0.03
FeO	20.56	21.12	20.86	17.67	19.02	20.31	19.67	19.18
MnO	0.24	0.25	0.24	0.21	0.23	0.24	0.23	0.23
MgO	41.04	39.82	40.68	42.96	42.27	41.13	41.86	41.37
CaO	0.28	0.31	0.29	0.29	0.29	0.28	0.29	0.28
NiO	0.36	0.34	0.34	0.38	0.36	0.35	0.34	0.35
Cr_2O_3	0.03	0.02	0.03	0.04	0.04	0.03	0.03	0.04
Summa	101.37	100.36	101.44	101.00	101.84	101.52	101.91	100.22

No	19	20	21	22	23	24	25	26	27	28
Component	Melt inclusions									
SiO_2	50.28	50.11	50.08	48.60	49.74	50.65	49.73	52.86	52.95	52.78
TiO_2	2.31	2.32	2.25	2.57	2.25	2.28	2.31	1.71	1.65	1.76
Al_2O_3	13.21	12.84	13.11	12.89	12.72	12.80	12.85	13.46	13.29	13.43
FeO	9.26	10.15	9.90	10.98	11.20	9.95	10.83	8.37	9.40	8.51
MnO	0.13	0.14	0.13	0.14	0.16	0.15	0.14	0.12	0.13	0.11
MgO	9.30	9.94	9.96	9.73	9.84	10.05	9.69	9.50	9.63	9.42
CaO	11.21	10.55	10.62	10.69	10.42	10.88	10.57	9.72	9.76	9.95
Na_2O	2.27	2.17	2.20	2.20	2.32	2.19	2.28	2.41	2.28	2.37
K_2O	0.33	0.36	0.34	0.37	0.34	0.33	0.32	0.63	0.59	0.61
P_2O_5	0.22	0.24	0.20	0.23	0.18	0.21	0.22	0.16	0.15	0.18
S, wt %	0.02	0.04	0.04	0.04	0.05	0.04	0.03	0.03	0.04	0.04
Cl, wt %	0.04	0.04	0.04	0.03	0.02	0.03	0.03	0.003	0.005	0.005
Summa	98.58	98.91	98.87	98.47	99.25	99.57	99.01	98.98	99.86	99.17
NiO	0.03	0.05	0.05	0.05	0.04	0.05	0.05	0.03	0.03	0.03
Cr_2O_3	0.23	0.19	0.09	0.11	0.04	0.13	0.15	0.16	0.09	0.18
TiO_2, wt %		2.32	2.26	2.72	2.14	2.24	2.34	1.65	1.60	1.74
H_2O, wt % мас.%		0.18	0.12	0.25	0.13	0.19	0.14	0.05	0.06	0.05
B		3.8	3.5	3.7	3.0	3.7	3.2	3.9	5.3	3.7
Li		4.5	3.6	5.3	3.2	4.8	5.0	7.0	7.7	7.3
Rb	7.51	7.06	6.87	8.12	6.59	6.95	6.65	20.46		19.81
Ba	73.7	75.7	67.8	81.0	68.7	69.1	72.6	115.4		114.5
Th	0.89	1.06	0.91	1.02	0.89	0.89	0.92	1.38		1.39
U	0.26	0.28	0.27	0.31	0.26	0.27	0.25	0.53		0.54
Nb	10.04	10.62	9.15	11.40	8.66	9.31	8.92	8.20		8.32
Ta	0.67	0.79	0.62	0.81	0.55	0.64	0.60	0.56		0.62
La	10.50	11.36	9.81	11.53	9.23	9.83	10.09	9.59		9.62

(continued)

Table 3.10 (continued)

No	19	20	21	22	23	24	25	26	27	28
Component	Melt inclusions									
Ce	28.24	29.27	24.73	29.92	24.07	25.91	26.29	23.42		24.17
Pb	1.22	1.46	1.37	1.54	1.29	1.37	1.18	3.68		3.47
Pr	4.19	4.40	3.83	4.53	3.58	3.82	3.89	3.36		3.58
Nd	21.14	21.72	19.35	22.10	17.59	19.27	19.17	16.37		17.89
Sr	303	309	287	329	282	292	282	258		279
Sm	5.93	6.44	5.30	6.04	5.21	5.51	5.44	4.50		4.65
Zr	148	160	138	166	132	137	134	120		131
Hf	3.97	4.22	3.65	4.40	3.69	3.68	3.55	3.25		3.59
Eu	1.90	2.03	1.77	2.00	1.78	1.93	1.79	1.53		1.66
Ti	14,202	13,724	13,508	16,659	13,326	14,398	13,649	10,309		10,927
Gd	6.63	6.88	6.11	6.73	5.61	5.93	6.25	4.80		5.48
Tb	1.03	1.00	0.91	1.02	0.94	0.96	0.98	0.81		0.80
Dy	5.79	5.93	5.47	6.05	5.81	5.56	5.46	4.81		4.95
Ho	1.03	1.11	0.98	1.10	1.03	0.98	1.03	0.84		0.92
Y	26.80	26.23	25.47	27.89	26.60	25.79	25.36	22.46		23.75
Er	2.70	2.54	2.48	2.87	2.86	2.66	2.50	2.21		2.61
Tm	0.27	0.39	0.34	0.34	0.34	0.31	0.33	0.30		0.31
Yb	2.01	2.01	1.96	2.26	2.34	2.01	1.94	1.76		1.82
Lu	0.27	0.26	0.27	0.29	0.26	0.25	0.28	0.25		0.27
	Ol-host									
Fo, mol %	80.55	80.41	80.23	77.60	77.90	80.32	78.97	79.93	79.63	80.17
SiO_2	39.35	39.04	39.54	38.98	39.32	39.18	39.13	38.83	39.29	39.21
TiO_2	0.01	0.01	0.01	0.01	0.01	0.01	0.01	0.01	0.008	0.01
Al_2O_3	0.02	0.03	0.03	0.02	0.02	0.02	0.02	0.02	0.02	0.02
FeO	18.38	18.40	18.66	20.89	20.72	18.52	19.77	18.89	19.18	18.73
MnO	0.21	0.22	0.22	0.24	0.24	0.22	0.23	0.23	0.24	0.23
MgO	42.70	42.37	42.50	40.60	40.98	42.41	41.66	42.20	42.14	42.24
CaO	0.29	0.29	0.29	0.30	0.29	0.29	0.29	0.26	0.26	0.27
NiO	0.37	0.37	0.38	0.33	0.36	0.36	0.35	0.34	0.34	0.34
Cr_2O_3	0.03	0.03	0.04	0.03	0.03	0.04	0.04	0.04	0.03	0.03
Summa	101.41	100.78	101.72	101.46	102.02	101.11	101.54	100.85	101.53	101.11

No	29	30	31	32	33	34	35	36	37
Component	Melt inclusions								
SiO_2	50.28	50.11	50.08	48.60	49.74	50.65	49.73	52.86	52.95
TiO_2	2.31	2.32	2.25	2.57	2.25	2.28	2.31	1.71	1.65
Al_2O_3	13.21	12.84	13.11	12.89	12.72	12.80	12.85	13.46	13.29
FeO	9.26	10.15	9.90	10.98	11.20	9.95	10.83	8.37	9.40
MnO	0.13	0.14	0.13	0.14	0.16	0.15	0.14	0.12	0.13
MgO	9.30	9.94	9.96	9.73	9.84	10.05	9.69	9.50	9.63
CaO	11.21	10.55	10.62	10.69	10.42	10.88	10.57	9.72	9.76
Na_2O	2.27	2.17	2.20	2.20	2.32	2.19	2.28	2.41	2.28
K_2O	0.33	0.36	0.34	0.37	0.34	0.33	0.32	0.63	0.59
P_2O_5	0.22	0.24	0.20	0.23	0.18	0.21	0.22	0.16	0.15
S, wt %	0.02	0.04	0.04	0.04	0.05	0.04	0.03	0.03	0.04
Cl, wt %	0.04	0.04	0.04	0.03	0.02	0.03	0.03	0.003	0.003
Summa	98.58	98.91	98.87	98.47	99.25	99.57	99.01	98.98	99.86
NiO	0.03	0.05	0.05	0.05	0.04	0.05	0.05	0.03	0.03
Cr_2O_3	0.23	0.19	0.09	0.11	0.04	0.13	0.15	0.16	0.09

(continued)

Table 3.10 (continued)

No	29	30	31	32	33	34	35	36	37
Component	Melt inclusions								
TiO_2, wt %		2.32	2.26	2.72	2.14	2.24	2.34	1.65	1.60
H_2O, wt %		0.18	0.12	0.25	0.13	0.19	0.14	0.05	0.06
B		3.8	3.5	3.7	3.0	3.7	3.2	3.9	5.3
Li		4.5	3.6	5.3	3.2	4.8	5.0	7.0	7.7
Rb	7.51	7.06	6.87	8.12	6.59	6.95	6.65	20.46	
Ba	73.7	75.7	67.8	81.0	68.7	69.1	72.6	115.4	
Th	0.89	1.06	0.91	1.02	0.89	0.89	0.92	1.38	
U	0.26	0.28	0.27	0.31	0.26	0.27	0.25	0.53	
Nb	10.04	10.62	9.15	11.40	8.66	9.31	8.92	8.20	
Ta	0.67	0.79	0.62	0.81	0.55	0.64	0.60	0.56	
La	10.50	11.36	9.81	11.53	9.23	9.83	10.09	9.59	
Ce	28.24	29.27	24.73	29.92	24.07	25.91	26.29	23.42	
Pb	1.22	1.46	1.37	1.54	1.29	1.37	1.18	3.68	
Pr	4.19	4.40	3.83	4.53	3.58	3.82	3.89	3.36	
Nd	21.14	21.72	19.35	22.10	17.59	19.27	19.17	16.37	
Sr	303	309	287	329	282	292	282	258	
Sm	5.93	6.44	5.30	6.04	5.21	5.51	5.44	4.50	
Zr	148	160	138	166	132	137	134	120	
Hf	3.97	4.22	3.65	4.40	3.69	3.68	3.55	3.25	
Eu	1.90	2.03	1.77	2.00	1.78	1.93	1.79	1.53	
Ti	14,202	13,724	13,508	16,659	13,326	14,398	13,649	10,309	
Gd	6.63	6.88	6.11	6.73	5.61	5.93	6.25	4.80	
Tb	1.03	1.00	0.91	1.02	0.94	0.96	0.98	0.81	
Dy	5.79	5.93	5.47	6.05	5.81	5.56	5.46	4.81	
Ho	1.03	1.11	0.98	1.10	1.03	0.98	1.03	0.84	
Y	26.80	26.23	25.47	27.89	26.60	25.79	25.36	22.46	
Er	2.70	2.54	2.48	2.87	2.86	2.66	2.50	2.21	
Tm	0.27	0.39	0.34	0.34	0.34	0.31	0.33	0.30	
Yb	2.01	2.01	1.96	2.26	2.34	2.01	1.94	1.76	
Lu	0.27	0.26	0.27	0.29	0.26	0.25	0.28	0.25	
	Ol-host								
Fo, mol %	80.55	80.41	80.23	77.60	77.90	80.32	78.97	79.93	79.63
SiO_2	39.35	39.04	39.54	38.98	39.32	39.18	39.13	38.83	39.29
TiO_2	0.01	0.01	0.01	0.01	0.01	0.01	0.01	0.01	0.00
Al_2O_3	0.02	0.03	0.03	0.02	0.02	0.02	0.02	0.02	0.02
FeO	18.38	18.40	18.66	20.89	20.72	18.52	19.77	18.89	19.18
MnO	0.21	0.22	0.22	0.24	0.24	0.22	0.23	0.23	0.24
MgO	42.70	42.37	42.50	40.60	40.98	42.41	41.66	42.20	42.14
CaO	0.29	0.29	0.29	0.30	0.29	0.29	0.29	0.26	0.26
NiO	0.37	0.37	0.38	0.33	0.36	0.36	0.35	0.34	0.34
Cr_2O_3	0.03	0.03	0.04	0.03	0.03	0.04	0.04	0.04	0.03
Summa	101.41	100.78	101.72	101.46	102.02	101.11	101.54	100.85	101.53

Note: (1) No inclusion: 1–21—sample CY-50, 22–27—обр. 4270/13, 28–39—KhS-51/130; (2) line subdivide group of elements determined by different methods (from the top to down): standard X-ray microanalysis, high-precious X-ray microanalysis, secondary ion mass spectrometry, mass spectrometry couple inductive with laser ablation. Analyses were carried out in Max-Planck Institute of Chemistry, Mainz, Germany, analyst D. V. Kuzmin; (3) oxides are given in %, elements – in ppm. After Sobolev et al. (2009)

Table 3.11 Composition of trapped melt in olivines from the picrate of the Gudchikhinsky Formation (wt %)

Component	CY50-3	CY50-5	CY50-6	CY50-7	CY50-8	CY50-9	CY50-10
SiO_2	50.33	50.10	50.02	49.72	50.19	50.14	50.43
TiO_2	2.27	2.12	2.12	2.58	2.18	2.18	2.33
Al_2O_3	12.88	12.76	12.37	12.60	12.67	12.76	12.85
Fe_2O_3	0.52	0.53	0.53	0.53	0.53	0.53	0.52
FeO	11.79	11.88	11.90	12.01	11.85	11.86	11.76
MnO	0.12	0.11	0.11	0.12	0.11	0.11	0.12
MgO	8.23	8.64	9.15	8.40	8.52	8.66	7.86
CaO	10.77	10.76	10.77	10.95	10.93	10.66	11.03
Na_2O	2.31	2.34	2.11	2.22	2.21	2.30	2.33
K_2O	0.37	0.38	0.47	0.43	0.43	0.37	0.35
P_2O_5	0.21	0.18	0.22	0.25	0.18	0.20	0.21
Cr_2O_3	0.18	0.18	0.21	0.17	0.18	0.20	0.18
NiO	0.02	0.03	0.03	0.02	0.03	0.03	0.02
T_{clc}.°C	1,208	1,220	1,231	1,212	1,215	1,220	1,197
Fo	79.62	80.4	81.24	79.82	80.15	80.43	78.86
NiO_{ol}	0.36	0.38	0.38	0.35	0.37	0.38	0.35
K_{cor}	0.99	0.95	0.94	0.97	0.96	0.95	1.01
S_{cor}	0.04	0.03	0.05	0.05	0.03	0.04	0.02
Cl_{cor}	0.04	0.09	0.19	0.05	0.11	0.07	0.04

Component	CY50-11	CY50-12	CY50-13	CY50-201	CY50-202	CY50-203	CY50-204
SiO_2	49.92	49.80	50.52	50.36	50.26	49.46	50.32
TiO_2	2.23	2.25	2.32	2.38	2.42	2.31	2.29
Al_2O_3	12.52	12.70	13.09	13.15	13.09	12.70	12.78
Fe_2O_3	0.53	0.53	0.52	0.52	0.53	0.54	0.52
FeO	11.93	11.99	11.73	11.75	11.83	12.11	11.79
MnO	0.11	0.11	0.13	0.13	0.13	0.10	0.13
MgO	9.00	8.72	7.45	7.11	7.41	9.25	8.35
CaO	10.83	10.92	10.90	11.59	11.43	10.64	10.73
Na_2O	2.20	2.23	2.49	2.26	2.16	2.15	2.28
K_2O	0.38	0.37	0.50	0.35	0.35	0.36	0.37
P_2O_5	0.19	0.20	0.21	0.20	0.22	0.22	0.22
Cr_2O_3	0.14	0.15	0.13	0.19	0.16	0.15	0.18
NiO	0.03	0.03	0.02	0.02	0.02	0.03	0.02
T_{clc}.°C	1,228	1,221	1,189	1,172	1,181	1,234	1,211
Fo	81.01	80.45	78.06	77.07	77.66	81.25	79.85
NiO ol	0.38	0.38	0.36	0.34	0.34	0.38	0.36
K_{cor}	0.93	0.95	1.01	1.03	1.01	0.95	0.99
S_{cor}	0.02	0.03	0.05	0.04	0.05	0.05	0.05
Cl_{cor}	0.07	0.05	0.13	0.03	0.04	0.06	0.04

Component	CY50-205	CY50-206	CY50-207	CY50-208	4270-1	4270-2	4270-3
SiO_2	50.45	50.15	50.48	49.82	50.07	50.03	49.29
TiO_2	2.36	2.36	2.20	2.23	2.32	2.25	2.69
Al_2O_3	12.95	12.77	12.96	12.73	12.82	13.08	13.48
Fe_2O_3	0.52	0.53	0.52	0.53	0.53	0.53	0.54
FeO	11.75	11.86	11.75	11.98	11.88	11.91	12.16
MnO	0.13	0.13	0.13	0.12	0.13	0.13	0.14
MgO	7.66	8.08	8.07	8.82	8.72	8.62	7.46
CaO	11.30	11.25	10.66	10.80	10.54	10.60	11.18
Na_2O	2.09	2.07	2.45	2.19	2.17	2.20	2.30
K_2O	0.36	0.36	0.35	0.32	0.36	0.34	0.39
P_2O_5	0.20	0.23	0.19	0.21	0.24	0.20	0.24

(continued)

Table 3.11 (continued)

Component	CY50-205	CY50-206	CY50-207	CY50-208	4270-1	4270-2	4270-3
Cr_2O_3	0.21	0.19	0.21	0.23	0.20	0.09	0.12
NiO	0.02	0.02	0.02	0.03	0.03	0.03	0.02
T_{clc}.°C	1,187	1,199	1,206	1,222	1,220	1,218	1,185
Fo	78.3	79.14	79.36	80.55	80.41	80.23	77.6
NiO_{ol}	0.35	0.34	0.35	0.37	0.37	0.38	0.33
K_{cor}	1.01	0.99	0.99	0.96	0.99	0.99	1.04
S_{cor}	0.02	0.06	0.03	0.01	0.03	0.03	0.03
Cl_{cor}	0.04	0.04	0.06	0.04	0.04	0.04	0.03

Component	4270-4	4270-5	4270-6	XC51-1	XC51-4	XC51-5	XC51-6
SiO_2	50.33	50.26	50.02	52.43	52.57	52.18	52.87
TiO_2	2.37	2.26	2.37	1.68	1.65	1.72	1.57
Al_2O_3	13.37	12.70	13.16	13.25	13.30	13.11	13.13
Fe_2O_3	0.52	0.53	0.53	0.49	0.49	0.50	0.48
FeO	11.79	11.83	11.91	11.05	11.01	11.14	10.89
MnO	0.16	0.14	0.14	0.12	0.14	0.12	0.13
MgO	7.43	8.63	7.98	8.06	7.92	8.24	7.83
CaO	10.96	10.79	10.83	9.57	9.77	9.71	9.91
Na_2O	2.44	2.17	2.34	2.37	2.28	2.31	2.24
K_2O	0.36	0.33	0.33	0.62	0.59	0.60	0.57
P_2O_5	0.19	0.21	0.23	0.16	0.15	0.18	0.15
Cr_2O_3	0.05	0.13	0.15	0.17	0.10	0.18	0.20
NiO	0.02	0.03	0.02	0.02	0.02	0.02	0.02
T_{clc}.°C	1,187	1,217	1,201	1,213	1,207	1,216	1,203
Fo	77.9	80.32	78.97	79.93	79.63	80.17	79.5
NiO ol	0.36	0.36	0.35	0.34	0.34	0.34	0.34
K_{cor}	1.05	0.99	1.02	0.98	1.00	0.97	1.00
S_{cor}	0.04	0.03	0.03	0.02	0.04	0.03	0.03
Cl_{cor}	0.03	0.03	0.03	0.003	0.003	0.005	0.003

Component	XC51-9	XC51-11	XC51-12	XC51-13	XC51-14	XC51-16	XC51-17	XC51-18
SiO_2	52.23	52.37	52.85	51.75	52.20	53.34	53.20	51.68
TiO_2	1.65	1.73	1.64	1.45	1.82	1.39	1.64	1.69
Al_2O_3	12.81	12.74	12.85	12.95	13.26	12.30	12.72	13.05
Fe_2O_3	0.49	0.49	0.49	0.51	0.50	0.48	0.48	0.50
FeO	11.12	11.08	10.91	11.36	11.13	10.73	10.78	11.32
MnO	0.12	0.13	0.14	0.13	0.15	0.11	0.14	0.11
MgO	8.59	8.27	8.03	9.39	7.29	9.20	8.05	8.53
CaO	9.87	10.04	10.09	8.93	10.42	9.34	9.79	9.87
Na_2O	2.22	2.22	2.11	2.69	2.26	2.17	2.22	2.28
K_2O	0.55	0.57	0.53	0.52	0.59	0.57	0.56	0.59
P_2O_5	0.16	0.15	0.16	0.15	0.19	0.14	0.19	0.15
Cr_2O_3	0.17	0.19	0.18	0.15	0.20	0.20	0.21	0.20
NiO	0.02	0.02	0.02	0.03	0.02	0.03	0.02	0.02
T_{clc}.°C	1,223	1,214	1,206	1,253	1,185	1,241	1,209	1,222
Fo	80.84	80.26	79.86	82.14	78.01	82.23	80.07	80.57
NiO_{ol}	0.35	0.35	0.34	0.37	0.32	0.38	0.35	0.36
K_{cor}	0.96	0.96	1.00	0.94	1.04	0.95	1.01	0.96
S_{cor}	0.03	0.03	0.04	0.03	0.03	0.03	0.05	0.02
Cl_{cor}	0.002	0.003	0.004	0.002	0.008	0.004	0.003	0.003

Note: *T*cls. *C* is the temperature of equilibrium with the host olivine. *Fo* (mol %) and NiO (wt %) are the compositional characteristics of the host olivine. K_{cor} is the correction factor for incompatible elements (see text for explanation). S_{cor} and Cl_{cor} are the corrected contents of sulfur and chlorine in the melt. After Sobolev et al. (2009)

buffer), which is related to the strongly reduced spinel composition. A monotonous increase in CaO and Al_2O_3 concentrations with decreasing MgO (Fig. 3.34) suggests that the melt evolution was controlled by olivine crystallization without coexisting plagioclase and pyroxene. This fact allowed us to calculate the composition of the parental melt by modeling reverse olivine fractionation.

The compositions of inclusions from samples CY-50 and 4270/13 form a single trend with respect to all elements (Fig. 3.34) and are similar to the compositions of basaltic glasses from Mauna Loa, Hawaii. Inclusions from sample KhS-51/130 are distinguished by high SiO_2 and K_2O; low CaO, TiO_2, and P_2O_5; and elevated Rb, Ba, U, Th, La, and Pb concentrations (Figs. 3.34–3.36); i.e., they are enriched in elements typical of silicic rocks of the continental crust (Rudnick 2002). The lithophile element spider diagrams of these inclusions (Fig. 3.36a) display distinct characteristic anomalies of the continental crust, i.e., positive for Rb, U, and Pb and negative for Ta, Nb, and Ti. These observations suggest that the melts from which the olivine of sample KhS-51/130 crystallized were significantly contaminated by silicic continental materials, for example, quartz sandstones. Indications of contamination (elevated concentrations of K, Cl, and B) were also observed in some inclusions from sample CY-50 (Fig. 3.37). However, in that case, the contaminant was different, i.e., enriched in Cl and K and poor in Si (most likely evaporite). The compositions of the melt inclusions in olivine from samples CY-50 and 4270/13 containing Cl<0.045 wt % show a very narrow range of all incompatible elements, i.e., identical in the two samples (Fig. 3.36b). No evidence for crustal contamination was observed. The composition of the inclusions is very similar to the composition of melts with similar Mg concentrations from Mauna Loa, Hawaii.

3.3.2.8 Volatile Components in Melts

The melt inclusions in olivine from sample CY-50 and the majority of inclusions in olivine from sample KhS-51/130 contain anomalously low H_2O concentrations (Fig. 3.37), which is most likely related to the near-surface degassing of crystallizing magmas. The inclusions in olivine from sample 4270/13 contain slightly higher H_2O concentrations (up to 0.25 wt %), which are comparable with estimates obtained from the concentration of an element with similar incompatibility, e.g., Ce (Dixon et al. 2002). Consequently, they can be considered to be the initial characteristics of undegassed magmas. This concentration is also similar to the minimum the concentrations in the glasses of Mauna Loa, Hawaii. The melt inclusions in olivine from sample KhS-51/130 have extremely low Cl concentrations, which, similar to H_2O, probably result from near-surface degassing (Fig. 3.37). The compositions of inclusions in olivine from sample CY-50 display a steep trend of Cl accumulation with increasing K_2O content. The amounts of Cl reach remarkably high levels. The compositions of inclusions from sample 4270/13 lie at the beginning of the Cl accumulation trend and are comparable in this respect to the most Cl-rich glasses of Mauna Loa, Hawaii. A characteristic feature of all the melt inclusions is an elevated boron content, which reaches maximum values in the most Cl-rich inclusions from sample SY-50. All of the inclusions contain low S concentrations, more than 2–3 times lower than the level of basaltic melt saturation in sulfide liquid. Owing to early degassing, the S content in low-pressure melts is not representative of parental magmas. However, the S content in an undegassed melt can be estimated from the amount of a nonvolatile element of similar incompatibility (Nb) using the data of Saal et al. (2002). Given a CO_2/Nb value of 240 reported by these authors, the initial CO_2 content can be estimated as ~0.25 wt %.

3.3.3 Basalts of the Morongovsky–Samoedsky Formations

When studying the geochemical features of the basalts in the Kharaelakh Trough, in borehole SG-32, the transitional rocks were established (Wooden et al. 1993; Fedorenko et al. 1996). They include the upper portion of the Nadezhdinsky Formation and the lower portion of the Morongovsky Formation. These units are characterized by geochemical parameters intermediate between typical representatives of these formations. On a spider diagram, we plotted data relating to the "transition" formations (Fig. 3.38). Unfortunately, not all elements were analyzed in the samples presented in the reference section of these papers. However, a pattern is visible: the series of samples exhibits a fairly continuous series of compositions. Based on this finding, the intermediate phase of trap magmatism was attributed to this origin.

As we studied several sections in the eastern part of the Noril'sk area, we noted a similar pattern in rare cases (e.g., the Noril'sk Trough, described above). However, in the eastern part of the region, this pattern changes significantly: the section can be clearly divided into specific formations, i.e., the Morongovsky and Nadezhdinsky Formations, and the contact between them can be traced clearly. For the reference section, we chose a cut on the right side of the Iken River (in Fig. 3.39), which incises the Nadezhdinsky Formation (including the lower subformation of tholeiitic basalts, porphyritic basalts; the middle subformation and 2 flow glomeroporphyric basalts traditionally assigned to the upper subformation of the Nadezhdinsky Formation) and the 200-m-thick Morongovsky Formation. The compositions of these rocks are shown in Table A4 in Appendix which clearly shows that rocks of the lower and upper portions of

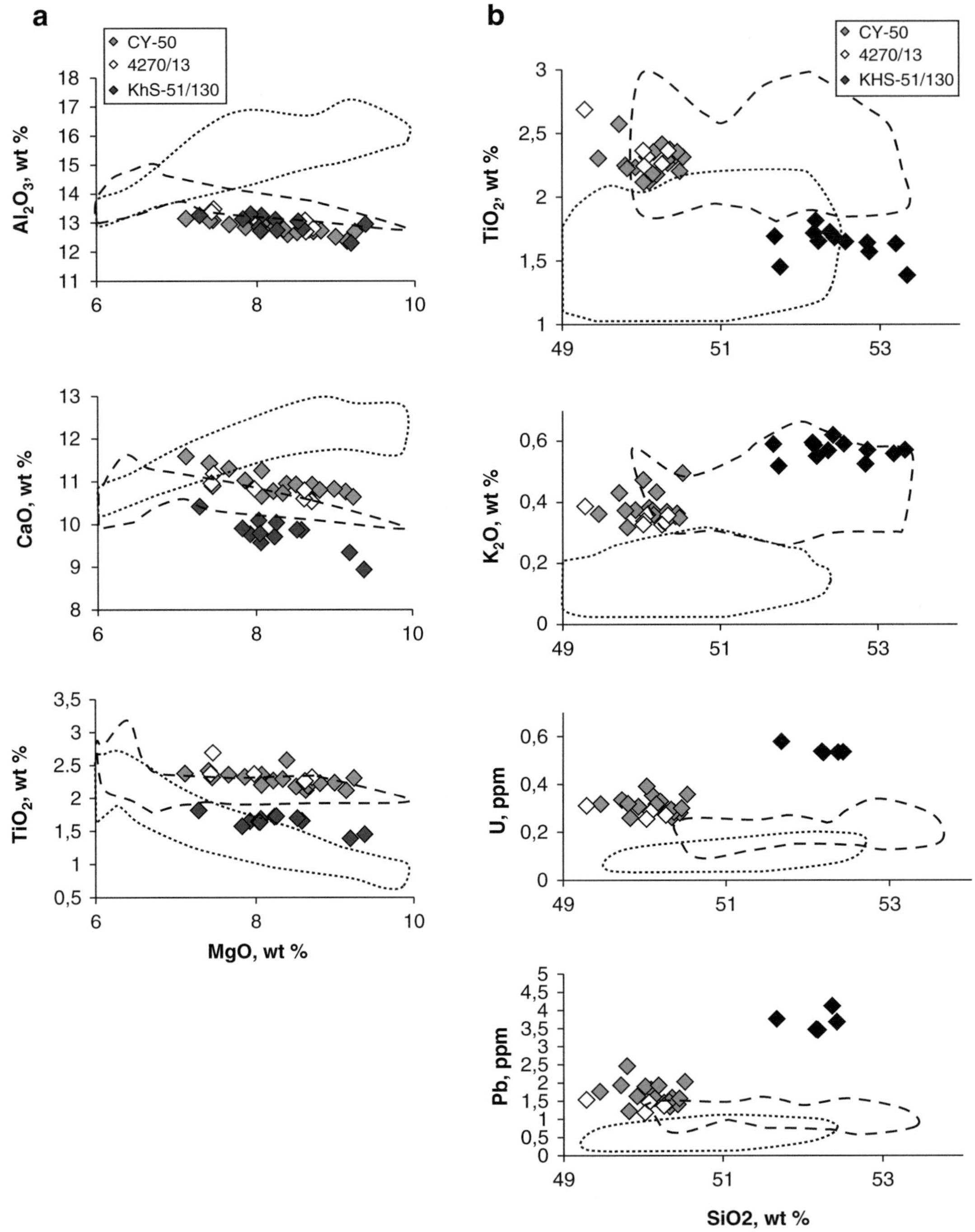

Fig. 3.34 Diagram of the compositions of trapped melts in olivine phenocrysts from the Gudchikhinsky picrites
All compositions were recalculated to equilibrium with the host olivine (see text). The *dotted line* encloses the compositional field of glasses from mid-ocean ridge basalts according to the PetDB database (http://www.petdb.org/petdbWeb/index.jsp); and the *dashed line* encloses the field of glass compositions from Mauna Loa Volcano, Hawaii, according to the GEOROC database (http://georoc.mpchmainz.gwdg.de/georoc/)
After Sobolev et al. (2009)

the Nadezhdinsky Formation differ markedly in their concentrations of copper. The copper content in the first unit varies from 15 to 30 ppm and that of the second unit reaches 90 ppm, in three times higher. However, the two units are identical in terms of their trace element composition and are distinct from the bottom of the Morongovsky Formation (Fig. 3.39). The rocks of intermediate rocks are similar to those in Fig. 3.38, which are missing in the eastern part of the region where the lava of the Morongovsky Formation displays its maximum lateral extent.

Thus, a comparison of the data obtained from these rocks and those of the Nadezhdinsky and Morongovsky Formations indicates that they are significantly different from each other and lack any mixed traits. Our studies of a number of key sections of different tectonic structures (Vologochansky Trough, northern periclinal framework of Khantaysko–Rybninsky swell, the western part of the Tunguska syneclise) indicated that the presence of rocks of

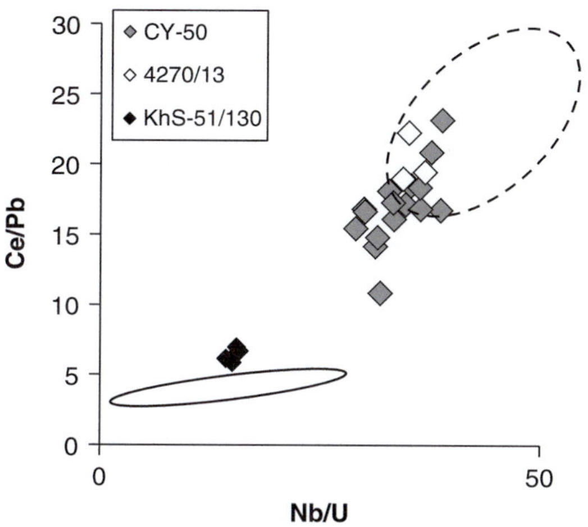

Fig. 3.35 Indicator trace element ratios in melt inclusions in olivine from the Gudchikhinsky picrites
The *dashed line* encloses the field of basalt compositions from ocean islands according to Hofmann (2002) and GEOROC database (http://georoc.mpch-mainz.gwdg.de/georoc/). The *solid line* encloses the compositional field of the crystalline rocks of the continental crust (Rudnick 2002)
After Sobolev et al. (2009)

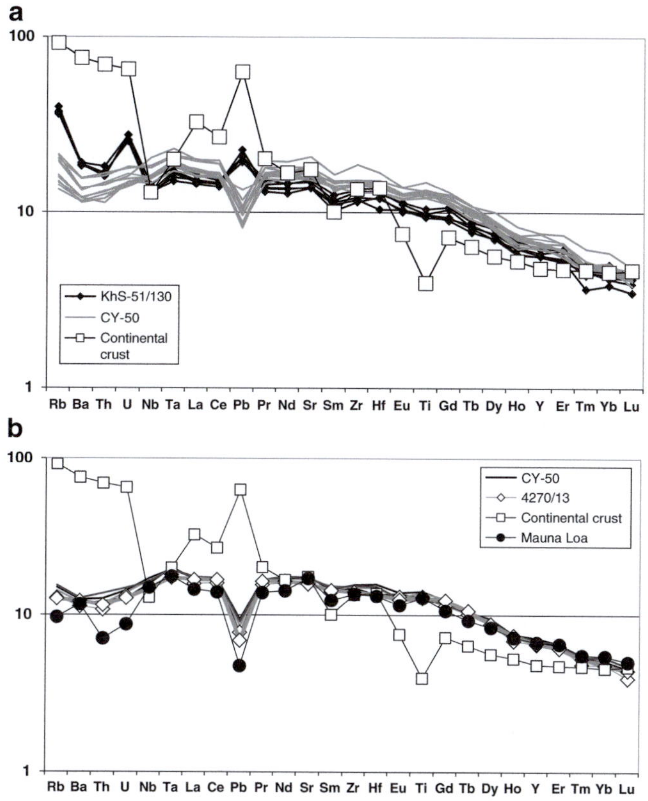

Fig. 3.36 Distribution patterns of incompatible lithophile elements in melt inclusions in olivine from the Gudchikhinsky picrites
Compositions of contaminated melts. (**b**) Compositions of weakly contaminated or uncontaminated melts. Cont. crust is the average composition of continental crust (Rudnick 2002). Also shown is the average composition of melt inclusions in olivine from Mauna Loa Volcano, Hawaii (Sobolev et al. 2005 and unpublished data)
After Sobolev et al. (2009)

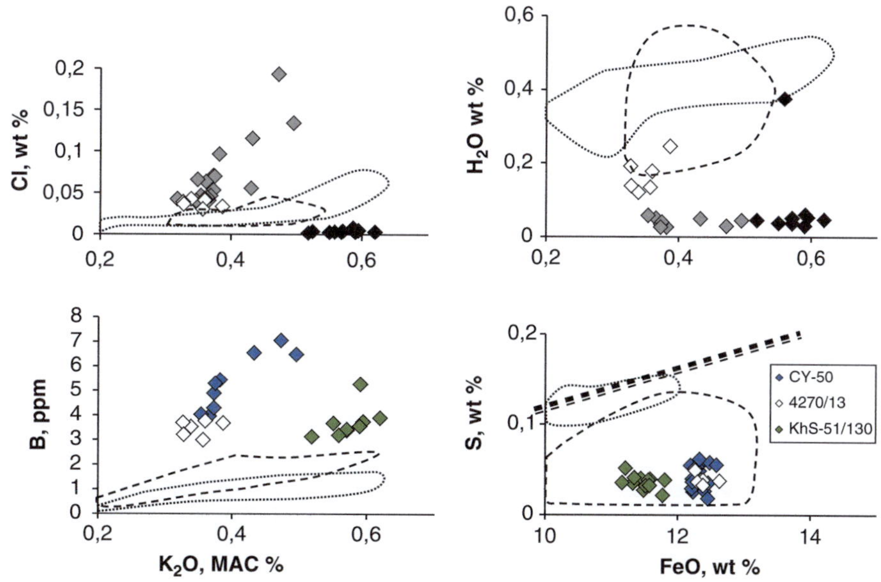

Fig. 3.37 Contents of volatile components in melts trapped in olivine phenocrysts from the Gudchikhinsky picrites
The *dashed line* encloses the field of glasses from Mauna Loa Volcano, Hawaii according to the GEOROC database (http://georoc.mpch-mainz.gwdg.de/georoc/), and the *dotted line* encloses the field of glasses from mid-ocean ridge basalts without evidence for seawater contamination according to the PetDB database (http://www.petdb.org/petdbWeb/index.jsp). The fields of glasses in the K_2O–B coordinates are shown according to the unpublished data of A.V. Sobolev from the investigation of melt inclusions in olivine. The *double dashed line* encloses the region of sulfur saturation according to the data of Mathez (1976)
After Sobolev et al. (2009)

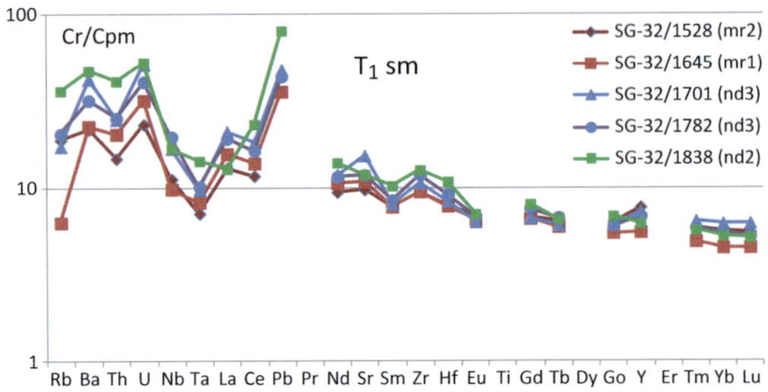

Fig. 3.38 Trace element patterns for volcanic rocks of the Upper Nadezhdinsky and Low Morongovsky Formations in the Kharaelakh Trough After Wooden et al. (1993)

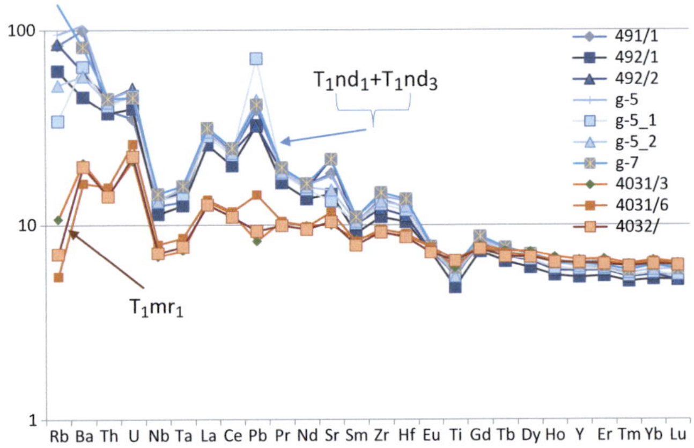

Fig. 3.39 Trace element patterns for volcanic rocks of the Upper Nadezhdinsky and Low Morongovsky Formations in the eastern part of the Noril'sk area

composition intermediate between certain rocks and the volcanics of the Nadezhdinsky and Morongovsky Formations is rare. Such rocks are more prevalent in the western part of the region. More often, the representative rocks from these suites have properties typical of these stratigraphic units. Additionally, these rocks sometimes evolve "anomalous" properties (enrichment of the spectrum) in the central parts of other formations (e.g., the Samoedsky), which cannot be attributed to a "transitional" stage of magmatism (between the OIB and intraplate basalts, Lightfoot et al. 1993; Wooden et al. 1993).

3.4 Evolution of Tuff–Lava Sequence

Based on a comparison of the sections of volcanic rocks from the Kharaelakh, Vologochansky, and Noril'sk Troughs, the eastern side of the Khantaysko–Rybninsky swell (the western part of the Tunguska syncline), and the lower stream of the Mikchangda River, we have developed certain conclusions. The thicknesses of almost all formations decrease by 20–25 % from the central part of troughs to its eastern side toward the western part of the Khantaysko–Rybninsky swell. However, in the eastern part, they thicken again (slightly in the case of the Ivakinsky and Syverminsky Formations, i.e., from 110 to 140 m, and substantially in the case of the Khakanchansky Formation and especially the Tuklonsky Formation, i.e., from 50 to 120 m). A completely different conclusion is developed regarding the rocks of the Gudchikhinsky Formation, whose dimensions are very different from those of the other formations: its thickness rapidly decreases from 450 m in the western Kharaelakh Trough to 15 m in the Lake Glubokoe region and completely wedges out on the Putorana Plateau. In the northern periclinal part of the Khantaysko–Rybninsky swell, the Gudchikhinsky Formation is represented by two flows of picrites (total thickness is less 25 m). The structure of the upper formations also changes; for example, in the upper subformation of the Morongovsky Formation, the pyroclastic rocks, which are developed within the Noril'sk trough, are replaced lavas (in the Mikchangdinsky area).

Thus, the structure of the volcanic rocks within different tectonic elements varies considerably: the thicknesses of the separate formations varies in relation to that of the lavas and tuffs. The observed lateral variability of the volcanic flows is most typical of the Gudchikhinsky, Nadezhdinsky, and Tuklonsky Formations. Each plicative structure is characterized by individual features of the geological structure, indicating a certain specificity of its formation on the background of the general development of the magmatism. In light of the above, the conclusion that all plicative structures in the Noril'sk region were formed after the traps (Geology and ore … 1994) and did not play a significant role in the formation of volcanic strata does not seem justified. A significant difference between the structure on the western and eastern sides of the Khantaysko–Rybninsky swell indicates its existence as a long-lived structure that formed before the presence of the first magmatic melts on the surface in the area. It is likely that the volcanic and intrusive complexes in different structural elements also had their own peculiarities, which could affect the formation of Pt–Cu–Ni mineralization. Analysis of this situation requires further detailed studies of the comparison of the geological structure and development of the various tectonic elements of the district. In general, the areal distribution and thicknesses of the individual formations differ significantly in the Noril'sk area. The thicknesses of the Ivakinsky and Syverminsky Formations are nearly equal. The first of these formations has a smaller lateral extent and is partly missing in the central part of the basin of the Vologochansky Trough (borehole OV-36). At this location, there was uplift of Paleozoic rocks during the deposition of the Ivakinsky Formation.

Isopach maps of the most important formations were developed for the Gudchikhinsky, Tuklonsky, and Nadezhdinsky Formations. These maps show increasing thicknesses in opposite directions: the Tuklonsky Formation thickens from west to east, and the Gudchikhinsky and the Nadezhdinsky Formations thicken from east to west. First but less extensively, the Morongovsky and Mokulaevsky Formations were deposited. The rocks overlying these volcanic rocks, unfortunately, were only developed very locally, i.e., within the central part of the Ikonsky Trough, and thus the patterns of their distribution in the Triassic are difficult to discern (Fig. 3.40).

The area (Fig. 3.41) has been well studied using geophysical methods (Dolgal' and Chekhovich 1998; Dolgal' 2001). The

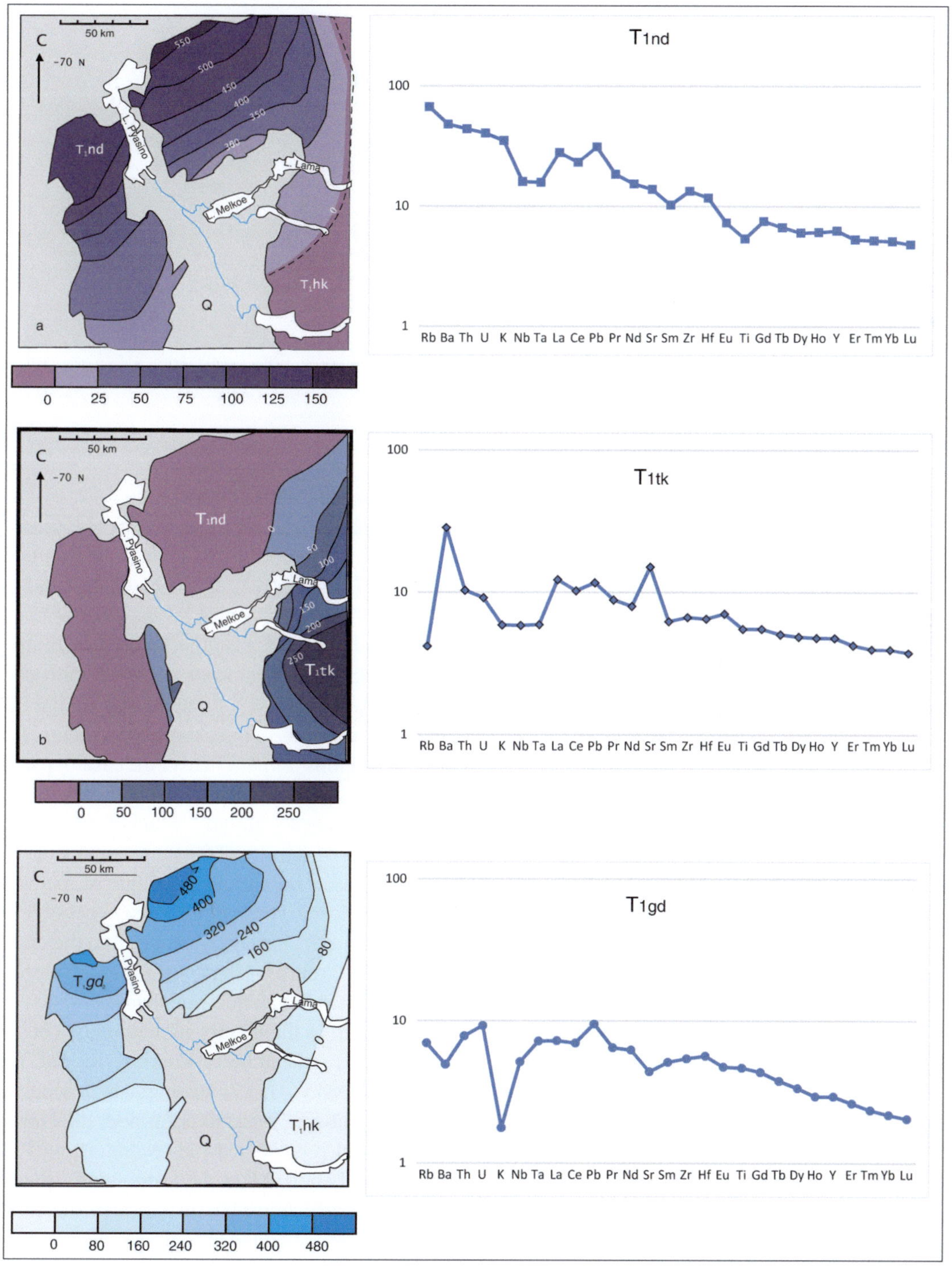

Fig. 3.40 Isopach maps for different formations

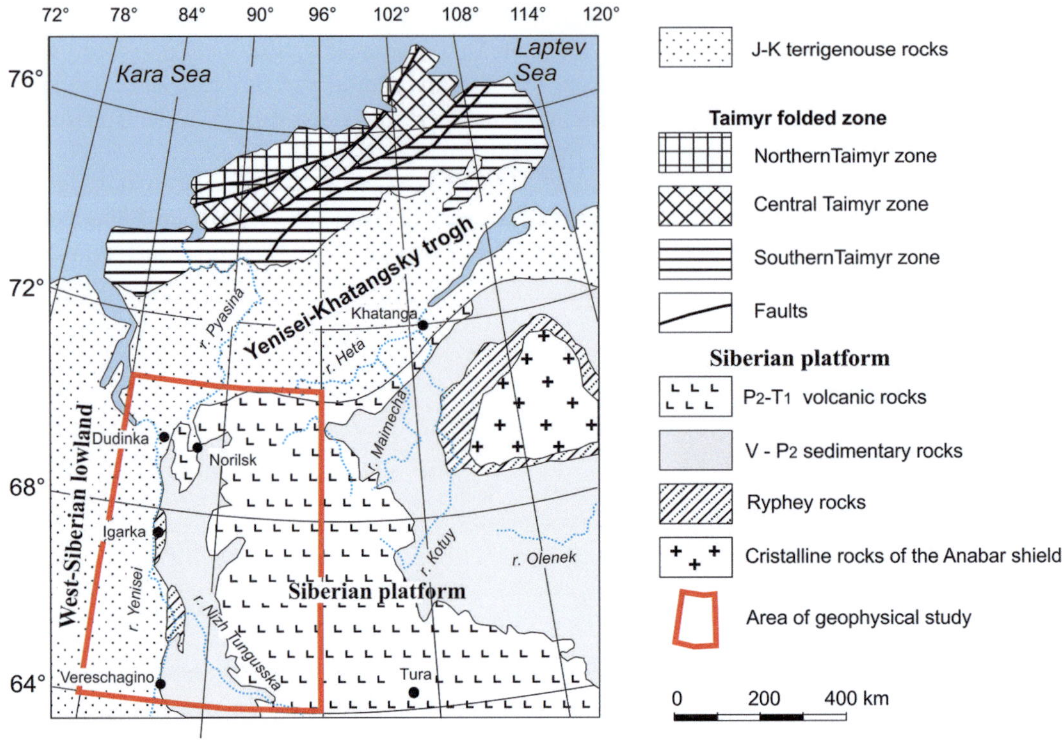

Fig. 3.41 Position of the studied area inside Siberian Platform (Shown in Fig. 3.42)

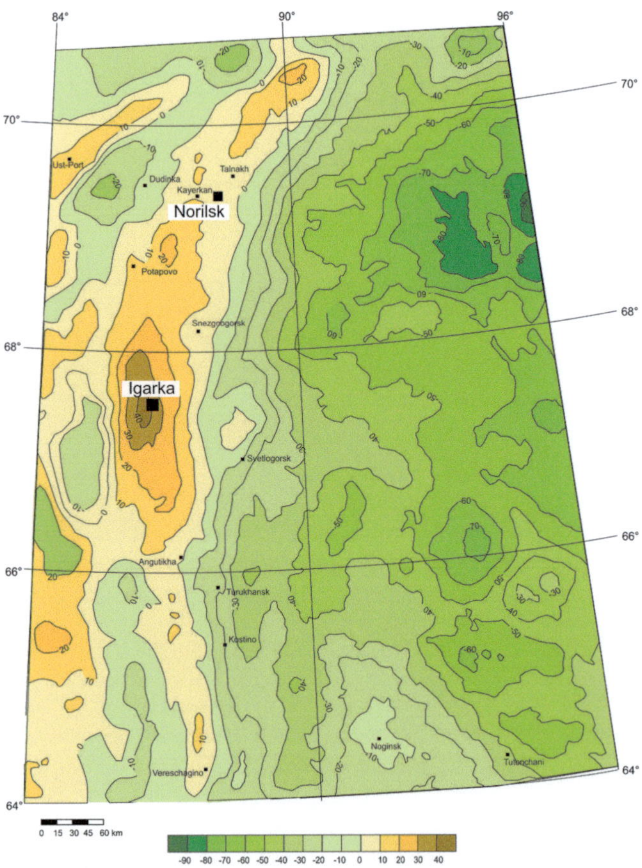

Fig. 3.42 Noril'sk-Igarka paleorift system on the gravity geophysical map. Scale in conventional units
After A. Dolgal' (2012)

Noril'sk–Igarka paleorift structure was suggested by N. Malich et al. (1988) and clearly stands out based on the gravity data (Fig. 3.42; Dolgal' 2012). We assessed the areal distribution of individual formations based on the gravity map (Fig. 3.43), which clearly shows that the lower formations (Ivakinsky, Syverminsky, Gudchikhinsky, and Nadezhdinsky) are distributed only within the rift structure and wedge out toward the Tunguska syncline—a typical platform region. In contrast, the upper formations primarily cover the eastern part of the region. It is logical to assume the existence of early formations in the Yenisei–Khatangsky trough, which formed at the stage of development of the rift (Triassic volcanics have been identified there based on various geophysical methods). Locating the upper volcanic rocks in this area is possible but not likely because their effusion likely occurred in the east area and did not reach the Yenisei–Khatangsky Trough (as shown in Fig. 3.43 as denoted by the question mark).

Periodicity in the history of magmatism has been emphasized by all researchers working in the region. Thus, M.N. Godlevsky (1959) distinguished four stages during which only lava or intrusions formed; the Noril'sk complex in this model represented the second stage. P. C. Lightfoot et al (1993) believed that the formation of the basalts of the Noril'sk region occurred in two primary stages, during which the basalts corresponded to those of oceanic islands and then intraplate volcanism occurred between which volcanics of the intermediate stage formed. A. Al'mukhamedov et al. (1999) identified rift and platform stages on the basis of

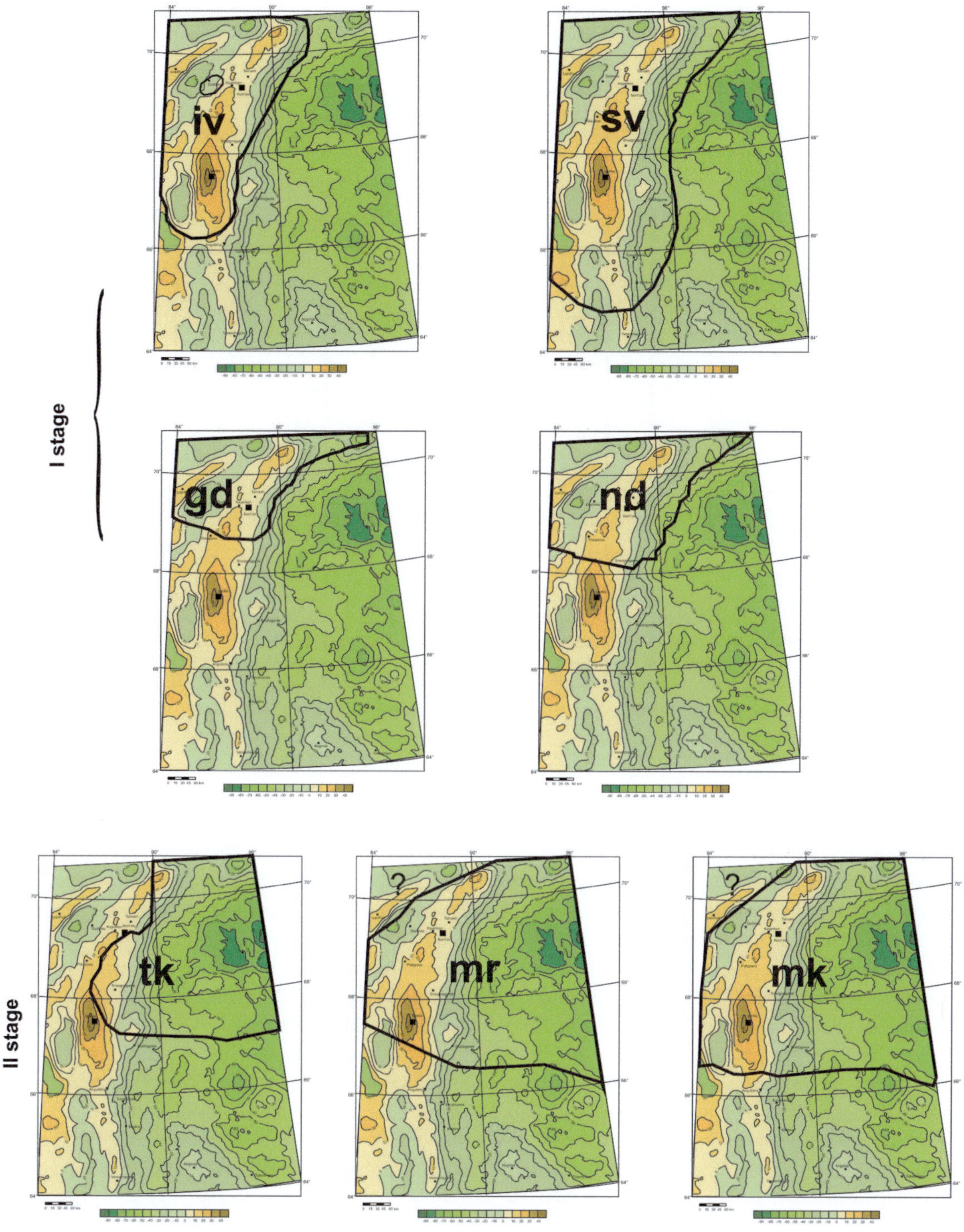

Fig. 3.43 Areas of distribution for main formations in NW part of the Siberian Platform After Krivolutskaya (2014)

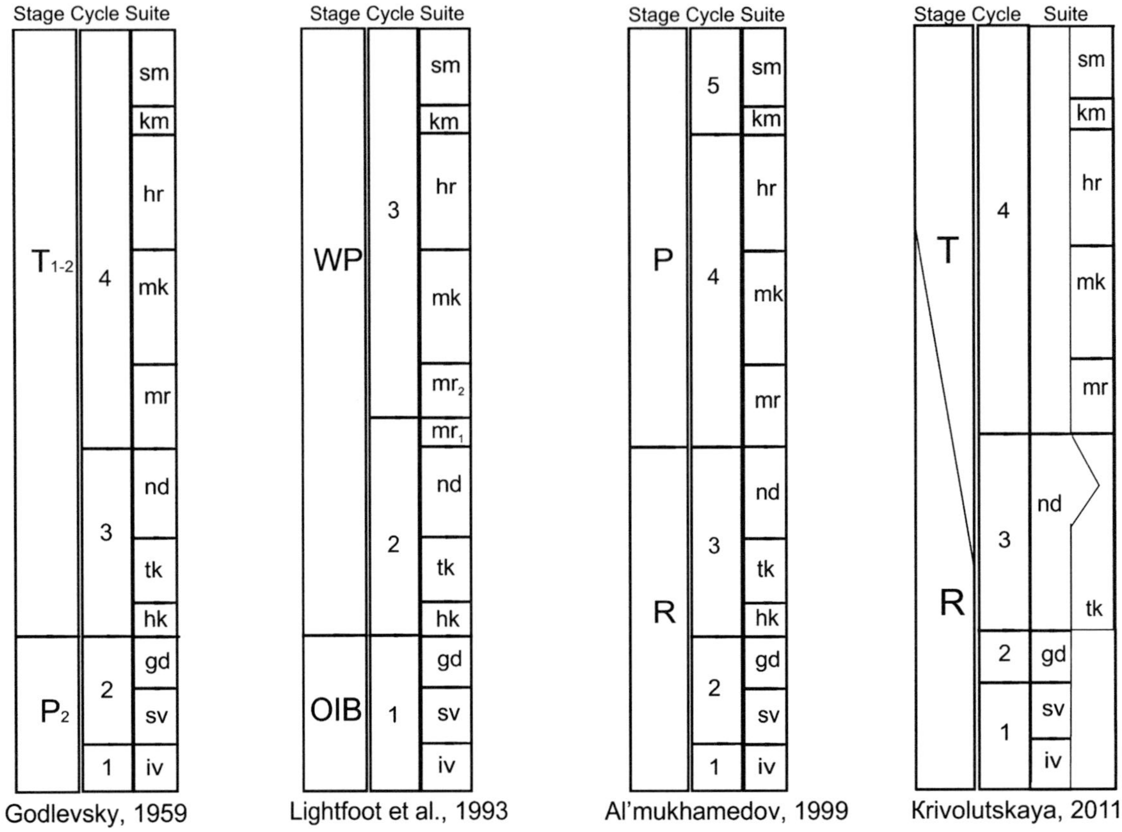

Fig. 3.44 Schemas of trap magmatism evolution in the Noril'sk region (According to different authors)

comparison of the structure and composition of volcanic formations in Western and Eastern Siberia. The model presented in this book is based on the data of the last researchers; however, its fundamental difference lies in the fact that the primary stages did not clearly replace one another in time but coexisted during a certain period of time in neighboring areas. In terms of the allocation of specific cycles, the differences are even greater. For example, these authors combined in one cycle the Syverminsky and Gudchikhinsky Formations, which have fundamentally different structures and areal distributions, although the cycle is separated from them by the Ivakinsky Formation, very close to Syverminsky Formation in terms of geochemistry and areal extent (Fig. 3.44). The Khakanchansky, Tuklonsky, and Nadezhdinsky Formations represent the next cycle. We have demonstrated the proximity of the Khakanchansky and Nadezhdinsky Formations (causing us to conclude that the former belongs to the early phase of the Nadezhdinsky magmatism) and the simultaneous eruption of the lava of the Tuklonsky Formation and the Khakanchansky (Nadezhdinsky) tuffs.

Based on the volcanism data, the history of the evolution of this area may be as follows. The first stage of magmatism was associated with continental rifting, during which time the melts of deep origin (with a high Gd/Yb ratio, which indicates the presence of garnet in the source) erupted. The centers of magmatic activity (Ivakinsky and Syverminsky Formations) were located adjacent to Yenisei–Khatangsky deflection areas, as is most clearly observed in the Gudchikhinsky Formation. Within the rifts, there later occurred a stage in the formation of basalts and tuffs of the Nadezhdinsky Formation that are significantly different from the lower ones: these have a low Gd/Yb ratio and enrichment in the spectrum HLRE elements. A similar development took place in analogous structures of the Western Siberian platform (Fig. 3.45, Masaitis 1983; Reichow et al. 2005), which consist of subalkaline rocks and basalts that are similar in composition to the Nadezhdinsky Formation and even acidic volcanic rocks (Al'mukhamedov et al. 1999, 2000, 2004).

At the end of the western rift magmatism, a new center of magmatism developed, the source of which was significantly different from that in the eastern part of the region at the time. This new center produced the tholeiitic and poikilophitic (sometimes picritic) basalts of the Tuklonsky Formation, and all parameters indicate that the products of eruption were very close to the primary stage of trap magmatism (Morongovsky–Samoedsky Formations). The center of the eruption of the second stage was most likely

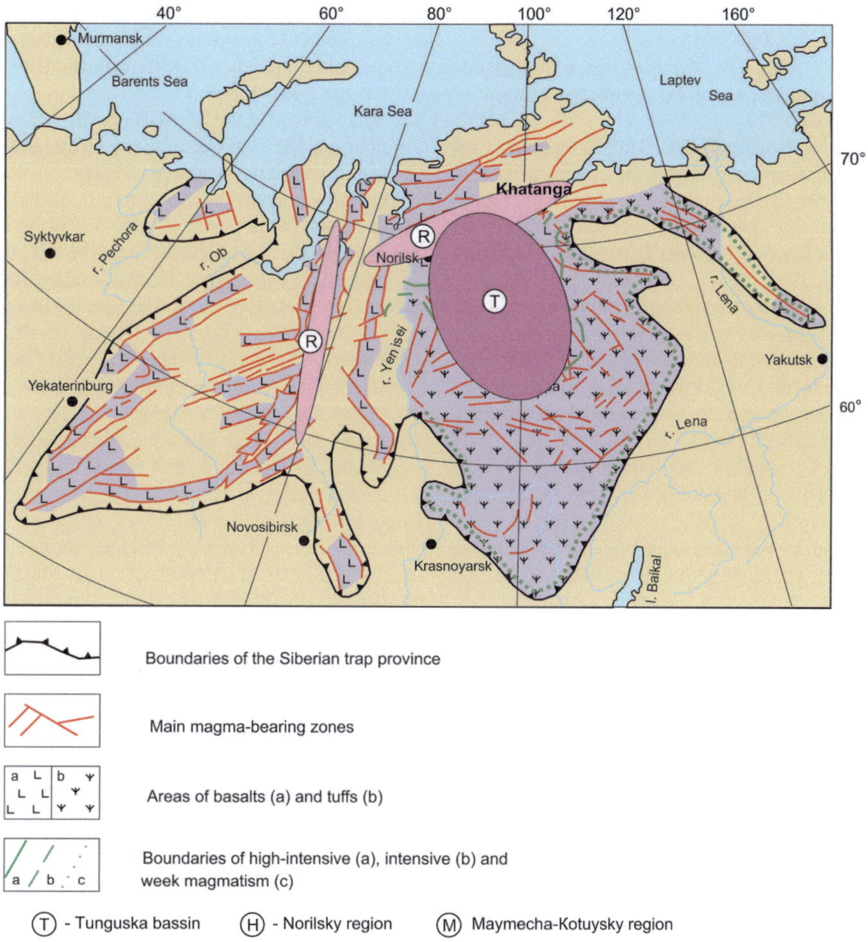

Fig. 3.45 Aureoles of distribution volcanic rocks of rift (*R*) and trap (*T*) stage within the Siberian trap province
Province boundaries are given after V. Masitis (1983)

located in the Putorana Plateau or in the central part of the Tunguska syneclise, judging by the increase in thickness of rocks in this direction. However, a great deal of the upper part of the effusions of this stage have been eroded away. Ore deposits are located inside paleorift zone as well (Dodin et al. 1971; Distler et al. 1984). Magmatic activity in deep faults continued after the primary stage, as indicated by alkaline rocks of Maimecha-Kotuy Province (Fedorenko and Czamanske 1997).

3.5 Conclusions

Based on the distribution of volcanic rocks in different tectonic structures of the area and their petrogeochemical compositions (major, rare, and radiogenic elements), the rocks of the tuff–lava sequences of the Noril'sk region were formed during two stages: rifting (iv–sv–gd–nd) and the actual trapping (tk–mr–sm), which were not only sequential but also simultaneous for a long period of time. The volcanic development occurred in four cycles (iv–sv, gd, nd–tk, and mr–sm).

References

Al'mukhamedov AI, Medvedev AY, Kirda NP (1999) Comparing analyses of geodynamics of Permian -Triassic magmatism. Russ Geol Geophys 40:1575–1587

Al'mukhamedov AI, Medvedev AY, Kirda NP (2000) Rhiolites as a part of Triassic volcano-sediment complex in West Siberia. Doklady Earth Sci 371(2):200–203

Al'mukhamedov AI, Medvedev AY, Zolotukhin VV (2004) Evolutions of Permo-Triassic basalts of the Siberian Platform in time and space. Petrology 12(4):339–353

Arndt NT, Czamanske GK, Wooden JL, Fedorenko VA (1993) Mantle and crustal contributions to continental flood volcanism. Tectonophysics 223:39–52

Arndt NT, Chauvel C, Czamanske G, Fedorenko VA (1998) Two mantle sources, two plumbing systems: tholeiitic and alkaline magmatism of the Maymecha River basin, Siberian flood volcanic province. Contrib Mineral Petrol 133:279–313

Balashov YuA, Nesterenko GV (1966) Distribution of rare elements in Siberian traps. Geochem Int, no 7, pp 854–860 (in Russian)

Beattie p (1993) Olivine–melt and orthopyroxene–melt equilibria. Contrib Mineral Petrol 115:103–111

Brügmann GE, Naldrett AJ, Asif M, Lightfoot PC, Gorbachev NS, Fedorenko VA (1993) Siderophile and chalcophile metals as tracers of the evolution of the Siberian trap in the Noril'sk region, Russia. Geochim Cosmochim Acta 57:2001–2018

Campbell IH, Griffiths RW (1992) The changing nature of mantle hotspots through time - implications for the chemical evolution of the mantle. J Geol 100(5):497–523

Campbell IH, Czamanske GK, Fedorenko VA, Hill JR, Stepanov VS (1992) Synchronism of the Siberian traps and the Permian-Triassic boundary. Science 354:1760–1763

Canil D (2002) Vanadium in peridotites, mantle redox and tectonic environments: archean to present. Earth Planet Sci Lett 195:75–90

Dalrymple GB, Czamanske GK, Fedorenko V et al (1995) A reconnaissance $^{40}Ar/^{39}Ar$ geochronologic study of ore-bearing and related rocks, Siberian Russia. Geochim Cosmochim Acta 59(10):2071–2083

Distler VV, Genkin AD, Duzhikov OA (1984) Sulfide petrology and genesis of copper-nickel ore deposits. In: Geology and metallogeny of copper deposits. Springer, Berlin, pp 111–123

Dixon JE, Leist L, Langmuir C, Schilling J-G (2002) Recycled dehydrated lithosphere observed in plume-influenced mid-ocean-ridge basalt. Nature 420:385–389

Dobretsov NL (1997) Permian-Triassic magmatism and sedimentation in Eurasia as reflection of superplume. Doklady Earth Sci 354:220–223

Dobretsov NL, Kirdyashkin AA, Kirdyashkin AG, Vernikovsky VA, Gladkov IN (2008) Modelling of thermochemical plumes and implications for the origin of the Siberian traps. Lithos 100:66–92

Dodin DA (1967) Traps petrology of East Kharaelakh. Dissertation on candidate of geology-mineralogical sciences. Leningrad, VSEGEI, 256 p (in Russian)

Dodin DA, Batuev BN, Mitenkov GA (1971) Atlas of the rocks and ores of the Noril'sk copper–nickel deposits. Nedra, Leningrad (in Russian)

Dolgal' AS (2001) Computer technologies for interpretation of geopotential fields during research of copper-nickel mineralization. Geophys J 23(2):106–112

Dolgal' AS (2012) Realization of V.N. Strakhov ideas in interpretation of geopotential fields. In: Academician VN (ed) Strakhov as geophysics and mathematic. Nauka, Moscow, pp 55–78 (in Russian)

Dolgal' AS, Chekhovich KM (1998) Complex interpretation of geopotential fields during prospecting of copper-nickel-platinum mineralization (Noril'sk region). Russ Geol Geophys 39(11):1615–1625

Dyuzhikov OA, Distler VV, Strunin BM et al (1988) Geology and ore potential of the Noril'sk ore district. Nauka, Moscow. Dyuzhikov OA, Distler VV, Strunin BM, Mkrtychyan AK, Sherman ML, Sluzhenikin SF, Lurye AM (1992) Geology and metallogeny of sulfide deposits Noril'sk region USSR (trans). Econ Geol Monogr. Ontario, Spec vol 1, 241 p

Elkins-Tanton LT (2005) Continental magmatism caused by lithospheric delamination. In: Foulger GR, Natland JH, Presnall DC, Anderson DL (eds) Plates, plumes and paradigms. Special paper. Geological Society of America, Boulder, pp 449–461

Fedorenko VA (1981) Petrochemical series of volcanic rocks in the Noril'sk district. Russ Geol Geophys 22:78–88 (in Russian)

Fedorenko VA, Czamanske G (1997) Results of new field and geochemical studies of the volcanic rocks of the Maymecha-Kotuy area, Siberian flood-basalt province, Russia. Int Geol Rev 39:479–531

Fedorenko VA, Kuligin VM, Vitozhents GCh et al (1989) Rare earth elements in magmatic rocks of the Noril'sk region. Russ Geol Geophys, no 8, pp 67–75

Fedorenko VA, Lightfoot PC, Naldrett AJ et al (1996) Petrogenesis of the Siberian flood-basalt sequence at Noril'sk, north central Siberia. Int Geol Rev 38:99–135

Fekiacova Z, Abouchami W, Galer SJG, Garcia MO, Hofmann AW (2007) Origin and temporal evolution of Ko'olau volcano, Hawaii: inferences from isotope data on the Ko'olau Scientific Drilling Project (KSDP), the Honolulu volcanics and ODP Site 843. Earth Planet Sci Lett 261:65–83

Ford CE, Russel DG, Craven JA, Fisk MR (1983) Olivine–liquid equilibria: temperature, pressure and composition dependence of crystal/liquid cation partition coefficients for Mg, Fe^{2+}, Ca and Mn. J Petrol 24:256–265

Geological map of the Norilsk ore district, scale 1:200000 (1994) In: Sherman ML (ed) Committee of Russian Federation on geology and use of mineral resources. Federacii Rossijskoj Committee on Geology and Nedra ispol'zovaniû, St. Petersburg (in Russian)

Geology and ore deposits of the Noril'sk region (1994) In: Distler VV, Kunilov VE (eds) Guidebook of the VII international platinum symposium Moscow-Noril'sk. Moskovsky Contact Press, Moscow

Godlevsky MN (1959) Traps and ore-bearing intrusions of the Norilsk district. Gosgeoltekhizdat, Moscow (in Russian)

Hofmann AW (1988) Chemical differentiation of the earth: the relationship between mantle, continental crust, and oceanic crust. Earth Planet Sci Lett 90:297–314

Hofmann AW (2002) Sampling mantle heterogeneity through oceanic basalts: isotopes and trace elements. In: Holland HD, Turekian KK (eds) Treatise on geochemistry: the mantle and core, vol 2. Elsevier, Amsterdam, pp 61–101

Howkesworth CJ, Lightfoot PC, Fedorenko VA et al (1995) Magma differentiation and mineralization in the Siberian continental flood basalts. Lithos 34:61–81

Ivanov AV, He H, Yan L, Ryabov VV, Shevko AY, Palesskii SV, Nikolaeva IV (2013) Siberian traps large igneous province: evidence for two flood basalt pulses around the Permo-Triassic boundary and in the Middle Triassic, and contemporaneous granitic magmatism. Earth Sci Rev 122:58–76

Kamo SL, Czamanske GK, Amelin Y et al (2003) Rapid eruption of Siberian flood-volcanic rocks and evidence for coincidence with the Permian-Triassic boundary and mass extinction at 251 Ma. Earth Planet Sci Lett 214:75–91

Keays RP, Lightfoot PC (2007) Siderophile and chalcophile metal variations in Tertiary picrites and basalts from West Greenland with implications for the sulphide history of continental flood basalt magmas. Miner Deposita 42:319–336

Keays RR, Lightfoot PC (2010) Crustal sulfur is required to form magmatic Ni-Cu sulfide deposits: evidence from chalcophile element signatures of Siberian and Deccan Trap basalts. Miner Deposita 45:241–257

Krivolutskaya NA (2011) The problem of subdivision of volcanic rocks of the trappean formation of the Norilsk region. Doklady Earth Sci 439(Part 2):1088–1092

Krivolutskaya NA (2014) Evolution of trap magmatism and processes producing Pt-Cu-Ni mineralization in the Noril'sk area. KMK, Moscow, 320 p (in Russian)

Krivolutskaya NA, Rudakova AV (2009) Structure and geochemical characteristics of trap rocks from the Noril'sk trough, Northwestern Siberian craton. Geochem Int 47(7):635–656

Krivolutskaya NA, Sobolev AV, Nistratov SV, Rudakova AV, Myzdrikova GYu (2009) Specific features of magmatism developement in different tectonic structures inside the Noril'sk area. In Magmatism and ore-forming processes. Moscow, IGEM 70–76 (in Russian)

Krivolutskaya NA, Sobolev AV, Mikhailov VN, Roshchina IA (2005) New data on the formational affiliation of picritic basalts of the Noril'sk district. Doklady Earth Sci 402(4):542–546

Krivolutskaya NA, Sobolev AV, Mikhailov VN, Plechova AA, Kostitsyn YA, Roschina IA, Fekiacova Z (2012) Parental melt of the Nadezhdinsky formation: geochemistry, petrology and connection with Cu-Ni deposits (Noril'sk area, Russia). Chem Geol 302-303:87–105

Li C, Ripley EM, Naldrett AJ (2009) A new genetic model for the giant Ni-Cu-PGE sulfide deposits associated with the Siberian flood basalts. Econ Geol 104:291–301

Lightfoot PC, Keays RR (1995) Siderophile and chalcophile metal variations in flood basalts from the Siberian trap, Noril'sk region: implications for the origin of the Ni-Cu-PGE sulfide ores. Econ Geol 100:439–462

Lightfoot PC, Naldrett AJ, Gorbachev NS (1990) Geochemistry of the Siberian trap of the Noril'sk area, USSR, with amplication for the relative contributions of crust and mantle to flood basalt magmatism. Contrib Mineral Petrol 104:631–644

Lightfoot PC, Howkesworth CJ, Hergt J et al (1993) Remobilisation of the continental lithosphere by a mantle plume: major-, trace-element, and Sr-, Nd-, and Pb-isotopic evidence from picritic and tholeiitic lavas of the Noril'sk District, Siberian Trap, Russia. Contrib Mineral Petrol 114:171–188

Lightfoot PC, Naldrett AJ, Gorbachev NS, Fedorenko VA, Howkesworth CJ, Hergt J, Doherty W (1994) Chemostratigraphy of Siberian trap lavas, Noril'sk district: implication for the source of flood basalt magmas and their associated Ni-Cu mineralization. Proceeding of the Sudbary—Noril'sk symposium, Ontario, spec vol 5, chap 22, pp 283-312

Lind EN, Shchsekoturov VV (1991) Paleomagnitic dating of trap volcanism in NW Siberian Platform. The fourth Paleomagnetism Congress, vol 2. SNIO USSR, Vladimir-Suzdal, pp 79–80

Malitch NS, Grinson AS, Tuganova EV, Chernyshev NM (1988) Rifting of the Siberian platform. In: 28th session of International Geological Congress. Tectonic processes. Nauka Press, Moscow, pp 184–193

Masaitis VL (1983) Permian and Triassic volcanism of Siberia: problems of dynamic reconstructions. Zap Vses Mineral O-va 112:412–425 (in Russian)

Mathez EA (1976) Sulfur solubility and magmatic sulfides in submarine basalt glass. J Geophys Res 81:4269–4276

Maurel C, Maurel P (1982) Etude experimentale de l'equilibre Fe^{2+}–Fe^{3+} dans les spinelles chromiferes et les liquides silicates basiques coexistants a 1 atm. C R Acad Sci, Paris. 285:209–215

Medvedev AYa (2004) Permian-Triassic volcanism of North Asian craton (West Siberian and Tunguska sineclise). Dissertation on doctor geology-mineralogical sciences, Irkutsk, 248 p (in Russian)

Naldrett AJ (1992) A model for the Ni-Cu-PGE ores of the Noril'sk region and its application to other areas of flood basalts. Econ Geol 87:1945–1962

Naldrett AJ (2004) Magmatic sulphide deposits: geology, geochemistry and exploration. Springer, Berlin/Heidelberg/New York

Nesterenko GV, Alfmukhamedov AI (1973) Geochemistry of differentiated traps. Nauka, Moscow (in Russian)

Nesterenko GV, Avilova NS, Smirnova NP (1964) Rare elements in traps of the Siberian Platform. Geokhimiya, no 10, pp 1015–1021

Nesterenko GV, Tikhonenkov PI, Romashova TV (1991) Basalts of plateau Putorana. Geokhimiya, no 10, pp 1419–1425

Polovinkina YuIr, (1966) Structures and textures of igneous and metamorphic rocks. Nedra, Moscow, Parts 1, 2 (in Russian)

Reichow MK, Saunders AD, White RV, Al'mukhamedov AI, Medvedev AY (2005) Geochemistry and petrogenesis of basalts from the West Siberian Bassin: an extension of Permo-Triassic Siberian traps, Russia. Lithos 79:425–452

Reichow MK, Pringle MS, Al'Mukhamedov AI et al (2009) The timing and extent of the eruption of the Siberian traps large igneous province: implications for the end-Permian environmental crisis. Earth Planet Sci Lett 277:9–20

Renne PR, Basu AR (1991) Rapid eruption of the Siberian Traps flood basalts at the Permo-Triassic boundary. Science 253:176–179

Roedder E (1984) Fluid inclusions. Miner Soc Am 14:644 p

Rudakova AV, Krivolutskaya NA (2009) Structural and textural specific features of rocks of trap rock association in the Noril'sk trough (NW of Siberian Platform). Univ Geol Bull 64:364–375

Rudnick AW (2002) Composition of the continental crust. In: Holland HD, Turekian KK (eds) Treatise on geochemistry, vol 3, The crust. Elsevier, Amsterdam, pp 1–64

Ryabov VV, Shevko AYa, Gora MP (2000) Igneous rocks of the Noril'sk district. V.1. Petrology of traps. Nonparel, Novosibirsk, vol 1, 2 (in Russian)

Ryabov VV, Shevko AYa, Gora MP (2014) Trap magmatism and ore formation in the Siberian Noril'sk region, V 1, 2. Springer

Saal AE, Hauri EH, Langmuir CH, Perfit MR (2002) Vapour undersaturation in primitive mid-ocean ridge basalt and the volatile content of earth's upper mantle. Nature 419:451–455

Sharma M (1997) Siberian traps. In: Mahoney JJ, Coffin MF (eds) Large igneous provinces: continental, oceanic, and planetary flood volcanism, vol 100, Geophysical monograph series. AGU, Washington, DC, pp 273–295

Sluzhenikin SF, Krivolutskaya NA, Rad'ko VA, Malich KN, Distler VV, Fedorenko VA (2014) In: Simonov ON (ed) Ultramafic-mafic intrusions, volcanic rocks and PGE-Cu-Ni sulfide deposits of the Noril'sk. Field trip guidebook. 12th international platinum symposium. IGG UB RAS, Yekaterinburg, 83 p

Sobolev AV, Danyushevsky LV (1994) Petrology and geochemistry of boninites from the north termination of the Tonga Trench: constraints on the generation conditions of primary high-Ca boninite magmas. J Petrol 35:1183–1213

Sobolev AV, Slutsky AB (1984) Composition and crystallization conditions of the parental melt of the Siberian meymechites in connection with the general problem of ultrabasic magmas. Geol Geofiz 12:97–110 (in Russian)

Sobolev AV, Hofmann AW, Sobolev SV, Nikogosian IK (2005) An olivine-free mantle source of Hawaiian shield basalts. Nature 434:590–597

Sobolev AV, Hofmann AW, Brügmann G et al (2008) A quantitative link between recycling and osmium isotopes. Science 321:536

Sobolev AV, Krivolutskaya NA, Kuzmin DV (2009) Petrology of the parental melts and mantle sources of Siberian trap magmatism. Petrology 17:253–286

Sobolev AV, Hofmann AW, Kuzmin DV, Yaxley GM, Arndt NT, Chung S-L, Danyushevsky LV, Elliott T, Frey FA, Garcia MO, Gurenko AA, Kamenetsky VS, Kerr AC, Krivolutskaya NA, Matvienkov VV, Nikogosian IK, Rocholl A, Suschevskaya NM, Teklay M (2007) Estimating the amount of recycled crust in sources of mantle derived melts. Science 316:412–417

Sobolev SV, Sobolev AV, Kuzmin DV, Krivolutskaya NA, Petrunin AG, Arndt NT, Radko VA, Vasilev YR (2011) Linking mantle plumes, large igneous provinces and environmental catastrophes. Nature 477:312–316

Staroseltsev VS (1989) Tectonic of lava plateau. Nedra, Moscow, 258 p (in Russian)

The reference legend for 1: 50 000 scale map, Noril'sk Group (1993) Ed. V. Lyul'ko, St-Petersburg (in Russian)

Vasil'ev YR, Zolotukhin VV (1975) Petrology of the ultrabasic rocks in the North Siberian Platform and problems of their genesis. Nauka, Novosibirsk, p 271 (in Russian)

Volkov ID (1963) Distribution of trace elements in magmatic rocks of the Noril'sk area. In: Problems of magmatism, metamorphism and ore-forming processes. рудоносности. Gosgeoltechizdat, Moscow, pp 29–38 (in Russian)

White RS, McKenzie D (1995) Mantle plumes and flood basalts. J Geophys Res Solid Earth 100(B9):17543–17585

White RV, Saunders AD (2005) Volcanism, impact and mass extinctions: incredible or credible coincidences. Lithos 79:299–316

Wooden JL, Czamanske GK, Bouse RM, King RJ, Siems DT (1993) Isotopic and trace-element constraints on mantle and crustal contributions to Siberian continental flood basalts, Norilsk area, Siberia. Geochim Cosmochim Acta 57:3677–3704

Yarmolyuk VV, Kovalenko VI, Kuzmin MI (2000) North Asia superplume in Phanerozoic: magmatism and deep geodynamics. Geotectonics 5:3–29

Zolotukhin VV, Vasil'ev YR, Dyuzhikov OA (1978) Diversity of traps and initial magmas: a case of the Siberian Platform. Nauka, Novosibirsk, p 289 (in Russian)

Zolotukhin VV, Vilensky AM, Vasil'ev YR et al (1984) Magnesium basic rocks of western part of the Siberian Platform and problems of nickel potential. Nauka, Novosibirsk, 208 p (in Russian)

Zolotukhin VV, Vilensky AM, Dyuzhikov OA (1986) Basalts of the Siberian Platform. Nauka, Novosibirsk (in Russian)

Intrusive Rocks

4

Intrusive rocks of the Noril'sk region were subdivided into several complexes according to their internal structure. Intrusions of basic–ultrabasic composition form three types, taking into account the distribution of rare elements in rocks—Dyumtaleysky, Low Talnakh, and Noril'sk close to Gudchikhinsky, Nadezhdinsky, and Morongovsky lavas. Main ore-bearing intrusions of the Noril'sk Complex are described based on 1–2 sections. They are as follows: Talnakh, Kharaelakh, Noril'sk 1, Maslovsky, Noril'sk 2, Chernogorsky, Zub-Marksheydersky, Bol'shaya Bar'ernaya, and Mikchangdinsky. They have similar distribution of rare elements in rocks which are very close to the spectra of the crust.

4.1 Problems in Distinguishing of the Intrusive Complexes in the Noril'sk Area and Distinctive Structural and Compositional Features of These Complexes

The intrusive rocks in the Noril'sk area mostly comprise gently dipping bodies (sills and chonoliths) that are subconformable with the host terrigenous sedimentary and volcanic rocks and occasionally form irregularly shaped bodies and dikes that cut across the horizontal host rocks of the massif. It is often difficult to correlate the intrusive rocks with the lavas because their contacts are rarely exposed. The intrusions are typically much lower stratigraphically than the tuff–lava sequence because of the lithologies of the rocks; the Devonian rocks are most favorable for hosting the intrusions.

The intrusive rocks in the Noril'sk area have been classified according to several schemes (Geology and ore deposits … 1994; Petrov et al. 2001; Ryabov et al. 2014). Surveys in the Noril'sk area resulted in official legend to State Geological Map 1:200,000 scale (Petrov et al. 2001) and reference materials in which all of the intrusive rocks in the area are classified into five major complexes (Table 4.1). We relied on this classification in our studies. The intrusions in the area include reliably identified alkaline massifs that are grouped into the Ergalakhsky and Pyasinsky Complexes, which can be correlated with the rocks of the Ivakinsky Formation. The rocks of the Daldykansky Complex can be distinguished from the other normal alkalinity ultrabasite–basite rocks; they intrude the basalts of the upper formations, are compositionally similar to them, and have elevated Ti concentrations. These rocks appear to be the youngest from the Mokulaevsky–Samoedsky episode of magmatism.

It is difficult to classify the rest of the numerous basite rocks into complexes or types because these rocks are fairly similar in composition, and their relationships with the lavas have not been observed and remain uncertain. The narrow time span of their emplacement (Kamo et al. 2003; Pavlov et al. 2007, 2011) prevents the application of isotopic geochemical techniques to classify and differentiate these rocks. Thus, numerous massifs in the Noril'sk area have been subdivided into complexes and subcomplexes based on analyses of their structural–textural and compositional features and their vertical distribution. These classifications rely largely on the degree of differentiation of the intrusive bodies. For example, the Noril'sk Complex comprises numerous (dozens to a few hundred) variably differentiated and mineralized massifs (forming Northern Siberian nickel-bearing area – Urvantsev 1974) with elevated MgO contents (8–16 wt %), which can be further subdivided into subtypes based on their internal structures (lower and upper varieties of the rocks), including the Noril'sk (picritic gabbro–dolerites–gabbro-diorites), Low Talnakh (picritic gabbro–dolerites–olivine gabbro–dolerites), Zubovsky (troctolites–gabbro–dolerites), and Kruglogorsky (olivine gabbro–dolerites–leucogabbro) subtypes. The weakly differentiated or undifferentiated intrusions are attributed to the Ogonersky Complex. However, the systematics of the massifs reflect the crystallization conditions of their magmas rather than their origin.

One of the important characteristics of the intrusive rocks is the distribution of trace elements and primarily the rare earth elements (REE), which provide insight into the

N.A. Krivolutskaya, *Siberian Traps and Pt-Cu-Ni Deposits in the Noril'sk Area*,
DOI 10.1007/978-3-319-17205-7_4

Table 4.1 Main intrusions of the Noril'sk area (studied by author)

Complex	Type	Ore potential (if any)	Intrusions
Daldykansky		Barren massifs	1. Daldykansky (prim. exposures)
			2. Massif on the Valek R. (prim. exposures)
Morongovsky		Insignificantly mineralized, barren	3. Morongovsky (prim. exposures)
Noril'sk	Noril'sk	1. Uniquely large deposits	4. Kharaelakh (TG-21, KZ-456, KCS-56),
			5. Talnakh (OUG-2)
			6. Noril'sk 1 (G-22, MS-31, MN-7, MR open pit)
		2. Medium-sized and small deposits	7. Chernogorsky (Ch-55, prim. exp.)
			8. Northern Maslovsky (OM-4),
			9. Southern Maslovsky (OM-24)
			10. Noril'sk 2 (MN-18, prim. exp.)
		3. Weakly mineralized or barren massifs	11. Mikchangdinsky (MD-48/1200),
			12. Massif, borehole MD-57/1200
			13. Massif, borehole MD-50/1000
	Zubovsky	1. Large deposits	14. Southern Pyasinsky (OV-36, NV-12)
			15. Vologochansky (OV-29)
		2. Small deposits	16. Zub-Marksheydersky (MP-34),
			17. Mt. Bol'shaya Bar'ernaya Massif (MP-38)
	Low Talnakh	Occurrences of ore mineralization and barren massifs	18. Low Talnakh (TG-31, ZF-15)
			19. Zelenogrivsky (F-233)
			20. Massifs in the Fokinsky Prospect (F-225/1220, F-230/790)
	Kruglogorsky	Weakly mineralized or barren	21. Lower sill Noril'sk 1 (MR open pit), MPB-2, massifs of borehole VT-2
Ogonersky			22. Ogonersky
			23. Massifs of borehole OV-36/466
Ergalakhsky			24. VT-2
?			25. Cherevkovsky (sampling site 4180)
			26–27. Massifs in the lower reaches of the Mikchangda R. (sampling sites 2309, 490)
			28. MZhK-1, 2
			29. MD-57/300
			30. Dikes at Lama Lake
Dyumtaleysky			Dyumtaleysky (S. Taimyr)

Note: Number of intrusive body corresponds to number in Fig. 4.1; ? – unknown complex; OM-4, MD-57/1200 et ctr. - drill holes and depth where intrusions were studied

composition of the parental magmas and their sources. The wide application of modern analytical techniques has resulted in the accumulation of geochemical data from the magmatic rocks in the Noril'sk area. However, most of these data pertain to the volcanic rocks, whereas the chemistries of the intrusive complexes are less well understood, which limits the ability to perform detailed comparative analyses of the ultrabasite–basite rocks in the Siberian Platform as a whole and in the Noril'sk area in particular.

Massifs are concentrated within several ore junctions (Fig. 4.1). We have studied numerous intrusive bodies (more than 30 massifs) in the Noril'sk area (Fig. 4.2). Here, we present extensive mineralogical and geochemical data for ore-bearing massifs mostly belonging Talnakh and Noril'sk ore junctions. These data allowed us to consider the systematics of the massifs from a new point of view to compare them to the volcanic rocks produced by several episodes of volcanism and to correlate them with the episodes of magmatism in the area.

During the fieldwork in the Noril'sk area, the author examined variably mineralized massifs (Fig. 4.2). The reference vertical section was prepared by studying borehole materials and natural exposures. The massifs are shown in the map of factual material in Fig. 4.2, and the numeric labels of the massifs in the map correspond to the numbers and names in Table 4.1. Some of these results were published in our earlier papers (Krivolutskaya et al. 2001, 2012).

In discussing the geochemistry of the mineralized massifs in the Noril'sk area, we obtained most of the characteristics of the mineralized ultrabasite–basite bodies from the author's data from Karelia and the Kola Peninsula (nineteen intrusive bodies and several layered intrusions). All of the analyses

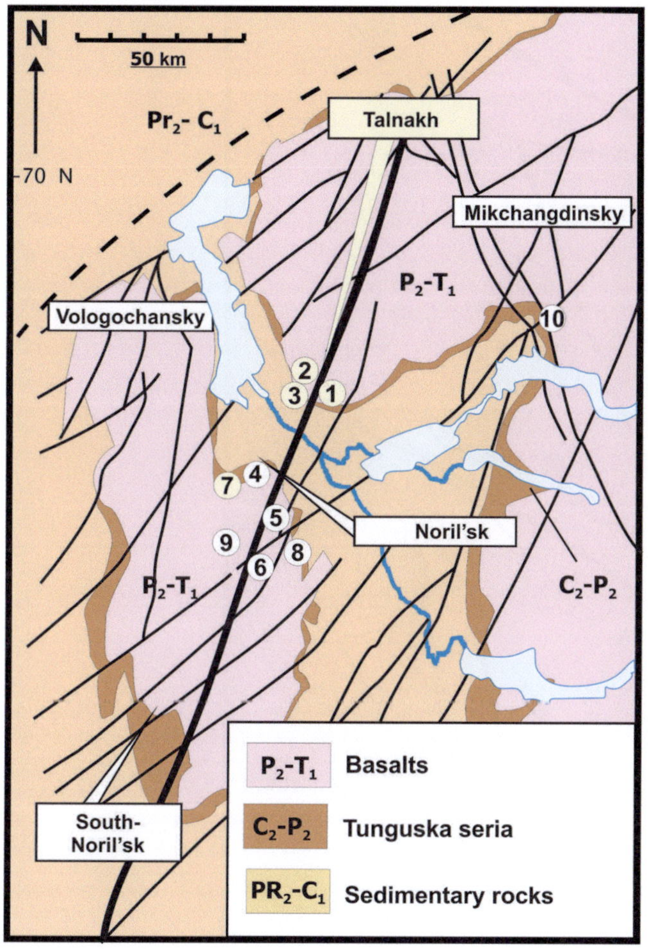

Fig. 4.1 Schematic geological map of the Noril'sk area with position of main ore junctions (in *rectangles*).
Intrusions described in this study, number: (1) Talnakh, (2) Kharaelakh, (3) Low Talnakh, (4) Noril'sk 1, (5) Noril'sk 2, (6) Chernogorsky, (7) Zub-Marksheydersky, (8) Bol'shaya Bar'ernaya, (9) Maslovsky, (10) Mikchangdinsky

were performed in the same laboratories within a short period of time, which indicates that the comparisons should be accurate.

4.2 Geochemical Features of the Intrusive Rocks the Noril'sk Area

We studied normal alkalinity basic–ultrabasic massifs, including the Ogonersky, Noril'sk, Morongovsky, and Daldykansky Complexes. The rocks of the Daldykansky Complex are considered to have been produced by the most strongly fractionated portion of the same parental melt that produced other bodies (TiO_2 = 1.8–2.0 wt %). The intrusive rocks of the other complexes are similar.

The massifs are listed in Table 4.1, as are the boreholes that were analyzed and the sampling sites in exposures. For example, the material recovered from borehole OV-36 was utilized to study a particular massif. If a borehole penetrated more than one intrusion, the depth intervals (in meters) from which the rocks from particular massifs were recovered are also specified. For example, "MD-50/1000, pr. exp." means that the samples were collected from borehole MD-50 from the depth 1000 m or the Medvezhy Creek open pit ("MC" indicates the mine; "IMS" indicates the space between Ugolny and Medvezhy Greek mines). Most of our analyses are presented in Appendix, and some data of the massifs are listed in the tables in this chapter.

The petrochemical parameters of the ultrabasite–basite rocks of all of the massifs are similar. The Harker diagrams show significant variations in the concentrations of the major components, which is typical of the most strongly differentiated massifs in the Noril'sk Complex, such as those classified as the Noril'sk type (massifs of the first group, which host unique large ore reserves). The concentrations of the major components in the rocks vary as follows (wt %): SiO_2, 40–55 %; MgO, 3–26 %; TiO_2, 0.2–3.9 %; FeO, 8–16 %; and Na_2O, 1.0–3.8 %. The rock compositions of the intrusive complexes (including the Daldykansky and Ogonersky Complexes) define a single trend in the MgO–TiO_2 and SiO_2 diagrams (Fig. 4.3).

However, the geochemistry of the massifs allows them to be classified more reliably. The normal alkalinity ultrabasite–basite massifs can be subdivided into three groups based on the concentrations and ratios of the trace elements (Fig. 4.4; Gd/Yb-La/Sm, see Fig. 4.5), the Dyumtaleysky, Low Talnakh, and Noril'sk groups, whose geochemical characteristics (including isotopic parameters) are analogous to those of the Gudchikhinsky, Nadezhdinsky, and Morongovsky formations, respectively. The first group consists of two massifs: the Dyumtaleysky massif in southern Taimyr and the sill exposed in the open pit at "Medvezhy Creek" (between the "Medvezhy Creek" and "Ugolny Creek" pits). The second group includes three massifs: the Low Talnakh and Zelenogrivsky massifs and the massif penetrated by borehole F-225 in the Fokinsky area (Southern Noril'sk group of mineral deposits, i.e., Southern Noril'sk ore junction, Fig. 4.1). All of the other massifs are similar to Noril'sk-type intrusions.

The compositional data of almost all of the massifs plot within the field of rocks of the Noril'sk Complex regardless of their internal structures or their ore potential. This is even more clear in the primitive mantle-normalized ratios of elements (Hofmann 1988), such as (Th/Ta)n–(Nb/Ta)n and (U/Nb)n–(Nb/La)n. In these diagrams, only the Dyumtaleysky-type rocks are clearly separated, and the other rocks (the Noril'sk and Low Talnakh types) cannot be distinguished from one another (see diagram Gd/Yb-La/Sm, Fig. 4.5b).

In this study, the massifs are compared regardless of their concentrations of base and precious metals because the

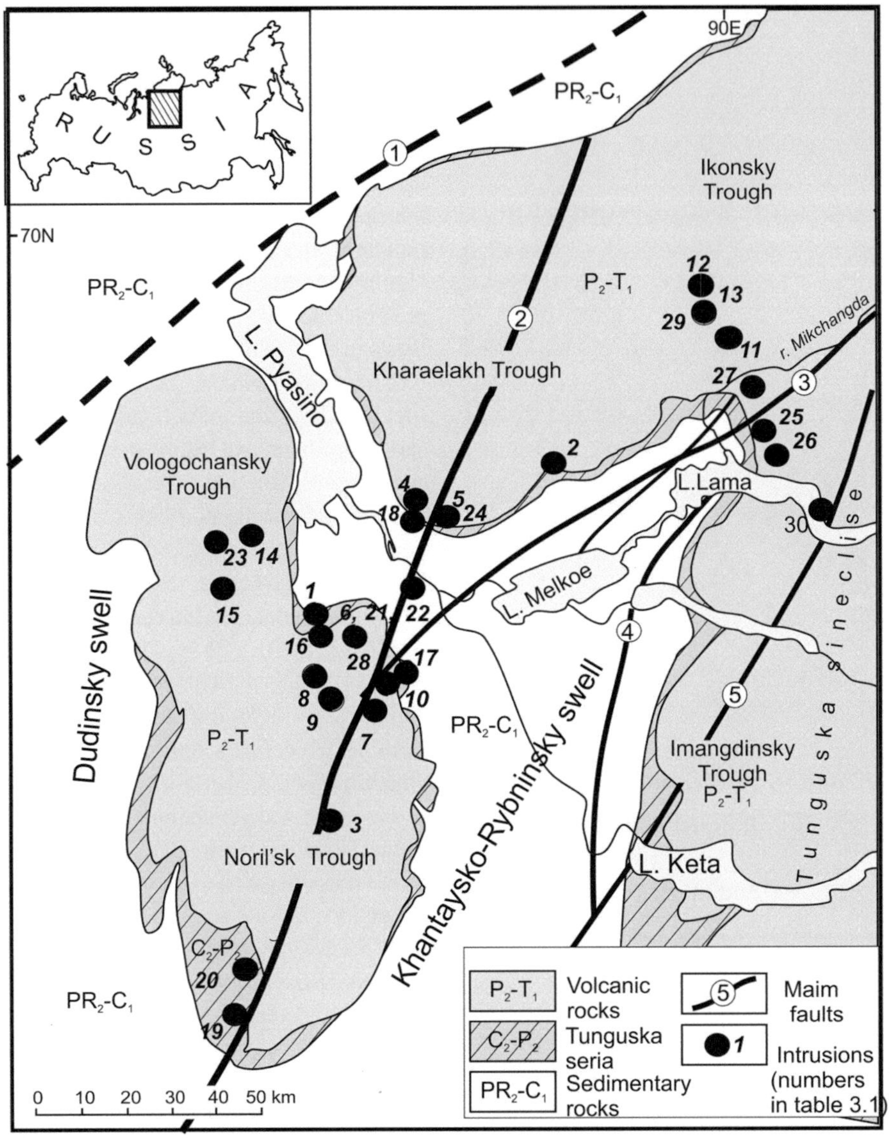

Fig. 4.2 Location map of intrusions studied in the Noril'sk area
The numbers of the massifs correspond to those in Table 4.1 and Fig. 3.1

concentrations of these metals are usually controlled by the occurrence of sulfides in the rocks (with sulfide segregations that are often less than a few micrometers across), which are difficult to analyze. This would require the use of specialized analytical techniques (to identify very low concentrations of PGE and Au on the order of a few ppb). This interesting avenue of research requires further investigation.

The most important conclusion of our research is that the geochemistry of the rocks is not correlated with the presence of sulfide mineralization, i.e., variably mineralized massifs can have similar geochemical characteristics, and their compositional data plot within common fields in all of the geochemical diagrams (Fig. 4.5b). We present data from the main ore-bearing intrusions from the Noril'sk area and mainly from Talnakh and Noril'sk ore junctions, below.

4.3 Inner Structures of the Intrusions of the Noril'sk Complex and Petrography of the Rocks

The dolerite textures of the rocks (Dodin et al. 1971; Ryabov et al. 2000), the zoning of the minerals, the occurrence of melt inclusions (which are often vitreous), and the fact that the rocks often contain recrystallized glass indicate that all of the ultrabasite–basite massifs in the Noril'sk area are subvolcanic bodies that crystallized near the surface. These features suggest that the melts cooled rapidly during the formation of the intrusive bodies. These rocks are compositionally diverse gabbro-dolerites, which are subvolcanic varieties of gabbro. The structural and textural features of the rocks in the Noril'sk area are different from those of Cu–Ni and PGE deposits worldwide, and thus it is not appropriate to classify them

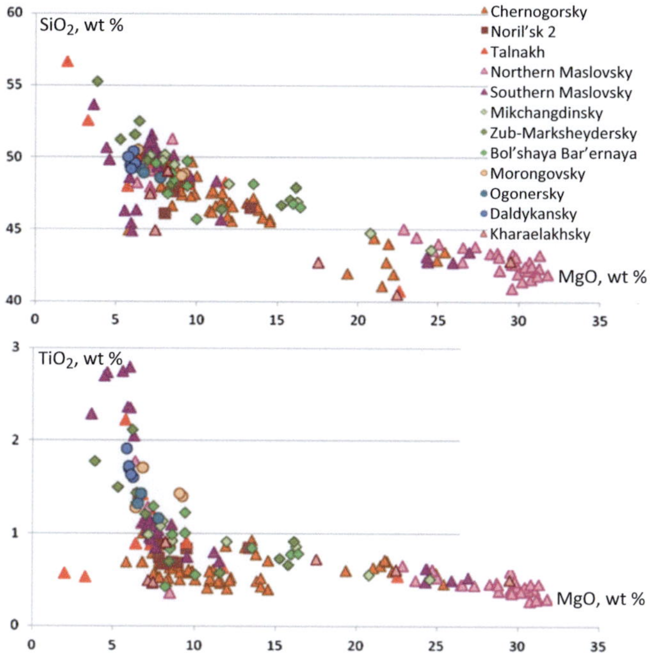

Fig. 4.3 Diagrams SiO_2–MgO and TiO_2–MgO diagrams for intrusive rocks of the Noril'sk area

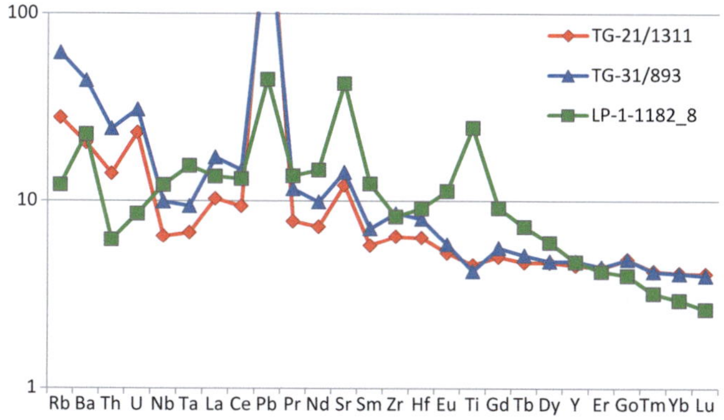

Fig. 4.4 Rare element patterns for major geochemical types of rocks of normal alkalinity
Green line – Dyumateysky massif, *blue line* – Low Talnakh massif, *red line* – Talnakh massif

according to a single scheme (Zen'ko and Czamanske 1994; Naldrett 2009). Herein, we use the terminology that is traditionally applied to local rocks, including gabbro-dolerites; picritic gabbro-dolerites, which contain >40 vol.% olivine; olivine gabbro-dolerites, which contain 10–40 % olivine; olivine-bearing gabbro-dolerites, which contain <10 vol.% olivine; and olivine-free, contact, and taxitic gabbro-dolerites rocks with ataxitic structures and variable compositions, which are mostly similar to leucogabbro and often contain olivine (up to 10 %), which is unevenly distributed in the rocks. For brevity, abridged names such as *picrite* and *taxite* are used in the vertical sections (e.g., Fig. 4.10).

All of the massifs consist of rocks that are similar in composition and fabric, and the proportions of the rocks vary vertically throughout the individual intrusions. The inner structures of the large massifs show lateral variability in which the frontal portions are enriched in leucocratic rock types, and the rear regions are more melanocratic because of the early crystallization of the melts at depth and the separation of plagioclase and olivine crystals (Likhachev 1965, 1977, 2015). These processes are reflected in the behavior of the major components (e.g., MgO and Al_2O_3) but only weakly affect the distribution of trace elements. This is important for further analyses of the geochemistry of the rocks from the massifs and for comparing their characteristics because the reference vertical sections are often located in different parts of the intrusive bodies, including the central and/or marginal portions from which the rock samples could be collected.

The major types of intrusive rocks in the Noril'sk area are described in detail elsewhere (Likhachev 1965; Dodin et al. 1971; Dodin and Batuev 1971; Naldrett 1992, 2004, 2009; Naldrett and Lightfoot 1999; Naldrett et al. 1992, 1995, 1996; Czamanske et al. 1995; Zolotukhin 1997; Ryabov et al. 2000, 2014; Likhachev 1996a, b; Turovtsev 2002; Li et al. 2003; Malitch et al. 2010, 2013 and others). Thus, we only briefly present the characteristics of the major varieties of rocks and focus more on the rocks that were documented and studied for the first time by the author in the Southern Maslovsky intrusion.

The clearly differentiated and variably mineralized rocks have attracted much attention. They are grouped together as the Noril'sk Complex and are further subdivided into types and subtypes (Table 4.1). For example, the variably mineralized and differentiated chonolith-shaped intrusions that are up to 12 km long, up to 2–4 km wide, and from 20–50 to 400 m thick with weighted mean Mg numbers of 10–12 % that host the large Noril'sk 1, Talnakh, and Kharaelakh deposits and other small deposits (Chernogorsky and Noril'sk 2; see Table 4.1) or are barren (massifs in the Fokinsky area penetrated by boreholes F-230 and F-225) are ascribed to the Noril'sk type. The Zubovsky type comprises massifs with similar weighted mean Mg numbers but are less clearly differentiated (Vologochansky and Southern Pyasinsky intrusions). The Low Talnakh type encompasses weakly differentiated highly magnesian (13–16 wt % MgO) intrusions. The Kruglogorsky type consists of massifs that are dominated by leucogabbro that forms individual bodies or parts of differentiated bodies (differentiation products of the parental magma). The rocks often form tongues on the peripheries of the Talnakh ore field, and the relationships between these thin vein-shaped bodies and the main intrusion are often unclear. As illustrative examples of the inner structures of the different types of massifs, several intrusions will be analyzed below.

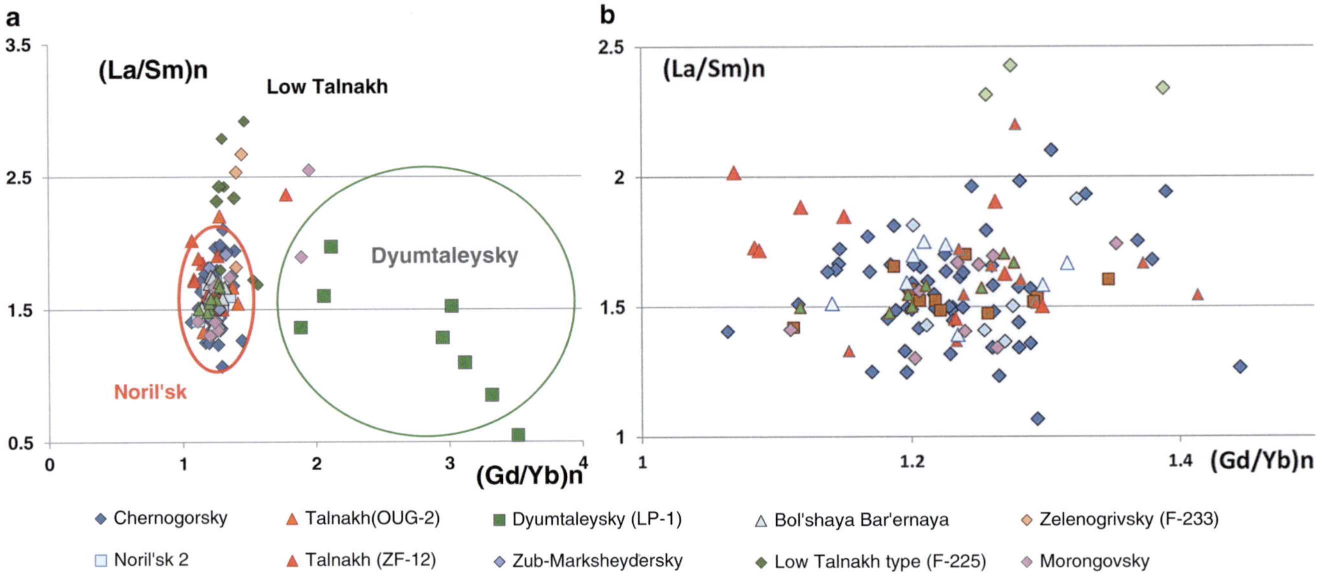

Fig. 4.5 (La/Sm)n–(Gd/Yb)n diagrams for intrusive rocks of (**a**) the Noril'sk area and separately of (**b**) the Noril'sk type
Red and *pink* symbols correspond to mineralized massifs (hosting mineral deposits), and *green* and *blue* symbols show weakly mineralized massifs (After Krivolutskaya 2014)

4.4 Massifs of the Talnakh Ore Junction

We begin the description of the Noril'sk deposits with the Talnakh and Oktyabr'skoe (Fig. 4.6), which are related to Talnakh and Kharaelakh intrusions and are the most economically important deposits despite their late discovery (Fig. 4.2). The end of production of the rich ores in the Noril'sk deposit in 1959 led to an intensive research (L. B. Bertosh, V.S. Staroseltsev, G.I. Kharchenko). Gabbro-dolerite outcrops were found in Western Kharaekh (Mt Otdel'naya) by V.F. Kravtsov, V.F. Nesterovsky and Yu. D. Kuznetsov in 1960, and the disseminated sulfide ores of the Talnakh deposit were discovered in borehole KZ-21 one month later. Oktyabr'skoe deposit was found in 1965 within the deepest Devonian sediments (Fig. 4.7). Several mines explore these ores (Fig. 4.8).

4.4.1 Talnakh Massif

Morphologically, this intrusion is a chonolith with a flat or, in places, a smooth convex roof, and a smooth concave floor. The massif can be traced for 20 km and is 0.5–2 km wide and up to 218 m thick (Fig. 4.9). The body dips 4–6° NNE, and its roof is in contact with the Permian Ivakinsky Formation in the south and the Lower Devonian Zubovsky Formation in the north. The geology and inner structure of the Talnakh intrusion were described in detail in several earlier publications (Sukhanova 1968; Urvantsev 1972; Dodin and Batuev 1971; Zolotukhin et al. 1975; Ryabov and Zolotukhin 1977; Distler et al. 1988; Dyuzhikov et al. 1988; Likhachev 1994, 1996a, b, 2006; Zen'ko and Czamanske 1994; Ryabov et al. 2000; Krivolutskaya et al. 2001; Krivolutskaya 2011; Sluzhenikin et al. 2014).

4.4.1.1 Inner Structure and Petrography

Similar to other Noril'sk-type massifs, the Talnakh intrusion is characterized by a layered inner structure with rocks that systematically alternate vertically. The succession of derivatives is normally visualized as a layered series with the following three units (listed in order from bottom to top): the Lower Gabbro Series (contact and taxitic gabbro-dolerites), the Main Layered Series (picritic, olivine–biotite, olivine, olivine-bearing, and olivine-free gabbro-dolerites), and the Upper Gabbro Series (gabbro-diorites, ferrogabbro, upper contact gabbro-dolerites, eruption breccias, upper picritic gabbro-dolerites, leucogabbro, and taxitic chromite-bearing gabbro). All of the units are typically present in the central part of the intrusion (Fig. 4.9), and its peripheral regions are devoid of picritic gabbro-dolerites; consequently, the inner structures of these areas are more homogeneous (Krivolutskaya et al. 2001).

The Main Layered Series is characterized by gradual transitions between the rocks, which are mainly caused by the systematic upward decrease in the *Ol* concentration. The most typical rocks (Fig. 4.10) of the Upper Gabbro Series are gabbro-diorites, ferrogabbro, and near-contact gabbro-dolerites, all of which are variably altered (Dodin et al. 1971).

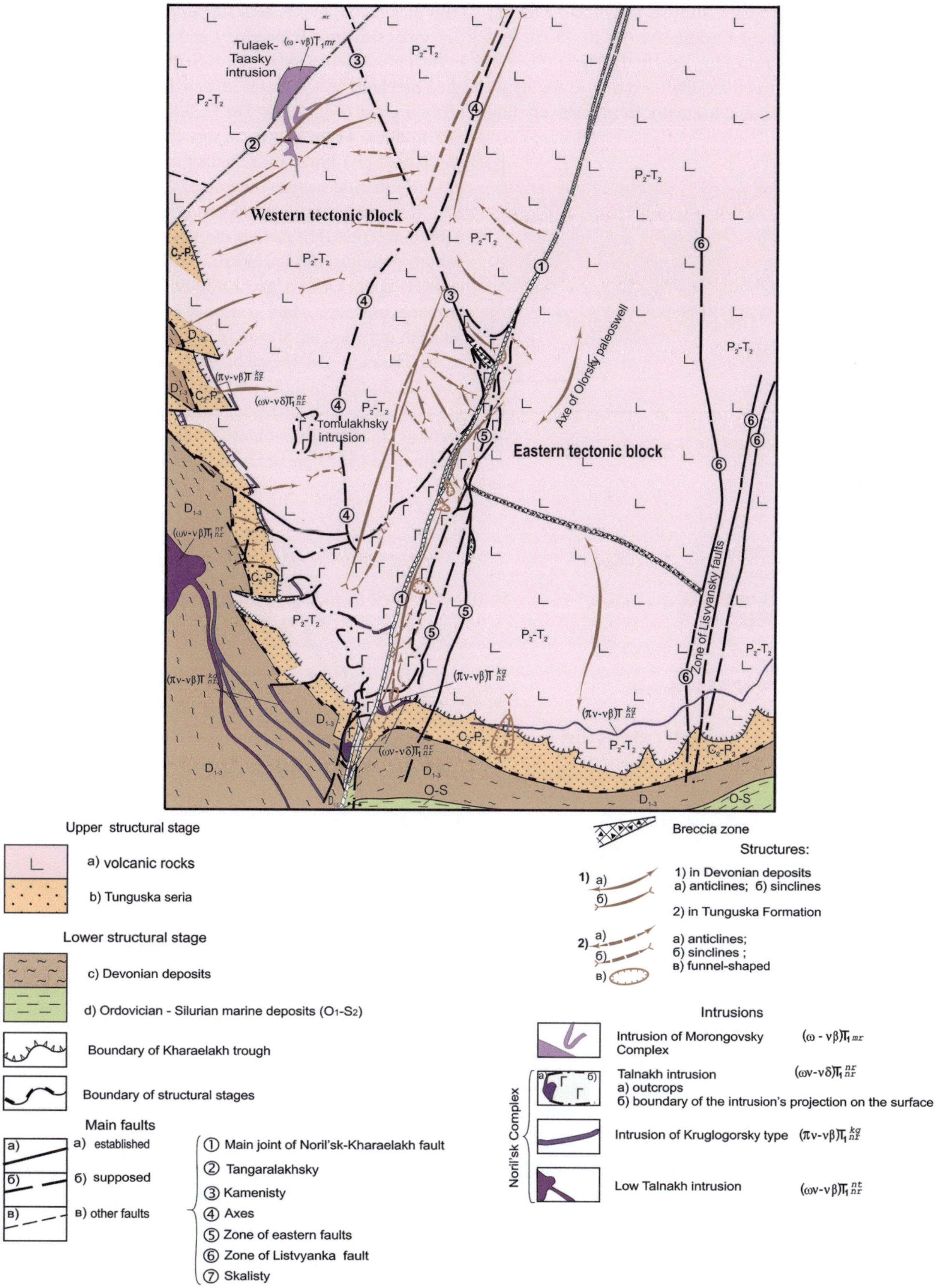

Fig. 4.6 Schematic geological map of the Talnakh ore junction Simplified from Sokolov et al. (unpublished)

The leucocratic gabbros do not form an individual layer but occur as lenses of different thicknesses (up to 25 m) and lengths (from a few meters to a few hundred meters). The *Ol*-rich rocks (picrite- and troctolite-like) are rare in the upper contact zone of the massif and normally form relatively thin (less than 0.5 m) lenses in the upper parts of the leucogabbro. The inner structure, mineralogy, and chemistry of the Talnakh intrusion are described below based on the data from reference borehole OUG- 2, which was drilled in the central part of the massif (Fig. 4.11). The penetrated intrusive rocks are 155 m thick. From bottom to top, the magmatic succession consists of (A) lower near-contact and taxitic gabbro-dolerites, (B) picritic gabbro-dolerites, (C) olivine gabbro-dolerites, (D) olivine-bearing gabbro-dolerites, (E) olivine-free gabbro-dolerites, (F) gabbro-diorites and prismatic–granular gabbro-dolerites, (G) leucocratic (sometimes taxitic) gabbro, and (H) upper contact gabbro-dolerites, quartz diorites, and eruption breccias. Main texture types for the rocks of the Talnakh intrusion are shown in Fig. 4.10.

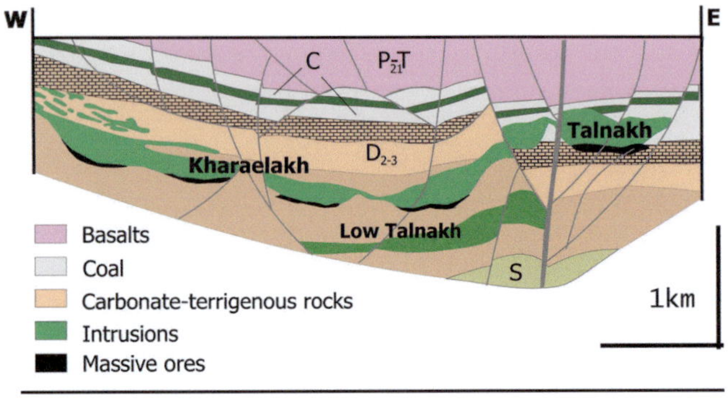

Fig. 4.7 Schematic cross section of the Talnakh ore junction After Dyuzhikov et al. (1988)

A. The near-contact gabbro-dolerites (1.5–2 m) form the lower and upper marginal zones. These are fine- and small-grained rocks with microdoleritic, ophitic, or poikilophitic textures in the groundmass, which sometimes contains plagioclase (An_{70}) phenocrysts. The average composition of

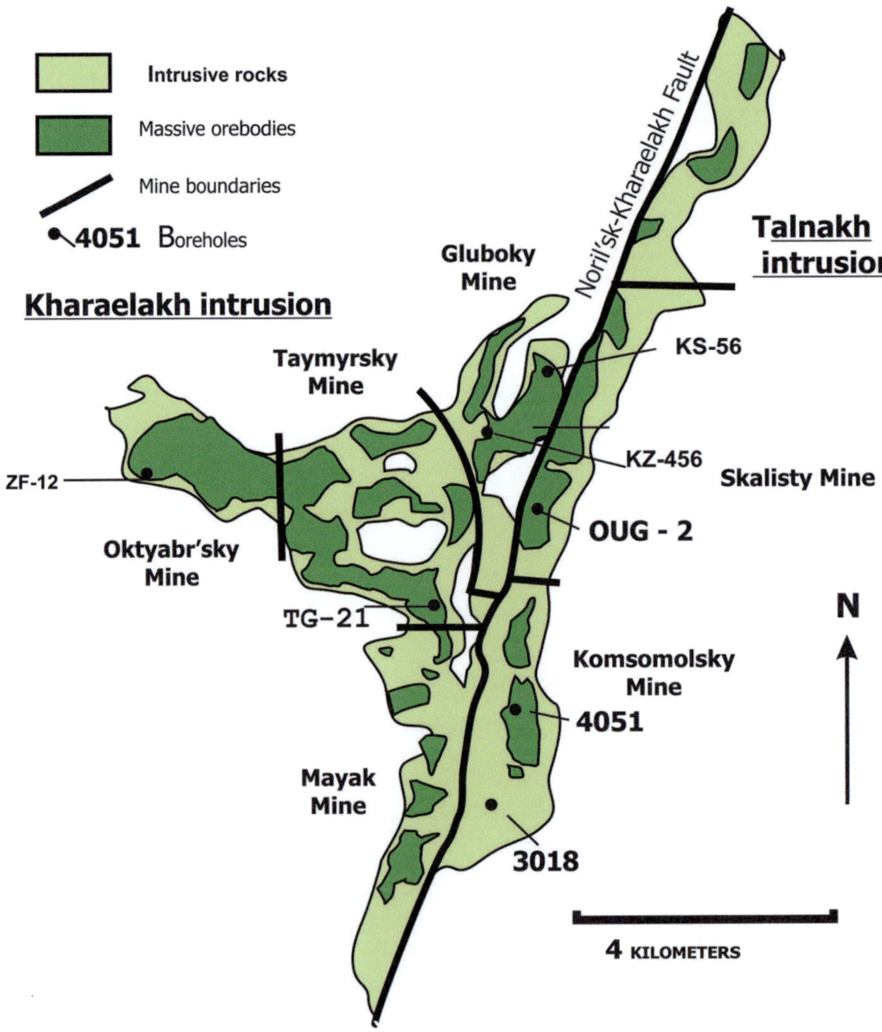

Fig. 4.8 Position of studied boreholes on the plan of Talnakh ore junction

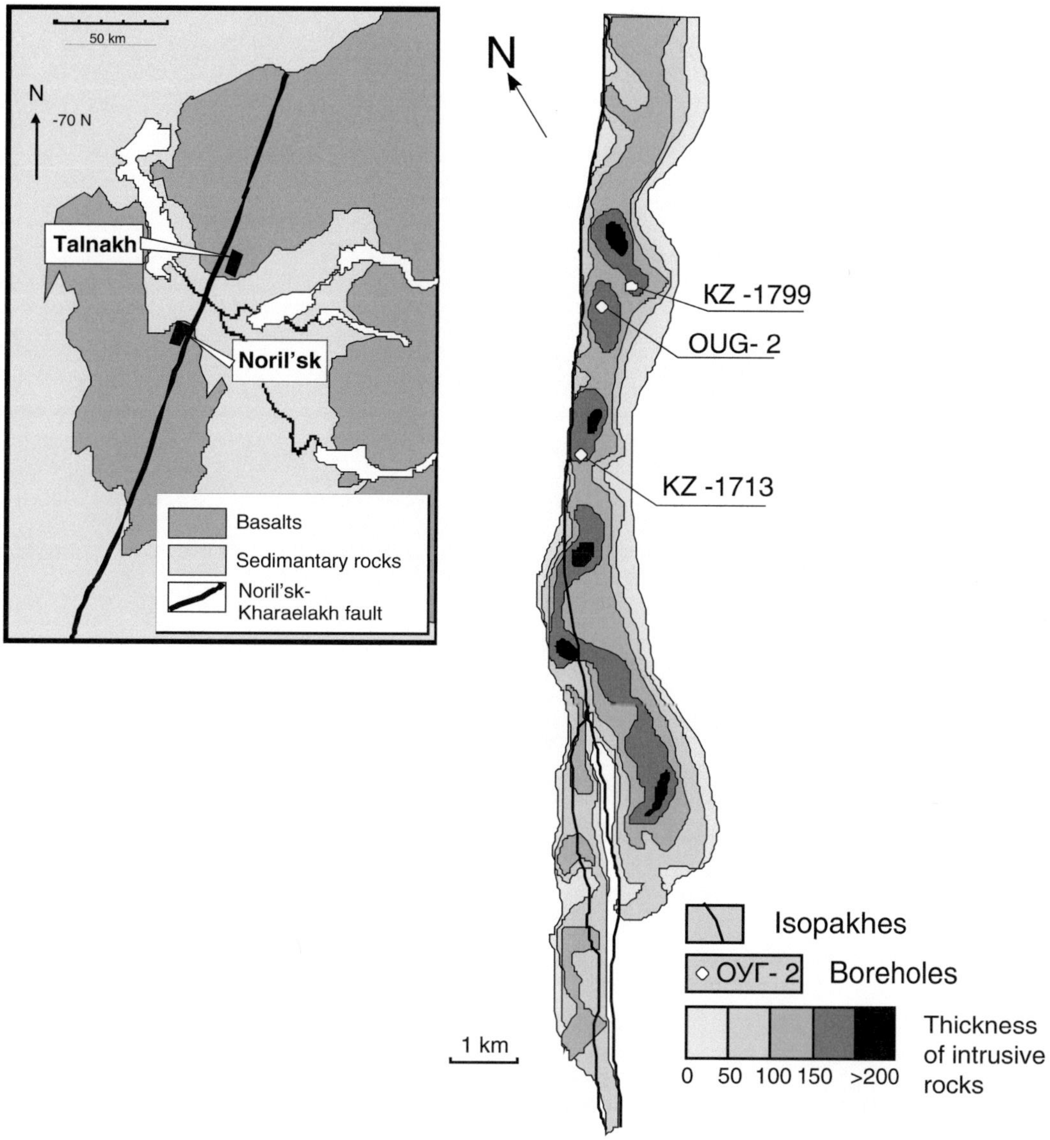

Fig. 4.9 Schematic map showing the setting of the Talnakh intrusion in the Noril'sk district and the morphology of the intrusion (projection onto a horizontal plane)
(1) Faults, (2) holes and their numbers, (3) gradation of the intrusion thickness, m (After Krivolutskaya et al. 2001)

the rocks is as follows: 50–60 vol.% *Pl*, 32–40 % *Cpx*, 5–10 % *Ol*, and 5–15 % ore minerals. The subordinate minerals are hornblende, biotite, apatite, and sphene. Olivine (Fo_{64}) and ore minerals occur only in the lower-contact gabbro-dolerites. The *taxitic gabbro-dolerites* (13 m) are light gray and white coarse-grained rocks with a clearly pronounced ataxitic structure that is caused by the contrasting grain sizes and uneven distribution of the mafic minerals. The mineralogical composition of the rocks varies within wide limits: 35–75 % *Pl* (including two generations, An_{51-66} and An_{76-83}), 10–30 % high-Ca *Cpx* (Fs_{10-12}), 5–25 % *Ol* (Fo_{62-82}), and up to 5 % *Opx* (Fs_{21-25}). The rocks are characterized by variable structures and textures with variations that are often observable even within a single thin section. The dominant textural pattern is controlled by large (3–5 mm) zoned tabular plagioclase grains (An_{66-57} and An_{83-76}), whose marginal parts are more sodic, up to An_{30}, with inclusions of anhedral clinopyroxene grains. The clinopyroxene (up to 5 mm) contains inclusions of small (<1 mm) tabular crystals, laths of plagioclase, and grains of olivine and thus causes the poikilophitic texture of the rock. The clinopyroxene is often zoned with green inner and brownish outer zones. Along with these coarse-grained patches, the rocks contain finer-grained zones (grains <1 mm) that consist of *Pl* laths (with occasional *Cpx* grains in between) with a clear ophitic texture. These rocks sometimes have a gabbroic texture, which is caused by roughly euhedral equant *Pl* and *Cpx* grains. The taxites also contain relatively small (up to 1–2 cm) lens-shaped segregations of fine-grained (0.1–0.3 mm) *Ol* and *Pl* (sometimes with an admixture of Al–Mg spinel),

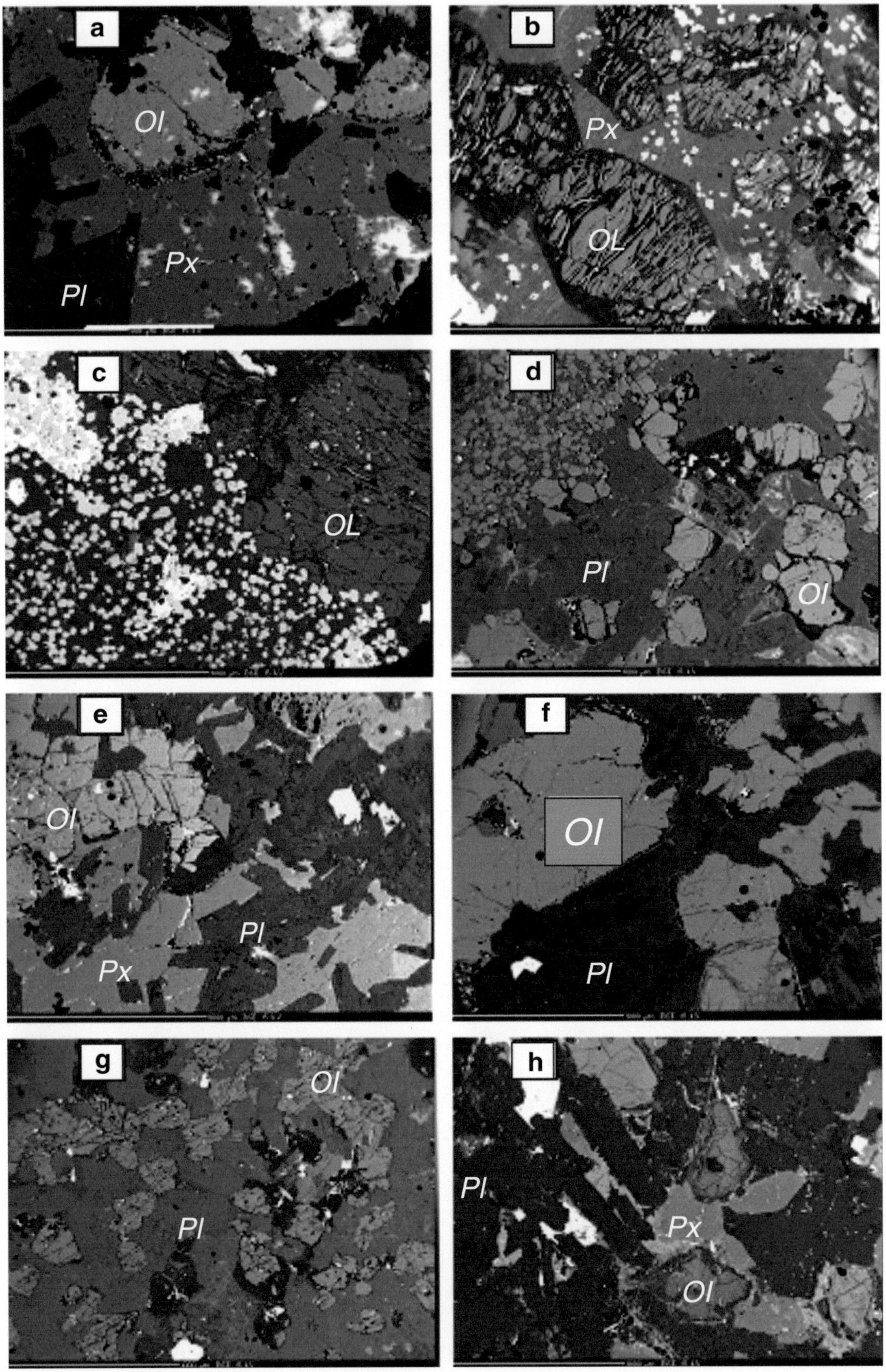

Fig. 4.10 Microphotos of main types of intrusive rocks (image in BSE)
Gabbro-dolerites: (**a**, **b**, **c**) are picritic, (**d**–**f**) are taxitic, (**g**–**h**) are contactic
Scale bar is 1 mm. *Ol* olivine, *Px* pyroxene, *Pl* plagioclase; white grains are spinel (After Krivolutskaya 2014)

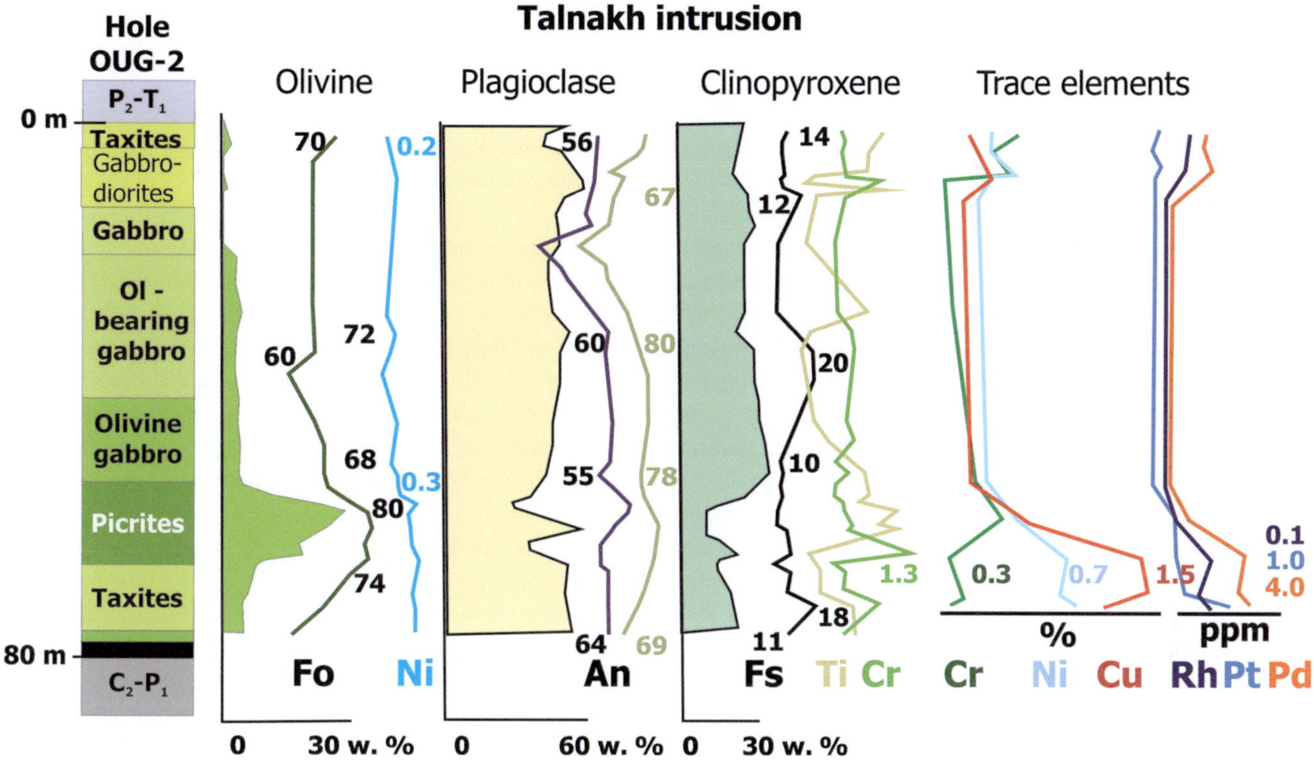

Fig. 4.11 Inner structure of the Talnakh intrusion, variations of amount, and composition of rock-forming minerals and metals' distribution (borehole OUG-2)
After Distler et al. (1999)

whose composition corresponds to troctolite. Olivine is most often present in these rocks as anhedral grains with inclusions of tabular plagioclase crystals. Oxides account for up to 5 vol.% of the rocks and include titanomagnetite and ilmenite. The taxitic gabbro-dolerites (along with picritic gabbro-dolerites) form a mineralized unit with sulfide concentrations of 10–15 vol.%. The ore minerals generally occur as large (up to 5 cm) anhedral pockets and schlieren. The secondary alterations of the taxites involve saussuritization of plagioclase, clinopyroxene replacement by hornblende, and biotite replacement by chlorite. The rocks usually grade smoothly into the overlying picrites.

B. The picritic gabbro-dolerites (14 m) form a unit with well-defined lower and, particularly, upper boundaries. They consist of dark gray fine- to medium-grained massive rocks in which *Ol* is more than 40 vol.% (sometimes up to 70 %). The picrites mostly have poikilophitic, poikilitic, and, in places, panidiomorphic–granular textures. The average mineralogical composition of the rocks is as follows: 25–30 % *Ol* (Fo_{72-78}), 30–35 % *Pl* (An_{53-66} and An_{72-84}), 30–35 % augite (Fs_{10-11}), up to 5 % *Opx* (Fs_{20-23}), and 5–7 % *Bt*. Apatite and spinel are the accessory minerals. The olivine is distributed relatively equally throughout the rock and occurs as euhedral grains 0.3–1 mm in size without notable zoning. The plagioclase forms large (up to 1.5 mm) tabular crystals (with euhedral olivine crystals) and anhedral grains with pronounced zoning. Another generation of plagioclase, which occurs as laths in clinopyroxene, has clear twins. The clinopyroxene (augite) occurs as oikocrysts that are 1–5 mm in size with inclusions of *Ol* grains and *Pl* platelets. Orthopyroxene is present in strongly subordinate amounts as euhedral prisms or, more rarely, oikocrysts with *Ol* inclusions. *Opx* sometimes develops as thin rims around *Ol*. Brown biotite forms lepidoblastic aggregates that often form near oxide and sulfide inclusions and replace clinopyroxene. The latter, in turn, is also replaced by hornblende. The olivine is typically replaced by minerals of the serpentine group (with replacement grades varying even within a single thin section) first along fractures (together with magnetite and pyrrhotite) and, later, in the form of aggregates of lizardite platelets. The oxides (titanomagnetite, ilmenite, Cr–magnetite, and chromite) account for 3–5 % by volume, and the sulfide contents reach 5–8 %. The latter minerals are present as fine-grained (<2 mm) interstitial crystals that are disseminated in the upper portion of the picrite layer and predominantly as ovoids up to 25 mm across in the middle and lower parts of the unit. The transition zone to the taxitic gabbro-dolerites contains segregations of large plagioclase grains, whose concentration increases downward. These segregations eventually merge, and the rock acquires a typical taxitic fabric.

C. The transition to the overlying *olivine gabbro-dolerites* (20 m) is relatively sharp and occurs within a 0.1- to 0.3-m-thick zone. The olivine gabbro-dolerites are gray, medium-grained, massive rocks that are characterized by poikilophitic and ophitic textures. Their *Ol* ($Fo_{69–75}$) concentration reaches 10–15 vol.%, *Pl* ($An_{57–60}$ and $An_{75–84}$) accounts for 40–50 vol.%, and *Cpx* ($Fs_{11–13}$) and *Opx* (Fs_{26}) are present in amounts of 25–30 % and up to 3 %, respectively. The boundary between these rocks and the picrites is marked by elevated contents of biotite (up to 6–7 %) and sulfides. The accessory minerals are apatite and sphene, and the secondary minerals are prehnite, carbonates, iddingsite, chlorite, and serpentine.

D and E. Olivine-bearing and olivine-free gabbro-dolerites (60 m) gradually replace the underlying unit due to a gradual systematic decrease in the content of *Ol* upsection to its complete disappearance and the replacement of the gabbrophitic, poikilophitic, and panidiomorphic–granular textures by prismatic-ophitic and prismatic–granular textures. The main rock-forming minerals are *Pl* ($An_{55–57}$ and $An_{64–77}$), *Cpx* ($Fs_{10–19}$), *Ol* ($Fo_{61–71}$), and *Opx*. The rocks in the upper part of the unit sometimes contain quartz. The rocks also often contain oxides and disseminated sulfides. The accessory and secondary minerals are the same as in the rocks of the underlying units.

F. The gabbro-diorites and prismatic–granular gabbro-dolerites (30 m) are medium- to coarse-grained greenish- or pinkish-gray rocks with prismatic–granular, ophitic textures that contain schlieren of pegmatoid gabbro-dolerites and thin layers of titanomagnetite gabbro. The rocks consist of 50–70 vol.% *Pl* ($An_{56–59}$) and 20–30 % augite ($Fs_{17–24}$). The mesostasis contains albite, microcline, quartz, biotite, chlorite, pumpellyite, orthite, apatite, zircon, and sphene. The secondary minerals are aegirine–augite, hornblende, chlorite, saponite, hydrotalcite, prehnite, and saussurite aggregates.

G. The *taxitic gabbro* (4 m) in the upper portion of the massif is an unequally grained leucogabbro with an ataxitic structure and consists of 60–70 % *Pl* ($An_{60–70}$), 20–25 % *Cpx* (Fs_{20}), hornblende, biotite, rare grains of quartz, micropegmatite, sphene, apatite, and carbonates. In places, the rocks in this unit have consistent textures and structures and form a common anorthite leucogabbro. The taxitic gabbro often hosts low-sulfide platinum mineralization. These rocks are enriched in chromite and contain aggregates of H_2O-, Cl-, and F-bearing minerals (Sluzhenikin et al. 1994).

H. Quartz diorites (13 m) compose the roof of the intrusion. These compositionally, texturally, and structurally variable rocks include clearly bounded shadow xenoliths of the host sand–siltstones of the Tunguska Group. The rocks consist of quartz, potassic feldspar, sodic plagioclase, micropegmatite, biotite, hornblende, relict clinopyroxene, apatite, and fragments of microquartzite, hornfelsed carbonaceous rocks, and graphitized coal, which form the eruption breccias.

4.4.1.2 Cryptic Layering of the Talnakh Intrusion

The compositional trends of the principal rock-forming minerals in the section of borehole OUG-2 are shown in Fig. 4.11. Tables 4.2, 4.3, 4.4, and 4.5 present representative analyses of the olivine, plagioclase, and clinopyroxene that were used to distinguish the units.

Olivine The compositional variations of this mineral in distinct zones of the massif are particularly significant. The maximum forsterite concentration ($Fo_{82.3}$) was detected in the upper part of the taxitic unit (Fig. 4.3, Table 4.2), although the overlying picritic gabbro-dolerites usually have higher-magnesium compositions. In the olivine gabbro-dolerites, *Fo* varies from 69 to 72 mol%, and the Mg# of *Ol* decreases to $Fo_{61.3}$ with the transition to *Ol*-bearing gabbro-dolerites. These variations are correlated with the NiO concentration of *Ol*, which ranges from 0.20–0.25 wt % in the picrites to 0.08–0.11 wt % in the *Ol*-bearing gabbro-dolerites. The MnO and CaO concentrations of the olivine increase upsection.

Plagioclase Two generations of this mineral can be distinguished: (1) large platelets of zoned plagioclase, whose composition varies from $An_{75–84}$ in the olivine, picritic, and taxitic gabbro-dolerites and the upper anorthite leucogabbro to $An_{56–39}$ in the gabbro-diorites, and (2) intercumulus plagioclase with polysynthetic and simple twins and less pronounced compositional variations, roughly from An_{62} to An_{47} (Fig. 4.3, Table 4.3). Both generations exhibit a common trend: the plagioclase becomes more sodic upsection until it forms of albite, which is widespread in the gabbro-diorites.

Clinopyroxene The concentration of *Cpx* systematically increases from the bottom to the top from 15–25 vol.% in the picrites to 40–45 vol.% in the rocks of the Upper Gabbro Series. The mineral can be subdivided into two varieties: (1) green and (2) brown, with the predominance of *Cpx* of the former type. Both varieties correspond to augite, although the green grains are more magnesian and chromic, and the brown grains are more titanic (Dobretsov et al. 1971; Zolotukhin et al. 1975; Czamanske et al. 1995). The compositions of the clinopyroxene from the rocks of the Upper Gabbro Series fall within the field of typical augite (Hess et al. 1941), and the augite from the Main Layered and Lower Gabbro Series are more magnesian and close to the endiopside field. The rims of the clinopyroxene grains are usually more Fe-rich than the cores. The composition of the clinopyroxene systematically varies over the vertical section of the massif. The concentration of the ferrosilite end-member increases upward from 10–11 mol% in the picrites and taxites to 24 mol% in the gabbro-diorites and then decreases to 13 mol% in the upper marginal gabbro-dolerites. The concentration of the enstatite component accordingly decreases upward from 53 to 35 mol%, whereas the concentration of

Table 4.2 Olivine composition from the Talnakh intrusion, wt %

Depth, m	SiO_2	TiO_2	FeO	MnO	MgO	CaO	NiO	Total	*Fo*, mol %
1,156.0	36.75	0.03	33.24	0.63	29.55	0.38	0.08	100.79	61.3
1,164.5	37.03	0.02	28.60	0.61	33.39	0.29	0.05	100.07	67.5
1,169.2	36.23	0.02	32.14	0.65	31.29	0.34	0.08	100.80	63.0
1,175.0	38.16	0.03	26.00	0.54	34.80	0.30	0.11	99.99	70.0
1,181.6	38.09	0.02	23.94	0.47	36.90	0.38	0.13	100.00	73.6
1,188.3	37.28	0.02	25.48	0.46	35.15	0.30	0.08	98.90	70.9
1,188.3	38.35	0.02	24.75	0.44	37.11	0.34	0.13	101.32	73.0
1,200.0	37.67	0.03	25.13	0.45	35.75	0.36	0.13	99.65	71.6
1,200.0	36.44	0.10	27.53	0.43	34.64	0.26	0.09	99.58	68.9
1,208.0	37.20	–	23.78	0.44	37.70	0.31	0.13	99.75	74.2
1,209.5	38.72	–	19.28	0.31	41.60	0.29	0.13	100.44	79.7
1,218.0	38.37	0.03	21.83	0.36	40.12	0.10	0.25	101.76	76.7
1,218.0	38.31	0.03	21.13	0.37	39.49	0.13	0.18	99.70	77.1
1,221.2	39.31	0.05	17.00	0.31	43.27	0.15	0.22	100.39	78.5
1,221.2	38.63	0.05	16.78	0.30	43.64	0.14	0.20	99.76	82.3
1,221.8	37.82	0	21.97	0.32	39.74	0.20	0.18	100.24	76.4
1,222.8	35.12	0.03	25.30	0.40	34.17	0.20	0.19	97.94	70.2
1,225.7	37.90	0.02	24.20	0.41	38.12	0.17	0.19	101.10	74.0
1,225.7	38.57	0.07	21.62	0.34	40.29	0.11	0.19	101.20	77.0
1,227.5	38.95	0.02	24.45	0.50	36.69	0.14	0.19	101.11	72.4
1,227.5	38.93	0.02	24.57	0.50	35.90	0.14	0.19	100.44	71.8

Note: Here and in Tables 4.3, 4.4, and 4.5: Depth, m is a depth in borehole OUG-2; analyses were carried out in IGEM RAS, analyst T. Golovanova. After Krivolutskaya et al. (2001)

Table 4.3 Plagioclase composition from the Talnakh intrusion, wt %

Depth, m zone	SiO_2	Al_2O_3	FeO	MgO	CaO	Na_2O	K_2O	Total	An mol.%
1085 G	50.89	28.95	1.80	0.36	14.55	3.14	0.17	99.46	70.2
1097 G	52.56	27.13	1.21	0.32	10.98	4.39	0.24	98.03	59.3
1097 G	57.00	26.25	0.75	0.05	9.82	5.82	0.34	100.11	47.3
1104.6 F	55.29	26.59	0.76	0.10	10.70	4.37	0.29	98.15	56.5
1123.5 F	54.99	27.20	0.93	0.10	11.55	5.09	0.22	100.15	55.2
1130.3 F	53.86	27.80	1.01	0.26	12.02	4.61	0.21	100.06	58.1
1130.3 F	53.19	28.42	0.82	0.12	12.44	4.77	0.16	99.98	58.6
1139.2 E	58.95	24.17	0.99	0.13	7.81	7.02	0.57	99.76	37.0
1139.2 E	53.09	27.59	0.95	0.08	11.63	4.83	0.27	98.51	56.5
1144.4 E	53.45	28.49	0.97	0.36	13.18	4.02	0.17	101.20	63.8
1144.4 E	56.0.8	27.13	0.76	0.07	9.90	5.78	0.30	100.08	57.4
1156.2 E	50.67	31.51	0.90	0.18	15.12	3.10	0.19	101.71	72.6
1156.2 E	56.23	26.76	1.13	0.13	9.89	5.65	0.40	100.83	48.2
1164.5 D	48.85	30.86	0.73	0.13	15.09	2.51	0.13	98.34	76.4
1169.2 D	48.88	31.36	0.93	0.33	15.71	2.52	0.10	99.87	77.1
1169.2 D	54.39	28.24	0.79	0.08	10.34	4.53	0.30	98.78	54.6
1175.0 D	52.32	29.54	1.00	0.32	13.72	3.49	0.17	100.63	68.0
1181.6 D	49.54	31.71	0.75	0.22	16.12	2.83	0.11	101.58	75.3
1188.3 D	50.69	28.37	0.79	0.18	15.58	2.45	0.08	98.17	77.4
1194.4 D	50.37	29.60	0.75	0.22	16.26	3.01	0.13	100.34	74.5
1200.0 C	53.82	27.76	1.07	0.45	10.84	4.72	0.29	100.02	57.2
1200.0 C	47.94	32.16	0.74	0.12	17.18	1.78	0.06	99.98	83.9
1204.8 C	49.88	30.22	0.82	0.25	15.86	2.63	0.12	99.83	76.3
1204.8 C	48.51	31.32	0.98	0.40	16.06	2.04	0.11	99.45	81.0
1208.0 C	51.93	28.26	1.00	0.40	12.40	4.45	0.25	98.77	59.7

(continued)

Table 4.3 (continued)

Depth, m zone	SiO_2	Al_2O_3	FeO	MgO	CaO	Na_2O	K_2O	Total	An mol.%
1208.0 C	53.41	27.54	1.48	0.40	10.67	5.03	0.18	98.80	53.2
1209.5 B	51.49	31.17	0.75	0.41	13.36	3.33	0.13	100.67	68.2
1218.0 **B**	50.05	30.86	0.72	0.05	14.27	3.03	0.16	99.22	71.7
1221.2 B	48.38	31.85	0.54	0.02	15.78	2.64	0.13	99.40	76.2
1221.2 A	50.87	30.13	0.53	0.00	13.56	3.60	0.27	98.99	66.3
1222.8 A	48.77	31.41	0.39	0.00	16.33	2.25	0.19	99.87	79.3
1222.8 A	48.85	31.41	0.51	0.05	16.34	2.14	0.19	99.56	79.9
1225.7 A	48.19	31.53	0.57	0.03	15.22	2.49	0.14	98.21	76.6
1227.5 A	51.30	29.98	0.52	0.12	15.66	1.73	0.17	99.99	82.7
1227.5 A	49.41	31.17	0.59	0.12	16.49	1.66	0.18	99.63	83.7
1230.0 A	54.74	30.00	0.51	0.10	13.21	2.74	0.31	101.65	71.2

Note: Depth, m is depth in borehole OUG-12; zone, see text. After Krivolutskaya et al. (2001)

Table 4.4 Pyroxene composition from the Talnakh intrusion, wt %

Depth, m; zone	SiO_2	TiO_2	Al_2O_3	Cr_2O_3	FeO	MnO	MgO	CaO	Na_2O	K_2O	Total	*Fs*	*En*	*Wo*	Mg#
1085.6H	51.08	1.00	3.80	0.23	8.02	0.19	14.99	21.14	0.31	0.02	100.78	13	43	44	76.5
1097.9G	50.42	0.80	2.12	0.04	11.54	0.39	13.65	20.23	0.36	0.01	99.56	19	39	42	67.1
1104.6 F	51.70	0.65	1.68	0.01	11.63	0.36	14.82	18.59	0.24	0	99.68	19	43	38	68.8
1116.5 F	50.14	0.57	1.83	0.03	13.58	0.74	12.12	19.70	0.30	0	99.01	24	35	41	60.2
1123.5E	50.94	0.74	2.25	0.03	11.32	0.32	14.96	19.18	0.23	0.01	99.98	18.4	42.4	39.2	69.7
1130.3E	51.64	0.70	2.31	0.07	10.53	0.36	15.02	19.56	0.12	0.01	100.32	17.5	42.6	39.9	71.1
1139.2E	52.62	0.70	2.04	0.01	11.39	0.40	15.37	18.83	0.27	0.04	101.67	18.7	43.2	38.1	69.9
1144.4E	52.19	0.62	2.36	0.10	8.25	0.22	15.72	20.37	0.11	0.01	99.95	13.5	44.8	41.7	76.8
1156.2D	51.31	0.65	2.40	0.09	8.78	0.28	15.60	19.75	0.20	0.01	99.07	14.6	44.7	40.7	75.4
1156.2D	52.62	0.42	2.25	0.19	6.99	0.23	16.81	20.34	0.15	0.01	100.01	11.4	47.4	41.2	80.6
1164.5D	52.34	0.43	2.89	0.45	6.20	0.25	16.08	20.72	0.15	0.01	99.52	10.3	46.6	43.1	81.6
1164.5D	52.64	0.52	3.02	0.57	6.22	0.26	16.56	20.68	0.12	0.02	100.61	10.4	47.2	42.4	82.0
1169.2D	52.53	0.40	1.89	0.20	6.77	0.26	16.98	19.95	0.11	0.01	99.10	11.2	48.2	40.6	81.1
1175.0C	51.04	0.52	2.51	0.16	6.02	0.18	15.92	21.59	0.11	0.01	98.06	9.9	45.6	44.4	82.1
1181.6C	52.37	0.70	2.33	0.01	7.48	0.18	16.11	20.63	0.18	–	99.99	12	46	42	78.9
1188.3C	52.02	0.62	2.51	0.10	8.10	0.22	16.30	20.40	0.34	–	100.61	13	46	41	77.7
1194.4C	52.70	0.65	2.99	0.42	7.45	0.19	15.97	19.87	0.24	–	100.48	12	46	42	78.7
1200.0C	51.42	0.43	2.74	0.76	6.45	0.15	15.72	20.33	0.28	0.01	98.29	11	46	43	80.9
1204.8B	51.93	0.52	2.66	0.77	6.02	0.14	16.98	21.21	0.31	–	100.54	10	48	42	83.0
1204.8B	50.67	0.88	2.17	0.32	9.31	0.26	15.22	20.15	–	–	99.78	15	43	42	73.9
1208.0B	51.83	0.47	2.53	1.21	6.22	0.18	16.33	21.38	–	–	100.35	10	46	44	82.0
1209.5B	53.03	0.80	2.49	0.92	6.09	0.13	16.33	21.18	0.31	–	101.30	10	47	43	82.3
1209.5B	53.44	0.61	2.64	1.13	6.23	0.19	17.22	19.85	–	0.04	101.31	10	49	41	82.6
1218.0B	52.02	0.54	3.09	0.96	6.43	0.09	16.83	19.58	0.24	0	99.97	10.5	48.7	40.7	82.2
1221.2A	52.06	0.47	2.17	0.63	6.46	0.19	17.31	18.30	0.28	0.02	98.00	10.сен	50.6	38.5	82.2
1227.5A	52.83	0.47	1.95	0.66	5.33	0.19	18.92	18.89	0.08	0	99.45	08.8	53.1	38.1	85.9
1227.5A	52.88	0.63	1.70	0.28	6.34	0.18	16.66	20.20	0.03	0	98.97	10.6	47.8	41.6	81.9
1227.5A	54.35	0.43	1.47	0.31	6.54	0.23	17.06	19.99	0.03	0	100.45	10.8	48.4	40.8	81.8
1230.0C	52.78	0.73	2.02	0.19	9.75	0.34	14.89	20.04	0.20	0.01	100.03	14.8	43.3	41.9	74.4
1188.3C	54.72	0.63	1.55	–	16.64	0.41	25.07	2.41	0.05	0.01	101.49	26	69	5	72.4
1213.7B	52.34	0.65	0.96	0.47	13.33	0.43	29.03	1.96	–	–	99.17	20	76	4	79.0
1222.8A	53.28	0.47	1.19	0.13	13.54	0.34	27.57	2.28	0.04	0	99.08	21	74	5	78.0
1222.8A	53.43	0.40	0.68	0.04	14.85	0.36	27.11	1.93	0.03	0	98.91	23	73	4	76.1
1225.7A	53.88	0.45	0.79	0	16.77	0.41	27.06	1.58	0	0.01	101.14	25	72	3	73.7
1230.0A	55.91	0.47	0.83	0.13	14.21	0.50	26.01	1.97	0	0	100.12	23	73	4	75.9

Table 4.5 Composition of biotite from the Talnakh intrusion, wt %

Depth, m; zone	SiO_2	TiO_2	Al_2O_3	Cr_2O_3	FeO	MnO	MgO	CaO	Na_2O	K_2O	Total	Mg#
1144 E	34.82	2.57	11.96	0.23	29.58	0.31	6.86	0.49	0.08	5.60	93.50	69.5
1164. D	38.95	2.87	12.89	0.07	27.82	0.22	6.38	0.25	0.18	9.33	99.96	70.1
1194 C	40.85	2.40	11.70	–	16.71	–	14.96	0.27	0.19	9.13	96.24	41.4
1218.B	38.76	6.99	11.78	0.15	9.45	0.05	18.47	0	0.47	8.40	95.77	32.5
1225 A	38.89	5.94	12.63	0.35	9.12	0.01	18.75	0.03	0.34	8.93	95.25	30.2

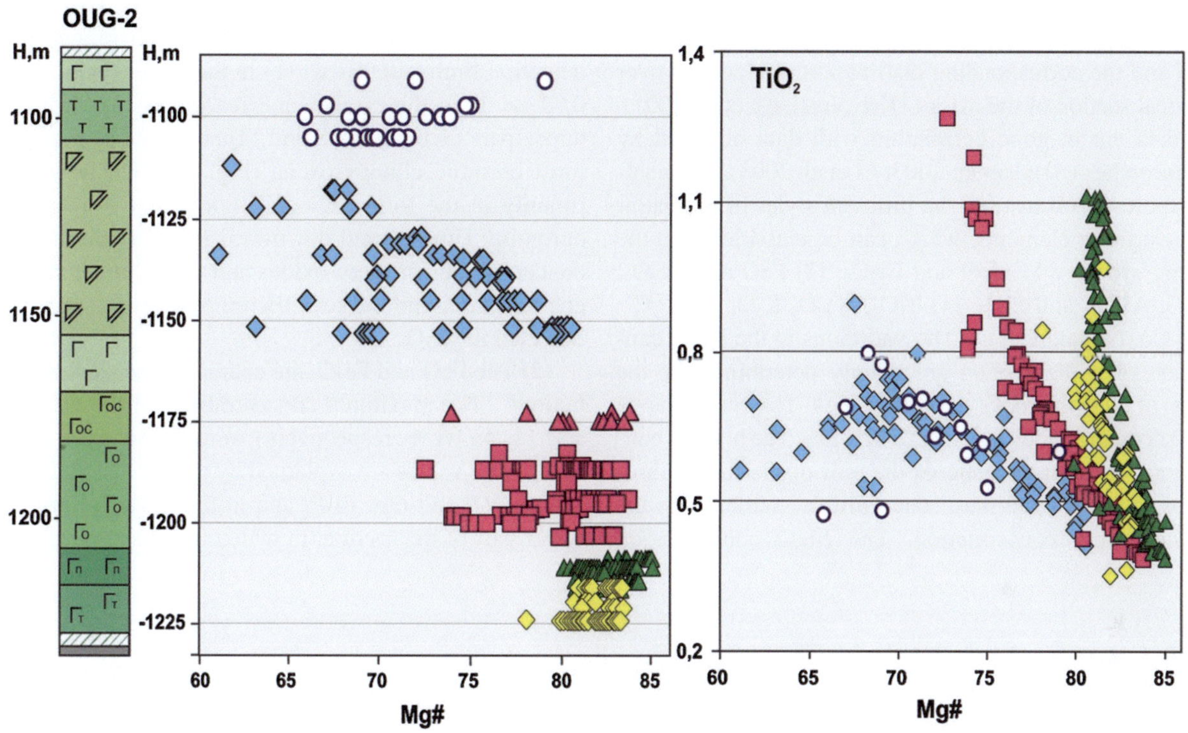

Fig. 4.12 Variations in the Mg# of pyroxene in the vertical section of the Talnakh intrusion (left-hand diagram) and a correlation between TiO_2 concentration and Mg# in various units of the massif
Gabbro-dolerites: T, taxitic; П, picritic; Гo, olivine; Гoc, olivine-bearing; Г, olivine-free. *Triangles* with a *double bar* indicate gabbro-diorites, *hatchwork marks* massive sulfide ores, *small triangles* with Г denote magmatic breccia (After Krivolutskaya et al. 2012)

the wollastonite end-member varies relatively little from 38 to 44 mol%. The Cr_2O_3 concentration of the pyroxenes reaches a maximum of 1.13–1.21 wt % in the uppermost third of the picrite unit and then decreases rapidly toward the taxitic gabbro-dolerites. The systematic decrease in the Cr_2O_3 content of the *Cpx* is also characteristic of the transition from the picrites to olivine and olivine-bearing gabbro-dolerites. The TiO_2 is characterized by smaller variations but is enriched consistently in the upper inner-contact zone (Table 4.4) and particularly in the gabbro-diorites. It is necessary emphasized (Krivolutskaya et al. 2012) that compositions of pyroxene form separate trends for each horizon (Fig. 4.12).

Orthopyroxene The most magnesian orthopyroxene, with 76 mol% *En*, was encountered in the picritic gabbro-dolerites. The Mg# of this mineral decreases up- and down-section, and its Fe # correspondingly increases (Fig. 4.11). The orthopyroxene from the picritic gabbro-dolerites is characterized by the highest concentrations of Cr_2O_3 (0.47 wt %), and the other components vary as follows: Al_2O_3 from 0.7 to 1.15 wt %, CaO from 1.6 to 2.4 wt %, MnO from 0.35 to 0.50 wt %, and TiO_2 from 0.4 to 0.65 wt % (Table 4.4).

Biotite The composition of biotite varies over a broad interval in the vertical section of the Talnakh Massif (Fig. 4.3, Table 4.5). The biotite of the olivine, picritic, and taxitic gabbro-dolerites is the highest in MgO (15–18 wt %), TiO_2 (6–7 wt %), and alkalis (8.4–9.1 wt %) at the lowest concentration of CaO. In the upper sections of the intrusion (upper taxites, leucogabbro, gabbro-diorites, and olivine-free and olivine-bearing gabbro-dolerites), this mineral has a relatively low-Mg# and is low in alkalis but high in CaO (up to 0.5 wt %) at a TiO_2 concentration of 2.4–2.6 wt %.

4.4.1.3 Major and Trace Element Chemistry of the Rocks

The chemistry of typical rocks from the Talnakh Massif (samples from borehole OUG-2) is shown in Tables 4.6 and 4.7 and the corresponding distribution of elements over the vertical section of the massif (Krivolutskaya et al. 2001). These data are in good correlation with data obtained by other researches (Arndt et al. 2003; Li et al. 2003). The analysis of these data illustrates the different styles of the major and ore-forming elements, which can be classified into the following groups: (1) MgO and Cr_2O_3; (2) FeO and Fe_2O_3; (3) CaO, Al_2O_3, and SiO_2; (4) TiO_2, Na_2O, K_2O, and P_2O_5; and (5) Co, Ni, and Cu. (1) The variations in the MgO concentration over the section are mainly determined by the amounts of *Ol* in the rocks and by the Mg#. The overall range of MgO concentrations is 5.11–22.41 wt %. The highest concentrations are characteristic of the unit of picritic gabbro-dolerites, whose uppermost one-third is marked by the maximum MgO concentration. The MgO concentration decreases upward and downward from this ultrabasic unit. It should also be noted that in some sections of the intrusion, *Ol*-enriched rocks are located in the upper inner-contact zone and form a second maximum in the MgO concentration. The distribution curve of Cr_2O_3 follows the variations in the MgO concentration. Significant differences in the concentrations of the two components are observed only in the upper part of the olivine gabbro-dolerites. The relative minimum in the Cr_2O_3 concentration in this layer is explained by the strong decrease in the Cr–magnetite content of the rock with a gradual upward decrease in the olivine content. The Cr_2O_3 concentration over this section of the massif ranges from the detection limit (<0.007 wt %) in the upper taxitic gabbro to 0.77 wt % in the picrites and reaches its maximum in the upper part of the picrite unit. These units also contain the most chromic clinopyroxene (Fig. 4.3) and large amounts (mainly in the form of segregations) of Cr–magnetite and chromite. Upward and downward from the picrite unit, the concentrations of these oxides and Cr_2O_3 in the clinopyroxene decrease, and the rock therefore becomes progressively depleted in Cr_2O_3.

(2) The FeO and Fe_2O_3 are characterized by similar distributions . The maximum concentrations of these oxides (7.6 and 11. 45 wt %, respectively) were encountered in the picrites; these values are explained by an increase in the proportions of *Ol* and *Opx* to *Pl* and in the gabbro-diorites of the upper part of the section, in which they have elevated (up to

Table 4.6 Composition of rocks from the Talnakh intrusion, wt %

Component	1104.6*	1130.3	1144.4	1169.2	1181.6	1188.3	1200	1208	1213.7	1218	1222.8	1230
SiO_2	46.43	47.21	49.66	49.01	46.79	46.57	46.69	40.49	37.66	38.96	41.49	43.34
TiO_2	2.03	1.47	1.15	0.89	0.71	0.83	0.72	0.6	0.4	0.44	0.55	0.54
Al_2O_3	12.38	13.45	14.51	15.99	17.32	16.78	17.32	8.02	7.09	7.75	11.22	14.01
Fe_2O_3	7.11	5.15	4.02	4.26	4.29	3.68	3.88	4.81	6.59	4.06	11.45	8.76
FeO	9.11	9.32	7.98	5.64	.94	7.04	6.06	10.79	12.52	12.06	7.06	6.75
MnO	0.23	0.24	0.23	0.18	0.15	0.18	0.16	0.22	0.18	0.23	0.15	0.15
MgO	5.11	6.28	6.13	7.11	8.86	10.43	10.05	22.41	18.77	19.13	7.15	6.64
CaO	9.25	10.01	10.77	12.21	10.98	9.32	10.84	5.03	05.01	4.96	5.53	9.22
Na_2O	3.58	2.32	2.05	2.18	1.98	1.92	1.81	0.87	0.88	0.62	2.22	1.94
K_2O	0.99	0.83	0.67	0.69	0.49	0.57	0.46	0.25	0.32	0.41	1.09	0.79
Cr_2O_3	<0.007	<0.007	<0.007	0.05	0.07	0.02	0.08	0.77	0.57	0.28	0.03	0.03
P_2O_5	0.27	0.25	0.25	0.15	0.14	0.17	0.15	0.12	0.11	0.11	0.15	0.14
H_2O-	0.14	0.43	0.24	0.2	0.38	0.26	0.2	0.14	0.22	0.78	0.22	0.24
H_2O+	1.67	1.29	1.59	1.51	2.35	1.75	1.89	4.49	2.96	6.05	2.12	1.79
Ni	0.01	0.01	0.01	0.01	0.28	0.35	0.04	0.17	0.53	0.45	0.89	0.59
Co	0.01	0.01	0.01	0.01	0.01	0.01	0.01	0.02	0.03	0.02	0.04	0.03
Cu	0.01	0.02	0.02	0.01	0.01	0.01	0.01	0.13	1.58	0.30	3.00	1.79
S	0.33	0.2	0.25	0.09	0.07	0.13	0.05	0.45	4.23	2.33	6.24	4.85
F	0.11	0.08	0.08	0.06	0.06	0.06	0.05	<0.05	<0.05	0.07	0.06	0.06
Total	99.68	98.66	100.06	100.25	99.88	100.44	100.48	100.04	100.47	100.09	101.66	101.66

Note: * - N sample (*first line*) is a depth (m) in borehole OUG-2. Analyses were carried out in IGEM RAS, analyst N. Malysheva. After Krivolutskaya et al. (2001)

Table 4.7 Concentrations of major and rare elements in the rocks from the Talnakh intrusion (OUG-2)

№ sample	1,085	1,097	1,116	1,130	1,151	1,173	1,181	1,186	1,191	1,200.5
SiO_2	50.25	49.48	48.24	48.11	49.96	49.48	46.57	49.78	49.56	46.69
TiO_2	1.37	1.988	0.63	1.47	1.26	0.88	0.81	0.89	1.42	0.72
Al_2O_3	14.66	14.13	16.28	13.45	14.16	16.71	17.32	19.42	13.35	17.32
FeO	11.7	14.04	9.82	13.96	12.12	9.14	9.8	8.10	12.34	9.56
MnO	0.24	0.264	0.18	0.24	0.21	0.19	0.15	0.17	0.39	0.16
MgO	6.4	5.3	11.73	6.28	7.29	7.27	8.87	6.39	6.73	10.05
CaO	10.82	10.03	10.97	10.1	12.19	13.20	10.98	12.42	11.62	10.84
Na_2O	1.83	1.00	1.53	2.32	2.22	2.50	1.98	2.45	2.84	1.81
K_2O	1.25	0.81	0.35	0.83	0.50	0.54	0.49	0.53	1.06	0.46
P_2O_5	0.11	0.12	0.07	0.25	0.13	0.10	0.14	0.09	0.17	0.15
Cr_2O_3	0.04	0.01	0.08	0.00	0.00	0.08	0.07	0.07	0.05	0.09
LOW	1.67	1.78	0.87	0.9	0.13	0.19	2.14	0.18	0.77	2.19
Total	100.34	98.95	100.76	97.91	100.18	100.28	99.32	100.49	100.31	100.04
S	0.28	0.52	0.91	0.2	0.08	0.06	0.076	0.04	0.69	0.051
Rb	40.8	n.a.	12.6	23.4	17.9	16.0	14.7	13.2	37.9	3.4
Ba	505	n.a.	181	219	147	169	154	185	217	130
Th	1.25	n.a.	1.03	1.36	1.58	0.40	0.82	1.11	0.82	0.69
U	0.56	n.a.	0.28	0.58	0.46	0.21	0.37	0.42	0.35	0.35
Nb	4.70	n.a.	2.38	4.94	4.22	2.36	2.78	3.74	4.90	2.85
Ta	0.33	n.a.	0.17	0.35	0.26	0.15	0.21	0.22	0.30	0.22
La	8.91	n.a.	3.8	9.87	7.5	5.4	5.41	7.1	8.1	4.87
Ce	20.74	n.a.	8.5	22.71	16.9	12.3	12.73	16.3	18.8	11.94
Pb	35.21	n.a.	35	86	80	44	80	359	2410	44
Pr	2.80	n.a.	1.17	3.05	2.32	1.66	1.74	2.19	2.45	1.60
Nd	12.7	n.a.	5.54	13.9	11.19	7.87	7.9	10.04	11.73	7.3
Sr	411	n.a.	233	250	242	262	268	275	288	230
Sm	3.48	n.a.	1.53	3.74	3.11	2.19	2.12	2.67	3.19	1.95
Zr	92.5	n.a.	40	99	74	67	61	59	70	59
Hf	2.41	n.a.	1.13	2.71	1.97	1.67	1.70	1.67	1.81	1.61
Eu	1.18	n.a.	0.58	1.41	1.02	0.79	0.80	0.97	1.09	0.75
Ti	7,850	n.a.	3,869	8,832	7,610	5,249	4,807	5,472	8,608	4,562
Gd	4.10	n.a.	2.04	4.48	3.77	2.68	2.53	3.14	4.00	2.29
Tb	0.70	n.a.	0.34	0.76	0.65	0.43	0.44	0.52	0.69	0.39
Dy	4.56	n.a.	2.22	5.01	4.19	2.77	2.85	3.34	4.27	2.61
Ho	0.98	n.a.	0.47	1.07	0.87	0.56	0.62	0.70	0.90	0.55
Y	28.3	n.a.	11.2	30.1	21.7	14.2	17.0	17.3	22.6	14.5
Er	2.76	n.a.	1.28	3.02	2.44	1.57	1.73	1.91	2.45	1.56
Tm	0.41	n.a.	0.19	0.46	0.33	0.22	0.26	0.27	0.35	0.24
Yb	2.60	n.a.	1.33	2.85	2.36	1.53	1.63	1.85	2.52	1.52
Lu	0.40	n.a.	0.20	0.44	0.36	0.23	0.25	0.28	0.37	0.23
Ni	92.9	n.a.	1,603	56	136	275	254	415	148	287
Cu	59.0	n.a.	3072	158	399	716	134	1,970	4,047	91
Zn	70.5	n.a.	151	186	168	230	61	143	559	68
Co	42.9	n.a.	105	49	59	83	47	61	117	52
V	301	n.a.	170	345	332	241	182	222	355	178

(continued)

Table 4.7 (continued)

№ sample	1,203	1,211	1,216	1,221	1,222	1,225	1,227	1,230	1,231	1,234	1,259
SiO_2	48.15	40.73	36.16	32.97	48.02	38.09	44.98	47.34	18.4	22.42	56.63
TiO_2	0.90	0.54	0.822	0.364	2.23	0.292	0.59	0.54	0.179	0.274	0.57
Al_2O_3	16.25	8.03	9.32	6.97	12.00	12.76	16.43	14.01	6.65	6.28	13.53
FeO	10.77	17.55	18.15	21.55	17.71	16.27	10.97	10.75	33.35	32.11	16.08
MnO	0.26	0.28	0.292	0.232	0.65	0.25	0.146	0.15	0.137	0.075	0.25
MgO	9.48	22.57	15.33	17.89	5.72	12.69	7.62	6.64	0.95	1.28	1.97
CaO	11.55	6.09	6.63	4.17	9.68	7.18	12.71	9.22	5.08	4.98	4.71
Na_2O	1.98	0.85	0	0	2.52	0	0.16	1.94	0	0.39	4.08
K_2O	0.30	0.23	0.24	0.32	0.89	0.5	0.3	0.79	0.15	0.35	1.53
P_2O_5	0.09	0.07	0.10	0.04	0.21	0.02	0.02	0.14	0.00	0.04	0.04
Cr_2O_3	0.09	0.92	0.81	0.09	0.07	0.14	0.08	0.03	0.10	0.07	0.00
LOW	0.51	4.32	4.38	6.32	1.42	3.8	2.13	2.01	9.74	7.91	2.02
Total	100.34	102.16	92.24	90.91	101.13	91.99	96.14	93.56	74.74	76.19	101.42
S	0.49	4.26	0.52	6.56	0.03	4.96	2.68	4.85	29.95	28.65	1.01
Rb	8.9	9.7	9.6	12.7	38.5	n.a.	n.a.	6.4	n.a.	n.a.	35.0
Ba	159	97	121	84	441	n.a.	n.a.	170	n.a.	n.a.	1911
Th	0.73	0.62	1.02	0.32	2.31	n.a.	n.a.	4.82	n.a.	n.a.	10.76
U	0.32	0.20	0.47	0.16	1.20	n.a.	n.a.	0.52	n.a.	n.a.	2.09
Nb	3.33	1.67	3.42	1.54	7.29	n.a.	n.a.	2.20	n.a.	n.a.	10.99
Ta	0.23	0.10	0.27	0.10	0.54	n.a.	n.a.	0.16	n.a.	n.a.	0.83
La	6.0	3.3	170.87	2.75	11.7	n.a.	n.a.	20.09	n.a.	n.a.	14.3
Ce	13.7	7.5	308.48	6.33	28.1	n.a.	n.a.	31.83	n.a.	n.a.	27.6
Pb	380	1,874	359	2,410	263	n.a.	n.a.	380	n.a.	n.a.	32
Pr	1.84	0.98	31.93	0.81	3.99	n.a.	n.a.	5.97	n.a.	n.a.	2.91
Nd	8.69	4.56	96.2	3.6	19.61	n.a.	n.a.	22.0	n.a.	n.a.	10.63
Sr	239	133	134	93	257	n.a.	n.a.	275	n.a.	n.a.	299
Sm	2.29	1.22	11.88	0.95	5.53	n.a.	n.a.	3.01	n.a.	n.a.	2.01
Zr	70	28	75	27	414	n.a.	n.a.	46	n.a.	n.a.	164
Hf	2.01	0.81	2.11	0.72	9.66	n.a.	n.a.	1.32	n.a.	n.a.	4.64
Eu	0.89	0.49	2.63	0.36	1.86	n.a.	n.a.	1.13	n.a.	n.a.	0.62
Ti	5,431	3,382	4,868	2,198	13,552	n.a.	n.a.	2,735	n.a.	n.a.	3,540
Gd	2.79	1.47	6.92	1.14	7.17	n.a.	n.a.	2.38	n.a.	n.a.	1.71
Tb	0.47	0.25	0.90	0.19	1.24	n.a.	n.a.	0.39	n.a.	n.a.	0.26
Dy	3.02	1.56	4.71	1.23	8.29	n.a.	n.a.	2.38	n.a.	n.a.	1.57
Ho	0.62	0.32	0.91	0.26	1.79	n.a.	n.a.	0.48	n.a.	n.a.	0.32
Y	15.4	8.1	30.8	7.7	43.7	n.a.	n.a.	16.1	n.a.	n.a.	8.2
Er	1.71	0.96	2.45	0.74	5.07	n.a.	n.a.	1.33	n.a.	n.a.	0.95
Tm	0.24	0.13	0.34	0.11	0.71	n.a.	n.a.	0.20	n.a.	n.a.	0.16
Yb	1.79	0.96	2.00	0.74	5.03	n.a.	n.a.	1.29	n.a.	n.a.	1.25
Lu	0.27	0.16	0.29	0.11	0.74	n.a.	n.a.	0.19	n.a.	n.a.	0.23
Ni	556	4,930	6,450	8,706	81	n.a.	n.a.	7097	n.a.	n.a.	2579
Cu	3,084	17,236	7,774	10,599	686	n.a.	n.a.	12,086	n.a.	n.a.	9,236
Zn	1,209	654	267	128	358	n.a.	n.a.	98	n.a.	n.a.	96
Co	96	250	241	303	82	n.a.	n.a.	208	n.a.	n.a.	195
V	220	203	201	87	610	n.a.	n.a.	196	n.a.	n.a.	118

Note: Oxides are given in wt %, elements in ppm. N sample is a depth (m) in borehole OUG-2. Here and in Tables 4.8, 4.9, 4.12, and 4.15 analyses were carried out in GEOKHI RAS, analysts I. Rostshina, T. Romashova (XRF), and rare elements were analyzed in IMGRE, analyst D. Zhuravlev (ICP-MS); n.a., element was not analyzed

10–15 %) concentrations of titanomagnetite. The lowest concentrations of the iron oxides are located in the contact zone between the picrites and the olivine gabbro-dolerites.

(3) The highest concentrations of CaO and Al_2O_3 (12.21 and 17.32 wt %, respectively) are spatially restricted to the olivine and olivine-bearing gabbro-dolerites, which is explained by the enrichment of plagioclase in these rocks and the high-Ca# (up to An_{84}) of this mineral. A similar characteristic was observed in the plagioclase-rich rocks, including the taxitic gabbro-dolerites and anorthite leucogabbro from the lower and upper units. The CaO and Al_2O_3 concentrations in the upper taxites decrease slightly, which is correlated with an increase in the albite concentration of the plagioclase. The concentrations of these elements in the picritic gabbro-dolerites decrease greatly (to 4.96 and 7.75 wt %, respectively) due to the overall decrease in the *Pl* content at the constant Ca#. The behavior of SiO_2 in the lower part of the intrusion is analogous to that of CaO and Al_2O_3, and, beginning with the olivine gabbro-dolerites, the silica content of the rocks remains nearly constant.

(4) Phosphorus, titanium, and alkalis are gradually enriched toward the roof of the intrusion, i.e., they exhibit trends opposite to those of CaO and Al_2O_3. All of the incompatible elements are characterized by slightly elevated concentrations near the bottom of the massif, with each element showing small local maxima and minima in the section. For example, an increase in the Na concentration in the upper part of the succession is related to a decrease in the Ca# of *Pl* and, to a lesser degree, secondary alteration of the rocks. The analogous behavior with Ti is explained by the appearance of titanomagnetite in the upper part of the sequence.

(5) The enrichment of the ore components Ni, Co, and Cu was mainly related to the sulfides, which controlled the maximum concentrations of these metals in the lower mineralized derivatives (picritic and taxitic gabbro-dolerites). Note that compared to Co and Cu, the Ni concentration is characterized by a smoother decrease over the section (Fig. 4.11), which may be explained by the relatively high Ni concentrations in the olivine and the monotonous variations in the content of this mineral. The near-contact taxitic to picritic gabbro-dolerites have the highest concentrations of MgO. These units are characterized by significant variations in the rock chemistry. The concentrations of the components vary as follows: SiO_2 varies by approximately 5 wt %, Al_2O_3 varies by approximately 10 wt %, MgO varies by approximately 15 wt %, and FeOtot and CaO vary by approximately 7 wt %. These vertical variations are systematic and may have been caused by the redistribution of crystals (mainly intratelluric) during the insignificant fractionation of the parental magmatic melt (Likhachev 1965; Dneprovskaya et al. 1987). As was described in the previous sections of this chapter, this characteristic is genetically informative and meets the conditions discussed above for the applicability of geochemical thermometric techniques.

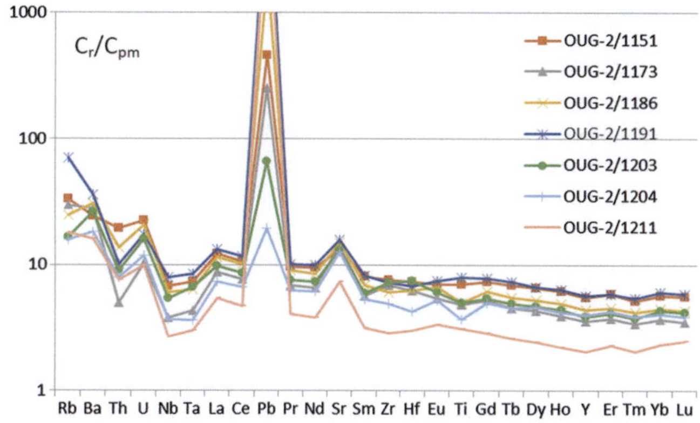

Fig. 4.13 Trace element patterns for intrusive rocks of the Talnakh intrusion (borehole OUG-2)

Considering the distribution range of rare elements normalized to the primitive mantle (Hofmann 1988; Fig. 4.13), the topology of the spectra of the Talnakh rocks are notably close to the volcanic rocks of the main stage of trap magmatism. They are especially similar to Morongovsky and Mokulaevsky Formations (see Chap. 3). They manifest the same characteristics: the presence of negative Ta–Nb and Ti and positive Pb and Sr anomalies, the same level of trace element accumulation, as well as the general inclination of the spectrum.

There is a significantly stronger lead anomaly in the rocks of the Talnakh rocks due to the presence of large amounts of sulfides in them. Some differences were established only in the values of the anomalies of titanium and strontium, which may reflect some variation in the mineral composition of rocks (more titanomagnetite and plagioclase in some gabbro-dolerite varieties). In general, the spectra are very similar to the spectra of the continental crust.

4.4.1.4 Sulfide Ores

Powerful massive orebodies (up to 22 m) are confined to the deflection at the base of the Talnakh intrusion (Likhachev 1997). They have a lenticular cross-sectional shape, but in plan they are round or have irregular form (Fig. 4.8). Their composition varies from north to south, with chalcopyrite–pyrrhotite species are replaced by chalcopyrite and substantially chalcopyrite–pentlandite which are typical high tenor of platinum group metals (up to 100 ppm). In the lower part of intrusion, in taxitic and picritic gabbro-dolerites, disseminated ores dominate.

Their mineral composition is similar to that of Kharaelakh intrusion, which will be described below. Low-sulfide PGE mineralization is rare in the upper part of the intrusive body; it takes place occasionally in the upper taxitic horizon. The characteristic of this type of mineralization is done for Noril'sk 1 intrusion where it is widespread.

4.4.2 Kharaelakh Massif

The Kharaelakh intrusion was discovered by V.A. Lul'ko and other geologists of the Noril'sk Exploration Expedition during 1:50,000-scale mapping with drilling boreholes (KZ-584 and T-56) in 1965. Disseminated ores were found initially, and massive ores were located later (the thickness of the first horizon was 11 m). The orebody was determined to be very large (up to 54 m) in 1967 using several boreholes, and this deposit was named Oktyabr'skoe. Up to now, many geologists support that the Kharaelakh intrusive body is a lower branch of the Talnakh intrusion and they regarded them as one massif. Indeed these intrusions are very similar in terms of their geochemistry and mineral composition. But they have some differences in sulfide tenor and type of ores. Thus, we describe them separately.

The Oktyabr'skoe deposit is located in the southwestern centroclinal part of the Kharaelakh Trough. The main fold structure is the Pyasinsky anticline. There are many intrusive bodies present inside this area. According to internal structure and chemical composition, the intrusions are subdivided into non-differentiated and differentiated intrusions (Based Legend for geological map… 1993) and are related to certain intrusive complexes: Ergalakhsky ($\tau\beta P_2er$), Ogonersky (βT_1og), Daldykansky ($\nu\beta T_{1\text{-}2}dl$), and Noril'sk, which include the subcomplexes of Low Talnakh $(\omega\nu\text{-}\nu\beta)T_1nr^{nt}$, Kruglogorsky $(\pi\nu-\nu\beta)T_1nr^{kg}$, and Noril'sk $(\omega\nu\text{-}\nu\delta)T_1nr^{nr}$

The intrusions differ in internal structure and the scale of sulfide mineralization.

The Kharaelakh intrusion belongs to the Noril'sk Complex. It is a thin horizontal body with 7-km long and 2-km wide. It has a low-angle inclination to the NE. The intrusion is located in terrigenous carbonate rocks of the Middle Devonian deposits and is associated with the contact of the Kureysky and Razvedochninsky Formations. To the north, it is located in the Zubovsky Formation (D_1). Many breccia rocks were found on the flanks of the intrusive body. Barren rocks are present across an area of 0.8–1.0 km^2 in the Kharaelakh intrusive body.

The internal structure of the intrusion in the eastern part is relatively homogeneous, whereas, in the west, the intrusion is split into a series of small apophyses and is heterogeneous both in the section and in plain view. In the upper central part of the intrusion, exocontact copper ores form thick horizons. In its lower part, disseminated ores present mostly in taxitic gabbro-dolerite (in picritic gabbro-dolerite as well), and in parts of the exocontact intrusion, rich in massive ores with a halo of vein-disseminated copper ores are observed. A distinctive feature of the Kharaelakh intrusion is the presence of massive sulfide ores at the top of the intrusive body; these ore mainly are located in the surrounding rocks.

4.4.2.1 The Structure of the Kharaelakh Massif

Kharaelakh intrusion is described in detail in numerous books and papers, such as D. Dodin and B. Batuev (1971), V. Zolotukhin et al. (1975), V. Ryabov (1992), V. Ryabov et al. (2000), A. Likhachev (1996a, b, 2006), and other researchers. Therefore, only a brief description of the two core-based cross sections investigated in this study are presented in detail. They are located in different parts of the deposit. One, borehole TG-21, located in the southern part of the intrusion, has been a reference for many years to geologists of different organizations and is therefore well understood (Fig. 4.8). However, the study of olivines using modern methods was performed for her first time. The second borehole, KZ-456, is much farther north. Here, different horizons of the intrusion are penetrated by the borehole. A third core is located in the western part of the field and is characterized by only one apophysis, which is rich in sulfide ores and is described in Chap. 7. The section of borehole TG-21 is not typical of the intrusion (Fig. 4.14) because it has a different structure: the upper part, constituting almost half of the section (approximately 50 m) and is composed of gabbro-diorite and gabbro (Table 4.8). Its characteristic feature is the presence of the distinct so-called ferrogabbro horizon, i.e., gabbros with a high content of coarse grains of titanomagnetite (1–1.5 cm) and a fine-grained mass. The lower part of the intrusion is enriched in olivine; therefore, the upper and lower parts have contrasting chemical compositions.

These features are evident in a series of Harker diagrams built according to X-ray fluorescence analysis. The difference between these rocks is reflected in the almost bimodal distribution of magnesium; the Mg field separates the enriched rocks that form the lower part of the array (featuring a MgO concentration range of 15–25 wt %) and the low-MgO rocks (featuring a MgO concentration range of 2–10 wt %). The upper part of the section is characterized by its specific features. The term on its breed is distinguished by higher TiO_2 contents (due to the previously mentioned presence of a high content of titanomagnetite). The corresponding points of these rocks form a separate field oriented along the *x*-axis (MgO). These patterns are manifested in the behavior of Na, P_2O_5, and, to a lesser extent, iron.

Rocks from the top endocontact are also enriched in phosphorus and depleted in Cr. The lower part of the section, composed of olivine (picritic and taxitic gabbro-dolerites) is

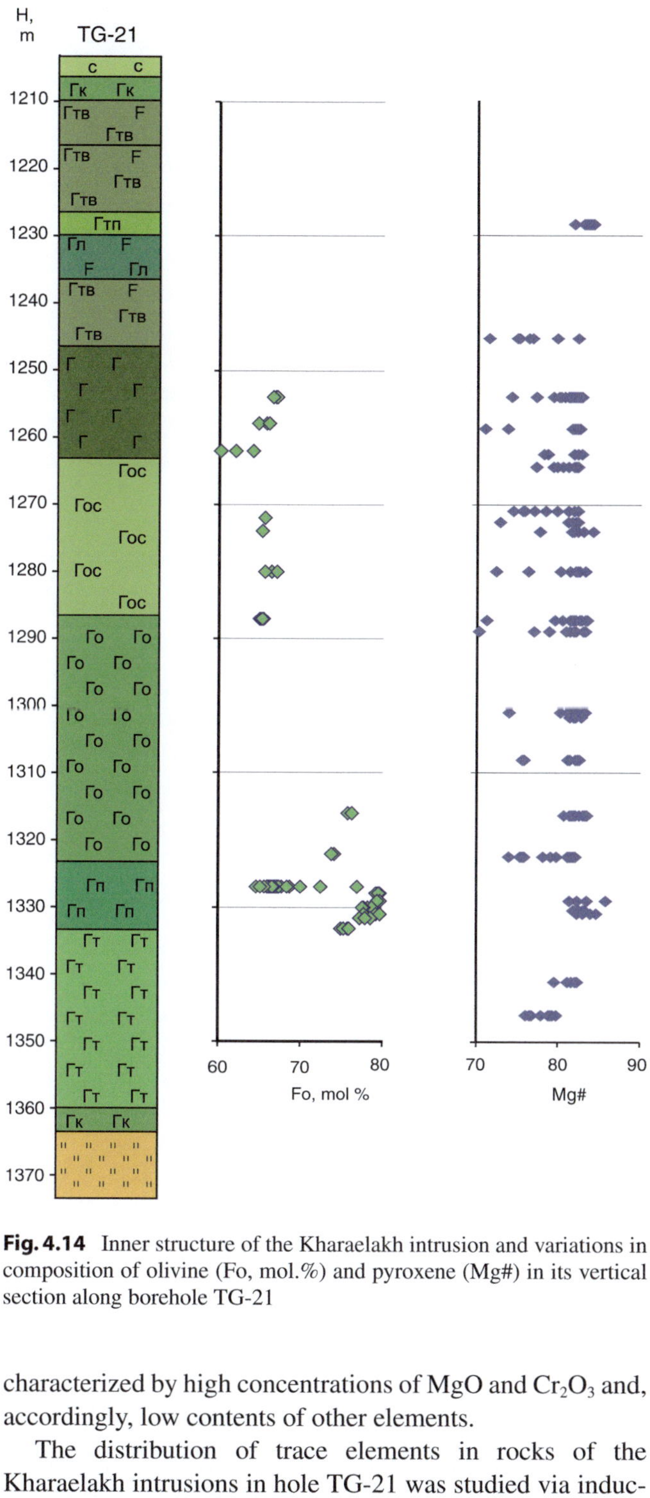

Fig. 4.14 Inner structure of the Kharaelakh intrusion and variations in composition of olivine (Fo, mol.%) and pyroxene (Mg#) in its vertical section along borehole TG-21

characterized by high concentrations of MgO and Cr_2O_3 and, accordingly, low contents of other elements.

The distribution of trace elements in rocks of the Kharaelakh intrusions in hole TG-21 was studied via inductively coupled plasma mass spectrometry (Table 4.8). The analysis results are shown in a spider diagram (Fig. 4.15). The typology of the spectra for all samples is almost identical, and they differ only in the content of the elements. The range of values for the content of heavy rare earths (and other elements) is determined by the amount of olivine in the rocks. The spectra corresponding to the lowest concentrations of trace elements are enriched in olivine (picritic and olivine gabbro-dolerite) because these rocks are depleted in clinopyroxene and plagioclase. The characteristic features of these spectra are that they have quite steep slopes from left to right (i.e., there is a significant enrichment in the light rare earth elements) and they have the presence of positive Th, U, and Pb anomalies and negative Ta and Nb anomalies. These characteristics demonstrate a significant role of crustal material in the initial melt.

The structure of the rocks in hole KZ-456. This borehole has a classical structure (Fig. 4.16). From the bottom to the top, we found the following textures/rock types: the contact, taxitic, picritic gabbro-dolerites, olivine-, olivine-bearing gabbro-dolerites, and olivine-free gabbro-dolerites. A distinctive feature of this section is the presence of a distinctive upper part (approximately 50 m), which is enriched in olivine and includes upper taxitic and upper picritic gabbro-dolerites.

A comparison of the studied sections shows that the lower, productive part of the section is thicker in hole TG-21. The total thickness of the taxitic and picritic gabbro-dolerites and partly mineralized olivine gabbro reaches almost 50 m, whereas it is less than 30 m in the borehole KZ-456. Moreover, these differences are due to differences in the thickness of the taxitic gabbro-dolerites (in contrast, the picritic gabbro-dolerite are approximately the same thickness, approximately 10 m). However, at the top of the borehole KZ-456, the taxitic gabbro-dolerite dominates, whereas this part of the section is missing in borehole TG-21.

Olivine. The variations in the composition of olivine from the rocks of the lower part of the Kharaelakh Massif in core TG-21 are shown in Figs. 4.17, 4.18, and 4.19 and are listed in Tables 4.9 and 4.10. The magnesium component in olivine in this section varies very significantly from 57 (in the olivine gabbro-dolerites) to 80 mol% Fo (in picritic gabbro-dolerites), with maximum values occurring in individual samples of the picritic gabbro-dolerites ($Fo_{74–80}$). In Fig. 4.17, the changes in the concentration of nickel in olivine exhibit a direct correlation with the magnesium content, whereas titanium is inversely proportional to the forsterite component in olivine.

The same trend is visible in the diagram TiO_2–Fo (Fig. 4.18), especially for the horizon of picritic gabbro-dolerite. Despite the considerable spread in values (0.01–0.04 wt %), Fe–olivine (Fo_{75}) on average has higher titanium content than the magnesium-rich olivine ($Fo_{78–80}$). The concentrations of nickel in olivine from olivine-bearing rocks do not exceed 0.1 wt % NiO, whereas the olivine from picritic gabbro-dolerite is enriched in NiO up to 25 wt %. In this case, there is a direct correlation between NiO and the iron content, especially in olivine ranging from 74 to 76 mol% Fo. It is likely due to the presence of sulfides in the rocks, which redistribute nickel and iron between the sulfides and silicate melts. Other components, such as Mn and Ca, are inversely proportional to the concentration of magnesium-rich olivine.

Table 4.8 Composition of the rocks from the Kharaelakh intrusion

N sample/component	1,210	1,218	1,240	1,272.8	1,287	1,311.8	1,316.4	1,333.6
SiO_2	52.12	54.55	52.92	47.81	51.17	45.58	51.04	39.9
TiO_2	1.25	1.02	1.2	0.89	1.01	0.57	0.96	0.49
Al_2O_3	18.45	23	22.08	16.66	18.79	17.66	18.56	7.56
FeO	11.83	8.45	6.12	9.52	8.53	10.53	10.21	18.17
MnO	0.34	0.23	0.14	0.17	0.16	0.17	0.17	0.23
MgO	5.08	4.07	3.67	8.28	6.65	7.67	8.22	22.72
CaO	0.7	7.03	14.37	13.1	12.48	9.65	9.81	5.16
Na_2O	3.04	1.26	1.62	1.76	1.12	1.32	1.54	0.34
K_2O	1.27	2.54	0.33	0.37	0.29	0.56	0.46	0.15
P_2O_5	0.2	0.05	0.05	0.12	0.07	0.1	0.12	0.12
Cr_2O_3	0.07	0.02	0.02	0.09	0.07	0.02	0.03	0.42
LOW	4.60	4.22	4.56	0.56	1.04	3	1.00	3.83
Total	99.0	99.4	100.0	99.3	99.1	99.3	99.2	99.1
Ti	6,540	2,660	2,825	4,755	4,785	5,885	4,830	2,515
V	196	90	104	189	191	216	171	127
Cr	529	180	171	704	529	188	221	2,700
Mn	2,110	1,460	890	1,100	1,060	1,150	1,080	1,520
Co	41	41	27	51	45	53	55	186
Ni	94	539	229	254	183	261	265	4070
Cu	1,300	893	708	2120	334	273	363	5060
Zn	669	281	470	1,019	376	299	327	530
Ga	24.4	20.1	18.7	17.6	19.6	17.4	18.0	9.0
Rb	61.8	86.6	16.7	9.3	12.7	17.6	16.3	5.3
Sr	108	317	127	252	325	255	252	88
Y	30.9	13.2	12.1	19.9	19.3	20.5	20.1	13.5
Zr	196.1	52.4	40.6	62.6	62.0	72.0	71.5	41.5
Nb	14.08	3.46	2.91	4.25	4.21	4.61	4.71	2.75
Mo	1.38	1.42	1.14	1.24	1.62	1.95	1.26	1.79
Cs	1.94	0.96	0.29	0.67	0.60	0.63	0.85	0.68
Ba	44	456	72	89	152	142	130	48
La	17.0	6.1	4.6	5.4	5.8	7.0	7.8	3.6
Ce	32.9	13.0	10.9	12.9	13.6	16.6	18.0	8.5
Pr	3.90	1.57	1.43	1.72	1.79	2.15	2.28	1.12
Nd	15.99	6.55	6.32	8.27	8.19	9.83	9.95	5.27
Sm	3.59	1.50	1.52	2.27	2.23	2.56	2.48	1.42
Eu	1.27	0.69	0.76	0.81	0.81	0.88	0.87	0.44
Gd	3.86	1.65	1.64	2.61	2.55	2.98	2.74	1.70
Tb	0.66	0.28	0.28	0.46	0.45	0.50	0.47	0.30
Dy	4.54	1.83	1.86	3.09	2.99	3.43	3.04	2.01
Ho	1.06	0.42	0.41	0.70	0.68	0.80	0.71	0.46
Er	2.93	1.16	1.11	1.80	1.79	2.10	1.88	1.26
Tm	0.47	0.17	0.16	0.27	0.26	0.31	0.28	0.19
Yb	3.03	1.14	1.03	1.72	1.68	2.02	1.76	1.21
Lu	0.48	0.17	0.16	0.25	0.25	0.30	0.27	0.19
Hf	5.03	1.30	1.01	1.67	1.59	1.96	1.85	1.10
Ta	0.81	0.67	1.46	0.53	0.64	0.63	0.46	0.15
W	0.78	0.86	0.78	0.47	0.75	0.73	0.49	0.35
Pb	36.5	35.3	45.3	31.7	31.4	33.1	32.0	33.4
Bi	0.05	0.15	0.16	0.05	0.13	0.10	0.10	0.32
Th	7.36	1.21	0.71	0.85	1.00	1.19	1.28	0.64
U	2.24	0.50	0.32	0.33	0.40	0.48	0.48	0.24

Note: Oxides are given in %, elements in ppm. No sample is the depth in the borehole TG-21 (m)

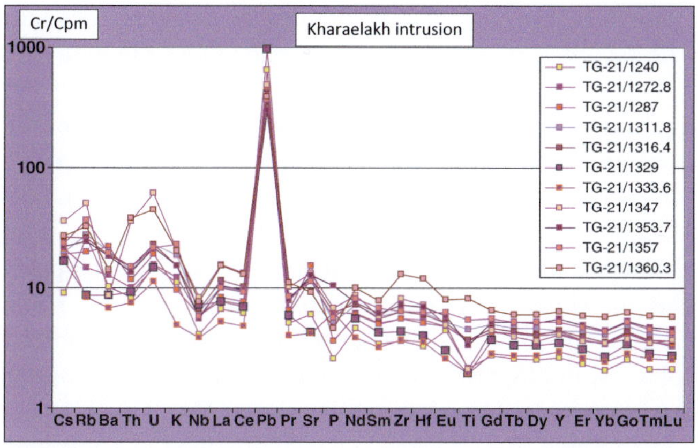

Fig. 4.15 Trace element patterns for intrusive rocks of the Kharaelakh intrusion (borehole TG-21)

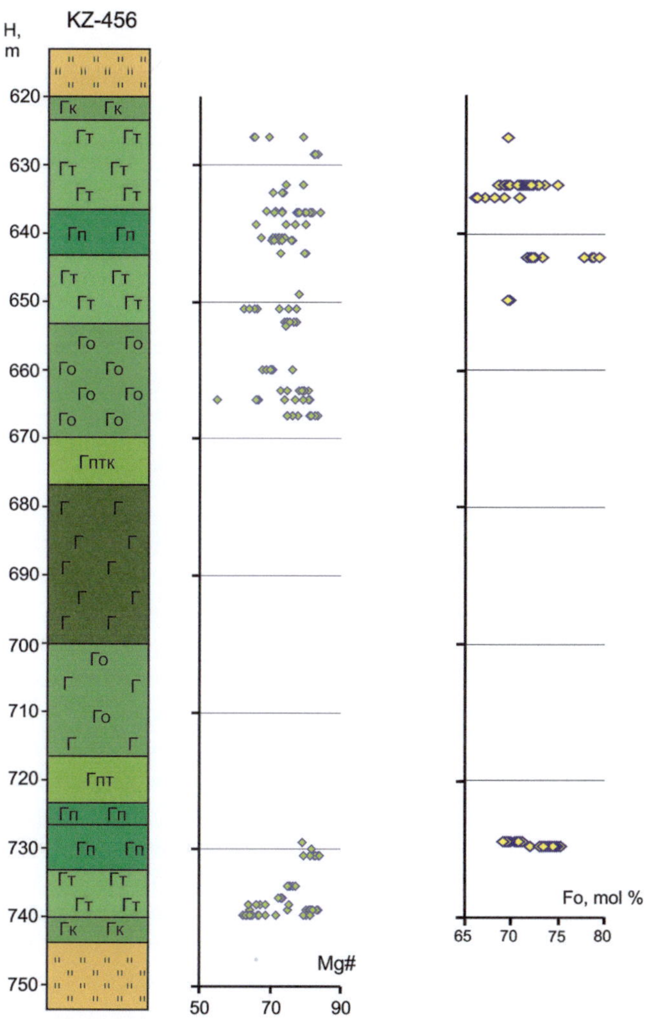

Fig. 4.16 Inner structure of the Kharaelakh intrusion and variations in composition of olivine (Fo, mol.%) and pyroxene (Mg#) in its vertical section along borehole KZ-456

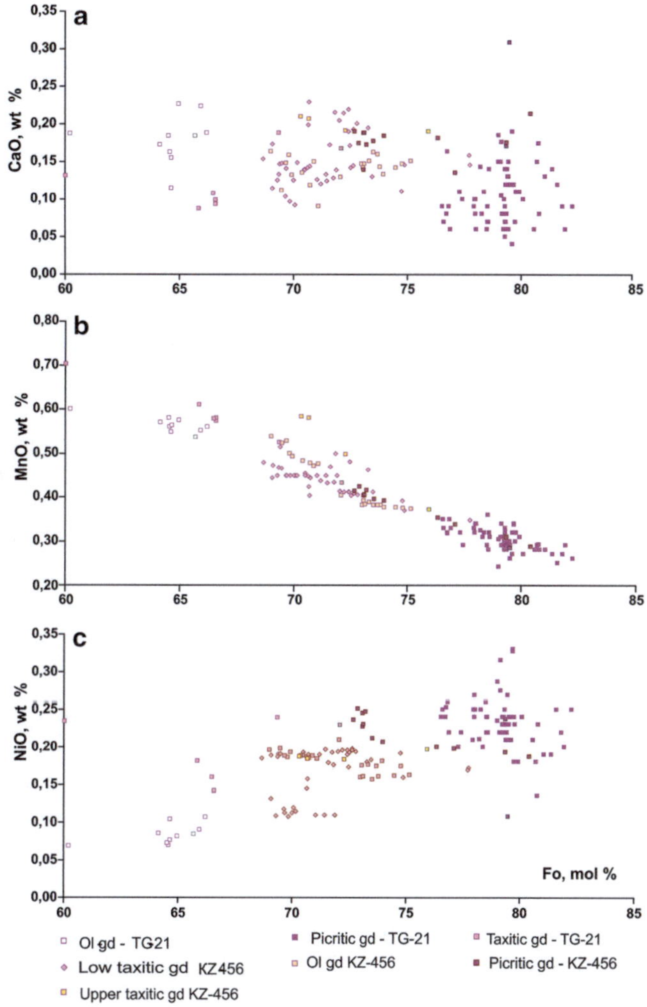

Fig. 4.17 CaO, MnO, NiO—*Fo* diagram for olivines from the Kharaelakh Massif

The composition of olivine was studied in more detail in the rocks from borehole KZ-456. Here, we have a rare opportunity to compare the composition of olivines from the upper and lower picritic and taxitic horizons (Fig. 4.17). First, the NiO contents are higher in the lower picrites (0.26 wt %) than in the upper ones (0.18 wt %). The NiO concentrations in olivine from the lower picritic gabbro-dolerites correlate with the iron content, as was noted for olivines from the same rocks in borehole TG-21. In the upper horizon of picrites, there is no such relationship due to a lack of essential sulfide minerals. However, the concentration of cobalt (for the same Mg–olivine, approximately 72 mol% Fo) is significantly higher in the upper picritic gabbro-dolerites (0.042 wt %) than in the lower picritic gabbro-dolerites (0.03 wt %). Olivines from the upper and lower taxitic gabbro-dolerites significantly differ in the content of MgO; olivine grains from the lower horizon are greatly enriched in Mg.

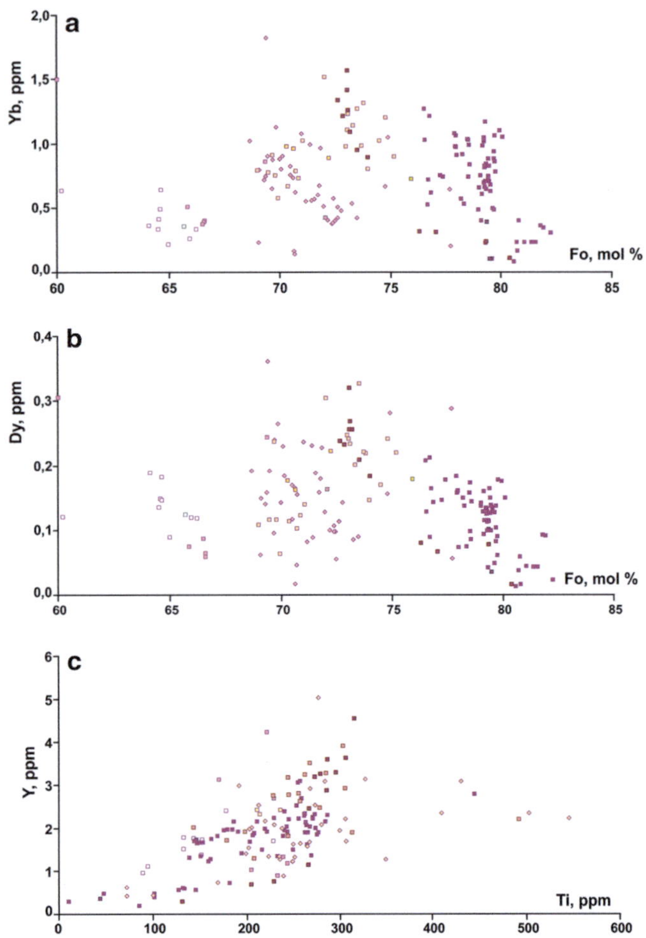

Fig. 4.18 Y, Yb—Fo and Y–Ti diagrams for olivine from the Kharaelakh Massif
Conventions are given in Fig. 4.17

The study of the distribution of trace elements (Y, Yb, Dy, Li, Na, Ti, V, and Cr) in olivines from rocks from the Kharaelakh Massif (boreholes KZ-456 and TG-21) was performed via secondary ion mass spectrometry (Table 4.10). In Fig. 4.18a–c, the contents of Y and HREE are shown to be well correlated with each other and exhibit similar patterns in the diagram. These patterns are distinct from the fields of picritic, taxitic, and olivine gabbro-dolerites. The points corresponding to olivines from picritic gabbro-dolerite form a trend line with a steep slope relative to the *x*-axis (Fo). The olivines from the olivine gabbro-dolerites form a shallower trend. The points of both the upper and lower taxitic gabbro-dolerites form a disorderly arrangement. The titanium and vanadium data from the picritic gabbro-dolerite form the most chaotic arrangement (Fig. 4.19). Titanium and, to a lesser extent, vanadium form on the diagrams two trend (not very clear manifestation): in high-Mg–olivine–Ti contents correlate with fayalite component in olivine and in low-Mg with forsterite. Perhaps, we see the appearance on Cr–Mt on the liquidus in the equilibrium with Fo_{75}. The distribution of chromium controlled by any patterns. Olivines from all varieties of rocks have high Cr contents, ranging from 200 to 1,600 ppm.

The sodium concentrations are not greater than 100 ppm. The maximum Li concentrations in olivines (1.77–13.35 ppm) are found in the upper taxitic and picritic gabbro-dolerites. In olivines from the gabbro-dolerites, these values reach only 8–9 ppm.

Pyroxenes. The changes in the composition of pyroxenes along the studied sections of the massif (Table 4.11) are shown in Figs. 4.8 and 4.15. The highest Mg# values (Mg#=Mg/Mg+Fe^{2+}) are typically observed in the pyroxenes from the picritic gabbro-dolerite. These pyroxenes are also enriched in NiO. The concentrations of Ti, in contrast, have an inverse correlation with the Mg#. Many elements demonstrate the distinct trends associated with the Mg# values in the picritic gabbro-dolerites (direct positive correlations with CaO, Cr_2O_3, and Al_2O_3 and inverse correlations with TiO_2 and SiO_2). In other horizons, the impurity elements behave erratically.

4.4.2.2 Ores of the Oktyabr'skoe Deposit Related to the Kharaelakh Intrusion

There are three main features of the Noril'sk deposits that differ them from other magmatic deposits in the world: (1) combination of low-sulfide PGE mineralization with Cu–Ni sulfide mineralization enriched in PGE, (2) high thickness of massive sulfide ores, and (3) unique mineralogical diversity of the ores compared to ores at other magmatic Pt and Cu–Ni deposits worldwide (including ore-forming minerals and rare minerals, especially PGM (and their huge size)).

The highest thickness of sulfide bodies is typical of the Oktyabr'skoe deposit (up to 54 m) and Talnakh (25 m). During the exploration of these deposits, many mines passed completely within sulfides (Fig. 4.20). The main body of the Oktyabr'skoe deposit has distinct zonation (Stekhin 1994; Torgashin 1994). Several orebodies consist of a new mineral, talnakhite, discovered here (Budko and Kulagov 1968) and other minerals of chalcopyrite group (putoronite, Ni–putoronite (Filimonova et al. 1974; 1980), and moikhukite (Murav'eva et al. 1972)) (Fig. 4.21). Thirty new minerals were discovered in ores (Genkin and Zvyagintsev 1962; Genkin et al. 1966, 1969; 1970; 1974; 1976; Evstigneeva et al. 1973, 1975; Filimonova et al. 1974, 1980; Kovalenker et al. 1974, 1976; Kulagov et al. 1969; Begizov et al. 1974), and 418 minerals were found in the Noril'sk deposits, with 174 mineral spices and 91 unknown tiny phases (May, 2011, unpublished data of E. Sereda).

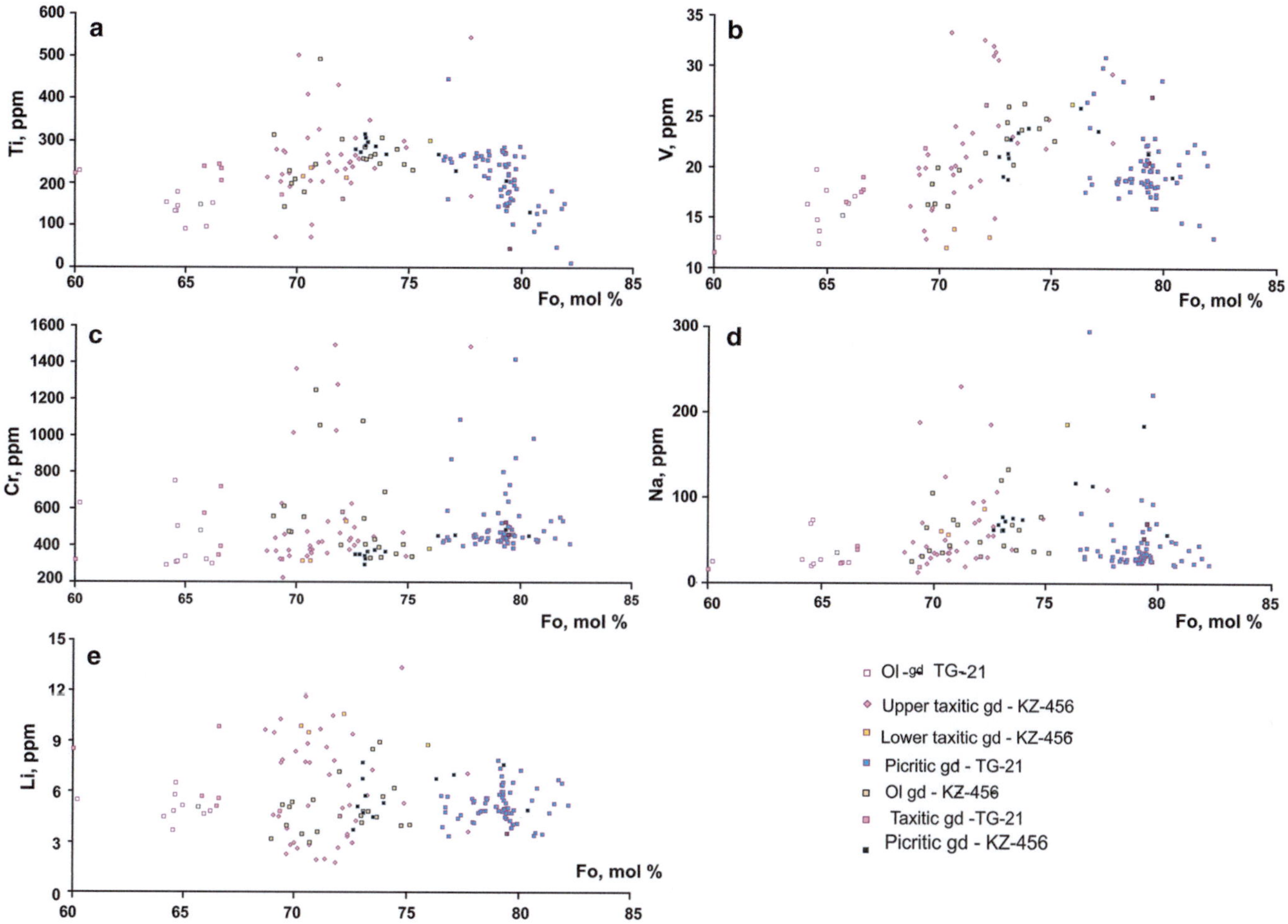

Fig. 4.19 Ti, V, Cr, Na, Li—*Fo* diagrams for olivine from the Kharaelakh Massif

The deposits of the Talnakh group have a highly diverse PGE mineralogical composition and host large grains and aggregates of PGM. The latter is particularly typical of sperrylite (Fig. 4.21k–n), whose crystals may be as large as 4–5 cm (along the cube edge). The veined ores, first of all, in their upper outer contact zone, often contain rims up to 4–5 mm thick consisting of large aggregates of Pd minerals. But mostly rare minerals have small size (Figs. 4.22 and 4.23).

The mineralogy and geochemistry of the Talnakh and Oktyabr'skoe deposits was examined in much detail (Genkin et al. 1981; Distler 1994a, b; Distler et al. 1988, Stekhin 1994, Naldrett et al. 1992, 1996; Sluzhenikin and Distler 1998; Likhachev 1994, 2006; Spiridonov and Gritsenko 2009; Malitch et al. 2010; Malich and Latypov 2011; Sluzhenikin and Mokhov 2015), but studies of new ore types of these deposits and the examination of previously poorly studied levels result in a continuous widening of the spectrum of interesting finds.

4.4.3 Low Talnakh Massif

The Low Talnakh intrusion has the form of an irregular body and a very complex morphology. This intrusion lies directly beneath the Talnakh and Kharaelakh intrusions. Its structural features have been described previously (Natorkhin et al. 1977; Turovtsev 2002; et ctr.). Usually, this intrusion is referred to as a barren intrusion in the literature, although sulfide-bearing rocks were encountered in a number of boreholes, and the content of nonferrous metals approached the conditional contents. The concentrations of noble metals are very low. Geologists of NorilskGeology, Ltd. regard the Low Talnakh intrusion as a specific subcomplex or type (Low Talnakh) inside the Noril'sk Complex. Nevertheless, it has many differences compared to the intrusions of the Noril'sk Complex, which will be shown later. This type of intrusion is characterized by higher average magnesium contents (13–15 wt % MgO) than the typical intrusions of the Noril'sk

Table 4.9 Olivine compositions from the Kharaelakh Massif, wt % (borehole TG-21)

№	N sample	Fo, mol.%	SiO_2	Al_2O_3	FeO	MnO	MgO	CaO	NiO	Total
1.	1,254	66.84	36.49	0.01	29.46	0.51	33.30	0.21	0.08	100.07
2.	1,254	67.10	36.66	0.01	29.27	0.50	33.48	0.24	0.08	100.26
3.	1,254	66.57	36.37	0.01	29.70	0.52	33.17	0.21	0.07	100.07
4.	1,258	66.07	37.17	0.00	30.19	0.52	32.98	0.19	0.08	101.15
5.	1,258	65.71	37.19	0.01	30.53	0.52	32.81	0.20	0.07	101.36
6.	1,258	64.81	36.90	0.01	31.27	0.53	32.30	0.20	0.07	101.30
7.	1,258	66.11	36.92	0.01	30.18	0.52	33.02	0.18	0.08	100.93
8.	1,262.5	64.11	37.39	0.01	31.62	0.52	31.68	0.21	0.08	101.53
9.	1,262.5	60.21	36.14	0.01	34.68	0.57	29.44	0.19	0.06	101.11
10.	1,262.5	62.01	36.48	0.01	33.33	0.55	30.52	0.20	0.07	101.18
11.	1,272.8	65.57	36.89	0.02	30.46	0.52	32.54	0.24	0.08	100.77
12.	1,274	65.28	37.42	0.02	30.91	0.53	32.59	0.20	0.07	101.75
13.	1,274	65.28	37.62	0.02	30.90	0.53	32.59	0.19	0.07	101.93
14.	1,280	66.41	37.63	0.01	29.96	0.51	33.23	0.21	0.08	101.65
15.	1,280	67.11	37.61	0.01	29.28	0.49	33.51	0.18	0.09	101.19
16.	1,280	65.62	37.59	0.02	30.56	0.52	32.71	0.18	0.07	101.67
17.	1,287.2	64.94	36.71	0.02	30.89	0.52	32.09	0.19	0.07	100.52
18.	1,287.2	65.30	37.07	0.02	30.84	0.52	32.55	0.23	0.08	101.32
19.	1,287.2	65.38	37.04	0.002	30.83	0.51	32.66	0.24	0.08	101.38
20.	1,287.2	65.14	37.04	0.01	31.04	0.52	32.53	0.24	0.08	101.48
21.	1,287.2	65.22	36.94	0.01	30.94	0.52	32.54	0.21	0.06	101.24
22.	1,316.4	75.79	38.39	0.02	22.51	0.35	39.53	0.16	0.14	101.12
23.	1,316.4	75.76	38.79	0.01	22.54	0.35	39.51	0.16	0.14	101.52
24.	1,316.4	75.80	38.82	0.01	22.48	0.35	39.50	0.16	0.14	101.48
25.	1,316.4	76.30	39.12	0.01	22.09	0.34	39.88	0.14	0.14	101.74
26.	1,316.4	76.26	38.71	0.01	22.04	0.35	39.70	0.14	0.14	101.10
27.	1,322.4	74.12	38.21	0.01	23.77	0.37	38.18	0.12	0.13	100.81
28.	1,322.4	73.80	38.12	0.01	24.04	0.37	37.99	0.13	0.13	100.80
29.	1,327	64.67	36.88	0.01	31.10	0.51	31.93	0.10	0.11	100.65
30.	1,327	66.27	37.12	0.01	30.06	0.51	33.13	0.14	0.11	101.10
31.	1,327	66.86	37.26	0.004	29.49	0.50	33.37	0.07	0.11	100.82
32.	1,327	66.37	37.08	0.008	29.98	0.51	33.18	0.13	0.10	101.00
33.	1,327	66.93	37.35	0.01	29.55	0.50	33.55	0.16	0.11	101.24
34.	1,327	66.80	37.20	0.01	29.62	0.50	33.42	0.13	0.10	101.00
35.	1,327	67.24	37.12	0.006	29.32	0.50	33.76	0.11	0.11	100.94
36.	1,327	67.75	37.51	0.005	28.96	0.49	34.13	0.10	0.11	101.32
37.	1,327	76.96	38.54	0.01	21.28	0.34	39.87	0.10	0.19	100.34
38.	1,327	66.81	37.08	0.01	29.66	0.51	33.49	0.15	0.11	101.03
39.	1,327	66.87	37.13	0.01	29.55	0.50	33.46	0.15	0.11	100.93
40.	1,327	68.66	36.79	0.00	26.77	0.42	32.89	0.09	0.10	97.08

Note: Here and in Tables 4.10, 4.11, 4.14, 4.18, 4.19, 4.27, 4.28, 4.29, 4.30, 4.31, 4.32, 4.33, and 4.34 analyses were carried out in Max Planck Institute of Chemistry, Mainz, Germany, analysts D. Kuzmin, N. Krivolutskaya; *N* sample, depth (m) in borehole TG-21

Table 4.10 Olivine composition from rocks of the Kharaelakh intrusion, wt %

N	N sample	Fo	SiO_2	CaO	FeO	MgO	NiO	MnO	Total
		mol %				wt %			
1	TG-1254.3	65.69	36.81	0.18	30.09	32.34	0.08	0.54	100.04
2	TG-1258.7	64.57	36.78	0.16	30.95	31.65	0.07	0.56	100.17
3	TG-1262.5	64.63	36.89	0.12	30.96	31.76	0.10	0.55	100.38
4	TG-1272	64.65	36.76	0.16	30.81	31.62	0.08	0.56	99.99
5	TG-1301	64.98	36.73	0.23	30.60	31.86	0.08	0.57	100.07
6	TG-1301a	65.93	36.82	0.22	29.98	32.57	0.09	0.55	100.23
7	TG-1306	64.13	36.28	0.17	31.35	31.46	0.09	0.57	99.92
8	TG-1311	60.23	36.51	0.19	34.04	28.93	0.07	0.60	100.34
9	TG-1339	72.07	37.70	0.17	25.19	36.48	0.23	0.43	100.20
10	TG-1354	65.86	36.98	0.09	30.19	32.69	0.18	0.61	100.74
11	TG-1354	66.59	37.31	0.09	29.61	33.13	0.14	0.57	100.85
12	TG-1354	66.59	37.28	0.10	29.66	33.19	0.14	0.58	100.95
13	TG-1354	66.50	36.99	0.11	29.46	32.83	0.16	0.58	100.13
14	TG-1361	69.35	37.39	0.19	27.58	35.03	0.24	0.52	100.95
15	kz-633-1	68.66	37.97	0.15	28.20	34.66	0.19	0.48	101.65
16	kz-633-5	73.46	38.60	0.13	24.34	37.79	0.18	0.41	101.45
17	kz-633-10	74.76	38.37	0.11	23.16	38.49	0.19	0.39	100.71
18	kz-633-11	70.48	38.16	0.14	26.82	35.92	0.19	0.45	101.68
19	kz-633-14	69.35	37.67	0.13	27.51	34.90	0.19	0.47	100.87
20	kz-633-16	69.08	37.69	0.12	27.65	34.64	0.19	0.47	100.76
21	kz-633-19	72.61	38.11	0.15	24.93	37.06	0.20	0.41	100.86
22	kz-633-21	71.16	37.94	0.12	26.09	36.11	0.19	0.44	100.89
23	kz-633-24	69.46	37.66	0.15	27.37	34.92	0.19	0.47	100.76
24	kz-729.66	75.17	39.20	0.15	23.27	39.51	0.16	0.37	102.66
25	kz-729.68	74.79	39.25	0.15	23.55	39.19	0.17	0.38	102.69
26	kz-729	74.49	38.34	0.14	23.36	38.27	0.16	0.38	100.65
27	kz-729.61	73.81	38.03	0.14	23.94	37.84	0.16	0.38	100.49
28	kz-729.66	73.51	37.20	0.16	23.68	36.85	0.16	0.38	98.43
29	kz-729.67	72.99	36.52	0.15	23.95	36.30	0.16	0.38	97.46
30	kz-731c-6	73.52	38.43	0.18	24.07	37.49	0.21	0.40	100.78
31	kz-731c-7	72.86	38.51	0.18	24.75	37.26	0.25	0.42	101.37
32	kz-731a-4	80.41	39.44	0.21	18.41	42.38	0.19	0.29	100.92
33	kz-731a5	76.34	38.08	0.18	21.61	39.10	0.20	0.35	99.52
34	kz-731a-6	77.10	37.69	0.14	21.03	39.73	0.20	0.34	99.13
35	kz-731a-8	79.36	39.22	0.18	19.36	41.74	0.19	0.31	101.00
36	kz-731a-9	75.95	39.65	0.19	22.94	40.64	0.20	0.37	103.99
37	kz-738-2	70.64	38.10	0.21	26.34	35.54	0.18	0.58	100.95
38	kz-738-5	70.30	38.10	0.21	26.69	35.44	0.18	0.58	101.21
39	kz-738-6	72.23	38.19	0.19	24.31	35.47	0.18	0.50	98.84

(continued)

Table 4.10 (continued)

N	N sample	Cr	Ti	V	Y	Yb	Dy	Na	Li
		ppm							
1	TG-1254.3	478	148	15	1.47	0.35	0.12	35	5.05
2	TG-1258.7	305	132	15	1.77	0.41	0.15	19	4.81
3	TG-1262.5	501	143	12	1.76	0.49	0.15	73	5.74
4	TG-1272	310	178	14	2.39	0.64	0.18	22	6.46
5	TG-1301	339	89	18	0.96	0.21	0.09	27	5.13
6	TG-1301a	322	95	16	1.11	0.26	0.12	23	4.62
7	TG-1306	289	152	16	1.72	0.36	0.19	27	4.47
8	TG-1311	626	228	13	1.69	0.64	0.12	24	5.45
9	TG-1339	579	161	26	1.86	0.42	0.16	31	4.48
10	TG-1354	572	239	16	1.32	0.51	0.07	23	5.66
11	TG-1354	718	205	19	1.03	0.40	0.06	43	5.55
12	TG-1354	394	233	18	0.89	0.38	0.06	38	9.85
13	TG-1354	344	243	17	1.18	0.37	0.09	987	5.08
14	TG-1361	323	170	22	3.12	0.86	0.24	19	4.79
15	kz-633-1	367	212	16	2.53	1.02	0.19	36	9.65
16	kz-633-5	449	235	22	1.27	0.42	0.09	39	7.25
17	kz-633-10	469	299	22	1.95	0.66	0.16	408	13.35
18	kz-633-11	343	305	20	2.21	0.82	0.12	124	11.65
19	kz-633-14	628	219	20	2.16	0.74	0.16	188	10.30
20	kz-633-16	371	280	19	2.08	0.80	0.15	545	9.50
21	kz-633-19	497	306	24	1.69	0.50	0.11	56	9.41
22	kz-633-21	421	202	21	1.54	0.55	0.08	230	9.72
23	kz-633-24	458	273	21	2.53	0.90	0.19	70	7.85
24	kz-729.6-6	336	229	23	2.67	0.90	0.22	35	4.00
25	kz-729.6-8	402	244	25	3.16	1.20	0.24	77	3.94
26	kz-729.6-10	355	278	24	2.46	1.02	0.17	37	6.18
27	kz-729.6-12	335	305	26	2.92	1.31	0.22	62	8.92
28	kz-729.6-16	431	268	46	3.50	1.26	0.33	68	8.47
29	kz-729.6-17	1078	257	23	2.62	0.98	0.25	121	4.55
30	kz-731c-26	373	286	23	2.86	0.95	0.21	75	4.47
31	kz-731c-27	350	272	19	3.18	1.21	0.23	68	5.09
32	kz-731a-14	451	130	19	0.28	0.11	0.02	57	4.88
33	kz-731a-15	453	266	26	1.14	0.31	0.08	117	6.72
34	kz-731a-16	453	229	24	0.76	0.31	0.07	113	6.99
35	kz-731a-18	488	205	21	0.69	0.24	0.08	184	7.55
36	kz-731a-19	380	299	26	1.98	0.72	0.18	186	8.76
37	kz-738-2	314	236	14	2.41	0.96	0.16	57	9.49
38	kz-738-5	315	214	12	2.32	0.97	0.18	61	9.89
39	kz-738-6	529	210	13	2.42	0.89	0.22	87	10.61

Note: N 1–8 Ol gabbro-dolerites, 9–15 – upper taxitic gabbro-dolerites, 16–25 – upper picrites, 26–31 – Ol gabbro-dolerites, 32–37 – low picrites, 38–41 – low taxitic gabbro-dolerites

Table 4.11 Representative analyses of clinopyroxenes from the Kharaelakh intrusion, wt % (borehole TG-21)

№	№ sample	MgO#	SiO_2	TiO_2	Al_2O_3	FeO	MnO	MgO	CaO	Na_2O	Cr_2O_3	Total
1.	1228.3	84.04	52.33	0.47	2.88	5.73	0.14	16.92	20.57	0.25	0.92	100.22
2.	1228.3	84.29	52.38	0.47	2.75	5.65	0.14	17.00	20.53	0.27	0.88	100.09
3.	1245.3	82.39	51.47	0.44	3.46	6.32	0.16	16.59	20.40	0.25	0.87	99.99
4.	1245.3	71.48	52.61	0.18	0.94	9.89	0.29	13.90	22.15	0.20	0.00	100.18
5.	1254	82.79	52.72	0.37	2.37	6.31	0.15	17.03	20.40	0.24	0.47	100.08
6.	1254	82.47	52.66	0.47	3.10	6.23	0.14	16.44	21.09	0.27	0.54	100.95
7.	1254	74.12	53.56	0.45	0.94	9.37	0.25	15.05	20.96	0.30	0.00	100.92
8.	1258.7	81.79	51.89	0.44	2.94	6.53	0.15	16.45	20.74	0.28	0.55	99.99
9.	1258.7	82.12	51.82	0.47	3.06	6.47	0.16	16.67	20.61	0.26	0.58	100.11
10.	1262.5	82.26	51.70	0.42	2.96	6.33	0.15	16.46	20.97	0.25	0.58	99.84
11.	1262.5	81.77	50.87	0.53	3.77	6.48	0.15	16.30	20.60	0.25	0.87	99.84
12.	1264.5	81.94	50.31	0.46	3.37	6.29	0.14	16.01	20.48	0.25	0.77	98.11
13.	1264.5	68.38	52.11	0.76	1.96	12.4	0.34	15.08	18.03	0.29	0.00	101.04
14.	1271	78.40	51.66	0.56	2.34	7.97	0.19	16.23	19.91	0.27	0.17	99.32
15.	1272.8	81.54	51.39	0.51	3.48	6.59	0.14	16.33	20.45	0.23	0.68	99.83
16.	1272.8	82.39	52.31	0.41	2.49	6.45	0.16	16.93	20.44	0.22	0.43	99.85
17.	1274	77.68	51.71	0.57	2.38	8.33	0.20	16.26	19.95	0.27	0.04	99.73
18.	1274	84.10	51.65	0.40	2.87	5.57	0.11	16.52	21.53	0.26	0.70	99.65
19.	1274	81.97	50.33	0.46	3.31	6.35	0.14	16.19	20.79	0.27	0.78	98.65
20.	1280	82.12	51.23	0.47	3.34	6.36	0.16	16.38	20.65	0.24	0.73	99.56
21.	1280	82.07	51.92	0.38	2.47	6.58	0.16	16.89	20.35	0.26	0.48	99.51
22.	1287.2	82.86	50.99	0.45	3.07	6.11	0.14	16.57	20.75	0.22	0.79	99.10
23.	1287.2	82.06	52.39	0.46	3.11	6.39	0.14	16.39	20.64	0.29	0.65	100.48
24.	1289	81.88	51.55	0.48	3.63	6.46	0.15	16.37	20.45	0.28	0.76	100.15
25.	1289	82.16	51.52	0.48	3.22	6.30	0.15	16.27	20.78	0.28	0.72	99.74
26.	1301	81.76	51.78	0.44	2.84	6.53	0.15	16.42	20.76	0.25	0.47	99.67
27.	1301.7	82.86	52.01	0.38	2.62	6.22	0.14	16.86	20.77	0.22	0.60	99.84
28.	1301.7	81.42	52.02	0.57	3.53	6.76	0.16	16.61	20.27	0.27	0.64	100.82
29.	1308	81.22	51.15	0.45	2.60	6.77	0.16	16.42	20.59	0.25	0.26	98.66
30.	1308	81.49	50.17	0.55	3.87	6.48	0.15	16.00	20.58	0.26	0.86	98.95
31.	1322.4	81.55	52.18	0.51	3.30	6.58	0.16	16.31	20.39	0.26	0.81	100.52
32.	1322.4	81.84	52.44	0.44	2.91	6.52	0.15	16.48	20.51	0.24	0.69	100.40
33.	1329	82.31	51.57	1.16	2.24	6.42	0.17	16.75	20.17	0.37	0.58	99.44
34.	1329	85.87	49.28	0.44	3.02	6.62	0.11	22.56	15.22	0.23	0.93	98.42
35.	1330.5	83.10	52.18	0.60	3.28	5.96	0.14	16.44	20.43	0.29	1.03	100.37
36.	1330.5	83.29	52.33	0.56	2.65	6.03	0.15	16.86	20.25	0.29	0.81	99.94
37.	1331	83.94	51.61	0.54	3.35	5.63	0.13	16.50	20.63	0.27	1.08	99.76
38.	1341	82.05	51.93	0.56	2.99	6.46	0.16	16.56	20.21	0.28	0.79	99.94
39.	1346	76.57	52.11	0.57	2.34	8.86	0.22	16.24	19.44	0.24	0.11	100.16

Note: N sample is the depth in borehole TG-21 (m)

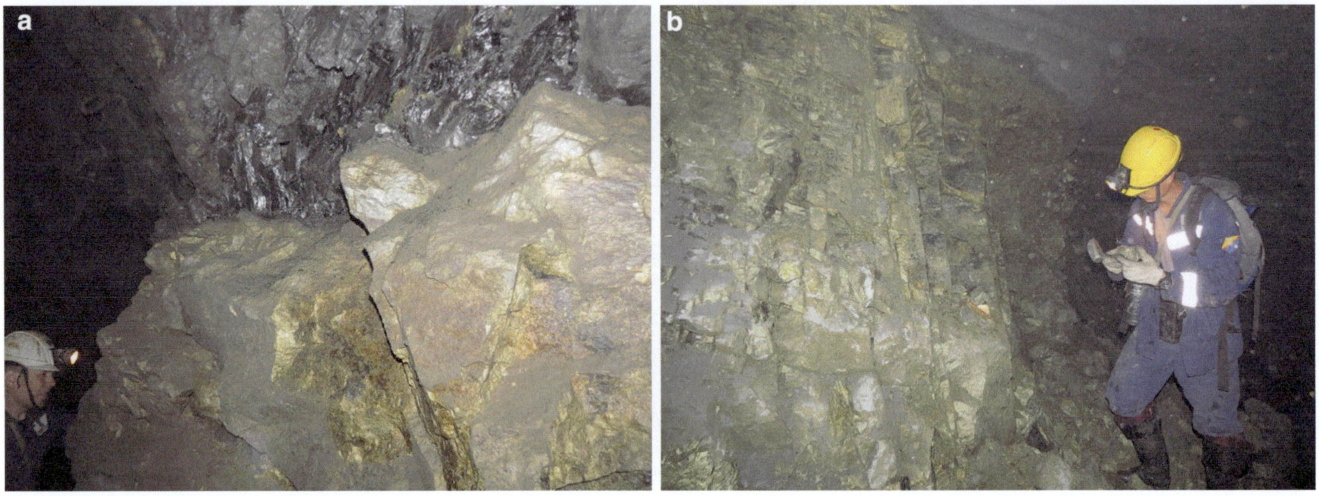

Fig. 4.20 Massive ores of the Talnakh deposit
(**a**) mine Skalisty, photo S. Belorustseva; (**b**) mine Mayak, photo I. Fomin

Complex, which usually have weighted MgO contents of 10–12 wt %. This is due to the development of thicker horizons of olivine, picritic, and troctolitic rocks not only on the bottom of the Low Talnakh-type massifs but also in the middle and even the upper parts (Fig. 4.24).

4.4.3.1 Internal Structure of the Low Talnakh Massif

The study of the Low Talnakh intrusion was conducted mainly in borehole TG-31, from which a complete section of the intrusion was available.

Figure 4.24, which shows a section of the Low Talnakh Massif through borehole TG-31, illustrates the differentiated structure of the intrusion. The lower part of the massif is composed of more melanocratic rocks than the upper part, where interlayer olivine-free gabbro-dolerites are present. A small troctolite up to 10 m thick is located at the bottom of the intrusion. Fine gabbro-dolerites, in which olivine was not detected, were found in the contact zone. The composition of the rocks of the Low Talnakh intrusion in borehole TG-31 differs significantly from the other massifs of the Noril'sk Complex; therefore, this body is a Low Talnakh-type intrusion.

As noted above, the main feature of the intrusion is the enrichment of Mg (Table 4.12). However, the most distinctive feature is the behavior of the trace elements in the rocks. Their trace element spectra normalized to the primitive mantle (Hofmann 1988) have dramatically different characteristics than those of other massifs of the Noril'sk Complex. First, they are characterized by high concentrations of LIRE elements and LREE, where the left part of the spectrum is higher than the right. These features distinguish the typological spectrum of the Low Talnakh intrusion (blue line in Fig. 4.25) from the Noril'sk Complex or Noril'sk-type intrusions, which are shown by the Kharaelakh rocks (red line). The right portions of the lines that correspond to the Low Talnakh intrusion are significantly lower than those of the other intrusion, i.e., the rocks of the Low Talnakh intrusion are depleted in heavy rare earth elements. The left parts of the spectra are close to each other. It is possible to characterize this difference using the La/Sm and Gd/Yb ratios.

Olivine. The main mineralogical feature of the rocks from the Low Talnakh intrusion are the changes in the olivine composition, which are shown in the data from borehole TG-31 in Fig. 4.24 and in Table 4.13. The magnesium component of the olivine in this section varies from 73 (in the olivine gabbro-dolerites) to 84 mol% Fo (in the picritic gabbro-dolerites), with values of individual samples ranging from Fo_{74} to Fo_{80}. One can see the variations in the olivine contents in the section as well as in the nickel and manganese; the latter is inversely proportional to the magnesium content, whereas the low concentration of nickel corresponds to a low magnesium content in the olivine.

The distributions of calcium, aluminum, and chromium are not dependent on the Mg in the olivine; they vary significantly for the same magnesium content.

The distribution of trace elements in olivines from the Low Talnakh intrusion in borehole TG-31 was investigated

Fig. 4.21 Ores and minerals from Oktyabr'skoe deposit
Ores: (**a**, **b**) chalcopyrite–talnakhite, (**c**) pentlandite–moikhukite, (**d**) moikhukite breccia, (**e**) millerite–bornite, (**f**) chalcopyrite–pyrrhotite, (**g**) galenite–moikhukite, (**h**) galenite–bornite ore. Samples from collection E. Sereda (Noril'sk), photos E. Sereda. Crystals: (**i**) pyrrhotite, (**j**) calcite on millerite, (**k**, **l**, **m**, **n**) sperrylite, (**o**) chalcopyrite, (**p**) pyrrhotite. Sample n data is from (http://rusmineral.ru), samples **i–m**, **o**, **p** are from collection E. Sereda. Photos **i–m**, **o** by M. Bogomolov, **p** by E. Sereda. Lower frame for photos, cm: g - 4, h - 1, i - 0.5, j - 0.3, k - 3, l - 1.5, m - 3, n - 5, o - 5

Fig. 4.21 (continued)

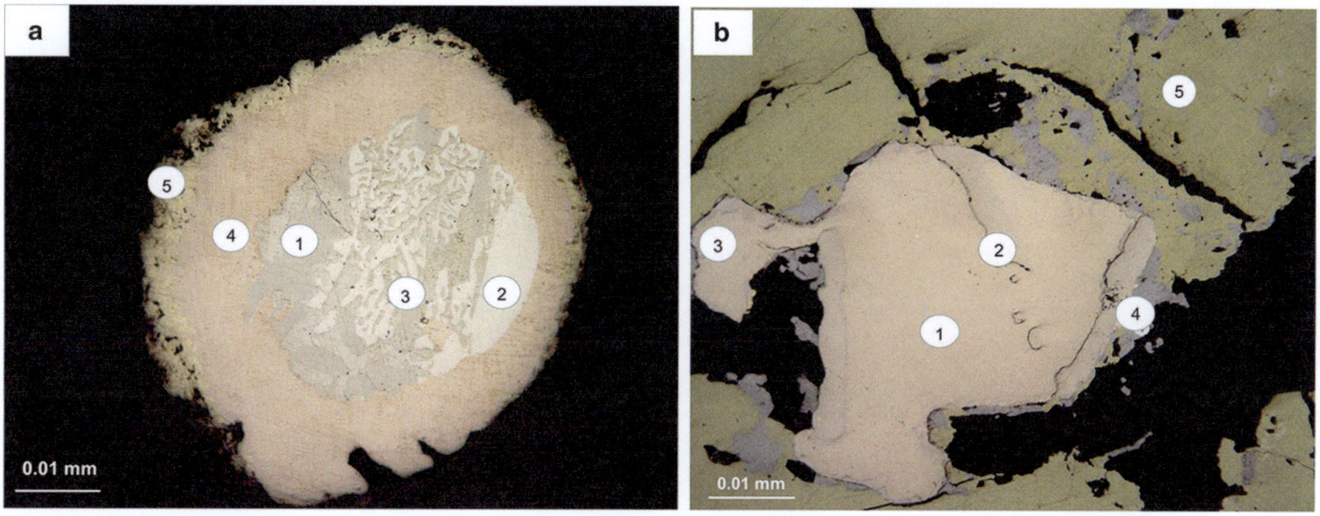

Fig. 4.22 Minerals of Au-Ag and Pd in sulfide ores
After: a – Sluzheninkin and Mokhov 2015, b – Dodin et al. 2009. (**a**) *1* – polyarite, *2* – zvyagintsevite, *3* – plumbopalladinite, *4* – Pt-Pd-Cu-Au, *5* – Au, Ag. (**b**) *1* – Pd (Bi, Te, Pb), *2* – Pd3 (Sb, As, Sn), *3* – Bi-moncheite, *4* – kustelite, *5* – chalcopyrite

by mass spectrometry with inductively coupled plasma. In the olivine, virtually all of the rare earth elements (e.g., Y, Yb, and Dy) have identical trends that are inversely proportional to Fo. Most of the other elements do not exhibit regular variations, although the vanadium concentration increases with increasing Mg concentration.

Pyroxene The variations in the pyroxene composition were studied along sections (Table 4.14) and are shown in Fig. 4.24. All of the studied elements form a single trend line, which is the same for the pyroxenes in the main rock types. Manganese, titanium, and sodium are inversely proportional to the Mg concentration, and calcium and aluminum increase with increasing magnesium content. The maximum concentrations of chromium are characteristic of most varieties of magnesium pyroxenes.

The distribution of trace elements in the pyroxenes of the Low Talnakh intrusion in borehole TG-31 was also investigated by mass spectrometry with inductively coupled plasma. Of the rare earth elements, the concentrations of cerium and dysprosium increase with increasing Mg# of the nickel in the pyroxenes, while zirconium and niobium have inverse relationships with the magnesium content.

4.5 Massifs of the Noril'sk Ore Junction

There are several intrusive bodies with economic PGE–Cu–Ni mineralization within the Noril'sk Trough; they are Noril'sk 1, Noril'sk 2, Bol'shaya Bar'ernaya, Chernogorsky, Zub-Marksheydersky, and Maslovsky. Their position in this structure are shown on Fig. 4.26. Despite their similar internal structure and geochemical features of rocks, the chemical composition of sulfide mineralization related to the intrusive bodies is different. The most obvious feature of ores is PGE tenor in sulfides. The highest PGE concentrations are typical of the Noril'sk 1 intrusion (average 10–12 ppm for disseminated ores, up to 2,000 ppm in massive ores). This intrusion contains economic low-sulfide mineralization as well.

4.5.1 Noril'sk 1 Massif

Noril'sk 1 was the first ore deposit to be discovered in the Noril'sk area. In 1865, K.P. Sotnikov organized the extraction of copper ores and copper smelting on the northern slope of the Rudnaya Mountain, which resulted in the

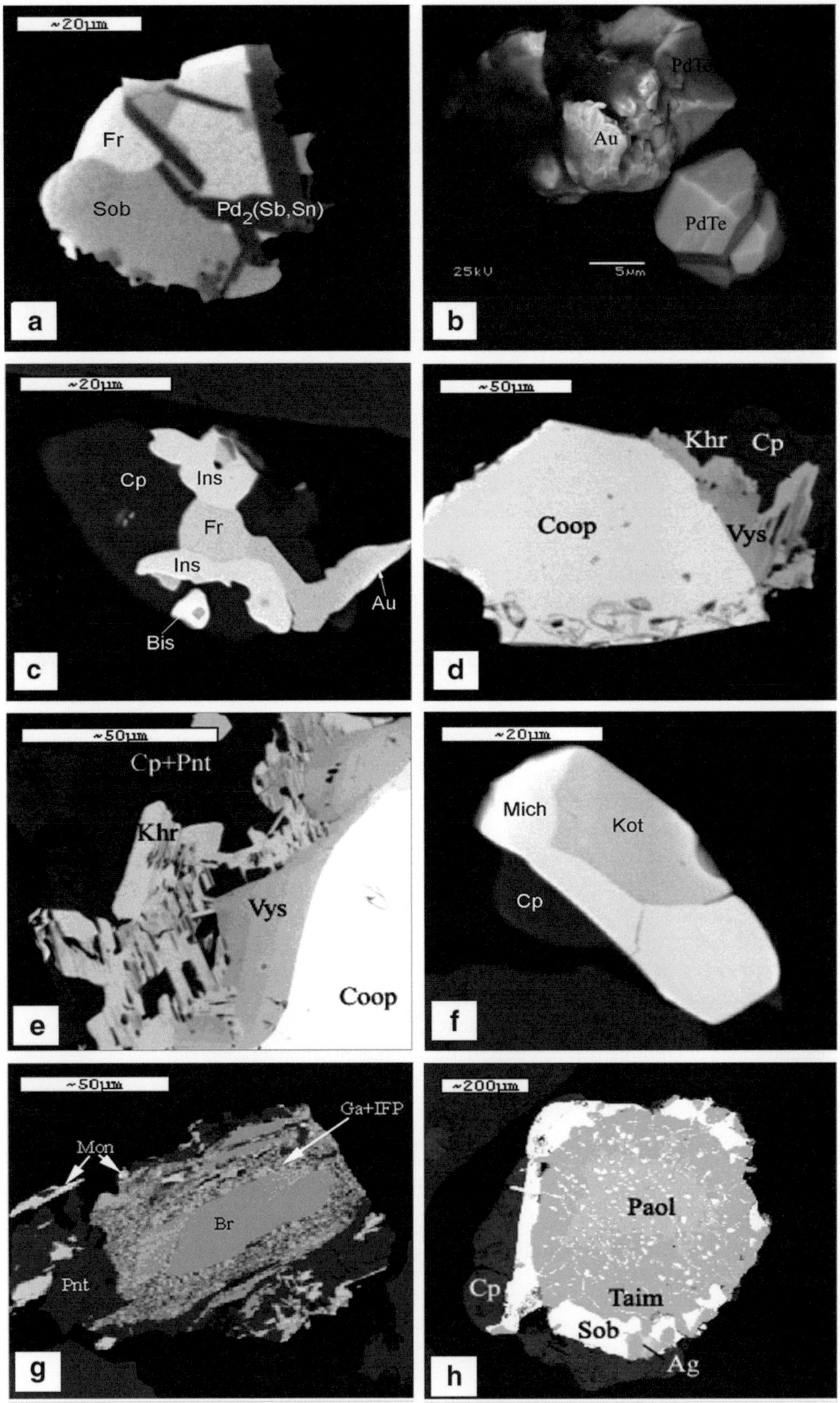

Fig. 4.23 Minerals of platinum group elements
Images in BSE fr, frudite; Kot, kotulskite; Mich, michenerite; Ins, insizvaite; Vys, vysotskite; Coop, cooperate; Bis, bismuthine; Cp, chalcopyrite; Paol, paolovite; Taim, taimyrite; Mon, moncheete; Pnt, pentlandite; Ga, galena; Sob, sobolevskite (After Kozyrev 2002)

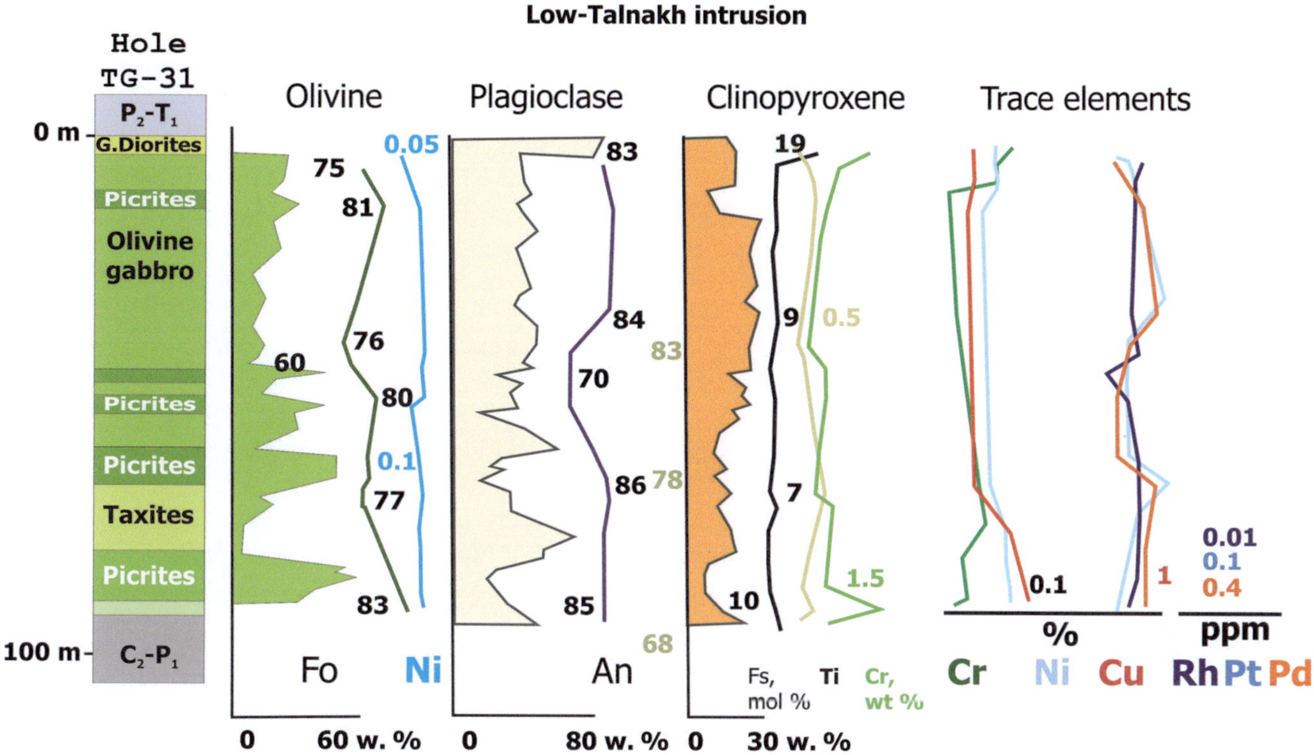

Fig. 4.24 Inner structure of the Low Talnakh intrusion, variations of amount, and composition of rock-forming minerals and the distribution of elements in its vertical section (borehole TG-31)
After Turovtsev (2002)

melting of 200 lb of blister copper. The ore deposit was first described in 1866 by the academician Schmidt. A.A. Sotnikov collected ores and coals from the Noril'sk region and published an article on the prospects for the exploitation of ores, coals, and graphite in connection with the development of the Northern Sea Route (Sotnikov 1919). In 1920, a large expedition was sent to explore this area under the direction of N.N. Urvantsev. Following the decision by the government of the USSR to construct the Noril'sk Mining and Metallurgical Combine (based on the approved reserves), Noril'sk 1 was expanded, and prospecting and exploration began in 1935. Since 1951, the "Coal Creek" and "Medvezhy Creek" mines and the "Zapolyarny" and underground mine have produced Cu–Ni–Pt ores.

The differentiated Noril'sk 1 intrusion outcrops at the surface along the northern and eastern slopes of Rudnaya Mountain (Godlevsky 1959). At this location, it consists of two branches that are sometimes separate and sometimes connected (Fig. 4.27). Both branches form a trough-shaped body with steep eastern and western sides. The maximum thickness is located along the axis of the inflection from gentle to steep bedding. The southern half of the intrusion is a tube-shaped body that is flattened in the vertical direction. The maximum thickness is in the central part of the intrusion, and the average thickness is approximately 135 m. The eastern and western branches of the northern half of the intrusion are up to 100–150 m thick, while the steep sides are 150–200 m thick. The intrusion plunges at 5-60° to the southwest.

The host rocks of the Noril'sk 1 intrusion are labradorite and titanium–augite basalts of Ivakinsky Formation and basalts, picrites of the Syverminsky and Gudchikhinsky Formations and sediments of the Tunguska Group.

The ore field can be divided into north and south parts based on the presence of the ore-bearing deposits. The north part is the richest; the veins and richly disseminated ores are represented by a single orebody with very high ore metal contents. The orebody contains ores in two layers.

The orebody is confined to the bottom of the intrusion and is associated with picritic, taxitic, and contact gabbro-dolerites and underlying basalts and sandstones. The ore types are disseminated in gabbro-dolerites, are

Table 4.12 Rock composition from the Low Talnakh intrusion (borehole TG-31)

N sample	798	817	831	836	853	868	878.6	893
SiO_2	45.1	45.02	46.4	42.45	45.77	49.88	49.85	45.26
TiO_2	0.46	0.44	0.46	0.50	0.54	0.94	0.96	0.43
Al_2O_3	10.56	10.7	11.23	7.7	9.3	15.05	14.31	9.86
FeO	12.82	12.2	11.53	15.57	13.7	10.62	10.8	12.66
MnO	0.23	0.18	0.22	0.24	0.23	0.23	0.25	0.28
MgO	19.77	21.42	19.32	24.13	20.8	7.74	7.58	18.05
CaO	7.2	6.34	8.04	5.93	7.05	10.66	10.29	7
Na_2O	0	0.2	0.01	0.12	0.12	1.41	1.6	0
K_2O	0.36	0.52	0.44	0.33	0.61	1.19	1.26	0.88
P_2O_5	0.06	0.07	0.05		0.09	0.1	0.11	0.04
Cr_2O_3	0.03	0.02	0.04		0.02	0.01	0.01	0.02
LOW	2.06	2.7	2.2	4.01	1.04	1.71	1.89	5.27
Total	98.65	99.81	99.94	100.99	99.27	99.54	98.91	99.77
Sc	18	18	17	16	19	18	20	31
Ti	2,395	2,500	2,065	2,280	2,470	2,480	2,855	5,455
V	108	108	96	103	114	102	126	216
Cr	138	129	303	127	207	226	108	86
Mn	1,410	1,210	1,140	1,250	1,380	1,550	1670	1,660
Co	105	71	89	110	88	151	105	45
Ni	10	316	463	746	677	1,089	594	106
Cu	754	206	622	663	614	1,880	503	242
Zn	227	324	307	160	353	1,081	306	260
Rb	13.2	14.2	13.1	20.3	15.1	12.5	20.7	39.1
Sr	172	274	191	143	208	152	145	299
Y	10.7	10.8	10.4	10.5	10.9	11.1	13.7	21.6
Zr	41.1	41.8	38.3	45.3	38.7	45.5	40.6	95.9
Nb	3.30	4.52	2.59	3.13	3.25	3.27	4.48	7.05
Mo	1.18	0.98	5.18	0.70	1.49	2.12	1.01	1.03
Cs	0.80	0.61	0.58	0.97	1.06	0.67	1.19	0.94
Ba	152	98	99	81	210	87	185	307
La	5.2	4.9	5.7	4.6	5.4	4.5	7.6	11.7
Ce	11.1	11.2	12.0	12.1	11.7	10.6	15.5	25.9
Pr	1.36	1.35	1.46	1.58	1.44	1.36	1.83	3.19
Nd	5.80	5.86	6.30	6.80	6.19	6.03	7.70	13.35
Sm	1.34	1.41	1.44	1.47	1.51	1.41	1.73	3.15
Eu	0.45	0.47	0.48	0.38	0.53	0.42	0.62	0.98
Gd	1.51	1.48	1.53	1.51	1.63	1.59	1.95	3.32
Tb	0.25	0.24	0.25	0.26	0.26	0.26	0.31	0.55
Dy	1.61	1.60	1.60	1.57	1.70	1.70	2.07	3.49
Ho	0.36	0.36	0.36	0.35	0.40	0.38	0.46	0.80
Er	0.97	0.97	0.96	0.92	1.03	1.04	1.21	2.12
Tm	0.15	0.14	0.14	0.13	0.15	0.15	0.18	0.31
Yb	0.93	0.88	0.89	0.87	0.98	0.95	1.13	2.01
Lu	0.14	0.14	0.13	0.13	0.14	0.14	0.17	0.29
Hf	1.04	1.01	0.93	1.07	1.06	1.18	1.10	2.47
Pb	30.4	34.0	10.2	17.4	42.8	87.2	49.3	21.1
Th	0.86	0.90	0.86	0.86	0.91	0.91	0.99	2.06
U	0.30	0.24	0.34	0.30	0.33	0.29	0.45	0.64

Note: N sample is depth in borehole TG-31, m. Oxides - in wt %, elements - in ppm

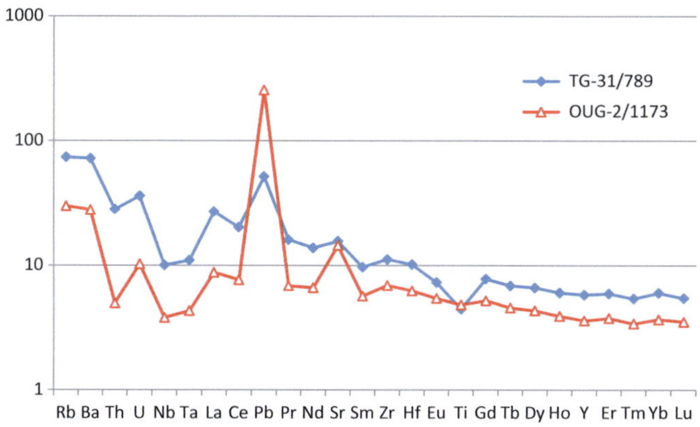

Fig. 4.25 Rare elements patterns for Low Talnakh and Talnakh intrusive rocks
Blue and *red* lines respectively

veinlet-disseminated (cuprous) in the underlying rocks (basalts and rocks of the Tunguska Group), and are present in veins. The average grades are 0.42 % nickel, 0.63 % copper, and 10 ppm PGE.

The ore minerals include pyrrhotite, chalcopyrite and pentlandite, cubanite, millerite, nickel pyrite, and other minerals.

4.5.1.1 Internal Structure of the Noril'sk 1 Intrusion

The massif was studied based on samples collected in the open-pit "Medvezhy Creek" (contact, olivine, picritic, and taxitic gabbro-dolerite, veins, and disseminated sulfide ores) and cores from boreholes G-22, MS-31, and MN-7 (the "Prirezka" field, Fig. 4.27). The first borehole is unique because it comprises the upper taxitic and picritic gabbro-dolerite with elevated PGE contents, in which fresh olivine was detected and analyzed. The upper picrites were also found in the other boreholes.

These boreholes were located in the marginal part of the Noril'sk 1 intrusion and are important for this study because the flanks of intrusive bodies are of special interest for prospecting for Cu–Ni mineralization in new intrusions. The results of the analyses of minerals obtained from the boreholes allow a comparison between the central and peripheral zones of the intrusion. The structures of the intrusion in boreholes MS-31 and G-22 are very similar (Figs. 4.28 and 4.29) because they are both located along the edge of the massif (Fig. 4.27). The intrusive body is 58.3 m and 60.1 m thick, respectively, in the two boreholes. In the vertical section of the intrusion, several classical horizons were distinguished from the base to the top, including the contact (2.2 and 1.8 m thick in MS-31 and G-22, respectively), taxitic (10.0 and 6.2 m thick), picritic (5 and 8 m thick), and olivine (0 and 10 m thick) gabbro-dolerites, olivine-bearing gabbro-dolerites (12 and 14 m thick), olivine-free gabbro-dolerites (3 and 6 m thick), leucogabbro (only in MS-31; 18 m thick), the upper picritic gabbro-dolerites (only present in G-22; 8 m thick), the upper taxitic gabbro-dolerites (in G-22; 8 m thick), and eruption breccias (in MS 31; 4 m thick).

These characteristics of the two sections show that the lower part of the section contains similar structures; the contact, taxitic, picritic, and olivine varieties have similar thicknesses (Fig. 4.30). Significant differences are found in the upper part of the section; a significant leucogabbro zone (18 m thick) and eruption breccia are present in borehole MS-31, and borehole G-22 contains the upper picritic and taxitic gabbro-dolerites at the top.

The petrochemical features of the rocks from the Noril'sk 1 intrusion (Table. 4.15) can be illustrated using Harker diagrams (Fig. 4.31). The compositions of the analyzed rocks form a continuous trend from high-Mg (26 wt %) to low-Mg (4 wt % MgO) differentiations. The most distinct trends are observed for Al_2O_3, CaO, SiO_2, and Na_2O. The distinct trends for the latter two elements indicate the appearance of plagioclase on the liquidus. Accumulation alkalines occur with increasing silica content in the melt and are caused by fractional crystallization. It is interesting to consider the behavior of iron and titanium in the rocks. The iron content decreases gradually from the high-Mg to low-Mg rocks. Titanium exhibits the opposite pattern, although its behavior is more complex. Two types of rocks are present: low-Ti and high-Mg rocks and moderately high-Ti and low-Mg rocks. The compositions of the contact gabbro-dolerites can be explained by an increasing content of titanomagnetite. They have a separate trend that is not associated with the bulk rock. The observed data for these samples demonstrate that the growth of alkalinity, phosphorus, and zirconium may represent the contact processes of the assimilation of the rocks by magma at the bottom of the intrusive body.

Several horizons in the different boreholes have similar compositions. Significant variations of many components are observed in the upper picritic and olivine gabbro-dolerites in the upper part of the intrusion in boreholes G-22 and MC ("Medvezhy Creek" open pit). This indicates that the rocks in the upper part of the massif are more heterogeneous than those in the lower horizons. The chromium content is extremely high (up to 8 wt %) in the picritic gabbro-dolerites and can be explained by accumulations of chromite and chrommagnetite. A comparison of the "upper" and "lower"

Table 4.13 Olivine compositions from rocks of the Low Talnakh intrusion

№	№ sample	Fo, Mol %	SiO_2	FeO	MnO	MgO	CaO	NiO	Summa	Cr	Yb	Dy	Y	V	Ti	Na	Li
1.	TG-31/878.6	76.66	38.6	21.4	0.28	39.5	0.19	0.03	100.10	539	0.38	0.12	1.51	22.4	104.3	46.8	6.08
2.	TG-31/817	76.86	37.9	21.5	0.28	40.1	0.14	0.07	100.04	422	0.20	0.08	0.92	24.0	122.6	39.6	5.40
3.	TG-31/862.5	71.81	37.0	25.7	0.34	36.8	0.12	0.11	100.04	366	0.67	0.12	1.88	17.4	118.2	26.6	5.54
4.	TG-31/862.5	73.35	37.3	24.4	0.36	37.7	0.16	0.13	100.11	448	0.46	0.14	1.82	19.0	114.6	35.3	6.07
5.	TG-31/878.6	74.75	37.0	23.1	0.33	38.4	0.09	0.05	98.98	1014	0.41	0.09	1.71	23.7	102.9	31.2	6.64
6.	TG-31/878.6	74.92	37.7	23.2	0.40	38.9	0.13	0.03	100.25	1320	0.44	0.13	1.88	24.5	123.0	248.3	6.42
7.	TG-31/843	78.74	41.0	19.7	0.33	38.9	0.13	0.01	100.09	398	0.35	0.09	1.27	23.8	146.8	25.6	6.80
8.	TG-31/810	79.51	41.5	19.1	0.31	39.4	0.17	0.17	100.61	381	0.28	0.08	1.11	25.7	137.3	27.3	7.03
9.	TG-31/875	80.61	42.4	18.2	0.32	39.2	0.16	0.10	100.37	322	0.09	0.03	0.51	18.8	98.5	119.3	5.54
10.	TG-31/820	80.23	41.5	18.2	0.28	39.1	0.12	0.10	99.25	448	0.17	0.05	0.60	25.3	125.1	26.6	5.13
11.	TG-31/858	75.51	39.1	22.6	0.34	38.3	0.24	0.13	100.68	1779	0.26	0.10	1.05	28.3	52.1	137.5	5.76
12.	TG-31/786	78.79	40.2	19.3	0.27	39.2	0.20	0.13	99.25	325	0.22	0.12	1.09	28.1	75.3	64.3	5.94

Note: Analyses were carried out: major elements in (wt %) GEOKHI RAS by EPMA, analyst N. Kononkova; rare elements in ppm Institute of Microelectronics, Yaroslavl, Russia, analysts S. Simakin

Table 4.14 Pyroxene composition from the Low Talnakh intrusion (borehole TG-31), wt %

№	Depth, m	MgO#	SiO_2	TiO_2	Al_2O_3	FeO	MnO	MgO	CaO	Na_2O	Cr_2O_3	Total
1.	820	83.68	52.51	0.91	2.02	6.09	0.16	17.52	20.65	0.24	0.02	100.15
2.	820	83.75	52.78	0.70	2.00	5.96	0.11	17.23	20.74	0.24	0.09	99.86
3.	820	83.32	52.44	0.87	1.96	6.14	0.17	17.20	20.62	0.26	0.08	99.76
4.	820	86.14	53.09	0.34	2.37	5.01	0.13	17.47	21.29	0.12	0.23	100.07
5.	820	85.91	52.70	0.37	2.39	5.08	0.09	17.37	21.25	0.19	0.23	99.71
6.	820	85.52	52.81	0.42	2.29	5.23	0.11	17.32	21.38	0.20	0.19	99.96
7.	875	85.84	52.53	0.41	2.51	5.04	0.12	17.13	21.20	0.17	0.29	99.43
8.	875	85.79	52.60	0.32	2.46	5.15	0.13	17.44	21.25	0.19	0.19	99.76
9.	875	86.21	52.89	0.34	2.16	4.95	0.14	17.36	21.41	0.15	0.17	99.60
10.	875	85.69	52.75	0.36	2.38	5.19	0.13	17.43	21.15	0.18	0.23	99.84
11.	875	85.66	52.53	0.36	2.35	5.18	0.13	17.36	21.08	0.16	0.20	99.38
12.	875	85.72	52.82	0.33	2.42	5.19	0.10	17.48	21.19	0.15	0.22	99.92
13.	876	84.61	52.85	0.38	1.90	5.60	0.15	17.27	21.07	0.21	0.04	99.54
14.	876	84.64	52.54	0.43	2.32	5.55	0.12	17.15	21.24	0.18	0.07	99.65
15.	876	81.41	52.19	0.66	1.97	6.85	0.17	16.83	20.32	0.22	0.01	99.26
16.	876	83.12	52.61	0.49	1.92	6.19	0.16	17.10	20.75	0.25	0.02	99.52
17.	876	82.72	52.11	0.53	2.40	6.32	0.17	16.97	20.42	0.18	0.03	99.16
18.	876	81.99	52.21	0.58	2.03	6.62	0.18	16.90	20.34	0.24	0.01	99.14
19.	876	84.78	52.89	0.35	1.90	5.53	0.15	17.28	21.00	0.17	0.05	99.34
20.	876	78.11	51.48	1.17	2.10	8.18	0.24	16.37	19.44	0.31	0.01	99.33
21.	876	80.91	51.95	0.76	2.29	7.04	0.18	16.74	20.25	0.29	0.01	99.54
22.	876	82.52	52.23	0.54	2.33	6.52	0.17	17.26	20.18	0.24	0.03	99.53
23.	876	81.94	52.53	0.60	2.07	6.74	0.18	17.15	20.28	0.13	0.01	99.72
24.	876	85.71	52.28	0.41	2.88	5.07	0.12	17.05	21.50	0.12	0.22	99.70

№	Depth, m	MgO#	SiO_2	TiO_2	Al_2O_3	FeO	MnO	MgO	CaO	Na_2O	Cr_2O_3	Сумма
25.	876	83.12	52.88	0.47	1.89	6.16	0.24	17.01	20.91	0.20	0.03	99.81
26.	876	83.54	52.64	0.48	2.14	5.99	0.16	17.05	20.91	0.23	0.05	99.67
27.	876	84.39	52.91	0.38	1.78	5.71	0.15	17.32	21.03	0.20	0.05	99.54
28.	876	84.28	52.72	0.41	2.21	5.68	0.14	17.08	21.02	0.18	0.06	99.53
29.	876	83.48	52.73	0.45	1.90	6.06	0.16	17.18	20.83	0.20	0.04	99.55
30.	876	84.66	53.06	0.36	1.77	5.60	0.13	17.33	21.06	0.21	0.05	99.57
31.	876	79.04	51.83	0.94	1.96	7.81	0.22	16.52	19.80	0.29	0.01	99.39
32.	876	80.39	52.05	0.78	2.07	7.28	0.19	16.74	20.22	0.25	0.01	99.60
33	876	84.49	52.95	0.38	1.83	5.66	0.14	17.30	20.98	0.18	0.06	99.50
34.	876	85.3	52.82	0.38	2.36	5.28	0.13	17.18	21.31	0.19	0.14	99.80
35.	876	83.3	52.53	0.46	2.19	6.16	0.16	17.24	20.57	0.23	0.05	99.60
36.	876	83.12	52.88	0.47	1.89	6.16	0.24	17.01	20.91	0.20	0.03	99.81
37.	876	83.54	52.64	0.48	2.14	5.99	0.16	17.05	20.91	0.23	0.05	99.67
38.	876	84.39	52.91	0.38	1.78	5.71	0.15	17.32	21.03	0.20	0.05	99.54
39.	876	84.28	52.72	0.41	2.21	5.68	0.14	17.08	21.02	0.18	0.06	99.53
40.	876	83.48	52.73	0.45	1.90	6.06	0.16	17.18	20.83	0.20	0.04	99.55
41.	876	84.66	53.06	0.36	1.77	5.60	0.13	17.33	21.06	0.21	0.05	99.57
42.	876	79.04	51.83	0.94	1.96	7.81	0.22	16.52	19.80	0.29	0.01	99.39
43.	876	80.39	52.05	0.78	2.07	7.28	0.19	16.74	20.22	0.25	0.01	99.60
44.	876	84.49	52.95	0.38	1.83	5.66	0.14	17.30	20.98	0.18	0.06	99.50
45.	876	85.3	52.82	0.38	2.36	5.28	0.13	17.18	21.31	0.19	0.14	99.80
46.	876	83.3	52.53	0.46	2.19	6.16	0.16	17.24	20.57	0.23	0.05	99.60

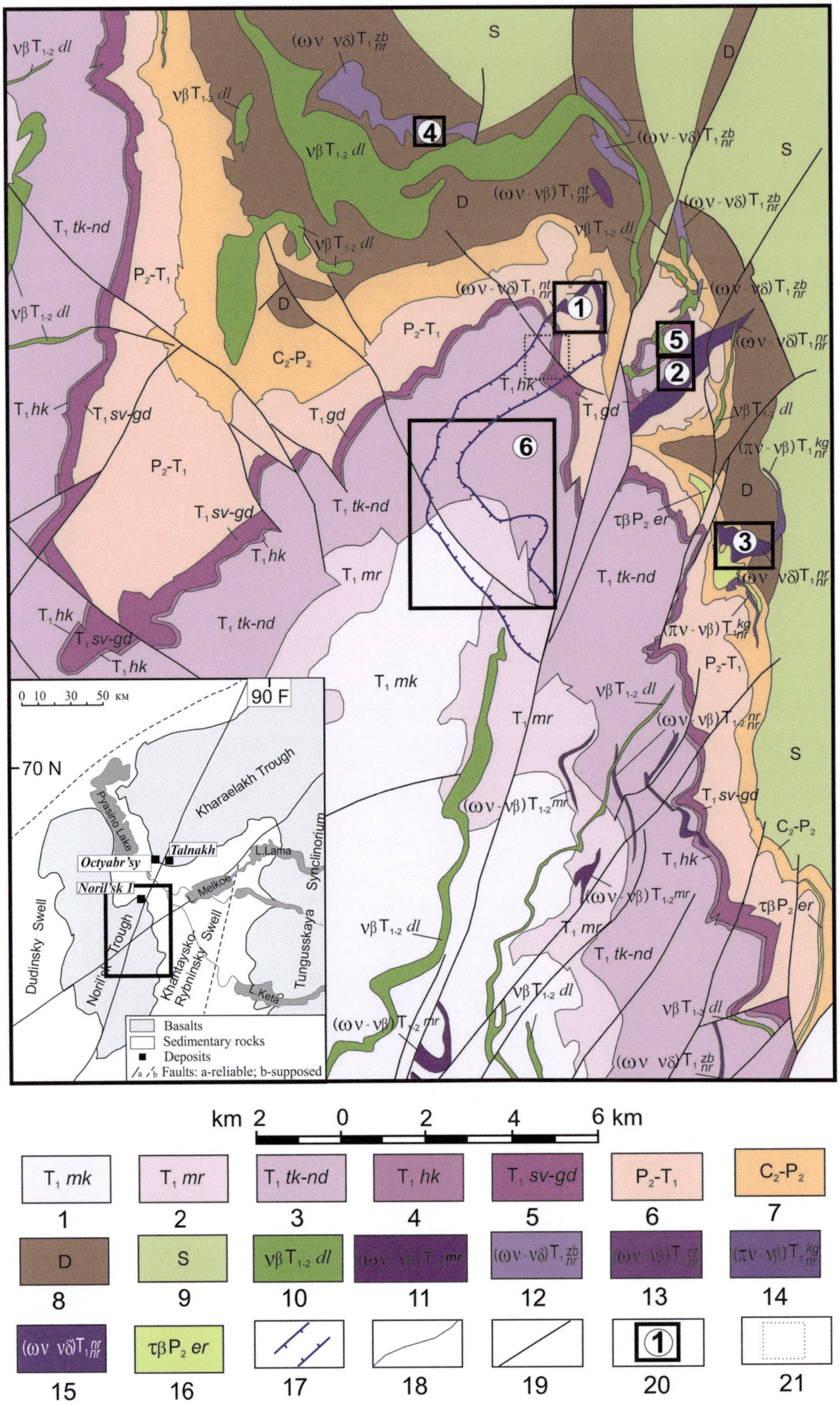

Fig. 4.26 Geological map of the Noril'sk Trough (Inset: its location in the Noril'sk area; Geology and ore deposits 1994)
The map was prepared by V.A. Teteryuk based on materials of NorilskGeology, Ltd. in 2003 and modified by the authors. (1–6) Basalts, formations: (1) Mokulaevsky, (2) Morongovsky, (3) Tuklonsky, (4) Gudchikhinsky, (5) Syverminsky, (6) Ivakinsky; (7) terrigenous deposits with coal of the Tunguska suit. (8–9) carbonate–terrigenous rocks: (8) Devonian, (9) Silurian. (10–16) gabbro-dolerites of intrusive complexes: (10) Daldykansky, (11) Morongovsky. (12–15) Noril'sk Complex, subcomplexes: (12) Zubovsky, (13) Lower Talnakh (14) Kruglogorsky, (15) Noril'sk; (16) trakhi-dolerites of Ergalakhsky Complex; (17) deep contour of the Noril'sk 1 intrusion; (18) geological boundaries, (19) faults. (20) mineral deposits: (1) Noril'sk 1, (2) Noril'sk 2, (3) Mount Chernaya, (4) Zub-Marksheydersky; (21) contour of the Maslovsky deposit

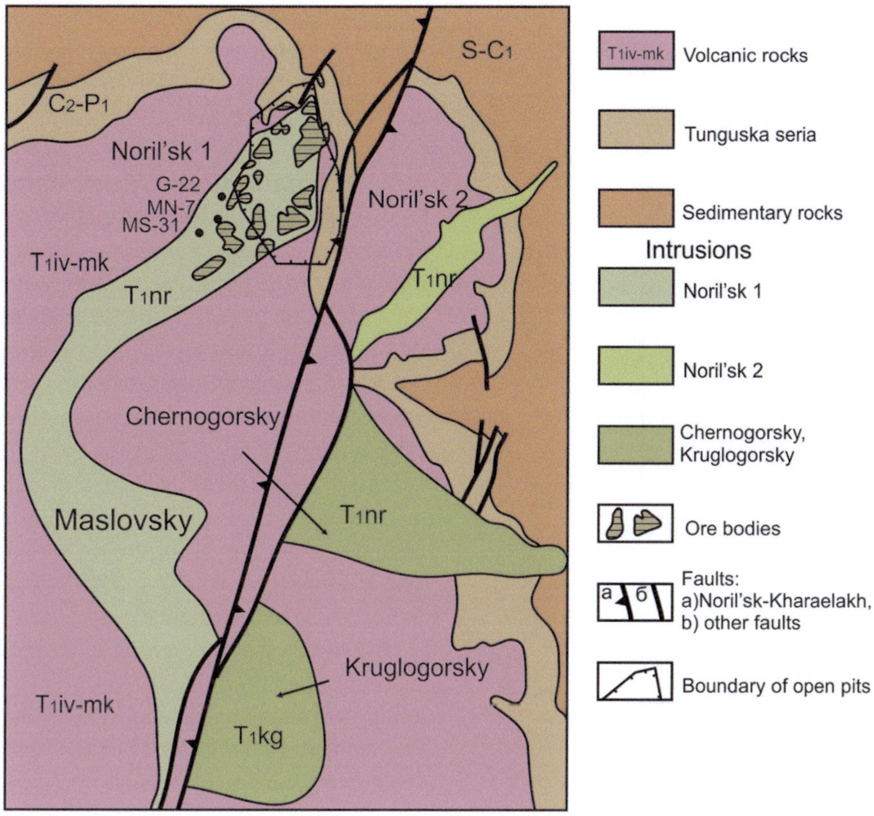

Fig. 4.27 Geological map of the Noril'sk 1 deposit with position of studied boreholes
After NorilskGeology, Ltd. data, with changes by the author

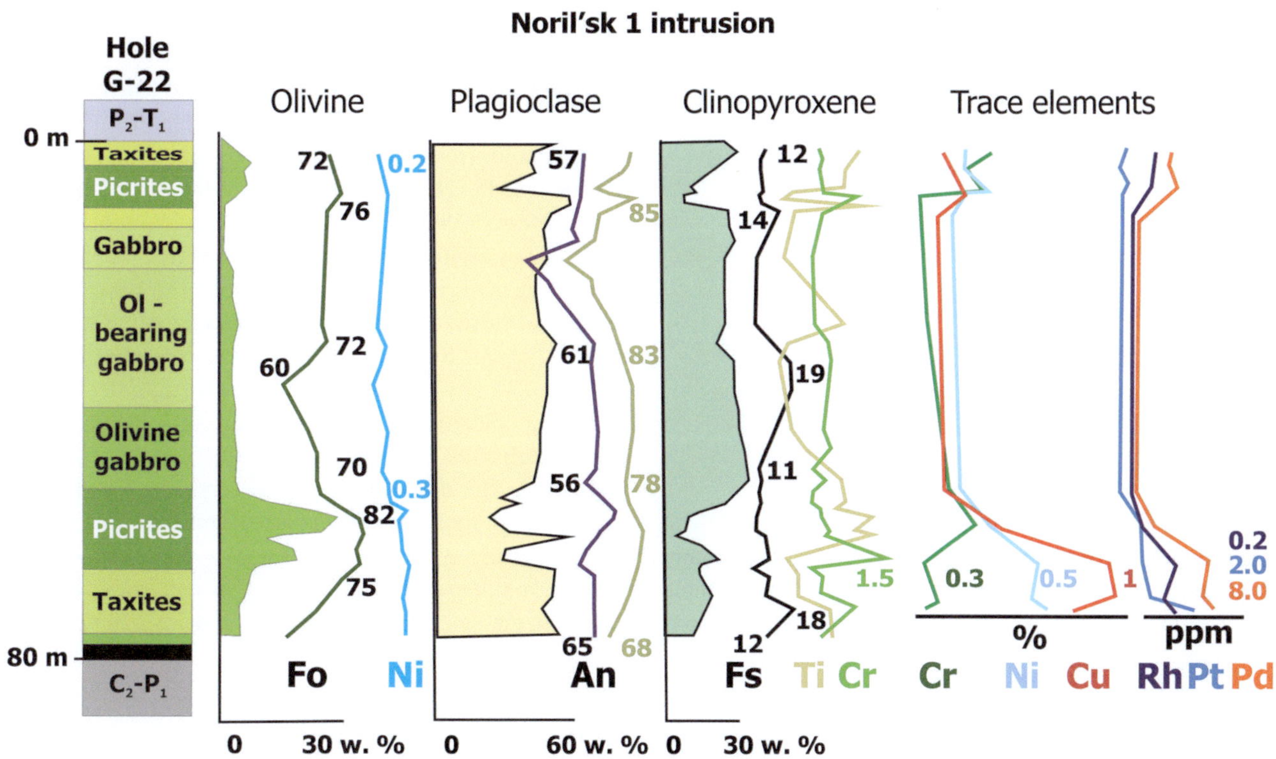

Fig. 4.28 Inner structure of the Noril'sk 1 intrusion, variations of amount, and composition of rock-forming minerals and metals' distribution (borehole G-22)
After Distler et al. (1999)

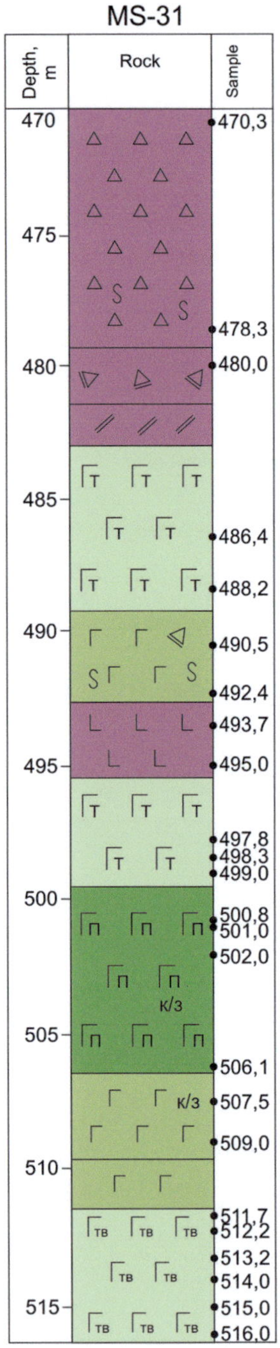

Fig. 4.29 Inner structure of the Noril'sk 1 intrusion (borehole MS-31)

picrites from the two boreholes indicates greater enrichment in the former rocks than in the latter. This observation cannot be extended to the entire massif because of the limited amount of material that was available for analysis. Because segregation of the chromite in the upper and lower parts of the intrusion appears irregularly, additional samples must be analyzed to develop conclusions about the chromium contents in these horizons.

4.5.1.2 Distribution of Trace Elements in the Rocks of the Noril'sk 1 Intrusion

The distribution spectra of trace elements (Table 4.15), normalized to the primitive mantle Hofmann 1988), provide indications of the origin of these rocks. It should be noted that all of the rocks have identical morphologies and differ only in the concentrations of the elements. The minimum concentration of rare elements is present in rocks that are enriched in olivine, such as the "upper" and "lower" picritic gabbro-dolerites (see Chap. 5), due to the low concentrations of most trace elements in this mineral. The maximum concentrations are typical of the taxitic upper gabbro-dolerites. The remaining rocks (i.e., olivine and olivine-free gabbro-dolerites) have intermediate concentrations.

All of the spectra have common characteristics, including low concentrations of trace elements, subhorizontal orientations and the resulting low $(La/Sm)_n$ and $(Gd/Yb)_n$ ratios that characterize the slopes of the right and left parts of the spectra, and positive anomalies of Pb, Sr, and U and negative anomalies of Ta–Nb and, partially, Ti. The presence of lead and tantalum–niobium anomalies is usually interpreted as a sign of crustal contamination of the melt, and strontium indicates the presence of significant amounts of plagioclase in the rocks. The Noril'sk 1 massif differs from the Low Talnakh type of intrusions with reduced Gd/Yb and La/Sm ratios.

4.5.1.3 Composition of Rock-Forming Minerals

The mineral composition data from the intrusion were consolidated from both boreholes: data from the main section of MS-31 were added to the data from the upper rocks of borehole G-22. This was of fundamental importance for studying the olivine composition because it is interesting to compare minerals from the upper picritic and taxitic horizons with similar rocks from the lower part of the section. Several samples were collected from the 120 and 135 horizons from the "Medvezhy Creek" mine, including picritic gabbro-dolerites and upper taxitic gabbro-dolerites.

Olivine. The morphology of the olivine grains has been described in detail by several researchers; however, the studies mainly examined the Talnakh intrusion. Comprehensive summaries of the distribution and composition of olivine in the igneous rocks of the region are given in several

Fig. 4.30 Microphotos of rocks from the Noril'sk 1 intrusion
Lower frame is 2 mm

publications (Ryabov and Zolotukhin 1977; Ryabov 1992; Zenko and Czamanske 1994; Czamansky et al. 1994; Ryabov et al. 2000; Likhachev 2015; Krivolutskaya et al. 2009a; b). The most logical classification of the morphological types of olivine was proposed by V.V. Ryabov (1992).

In this study of the samples from boreholes MS-31 and G-22, we identified the following types of olivine sizes and morphologies (Fig. 4.30): (1) large (1.2–1.4 mm) and very large (up to 3 mm) subidiomorphic grains (in picritic olivine gabbro-dolerites and rarely in olivine gabbro-dolerites), (2) average size grains with irregular shapes (0.4–0.6 mm; in picritic and olivine gabbro-dolerites), (3) small rounded grains (average 0.3 mm; located in the plagioclase of the picritic gabbro-dolerites), (4) medium and small interstitial grains in the taxitic gabbro-dolerites (0.5–0.2 mm; in both the lower and upper horizons), and (5) granular olivine grains, which occur as rounded, irregular clusters in picritic and other gabbro-dolerites (less than 0.2 mm, Fig. 4.30d).

The olivine composition data that were obtained from different grains are shown in Table 4.16 and Fig. 4.31 for rocks from borehole MS-31. The olivine data from the Noril'sk 1 intrusion in boreholes MS-31 and G-22 are summarized together. The composition of the olivine varies greatly from Fo_{43} to Fo_{83}. The most magnesium-rich olivine is concentrated in the picritic gabbro-dolerites, and the nickel and forsterite contents of the olivine are direct correlated. Variations of the nickel content were observed in the contact,

Table 4.15 Representative analyses of the rocks from the Noril'sk 1 intrusion (borehole G-22)

N sample	G22/64.4	G22/70.5	G22/71.2	G22/78.0	G22/99.5	G22/102.7
			Component			
SiO_2	42.06	47.42	46.5	49.67	47.4	49.34
TiO_2	0.41	0.8	0.83	0.85	0.95	1.84
Al_2O_3	8.7	13.27	13.74	16.34	15.28	13.58
FeO	14.13	11.82	11.81	9.8	10.6	13.46
MnO	0.2	0.2	0.2	0.19	0.16	0.17
MgO	22.58	15.05	13.75	8.92	7.62	4.89
CaO	5.33	8.62	9.23	11.22	12.52	9.76
Na_2O	0.01	0.3	0.43	0.98	2.33	4.5
K_2O	0.08	0.39	0.42	0.42	0.53	0.8
P_2O_5	0.06	0.08	0.09	0.08	0.14	0.24
Cr_2O_3	0.08	0.4	0.32	0.06	0.05	0.06
LOW	6.06	1.26	1.8	1.1	2.12	1.79
Total	99.7	99.61	99.12	99.63	99.69	100.43
Sc	21.7	30.1	31.5	34.4	43.1	44.3
Ti	3,630	4,935	4,660	4,565	4,730	5,390
V	189	223	215	190	187	209
Cr	7,190	3,220	2,320	395	780	491
Mn	1,250	1,500	1,340	1,180	968	1,140
Co	143	86	78	61	51	50
Ni	2,875	692	557	295	210	164
Cu	2,590	659	132	277	81	91
Zn	94	506	98	332	64	80
Rb	8	16	15	19	25	22
Sr	125	209	226	257	295	461
Y	17.5	20.6	20.6	20.9	19.2	25.2
Zr	67.4	64.2	62.5	65.8	57.5	77.6
Nb	4.13	4.56	3.77	4.30	3.55	4.78
Mo	1.06	2.65	1.66	1.88	1.42	1.31
Cs	0.77	1.01	0.90	1.05	0.49	0.46
Ba	71	122	121	105	97	115
La	4.74	5.07	5.25	5.74	5.00	6.82
Ce	11.3	12.3	12.5	13.6	12.0	16.5
Pr	1.46	1.63	1.67	1.74	1.64	2.20
Nd	6.72	7.57	7.84	8.11	7.46	10.15
Sm	1.83	2.08	2.16	2.30	2.14	2.86
Eu	0.58	0.71	0.76	0.82	0.74	0.91
Gd	2.20	2.46	2.58	2.66	2.51	3.30
Tb	0.37	0.42	0.45	0.46	0.43	0.58
Dy	2.54	2.80	3.02	3.15	2.91	3.83
Ho	0.59	0.65	0.70	0.72	0.67	0.87
Er	1.60	1.75	1.89	1.90	1.76	2.36
Tm	0.24	0.26	0.27	0.29	0.26	0.35
Yb	1.50	1.67	1.76	1.84	1.69	2.15
Lu	0.22	0.25	0.25	0.27	0.25	0.32
Hf	1.76	1.68	1.63	1.72	1.54	2.08
W	0.47	0.75	0.50	1.07	0.61	0.58
Pb	9.80	43.38	3.80	29.75	2.36	2.70
Bi	0.35	0.09	0.05	0.11	0.04	0.04
Th	1.05	1.02	0.95	1.08	0.95	1.24
U	0.44	0.39	0.36	0.41	0.35	0.45

Note: Analyses were carried out by XRF in GEOKHI (analyst I. Roshchina) and by ICP-MS in IMGRE (analyst D. Zhuravlev)

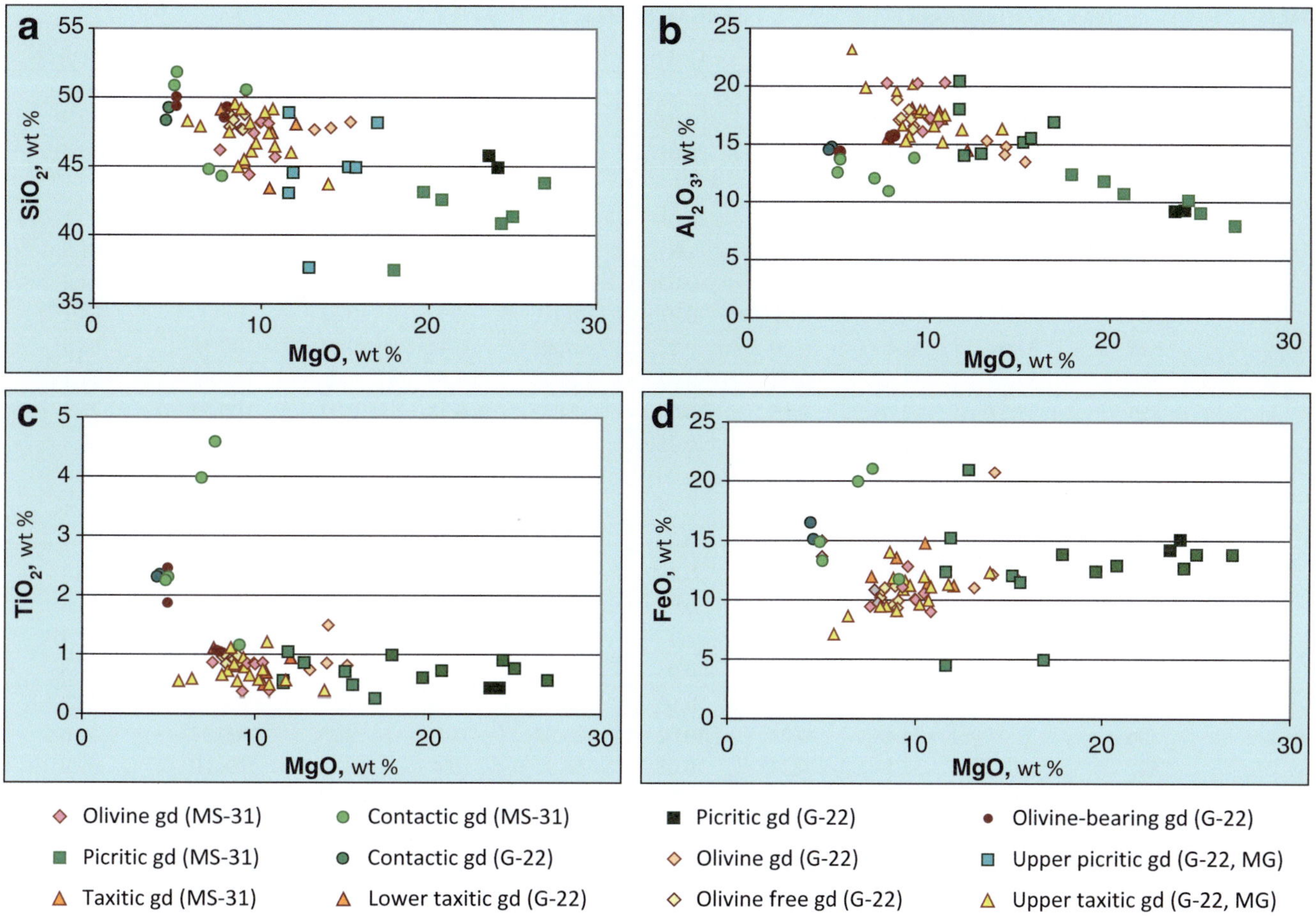

Fig. 4.31 Harker diagrams for rocks of the Noril'sk 1 intrusion

taxitic, and picritic gabbro-dolerites in the lowermost part of the section due to the presence of sulfide. The clear correlation of nickel in the rocks with the nickel content of the olivine in the picritic gabbro-dolerites is logical because these rocks are known to contain disseminated mineralization. The presence of a very ferruginous olivine in the contact rock varieties is also interesting, although according to the model structure, it should have a significantly more magnesium-rich composition (70–72 mol% forsterite).

It is interesting to compare the compositions of the olivine from different horizons of the intrusion and especially those from the upper and lower taxitic and picritic gabbro-dolerite horizons, because the mechanism for the formation of olivine-rich rocks near the top of the intrusion is a topic of discussion and has not been resolved. As is clearly shown in Fig. 4.31 (top right), the olivine from the lower horizons (picrite and taxitic gabbro-dolerites) is more magnesium-rich and contains high concentrations of nickel (0.2–0.35 wt % NiO in the picritic gabbro-dolerites and 0.16–0.32 wt % in the taxitic gabbro-dolerites) compared to the olivine from the upper horizons (0.16–0.22 wt % in the picritic gabbro-dolerites and 0.09-0.12 wt % in the taxitic gabbro-dolerites). The presence of high-magnesium and low-Ni (0.07–0.16 wt % NiO) olivine in the lower taxitic gabbro-dolerites is noteworthy. The calcium contents in the olivine are similar; the average CaO concentrations range from 0.08 to 0.16 wt % for the olivine in all of the horizons, with the exception of the lower taxitic gabbro-dolerites, where the olivine content can exceed 0.3 wt %. Surprisingly, the CaO composition is not dependent on the concentrations of the main components. The low calcium concentrations of the olivines in many of the intrusions in the Noril'sk region remain poorly explained. Because the CaO content in olivine depends on pressure, the crystallization of intratelluric phenocrysts can be explained by the deep conditions during their formation, especially those in the picritic gabbro-dolerite.

Table 4.16 Olivine composition from rocks of the Noril'sk 1 intrusion (borehole MS-31, wt %)

N	Depth, m	SiO_2	MgO	CaO	Cr_2O_3	FeO	MnO	NiO	Total	Fo, mol %
1.	512	38.75	42.71	0.22	0.02	18.29	0.3	0.08	100.37	80.6
2.	512	38.86	42.45	0.2	0.03	18.6	0.3	0.08	100.52	80.3
3.	512	38.71	41.83	0.22	0.05	18.97	0.29	0.07	100.14	79.7
4.	512	38.92	42.08	0.24	0.03	18.97	0.29	0.1	100.63	79.8
5.	512	38.97	42.76	0.19	0.01	18.5	0.29	0.1	100.83	80.5
6.	512	38.78	43.26	0.17	0	17.69	0.28	0.11	100.28	81.3
7.	522	38.68	43.46	0.17	0.02	17.8	0.28	0.09	100.49	81.3
8.	524	36.55	31.88	0.26	0.01	31.05	0.5	0.23	100.5	64.7
9.	525	37.04	34.46	0.26	0.02	27.88	0.44	0.24	100.34	68.8
10.	525	37.44	36.06	0.22	0.02	26.1	0.39	0.2	100.42	71.1
11.	525	37.57	37.76	0.15	0.04	24.5	0.38	0.22	100.62	73.3
12.	525	37.07	34.39	0.15	0.02	28.35	0.43	0.2	100.63	68.4
13.	525	37.23	36.17	0.19	0.06	25.85	0.42	0.24	100.16	71.4
14.	525	37.72	38.29	0.17	0.03	23.81	0.37	0.22	100.61	74.1
15.	525	37.65	37.88	0.18	0.01	23.89	0.38	0.2	100.18	73.9
16	525	37.23	37.25	0.17	0.29	23.54	0.39	0.19	99.05	73.8
17.	525	38.17	38.51	0.17	0.05	23.6	0.37	0.2	101.07	74.4
18.	506	36.42	33.72	0.2	0	29.06	0.46	0.25	100.11	67.4
19.	506	37.17	36.34	0.23	0.03	26.07	0.4	0.21	100.45	71.3
20.	506	36.95	36.13	0.17	0.03	26.07	0.39	0.2	99.94	71.2
21.	506	37.07	36.48	0.21	0.02	25.83	0.41	0.25	100.27	71.6
22.	522	36.87	35.13	0.21	0.03	26.95	0.4	0.24	99.83	69.9
23.	522	37.01	35.97	0.25	0.04	26.18	0.39	0.24	100.08	71.0
24.	522	36.85	33.4	0.19	0.01	29.37	0.46	0.19	100.48	67.0
25	522	36.82	34.49	0.18	0.03	27.85	0.43	0.24	100.04	68.8
26.	522	36.96	34.86	0.16	0.02	27.58	0.43	0.21	100.22	69.2
27.	522	37.19	35.03	0.16	0.01	27.53	0.45	0.19	100.56	69.4
28.	522	37.12	34.9	0.13	0.01	27.51	0.43	0.24	100.34	69.3
29.	515	37.23	35.26	0.15	0.01	27.21	0.43	0.22	100.51	69.8
30.	515	37.02	37.59	0.18	0.17	23.43	0.36	0.22	98.98	74.1
31.	515	37.61	38.46	0.18	0.01	23.68	0.36	0.22	100.52	74.3
32.	515	37.24	38.2	0.19	0.05	23.95	0.38	0.18	100.19	74.0
33.	515	37.56	38.62	0.18	0.03	23.16	0.36	0.19	100.1	74.8
34.	515	37.58	38.49	0.16	0.04	23.26	0.36	0.18	100.07	74.7
35.	515	37.58	38.46	0.18	0.03	23.24	0.36	0.21	100.07	74.7
36.	515	38.39	40.59	0.17	0.04	20.06	0.32	0.31	99.89	78.3
37.	515	37.9	40.28	0.18	0.04	20.69	0.35	0.24	99.68	77.6
38.	515	37.99	40.25	0.16	0.04	20.56	0.34	0.27	99.62	77.7
39.	515	38.05	40.27	0.21	0.03	20.76	0.34	0.27	99.94	77.6

Note: Analyses were carried out in GEOKHI RAS, analyst N. Kononkova

However, it is strange that this also applies to other types of rocks. Olivines in subvolcanic formations usually have average CaO contents of 0.3 wt %, but only some grains from the lower taxitic gabbro-dolerite have similar calcium concentrations. This provides new information for interpreting their origin. The complexity is that the calcium content of olivine is not a direct function of the pressure; the process is more complicated because the occurrence of olivine in a mineral affects the melt composition from which the olivine is crystallized. The Mn concentration of olivine is directly correlated with the iron content.

Interesting data of the distribution of trace elements in olivines (Table 4.17) were obtained by ion microprobe (S.G. Simakin, Institute of Microelectronic Russian Academy of Sciences, Yaroslavl', Russia). The heavy rare earths and yttrium are well correlated with the fayalite component in

Table 4.17 Representative analyses of olivines from picritic gabbro-dolerites of the Noril'sk 1 intrusion ("Medvezhy Creek" open pit)

Component	1	2	3	4	5	6	7	8	9
SiO_2	38.93	38.96	38.62	38.86	38.78	38.89	38.74	38.88	38.52
FeO	18.64	19.02	19.62	18.37	18.76	19.51	19.11	19.26	20.36
MnO	0.29	0.30	0.30	0.29	0.30	0.30	0.30	0.30	0.31
MgO	42.23	41.95	41.48	42.39	42.07	41.53	41.60	41.72	40.63
CaO	0.11	0.12	0.12	0.11	0.12	0.12	0.12	0.12	0.24
NiO	0.23	0.21	0.30	0.24	0.22	0.24	0.26	0.24	0.21
Cr_2O_3	0.03	0.02	0.01	0.01	0.01	0.01	0.01	0.01	0.02
Total	100.51	100.63	100.49	100.31	100.31	100.66	100.18	100.58	100.36
Fo	80.2	79.7	79.0	80.4	80.0	79.1	79.5	79.4	78.1
Y	1.11	1.72	1.66	1.23	1.42	1.98	1.77	1.71	0.68
Dy	0.13	0.16	0.15	0.12	0.13	0.16	0.16	0.19	0.07
Ho	0.05	0.06	0.07	0.04	0.04	0.07	0.07	0.06	0.02
Er	0.18	0.31	0.31	0.20	0.24	0.34	0.35	0.31	0.10
Tm	0.04	0.07	0.07	0.05	0.06	0.07	0.07	0.07	0.03
Yb	0.44	0.63	0.62	0.47	0.53	0.63	0.65	0.67	0.20
Lu	0.07	0.14	0.11	0.09	0.11	0.13	0.13	0.12	0.04
Ti	140	185	221	190	178	211	187	227	89
Zn	121	130	145	126	127	153	143	161	173
Sc	9.4	11.3	9.4	9.0	9.6	10.3	10.6	9.3	11.0
Al	135	123	94	107	114	120	107	108	181
Zr	0.26	0.32	0.35	0.27	0.25	0.30	0.27	0.27	0.08
Co	192	204	197	202	200	227	211	240	233
Ge	0.87	0.78	0.96	0.93	0.86	1.14	1.02	1.12	1.13
V	7.31	7.36	6.27	6.57	6.05	7.55	6.77	8.17	23.20
Li	6.51	5.96	5.88	6.51	5.22	6.20	4.87	6.84	10.06

Note: Major components (wt %) were analyzed by EPMA in GEOKHI RAS, analyst N. Kononkova; rare elements (ppm) were determined by SIMS in Yaroslavl', analyst S. Simakin

the olivine. The maximum slope of the trend line is typical of picritic gabbro-dolerites, and the trends of the rare earths in olivine are significantly flatter for the taxitic gabbro-dolerites and leucogabbro. The olivines from the contact gabbro-dolerites form a separate trend that is characterized by the highest concentrations of rare earth elements (Fig. 4.32).

Figure 4.33 shows the distributions of Y, Yb, and Dy in grains of different sizes. Diagrams a–d show that the morphological features of the mineral do not influence the distribution of these elements, i.e., large, medium, and small grains form one trend. Only points that represent granular high-Mg grains of olivine form an isolated compact field with very low contents of rare earths and yttrium. The concentration of Cr shows some dependence on the grain size. The maximum element contents typically occur in medium and large grains, and small and granular grains generally have lower Cr contents. The distributions of Ti, V, and Li are clearly distinguished only in the contact gabbro-dolerites, which are characterized by a direct correlation between the vanadium and forsterite contents and an inverse correlation between titanium and lithium. Sodium does not have any relationship with the compositions of the main components in the olivine and their affiliations with the types of rocks.

The rare earth elements measured in the olivines can be divided into groups according to their correlations with the fundamental characteristic of the olivine, its forsterite component, as follows:

1. A group of rare earth elements (Y, Yb, Dy, Lu, Ho, Er, and Tm) and Ti are negatively correlated with Mg in olivine, but the elements are well correlated with each other. The light rare earths have very low concentrations in the olivines, so they cannot be measured using SIMS or LA; however, it is possible that with further development of the method (full involvement of the evaporated material in the mass spectrometer), this problem will be solved. Increasing the diameter of the laser to 160 μm (in the presence of large grains of pure olivine that are 200–250 μm in size) might lead to the identification of several more rare earth elements.

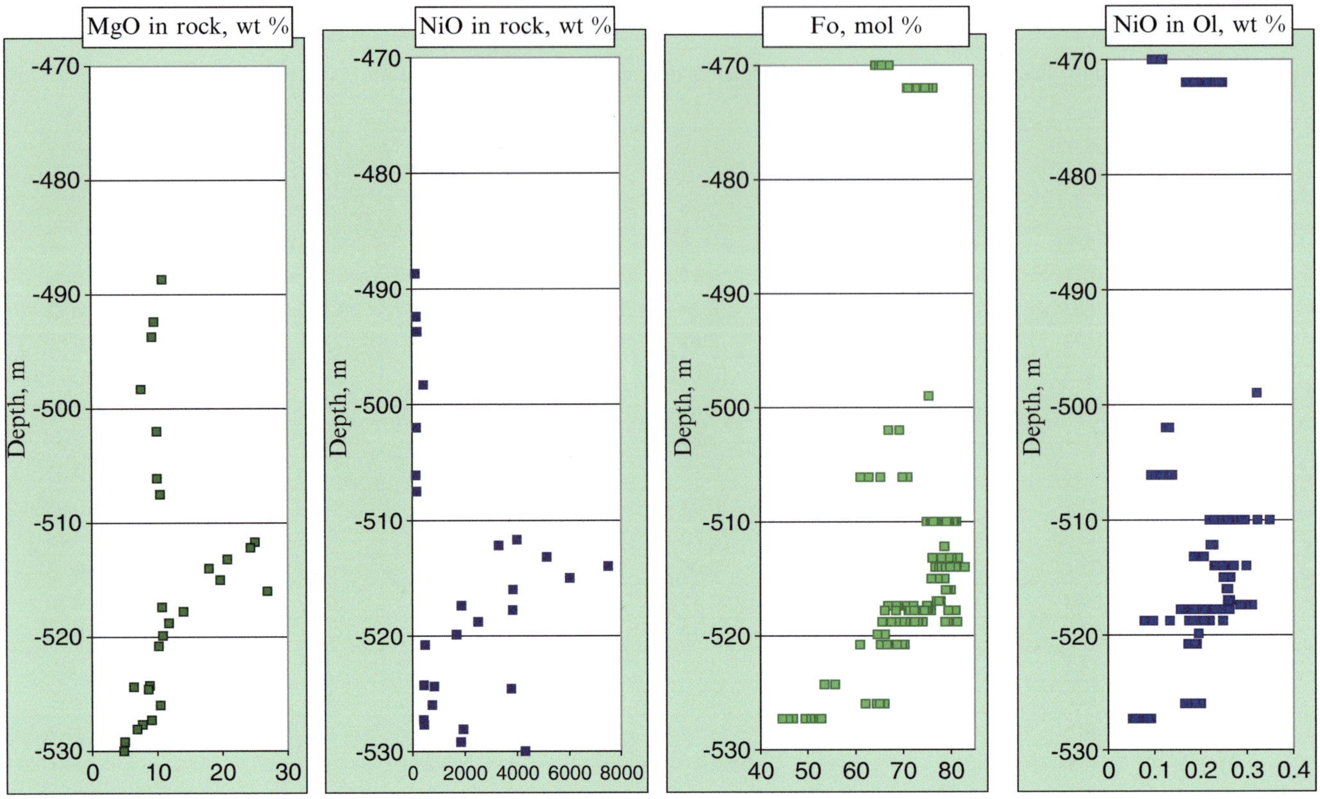

Fig. 4.32 Variations of MgO and NiO contents in rocks and Fo and NiO in olivine in borehole MS-31

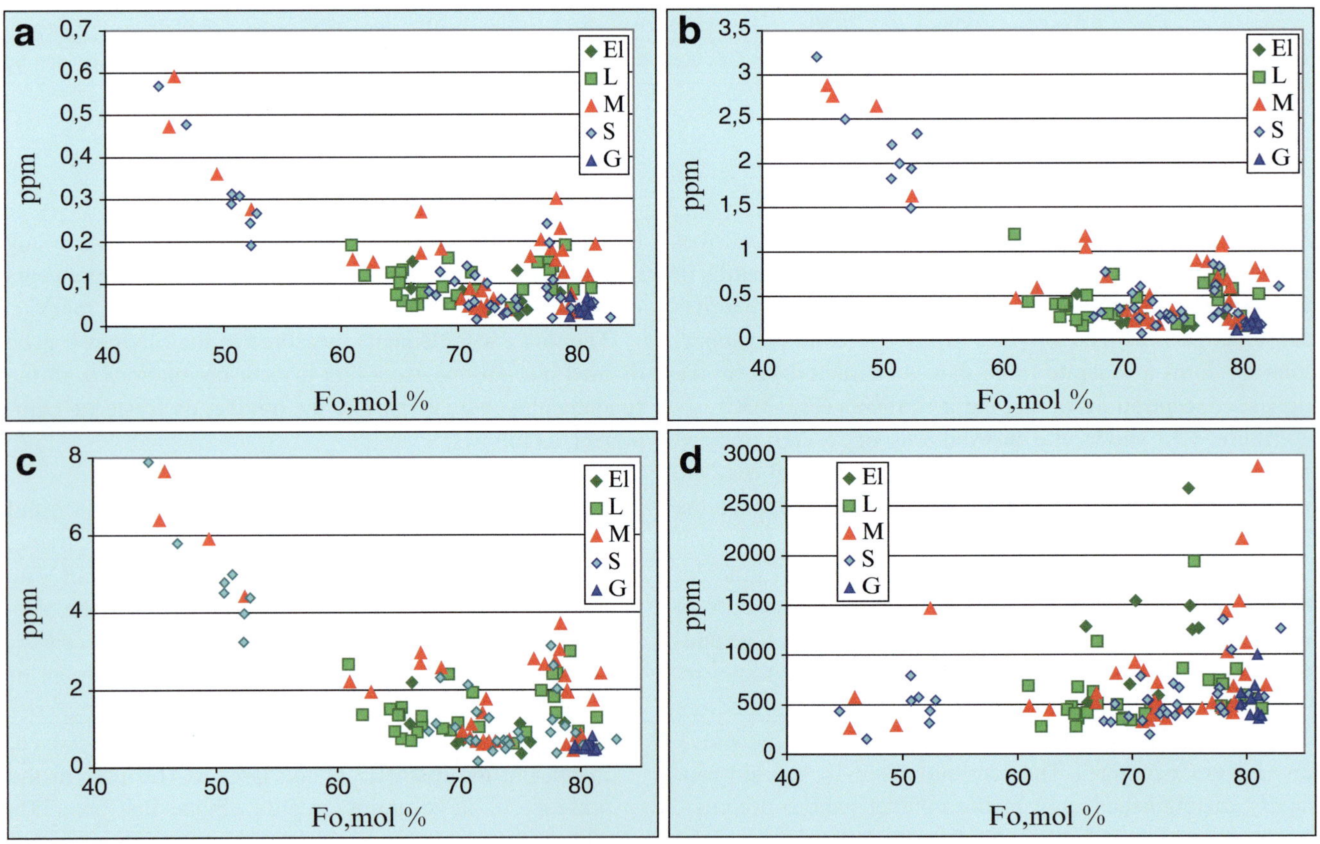

Fig. 4.33 Diagrams Dy (**a**), Yb (**b**), Y (**c**), Cr (**d**) – Fo for olivines of different size from Noril'sk 1 intrusion
Grains, size: El, extra large; l, large; M, middle; S, small; G, granulated

2. A group of nonferrous metals (Cu, Ni, Zn, Co) shows a weak negative correlation with forsterite in the olivine. This is somewhat unusual for nickel from picrite horizons that are enriched in disseminated sulfide and indicates that Ni was redistributed between the olivine and the sulfide phase. The same characteristic occurred with cobalt.
3. Ca and V are independent of the main components of the olivine.
4. Sc, Ge, and Sr have large spreads and show no patterns with respect to the main components of the olivine. These patterns are probably determined by their low concentrations; therefore, the results are not reliable. Li and Zr are in the same group, although they have higher concentrations. However, their distribution patterns were also detected.

Pyroxene. We also studied pyroxene because of the large variations in the compositions of its main components as well as the presence of large amounts of impurities, which can significantly change the concentration depending on the crystallization conditions. However, because pyroxene is a very late phase in the formation of the rocks of the Noril'sk 1 massif it contains information about the processes that occurred only within the intrusive chamber. In this respect, the data are complementary to those obtained from olivine, which reflect the earliest stages of the formation of the rocks. The pyroxene compositions help us to understand the peculiarities of the complete crystallization process of the gabbro-dolerites. We do not know the stage that mineralization occurred within the massif. The composition of the melt does not provide information about the ore components if the magma only mechanically transported the mass of sulfide from the source to the crystallization chamber. This is why it is important to examine all of the stages of massif formation in detail.

Pyroxenes from the sequence in the boreholes discussed above (MS-31, G-22, and MG) were studied (Table. 4.18). In all of the samples, the Mg# vary from 58–84, and the predominant pyroxenes from picritic gabbro-dolerites have Mg# of 80–84. Low differences in the magnesium concentrations were present in the pyroxenes of the lower picritic gabbro-dolerites (Mg# of 81–84), while the upper picritic gabbro-dolerites are characterized by the presence of three generations of pyroxene: (1) moderate magnesium minerals (Mg# 72–76); (2) high-Mg minerals (Mg# 80–82), which are found only in high-Mg rocks; and (3) low-Mg minerals (Mg# 60–64). The upper taxitic and olivine gabbro-dolerites contain pyroxenes with nearly identical compositions (Mg# of 80–82 and 72–83, respectively), while the pyroxenes in the lower taxitic gabbro-dolerites had Mg# of 78–79. The contact gabbro-dolerites had typical pyroxene Mg# of 63–73, and most had Mg# of 65–66.

Profiles of large poikilitic pyroxene grains were studied to determine how their compositions vary from the center to the periphery. No significant changes in Mg# were observed; the Mg# varied slightly from 78.3 at the center to 77.5 at the periphery. These changes in the overall variability of grains occur within individual horizons, not across the entire massif.

The main patterns of the impurity elements in the pyroxenes, including Cr, Ti, Na, and Mn, were analyzed by electron probe. The concentrations of the elements, which were substantially lower than those from the microprobe analysis, were studied by laser ablation and will be provided later.

The clearest dependence of the Mg–pyroxene is observed for chromium. Fig. 4.34a demonstrates the hyperbolic dependence of Cr_2O_3 content in pyroxenes from different rock types of the Noril'sk 1 intrusion. The maximum enrichment in Cr is observed in pyroxenes from the lower picritic gabbro-dolerites. The pyroxenes from the upper picritic and taxitic gabbro-dolerites have large variations of the Cr_2O_3 contents from 0.1 to 1.2 wt %. Chromium is almost completely absent in the pyroxenes from the lower-contact gabbro-dolerites. The opposite characteristic is observed for titanium (Fig. 4.34b). Its concentration increases with the iron content of the pyroxene; thus, the most depleted pyroxenes are in the lower and upper picritic gabbro-dolerites (Fig. 4.35). In general, the TiO_2 concentrations in the pyroxenes are not as high as the chromium concentrations. Nevertheless, it is clear that the most Ti-rich pyroxenes are in the contact gabbro-dolerite. The sodium and manganese contents in the pyroxenes are very similar.

A comparison of the compositions of the olivines and pyroxenes from the same horizons in the central and marginal parts of the Noril'sk 1 massif (from the boreholes and the open pit mine) demonstrates their similar compositions.

Plagioclase. Studying the composition of plagioclase is a laborious process because the size and morphology of the grains of the mineral vary greatly in each horizon of the massif. Obtaining an objective picture requires having statistics that reflect the composition of the grains from each generation (Table 4.19). In addition, plagioclase generally has a zonal structure, so comparisons between grains are difficult. Table 4.19 presents data from the individual grains; in the last column, the letters "l," "m," and "s" denote the large (>0.5 mm), middle (0.5–0.2 mm), and small (<0.2 mm) grains, respectively. The data from the central portion of the grain are given in Table 4.19. The plagioclase composition can vary from 82.96 to 69.50 mol% An.

Table 4.18 Representative analyses of pyroxenes from the Noril'sk 1 intrusion (boreholes MS-31. MN-7), wt %

	N sample	MgO#	SiO_2	TiO_2	Al_2O_3	FeO	MnO	MgO	CaO	Na_2O	K_2O	Cr_2O_3	Total
1.	MN-7/463.6	78.01	38.34	0.03	0.02	20.61	0.31	41.01	0.14	0.01	0.000	0.01	100.50
2.	MN-7/463.6	86.08	53.34	0.14	0.55	4.77	0.18	16.55	22.63	0.37	0.009	0.52	99.08
3.	MN-7/463.6	81.73	50.88	0.68	2.87	6.61	0.17	16.58	19.85	0.29	0.000	0.96	98.91
4.	MN-7/469.7	61.72	48.64	0.31	7.02	15.16	0.29	13.71	11.78	1.28	0.07	0.004	98.28
5.	MN-7/469.7	74.32	53.16	0.66	1.09	16.09	0.35	26.12	1.99	0.04	0.000	0.01	99.53
6.	MN-7/469.7	74.26	53.47	0.53	0.92	16.38	0.35	26.51	1.63	0.05	0.000	0.01	99.86
7.	MN-7/469.7	82.01	51.58	0.55	2.54	6.60	0.15	16.88	20.13	0.25	0.001	0.42	99.13
8.	MN-7/471.6	25.17	47.52	0.03	34.28	0.53	0.00	0.10	16.04	2.26	0.07	0.000	100.86
9.	MN-7/471.6	24.82	47.32	0.02	34.79	0.54	0.00	0.10	16.51	1.97	0.05	0.009	101.34
10.	MN-7/471.6	67.51	37.00	0.03	0.01	29.37	0.42	34.23	0.21	0.00	0.005	0.004	101.30
11.	MN-7/471.6	73.18	37.44	0.02	0.04	24.82	0.34	37.99	0.18	0.01	0.001	0.02	100.89
12.	MN-7/471.6	82.70	51.90	0.41	2.31	6.18	0.15	16.57	20.65	0.24	0.000	0.67	99.09
13.	MH-7/472.7	73.92	50.64	0.87	2.09	9.88	0.24	15.71	19.01	0.25	0.000	0.02	98.74
14.	MN-7/472.7	72.78	51.08	0.90	1.85	10.49	0.27	15.73	18.72	0.27	0.004	0.01	99.34
15.	MN-7/472.7	73.14	51.84	0.87	1.88	10.37	0.26	15.84	18.84	0.24	0.000	0.02	100.16
16.	MN-7/474.7	71.06	50.94	0.77	2.54	10.68	0.25	14.71	19.38	0.25	0.000	0.08	99.62
17.	MN-7/475.2	69.89	51.36	0.76	1.79	11.45	0.33	14.91	18.80	0.29	0.000	0.01	99.71
18.	MN-7/475.2	66.34	51.41	0.66	1.66	12.81	0.40	14.16	18.52	0.30	0.000	0.00	99.94
19.	MN-7/475.2	69.74	50.79	1.00	2.55	11.31	0.30	14.62	19.22	0.32	0.002	0.009	100.14
20.	MN-7/475.2	69.10	51.07	0.90	2.19	11.51	0.31	14.44	19.42	0.30	0.004	0.01	100.18
21.	MN-7/475.2	68.50	50.33	1.06	2.67	11.82	0.29	14.42	18.96	0.31	0.04	0.005	99.94
22.	MN-7/475.2	68.85	51.43	0.73	1.85	11.84	0.34	14.68	18.81	0.30	0.000	0.008	100.02
23.	MN-7/475.4	67.25	50.63	0.82	1.94	12.20	0.35	14.05	19.30	0.30	0.005	0.009	99.61
24.	MN-7/475.4	68.91	49.83	0.86	2.19	11.46	0.30	14.25	19.40	0.29	0.000	0.005	98.60
25.	MN-7/475.4	66.31	50.33	1.14	2.46	12.42	0.37	13.71	19.21	0.34	0.001	0.02	100.01
26.	MN-7/475.4	66.10	51.64	0.63	1.55	12.95	0.40	14.16	18.44	0.29	0.01	0.00	100.09
27.	MN-7/475.4	65.76	50.17	0.82	2.17	12.68	0.37	13.66	19.01	0.34	0.000	0.02	99.26
28.	MS-31/486.4	89.91	44.54	0.00	26.49	0.01	0.00	0.05	27.04	0.00	0.004	0.01	98.16
29.	MS-31/486.4	81.68	44.48	0.01	26.37	0.02	0.00	0.05	26.89	0.00	0.000	0.02	97.85
30.	MS-31/486.4	66.20	46.87	0.11	16.81	4.77	0.15	5.24	23.98	0.08	0.007	0.04	98.06
31.	MS-31/486.4	74.82	44.70	0.01	26.36	0.03	0.00	0.05	26.96	0.01	0.000	0.003	98.13
32.	MS-31/486.4	67.06	53.60	0.14	0.76	11.76	0.27	13.43	20.90	0.12	0.002	0.17	101.17
33.	MS-31/486.4	77.22	50.59	0.51	4.64	7.18	0.20	13.65	20.84	0.18	0.005	0.35	98.15

Note: Analyses were carried out in GEOKHI RAS, analyst N. Kononkova

The composition of a grain does not always depend on its size. This is clearly observed in the picritic gabbro-dolerites (Table 4.19; "size" indicates the grain size). The compositions of the plagioclases from different horizons vary from An_{50} to An_{90} and are dominated by $An_{80–85}$ (Fig. 4.36). The most basic plagioclases were found in the picritic gabbro-dolerites and partially in the olivine gabbro-dolerites. The picritic gabbro-dolerites contain a nearly continuous series of compositions from An_{62} to An_{88}, and the olivine gabbro-dolerites have a bimodal distribution with peaks at $An_{75–90}$ and $An_{64–66}$. A wide range of plagioclase compositions ($An_{62–88}$) is observed in the lower taxitic gabbro-dolerites and leucogabbro; the compositions are fairly evenly distributed and are not dominated by any compositions. The contact gabbro-dolerites contain the most acid plagioclase ($An_{54–70}$ with a peak at $An_{60–64}$).

Table 4.19 shows the most interesting plagioclase data, which are from the upper picrite and taxitic horizons in G-22. These horizons are characterized by wider scatter ($An_{55–80}$ in the upper taxites and $An_{55–90}$ in the picrites) than the lower analogous rocks. Fe was the only trace element found in the plagioclase (by electron microprobe), but its distribution was not established. The FeO content has a wide variability of 0.4 to 1 wt % in nearly all of the horizons.

Chrome Spinels. Minerals from the spinel class were only detected in the upper and lower picrites They are located

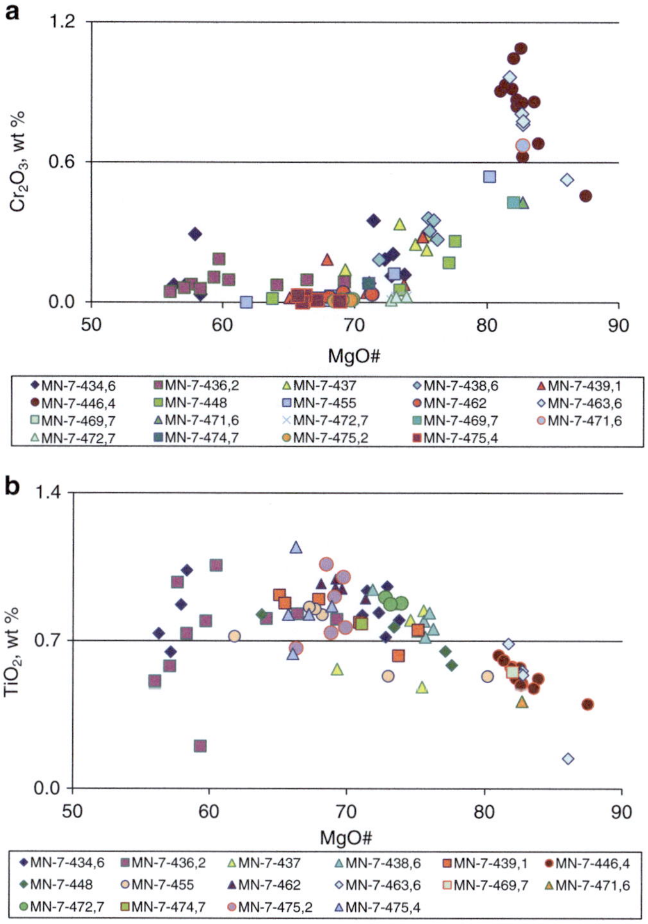

Fig. 4.34 Diagrams Cr, Ti–Mg# for pyroxenes from the Noril'sk 1 intrusion (borehole MN-7)
No sample = MN-7-448 – No borehole and depth in m

within the main rock-forming silicates: olivine (Table 4.20), pyroxene, and plagioclase (Table 4.21). The smallest grains (5–10 microns) are found in the olivine from the lower and upper picritic gabbro-dolerites. Their distinguishing feature is their similar and elevated Cr_2O_3 contents. Spinels from the olivines of the lower picritic gabbro-dolerite differ from those in the upper picritic gabbro-dolerite in their higher-magnesium contents, high zinc concentrations (up to 0.25 wt %), and reduced titanium contents (2–4 wt %). The spinel grains from plagioclase and pyroxene in these horizons are characterized by higher-Mg contents (Fig. 4.37a–c).

Spinels from different massifs of the Noril'sk area were studied by S. Barnes and V. Kunilov (2000). They described unusual high-Ti spinel in high-Mg rocks and explained this fact by long time presence of crystals in open magmatic system.

All of the components of the spinels have similar compositions with the exception of zinc, which has significantly lower contents in the spinel in olivine grains in the lower picritic gabbro-dolerite than in the upper spinel of the

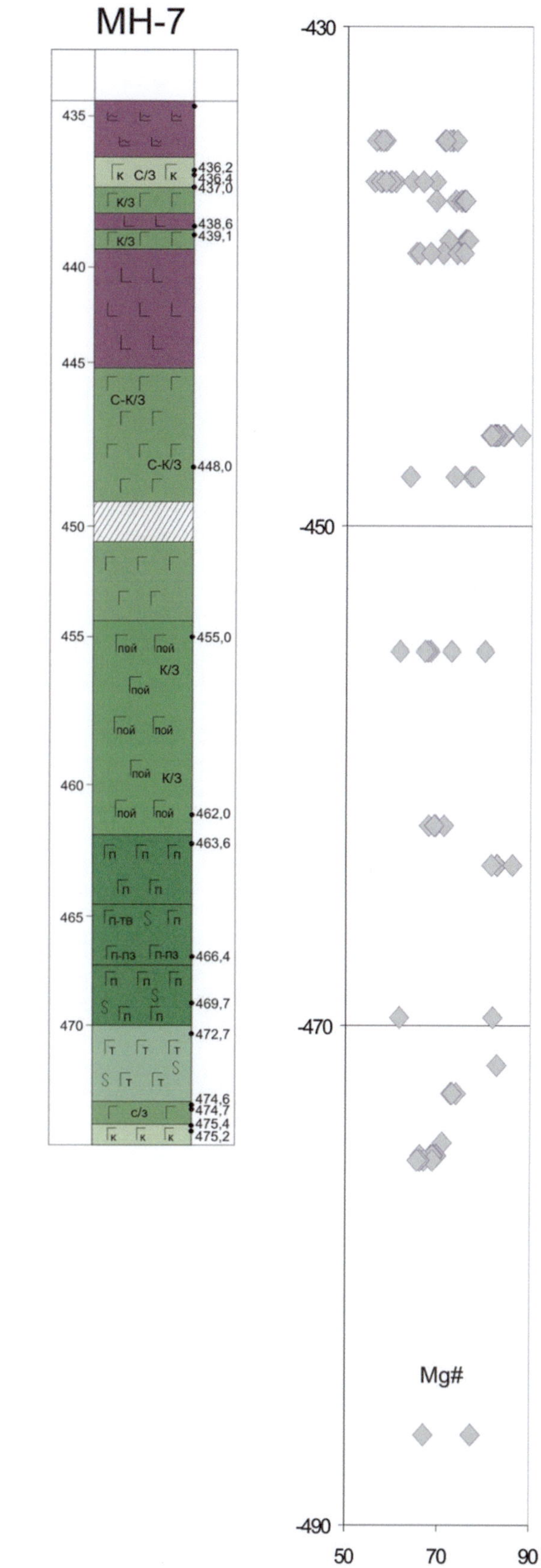

Fig. 4.35 Pyroxene variation compositions throw vertical section of the Noril'sk 1 intrusion (borehole MN-7)

Table 4.19 Plagioclase composition from the rocks of the Noril'sk 1 intrusion, wt %

№ п/п	Depth, m	Rock	Na_2O	SiO_2	CaO	FeO	Al_2O_3	MgO	K_2O	Total	An, mol%	Size
1.	G – 22	Upper taxitic gabbro-dolerites	2.46	48.76	15.14	0.87	30.66	0.02	0.13	98.10	77.33	s
2.	108		1.90	47.59	15.93	0.71	30.96	0.04	0.12	97.28	82.26	m
3.	108		1.72	47.58	16.24	0.66	31.23	0.00	0.09	97.62	83.95	m
4.	108		2.33	49.38	14.97	0.86	30.42	0.00	0.12	98.17	78.07	s
5.	108		3.18	51.97	13.05	0.93	28.60	0.06	0.19	98.06	69.45	m
6.	108		3.47	51.77	12.72	0.92	28.28	0.08	0.23	97.56	66.98	m
7.	108		2.80	50.53	13.94	0.85	29.86	0.00	0.18	98.26	73.37	m
8.	108		2.73	49.97	14.25	0.68	29.99	0.00	0.18	97.86	74.29	m
9.	108		4.16	53.91	11.10	0.78	27.06	0.02	0.29	97.46	59.59	m
10	108		2.30	49.00	15.07	0.68	30.53	0.01	0.12	97.74	78.38	l
11.	108		1.98	48.12	15.83	0.72	31.30	0.00	0.10	98.06	81.57	s
12.	108		1.92	47.35	15.54	0.79	30.87	0.00	0.13	96.75	81.78	s
13.	108		1.80	47.57	16.29	0.93	31.56	0.04	0.08	98.28	83.37	s
14.	108		3.77	53.33	12.04	0.83	28.15	0.01	0.20	98.41	63.88	m
15.	108		2.13	48.26	15.94	0.93	31.32	0.01	0.10	98.76	80.55	s
16.	108		2.59	49.54	14.72	1.36	30.44	0.03	0.17	98.89	75.86	s
17.	108		3.15	51.71	12.92	1.55	28.87	0.06	0.29	98.70	69.40	s
18.	108		3.11	51.43	13.12	1.00	28.95	0.25	0.21	98.18	70.03	l
19.	108		1.96	47.68	16.13	0.80	31.49	0.05	0.07	98.23	81.99	m
20.	108		2.36	48.98	15.18	0.80	30.83	0.02	0.08	98.32	78.07	m
21.	108		2.52	49.50	14.58	0.83	30.08	0.00	0.13	97.74	76.22	m
22.	108		2.60	49.78	14.24	0.83	29.94	0.06	0.14	97.63	75.17	s
23.	108		4.21	54.10	11.08	0.71	27.20	0.04	0.29	97.71	59.29	m
24.	108		1.64	47.01	16.59	0.72	31.86	0.03	0.08	97.97	84.86	s
25.	108		3.39	51.62	13.02	0.80	28.78	0.05	0.22	97.94	68.00	m
26.	108		2.56	49.38	14.96	0.76	30.41	0.11	0.14	98.37	76.42	m
27.	108		2.77	49.58	14.39	0.68	29.82	0.05	0.13	97.53	74.20	m
28.	107	Upper picritic gabbro-dolerites	2.42	49.63	14.85	0.95	30.57	0.06	0.15	98.86	77.24	l
29.	107		1.58	47.31	16.54	0.78	31.93	0.06	0.08	98.34	85.29	l
30.	107		3.67	52.57	12.53	0.85	28.62	0.07	0.34	98.69	65.40	m
31.	107		2.05	48.37	15.86	0.68	31.55	0.08	0.16	98.79	81.03	s
32.	107		1.56	46.88	16.71	0.74	32.17	0.04	0.08	98.22	85.59	l
33.	107		2.11	48.85	15.53	0.69	31.41	0.05	0.14	98.86	80.26	m
34.	107		3.17	51.26	13.07	0.61	29.19	0.05	0.21	97.66	69.50	l
35.	107		1.83	48.03	16.11	0.77	31.47	0.07	0.11	98.43	82.96	l
36.	107		3.50	52.67	12.69	0.75	28.34	0.07	0.23	98.30	66.74	l
37.	107		2.23	48.71	15.63	0.62	31.27	0.03	0.11	98.65	79.49	l
38.	107		2.25	49.45	15.16	0.79	30.72	0.09	0.19	98.70	78.85	l
39.	107		2.10	48.41	15.41	0.67	31.30	0.05	0.11	98.14	80.27	l
40.	107		2.24	49.05	15.42	0.66	31.12	0.11	0.09	98.73	79.23	l
41.	107		2.31	48.94	15.38	0.62	31.03	0.09	0.13	98.59	78.65	s
42.	107		2.34	49.34	15.13	0.73	31.11	0.07	0.09	98.87	78.15	l
43.	107		2.41	49.39	14.91	0.68	30.83	0.08	0.11	98.53	77.39	l
44.	107		2.31	49.10	14.99	0.77	31.00	0.07	0.11	98.40	78.23	l
45.	107		3.50	52.25	12.57	0.75	28.76	0.06	0.25	98.24	66.51	m
46.	107		1.05	9.43	10.65	0.48	7.69	0.00	0.04	29.43	84.83	m
47.	107		2.24	48.71	15.64	0.81	30.94	0.05	0.12	98.55	79.43	m
48.	107		3.05	50.98	13.27	0.78	28.78	0.06	0.19	97.18	70.63	s
49.	107		2.84	50.97	13.77	0.81	29.27	0.07	0.18	98.00	72.84	s
50.	107		3.19	51.46	13.38	0.79	28.82	0.08	0.18	97.98	69.87	s

(continued)

Table 4.19 (continued)

№ п/п	Depth, m	Rock	Na_2O	SiO_2	CaO	FeO	Al_2O_3	MgO	K_2O	Total	An, mol%	Size
51.	107		2.33	48.95	15.39	0.72	31.18	0.01	0.12	98.80	78.53	s
52.	107		2.30	48.99	15.06	0.80	30.55	0.06	0.14	98.00	78.35	s
53.	107		3.28	52.30	12.53	0.83	28.58	0.06	0.26	97.90	67.88	l
54.	107		3.17	51.76	12.79	0.84	28.81	0.04	0.21	97.69	69.08	l
55.	107	Upper picritic gabbro-dolerites	1.59	10.25	10.12	0.52	7.75	0.00	0.09	30.41	77.85	s
56.	107		2.39	49.52	14.71	0.67	30.39	0.08	0.12	98.00	77.27	l
57.	107		2.21	49.46	15.07	0.81	30.82	0.04	0.08	98.55	79.06	l
58.	107		2.07	48.91	15.25	0.72	30.84	0.05	0.13	98.06	80.32	s
59.	107		1.67	47.29	16.27	0.87	31.56	0.28	0.10	98.12	84.37	s
60.	107		3.63	52.72	11.98	0.64	28.31	0.10	0.31	97.82	64.64	m
61.	107		3.85	53.26	11.69	0.81	27.79	0.05	0.30	97.82	62.71	m
62.	107		2.22	48.69	15.18	0.69	30.97	0.05	0.11	97.99	79.12	s
63.	107		2.10	47.52	15.32	0.78	31.24	0.08	0.11	97.18	80.14	m
64.	107		2.21	47.68	15.39	0.78	31.26	0.04	0.12	97.54	79.38	m
65.	107		2.17	49.12	15.11	0.70	30.89	0.04	0.14	98.25	79.40	m
66.	107		1.75	47.47	16.49	0.77	31.90	0.05	0.10	98.62	83.94	l
67.	107		2.00	48.02	15.97	0.74	31.33	0.06	0.10	98.25	81.56	l
68.	107		1.90	48.20	15.95	0.77	31.55	0.06	0.10	98.64	82.25	l
69.	107		2.10	48.31	15.53	0.78	31.14	0.00	0.11	98.05	80.36	l
70.	107		2.45	48.97	15.29	0.70	30.73	0.08	0.08	98.34	77.55	l
71.	107		1.89	47.91	16.05	0.72	31.44	0.04	0.08	98.17	82.42	l
72.	107		2.20	48.73	15.39	0.82	30.89	0.06	0.13	98.29	79.44	l
73.	107		2.00	48.17	15.78	0.77	31.42	0.08	0.12	98.49	81.39	l
74.	107		1.78	47.68	16.01	0.73	31.31	0.05	0.11	97.69	83.30	l
75.	107		2.47	49.05	15.04	0.74	30.87	0.09	0.14	98.42	77.10	l
76.	107		2.19	49.13	15.39	0.75	30.63	0.09	0.10	98.35	79.53	l
77.	107		3.59	52.78	12.25	0.64	28.41	0.07	0.29	98.16	65.39	l
78.	107		2.61	49.58	14.62	0.90	30.18	0.06	0.18	98.15	75.60	l
79.	107		2.40	49.03	15.24	0.76	30.69	0.06	0.15	98.37	77.84	l
80.	107		1.91	47.68	16.32	0.67	31.76	0.05	0.09	98.56	82.54	l
81.	107	Upper picritic gabbro-dolerites	2.49	48.65	15.18	0.78	30.55	0.08	0.09	97.89	77.11	s
82.	107		1.62	46.75	16.44	0.76	31.23	0.04	0.06	96.99	84.91	s
83.	107		2.64	49.61	14.26	0.68	29.62	0.08	0.12	97.10	74.94	l
84.	107		1.69	47.11	16.40	0.87	32.04	0.02	0.11	98.30	84.29	s
85.	107		2.11	48.71	15.32	0.77	30.60	0.06	0.17	97.78	80.11	s
86.	107		4.15	53.62	10.46	0.71	26.69	0.08	0.34	96.07	58.25	l
87.	107		2.16	48.04	15.08	0.58	30.52	0.04	0.11	96.66	79.44	l
88.	107		4.06	52.90	11.91	0.75	27.66	0.03	0.21	97.75	61.86	m
89.	107		2.24	48.63	15.04	0.82	30.09	0.04	0.06	97.03	78.77	m
90.	107		2.12	48.02	15.53	0.72	30.72	0.00	0.13	97.28	80.25	m
91.	107		1.85	47.20	15.89	0.66	31.27	0.06	0.11	97.12	82.65	m
92.	107		1.72	46.92	16.25	0.66	31.47	0.04	0.08	97.18	83.97	m
93.	107		1.95	47.51	15.97	0.73	31.17	0.04	0.13	97.50	81.96	s
94.	107		2.33	49.13	15.11	0.69	30.24	0.04	0.12	97.72	78.22	s
95.	107		2.32	48.36	15.24	0.78	30.65	0.06	0.13	97.69	78.41	s
96.	107		3.10	50.51	13.02	0.75	29.06	0.07	0.22	97.08	69.93	l
97.	107		4.31	54.40	10.55	0.68	26.75	0.10	0.41	97.35	57.56	l
98.	107		3.67	52.04	12.36	0.76	27.95	0.06	0.27	97.17	65.11	l

(continued)

Table 4.19 (continued)

№ п/п	Depth, m	Rock	Na_2O	SiO_2	CaO	FeO	Al_2O_3	MgO	K_2O	Total	An, mol%	Size
99.	107		3.10	51.24	13.20	0.72	28.58	0.09	0.21	97.23	70.18	l
100.	107		2.22	49.22	15.02	0.70	30.33	0.09	0.11	97.76	78.92	l
101.	107		4.41	54.10	10.44	0.78	26.86	0.06	0.40	97.24	56.73	s
102.	107		4.01	53.10	11.46	0.79	27.51	0.10	0.32	97.37	61.29	m
103.	MS31/499	Olivine	1.75	47.43	16.00	0.68	31.35	0.12	0.07	97.49	83.48	l
104.	499	Gabbro-	1.49	46.87	16.94	0.53	32.20	0.10	0.03	98.20	86.25	l
105.	499	dolerites	3.62	52.68	11.84	0.70	27.73	0.09	0.27	97.07	64.44	s
106.	499		1.20	45.97	17.42	0.59	32.83	0.11	0.04	98.20	88.95	l
107.	499		1.66	47.24	16.60	0.63	31.86	0.07	0.05	98.18	84.72	l
108.	502	Olivine	2.28	48.54	15.39	0.60	31.23	0.01	0.10	98.23	78.87	s
109.	502	gabbro-	3.13	51.49	13.05	0.98	28.67	0.21	0.21	97.81	69.73	l
110.	502	dolerites	2.03	48.09	15.51	0.81	30.61	0.04	0.11	97.24	80.90	m
111.	502		4.31	54.33	10.88	0.52	27.12	0.05	0.28	97.61	58.27	m
112.	506,1		1.66	47.44	16.20	0.81	31.69	0.02	0.07	97.96	84.37	m
113.	506,1		3.45	51.77	12.47	1.02	27.93	0.49	0.25	97.49	66.65	m
114.	506,1		4.15	54.07	10.73	0.68	27.00	0.05	0.39	97.17	58.85	m
115.	506,1		1.88	47.64	16.33	0.64	31.75	0.05	0.06	98.41	82.74	m
116.	506,1		3.42	52.12	12.55	0.70	28.64	0.09	0.17	97.72	67.01	m
117.	506,1		3.35	52.03	12.79	0.72	28.62	0.06	0.21	97.90	67.86	m
118.	506,1		1.52	46.95	16.66	0.75	32.19	0.01	0.10	98.20	85.84	m
119.	506,1		1.56	46.71	16.92	0.74	32.22	0.01	0.10	98.30	85.71	m
120.	507,5		1.26	46.21	17.18	0.63	32.60	0.02	0.06	97.99	88.26	m
121.	507,5		1.73	47.56	16.39	0.95	31.35	0.06	0.09	98.16	83.97	s
122.	G22-64	Picritic	1.98	48.14	15.72	0.57	31.47	0.02	0.06	98.02	81.48	l
123.	64	gabbro-	1.70	47.31	16.39	0.91	32.22	0.01	0.07	98.68	84.25	l
124.	64	dolerites	2.16	48.27	15.12	1.24	30.58	0.08	0.12	97.65	79.48	l
125.	64		1.58	47.02	16.91	0.73	32.00	0.06	0.07	98.44	85.52	
126.	64		1.61	47.10	16.44	0.77	31.67	0.03	0.06	97.74	84.98	l
127.	64		1.37	45.10	15.04	4.32	30.52	0.64	0.03	97.40	85.87	l
128.	64		1.76	47.14	16.45	0.74	31.64	0.03	0.07	97.87	83.80	l
129.	64		1.50	46.05	16.14	2.25	31.30	0.07	0.04	97.49	85.63	l
130.	64		1.89	47.38	16.24	0.85	31.83	0.02	0.08	98.31	82.64	l
131.	64		2.29	49.00	15.03	0.79	30.69	0.05	0.10	97.97	78.39	s
132.	64		1.83	47.34	16.19	0.86	31.61	0.00	0.06	97.94	83.02	s
133.	64		1.57	47.43	16.64	0.73	31.91	0.04	0.06	98.40	85.44	m

(continued)

Table 4.19 (continued)

№	Depth, m	Rock	Na_2O	SiO_2	CaO	FeO	Al_2O_3	MgO	K_2O	Total	An	Зерно
134.	MS31/ 512.2	Picritic gabbro-	2.22	48.01	15.08	0.99	30.19	0.10	0.12	96.72	79.01	m
135.	512.2	dolerites	2.15	47.94	15.39	0.75	30.90	0.06	0.09	97.36	79.86	m
136.	512.2		2.26	48.80	15.13	0.68	30.60	0.05	0.11	97.64	78.72	m
137.	512.2		3.10	49.86	12.58	2.08	28.96	0.26	0.22	97.41	69.21	s
138.	513.2		2.51	49.25	14.86	0.69	30.68	0.13	0.12	98.33	76.62	l
139.	513.2		1.65	46.98	16.66	0.49	32.21	0.07	0.04	98.11	84.81	m
140.	514		1.68	47.67	16.24	0.62	31.71	0.00	0.06	98.07	84.24	m
141.	514		1.43	45.97	16.96	0.47	32.38	0.00	0.07	97.32	86.80	l
142.	514		1.43	46.64	16.94	0.52	32.47	0.00	0.08	98.13	86.74	m
143.	514		1.36	46.26	17.36	0.42	32.85	0.02	0.07	98.39	87.59	m
144.	514		1.63	47.05	16.70	0.47	32.11	0.03	0.09	98.12	85.04	s
145.	514		1.49	46.12	17.01	0.53	32.71	0.00	0.08	97.99	86.30	m
146.	514		1.66	46.89	16.52	0.73	31.87	0.02	0.08	97.80	84.67	m
147.	514		1.79	46.74	15.78	1.61	31.09	0.11	0.08	97.25	82.99	m
148.	514		3.80	52.76	12.06	0.69	28.04	0.08	0.23	97.70	63.72	l
149.	515		1.90	47.74	16.14	0.59	31.67	0.07	0.06	98.21	82.43	l
150.	515		2.04	48.67	15.36	0.74	30.69	0.05	0.10	97.72	80.63	l
151.	515		1.60	46.96	16.64	0.71	32.01	0.05	0.08	98.09	85.16	l
152.	515		1.96	48.02	15.88	0.62	31.41	0.03	0.09	98.07	81.79	k
153.	515		1.58	47.40	16.53	0.56	31.50	0.03	0.08	97.71	85.30	m
154.	516		2.98	50.80	13.51	0.83	29.56	0.09	0.18	98.07	71.47	m
155.	516		3.48	52.13	12.34	0.71	28.20	0.12	0.24	97.30	66.26	m
156.	516		2.73	50.11	14.34	0.71	29.79	0.10	0.14	97.95	74.39	l
157.	517.8	Taxitic	1.65	47.31	16.48	0.67	31.94	0.08	0.08	98.29	84.68	m
158.	517.4	gabbro-dol.	1.89	47.51	16.17	0.66	31.30	0.04	0.10	97.73	82.54	l

(continued)

Table 4.19 (continued)

№	Depth, m	Rock	Na_2O	SiO_2	CaO	FeO	Al_2O_3	MgO	K_2O	Total	An, mol%	Size
159.	MS31/517.4	Taxitic gabbro-dolerites	1.81	47.50	16.14	0.61	31.43	0.05	0.06	97.63	83.18	l
160.	517.4		1.73	47.36	16.59	0.58	31.78	0.07	0.04	98.24	84.14	l
161.	517.4		3.09	51.18	13.41	0.62	28.85	0.06	0.23	97.53	70.62	m
162.	517.4		2.04	48.21	15.60	0.64	31.00	0.09	0.10	97.74	80.85	s
163.	517.4		1.64	47.43	16.78	0.75	32.21	0.08	0.05	98.99	85.01	m
164.	517.8		2.11	47.98	15.81	0.65	31.23	0.11	0.08	98.02	80.60	l
165.	517.8		1.67	46.57	16.61	0.65	31.68	0.06	0.08	97.33	84.64	l
166.	517.8		2.12	47.65	16.04	0.51	31.48	0.01	0.12	97.98	80.71	m
167.	517.8		1.66	46.66	16.69	0.54	32.08	0.02	0.08	97.83	84.76	l
168.	517.8		1.37	46.29	16.86	0.62	31.94	0.02	0.08	97.23	87.23	l
169.	517.8		1.84	47.58	16.18	0.66	32.09	0.10	0.11	98.60	82.94	m
170.	517.8		1.72	47.19	16.72	0.62	32.16	0.03	0.09	98.57	84.36	s
171.	518.8		1.73	47.13	16.39	0.58	31.61	0.03	0.08	97.60	83.96	l
172.	518.8		2.26	48.92	15.23	0.60	30.96	0.09	0.06	98.17	78.88	s
173.	518.8		2.03	47.82	16.00	0.58	31.30	0.08	0.06	97.89	81.38	s
174.	518.8		2.13	48.43	14.68	0.68	31.15	0.16	0.52	97.81	79.21	s
175.	518.8		2.78	50.37	14.24	0.70	29.99	0.07	0.12	98.29	73.94	l
176.	518.8		2.55	49.09	14.91	0.73	30.46	0.11	0.10	97.99	76.38	l
177.	519.9		2.06	47.96	15.51	0.68	30.90	0.09	0.06	97.30	80.61	l
178.	519.9		2.50	49.06	15.12	0.66	30.71	0.05	0.11	98.25	77.02	m
179.	519.9		2.43	49.40	14.84	0.72	30.58	0.01	0.10	98.13	77.13	s
180.	519.9		2.30	49.03	14.94	0.71	30.86	0.07	0.08	98.07	78.26	m
181.	519.9		2.40	48.65	15.32	0.65	30.86	0.11	0.07	98.10	77.97	l.
182.	519.9		2.21	48.84	15.35	0.66	30.88	0.08	0.07	98.12	79.38	l
183.	519.9		2.27	49.10	14.87	0.74	30.54	0.09	0.10	97.82	78.41	s
184.	MS31/ 520.8	Leuco-gabbro	2.66	49.27	14.63	0.71	30.27	0.06	0.11	97.78	75.26	l
185.	520.8		2.04	48.03	15.43	0.64	31.43	0.05	0.07	97.76	80.71	l
186	526		3.00	51.10	13.33	0.70	28.69	0.08	0.16	97.18	71.08	s
187.	526		3.19	52.02	12.79	0.79	28.38	0.08	0.20	97.59	68.91	m
188.	526		3.16	51.91	12.92	0.64	28.90	0.09	0.23	97.91	69.38	m
189.	526		3.95	52.97	11.90	0.65	27.52	0.03	0.29	97.42	62.51	l
190.	526		3.70	53.26	12.04	0.87	28.15	0.09	0.31	98.49	64.29	m
191.	526		2.72	50.49	14.26	0.72	29.83	0.08	0.21	98.42	74.34	m
192.	526		2.97	51.50	13.29	0.72	29.00	0.15	0.20	97.97	71.21	m
193.	527.7	Contactic gabbro-dolerites	4.17	53.42	11.31	0.69	27.91	0.01	0.35	98.02	60.00	m
194.	527.7		4.16	53.57	11.20	0.53	27.45	0.00	0.35	97.38	59.81	m
195.	527.7		4.29	54.22	10.75	0.62	27.12	0.08	0.41	97.58	58.10	s
196.	527.7		4.58	54.58	10.26	0.89	26.69	0.02	0.38	97.59	55.35	s
197.	527.7		4.64	55.10	10.04	0.62	26.48	0.03	0.48	97.50	54.49	m
198.	527.3		3.59	51.79	12.19	0.73	28.14	0.00	0.31	96.82	65.26	s
199.	527.7		4.27	53.57	10.82	0.68	27.04	0.06	0.43	96.94	58.36	l
200.	527.7		4.20	53.33	11.65	0.60	28.11	0.00	0.35	98.33	60.57	l
201.	527.7		3.26	50.95	13.12	0.65	29.07	0.01	0.28	97.41	69.01	l

(continued)

Table 4.19 (continued)

№	Depth, m	Rock	Na_2O	SiO_2	CaO	FeO	Al_2O_3	MgO	K_2O	Total	An, mol%	Size
202.	528.1		3.88	52.88	11.71	0.65	28.00	0.06	0.36	97.64	62.57	l
203.	528.1		3.85	53.07	11.49	0.64	27.90	0.02	0.29	97.30	62.30	m
204.	528.1		3.97	53.06	11.07	0.53	27.45	0.01	0.29	96.47	60.71	s
205.	528.1		3.96	52.83	11.68	0.66	28.01	0.00	0.38	97.65	61.99	l
206.	528.1		3.85	52.66	11.70	0.58	27.66	0.00	0.35	96.88	62.74	s
207.	528.1		3.91	53.29	11.68	0.49	27.98	0.01	0.31	97.81	62.29	l
208.	528.1		4.15	53.40	11.54	0.61	27.99	0.00	0.26	98.08	60.59	s
209.	528.1		3.49	52.18	12.26	0.70	28.55	0.07	0.32	97.70	66.04	l

Note: Size of Pl grain
l, large (>0.5 mm); m, middle (0.2–0.5 mm); S, small (<0.2 mm). Analyses were carried out in GEOKHI RAS, analyst N. Kononkova

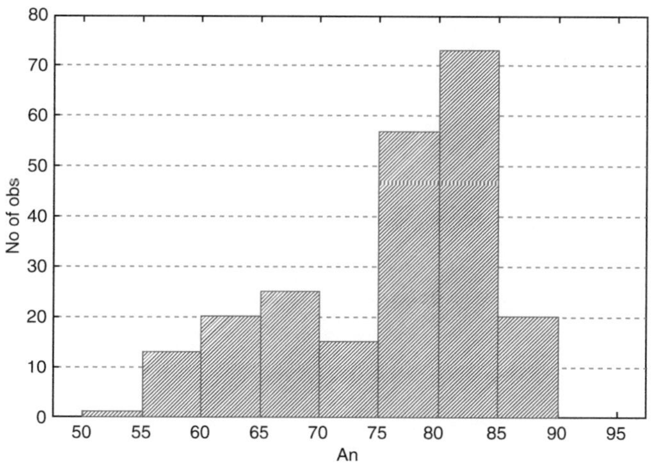

Fig. 4.36 Histogram for plagioclase from the rocks of the Noril'sk 1 intrusion

plagioclase and pyroxene from the same horizons. The high zinc contents suggest significant crystallization of this mineral; the highest concentrations in the olivine inclusions in the picrites from the lower horizons are additional evidence of the intratelluric nature of most of the olivine grains (as we have shown with other methods; Krivolutskaya and Sobolev 2001; Krivolutskaya et al. 2001).

4.5.1.4 Ores of the Noril'sk 1 Deposit

The Noril'sk sulfide copper–nickel deposits with well-known individual features of the distribution of platinum metals and ore types (Godlevsky and Shumskaya 1960; Genkin 1968; Genkin et al. 1981; Distler 1994a, b; Distler et al. 1996, 1999; Sluzhenikin and Mokhov 2015) are good examples. In these deposits, in addition to the presence of a large volume of massive magmatic sulfides, low-sulfide but PGE-rich ores were established. These ores, referred to as ores of the low-sulfide horizon, occur in the upper endocontact zone of the layered intrusions (Sluzhenikin et al. 1994). The contrast distribution of the PGE is typical of two main ore types of the Noril'sk deposits: the disseminated ores in the layered intrusions and massive sulfide ores associated with the same intrusions. As in the above examples, the massive ores of the Noril'sk deposits are relatively rich in the PGE content compared to the disseminated ores of the layered intrusions, where, taking into account only total sulfide volume, the concentration of the PGE is 5–10 times higher (Distler 1994a).

The mineralogy of these ores is still studied much less thoroughly than that of ores at the Talnakh deposits. The main reasons for this were that the discovery of the latter deposits coincided with the progressively broader application of analytical techniques of high spatial resolution (electron microscopy and X-ray microprobe analysis), which were utilized to examine the unique ores of the Talnakh and Oktyabr'skoe deposits (discovered in the early 1960s).

During the long period of the study of platinum mineralization of the Noril'sk 1 deposit, much data have been accumulated on mineralogy of platinum minerals, occurring mainly in veins of the massive sulfide ores (Genkin 1968; Genkin et al. 1981). There is considerably less data on the disseminated platinum mineralization (Genkin 1968; Razin and

Table 4.20 Spinel composition from olivines of the picritic gabbro-dolerites of the Noril'sk 1intrusion, wt %

Component	1	2	3	4	5	6	7	8	9
SiO_2	0.39	0.29	0.32	0.38	0.42	0.35	0.47	0.64	0.36
TiO_2	3.34	3.95	3.92	3.96	3.66	2.43	2.56	4.24	4.19
Al_2O_3	9.51	9.33	8.91	8.87	6.70	10.75	10.59	6.68	9.22
Cr_2O_3	35.67	34.13	32.45	32.38	29.37	39.34	33.54	25.88	32.18
V_2O_3	0.27	0.18	0.22	0.22	0.37	0.25	0.25	0.38	0.18
FeOtot	43.16	44.19	45.82	45.98	50.53	39.33	42.92	52.58	45.45
MnO	0.42	0.37	0.38	0.40	0.41	0.39	0.38	0.43	0.43
MgO	4.57	4.64	4.44	4.36	3.55	4.99	4.68	3.58	4.20
NiO	0.20	0.18	0.23	0.22	0.21	0.15	0.19	0.26	0.18
ZnO	0.19	0.17	0.20	0.21	0.22	0.24	0.19	0.22	0.23
Total	97.74	97.43	96.90	96.98	95.41	98.24	95.78	94.90	96.63
FeO	28.86	29.14	29.16	29.41	29.66	27.68	27.72	30.22	29.75
Fe_2O_3	15.89	16.72	18.51	18.42	23.19	12.94	16.90	24.85	17.44
New total	99.33	99.11	98.74	98.82	97.74	99.52	97.47	97.39	98.36
Fo	80.23	80.00	79.61	79.61	79.46	80.39	79.76	79.23	79.70
SiO_2	39.22	39.28	39.25	39.25	39.16	39.33	39.14	39.06	39.21
FeO	18.74	18.86	19.23	19.23	19.32	18.54	18.99	19.47	19.16
MnO	0.32	0.28	0.32	0.32	0.34	0.27	0.34	0.31	0.30
MgO	42.68	42.35	42.15	42.15	41.95	42.68	42.00	41.69	42.21
CaO	0.10	0.08	0.08	0.08	0.08	0.08	0.08	0.09	0.10
NiO	0.24	0.20	0.27	0.27	0.23	0.21	0.26	0.25	0.23
Cr_2O_3	0.01	0.01	0.00	0.00	0.01	0.02	0.01	0.01	0.01
Total	101.36	101.09	101.31	101.31	101.11	101.16	100.81	100.88	101.21
+QFM clc	2.15	2.24	2.40	2.40	2.84	1.73	2.21	3.00	2.35
Fe^{2+}/Fe^{3+}	2.09	2.01	1.81	1.84	1.47	2.47	1.89	1.40	1.96
Fe^{2+}/Fe^{3+} m	7.05	6.68	5.85	5.96	4.45	8.74	6.17	4.17	6.49

Note: Fe^{2+}/Fe^{3+} ratio were calculated based on spinel stoichiometry; fO_2 and Fe^{2+}/Fe^{3+} ratio in melt (Fe^{2+}/Fe^{3+} m) were calculated from Ballhause et al. (1990). Analyses were carried out in GEOKHI RAS, analyst N. Kononkova

Borishanskaya 1970). Special work was carried out for disseminated ores of the Noril'sk 1 intrusion with participation of the author (Distler et al. 1999). These results are given below.

We studied a continuous cross section in the Medvezhy Creek open pit. The cross section represents the full thickness of the horizon containing disseminated ore mineralization. Over the entire section, we mapped the zoning of the disseminated mineralization and took ore hand samples. From zones with varying mineral composition, we took large volume samples up to 100 kg in weight. In addition to the standard procedure of studying mineral composition, we made quantitative gravity separation of the PGM using the "ppm-mineralogy" procedure (Knauf 1996) with the subsequent determination of the chemical composition of phases with an electron microprobe. The scanning electron microscope ISM-5300 with an energy dispersing spectrometer Link ISIS was used for the direct determination of chemical composition of phases and relative proportions of PGM found in gravity concentrates. Data on distribution of the PGM in the sieve size fractions −250 + 125, −125 + 50 and less than 50 pm turned out to be very informative.

The solid solution PGE content in the ore-forming sulfides was determined using the proton microanalyzer (e.g., Cabri et al. 1985; Cabri 1988) at the Guelph Scanning Proton Microscope Laboratory, Guelph, Canada. Contents of the PGE and some other elements in pyrrhotite, pentlandite, chalcopyrite, talnakhite, and millerite were determined from hand samples, as well as from a fraction of the bulk samples, which were mounted in araldite and then polished. The sensitivity limits for various minerals out of the main sulfides studied were as follows (ppm): Pt, 24–37; Pd, 2–3.8; Rh, 2.6–5.3; Ru, 2.8–3.4; Cu, 25–38; Zn,5–160; Mo, 2.3–3; Cd, 5; As, 3–9; Se, 2.4; Te,7–11; and Ag, 3.9–5.3.

The mineral zoning of the disseminated sulfide mineralization in these intrusions was described for the first time in the publication (Distler et al. 1979). The zoning and cryptic layering of ore mineralization were shown to be fundamental characteristics of ore-bearing massifs by Distler et al. (1979).

It was shown that the zoning and cryptic layering of the disseminated ore mineralization are related to a regular distribution of three types of parageneses of the main ore-forming sulfides in the orebody section: (l) low-sulfur

Table 4.21 Spinel composition from olivine, plagioclase, and pyroxene of rocks from the Noril'sk 1 intrusion, wt %

№	Sample	SiO_2	TiO_2	Al_2O_3	Cr_2O_3	V_2O_3	FeOtot	MnO	MgO	NiO	ZnO	Total
1.	64	0.35	2.43	7.84	39.34	0.25	39.33	0.39	4.99	0.15	0.24	98.24
2.	64	0.47	2.56	7.84	33.54	0.25	42.92	0.38	4.68	0.19	0.19	95.78
3.	64	0.64	4.24	7.84	25.88	0.38	52.58	0.43	3.58	0.26	0.22	94.90
4.	64	0.36	4.19	7.84	32.18	0.18	45.45	0.43	4.20	0.18	0.23	96.63
5.	64	38.99	0.04	7.84	0.01	0.01	19.20	0.31	42.27	0.26	0.01	101.24
6.	64	0.47	1.28	7.84	31.52	0.64	51.62	0.45	3.03	0.20	0.25	94.83
7.	64	1.18	1.42	7.84	31.30	0.71	50.81	0.42	3.97	0.19	0.25	95.86
8.	107-sp1	0.00	3.99	7.84	32.75	0.27	44.70	0.38	5.78	0.15	0.11	99.50
9.	107-sp2	0.02	4.51	7.84	31.46	0.26	46.41	0.40	5.43	0.16	0.12	99.29
10.	107-sp3	0.03	3.48	7.84	33.26	0.30	42.08	0.36	6.99	0.15	0.11	99.72
11.	107-sp4	0.02	3.41	7.84	33.38	0.32	41.62	0.37	7.04	0.12	0.14	99.50
12.	107-sp5	0.03	3.54	7.84	32.96	0.30	42.51	0.36	6.98	0.13	0.10	99.66
13.	107-sp6	0.01	4.10	7.84	31.82	0.32	44.73	0.38	5.78	0.16	0.12	98.59
14.	107-sp7	0.02	3.07	7.84	33.99	0.32	39.91	0.36	7.14	0.15	0.07	98.60
15.	107-sp8	0.05	3.16	7.84	34.46	0.33	40.02	0.36	7.34	0.14	0.09	99.54
16.	107-9sp	0.00	4.74	7.84	28.31	0.27	53.39	0.43	3.45	0.18	0.16	97.96
17.	107-14	0.03	11.23	7.84	14.71	0.00	64.00	0.43	3.29	0.23	0.08	98.80
18.	107-15	0.01	11.44	7.84	14.52	0.00	63.44	0.44	3.47	0.23	0.09	98.40
19.	107-1	0.03	10.18	7.84	16.43	0.00	63.36	0.42	2.82	0.22	0.13	97.99
20.	107-Sp1	0.10	2.92	7.84	28.51	0.46	52.32	0.46	2.98	0.17	0.16	96.45
21.	107-Sp30	0.04	1.01	7.84	32.18	0.18	35.18	0.32	8.33	0.09	0.15	99.14
22.	107-Sp32	0.03	1.63	7.84	32.13	0.18	38.36	0.35	7.13	0.10	0.18	99.48
23.	107-Sp34	0.01	6.89	7.84	23.89	0.02	55.59	0.44	3.28	0.15	0.11	97.83
24.	107-Sp38	0.00	9.34	7.84	15.78	0.06	63.37	0.44	2.88	0.17	0.09	97.98
25.	107-Sp40	0.00	13.23	7.84	10.76	0.00	66.51	2.19	0.99	0.18	0.13	97.97
26.	107-10sp	0.04	10.43	7.84	18.74	0.03	60.32	0.43	3.29	0.21	0.11	98.65
27.	107-Sp18	0.04	8.82	7.84	14.90	0.23	65.19	0.38	2.94	0.26	0.09	98.23
28.	107-Sp20	0.02	11.89	7.84	15.02	0.00	62.32	0.41	4.18	0.26	0.08	98.93
29.	107-Sp21	0.00	12.34	7.84	15.11	0.00	61.38	0.42	4.39	0.25	0.08	98.53
30.	107- inPx-2	0.03	1.82	7.84	36.03	0.28	34.76	0.32	8.53	0.13	0.11	99.53
31.	107- inPx-3	0.06	1.78	7.84	36.02	0.27	34.50	0.33	8.45	0.14	0.12	99.11
32.	107- inPx-4	0.02	1.83	7.84	35.59	0.26	35.61	0.33	7.86	0.12	0.14	98.78
33.	107- inPx-5	0.64	1.84	7.84	34.37	0.27	35.76	0.31	9.46	0.13	0.07	98.90
34.	107- inPx-6	0.03	1.95	7.84	36.43	0.28	32.87	0.28	9.72	0.13	0.07	99.04

Note: N sample = borehole G-22/depth, m; NN (1–7) spinel in Ol (Fo_{79}), (8–26) spinel in Pl, (27–34) spinel in Px. Analyses were carried out in GEOKHI RAS, analyst N. Kononkova

paragenesis, (2) intermediate paragenesis, and (3) high-sulfur paragenesis. Each of these parageneses is considered to be a product of changes in sulfur activity in sulfide magmatic liquids coexisting with silicate melts of various compositions. These regularities correspond to the zoning and cryptic layering of the ore mineralization established in this section of the Noril'sk 1 intrusion.

Zone 1 is confined to the upper part of the orebody, which coincides with the uppermost part of the picrites next to the olivine gabbro-dolerites. This zone contains the least sulfides (no more than 3 vol.%). Sulfides form only small (no more than 2 mm) interstitial aggregates. Sulfides are represented by a low-sulfur assemblage: troilite + ferrigenous pentlandite + talnakhite + chalcopyrite + cubanite. Cubanite forms only platelike aggregates. Pentlandite of this assemblage has the highest ferrigenous (35.67 wt %).

Zone 2 includes the middle part of the picrite horizon. Here, sulfides make up to 5 vol.%. They occur as small interstitial ovoids and larger (to 2 cm) ovoids. Sulfides are represented by the assemblage of an intermediate sulfur content: hexagonal pyrrhotite + moderately ferruginous pentlandite + chalcopyrite. Pyrrhotite contains the least nickel. Pentlandite occurs in two varieties: (1) as rims around pyrrhotite and (2) as flame-like aggregates in pyrrhotite. A bulk sample H-l was taken from the zones 1 and 2 (Distler et al. 1999).

Zone 3 corresponds to the lowest part of the picritic gabbro-dolerite horizon and has maximum sulfide contents (to 8 vol.%) of the high-sulfur paragenesis. Pyrrhotite is

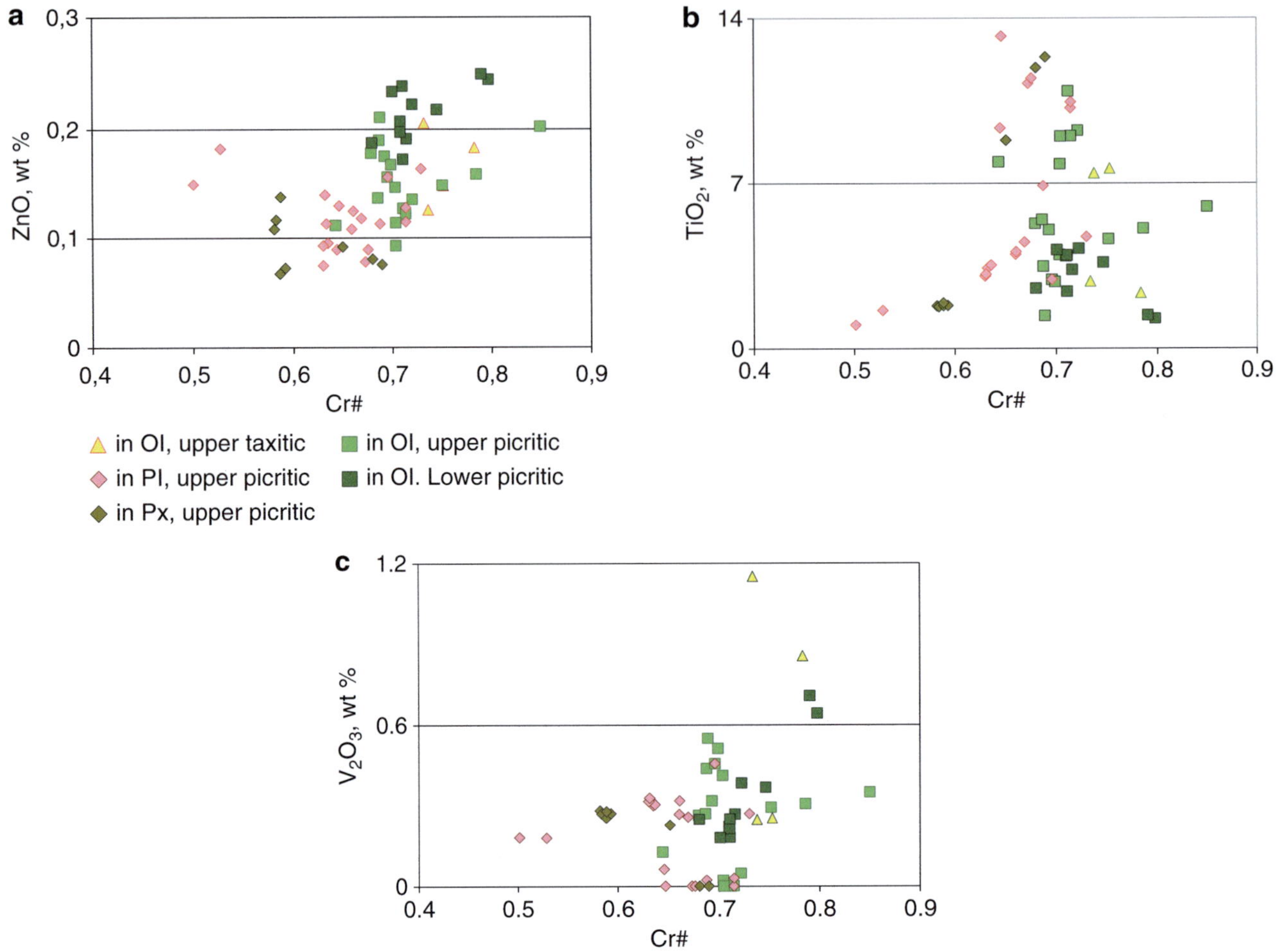

Fig. 4.37 Zn, Ti, V–Cr # diagrams for spinels from the Noril'sk 1 intrusion

represented by a monoclinic modification with nickel concentration reaching 2.5 wt %. As compared to pentlandite of the previous zone, the nickel content is considerably higher (to 40 wt %). At the base of the picritic horizon, pyrrhotite is extensively replaced by nickel-containing pyrite and pentlandite is replaced by millerite. The sample H-2 was taken from this part of the section.

Zone 4 includes the horizon of taxitic gabbro-dolerites containing large xenomorphic aggregates of sulfides of the low-sulfur assemblage. Here, monoclinic pyrrhotite and pentlandite with very high nickel contents are the major minerals. Extensive autometamorphism of primary ores is typical of this section of the disseminated ores and the taxitic horizon. Nickel-containing pyrite and violarite are abundant. Pyrite replaces pyrrhotite and millerite, and violarite completely replaces pentlandite. The H-3 sample represents the entire section of the disseminated ores in the taxitic gabbro-dolerites.

The unique mineralogy of the PGE in the Noril'sk deposits comprises about 60 species and varieties of PGM, the majority of which are characterized by and recognized to be established species. However, some of these minerals were studied mainly by determination of their chemical compositions (Table 4.22). Hence, these minerals should be subjected to exact determination of their structural parameters. All the PGM belong to the following main types of chemical compounds:

- Natural alloys and ordered and non-ordered metallic solid solutions
- Intermetallic compounds of variable composition with a wide range of isomorphic replacements of the ligands
- Chalcogenide compounds including arsenides, antimonides, selenides, and simple and complex sulfosalts

Palladium forms the largest quantity of mineral species comprising all the above classes, whereas platinum only comprises 11 mineral species, though trace concentrations of platinum from a few units to tens percent are present in some palladium minerals. In addition, continuous series of solid solutions occur between palladium and platinum end-members

Table 4.22 Minerals of platinum group from the ores of the Noril'sk 1 deposit

Mineral	Formula	Mineral	Formula
Isoferroplatinum	Pt_3Fe		(Au,Pd)Cu
Tetraferroplatinum	PtFe	Froodite	$PdBi_2$
Tulameenite	Pt_3CuFe	Maslovite	PtTeBi
	Pt_2Fe	Michenerite	PdTeBi
Atokite	Pd_3Sn	Urvantsevite	$Pd(BiPb)_2$
Rustenbergite	Pt_3Sn	Geversite	$PtSb_2$
Zvyagintsevite	Pd_3Pb		Pd_5As_2
Taimyrite	$(Pd,Cu, Pt)_3Sn$		$(Pd, Ni)_5As_2$
Paolovite	Pd_2Sn	Palladoarsenide	Pd_2As
	$Pd_2(Sn, Sb)$	Mayakite	PdNiAs
	$Pd_2(Sn, As)$		Ni_6Pd_2As
	$Pd_2(As, Sb)$	Stibiopalladinite	Pd_5Sb_2
	$(Pd, Ni)_2As$	Isomertieite	$Pd_{11}(AsSb)_5$
	$(Pd, Ni)_2(Sn, As)$	Mertieite	$(Pd, Pt)_8(As, Sb)_3$
	$(Pd, Ni)_2((AS,Sb)$		
Stannopalladinite	Pd5Sn2Cu	Guanqlinite	Pd_3As
Cabriite	Pd_2SnCu		$Pd_3(As,Te)$
Niggliite	PtSn	Palarstanide	Pd_5AsSn
Plumbopalladinite	Pd_3Pb_2	Telergpalite	$(Pd,Ag)_4Te$
Sobolevskite	PdBi	Sopcheite	Ag_4Pd_3Te
Kotulskite	PdTe		Ag_2PdS
Sudburite	PdSb	Kharaelakhite	$(Pt,Cu,Pb,Fe,Ni)_9S_8$
Polarite	Pd (Pb, Bi)	Cooperite	PtS
	Pd(Bi, Te, Pb)	Braggite	(Pt,Pd,Ni)S
Moncheite	$PtTe_2$	Vysotskite	(Pd,Ni)S
Merenskyite	$PdTe_2$	Sperrylite	$PtAs_2$
Insizwaite	$PtBi_2$	Hollingworthite	(Rh,Pt,Ir)AsS

Note: Table after Distler et al. (1999)

for some mineral species. Discrete mineral phases of rhodium, iridium, ruthenium, and osmium are essentially absent.

The PGM distribution in copper–nickel ore types is quite uneven. In massive ores, there is a differentiation in composition of PMG and in their abundance between the pyrrhotite ores and ores enriched in copper (cubanite, talnakhite, and chalcopyrite). As compared to other types and varieties, pyrrhotite massive ores have the simplest PGM composition and are characterized by some rare phases. Sperrylite is the major mineral of the pyrrhotite ores, and a large number of PGM were found in varieties of the pyrrhotite ores enriched in copper. It should be noted that in spite of a high-sulfur activity responsible for the formation of high-sulfur parageneses of the main ore-forming sulfides, the pyrrhotite ores do not contain sulfides of platinum metals; here, intermetallides and alloys are mainly present.

In the Noril'sk deposits, an increase in the concentrations of PGE in ores rich in copper (cubanite, talnakhite, chalcopyrite) results in an increase of the number of types of palladium minerals, whereas variability of platinum species changes insignificantly. The mineralogy of PGE of the stringer-disseminated ores differs from that of the massive ores. In the former, most of minerals of the intermetallic compound class, so typical of the massive ores, are essentially absent. The major PGM of this ore type is represented by platinum sulfides, which coexist with metallic solid solutions of the Pt–Fe system. Tellurides, tellurobismuthides, and bismuthides as well as the group of selenium-bearing phases are also present.

The disseminated ores are characterized by a considerable variability of PGM. These ores contain more than 20 PGM, only second in number of species to the massive ores, which are very rich in copper. Most of the PGM, known in other ore types, were found in the disseminated ores. This is the main peculiarity of the mineralogy of the PGM in the disseminated ores. Thus, compared to the rich ores, the disseminated ores contain all the PGM encountered in the pyrrhotite ores as well as the majority of the PGM associated with the talnakhite and cubanite ores, which do not occur in the pyrrhotite ores. The disseminated ores also contain most of the PGM found in the stringer-disseminated ores except for some very rare minerals of this group such as sulfoselenides and telluroselenides. However, the abundance of minerals of each of the PGE is quite different.

In the disseminated ores of the Noril'sk 1 deposit, the main PGM are alloys of platinum and iron, palladium rustenbergite, platinian atokite, and numerous phases of the general composition: (Pd, Pt)$_5$(Sb,Sn,As,Pb)$_2$ and (Pt Pd)$_3$(Sb,Sn,As,Pb) with a wide degree of isomorphism of platinum and palladium as well as ligands.

Alloys of platinum and iron can constitute up to half of the PGM grains in ores of the lower part of the picrite horizon. The quantity of these alloys abruptly decreases in the upper part of this horizon and in the taxitic gabbro-dolerites. These alloys correspond to three compositions: isofenoplatinum Pt_3Fe, tetraferroplatinum PtFe, and a mineral with a composition close to Pt_2Fe. According to some researchers, the latter phase is a solid solution of the two first compositions. It should be noted, that in the picrites, the iron–platinum alloys are represented exclusively by isoferroplatinum and that the Pt_2Fe phase appears in the fine fraction in the lower part of the picrites. In the taxitic gabbro-dolerites, alloys of platinum and iron are represented by all three compositions with a relatively more significant role of phases with higher iron content. In addition to platinum and iron, copper and nickel are constant components of the iron–platinum alloys.

The content of copper and nickel regularly increases from isoferroplatinum to tetraferroplatinum. The copper and nickel concentrations are as follows: in isoferroplatinum, Cu 0.13–O.17 wt % and Ni 0.174.81 wt %; in Pt_2Fe, Cu 0.57–1.58 wt % and Ni 0.51–.97 wt %; and in tetraferroplatinum, Cu 2.40–3.07 wt % and Ni 2.26–3.88 wt %. Aggregates of iron–platinum alloys often form crystallographic facets and cubic crystals. Palladium rustenbergite and platinum atokite form well-shaped crystals with growth facets. In coarse sample sizes, these minerals are practically devoid of other components except tin. Together with iron–platinum alloys, they are the principal minerals of the disseminated ores, and in the taxitic gabbro-dolerites, they even prevail over iron–platinum alloys. It should be noted that in the 125–250 pm size fraction, this mineral group is represented by platinum members of the isomorphic series Pt_3Sn-Pd_3Sn.

In large crystals (50–125 μm class), the isomorphism of Pt and Pd in these minerals is more widely pronounced, whereas among small grains (size fraction less than 50 pm), platinian atokite is a major mineral. Admixture of gold reaching 6 wt % is constantly present in rustenbergite and especially in atokite.

Minerals of the composition A_3B and A_5B_2, where A = Pd and Pt, B = Sb, Sn, As, and Pb, form predominantly aggregates of less than 50 μm but usually of 3–20 μm in size. In this size fraction, palladium minerals are the major PGM. Isomorphism of platinum and palladium is limited. As a rule, platinum content does not exceed 15 wt %. Sb, Sn, As, and Pb show a widely pronounced isomorphism. Tellurium and bismuth are often present in these minerals. Among phases of this composition are the following PGM: stibiopalladinite Pd_5Sb_2, zvyagintsevite Pd_3Pb, Pd_3As, and Pd_3(As,Te).

Unlike the disseminated ores of the Talnakh ore field, sperrylite, less than 0.1 mm in size, is abundant, but is not a major mineral in the disseminated ores of the Noril'sk 1 deposit. Cooperite is the other platinum mineral, subordinate to iron–platinum alloys. Cooperite is present mainly in the picrites. It forms grains of 50–125 μm in size. Of palladium minerals in finest size fraction, minerals of the system Pd–Bi–Te with a large degree of isomorphism of bismuth and tellurium are constantly noted. A tellurian kotulskite is typical of the picrites, whereas sobolevskite, bismuth-bearing variety, occurs in taxitic gabbro-dolerites. Admixtures of Pt, As, Pb, and Sb are constantly present in small amounts. Kotulskite and sobolevskite occur not only as irregular grains but also as well-shaped euhedral crystals.

The PGM in the disseminated ores form irregular monomineral aggregates and rarely form intergrows. Rustenbergite–atokite and iron–platinum alloys are common minerals of these intergrowths. The latter forms rims around platinum and palladium stannides. The central parts of stannide aggregates are higher in platinum than their peripheries. The other component of these intergrowths is stannopalladinite, which forms rims around rustenbergite and atokite. Often sperrylite, kotulskite, and sobolevskite form intergrowths with other PGM; this value is 10–15 %. Similar estimates were obtained for the PGE of ore mineralization in the Talnakh intrusion (Sluzhenikin and Distler 1998). Distribution of PGE in ore-forming minerals from disseminated ores of the Noril'sk 1 deposit ("Medvezhy Creek" open pit) was studied by S.-J. Barnes and coauthors (2006).

4.5.1.5 Low-Sulfide Mineralization

A new type of platinum mineralization for Noril'sk deposits—low sulfide—opened in the early 1980s (Ryabov et al. 1982; Ryabov 1984) although the first elevated contents of platinum group metals were established in the 1960s (Ginzburg and Rogover 1960; Smirnov 1966). Its commercial value was shown by V.V. Distler and Dyuzhikov (1988). S. Sluzhenikin and coauthors studied (1994) the mineral composition of platinum group elements from low-sulfide mineralization.

In terms of intrusions of platinum horizons are located inside the upper exocontact and confined to taxitic gabbro-dolerites with higher contents of chromite. They are separated from the main ore horizons (sulfide ores) by barren rocks of 100 m thickness. The most rich and mellow along strike is low-sulfide platinum horizon in the Noril'sk 1 intrusion. Petrographically it is heterogeneous. Rocks, enriched in PGE, are eruptive breccia, taxitic leucogabbro, and picritic gabbro-dolerites. Sulfides consist not more than 3 % of the volume of rocks. The ratio Ni:Cu = 1:1, metal contents are less 0.3–0.5 %. Monoclinic pyrrhotite dominate among sulfides, and chalcopyrite and pentlandite are of secondary importance. Ordinary PGE contents comprise 3–20 ppm, sometimes reaching 70–80 ppm. Ratio Pt–Pd = 0.3–0.4. Platinum group elements form solid solutions in the ore-forming sulfides, arsenides, and independent mineral species (their composition is shown in Table 4.23 after S. Sluzhenikin et al. 1994).

4.5.2 Noril'sk 2 Massif

The Noril'sk 2 intrusion was discovered in the summer of 1926. It is located on the northeastern slope of the plateau Noril'sk, 7 km from the city of Noril'sk, on the right side of the Noril'sk-Kharaelakh fault (Fig. 4.26). The deposit is associated with the similarly named differentiated intrusions, which abruptly cut Devonian sediments, the Tunguska Group, and a tuff–lava sequence. The intrusion is a lenticular body approximately 1 km wide and 7 km long. The intrusion is a complex body with a steep inclination (10–65°) and a generally northeast strike (400). The thickness of intrusive body is 100–370 m (Fig. 4.38). Sulfide mineralization is concentrated mainly in the picritic and taxitic gabbro-dolerites at the base of the intrusion. The disseminated mineralization is very uneven, and the thickness of the orebodies does not exceed 20 m. Sulfides in the form of lenses and irregularly shaped bodies exist at the bottom and in the middle part of the intrusion. The sulfide content reaches 4–6 %.

This massif forms the Noril'sk Complex, a Noril'sk-type intrusion, similar to the Talnakh and Kharaelakh intrusions. However, the massif lacks a large amount of sulfides. This difference is why it is especially interesting to compare the geochemical and mineralogical features of these intrusions. A cross section of the intrusion was studied using the core from borehole MN-18 (Fig. 4.39). The intrusion is not contrastingly differentiated. Its main volume (120 m) is composed of olivine gabbro-dolerites with olivine and taxitic gabbro-dolerites at the bottom (27 and 22 m, respectively). The upper and lower zones, approximately 20 m, represent the contact, taxitic gabbro-dolerites, and gabbro-diorites.

Olivines in the section of the massif were thoroughly studied. They have high FeO content and contain 55–75 mol% forsterite component (Table 4.24). The olivines also contain high concentrations of calcium. The range of variation in CaO content is large and reaches significant values, ranging from 0.15 to 0.30 wt % . Therefore, olivines from the Noril'sk 2 intrusion are significantly different from the olivines of other Noril'sk Complex massifs. The olivine data plotted on an Fo–CaO diagram form a continuous field of compositions, devoid of any rules or divisions. Aluminum also behaves erratically in the mineral, even though its content is very small (0–0.03 wt %). A similar pattern is observed for titanium—no regularities in the behavior of this element were detected. The behavior of Cr in olivine from Noril'sk 2 rocks differs from that in the Talnakh intrusion. Although Cr concentration in magnesium-rich olivines increases with increasing proportions of Fo, this is not the usual observed hyperbolic curve for this intrusion. Generally, the chromium concentrations in the Noril'sk 2 olivines are also lower than in the other massifs. CaO exhibits a conventional directly proportional relationship to the iron content of the olivine. This is even more obvious for Mn, which forms a distinct increasing trend with the increasing fayalite component in the olivine. The behavior of nickel in the olivine is especially noteworthy. Nickel varies commonly in the upper part of the section, forming a series of subparallel trends (for selected models) that reflect the growth of NiO with the increasing forsterite component in the olivine.

Although the NiO content in magnesium-rich olivines is typically higher (up to 0.2 wt %), the concentrations of nickel in olivine are stable and subparallel to the *x*-axis line. This indicates that the olivine was in equilibrium with a small amount of sulfide. Larger quantities of the sulfide phase in the volume of rock would force the line to tilt about a horizontal axis by increasing the nickel content in association with the increasing iron content of the mineral.

Rare elements were investigated using ion microprobe in 3 samples from borehole MP-18 taken from different parts of the intrusion. Each sample features its own characteristic distribution of rare earth elements and yttrium in olivine. The most magnesium-rich olivines (Fo = 66–74) were observed in sample 291, 6 m. Although these are much more iron-rich olivines than are usually observed in the picritic gabbro-dolerites of the different massifs, the yttrium trend is very similar to those of the ore-bearing massifs. The yttrium concentration is also high, reaching 2.5 ppm for samples taken from the section above (MN-279 and MN-264), which produces subparallel trends. These samples differ in the distribution of nickel: The first sample (291, 6 m) has an average concentration of 0.15 wt %. Typical trends are also observed for the other two samples, depending on the MgO in olivine. Titanium concentrations in the studied grains are distributed

Table 4.23 PGE mineral composition from the horizon of the low-sulfide mineralization (Noril'sk 1 intrusion), wt %

№	MIneral	Pt	Pd	Rh	Fe	Ni	Cu	Sn	Sb	As	Pb	Te	Bi	S	Total
	Kotulskite		44.00					0.13				53			98.03
	Sn–Kotulskite	0.41	52.16					13.68				34.22			100.47
	Sn–Kotulskite	2.54	49.69					13.55	0.07			32.94			98.79
	Pb–atokite	2.31	66.89					18.56			7.89		2.16		97.81
	Pb–atokite	0.44	65.4					13.73			19.01		0.03		98.61
	Atokite	0.36	70.34					24.72			1.58		1.5		98.5
	Atokite	5.23	66.59					24.08	0.28			1.72	1.06		99.68
	Pt–Atokite	0.23	50.08					24.99							98.07
	Pt–Atokite	28.08	46.36		1.62	0.26	0.63	21.02	0.2	0.03		0.04	0.1		98.34
	Zvyagintsevite	0.47	58.24								40.05		1.82		100.58
	Vysotskite	2.53	63.75			9.67								23.97	99.92
	Vysotskite		70.76		0.23	6.38			0.28				0.05	22.42	100.12
	Cuperite	32.69	52.61			0.51	0.09							13.43	99.33
	Pd3(Sn, As)	7.8	65.2					19.39		6.45		0.09			98.93
	Pd3(Sn, As)	4.21	68.45					19.26		7.49		0.14			99.55
	Pd3(Sn, As)	5.52	68.81					19.28		7.91		0.4			101.92
	Stibiopalladinite		53.48					0.76	43.52	1.22		1			99.98
	Sperrylite	57.25		0.22						42.51					99.98
	Sperrylite	57.94	0.81	0.15		0.45				40.65				0.18	100.18
	Sperrylite	57.08		0.3						40.32					98.42
	Sperrylite	56.34								42.91					99.25
	Sperrylite	56.56		0.06						41.64					98.26
	Sperrylite	57.34		0.45			0.16			42.5					100.45
	Sperrylite	57.26		0.21						40.53					98
	Sperrylite	55.93								43.25				0.44	99.62
	Pt2Fe	36.66	0.8		11.43	0.73	0.42								100.04
	Pt2Fe	83.23	1.83	0.18	11.06	0.99	1.38								98.67
	Pt2Fe	83.5	2.95	0.33	11.04	0.66	0.49								98.97
	Pt2Fe	86.08	2.14	0.48	10.36	0.63	0.83								100.52
	Pd3As	0.65	78.67		0.97	0.2	0.6	0.13		18.54				0.1	99.86
	Tetraferroplatinum	76.99	0.83	0.25	15.39	3.49	2.83								99.78
	Tetraferroplatinum	78.62	1.07		13.97	3.17	2.92								99.75
	Pd2SnAs	0.27	67.83					18.02	3.82	10.39					100.79
	Pd2AsSb	0.36	68.56					1.44	16.03	10.53	0.3	1.99			99.21
	Pd5(Sb,As)2		71.19						26.52	3.14				0.08	100.93
	Moncheite	37.85	5.27							0.71		45.61	7.73		97.17
	Merenskyite		9.58									48.56	21.86		100
	(PdNi)2As	0.41	48.82			17.89				32.72					99.89
	Mayakite		38.1			24.3				38.43					100.83

Note: Table after Sluzhenikin et al. (1994)

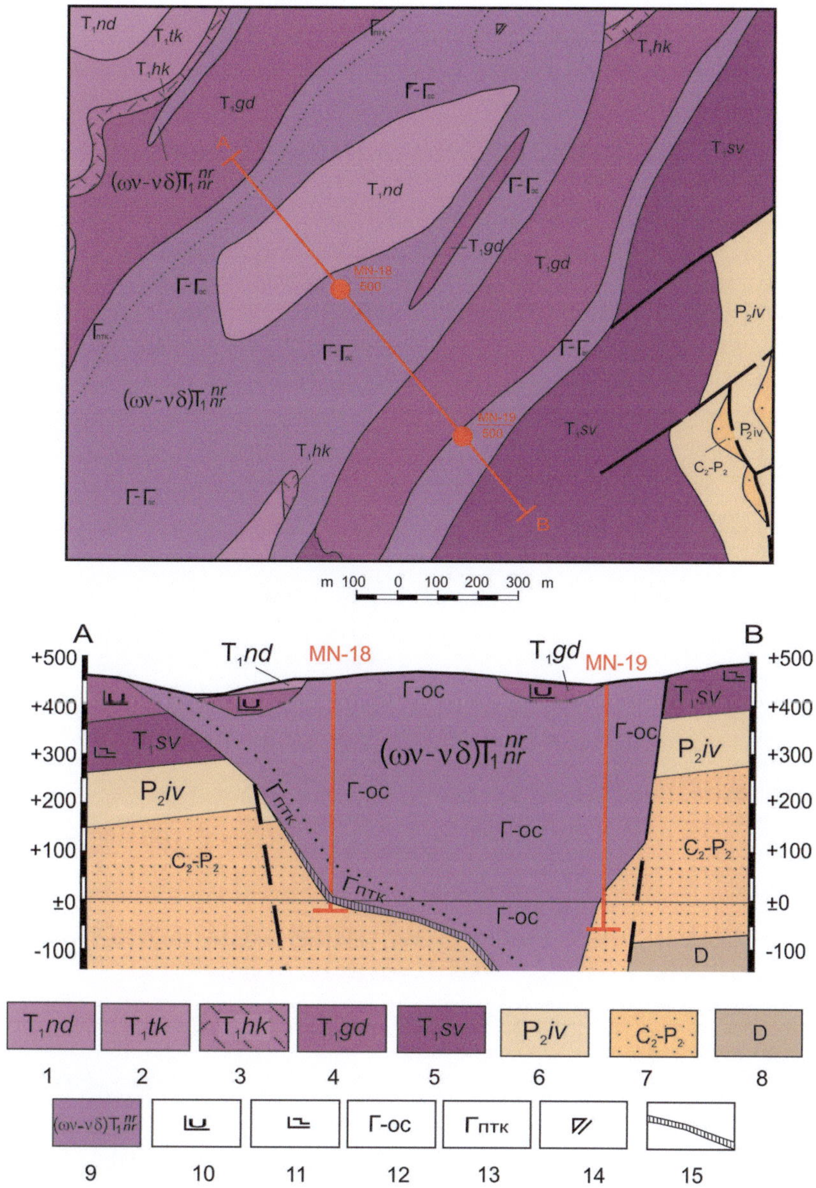

Fig. 4.38 Geological map and cross section of the Noril'sk 2 intrusion After NorilskGeology, Ltd. data., changed by the author

randomly, although the concentrations are substantially higher in the enriched olivines from sample 291.6 m. The concentrations of Zn and Mn are typical of the normal distribution patterns. The behavior of vanadium in the samples from the upper horizons of the two samples depends on magnesium; however, at the bottom of the sample (291 m), there is a random distribution of V in the mineral, and the highest concentration is observed (up to 29 ppm). The aluminum content, in contrast, is lowest at the bottom of the sample.

Pyroxene. The variations in the composition of pyroxenes in the each sample are very similar, and their magnesium content varies from 60 to 85 (Fig. 4.39). However, the behaviors of the trace elements in pyroxenes indicate the presence of certain behavioral regularities compared to olivine (Table 4.25). The differences are primarily related

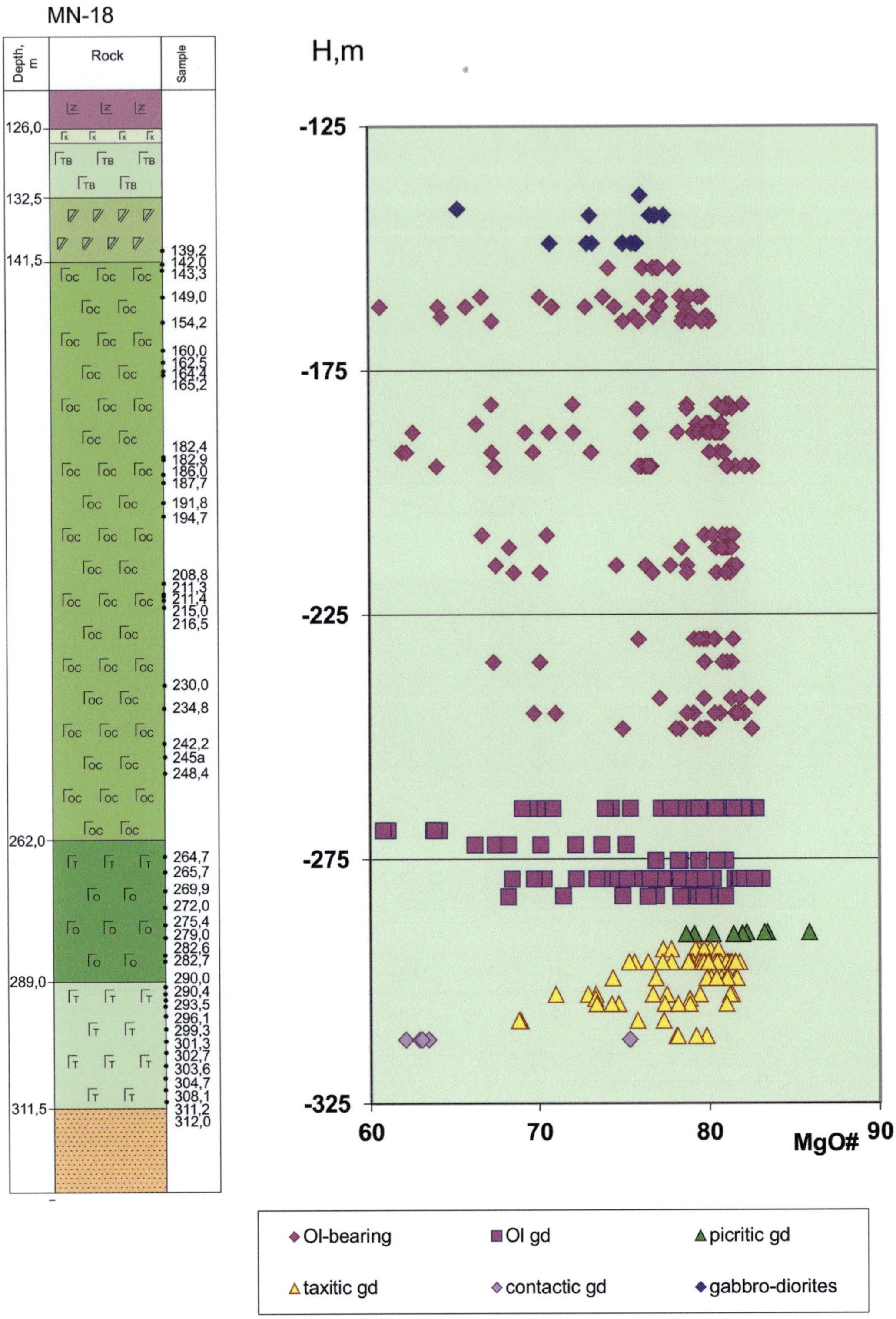

Fig. 4.39 Inner structure of the Noril'sk 2 intrusion and pyroxene compositions variations throw borehole MN-18

Table 4.24 Representative analyses of olivines from the Noril'sk 2 intrusion

N	Fo, mol %	SiO_2	CaO	FeO	MgO	Cr_2O_3	NiO	MnO	Al_2O_3	Total	Ti	V	Y	Yb	Dy
1.	73.18	38.12	0.18	24.58	37.63	0.01	0.20	0.38	0.02	101.15	126.90	24.77	1.43	0.46	0.15
2.	75.46	38.47	0.24	22.62	39.04	0.02	0.22	0.36	0.03	101.01	71.60	25.11	0.74	0.23	0.08
3.	73.45	38.38	0.23	24.21	37.57	0.00	0.26	0.37	0.03	101.07	113.27	29.45	1.32	0.45	0.16
4.	76.46	38.84	0.24	21.81	39.74	0.01	0.20	0.34	0.02	101.22	118.72	12.63	0.72	0.26	0.09
5.	77.31	39.10	0.23	21.08	40.30	0.01	0.20	0.34	0.02	101.32	104.41	22.18	0.65	0.22	0.09
6.	78.34	39.00	0.29	20.26	41.12	0.03	0.20	0.31	0.03	101.26	61.26	16.56	0.24	0.08	0.03
7.	77.81	38.95	0.26	20.60	40.52	0.02	0.20	0.32	0.03	100.92	82.43	21.55	0.60	0.20	0.09
8.	76.09	38.68	0.24	22.05	39.35	0.02	0.21	0.34	0.02	100.95	83.34	23.06	0.63	0.23	0.08
9.	75.63	38.39	0.22	22.55	39.25	0.01	0.20	0.35	0.02	101.03	79.92	25.25	0.90	0.28	0.11
10.	78.10	38.91	0.28	20.33	40.68	0.03	0.21	0.31	0.03	100.81	73.84	22.87	0.31	0.10	0.04
11.	77.66	38.59	0.18	20.84	40.65	0.03	0.19	0.33	0.02	100.84	127.59	23.47	1.00	0.26	0.11
12.	78.62	38.20	0.30	19.93	41.12	0.03	0.20	0.30	0.03	100.13	59.98	15.69	0.18	0.07	0.01
13.	76.10	38.14	0.16	22.10	39.48	0.00	0.24	0.34	0.01	100.49	149.52	23.75	1.24	0.49	0.13
14.	76.07	38.37	0.25	22.17	39.56	0.01	0.21	0.35	0.02	100.97	89.87	31.78	1.00	0.31	0.11
15.	77.78	38.53	0.25	20.70	40.64	0.02	0.18	0.32	0.02	100.69	100.12	23.02	0.68	0.21	0.08
16.	77.20	38.30	0.26	21.19	40.25	0.02	0.19	0.33	0.03	100.60	71.84	27.23	0.65	0.23	0.06
17.	74.97	38.15	0.16	23.12	38.85	0.02	0.20	0.36	0.01	100.89	151.45	18.35	1.11	0.38	0.14

Note: Major components (wt %) were determined by EPMA in GEOKHI RAS (analyst N. Kononkova), rare elements (ppm) were analyzed by SIMS in Institute of Microelectronic, Yaroslavl (analyst S. Simakin)

to the Ti and Ca compositional fields; these elements are very similar to those described for the pyroxenes from rocks penetrated by the hole KZ-456 (Talnakh intrusion). They also form "ascending and descending" trends, respectively, in the most magnesium pyroxenes. More diffuse patterns are observed for sodium. Al does not show a clear dependence on the Mg# in pyroxene; however, for more magnesium-rich members of the series, the range of Al concentrations increases sharply compared with the whole rock composition. Therefore, for pyroxenes with Mg# 80–82, the aluminum content varies from 2 to 4 wt %, and for iron pyroxene (Mg# = 60–70), the Al concentration varies from 1.5 to 3 wt %. Clear trend lines exist for Mn (ris. 3.2.9) (the Mg# is negatively correlated with Mn in the olivine) and chromium in pyroxene (Fig. 4.40).

4.5.3 Chernogorsky Massif

The geological study of the Chernogorsky intrusion began (Fig. 4.26) in 1940, when geologists V. Domarev and A. Koreshkov created a 1:50,000-scale geological map of the area to the east of the Noril'sk 1 intrusion and to the south of the Noril'sk 2 intrusion. On Mount Chernaya, a differentiated intrusion was found. The intrusion's morphology was then determined (Yakovleva 1947). In the marginal zones, small sulfide grains were found. In the years 1955–1956, a geological survey was conducted at a scale of 1:10,000. The resulting map provided detailed subdivisions (on the formation level) of effusive rocks and refined the stratigraphic position of the rocks of the Tunguska Group and their relationship with Devonian deposits (Fig. 4.41).

The Chernogorsky intrusion is located on the eastern side of the Noril'sk Trough and crops out at the surface in the middle of the mountain, on the northern and northeastern slopes, and on the summit. The intrusion is discordantly overlain and plunges to the west, but due to the difficult hypsometry, its bottom is unknown. The intrusion is located at the interformational contact between Lower Devonian sediments and the Tunguska Group. In general, the intrusion gradually sinks to the northwest. The angle of dip varies from 4–6 to 8–10 degrees. The hypsometry suggests that the base of the intrusion is at an elevation of 265 m in the east, −220 m in the west, and in between these values in the north (−200 to 240 m).

The intrusion is shaped like a language, and the long axis is oriented east–west. The intrusion is 4,250 m long and, at its center, 600–800 m wide. The deposit is divided into eastern and western parts. The eastern sector of the ore-bearing intrusion crops out at the surface, unlike the western sector, which is covered by deposits of the Tunguska Group and tuff–lavas. The boundary between the sectors is marked by the Zvonky stream. The eastern

Table 4.25 Representative analyses of pyroxenes from the rocks of the Noril'sk 2 intrusion, wt %

№	Depth, m	MgO#	SiO2	TiO_2	Al_2O_3	FeO	MnO	MgO	CaO	Na_2O	K_2O	Cr_2O_3	Total
1.	139.2	76.04	51.68	0.58	2.54	8.97	0.21	15.97	19.48	0.27	0.00	0.07	99.77
2.	142	65.30	50.24	0.78	2.04	13.33	0.31	14.07	17.98	0.28	0.001	0.01	99.07
3.	143.3	73.10	51.19	0.60	2.18	10.01	0.23	15.26	19.39	0.25	0.00	0.04	99.18
4.	149	75.53	52.41	0.45	1.99	9.91	0.25	17.16	17.61	0.21	0.00	0.08	100.09
5.	149	72.93	50.85	0.62	2.23	10.20	0.25	15.41	19.28	0.24	0.00	0.04	99.14
6.	154	77.12	51.21	0.53	2.23	8.58	0.20	16.22	19.83	0.22	0.00	0.06	99.08
7.	154	74.17	51.39	0.58	2.14	9.75	0.23	15.70	19.33	0.24	0.00	0.03	99.39
8.	160	76.24	52.76	0.53	2.12	8.91	0.21	16.04	19.98	0.24	0.00	0.02	100.84
9.	162	70.82	51.71	0.65	2.17	11.02	0.26	15.00	18.84	0.26	0.00	0.00	99.91
10.	162	77.24	51.61	0.52	2.11	8.40	0.20	15.99	19.98	0.22	0.001	0.05	99.10
11.	164	78.57	51.78	0.53	2.23	7.94	0.20	16.33	20.27	0.23	0.002	0.08	99.60
12.	164	79.88	52.24	0.47	2.26	7.38	0.18	16.43	20.33	0.21	0.00	0.16	99.68
13.	165	80.09	52.08	0.47	2.25	7.31	0.17	16.49	20.36	0.24	0.007	0.16	99.56
14.	182	67.24	50.16	0.78	2.06	12.69	0.31	14.61	18.02	0.33	0.003	0.00	98.98
15.	182.9	78.76	51.89	0.49	2.28	7.84	0.18	16.31	20.18	0.22	0.004	0.10	99.51
16.	182.9	75.84	51.91	0.56	2.23	9.04	0.22	15.92	19.46	0.24	0.00	0.02	99.62
17.	186	79.88	51.89	0.49	2.62	7.31	0.17	16.28	20.33	0.23	0.00	0.26	99.59
18.	187.7	80.68	52.45	0.44	2.33	7.07	0.16	16.56	20.34	0.23	0.003	0.23	99.83
19.	191.8	80.93	51.72	0.46	2.43	6.99	0.16	16.64	20.45	0.23	0.00	0.35	99.43
20.	191.8	62.21	50.46	0.72	1.62	13.96	0.42	12.89	19.05	0.25	0.00	0.03	99.42
21.	194.7	81.16	52.00	0.47	2.78	6.79	0.16	16.41	20.37	0.22	0.00	0.49	99.73
22.	194.7	81.59	51.72	0.47	3.17	6.46	0.15	16.06	20.39	0.25	0.00	0.88	99.57
23.	208.8	81.45	51.51	0.42	2.53	6.69	0.17	16.48	20.51	0.23	0.001	0.53	99.09
24.	208.8	80.27	52.09	0.46	2.39	7.17	0.16	16.36	20.43	0.24	0.006	0.24	99.56
25.	211.3	68.29	50.44	0.88	1.81	12.27	0.32	14.82	18.15	0.26	0.003	0.00	98.97
26.	211.4	81.16	51.46	0.41	2.10	6.94	0.17	16.77	20.18	0.22	0.00	0.28	98.54
27.	230.1	81.43	52.72	0.44	2.33	6.85	0.17	16.85	20.22	0.24	0.00	0.37	100.20
28.	230.1	79.17	51.43	0.65	3.48	7.61	0.17	16.22	19.71	0.23	0.001	0.36	99.87
29.	234.8	81.36	51.84	0.47	2.76	6.72	0.16	16.45	20.39	0.23	0.00	0.65	99.68
30.	242.2	79.72	53.32	0.48	2.25	7.55	0.18	16.65	20.30	0.22	0.00	0.11	101.08
31.	242.2	82.88	53.18	0.38	2.23	6.33	0.16	17.19	20.55	0.23	0.003	0.56	100.83
32.	242.2	81.86	51.89	0.43	2.71	6.53	0.15	16.53	20.73	0.24	0.00	0.61	99.85
33.	242.2	81.86	52.91	0.42	2.40	6.65	0.16	16.83	20.43	0.23	0.00	0.60	100.65

Note: N sample is the depth (m) in borehole MN-18, Analyses were carried out in MPI, analysts D. Kuzmin and N. Krivolutskaya

section has experienced the most prospecting for economic mineralization, whereas the western section is less explored.

In cross section, the intrusive body has the form of asymmetrical flattened cylinder with steep discordant northern and southern sides with slopes that are subconcordant. The intrusion farther north is split into a number of apophyses, and the underlying rocks are Lower Devonian sediments. There are three types of ore: disseminated, vein, and oxidized ores. All three ore types are found in the lower horizons of the differentiated intrusion of gabbro-dolerites and in the exocontact zone, occupying a very definite place in the vertical section of the deposit.

Disseminated ores, constituting the bulk of the ores, are confined to the lower differentiates of the intrusion. The impregnation sulfides usually appear only in the picritic gabbro-dolerites. Disseminated ores in this intrusion are represented by small drop-shaped sulfides 1–1.5 cm in diameter. In the taxitic gabbro-dolerites, large xenomorphic sulfide aggregates appear. This type of ore (in the taxitic gabbro-dolerites) represents most of the mass of disseminated ores.

Ore veins are present in the eastern, northern, and northwestern parts of the eastern section of the massif, as well as in the underlying rocks. In the intrusion, they are usually separate thin veinlets 3–4 cm wide and rarely up to 20–25 cm.

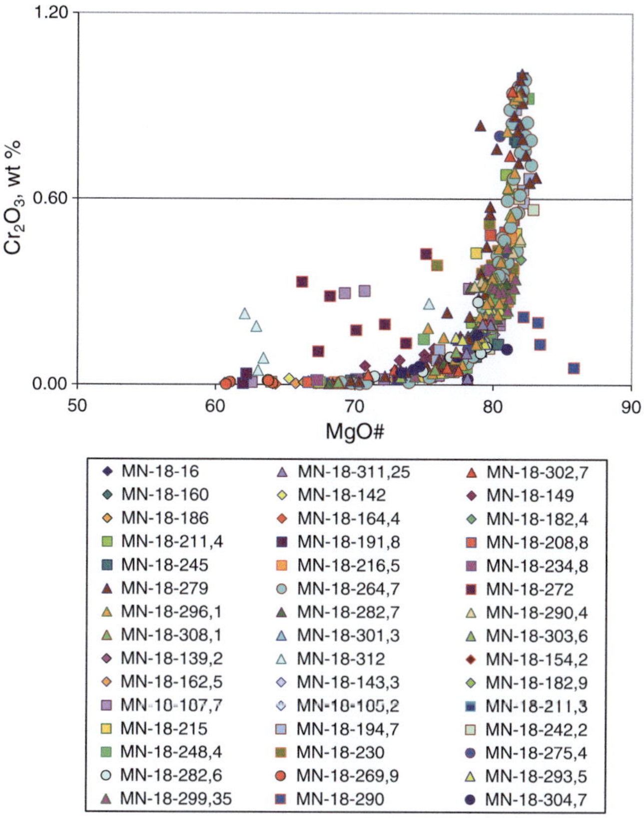

Fig. 4.40 Cr_2O_3–Mg# diagram for pyroxenes from the Noril'sk 2 intrusion

The distinctive feature of the chemical composition of the ore deposits are elevated levels of (often a few tens of ppm) PGE. Rich cuprous ore exists in the bedrock. The metal content in the western part of the deposit is slightly higher than the eastern part.

Oxidized ore mostly developed in the eastern part of the eroded area, where the disseminated ores are located beneath Quaternary sediments. The western border of oxidized ores has not been reliably established. They are linked via a secondary enrichment zone, in which the nickel content is lower but the contents of copper and platinum group metals are higher. Thus, where the content of nickel is 0.03 %, the copper and platinum metals contents reach 0.86 % and 14.65 ppm, respectively.

In 2000, the Chernogorsky massif was included in a number of promising objects for the detection of low-sulfide platinum horizons in the Upper Gabbro Series. Therefore, the distribution of low-sulfide mineralization in the Upper Gabbro Series was studied in the southwestern part of the eastern section and in Tunguska Group on the left bank of the Zvonky stream. The mineralization's thickness ranges from 8.0 to 29.6 m, which is approximately 10 % of the thickness of the intrusion. The sulfide concentration, represented by chalcopyrite, pyrite, and pyrrhotite, is 1 %. The contents of base metals are low, and the concentration of PGE reaches 2 ppm. The remaining sections of the massif's Upper Gabbro Series are completely eroded.

4.5.3.1 Internal Structure of the Chernogorsky Intrusion

The internal structure of the Chernogorsky intrusion was studied in borehole Ch-55 and MP-2 bis, where the intrusion is the thickest (120 m). In this section, the following horizons (from bottom to top) were distinguished: taxitic, picritic gabbro-dolerite, olivine gabbro-dolerites, and contact gabbro-dolerites. In this core, Kruglogorsky-type rocks (Noril'sk Complex) are present in the interval 180–230 m.

The compositions of the rock-forming minerals were studied only in the lower part of the massif, i.e., in the picritic and partly taxitic gabbro-dolerites from the borehole and from the surface outcrops.

The composition of the picritic gabbro-dolerite was studied in detail, and the content of the main components was defined by microprobe analysis obtained by sintering powder samples of rocks at a temperature of 1,500 °C (33 samples). The concentration of trace elements was determined by inductively coupled plasma mass spectrometry (laser-sampling type). The content of the main elements varied in the following ranges (wt %): SiO_2, 44–47; TiO_2, 0.5–0.6; Al_2O_3, 8.6–11.6; MgO, 17.6–23.2; CaO, 6.4–9.3; Na_2O, 0.4–0.9; KO_2, 0.4–0.7; and Cr_2O_3, 0.06–0.08. Representative analyses are given in Table 4.26.

The spectra of the distribution of trace elements are very typical of the Noril'sk-type intrusions. They are characterized by enrichment in light incompatible elements and by the presence of distinct negative Ta, Nb, and Ti anomalies and positive U, Pb, and Sr anomalies. Additionally, the curve features a steeper slope on the right side of the spectrum (Fig. 4.42).

Olivine. Variations in the olivine compositions of the picritic gabbro-dolerites rocks are shown in Fig. 4.43 and listed in Tables 4.27 and 4.28. All rocks of the lower part of the Chernogorsky intrusion are designated "picritic gabbro-dolerite," and the rocks of the Kruglogorsky-type massif are designated "taxitic gabbro-dolerite." The studied olivines from the rocks of the Chernogorsky intrusion differ significantly and feature a large range of MgO concentrations relative to the olivines from the underlying massif. The forsterite component has a range of 59–82 mol%

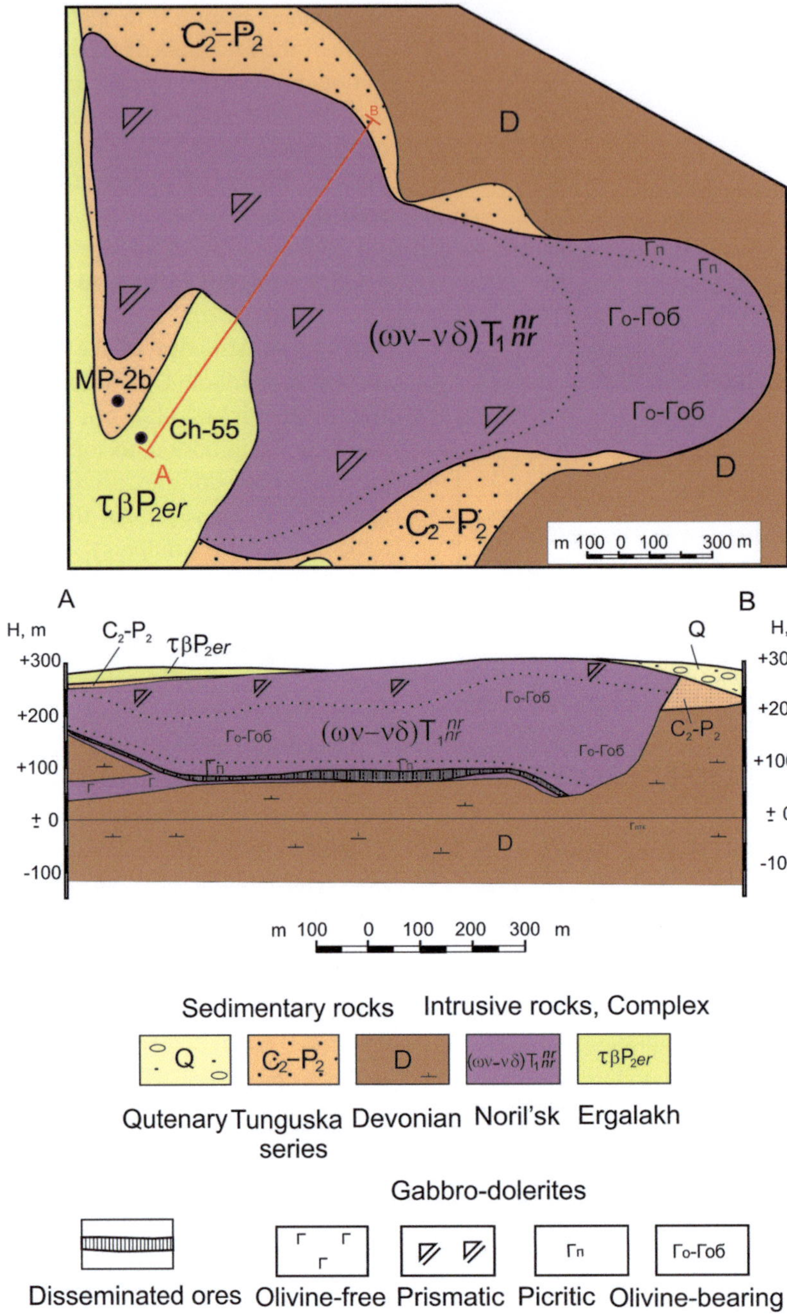

Fig. 4.41 Geological map and cross section of the Chernogorsky intrusion
After NorilskGeology, Ltd. data, with changes by the author

Fo, whereas olivines from the Kruglogorsky massif have a narrower range of 54–70 mol% Fo. In general, all the studied olivines show a distinct correlation between the major components: higher SiO_2 content correlates with higher concentrations of MgO, and Mn, Ti, Co, and Fe are also directly correlated. However, the behaviors of other trace elements in olivines from these intrusions differ. The olivines from the Chernogorsky intrusion are characterized by correlations between the contents of TiO_2 with increasing FeO and Al_2O_3 with increasing MgO, whereas no such obvious variations are observed in the Kruglogorsky olivines. There are weak correlations between the elements within the selected hori-

Table 4.26 Representative analyses of rocks from the Chernogorsky intrusion (borehole Ch-55)

Depth, m	202.4	207.6	215.8	219.3	246.1	255.5	258.4	268.2	269.5
Component	1	2	3	4	5	6	7	8	9
SiO_2	48.00	47.80	47.88	46.60	42.83	42.73	41.92	45.88	46.82
TiO_2	0.69	0.68	0.70	0.61	0.57	0.72	0.60	0.86	0.85
Al_2O_3	20.82	20.33	19.96	17.56	8.68	9.43	10.69	15.18	12.16
FeO	8.19	8.53	8.82	10.32	15.70	15.64	17.58	14.05	9.45
MnO	0.16	0.13	0.17	0.09	0.24	0.26	0.20	0.23	0.34
MgO	7.79	8.20	8.45	11.16	24.93	21.75	19.33	11.91	13.10
CaO	12.34	12.02	11.12	11.00	5.99	6.32	7.10	9.44	15.94
Na_2O	1.78	1.78	2.03	1.57	0.75	1.04	0.88	1.47	0.40
K_2O	0.54	0.49	0.70	0.52	0.25	0.35	0.30	0.47	0.61
P_2O_5	0.09	0.09	0.08	0.09	0.06	0.09	0.06	0.10	0.13
Total	100.3	100.05	99.91	99.53	100.00	98.33	98.67	99.59	99.80
Rb	10.7	11.4	16.7	13.9	5.7	7.9	7.8	9.6	13.8
Ba	163	104	177	93	103	115	129	191	162
Th	0.78	0.59	0.69	0.39	0.61	0.61	0.43	0.59	0.74
U	0.31	0.25	0.38	0.15	0.31	0.24	0.20	0.25	0.32
Nb	3.01	2.14	3.30	1.94	1.79	2.04	1.63	2.72	2.30
Ta	0.18	0.13	0.19	0.12	0.10	0.13	0.10	0.16	0.17
La	6.90	4.13	5.87	3.35	3.94	3.95	3.17	4.08	7.36
Ce	14.36	9.61	13.80	7.93	9.06	8.70	6.98	9.31	19.28
Pb	2.51	1.63	5.34	3.72	0.41	10.66	7.06	4.12	62.71
Pr	1.79	1.32	1.81	1.11	1.21	1.18	0.99	1.30	2.77
Nd	8.00	6.16	8.20	5.30	5.41	5.58	4.59	6.35	13.60
Sr	405	356	351	332	182	138	138	387	267
Sm	2.07	1.68	2.24	1.41	1.44	1.51	1.32	1.77	3.67
Zr	58.0	42.2	80.8	33.6	44.9	44.6	27.9	57.3	86.8
Hf	1.55	1.10	2.23	0.87	1.08	1.26	0.77	1.36	2.91
Eu	0.74	0.66	0.75	0.57	0.54	0.56	0.54	0.74	1.18
Ti	4117	3999	4336	3688	3634	4282	3603	5035	5127
Gd	2.44	2.01	2.63	1.69	1.73	1.84	1.59	2.28	4.16
Tb	0.41	0.34	0.43	0.29	0.29	0.32	0.27	0.36	0.67
Dy	2.66	2.28	2.91	1.91	2.00	2.15	1.83	2.48	4.35
Ho	0.54	0.47	0.62	0.40	0.41	0.45	0.38	0.52	0.87
Y	14.7	12.5	16.1	10.8	11.3	12.1	10.0	14.5	24.2
Er	1.57	1.33	1.77	1.18	1.21	1.29	1.11	1.54	2.52
Tm	0.22	0.20	0.24	0.17	0.17	0.19	0.15	0.23	0.34
Yb	1.51	1.34	1.76	1.14	1.22	1.30	1.15	1.56	2.33
Lu	0.22	0.20	0.25	0.18	0.18	0.20	0.17	0.24	0.35
Ni	191	216	199	309	1545	2956	3573	1551	227
Cu	172	102	134	182	511	4112	4837	1539	728
Zn	94	79	120	84	28	1179	591	206	853
Mn	1150	1132	1279	1206	1820	1674	1641	1634	2502
Sc	22	23	22	25	23	25	22	26	30
Co	58	42	81	34	45	45	28	57	87

Note: Analyses were carried out in Max Planck Institute of Chemistry, Mainz, Germany. Analysts B. Stoll, D. Kuzmin, N. Krivolutskaya zons of rocks, i.e., the olivine and olivine gabbro-dolerite (in the lower fraction of ferruginous olivine—$Fo_{54–60}$—relative to the higher fraction, $Fo_{60–70}$). In the olivine of the Chernogorsky intrusion, the NiO and CaO concentrations are slightly higher than those in the olivines of the Kruglogorsky intrusion with the same MgO, although they form common trend lines (Fig. 4.43).

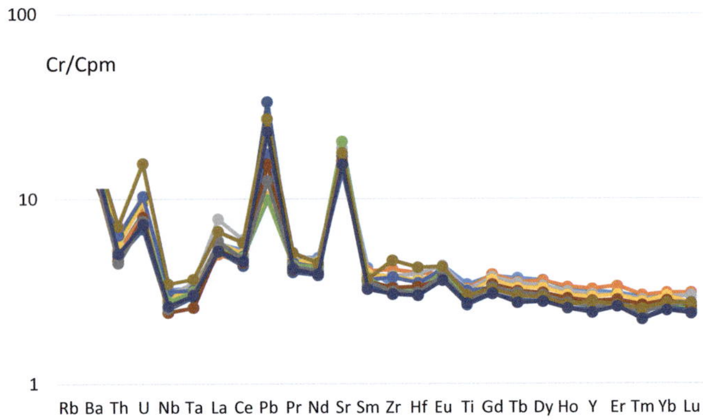

Fig. 4.42 Trace element patterns for intrusive rocks of the Chernogorsky intrusion

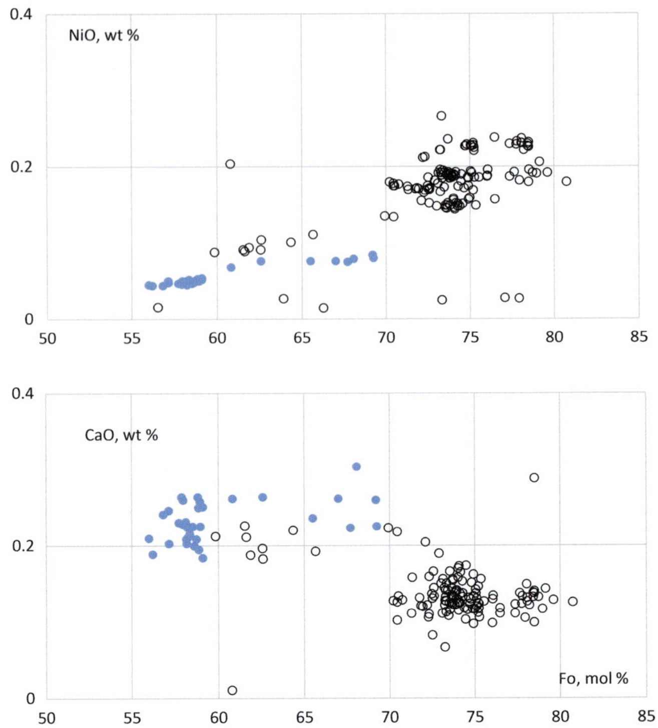

Fig. 4.43 NiO, CaO–Fo diagrams for olivines of the Chernogorsky (*black circles*) and Kruglogorsky (*blue circles*) intrusions

Table 4.27 Representative analyses of olivines from the picritic gabbro-dolerites of the Chernogorsky intrusion, wt %

	N Fo	SiO_2	CaO	FeO	MgO	Cr_2O_3	NiO	MnO	Total
1.	78.27	38.16	0.09	20.12	40.67	0.01	0.20	0.32	99.66
2.	78.20	38.60	0.11	20.24	40.72	0.01	0.22	0.32	100.31
3.	79.98	38.85	0.10	18.59	41.68	0.01	0.21	0.29	99.85
4.	78.35	38.49	0.07	20.07	40.75	0.00	0.23	0.31	100.03
5.	80.31	38.85	0.09	18.38	42.07	0.01	0.12	0.29	99.93
6.	78.78	38.53	0.08	19.71	41.05	0.01	0.21	0.31	99.99
7.	78.54	38.55	0.10	19.88	40.83	0.01	0.21	0.32	99.99
8.	78.50	38.40	0.10	19.92	40.82	0.01	0.22	0.32	99.89
9.	78.21	38.14	0.09	20.15	40.57	0.01	0.18	0.32	99.57
10.	78.33	38.69	0.12	20.14	40.84	0.01	0.11	0.32	100.32
11.	77.98	38.48	0.09	20.45	40.63	0.00	0.22	0.32	100.29
12.	78.69	38.74	0.09	19.83	41.08	0.01	0.20	0.32	100.36
13.	78.18	38.39	0.13	20.19	40.58	0.01	0.19	0.32	99.91
14.	78.86	38.18	0.10	19.63	41.08	0.01	0.22	0.32	99.64
15.	78.60	38.47	0.11	19.91	41.05	0.01	0.22	0.32	100.20
16.	78.40	38.12	0.09	20.03	40.81	0.01	0.23	0.32	99.70
17.	78.57	38.53	0.12	19.96	41.06	0.01	0.21	0.32	100.31
18.	78.43	38.36	0.07	19.97	40.74	0.00	0.21	0.32	99.77
19.	78.30	38.78	0.08	20.14	40.76	0.01	0.24	0.32	100.43
20.	78.06	38.84	0.1	20.35	40.63	0.00	0.19	0.32	100.52
21.	78.81	38.55	0.08	19.71	41.13	0.01	0.16	0.32	100.05
22.	79.07	38.78	0.09	19.49	41.31	0.01	0.21	0.32	100.31
23.	78.57	38.83	0.08	19.89	40.94	0.01	0.19	0.32	100.37
24.	78.80	38.73	0.07	19.76	41.23	0.01	0.20	0.32	100.41

(continued)

Table 4.27 (continued)

	N Fo	SiO_2	CaO	FeO	MgO	Cr_2O_3	NiO	MnO	Total
25.	78.59	38.73	0.08	19.91	41.01	0.006	0.22	0.32	100.38
26.	78.67	38.84	0.08	19.84	41.04	0.007	0.23	0.32	100.45
27.	78.88	38.72	0.07	19.70	41.30	0.007	0.21	0.32	100.42
28.	78.42	38.61	0.10	20.01	40.81	0.01	0.23	0.32	100.19
29.	78.50	38.62	0.11	19.96	40.89	0.008	0.23	0.32	100.23
30.	78.24	38.43	0.09	20.24	40.84	0.005	0.19	0.33	100.20
31.	78.78	38.94	0.07	19.78	41.20	0.009	0.22	0.32	100.62
32.	78.54	38.85	0.08	19.98	41.03	0.01	0.22	0.32	100.58
33.	78.48	38.72	0.10	20.07	41.06	0.01	0.18	0.33	100.55
34.	78.14	38.33	0.09	20.37	40.85	0.01	0.17	0.32	100.24
35.	78.50	38.85	0.11	20.06	41.10	0.01	0.24	0.32	100.78
36.	78.50	38.83	0.09	20.09	41.17	0.01	0.21	0.32	100.81
37.	78.46	38.46	0.07	20.06	40.99	0.01	0.24	0.32	100.24
38.	78.70	38.72	0.10	19.82	41.09	0.01	0.22	0.32	100.37
39.	78.50	38.66	0.09	19.98	40.92	0.01	0.23	0.32	100.31
40.	78.44	38.70	0.09	20.06	40.95	0.009	0.22	0.32	100.44
41.	78.54	39.05	0.11	20.02	41.11	0.01	0.16	0.33	100.87
42.	78.67	38.80	0.10	19.91	41.20	0.01	0.21	0.32	100.64
43.	78.88	39.10	0.08	19.71	41.31	0.01	0.20	0.31	100.81
44.	79.68	39.12	0.09	19.01	41.84	0.01	0.20	0.31	100.67
45.	78.31	38.88	0.09	20.20	40.92	0.009	0.19	0.32	100.71
46.	80.38	39.22	0.11	18.39	42.26	0.01	0.21	0.29	100.59
47.	78.39	38.92	0.09	20.11	40.92	0.007	0.22	0.32	100.67
48.	78.84	38.95	0.07	19.77	41.33	0.01	0.20	0.32	100.74
49.	78.56	39.00	0.07	20.00	41.12	0.006	0.24	0.32	100.85
50.	78.70	38.85	0.08	19.85	41.15	0.01	0.23	0.32	100.57
51.	78.39	38.44	0.09	20.11	40.94	0.008	0.21	0.32	100.22
52.	81.33	39.09	0.09	17.58	42.97	0.01	0.21	0.28	100.32

Note: Here and in Tables 4.28, 4.29 analyses were carried out in MPI, analysts D. Kuzmin and N. Krivolutskaya

Pyroxenes. Due to unfortunate technical reasons, the pyroxene compositions have not been studied uniformly throughout the section. However, the analysis was performed for the main types of rocks of both intrusions. For the Chernogorsky intrusion, clinopyroxenes from the taxitic and taxitic-like gabbro-dolerites from upper zone and picritic gabbro-dolerites from the bottom were investigated (Table 4.29).

For the Kruglogorsky massif, pyroxenes from olivine gabbro-dolerite were studied (Table 4.29). In the first of these samples, there is a very wide range in magnesium content (Mg#: 66–82), and the samples are similar in composition to the Kruglogorsky massif, which is characterized by similar variations in the composition of clinopyroxene but with a lower maximum magnesium content (Mg# of 77, not 82). In contrast, the values of the clinopyroxene from the picritic gabbro-dolerite of the Chernogorsky intrusion are 77–84. The clinopyroxenes are also characterized by low-titanium values (0.4–0.6 wt %), whereas the higher contents of Ti in the Kruglogorsky intrusion have values of 0.6–1.0 wt %.

Table 4.28 Representative analyses of olivine from Chernogorsky (borehole MP-2b), wt %

N	Depth, m	Fo	SiO_2	CaO	FeO	MgO	NiO	MnO	Total
1.	133	71.32	37.84	0.11	25.98	36.21	0.17	0.39	100.71
2.	133	70.5	37.62	0.10	26.67	35.71	0.17	0.39	100.68
3.	133	70.2	37.69	0.12	27.05	35.82	0.18	0.39	101.26
4.	133	71.9	37.43	0.12	25.36	36.35	0.17	0.38	99.81
5.	133	72.6	37.86	0.14	25.00	37.20	0.17	0.38	100.76
6.	133	72.6	37.82	0.13	24.99	37.20	0.17	0.38	100.71
7.	133	72.5	37.87	0.13	25.05	37.12	0.17	0.38	100.74
8.	133	72.3	37.74	0.10	25.21	36.91	0.16	0.38	100.53
9.	133	72.3	37.62	0.11	25.04	36.75	0.16	0.38	100.08
10.	133	70.5	37.82	0.21	26.80	35.89	0.13	0.40	101.29
11.	133	69.9	37.80	0.22	27.26	35.61	0.13	0.41	101.46
12.	134	78.1	38.74	0.14	20.52	40.98	0.23	0.32	100.96
13.	134	78.1	38.78	0.13	20.42	40.84	0.23	0.32	100.75
14.	134	78.5	38.73	0.13	20.04	41.07	0.22	0.31	100.53
15.	134	77.4	38.66	0.12	21.01	40.31	0.23	0.33	100.69
16.	134	78.1	38.59	0.12	20.36	40.73	0.23	0.31	100.38
17.	135	73.2	37.86	0.14	24.52	37.64	0.22	0.37	100.77
18.	137	74.3	38.53	0.13	23.83	38.67	0.17	0.36	101.72
19.	137	73.7	38.23	0.13	24.15	37.96	0.15	0.38	101.01
20.	137	73.6	38.35	0.13	24.21	37.92	0.14	0.37	101.15
21.	137	73.8	38.40	0.13	24.13	38.22	0.14	0.37	101.42
22.	137	74.9	38.35	0.13	23.11	38.76	0.15	0.36	100.88
23.	137	74.1	38.37	0.12	23.89	38.37	0.14	0.37	101.29
24.	137	75.1	38.47	0.14	23.01	38.86	0.19	0.35	101.05
25.	137	73.2	38.23	0.13	24.69	37.84	0.15	0.38	101.44
26.	137	73.0	38.18	0.15	24.86	37.71	0.15	0.38	101.45
27.	137	74.6	38.51	0.11	23.48	38.72	0.14	0.36	101.35
28.	139	65.7	37.26	0.19	30.68	32.95	0.11	0.47	101.68
29.	139	64.4	36.99	0.22	31.62	32.05	0.10	0.48	101.49
30.	140	74.0	38.14	0.12	23.85	38.19	0.19	0.37	100.86
31.	140	73.4	38.40	0.14	24.54	38.05	0.19	0.37	101.71
32.	140	73.7	38.35	0.13	24.17	37.95	0.19	0.37	101.18
33.	140	73.1	38.26	0.15	24.72	37.79	0.19	0.37	101.51

Table 4.29 Representative analyses of pyroxenes from the Kruglogorsky intrusions (borehole MP-2b), wt %

№	Depth, m	MgO#	SiO_2	TiO_2	Al_2O_3	FeO	MnO	MgO	CaO	Na_2O	Cr_2O_3	Total
1.	198	71.82	51.31	0.78	2.43	10.60	0.25	15.15	18.89	0.29	0.03	99.76
2.	198	65.74	51.12	0.90	1.67	13.89	0.33	14.95	17.07	0.26	0.00	100.23
3.	198	76.26	51.29	0.64	2.87	8.74	0.21	15.75	19.53	0.31	0.23	99.62
4.	198	66.10	51.47	0.84	1.79	13.42	0.31	14.68	17.63	0.30	0.01	100.49
5.	198	67.29	51.23	0.85	1.95	12.75	0.29	14.71	18.19	0.27	0.03	100.31
6.	200	68.28	51.09	0.82	2.26	11.92	0.28	14.39	18.87	0.32	0.02	99.99
7.	200	64.07	50.77	0.95	2.28	13.82	0.33	13.82	18.08	0.29	0.01	100.39
8.	201	67.35	51.10	0.87	2.23	12.55	0.29	14.52	18.24	0.23	0.01	100.09
9.	201	70.16	51.13	0.84	2.36	11.32	0.27	14.93	18.68	0.21	0.03	99.81
10.	206	66.92	51.07	0.88	2.38	13.21	0.33	14.99	17.06	0.17	0.01	100.13
11.	207	62.83	50.67	0.95	1.97	14.27	0.34	13.53	17.86	0.16	0.01	99.78
12.	211	77.77	51.22	0.60	2.90	8.11	0.18	15.91	19.70	0.36	0.31	99.33
13.	211	77.29	51.55	0.59	2.59	8.44	0.19	16.11	19.44	0.19	0.24	99.39
14.	215	77.38	51.84	0.58	2.39	8.39	0.20	16.10	19.71	0.34	0.15	99.75
15.	215	65.88	51.61	0.80	1.36	14.33	0.37	15.52	15.87	0.21	0.00	100.07

4.5.4 Zub-Marksheydersky Massif

The first information on the differentiated intrusions of Zub-Marksheydersky (Fig. 4.26) was obtained by Yu. Sherman in 1940 during 1:50,000-scale mapping for the Geological Survey between 1940 and 1943. In this intrusion 8 boreholes were drilled (along the strike of the intrusive body) and disseminated ores were found. The host rocks of this massif in the western part of the field are marls and mudstones, and the center is composed of Middle Devonian mudstones and Lower Devonian gypsum-bearing rocks. These formations all have a common gentle dip to the southwest (Chernova 1961). The intrusion has the shape of a chonolith and is entirely surrounded by Devonian rocks (Fig. 4.44). Despite the irregular shape of the intrusion and its high degree of bottom roughness, the intrusion is generally inclined toward the southwest. The intrusion extends toward the west–northwest and has a length of approximately 5 km; its width ranges from 100 to 2,000 m. The average thickness of the intrusion is 100 m.

4.5.4.1 The Internal Structure of the Zub-Marksheydersky Intrusion

The internal structure of the intrusion features peculiar, widespread hybridization of surrounding rocks, including Lower and Middle Devonian recycled carbonates, terrigenous carbonate rocks, marl rocks, and gypsum–anhydrite rocks. This assimilation process is most developed in the upper part of the intrusion, but the hybrid rocks and xenoliths of the host rocks are found not only in areas of the upper and lower exocontact but also in the center body of the intrusion. Two main types of ores were found: (1) disseminated ore, which is confined to the lower differentiated intrusions, and (2) veins, which are localized in the form of small veins and lenses in the underlying rocks. The orebody is in the lower portion of the intrusion and generally has the form of layer; in places, the orebody features somewhat complicated pinches and bulges. The distribution of the mineralization in the deposit area is vertically irregular. The thickness of the sulfide dissemination zone throughout the field ranges from 10 to 20 m. The main ore minerals are pyrrhotite, chalcopyrite, pentlandite, cubanite, pyrite, and sphalerite, as well as less commonly millerite and nickel pyrite. The structure and composition of the intrusive rocks were studied in borehole MP-34 (Fig. 4.45). In the upper part of the section, gabbro-diorite and olivine-free gabbro dominate. These rocks have experienced heavy secondary alteration, including feldspathization, silicification, and so on. The rest of the intrusion has a classical structure (gabbro-dolerites): olivine-bearing, olivine gabbro-dolerites, picritic, taxitic, and contactic.

Olivine The composition of olivine has been studied mainly in the lower part of the Zub-Marksheydersky massif because the upper part of the massif is too altered (Table 4.30). Variations in the olivine composition in this section are shown in Fig. 4.45. It should be noted that the MgO concentration in the minerals varies widely. For example, the Fo proportion in olivine varies from 58.6 to 78.59 mol%, and most of the compositions fall within the range of 70–75 mol%. According to the content of the impurity elements, in particular titanium, there does not appear to be any difference between the olivines from picritic, taxitic, olivine, and olivine-bearing gabbro-dolerites. The CaO contents of the aforementioned olivine horizons are also similar and vary greatly from 0.1 to 0.3 wt %. Most of the studied grains contain 0.15 to 0.20 wt % CaO which is, in general, higher than in the olivines from intrusions with large deposits. The MnO content in olivine exhibits the usual dependence on the iron content (Table 4.30). The lowest NiO concentrations in olivine were found in the upper part of the section (Fig. 4.46). This observation highlights the inverse relationship between nickel and forsterite in picritic gabbro-dolerites due to the presence of sulfides in the rocks.

Pyroxene The composition of pyroxenes has also been studied, mainly in the lower section of the intrusive body (Table 4.31). However, this pattern is not due to its absence in the rocks (as with olivine) but is rather due to the high degree of alteration resulting from hydrothermal processes, especially at the top of the section. The most magnesium-rich clinopyroxenes are at the top of the picrite horizon (Mg#=82.4). Variations in the composition of pyroxenes are found in all horizons, but they are especially characteristic for the upper part of the horizon of picritic and olivine gabbro-dolerites. In both cases, the difference is Mg# 25. The most distinct changes in the content of the impurity elements depend on the composition of the minerals' titanium and manganese characteristics (Fig. 4.47). These elements correlate directly with the iron, whereas the chromium concentration increases with increasing Mg#. The distribution of aluminum, calcium, and sodium in the pyroxenes is not related to the major elements.

The compositions of orthopyroxene were studied less thoroughly due to its limited availability in the rocks. The analytical data are likely insufficient for a clear picture of the behavior of trace elements in the orthopyroxene of the Zub-Marksheydersky intrusion. Except for the manganese and chromium, no elements exhibit patterns in their distributions.

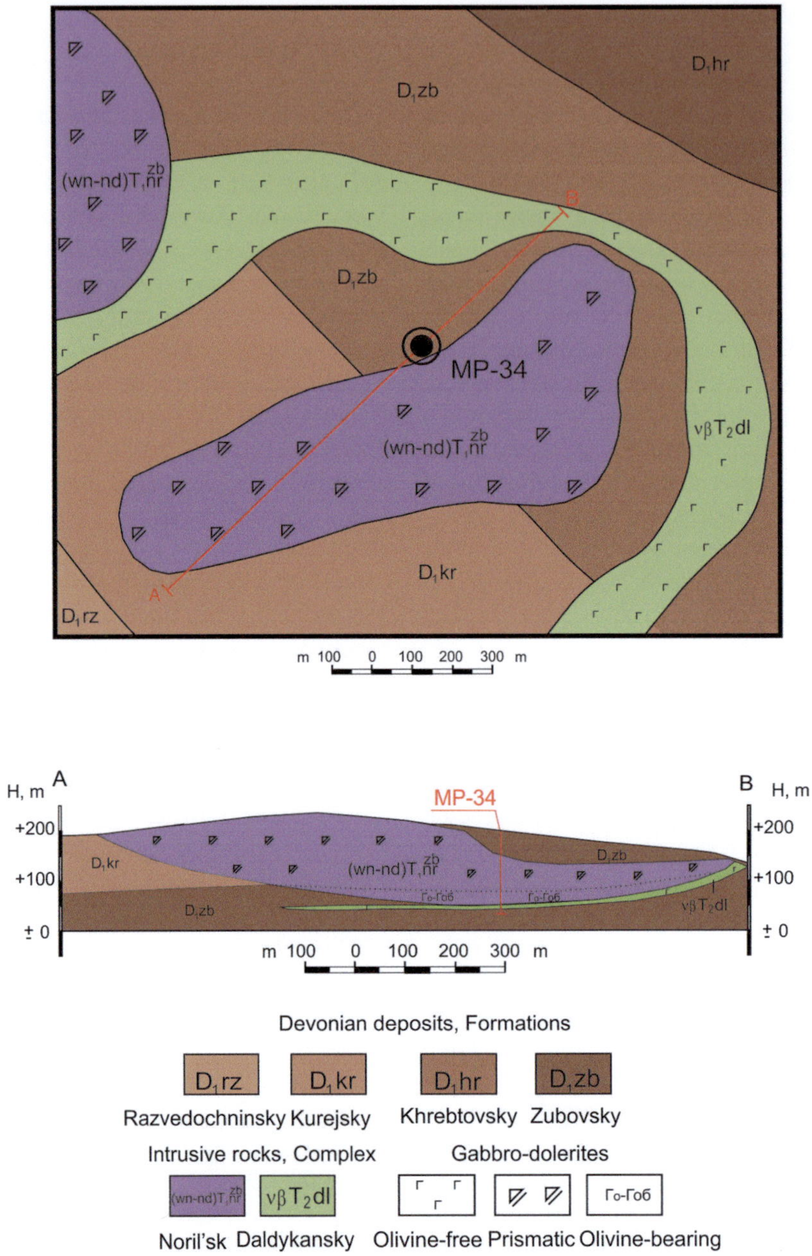

Fig. 4.44 Geological map and cross section of the Zub-Marksheydersky intrusion
After NorilskGeology, Ltd. data, changed by the author

4.5.5 Bol'shaya Bar'ernaya Massif

The intrusion in the mountain Bol'shaya Bar'ernaya is located subparallel to and approximately 300–600 m to northwest of the Noril'sk 2 massif (Fig. 4.26). This intrusion is thin (maximum thickness 100 m) and has very steep inclination to the NW. Its shape resembles a dyke and is approximately 3 km long. Geologists of NorilskGeology, Ltd. company regard it as a Noril'sk-type intrusion of the Noril'sk Complex (Fig. 4.48).

Samples from the Bol'shaya Bar'ernaya massif were collected from the core of borehole MP-38, in which the intrusive rocks form a section 55 m thick. This section consists mostly of olivine and olivine-bearing gabbro-dolerites (Fig. 4.49).

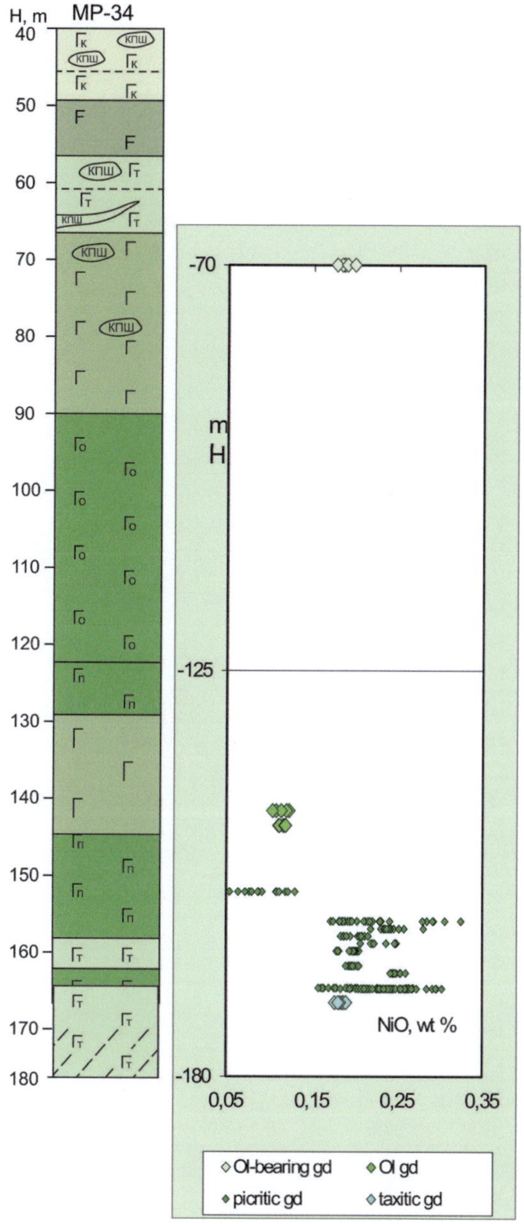

Fig. 4.45 Inner structure of the Zub-Marksheydersky intrusion and olivine compositions variations throw borehole MP-34

Olivine The composition of the olivine from the studied rocks varies quite significantly, and the forsterite component ranges from 73 to 79 mol% (Table 4.32). Calcium has weak tendency to increase in more magnesium-rich olivine, although its concentrations in general are sufficiently constant, ranging from 0.11 to 0.22 wt % CaO. For Fo_{79}, the CaO concentration is 0.25 wt %. The concentrations of nickel are characterized by surprising constancy (0.25 wt %) and are independent of the major elements, which is apparently caused by the admixture of sulfides. Manganese uniformly increases with fayalite from 0.3 to 0.38 wt %. The Fe/Mn ratio varies irregularly from 64 to 66. Ti and Al form sufficiently clear multidirectional trends; the former is correlated with Fe, and the latter is correlated with Mg.

Pyroxene The variations in the composition of pyroxenes in the intrusion of the Bol'shaya Bar'ernaya are significant, reaching 20 with respect to magnesium content, especially in the lowermost part of the section (Table 4.33). The pyroxenes are characterized by very monotonous behavior of rare elements, which are almost independent of the composition of the rocks. However, in the distribution of titanium in pyroxene, one can see two trends—one in picritic gabbro-dolerites and the other in all other rock types. However, this division is not very clear, especially in the area of magnesium-rich pyroxenes (Mg# greater than 78). The slopes of the trend are also very similar in Fe–pyroxenes. Based on sodium, which typically mirrors the behavior of titanium, the compositions of the pyroxenes can generally be separated into different types, in which sodium content increases with increasing iron content in pyroxene. The behavior of manganese closely resembles that of sodium in pyroxene. The distribution of the aluminum in the pyroxenes does not depend on the content of the major components, i.e., iron and magnesium. A larger range of Al concentrations were found in pyroxenes (1.8–4.0 wt %), whereas the pyroxenes from picritic and picritic-like gabbro-dolerites feature range from 2.0 to 3.5 wt %. The distribution of chromium in pyroxenes from Bol'shaya Bar'ernaya resembles the distribution in pyroxenes from the Zub-Marksheydersky intrusion.

4.5.6 Maslovsky Massifs

Massive and disseminated PGE–Cu–Ni sulfide mineralization is related to the Maslovsky intrusion that is thought to be a southwestern branch of the Noril'sk 1 intrusion based on data from boreholes drilled to the south of this massif. The deposit was found in the 1960s by G.D. Maslov, who previously has discovered the Talnakh deposit (unpublished data). Geologists from the NorilskGeology, Ltd., the Exploration Division of Noril'sk Nickel, have recently carried out exploration and acquired new data that not only shed light on the inner structure of this deposit but also provide a set of key features that help to understand better the genesis of the Noril'sk deposits.

Table 4.30 Representative analyses of olivine from the Zub-Marksheydersky intrusion (borehole MP-34), wt %

N	Depth, m	Fo, mol %	SiO_2	FeO	MnO	MgO	CaO	NiO	Total
1.	70	72.56	38.63	25.31	0.40	37.54	0.09	0.19	102.18
2.	70	71.95	38.80	25.77	0.41	37.07	0.11	0.19	102.45
3.	144	67.97	37.95	29.08	0.45	34.62	0.16	0.11	102.45
4.	144	69.29	38.20	28.01	0.43	35.44	0.15	0.12	102.43
5.	146	70.25	38.26	26.99	0.42	35.75	0.15	0.11	101.75
6.	149	71.24	37.74	26.01	0.40	36.13	0.17	0.12	100.63
7.	149	71.37	38.02	25.96	0.40	36.30	0.17	0.12	101.06
8.	155	62.66	36.97	32.96	0.55	31.02	0.25	0.08	101.90
9.	155	65.56	37.42	30.71	0.47	32.79	0.28	0.09	101.83
10.	159	72.19	38.07	25.40	0.40	36.98	0.15	0.19	101.25
11.	159	69.98	37.85	27.22	0.42	35.59	0.22	0.17	101.54
12.	159	73.28	38.13	24.55	0.39	37.76	0.12	0.20	101.23
13.	160	72.10	38.32	25.46	0.40	36.90	0.22	0.22	101.59
14.	160	70.49	37.89	26.83	0.42	35.95	0.18	0.25	101.60
15.	161	75.34	38.58	22.73	0.35	38.95	0.17	0.18	101.05
16.	161	75.60	38.70	22.27	0.36	38.70	0.27	0.23	100.60
17.	162	74.64	38.57	23.07	0.38	38.09	0.23	0.25	100.66
18.	162	73.61	38.14	23.89	0.39	37.37	0.20	0.24	100.33
19.	163	70.35	37.94	26.67	0.43	35.50	0.18	0.20	101.00
20.	163	71.41	38.10	25.75	0.42	36.07	0.16	0.20	100.77
21.	165	69.51	36.56	26.72	0.42	34.17	0.17	0.18	98.30
22.	165	68.71	36.48	27.43	0.43	33.79	0.21	0.18	98.60
23.	166	75.78	38.77	22.44	0.34	39.37	0.22	0.25	101.46
24.	166	74.44	38.68	23.58	0.36	38.51	0.16	0.25	101.62
25.	166	75.00	38.75	23.09	0.36	38.85	0.18	0.25	101.56
26.	166	75.56	38.70	22.65	0.35	39.28	0.19	0.24	101.48
27.	168	68.95	37.74	28.06	0.46	34.95	0.16	0.17	101.62
28.	168	67.37	37.27	29.13	0.48	33.73	0.17	0.20	101.05
29.	168	66.67	37.32	29.89	0.49	33.53	0.13	0.20	101.63
30.	169	72.32	37.99	25.37	0.38	37.18	0.13	0.26	101.39
31.	169	70.12	37.65	27.13	0.42	35.71	0.14	0.25	101.38
32.	169	71.97	37.91	25.53	0.39	36.77	0.18	0.26	101.10
33.	169	72.97	38.11	24.83	0.38	37.58	0.15	0.25	101.36
34.	169	69.21	37.62	27.89	0.46	35.17	0.15	0.29	101.64
35.	169	72.43	38.03	25.30	0.39	37.29	0.17	0.25	101.52
36.	169	70.60	37.71	26.73	0.43	36.01	0.17	0.25	101.36
37.	170	70.36	38.15	26.98	0.46	35.92	0.15	0.18	101.91
38.	170	68.92	37.92	28.09	0.47	34.93	0.13	0.19	101.80

Note: Here and in Table 4.31 analyses were carried out in GEOKHI RAS, analyst N. Kononkova

The Southern Maslovsky intrusion differs from the other ore-bearing intrusions of the Noril'sk region in cutting at a high level into the stratigraphic section of the flood basalts. The intrusions of the Noril'sk Complex carrying the economic ore mineralization are commonly hosted in Devonian–Permian terrigenous rocks that can be seen at the Zapolyarny, Mayak, Komsomolsky, Oktyabr'skoe, and Taymyrsky Mines. More rarely, they are located in the lower volcanic formations; this is seen in the Medvezhy Creek Mine as example. The Southern Maslovsky intrusion cuts across the rocks of the lower part of the Nadezhdinsky Formation (Krivolutskaya and Rudakova 2009), which belongs to the intermediate tuff–lava unit of the trap sequence (Geology and ore deposits 1994).

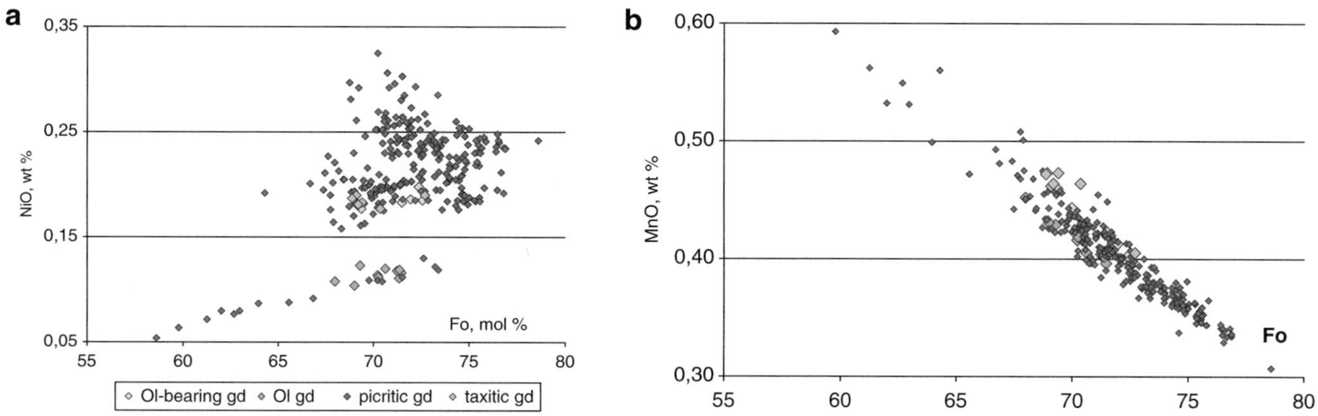

Fig. 4.46 NiO, MnO–Fo diagrams for olivines of the Zub-Marksheydersky intrusion

Table 4.31 Representative analyses of pyroxenes from the Zub-Marksheydersky intrusion (borehole MP-34), wt %

N	Depth, m	Mg#	SiO_2	TiO_2	Al_2O_3	FeO	MnO	MgO	CaO	Na2O	Cr_2O_3	Total
1.	70	78.44	52.26	0.51	2.81	8.19	0.20	16.71	19.57	0.26	0.18	100.75
2.	70	78.98	51.43	0.62	3.98	7.66	0.19	16.14	19.93	0.28	0.45	100.72
3.	146	74.28	52.35	0.12	0.78	9.05	0.33	14.66	22.25	0.10	0.00	99.63
4.	146	65.71	51.28	0.25	1.69	11.85	0.40	12.74	21.31	0.42	0.00	99.95
5.	150	75.36	51.05	0.66	3.42	9.16	0.23	15.72	19.36	0.28	0.35	100.27
6.	150	75.30	51.65	0.65	3.23	9.38	0.25	16.04	19.04	0.29	0.33	100.89
7.	155	75.85	51.09	0.65	3.73	8.93	0.22	15.74	19.24	0.25	0.38	100.26
8.	155	74.76	51.38	0.64	3.36	9.62	0.25	15.98	18.73	0.25	0.22	100.46
9.	159	80.05	52.63	0.50	2.85	7.49	0.19	16.86	19.87	0.20	0.32	100.95
10.	159	80.95	52.80	0.43	2.72	7.19	0.18	17.14	19.76	0.21	0.43	100.89
11.	160	81.07	52.05	0.60	2.60	7.11	0.17	17.08	19.74	0.28	0.31	99.94
12.	160	80.56	51.49	0.67	2.66	7.40	0.19	17.20	19.43	0.26	0.29	99.59
13.	160	79.96	51.70	0.72	2.54	7.78	0.20	17.41	19.06	0.26	0.23	99.89
14.	161	79.55	52.01	0.74	2.99	7.62	0.20	16.62	19.58	0.25	0.30	100.34
15.	161	76.30	50.81	1.52	3.33	8.89	0.22	16.06	18.95	0.31	0.11	100.24
16.	162	79.88	51.19	0.58	3.58	7.25	0.18	16.15	19.95	0.24	0.67	99.84
17.	162	78.99	50.71	0.66	3.32	7.73	0.19	16.30	19.53	0.24	0.43	99.14
18.	162	80.21	50.81	0.56	3.38	7.18	0.18	16.32	19.92	0.22	0.58	99.19
19.	163	77.24	51.75	0.76	2.41	8.50	0.21	16.18	19.66	0.29	0.30	100.07
20.	163	79.38	51.71	0.66	2.37	7.69	0.19	16.61	19.96	0.26	0.27	99.70
21.	163	78.43	50.53	0.67	3.26	7.97	0.20	16.27	19.52	0.22	0.31	98.99
22.	165	75.16	50.47	1.04	2.98	9.31	0.25	15.80	18.79	0.32	0.24	99.23
23.	165	74.73	50.51	1.04	3.04	9.65	0.26	16.01	18.29	0.34	0.25	99.43
24.	166	81.46	51.91	0.65	3.20	6.70	0.16	16.52	20.52	0.24	0.47	100.37
25.	166	81.17	51.54	0.69	3.09	6.81	0.16	16.46	20.40	0.26	0.41	99.80
26.	168	79.88	52.02	0.53	2.66	7.35	0.18	16.36	20.33	0.26	0.32	100.02
27.	168	76.06	52.72	0.65	2.15	9.08	0.23	16.18	19.51	0.30	0.03	100.84
28.	168	73.34	50.76	1.21	2.37	10.21	0.27	15.76	18.77	0.32	0.02	99.68
29.	170	69.84	54.35	0.27	1.79	12.64	0.27	16.42	11.82	0.37	0.01	98.08
30.	170	75.72	51.08	0.63	3.34	8.75	0.31	15.30	19.94	0.31	0.25	99.92
31.	171	75.16	54.11	0.43	1.93	10.62	0.50	18.02	11.15	0.61	0.00	97.50
32.	171	78.85	50.49	0.69	3.90	7.50	0.18	15.68	20.47	0.27	0.16	99.35
33.	171	80.73	51.93	0.49	2.62	6.99	0.18	16.43	20.79	0.28	0.13	99.84
34.	171	79.74	51.19	0.60	3.34	7.20	0.16	15.89	20.87	0.28	0.12	99.65

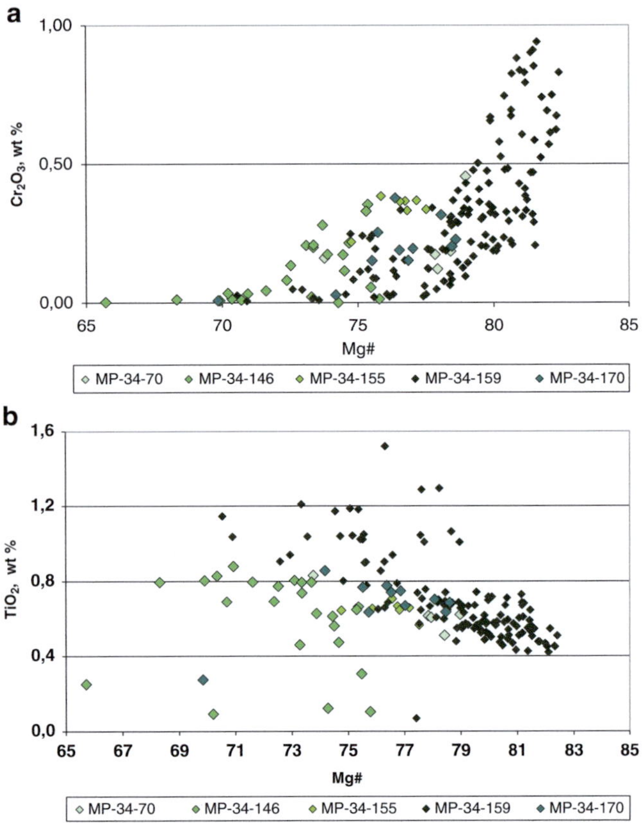

Fig. 4.47 TiO_2, Cr_2O_3–Mg# diagrams for pyroxenes from the Zub-Marksheydersky intrusion

These geological positions of the Southern Maslovsky intrusion help to resolve some of the poorly understood aspects of genesis of the Noril'sk deposits, viz., (1) to establish the temporal relationships between the ore-bearing intrusions and lavas and (2) to estimate the role of surrounding rocks' contamination by magma in the formation of sulfide mineralization.

The Maslovsky deposit has been explored with numerous boreholes, two of which (OM-4 and OM-24) were drilled through the thickest part of the intrusions (270 m in the north and 420 in the south) and, thus, they were selected for a more detailed examination for this study. Samples from boreholes OM-6 and OM-25 were also included in our research, and their data have been published in Krivolutskaya et al. 2012; Krivolutskaya and Rudakova (2009), Rudakova and Krivolutskaya (2009).

Methods. Two hundred gram samples of the rock were powdered, and 40 g were baked to glass using an Ir heater (Stoll et al. 2008). Major and trace elements were determined by EPMA on a *Jeol JXA 8200 SuperProbe* electron probe at the Max Planck Institute for Chemistry in Mainz, Germany. The major element compositions of the glasses were analyzed at an accelerating voltage of 15 kV and a beam current of 12 nA, with the reference sample being natural basaltic glass USNM111240/52 (VG2) (Jarosevich et al. 1980). A relative standard deviation (RSD) of 1–2 % was achieved during the analysis.

The composition of olivine was determined at an accelerating voltage of 20 kV and a beam current of 300 nA, following a special procedure which allows 20–30 ppm (two-sigma error) precision and an accuracy for Ni, Ca, Mn, Al, Ti, Cr, and Co at 0.02 mol% for the forsterite component in olivine (Sobolev et al. 2007). The composition of pyroxenes was analyzed at an accelerating voltage of 20 kV and a beam current of 80 nA. Sulfide compositions were determined on *SX 100 Cameca* at the Vernadsky Institute of Geochemistry RAS (Moscow, Russia) at an accelerating voltage of 15 kV and a beam current of 40 nA.

LA-ICP-MS was applied to determine trace elements in glasses and minerals on *ELEMENT-2* (*Thermo Scientific*, GB*)* mass spectrometer with an *UP-213 New Wave Research* solid-phase laser (GB) at the Max Planck Institute for Chemistry, with reference to the KL-2G and NIST 612 standard samples of basaltic glass (Jochum et al. 2000) and the GeoReM database (http://georem.mpch-mainz.gwdg.de), and Ca as the reference element for glasses and Si for minerals. The typical laser beam diameter was 160 μm for glasses and 60–80 μm for minerals, and the ablation time was 160 and 60–80 s, respectively. Element abundances were determined with two-sigma errors of no higher than 5 and 10 % for concentrations >1 ppm and <0.1 ppm, respectively.

Precious metals (Au and PGE) were determined at the Laboratory of the Geochemistry of PGE of the Semenenko Institute of Geochemistry, Mineralogy and Ore Formation of the National Academy of Sciences of Ukraine, by assaying with the use of atomic emission in 35-g samples (Kursky et al. 1992). Check analyses were made with 80–100 g of material. Simultaneously reference samples were prepared (RSP-1 and a set of original synthetic samples) whose precious metal concentrations were in the range of 0.003–20.0 ppm. The preparation of the samples involved their saturation with lead acetate solution with the addition of 10 mg Ag in the form of silver nitrate solution. The dried samples were mixed following conventional techniques, with a slight excess of the reducer. The reducing melting was carried out in chamotte refractory crucibles at a temperature of 1150°C. The lead bullions thus obtained were purified to get rid of minor interfering admixtures (first of all, Ni, Cu, Se, and Te) by means of scorifying dish melting. The purified lead bullions were cupellated at magnesite drips at a temperature of 1,050 °C until a silver–lead bead 50–100 mg in mass was obtained. The crushed material of the beads was analyzed in an electrode crater by arc burning,

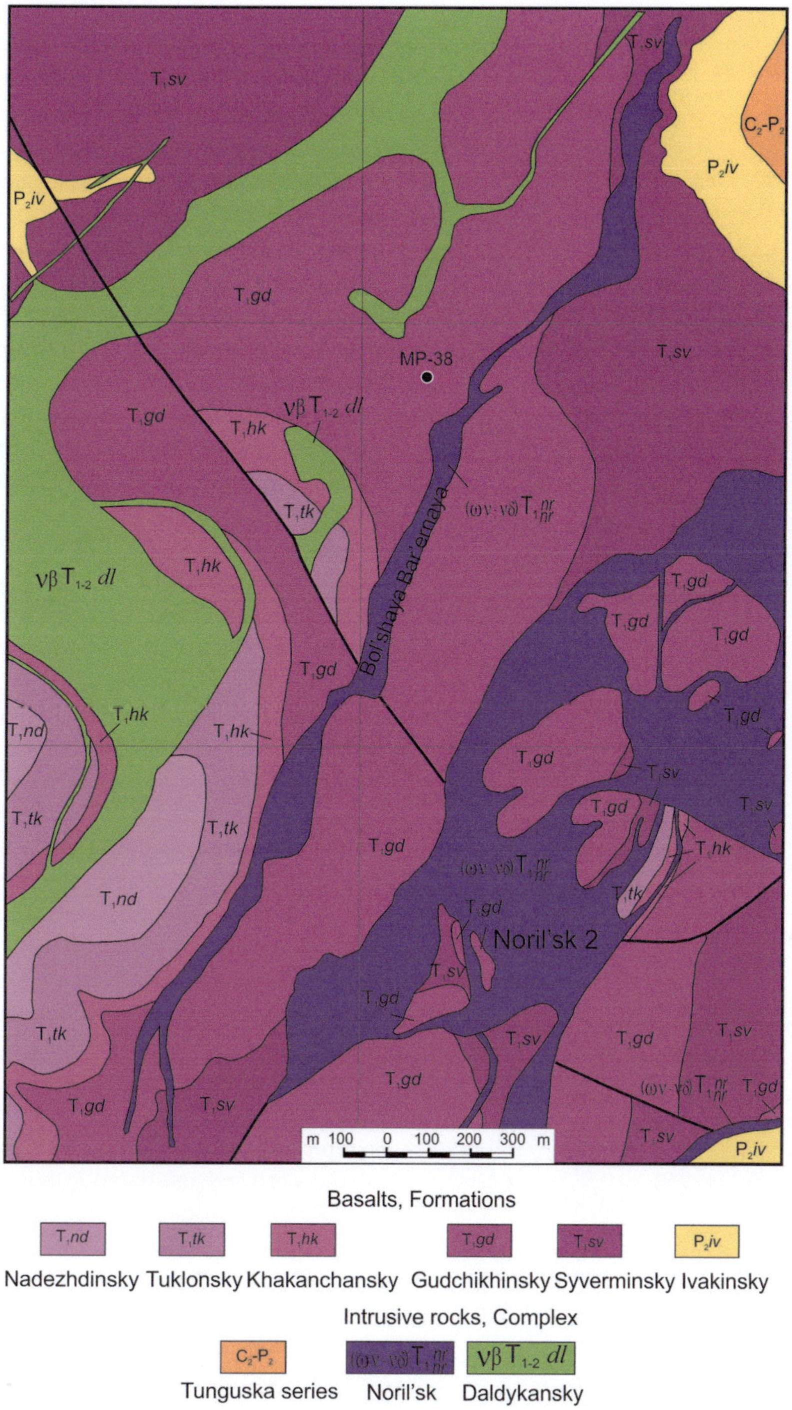

Fig. 4.48 Geological map and cross section of the Bol'shaya Bar'ernaya intrusion After NorilskGeology, Ltd. data, changed by the author

using a STE-1 spectrograph; the intensity of spectral lines was evaluated on MD-100. The detection limits were for Pt, 0.003; Pd, 0.001; Rh, 0.005–0.01; Ir, 0.03; Os, 0.1;Ru, 0.1 ppm, standard deviation for Pt, Pd, and Rh ≤15 %, for Ir, Ru, and Os = 30–35 %.

The chemical composition of sulfide and PGE minerals was analyzed on a Cameca SX 100 X-ray microprobe at the Vernadsky Institute (analyst N. Kononkova). The standards were minerals and synthetic oxides whose characteristics were close to those of the samples to be analyzed. The

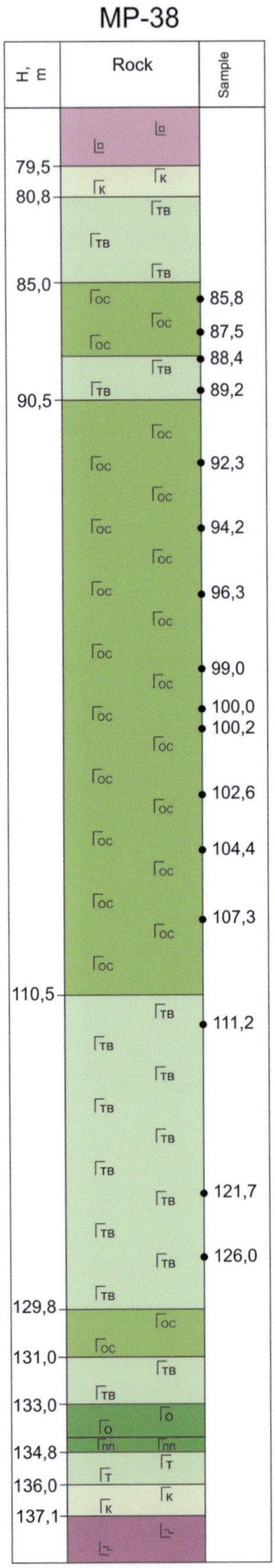

Fig. 4.49 Inner structure of the Bol'shaya Bar'ernaya intrusion, borehole MP-38
After NorilskGeology, Ltd. data, with changes by the author

detection limit of usual analyses (which were conducted at an accelerating voltage of 15 kV and a beam current of 40 nA) was 0.02 wt % for elements from Mg to Zn and 0.06 wt % for elements from F to Na. The spatial resolution of the technique was 2 μm. In order to more accurately analyze sulfides for As and Se, the counting time was increased to 200 s, the analyses were made at an accelerating voltage of 20 kV and a beam current of 200 nA, and the detection limit was 50 ppm. The standard for the calibration of the intensities and coordinates of the peaks and for determining background PGE concentrations were Astimex-certified standard samples of pure metals (Os, Ir, Rh, Pd, and Pt). Sulfur was calibrated against a standard of an Astimex natural pentlandite crystal $(Ni, Fe)_9S_8$. The standard for other elements were natural minerals and synthetic standard provided by Cameca. In calibrating peaks against standard, the following parameters were used: the accelerating voltage was 20 kV, the beam current was 20 nA, the counting time was 15 s, the background counting time was 5 s, and the beam was 1 μm in diameter.

The analyses for S isotopic composition were conducted at the Department for Isotopic Geochemistry of the Central Institute of Geological Exploration for Base and Precious Metals (TsNIGRI), following conventional methods (Ustinov and Grinenko 1965). Sulfur from sulfides was converted to SO_2 by its reaction with CuO at 760 °C in vacuum and the subsequent purification of the gas and analysis of the S isotopic composition on a MI-1201 mass spectrometer. The results were recalculated using the isotopic ratio of the CDT meteoritic standard. The laboratory reference standard samples were MSA-Pyrite (from TsNIGRI) with $\delta^{34}S = +0.7‰$ and Marine Sulfate (from GEOKHI) with $\delta^{34}S = +20.1‰$. Analyses of these standards at the beginning and immediately before the termination of analysis of our group of samples of the pyrrhotite–chalcopyrite ores corresponded to the certified tabulated value within 0.1‰ ($\delta^{34}S + 0.7$ and +20.1‰ respectively), which testified to the absence of *systematic analytical errors* during the analysis of the isotopic ratio. *Random analytical errors* (which involve the errors during sample preparation and measurement errors) were evaluated by multiple replicate analyses of certified standard samples as ±0.2 %. Hence, any differences >0.5‰ between the analyses of the two samples should be considered significant.

Sr–Nd–Pb isotope analyses were carried out on 100-mg rock powders that were dissolved in 5:1 mixture of concentrated HF and HNO_3. The element chromatography followed the methods outlined in Hoernle and Tilton (1991). Sr–Nd isotopic ratios were determined on a TRITON thermal ionization mass spectrometer (TIMS) and Pb isotope ratios on a MAT 2 thermal ionization mass spectrometer at IFM-GEOMAR. Both instruments operate in static multi-collection mode. Sr and Nd isotopic ratios are normalized within run to $^{86}Sr/^{88}Sr = 0.1194$

Table 4.32 Representative analyses of olivines from the Bol'shaya Bar'ernaya intrusion (borehole MP-38), wt %

N	Depth, m	Fo, mol %	SiO_2	FeO	MnO	MgO	CaO	NiO	Total
1.	121.7	74.70	38.22	23.21	0.35	38.44	0.11	0.22	100.66
2.	121.7	74.83	38.41	23.21	0.36	38.71	0.12	0.22	101.13
3.	121.7	75.79	38.59	22.36	0.34	39.26	0.14	0.22	101.02
4.	121.7	73.98	38.00	23.86	0.36	38.05	0.16	0.22	100.75
5.	121.7	76.04	38.35	22.15	0.34	39.43	0.17	0.23	100.76
6.	121.7	75.83	38.58	22.51	0.34	39.61	0.20	0.23	101.57
7.	121.7	75.08	38.36	23.02	0.35	38.90	0.20	0.23	101.16
8.	121.7	77.91	38.93	20.65	0.31	40.85	0.18	0.24	101.28
9.	121.7	76.50	38.71	21.90	0.33	39.98	0.19	0.24	101.46
10.	121.7	75.52	38.72	22.84	0.35	39.52	0.22	0.25	101.99
11.	121.7	75.12	38.64	23.10	0.35	39.11	0.20	0.24	101.73
12.	121.7	77.22	38.48	21.11	0.32	40.14	0.20	0.24	100.58
13.	121.7	76.83	38.17	21.40	0.33	39.81	0.18	0.24	100.23
14.	121.7	78.77	38.73	19.74	0.30	41.08	0.25	0.24	100.46
15.	121.7	77.79	38.66	20.62	0.31	40.51	0.24	0.24	100.71
16.	121.7	76.88	38.47	21.37	0.33	39.86	0.15	0.24	100.52
17.	121.7	76.35	38.33	21.83	0.33	39.52	0.15	0.24	100.51
18.	121.7	75.88	38.25	22.34	0.34	39.41	0.16	0.23	100.84
19.	121.7	75.11	38.15	22.84	0.35	38.66	0.13	0.24	100.45
20.	121.7	74.79	37.89	23.14	0.36	38.51	0.21	0.23	100.41
21.	121.7	74.01	38.13	23.85	0.36	38.10	0.16	0.22	100.92
22.	121.7	75.13	38.22	22.91	0.35	38.81	0.15	0.24	100.77
23.	121.7	75.19	38.15	22.82	0.35	38.79	0.16	0.24	100.58
24.	121.7	75.04	37.91	22.95	0.35	38.70	0.15	0.24	100.40
25.	121.7	74.91	37.91	23.01	0.35	38.54	0.17	0.25	100.32
26.	121.7	74.53	38.23	23.43	0.36	38.45	0.16	0.28	101.01
27.	121.7	75.20	38.35	22.84	0.35	38.84	0.16	0.26	100.89
28.	121.7	75.06	38.44	23.00	0.35	38.82	0.16	0.25	101.11

Note: Here and in Table 4.33 analyses were carried out in MPI, analysts D. Kuzmin and N. Krivolutskaya

Table 4.33 Representative analyses of pyroxenes from the Bol'shaya Bar'ernaya intrusion (borehole MP-38), wt %

№	Depth, m	Mg#	SiO_2	TiO_2	Al_2O_3	FeO	MnO	MgO	CaO	Na_2O	K_2O	Cr_2O_3	Total
1.	82	78.76	50.41	0.50	2.59	7.99	0.21	16.62	19.52	0.23	0.00	0.21	98.29
2.	82	78.90	51.46	0.52	2.71	7.85	0.21	16.47	19.72	0.22	0.00	0.27	99.43
3.	82.8	74.27	49.77	0.79	3.84	9.15	0.21	14.81	20.10	0.32	0.00	0.29	99.28
4.	82.8	71.83	49.65	0.90	3.64	10.05	0.23	14.38	19.91	0.31	0.00	0.12	99.20
5.	82.8	74.41	49.83	0.90	4.07	9.30	0.21	15.16	19.58	0.28	0.00	0.33	99.66
6.	85.8	78.04	52.59	0.53	1.79	8.28	0.21	16.50	19.72	0.24	0.00	0.21	100.08
7.	87.5	74.48	51.86	0.73	2.30	9.66	0.24	15.81	19.27	0.27	0.00	0.18	100.32
8.	88.4	72.78	51.55	0.83	2.80	10.22	0.24	15.33	19.00	0.25	0.00	0.12	100.37
9.	88.4	74.01	51.92	0.78	2.34	9.86	0.24	15.75	19.08	0.25	0.004	0.14	100.40
10.	88.4	74.44	51.57	0.63	2.54	9.38	0.25	15.32	19.71	0.33	0.001	0.32	100.07
11.	96.3	76.55	51.96	0.67	2.53	8.78	0.21	16.08	19.54	0.25	0.004	0.13	100.16
12.	96.3	76.21	52.14	0.68	2.46	8.96	0.21	16.10	19.44	0.27	0.005	0.10	100.40
13.	99	79.75	52.26	0.48	2.25	7.44	0.18	16.43	19.99	0.24	0.000	0.44	99.71
14.	99	79.96	51.16	0.57	3.06	7.24	0.18	16.20	20.08	0.26	0.000	0.67	99.44
15.	100	80.25	52.44	0.58	3.33	7.27	0.18	16.57	19.73	0.25	0.002	0.73	101.10
16.	100	74.72	52.35	0.65	2.75	9.83	0.26	16.30	18.35	0.24	0.002	0.09	100.85
17.	100	81.01	53.24	0.55	3.02	6.86	0.16	16.41	20.61	0.27	0.004	0.75	101.91
18.	100	78.85	52.50	0.61	2.65	7.81	0.18	16.33	20.07	0.26	0.002	0.35	100.79

(continued)

Table 4.33 (continued)

№	Depth, m	Mg#	SiO_2	TiO_2	Al_2O_3	FeO	MnO	MgO	CaO	Na_2O	K_2O	Cr_2O_3	Total
19.	100.2	77.66	52.57	0.58	2.10	8.39	0.20	16.36	19.85	0.24	0.002	0.07	100.37
20.	89.2	78.75	52.52	0.65	2.41	8.00	0.19	16.63	19.58	0.21	0.006	0.22	100.44
21.	100.2	81.17	52.41	0.44	2.18	6.96	0.16	16.83	20.31	0.25	0.005	0.43	100.00
22.	100.2	80.14	52.77	0.48	2.06	7.42	0.18	16.79	20.22	0.25	0.000	0.21	100.39
23.	100.2	78.20	51.87	0.62	2.58	8.12	0.19	16.34	19.91	0.24	0.000	0.13	100.02
24.	100.2	75.09	51.90	0.75	2.42	9.43	0.23	15.94	19.28	0.29	0.000	0.04	100.28
25.	100.2	81.81	52.03	0.42	2.14	6.73	0.18	16.98	20.11	0.23	0.000	0.54	99.38
26.	100.2	80.52	50.83	0.53	3.09	7.08	0.18	16.41	20.13	0.26	0.001	0.74	99.26
27.	100.2	76.85	51.85	0.62	2.17	8.69	0.22	16.18	19.86	0.25	0.002	0.04	99.89
28.	104.4	78.97	52.09	0.52	2.03	7.93	0.19	16.70	19.97	0.26	0.004	0.12	99.84
29.	104.4	78.19	52.72	0.56	2.01	8.23	0.21	16.55	19.93	0.27	0.001	0.09	100.58
30.	107.3	75.11	52.14	0.76	2.54	9.31	0.23	15.76	19.47	0.30	0.000	0.06	100.60
31.	107.3	78.47	52.56	0.55	2.02	8.05	0.20	16.46	19.97	0.26	0.001	0.10	100.18
32.	107.3	80.09	52.15	0.48	2.20	7.35	0.18	16.58	20.19	0.25	0.000	0.31	99.70
33.	111.2	81.60	50.75	0.57	3.15	6.56	0.15	16.32	20.55	0.27	0.003	0.88	99.22
34.	111.2	78.13	52.02	0.72	2.53	8.12	0.19	16.27	20.01	0.27	0.000	0.29	100.44
35.	121.7	80.01	51.89	0.70	1.86	7.65	0.19	17.17	19.54	0.24	0.000	0.09	99.35
36.	121.7	78.75	52.13	1.20	2.11	7.83	0.20	16.28	20.17	0.36	0.000	0.01	100.31
37.	121.7	79.41	51.93	0.85	2.56	7.70	0.19	16.66	19.71	0.23	0.000	0.09	99.95
38.	121.7	80.72	51.57	0.67	2.69	7.03	0.17	16.51	20.24	0.23	0.003	0.25	99.38
39.	121.7	81.75	52.50	0.54	2.08	6.76	0.17	16.99	20.34	0.22	0.004	0.22	99.85
40.	121.7	81.31	52.78	0.54	1.95	6.99	0.18	17.06	20.20	0.24	0.000	0.16	100.12
41.	121.7	80.90	52.50	0.55	1.90	7.18	0.18	17.06	20.06	0.25	0.005	0.12	99.83
42.	126	81.89	51.72	0.58	3.20	6.38	0.15	16.18	20.65	0.24	0.001	0.83	99.95
43.	126	82.23	52.39	0.59	3.28	6.34	0.20	16.45	20.55	0.26	0.001	0.83	100.90
44.	128.4	77.84	49.81	1.48	3.32	8.16	0.22	16.08	19.39	0.31	0.00	0.07	98.84
45.	131.8	69.88	52.07	0.26	1.51	10.91	0.33	14.19	20.38	0.22	0.01	0.04	99.91
46.	135	73.33	51.29	0.64	2.79	10.00	0.25	15.43	19.40	0.26	0.00	0.04	100.09
47.	135	82.30	51.05	0.43	3.31	6.30	0.16	16.43	20.78	0.23	0.00	0.57	99.27
48.	135.3	69.47	50.51	0.53	2.50	11.89	0.34	15.17	18.15	0.34	0.00	0.01	99.43
49.	135.3	75.42	52.08	0.56	2.47	9.62	0.25	16.55	18.76	0.22	0.00	0.08	100.59
50.	135.5	81.47	51.17	0.51	3.13	6.66	0.17	16.43	20.36	0.22	0.00	0.52	99.18
51.	135.5	82.59	51.42	0.42	2.81	6.20	0.15	16.49	20.69	0.24	0.00	0.68	99.10

and $^{146}Nd/^{144}Nd=0.7219$, respectively, and all errors are reported as 2σ. Over the course of the study normalized NBS-987 gave $^{87}Sr/^{86}Sr=0.710250\pm0.000008$ (N=18) and $^{143}Nd/^{144}Nd=0.511847\pm0.00006$ (N=9) was obtained for La Jolla and $^{143}Nd/^{144}Nd=0.511712\pm0.00006$ (N=8) for our in-house monitor Spex. The long-term average values for NSB-981 (N=47) are $^{206}Pb/^{204}Pb=16.900\pm0.007$, $^{207}Pb/^{204}Pb=15.437\pm0.009$, and $^{208}Pb/^{204}Pb=36.528\pm0.027$ and correlated to the NSB 981 values given in Todt et al. (1996). Total chemistry blanks were <100 pg for Sr, Nd, and Pb and thus considered negligible.

4.5.6.1 Geological Background

Numerous intrusions of ultrabasic–basic composition are widespread in the Noril'sk region (Fig. 4.2), while differentiated massifs stand apart from non-differentiated and were united into the Noril'sk Intrusive Complex. The most significant of the currently producing mineral deposits with massive ores (Talnakh and Oktyabr'skoe) are located inside Kharaelakh Trough in Devonian sediments; so genetic models for all Noril'sk deposits have been developed based on the location of these deposits under basalts in Devonian sediments (Likhachev 1965; Dyuzhikov et al. 1988; Rad'ko 1991; Naldrett 1992; Li et al. 2009). At the same time, the Noril'sk Trough hosts numerous smaller sulfide deposits (Noril'sk 1, Mountain Chernaya, Noril'sk 2, and Zub-Marksheydersky; after Korovyakov et al. 1963) (Fig. 4.26) which all are situated at the higher level of the stratigraphic sequence: they penetrate Tunguska sandstones and coals and, partially, Ivakinsky, Syverminsky, and Gudchikhinsky lavas. These intrusions contain disseminated sulfide mineralization, as well as numerous veins of massive sulfides a few meters thick in the lower parts of the intrusions. The veins have sharp boundaries with the host rocks, and they consist of chalcopyrite, pyrrhotite, and pentlandite in general, but there are many pure chalcopyrite veins. Their thickness changes from 0.1 up to 2–3 m, and the length is a few dozen meters.

The Maslovsky deposit is located to the south of the Noril'sk 1 deposit (Fig. 4.26). The morphology of the intrusion has been developed from an assessment of geophysical and drilling data. The intrusive rocks extend for a distance of 6 km and have a width from 1 to 3.5 km and an average thickness of 150 m. The isopach map in Fig. 4.50 shows that this body has two differently oriented funnel-shaped branches in which the thickness of the intrusive rocks reaches 270 and 400 m. This provides evidence to suggest that one large intrusion representing Maslovsky deposit comprises two massifs which are products of two different intrusive events. In order to test this hypothesis, detailed mineralogical and geochemical studies were conducted on materials recovered by boreholes OM-4 and OM-24, and the results are compared with data acquired from other boreholes.

4.5.6.2 Geology, Stratigraphy, and Geochemistry of the Maslovsky Deposit

The intrusive bodies in the northern and southern parts of the deposit have different stratigraphic positions in the host rocks which are dominated by volcanics of the trap association (Fig. 4.51). We have subdivided the tuff–lava sequence in this portion of the Noril'sk Trough based on materials recovered by the borehole OM-6, namely, in terms of the distinctive fabric and compositional features of the rocks (Krivolutskaya and Rudakova 2009). These rocks belong to the Ivakinsky, Syverminsky, Gudchikhinsky, Khakanchansky, Tuklonsky, Nadezhdinsky, and Morongovsky Formations. It was determined that the Northern Maslovsky intrusion is hosted in terrigenous–carbonaceous rocks of the Tunguska Group (C_2–P_1) and in the trachybasalts of the Ivakinsky Formation and tholeiitic basalts of the Syverminsky Formation, at a depth of 832–1,100 m. Likewise, it was demonstrated that the Southern Maslovsky intrusion cuts across the basalts of the Syverminsky–Nadezhdinsky Formations (borehole OM-24). The position of peripheral sill of the Southern Maslovsky intrusion inside the Low Nadezhdinsky basalts is shown in Fig. 4.52 (after Krivolutskaya and Rudakova 2009). It has been shown that the volcanic rocks of the Nadezhdinsky Formation have a very high La/Sm ratio and are depleted in Cu (e.g., Lightfoot et al. 1990, 1993, 1994; Naldrett 1992). This feature is also evident in the volcanic rocks from this study. Elevated La/Sm ratios were also found in the upper contact gabbro-dolerites of the Southern Maslovsky intrusion which reflects the contamination of Low Nadezhdinsky basalts by Maslovsky magma just in thin marginal zone.

A section through the Northern intrusive body is shown in Fig. 4.53. This body comprises from base to top different varieties of gabbro-dolerites: olivine, picritic-like, picritic, olivine-bearing, olivine-free, and other rocks, as troctolites and gabbro-diorites. This section is characterized by the thick picritic horizon and the absence of taxitic gabbro-dolerites. As a result, high-Mg rocks dominate in this part of the intrusion and account for two-thirds of the vertical section (180 m), whereas olivine-free rocks have a smaller thickness (55 m).

The Southern Maslovsky intrusion has a different inner structure (Fig. 4.54). The upper part of the massif consists of a thick gabbro-diorites and ferrogabbro horizons (gabbro with a titanomagnetite concentration of >5 wt %), which accounts for one-third of the vertical section. The thickness of the high-Mg rocks (picritic gabbro-dolerites) is much less (15 m), but this vertical section contains 34 m of taxitic gabbro-dolerites. The internal structure of the Southern Maslovsky intrusion is very similar to the one of the Talnakh intrusion in borehole OUG-2 (Fig. 4.11).

The heterogeneity in the internal structure of the Southern Maslovsky intrusion is accompanied by a wide variation in the geochemical characteristics of the rocks in the vertical section (Appendix Table A1; Krivolutskaya et al. 2012, Online Resource 1 footnote that this is available online). The Northern intrusion is characterized by a gradual upward increase in the Mg# of the rocks, whereas an increase in the Southern intrusion is stepwise. Most of the rocks have low Ti (TiO_2 <1 wt %), except for the ferrogabbro in the upper zone of the Southern intrusion which contains up to 2,79 wt % TiO_2. All these rocks plot on a normal Fe differentiation trend of gabbroic rocks (reaching up to 20 wt % FeO) with no more than 4 wt % alkalis and with Na dominated over K. The magnitude of the variation in the abundances of the major oxides is the same in rocks of the Southern and Northern intrusions, although the weighted mean compositions of the bodies calculated for the examined vertical sections significantly differ.

The REE patterns of rocks from both intrusions of the deposit are similar (Fig. 4.55) and correspond to those of other massifs of the Noril'sk Complex, including the Noril'sk 1 intrusion, as it has been previously demonstrated (Krivolutskaya and Rudakova 2009). They typically exhibit negative Ta–Nb and positive Th–U and Pb anomalies (2 RSD −2–3 % for these elements—it is enough to resolve these differences), and their patterns are only slightly tilted toward HREE, testifying to the fact that the magma source contained no garnet. Compositionally similar rocks (picritic gabbro-dolerites and troctolites) have the comparable ranges in the trace element concentrations. The presence of

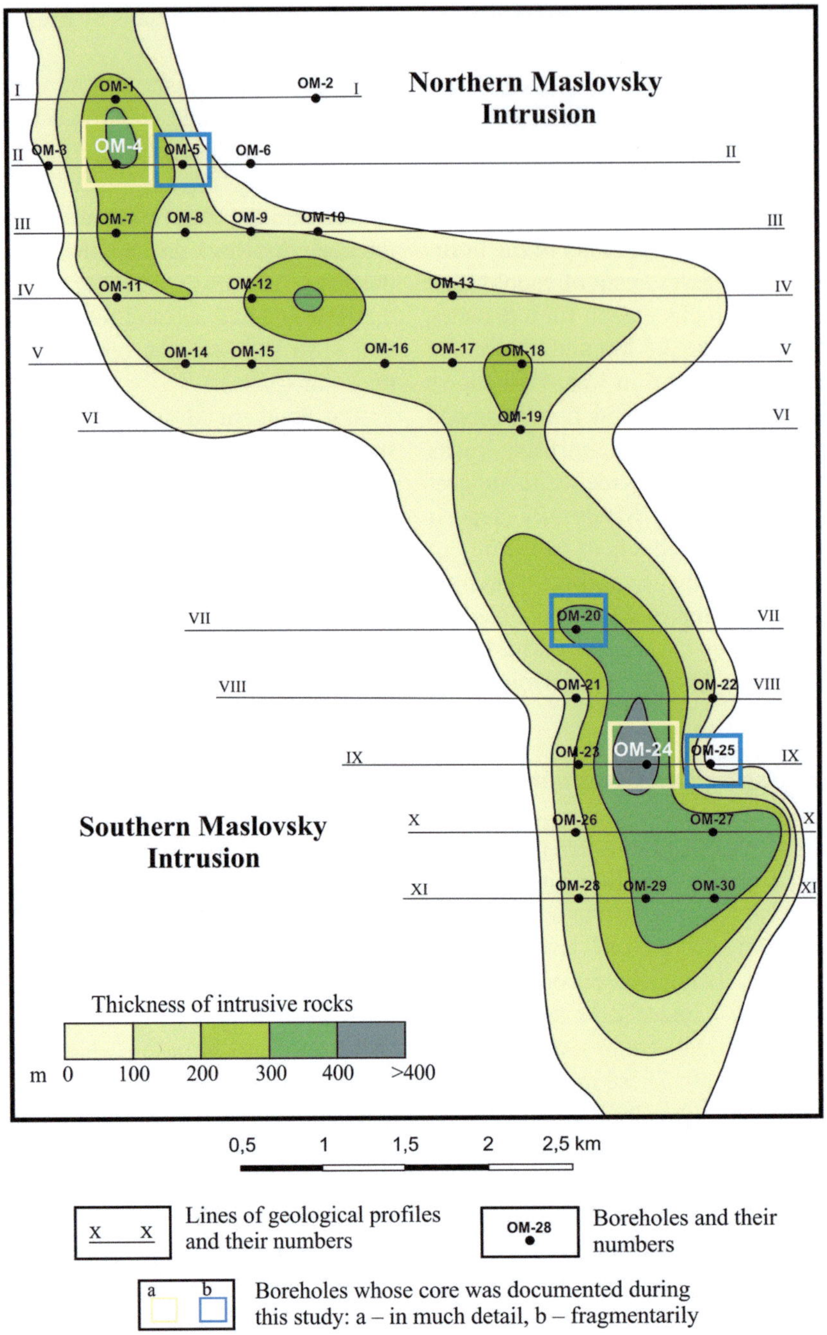

Fig. 4.50 Isopach map of the Maslovsky intrusion
Here and in Fig. 4.51—after materials NorilskGeology, Ltd. modified by the authors

gabbro-diorites in the cross section of the Northern Maslovsky intrusion is reflected in patterns with elevated trace element concentrations. A closer examination of the data reveals that there are differences in ratios of elements between intrusions. For example, the Ta–Nb ratio (Fig. 4.56) of silicate rocks from the intrusions that host the Maslovsky deposit plot in different fields, with the Southern intrusion having the higher values of Ta–Nb. The entire rock composition of both the Maslovsky intrusions is identical and is similar to those of the Talnakh Massif (Fig. 4.56).

4.5.6.3 Mineral Chemistry at the Maslovsky Deposit

Pyroxene is present in most samples from the profiles along with plagioclase, but it provides more valuable information due to its range in elemental composition.

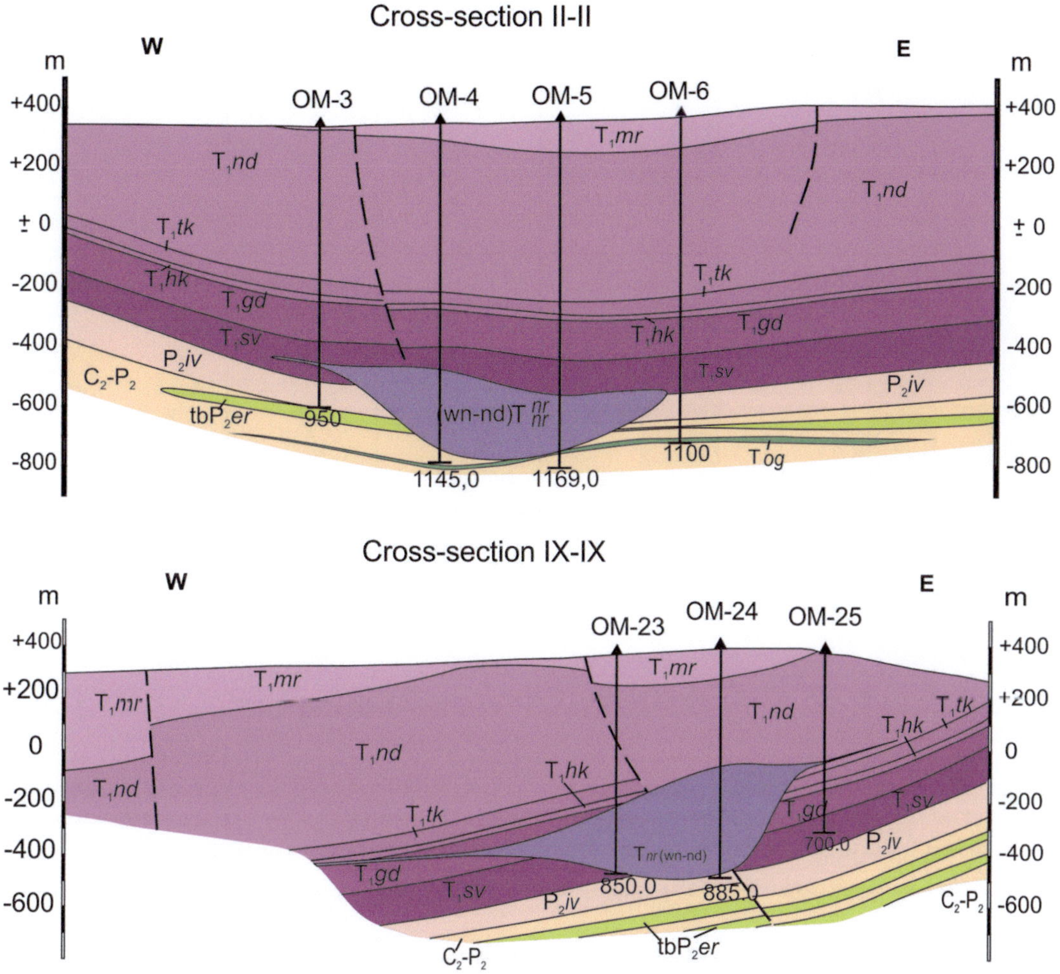

Fig. 4.51 Geological profiles through the northern (line II-II) and southern (line IX-IX) parts of the Maslovsky deposit See Fig. 4.26 for symbol explanations. *Dashed lines* show faults. Legend see Fig. 4.26

Pyroxene is usually one of the latest minerals in the rocks of Noril'sk Intrusive Complex, with rocks in the central portions of these massifs having a poikilitic texture formed by plagioclase laths in large round pyroxene grains. In olivine-bearing varieties, particularly in picritic gabbro-dolerites, it is intercumulus subhedral olivine grains and plagioclase laths. A very important characteristic of pyroxene is its Mg number (Mg# = Mg/Mg + Fe in atom), which varies depending on the Mg# of its host rocks (Table 4.33; Appendix Table A3 and in Krivolutskaya et al. 2012, Online Resource 2). The higher values of Mg# in pyroxene were detected in the picritic gabbro-dolerites (up to 85), and the lower ones (to 62) were found in gabbro-diorites in the upper parts of the intrusions. The most indicative minor elements in pyroxene are Ti, Cr, Na, and Al (Krivolutskaya et al. 2012, Online Resource 3). The TiO_2 concentrations have strong negative correlation with the Mg# in pyroxene). In the TiO_2–Mg# diagram (Figs. 4.53 and 4.54), the compositions of pyroxene define separate trends for compositionally distinct horizons of the intrusions (picritic, taxitic, olivine-bearing, and others gabbro-dolerites). The steepest trends, i.e., those characterized by the strong Ti enrichment in the clinopyroxene during crystallization of the melt, are typical of the high-Mg rocks: picritic and taxitic gabbro-dolerites. Other rock types yield more gently sloping trends relative to the Mg# axis that correspond to a weak increase in the concentration of this element with a decreasing Mg# of the host mineral. The relationship for Al is reversed: the Al concentrations directly correlate with the Mg# of pyroxene. The mineralization related to the intrusion is likely linked to the differentiation of minor elements. It is indicated, for example, by Ti in pyroxene

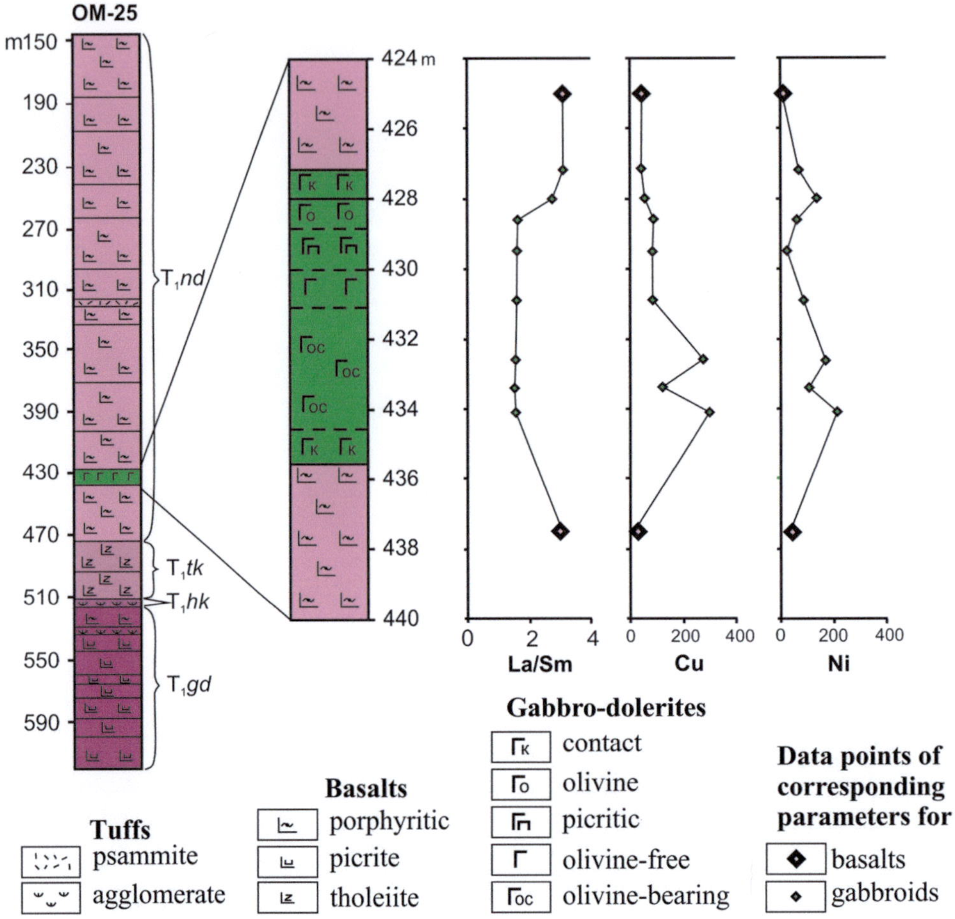

Fig. 4.52 Setting of the eastern peripheral sill of the Southern Maslovsky intrusion among basalts of the lower subformation of the Nadezhdinsky Formation (borehole OM-25)
After Krivolutskaya and Rudakova (2009)

which shows the more obvious trends and the more significant enrichment within the richer mineralized zone. The similar dependence was found for the Talnakh intrusion (Krivolutskaya 2014).

Olivine is one of the predominant rock-forming minerals of the Maslovsky intrusions, and its role is particularly important in mineralized horizons, which contain up to 60–70 vol.% of this mineral. The morphology of olivine grains varies in the vertical section through the intrusion. The picritic gabbro-dolerites and troctolites are dominated by subhedral crystals up to 5–7 mm across and also contain medium-sized and small (0.5 and 0.2 mm, respectively) olivine crystals included in plagioclase and pyroxene, whereas the olivine-bearing varieties are dominated by anhedral olivine grains up to 1 mm in grain size. The composition of the mineral systematically varies from high-Mg (Fo_{82}) at the bottom of the vertical section to Fo_{52} in its upper parts (Table 4.34, Appendix Table A2; Krivolutskaya et al. 2012, Online Resource 4). The most indicative minor element in olivine is Ni, which concentration is higher in the units containing disseminated sulfides. The behavior of Ni in olivine is different for picritic gabbro-dolerite in the Northern and Southern intrusions. An unusual inverse correlation between Ni and Fo similar to that of the Noril'sk 1 deposit was found in olivine from the Northern intrusion (Fig. 4.57), whereas Ni increases with increasing Fo in olivine from the Southern intrusion. The correlation between Ni and Fo indicates olivine fractionation with interaction of the magma, whereas the inverse correlation of Ni and Fo in olivine may argue for an exchange equilibrium with a sulfide melt. Chemistry of olivine from the Northern and Southern Maslovsky intrusions is also different in terms of content of other minor elements, including Cr and Ca, and particularly in terms of abundances of heavy REE and Y, as it is shown in Fig. 4.57 for high-Mg olivine from picritic gabbro-dolerites.

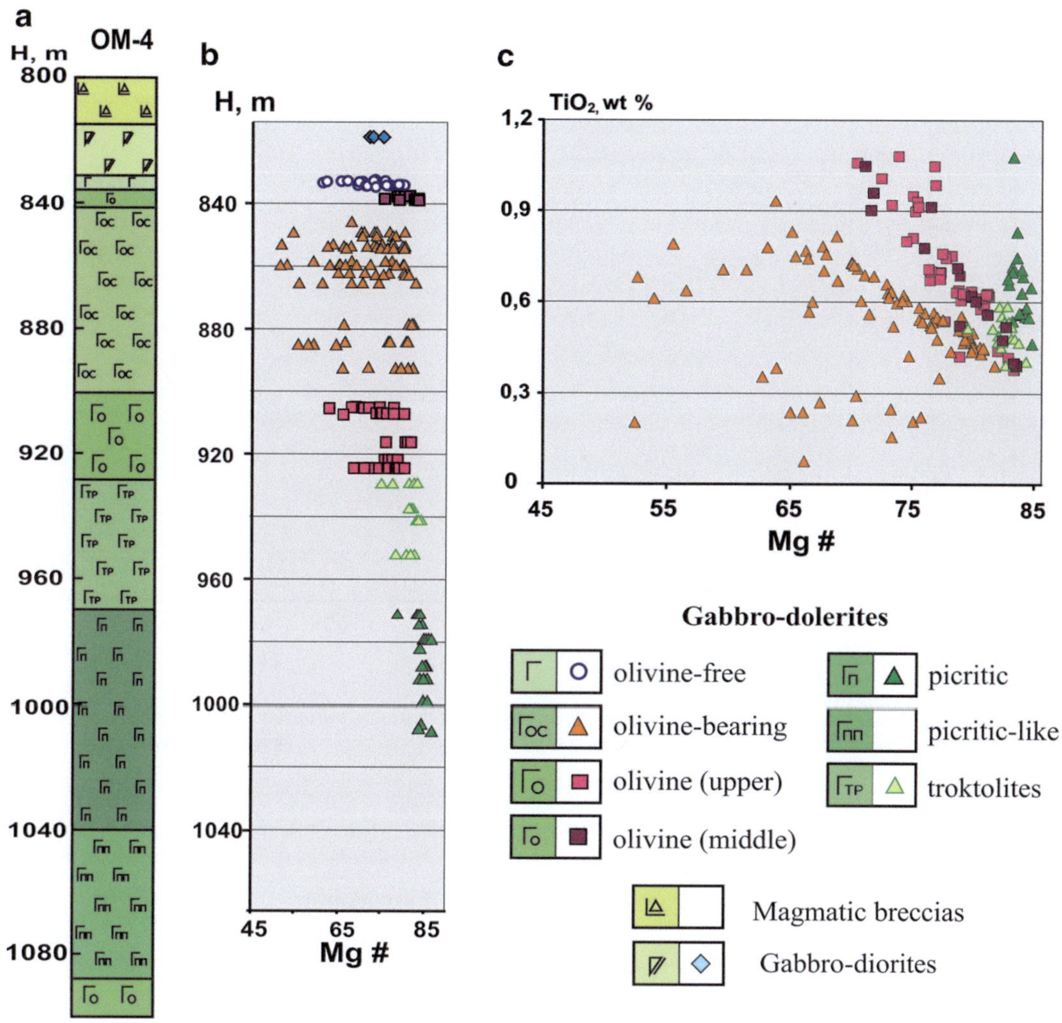

Fig. 4.53 (**a**) Vertical section of the Northern Maslovsky intrusion (based on borehole OM-4); (**b**) variations in the pyroxene composition in this vertical section and (**c**) variations in the TiO_2 concentration in clinopyroxene from various units
After Krivolutskaya et al. (2012)

Very high concentrations of Y, Yb, and Dy were found in olivines. There is a strong negative correlation between Fo and HREE contents in olivines. It assumes that the HREE enter olivine as a structural isomorphic admixture that was also verified by a high-resolution TEM study which confirms an absence of individual mineral phases of REE in olivine. Each magmatic body (the Southern and Northern Maslovsky intrusions and Noril'sk 1 intrusion) has its own trend line on a diagram Fo–Y (Fig. 4.57). Two last of the abovementioned intrusions are characterized by very steep slope on the diagram that is typical of olivine from picritic gabbro-dolerites with high-tenor mineralization, whereas the first one shows shallower slope that seems to be a characteristic of olivine from the Southern Maslovsky intrusion with poor mineralization.

Our data on the geochemistry of the rocks and the composition of their rock-forming minerals indicate that the Northern and Southern intrusions were formed by different magmas, whose parental melts crystallized differently that is confirmed by the different olivine compositions from the two bodies. The Northern intrusive body is a continuation of the Noril'sk 1 Massif, whereas the Southern body is likely a separate one.

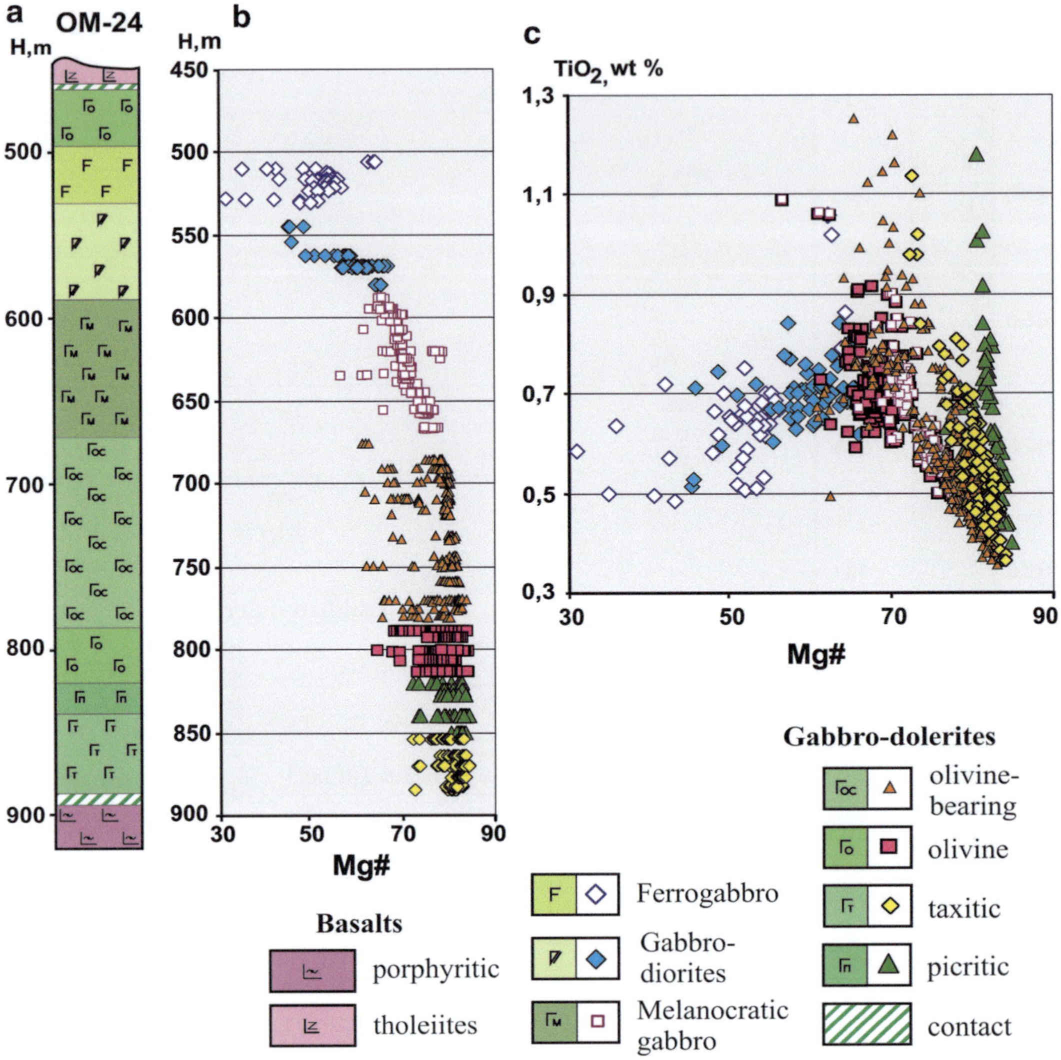

Fig. 4.54 (**a**) Vertical section of the Southern Maslovsky intrusion (based on borehole OM-24); (**b**) variations in the pyroxene composition in this vertical section and (**c**) variations in the TiO_2 concentration in clinopyroxene from various units
After Krivolutskaya et al. (2012)

4.5.6.4 Mineralization at the Maslovsky Deposit

In the disseminated ores of the Maslovsky deposit, sulfide minerals occur as thin veinlets, pockets, "droplets," and fine interstitial dissemination between rock-forming minerals. The former two morphological types of aggregates of ore minerals are more typical of the northern part of the Maslovsky deposit. The content of ore mineral usually does not exceed 12–15 % of the rock volume. Sulfides are distributed unevenly through the orebody, whose boundaries are drawn according to sampling and assaying results, although these boundaries generally coincide with the units of picritic and taxitic gabbro-dolerites. The morphology of aggregate of ore minerals is correlated with the petrography of the rocks (Krivolutskaya et al. 2011): the taxitic gabbro-dolerites contain pockets and irregularly shaped grains of sulfides, with the largest of them spatially related to pegmatoid domains (Fig. 4.58a, c). The upper part of the "droplets" often contains recrystallized glass or crystals of biotite, amphibole, and other minerals with volatile components (Fig. 4.58b), while the picrites are dominated by small droplet-shaped (Fig. 4.58d, g, h) and interstitial ore material (Fig. 4.58i), which is evenly distributed over the rock volume. Pyrrhotite veinlets (Fig. 4.58f) are very often in the northern part of the deposit.

The massive vein ores are dominated by chalcopyrite and contain minor amounts of pyrrhotite and pentlandite. These ores typically have reticulate (Fig. 4.59a), rimmed (Fig. 4.59b), porphyritic (Fig. 4.59e), and other textures. The former are caused by the occurrence of chalcopyrite rims around pyrrhotite crystals and the latter are large pentlandite grains in chalcopyrite groundmass.

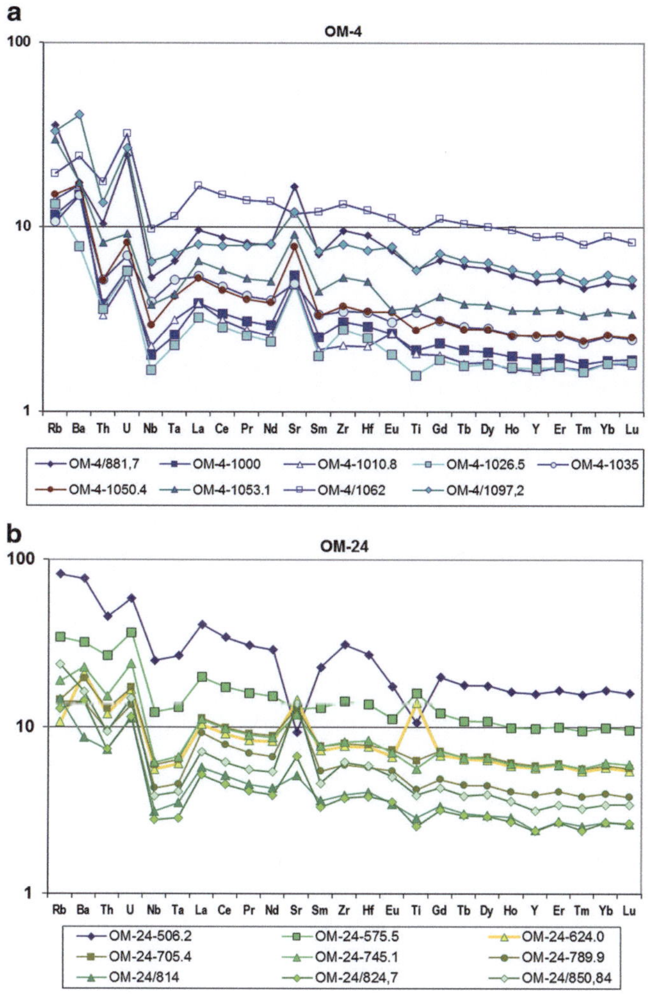

Fig. 4.55 Primitive mantle-normalized (Hofmann 1988) trace element patterns of intrusive rocks from the Northern (**a**) and Southern (**b**) parts of the Maslovsky deposit
After Krivolutskaya et al. (2012)

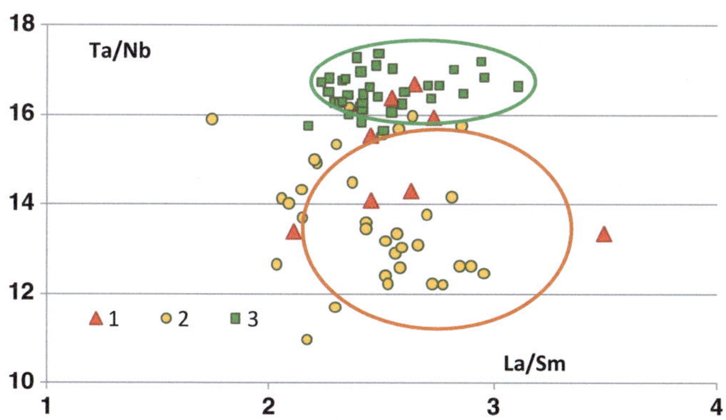

Fig. 4.56 Diagram Ta–Nb–La/Sm for the rocks of the Northern (OM-4, 3) and Southern (OM-24, 2) Maslovsky intrusions and Noril'sk 1 (1)
After Krivolutskaya et al. (2012)

The mineral potential of different parts of the deposit varies, with the richest ores restricted to the Northern Maslovsky intrusion. Compared to the Southern Maslovsky intrusion, the picritic gabbro-dolerites contain more abundant sulfide veins commonly 1–3 cm thick and about 1–3 m long. The rocks are also distinguished by the presence of large drops of magmatic sulfides in the picritic gabbro-dolerites (up to 3–4 cm in size). At the contacts of the intrusions with the rocks of the Tunguska Group, many boreholes penetrated massive sulfide ores up to a few m thick. We report analytical data on disseminated sulfides from the Northern Maslovsky intrusion in Table 4.35.

These ores comprise of chalcopyrite, pyrrhotite, and pentlandite and contain subordinate and minor amounts of cubanite, bornite, and millerite and a diverse assemblage of rare precious metal minerals. The mineral composition of ores in the northern part of the Maslovsky deposit differs from that in the southern part: in the former instance, the ores contain cubanite, while ores in the southern part more commonly bear bornite and millerite. Another distinctive feature of the northern part is the occurrence of large (up to 3–5 mm) accumulations of Pt and Pd minerals, in which numerous mineral species were identified, whereas the ores in the southern part of the deposit contain predominantly Au–Ag alloys, with grains and aggregates of precious metals only rarely being larger than 20 μm (because of their very small size, many phases were not identified).

Among the sulfides examined in the course of our study, the most interesting mineral is *pentlandite*, because its composition notably varies not only within discrete orebodies but also between various ore types of the deposit. A noteworthy compositional feature of this mineral is that its isomorphic series includes two end-members: Ni-poor pentlandite with elevated Fe and low-Ni concentrations (41 wt % FeO and 26 wt % NiO) and Ni-rich pentlandite with Ni strongly dominated over Fe (24 wt % FeO and 42 wt % NiO). Another interesting aspect of pentlandite composition is the variations in its Co concentration. The samples of disseminated ores from the Maslovsky deposit contain pentlandite with low Co contents, usually less than 1 wt % (we managed to find only one grain with 1.84 wt % Co: see Table 4.35 analysis 5). Our data show that the composition of pentlandite in disseminated ores from the Maslovsky deposit is similar to the composition of Noril'sk 1pentlandite with Co concentrations of 0.6–1.4 wt % (Genkin et al. 1981). It is also worth mentioning that the Ni-rich pentlandite persistently contains Cu, in which concentrations are occasionally higher than 3 wt %.

The *pyrrhotite* varies little in its composition, with Fe concentrations of 60–62 wt % and S concentrations of 37–38 wt %. The most typical minor element in the pyrrhotite is Ni, which was pervasively found in the examined grains of this mineral from disseminated ores of the

Table 4.34 Representative analyses of olivines from the Maslovsky deposit

№	1	2	3	4	5	6	7	8	9	10	11	12	13	14	15
№ sample	1010.8	1015.0	1026.5	1026.5	1030.6	1035.1	1039.4	1050.4	1023.5	807.0	811.0	811.0	811.0	814.5	814.5
Fo. mol.%	79.83	79.42	79.89	79.57	79.07	78.04	79.05	76.79	79.73	77.85	78.38	77.90	77.81	80.54	79.33
SiO_2	39.15	39.12	39.12	39.04	39.18	38.92	39.32	38.94	38.46	38.70	38.44	38.48	38.43	38.92	38.92
TiO_2	0.03	0.03	0.04	0.04	0.03	0.04	0.02	0.02	0.04	0.01	0.02	0.02	0.02	0.01	0.02
Al_2O_3	0.02	0.02	0.01	0.02	0.01	0.01	0.02	0.03	0.01	0.03	0.03	0.03	0.02	0.03	0.02
FeO	18.94	19.32	18.90	19.17	19.70	20.60	19.79	21.56	18.88	20.60	20.02	20.52	20.53	18.17	19.30
MnO	0.29	0.29	0.28	0.29	0.30	0.30	0.30	0.32	0.30	0.31	0.32	0.32	0.33	0.29	0.31
MgO	42.04	41.83	42.13	41.87	41.74	41.07	41.89	40.00	41.66	40.61	40.70	40.56	40.38	42.18	41.54
CaO	0.15	0.12	0.14	0.13	0.15	0.11	0.15	0.19	0.12	0.30	0.23	0.25	0.22	0.26	0.18
NiO	0.24	0.28	0.24	0.25	0.23	0.27	0.24	0.27	0.22	0.19	0.19	0.19	0.18	0.22	0.21
CoO	0.03	0.02	0.03	0.03	0.03	0.02	0.03	0.03	0.03	0.03	0.03	0.03	0.03	0.03	0.02
Cr_2O_3	0.01	0.01	0.02	0.02	0.01	0.01	0.01	0.00	0.02	0.03	0.03	0.02	0.02	0.03	0.02
Total	100.88	101.04	100.91	100.84	101.37	101.35	101.77	101.36	99.78	100.81	100.00	100.41	100.15	100.14	100.54
Y	1.33	1.87	0.65	2.11	1.53	1.72	0.80	1.31	2.12	0.22	0.65	0.52	0.87	0.27	0.95
Dy	0.12	0.17	0.05	0.17	0.14	0.17	0.08	0.13	0.21	0.01	0.07	0.05	0.07	0.03	0.10
Ho	0.05	0.06	0.02	0.06	0.05	0.05	0.03	0.05	0.07	0.01	0.03	0.02	0.03	0.01	0.03
Er	0.22	0.31	0.11	0.30	0.24	0.27	0.12	0.23	0.34	0.04	0.11	0.08	0.14	0.04	0.16
Tm	0.05	0.07	0.02	0.07	0.05	0.07	0.03	0.05	0.08	0.01	0.02	0.02	0.03	0.01	0.04
Yb	0.49	0.66	0.22	0.62	0.51	0.63	0.27	0.43	0.75	0.07	0.20	0.16	0.26	0.12	0.31
Lu	0.09	0.12	0.05	0.10	0.09	0.11	0.05	0.09	0.13	0.01	0.05	0.03	0.05	0.02	0.06
Ti	159.63	219.81	213.15	225.16	196.13	208.86	136.05	130.00	205.01	51.20	67.69	64.34	75.41	73.53	170.04
Cu	1.04	0.82	0.78	0.89	0.76	0.49	1.02	1.18	0.87	1.37	0.76	0.91	0.83	1.08	0.72
Zn	123.13	134.44	152.84	157.78	132.38	141.64	138.53	157.72	128.45	155.55	130.04	136.05	137.61	123.18	132.15
Sc	9.39	9.68	7.52	10.03	8.52	8.74	8.82	10.77	10.74	9.24	9.65	10.51	10.76	8.73	9.93
Al	79.98	79.63	140.50	82.30	101.52	78.74	118.32	144.23	85.03	174.77	146.96	151.07	148.46	168.03	124.46
Zr	0.13	0.19	0.25	0.18	0.12	0.18	0.10	0.07	0.26	0.03	0.06	0.06	0.07	0.05	0.15
Co	185.11	177.10	192.10	190.07	183.35	185.00	208.05	205.68	188.40	197.74	181.51	182.93	184.38	182.82	185.53
Ge	0.87	0.85	0.80	0.80	0.82	0.84	0.83	0.80	0.89	0.80	0.92	0.67	0.80	0.74	0.88
V	10.79	15.54	13.43	17.78	16.54	26.15	20.05	41.48	7.13	14.03	9.32	11.48	10.88	11.54	10.16
Li	4.19	4.46	2.50	4.45	6.28	5.34	4.88	5.28	6.51	4.26	5.97	5.15	6.00	3.59	7.15

Note: Olivines (1–9) Northern and (10–15) Southern Maslovsky deposits (N sample, depth (m) in boreholes OM-4 and OM-24 accordingly). Analyses were carried out in MPI, analysts D. Kuzmin and N. Krivolutskaya

Maslovsky deposit. The highest concentrations of Ni were detected in troilite (up to 2.65 wt %, Table 4.35). Fe and Ni in the pyrrhotite negatively correlate. The pyrrhotite is poor in Co and contains no more than 0.14 wt % of this element. Cu is an atypical minor element in pyrrhotite as a whole and was found in this mineral only near a contact with chalcopyrite (see, e.g., analyses 4 and 13 in Table 4.35).

The chalcopyrite has a narrow compositional range with practically constant proportions of the major elements: S, Cu, and Fe. The mineral is completely "pure" compared to other ore minerals and contains no concentrations of minor

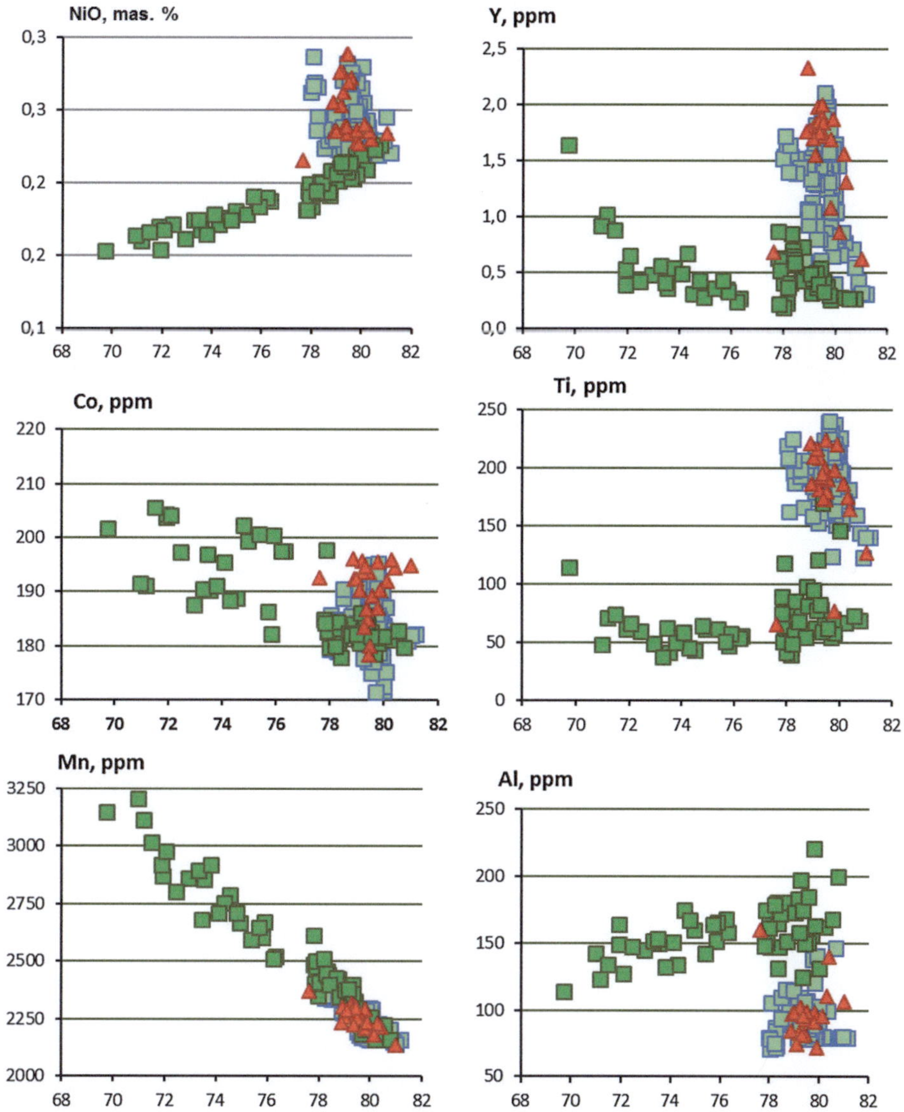

Fig. 4.57 11 *Fo*—NiO, Y, TiO_2, Ti, Al, Zr, Mn, and Cu diagrams for olivine from picritic gabbro-dolerites in the Northern and Southern Maslovsky and Noril'sk 1 intrusions
After Krivolutskaya et al. (2012)

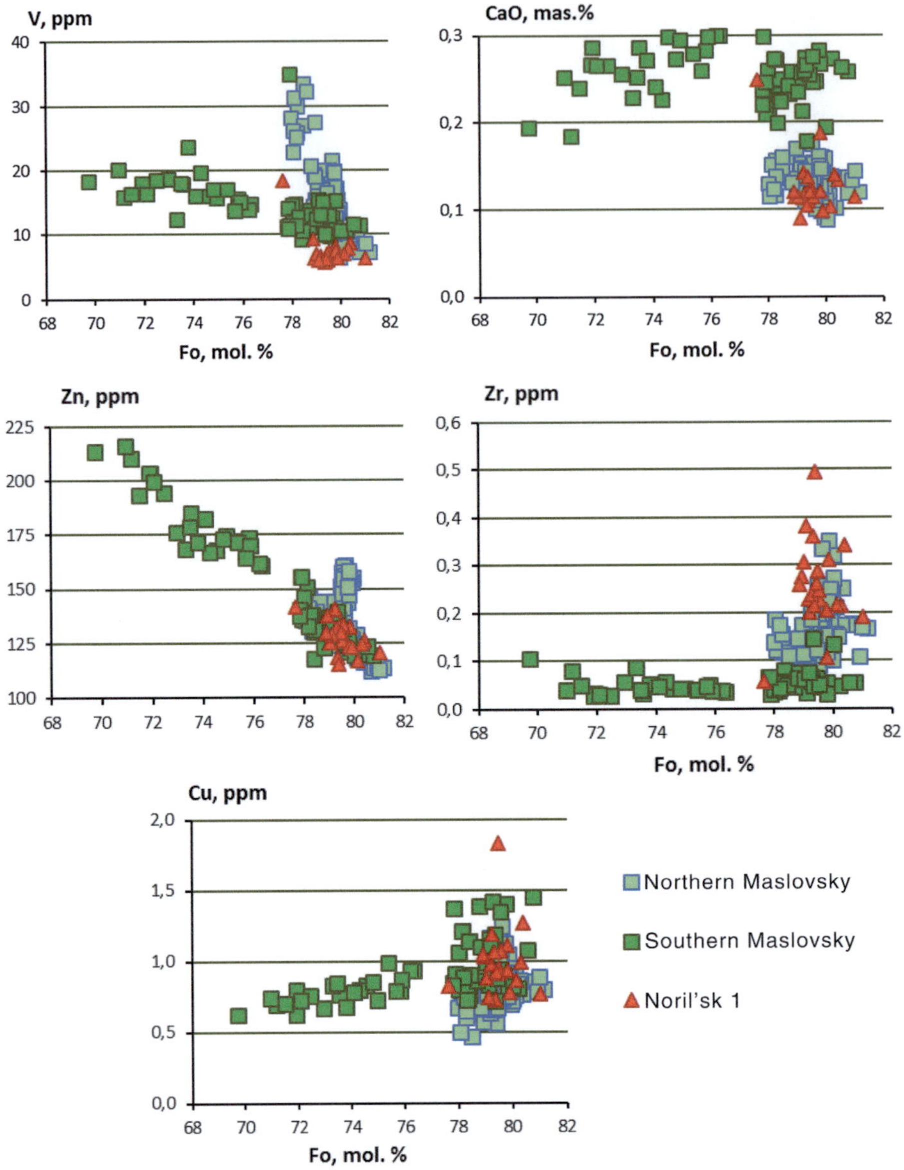

Fig. 4.57 (continued)

elements in higher than 0.07 wt %. However, this mineral ubiquitously contains Sb and As (a few hundred ppm) and traces of Ni and Co in concentrations close to the detection limits. Elevated concentrations of minor elements were found in the margins of chalcopyrite grains, at their contacts with pentlandite and other Co- and Ni-bearing minerals.

Millerite usually replaces chalcopyrite in disseminated ores (Fig. 4.59g) hosted in the taxite unit. The composition of the mineral was analyzed in three grains (Table 4.2) and changes very insignificantly. In addition to major components (Ni and S), it contains minor concentrations of Co (a little bit more than 1 wt %), Cu and Fe (up to 0.5 and 0.7 wt %, respectively), As, and Se.

Bornite is a fairly ubiquitous mineral of the disseminated ores of the Noril'sk deposits and is usually hosted in taxitic gabbro-dolerites and was found in picritic gabbro-dolerites

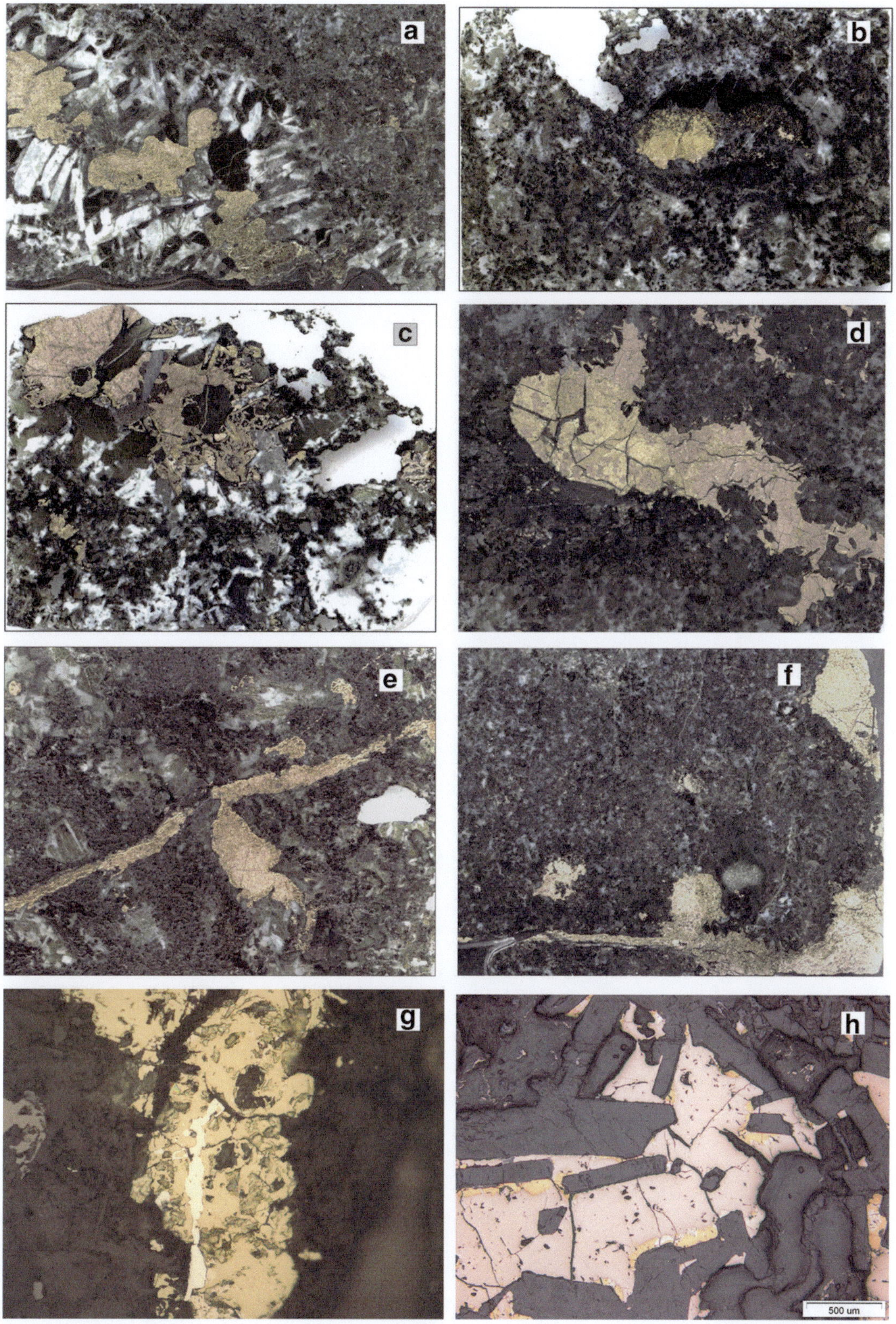

Fig. 4.58 Structures of ores at the Maslovsky deposit
(**a**) chalcopyrite–pyrrhotite aggregates in a pegmatoid segregation in taxitic gabbro-dolerite (sample OM-24/843.4), (**b**) droplet-shaped chalcopyrite segregation with a biotite rim in taxitic gabbro-dolerite (sample OM-24/838.7), (**c**) large pyrrhotite grains in taxitic gabbro-dolerite (sample OM-24/852.7), (**d**) chalcopyrite–pyrrhotite pocket in olivine gabbro-dolerite (sample OM-24/853.9), (**e**) pyrrhotite veinlet in olivine gabbro-dolerite (sample OM-24/843.6), (**f**) disseminated pyrrhotite in picrite gabbro-dolerites (sample OM-4/1006.6), (**g**) zvyagintsevite veinlet in chalcopyrite (sample OM-4/1052), (**h**) sideronitic texture of olivine gabbro-dolerite (sample OM-24/800,1 reflected light). a–h: polished samples, natural size (After Krivolutskaya et al. 2012)

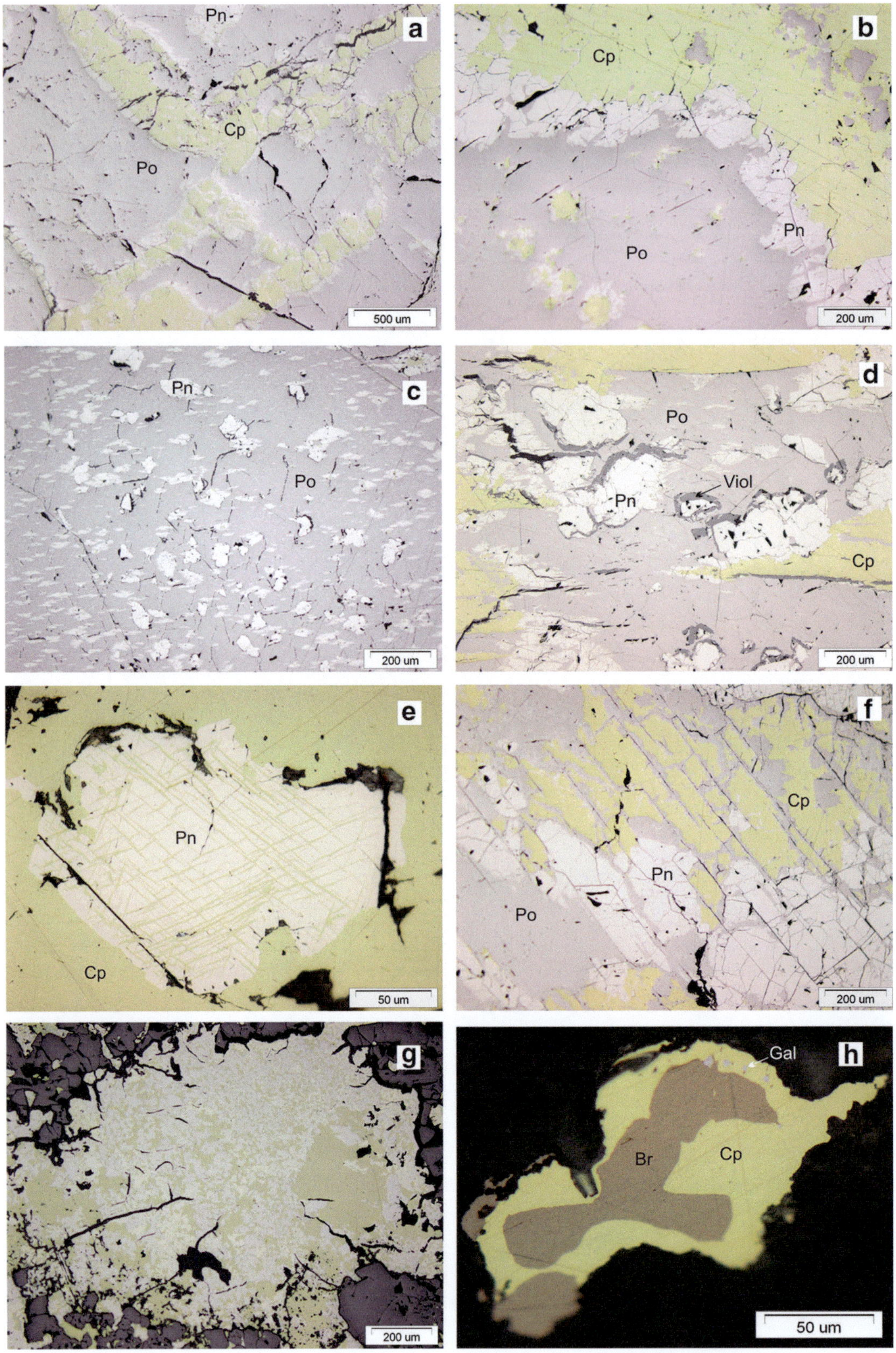

Fig. 4.59 Ore textures at the Maslovsky deposit (reflected light)
(**a**) reticulate, (**b**) rimmed, (**c**) exsolution textures of pentlandite solid solution in pyrrhotite, (**d**) porphyric, (**e**) porphyritic, (**f**) replacement, (**g**) myrmekite-like, (**h**) disseminated (After Krivolutskaya et al. 2012)

Table 4.35 Composition of main ore-forming minerals from the Northern Maslovsky intrusion, wt %

N	№ sample	Mineral	S	Fe	Cu	Co	As	Sb	Ni	Pd	Total
1.	OM-1/990,8	Cubanite	35.06	42.48	22.96	0.01	0.05	0.01	0.05	0.00	100.62
2.	OM-1/990,8	Chalcopyrite	35.03	31.66	34.07	0.04	0.04	0.02	0.05	0.00	100.91
3.	OM-1/1034,2	Pyrrhotite	37.92	62.5	0.01	0.03	0.04	0.01	0.47	0.04	101.02
4.	OM-1/1034,2	Pyrrhotite	37.25	62.41	0.21	0.01	0.04	0.01	0.46	0.00	100.39
5.	OM-1/1034,2	Pentlandite	31.83	31.62	0.01	1.84	0.03	0.01	34.81	0.00	100.15
6.	OM-5/1111,1	Pyrrhotite	38.74	60.36	0.02	0.12	0.02	0.03	1.58	0.00	100.87
7.	OM-5/1111,1	Pyrrhotite	38.06	60.27	0.03	0.08	0.05	0.03	1.61	0.03	100.16
8.	OM-12/1121,9	Bornite	24.59	8.98	68.03	0.04	0.02	0.02	0.01	0.06	101.75
9.	OM-5/1042,9	Chalcopyrite	34.08	31.44	34.34	0.02	0.06	0.02	0.04	0.00	100.00
10.	OM-5/1042,9	Pentlandite	31.80	24.61	0.36	0.88	0.03	0.01	42.47	0.00	100.16
11.	OM-12/1132,9	Cubanite	33.75	42.3	20.85	0.12	0.04	0.02	4.31	0.02	101.41
12.	OM-5/1093,0	Chalcopyrite	32.67	31.51	34.94	0.06	0.06	0.03	0.04	0.00	99.31
13.	OM-5/1093,0	Troilite	36.21	60.81	0.28	0.14	0.03	0.03	2.65	0.07	100.22
14.	OM-5/1093,0	Pyrrhotite	36.55	62.31	0.06	0.09	0.09	0.04	0.81	0.00	99.95
15.	OM-5/1093,0	Pentlandite	32.76	34.31	0.01	0.72	0.01	0.04	33.01	0.01	100.87
16.	OM-5/1093,0	Pentlandite	32.31	34.49	0.04	0.7	0.03	0.02	33.01	0.06	100.66
17.	OM-5/1111,1	Pyrite	48.61	46.56	0.07	0.46	0.05	0.06	2.39	0.00	98.20
18.	OM-12/1121,9	Bornite	24.67	9.4	68.05	0.03	0.02	0.03	0.01	0.00	102.21
19.	OM-9/1116,7	Pentlandite	30.99	34.35	0.03	0.81	0.06	0.03	33.24	0.00	99.51
20.	OM-12/1132,9	Cubanite	33.83	42.25	24.5	0.02	0.02	0.01	0.06	0.00	100.69
21.	OM-9/1116,7	Pyrrhotite	37.18	62.61	0.02	0.07	0.08	0.00	0.1	0.00	100.06
22.	OM-9/1116,7	Cubanite	34.45	41.74	22.12	0.07	0.01	0.04	0.28	0.00	98.71
23.	OM-9/1123,7	Chalcopyrite	33.1	32.48	34.28	0.02	0.03	0.02	0.03	0.00	99.96
24.	OM-9/1108,4	Pyrrhotite	36.14	62.79	0.06	0.03	0.04	0.03	0.27	0.04	99.40

Note: Analyses were carried out in GEOKHI RAS, N. Kononova (After Krivolutskaya et al. 2012)

in the southern part of the Maslovsky deposit, in which this mineral usually occurs in association with chalcopyrite and galena (Figs. 4.59h and 4.60). The mineral contains elevated As concentrations (up to 0.21 wt %).

Precious Metals The disseminated ores from the northern and southern parts of the deposit differ in concentrations of valuable components and mineralogy. The distribution of Ni, Cu, and precious metals (Pt, Pd, Rh, and Au) in rocks and disseminated ores from the Maslovsky deposit was examined in detail in its southern part, in the vertical section of Hole OM-24 (Table 4.36, Fig. 4.61).

Practically all samples, especially those from the upper portion of the intrusion, contain more Cu than Ni. In the uppermost part of the intrusion (to a depth of 600 m), the Cu and Ni concentrations are of the order of a few dozen ppm, and the Cu concentrations increase in the central part to a few hundred ppm (and reach a maximum of 606 ppm), whereas the Ni contents remain practically unchanging. Starting at a depth of 733 m, i.e., at the transition from the olivine-bearing to olivine gabbro-dolerites, the Ni to Cu ratio increases and becomes greater than 1.

The Pt and Au concentrations in the upper part of the vertical section (in the barren units of the massif, which consist of olivine-bearing and olivine gabbro-dolerites) are similar and vary from 10 to 30 ppb on average, whereas the Pd concentrations are much higher and reach a few hundred ppb in some samples. Because of this, the Pt–Pd ratio of the rocks broadly varies from 1.06 to 29.0, averaging at 3–5. An increase in the Pd concentration (as well as, to a certain extent, in the concentrations of Au and Pt) was detected in the uppermost near-roof part of the intrusion (at depths of 512–518 m, Hole OM-24). It is interesting that the Pt–Pd ratio remarkably varies within the mineralized interval (812–861 m): it becomes smaller than 1 in the disseminated ores, particularly those related to picrite gabbro-dolerites, i.e., the concentrations of Pt become greater than Pd, and this ratio is in some instances (first of all, in the taxitic gabbro-dolerites) slightly greater than 1. The increase in the Pt concentration in the rocks is associated with an increase in the Rh concentration, which reaches a maximum in the lower part of the orebody. Conversely, the highest Au concentrations are typical of the upper zones of the disseminated ores, which are hosted in picrites.

The distribution of base and precious metals in the northern part of the Maslovsky deposit was examined solely in the lower part of the intrusion (in Hole OM-4, Table 4.36), in the picritic and olivine gabbro-dolerites, which account

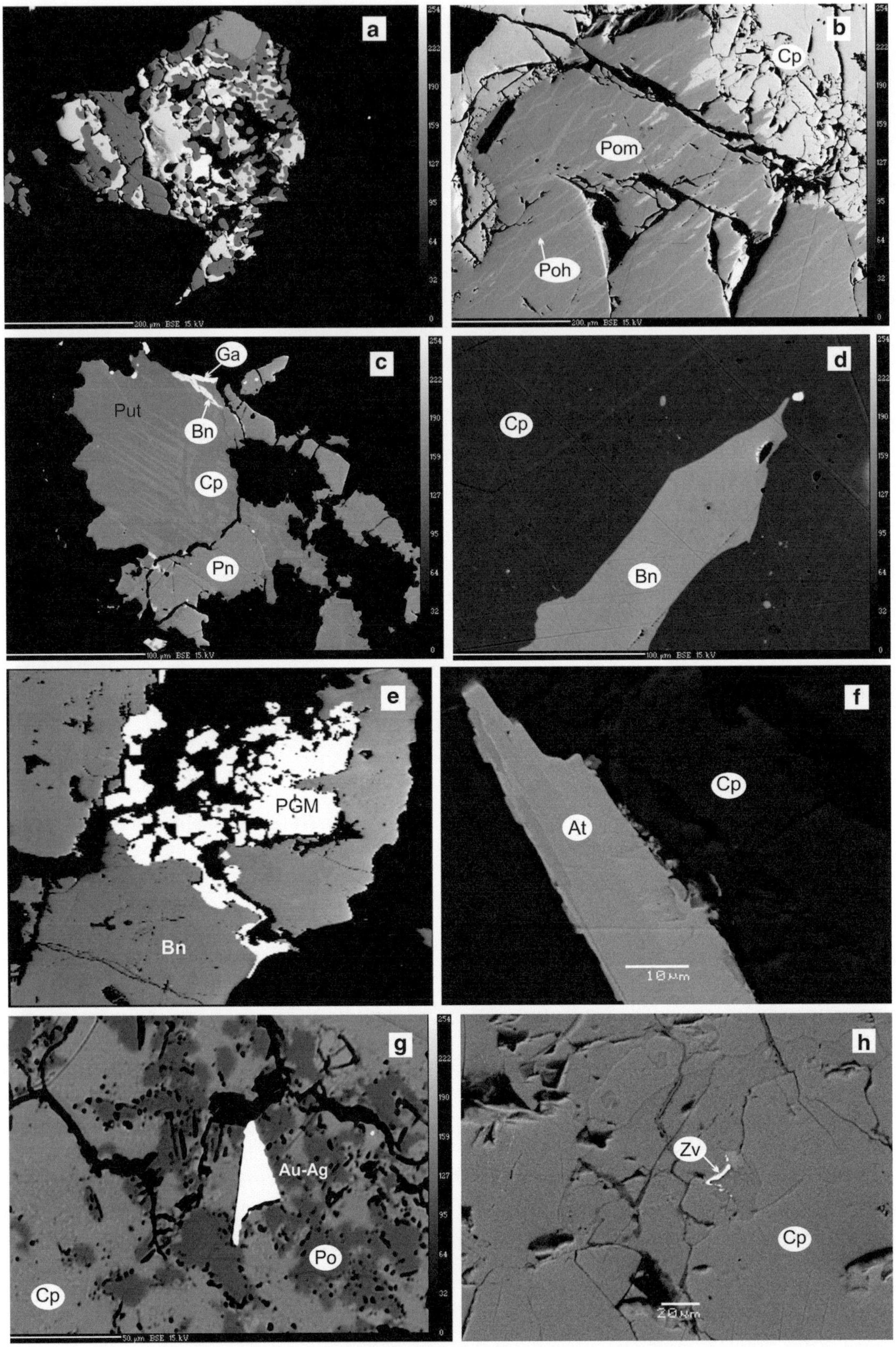

Fig. 4.60 BSE images (Cameca SX 100) of ore minerals at the Maslovsky deposit
(**a**) myrmekite-like aggregates of sulfides with Fe oxides, (**b**) aggregate of monoclinic and hexagonal pyrrhotite, (**c**) aggregate of tetragonal chalcopyrite and Ni–putoranite, (**d**) bornite grain in chalcopyrite, (**e**) PGM in bornite, (**f**) atokite crystal in chalcopyrite, (**g**) native Ag grain in a chalcopyrite–pyrrhotite aggregate, (**h**) zvyagintsevite veinlets in chalcopyrite Cp (After Krivolutskaya et al. 2011)

Table 4.36 PGE, Cu, and Ni concentrations (ppm) in the rocks and ores of the Maslovsky deposit

№	№ sample	Au	Pt	Pd	Rh	Pt+Pd+Rh	Pt–Pd	Ni	Cu
1.	OM-4/1007.5	1.15	0.09	0.37		0.46	4.30	1,636	203
2.	OM-4/1008.2	0.28	0.16	1.20	0.03	1.39	7.50	2,242	975
3.	OM-4/1009.8	0.32	0.20	0.18		0.38	0.90	1,808	328
4.	OM-4/1011.7	0.81	0.05	0.48		0.53	10.67	1,441	90
5.	OM-4/1013.2	0.13	0.24	0.18		0.42	0.75	3,367	1,932
6.	OM-4/1014.0	0.13	0.36	0.62	0.03	1.01	1.72	3,175	1,819
7.	OM-4/1016.0	0.07	0.12	0.65		0.77	5.65	3,180	2,142
8.	OM-4/1017.0	0.05	0.10	0.68		0.78	7.16	2,315	1,158
9.	OM-4/1022.0	0.10	0.18	0.85		1.03	4.72	3,169	4,755
10.	OM-4/1028.5.	0.07	0.20	0.57		0.77	2.85	2,558	3,320
11.	OM-4/1029.3	0.52	5.50	22.00	1.9	29.40	4.00	44,587	235,987
12.	OM-4/1031.4	0.11	0.09	0.75		0.84	8.82	1,517	481
13.	OM-4/1032.3	0.14	0.11	0.48		0.59	4.36	1,577	301
14.	OM-4/1032.7	0.02	0.25	0.33	0.02	0.60	1.32	2,251	1,107
15.	OM-4/1037.0	0.94	0.09	0.54	0.02	0.65	6.00	2,859	981
16.	OM-4/1037.4	0.13	0.38	0.57		0.95	1.50	1,768	178
17.	OM-4/1042.4	0.16	0.43	0.56		0.99	1.30	3,019	1,740
18.	OM-4/1045.0	0.19	0.62	0.36		0.98	0.58	1,950	848
19.	OM-4/1047.5	0.06	0.44	0.83		1.27	1.89	1,790	835
20.	OM-4/ 1049.0	0.04	0.22	0.46	0.24	0.92	2.09	3,259	1,536
21.	OM-4/1053.1	0.06	0.07	0.65		0.72	9.03	1,676	726
22.	OM-4/ 1060.0	0.05	0.43	0.64		1.07	1.49	159	1,021
23.	OM-4/1068.0	0.08	0.66	0.88		1.54	1.33	1,003	956
24.	OM-4/1073.0	0.03	0.07	0.42		0.49		2,256	3,548
25.	OM-4/1082.0	2.03	5.06	15.24		20.30	3.01	4,675	9,704
26.	OM-4/1086.0	0.05	0.14	0.76		0.90	5.63	190	2,297
27.	OM-4/1096.0	0.01	0.06	0.38		0.44	6.91	190	2,603
28.	OM-24/512.0	0.84	0.12	0.84		0.96	7.00	13	18
29.	OM-24/513.2	0.13	0.06	0.63		0.69	10.50	10	17
30.	OM-24/518.1	0.03	0.06	0.64		0.70	11.64	21	22
31.	OM-24/529.5	0.01	0.01	0.06		0.07	12.40	33	79
32.	OM-24/544.8	0.04	0.09	0.13		0.22	1.44	9	19
33.	OM-24/554.6	0.02	0.11	0.16		0.27	1.45	10	21
34.	OM-24/575.9	0.01	0.01	0.12		0.13	24.00	52	46
35.	OM-24/584.5	0.02	0.08	0.09		0.17	1.06	26	19
36.	OM-24/597.4	0.01	0.06	0.10		0.16	1.58	15	91
37.	OM-24/608.0	0.03	0.02	0.10		0.12	5.00	45	170
38.	OM-24/616.5	0.02	0.03	0.09		0.12	3.00	55	308
39.	OM-24/624.0	0.01	0.01	0.15		0.15	29.00	126	606
40.	OM-24/637.8	0.13	0.03	0.44		0.47	17.60	135	400
41.	OM-24/647.2	0.01	0.03	0.15		0.18	5.07	67	133
42.	OM-24/650.8	0.02	0.01	0.09		0.10	9.00	75	238
43.	OM-24/655.8	0.02	0.04	0.12		0.16	3.00	100	222
44.	OM-24/666.6	0.02	0.09	0.45		0.54	5.00	135	157
45.	OM-24/676.7	0.06	0.10	0.32		0.42	3.20	91	111
46.	OM-24/688.2	0.02	0.03	0.10		0.13	3.33	78	191
47.	OM-24/696.7	0.05	0.05	0.07		0.12	1.30	87	186
48.	OM-24/705.4	0.00	0.02	0.04		0.06	2.24	98	169
49.	OM-24/717.9	0.03	0.03	0.13		0.16	5.20	112	174
50.	OM-24/733.0	0.03	0.02	0.09		0.11	6.27	103	100

(continued)

Table 4.36 (continued)

№	№ sample	Au	Pt	Pd	Rh	Pt + Pd + Rh	Pt–Pd	Ni	Cu
51.	OM-24/745.1	0.02	0.03	0.05		0.08	2.08	109	87
52.	OM-24/758.6	0.01	0.03	0.05		0.08	1.60	117	85
53.	OM-24/769.8	0.01	0.02	0.04		0.06	2.50	187	103
54.	OM-24/789.9	0.01	0.02	0.07		0.09	3.40	n/a	n/a
55.	OM-24/803.3	0.02	0.02	0.05		0.07	3.60	n/a	n/a
56.	OM-24/813.0	0.36	2.10	1.85	0.05	4.00	0.88	n/a	n/a
57.	OM-24/832.0	0.19	1.10	0.68	0.08	1.86	0.62	n/a	n/a
58.	OM-24/837.0	0.18	1.15	0.76	0.84	2.75	0.66	n/a	n/a
59.	OM-24/842.0	0.11	0.87	0.81	0.02	1.70	0.93	n/a	n/a
60.	OM-24/849.0	0.11	0.50	0.72	0.65	1.87	1.44	n/a	n/a
61.	OM-24/851.0	0.10	0.78	0.87	0.6	2.25	1.12	n/a	n/a
62.	OM-24/858.0	0.08	0.55	0.50	0.03	1.08	0.91	n/a	n/a
63.	OM-24/860.0	0.21	2.00	2.72	0.94	5.66	1.36	n/a	n/a
64.	OM-24/ 861.0	0.08	0.52	0.84	0.15	1.51	1.62	n/a	n/a
66.	OM-24/869.0	0.04	0.09	0.64		0.73	7.11	n/a	n/a

Note: (1) Empty cell—element concentration is under detection limited, (2) here and Tables 2–5—*N* sample = *N* borehole/depth (m), (3) Samples OM-4 northern and OM-24 southern parts of the Maslovsky deposit, (4) n/a, element was not analyzed, Analyses were carried out in the Semenenko Institute of Geochemistry, Mineralogy and Ore Formation, National Academy of Sciences of Ukraine, analyst A. Yushin (After Krivolutskaya et al. 2011)

for one-third of the vertical section of the massif. The contents of Cu and, particularly, Ni are there remarkably higher than in the southern part of the deposit (Fig. 4.62) for the following two reasons: the occurrence of a thick unit of picritic gabbro-dolerites containing up to 60 vol.% olivine with an average Ni concentration of 0.2 wt % and the occurrence of fine sulfide dissemination in the upper part of this unit. The lower part of the picrites unit contains large crystals and veinlets of sulfide minerals, and hence, the concentrations of base metals notably increase. Their maximum values were detected in sample OM-4/1029, which consists of massive sulfides. In the lowermost part of the vertical section, in which picrite gabbro-dolerites are underlain by olivine gabbro-dolerites, Cu dominates over Ni.

The contents of precious metals in this portion of the vertical section are much higher than in the barren rocks, whose analyses are reported here for samples from Hole OM-24. For example, the concentrations Pd > Pt > Au and the Pt–Pd ratios are everywhere greater than 1 (except only for two analyses, in which this ratio is 0.9 and 0.75). The ores were determined to be rich in precious metals, whose total concentrations (Au + Pt + Pd + Rh) reach 30.0 and 22.3 ppm (in samples OM-4/1029.3 and OM-4/1082, respectively).

Minerals containing *precious metals* were found mostly in the Northern intrusion. These minerals are usually restricted to the selvages of thin sulfide veinlets in picritic gabbro-dolerites. More rarely single PGM grains occur inside sulfide drops or rock-forming minerals. The identified phases are native metals (Au, Ag, and Pd), intermetallic Sn–Pd–Pt–Bi–Pb compounds, and Pt–Fe alloys.

The largest aggregates of precious metal minerals reach 3–4 mm in size and are situated at the contacts between sulfides and silicates. Some of the phases are highly heterogeneous and some are zoned. The aggregate includes Pt and Pd compounds (Fig. 4.63): tetraferroplatinum, Pt with a Pd admixture, and Pt and Pd compounds with Sn and Te (including maslovite). Sperrylite may contain an inclusion of native gold (Fig. 4.64f). Another sperrylite grain is anhedral and is located at a sulfide–silicate contact. More rarely sperrylite occurs in association with other precious metal minerals. It coexists with Pt compound and Fe–Ni alloy with up to 11 wt % Cu. In all associations sperrylite contains relatively high concentrations of Fe (particularly in the intergrowth with Fe–Ni alloy up to 3.2 wt %), Pd (up to 6 wt %), Ni (up to 0.25 wt %), Cu (up to 6 %), and S (up to 0.96 %) and a very low concentration of Te (no more than 0.17 wt %). Native Ag is rather common in bornite (Fig. 4.65). Sperrylite, in contrast to Sn, Te, and Sb intermetallides, commonly occurs as a single crystals. One such crystal inside chalcopyrite grains in association with pyrite is shown in Fig. 4.66a.

The Sb–Sn–Bi system (Fig. 4.66b–f) was determined to contain exsolution products of a solid solution: the matrix of the grain is dominated by Sn, Sb, and Pd, whereas the lamellae contain Bi at a much lower concentrations (which

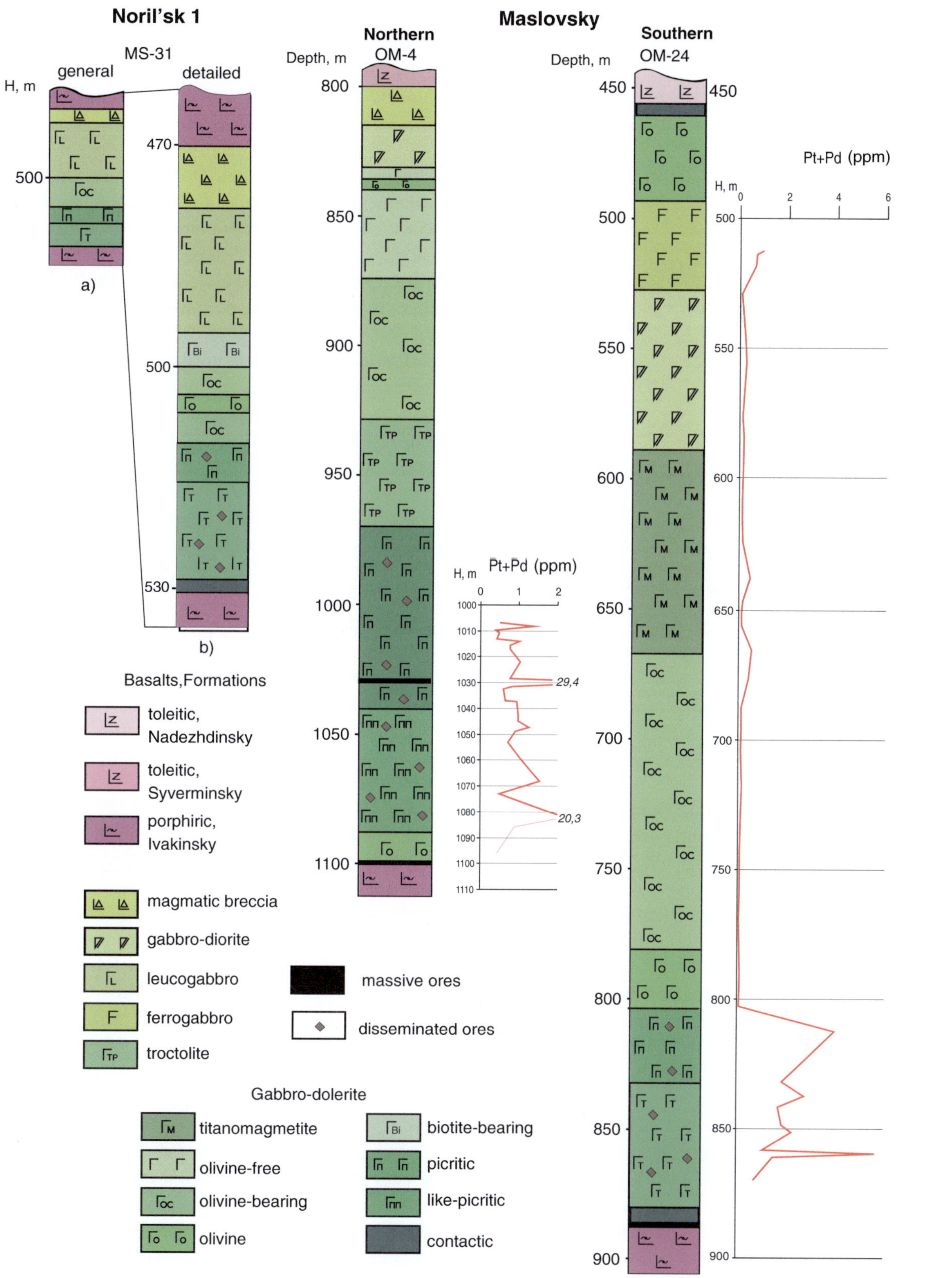

Fig. 4.61 Inner structure of intrusive bodies at the Maslovsky and Noril'sk 1 deposits and the distribution of PGE (Data on reference boreholes) After Krivolutskaya et al. (2011)

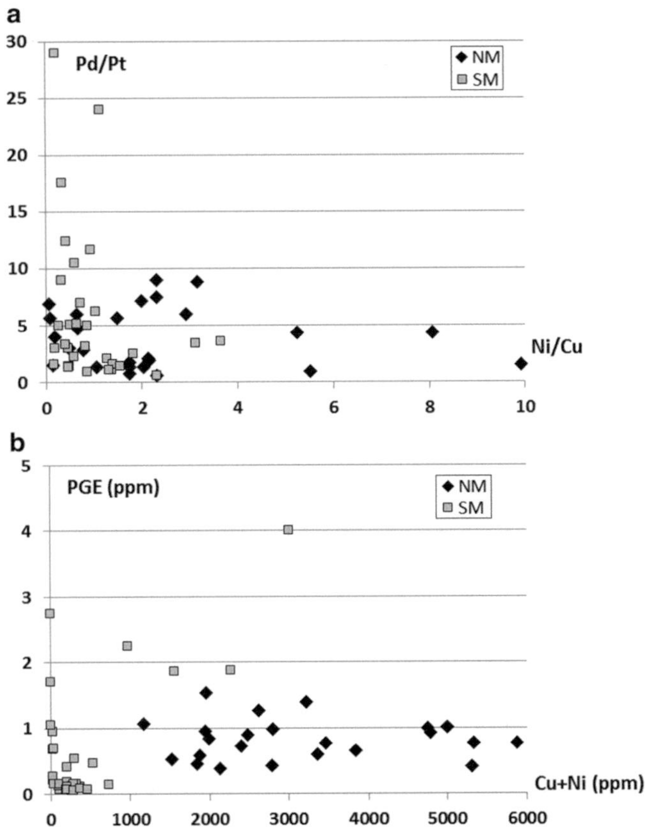

Fig. 4.62 Pt–Pd *versus* Cu (ppm) diagram for disseminated ores of the Maslovsky deposit
NM – Northern Maslovsky intrusion, SM – Southern Maslovsky intrusion. After Krivolutskaya et al. (2011)

decreases from 12.23 to 4.9 wt %) and Pd (decreases from 68.63 to 58.66 wt %). The system typically contains practically no Pt, whose very low concentration (0.18 wt %) was identified only in the lamellae.

An interesting find was that of Pd (98 and 93 wt % Pd) containing practically no Pt (0.21 wt %) (Fig. 4.66g, h), although Pd should contain 18 % Pt and have the formula (Pd,Pt) according to the classification of native metals (Table 4.37, Fig. 4.67e–h). A composition close to that was determined in one Pd grain (which contained up to 25 wt % Pt; no. 3 in Table 4.37). Pt-free Pd contains a minor Cu admixture (5 wt %) and contains Co and Fe (0.17 and 1.25 wt %, respectively). An elongated grain of irregular morphology of this mineral occurs at a contact between sulfide minerals and silicates (Fig. 4.66h), as also do most precious metal minerals.

Zvyagintsevite is one of the most widely spread minerals in the PGM ores of the Maslovsky deposit. It is most commonly contained in disseminated and massive ores in the northern part of the deposit and is more rare in its southern part. At the same time, this is the predominant PGM mineral in ores at the Noril'sk 1 deposit (Vein 1, Table 4.38). Its morphology varies insignificantly in different ore types. While the disseminated ores contain its thin veinlets among sulfides (Fig. 4.63g, h), rims around sulfides (Figs. 4.60h and 4.63f, g, h), as well as this mineral in the selvages of thin sulfide veinlets in rocks, the massive ores contain platelets of this mineral and its veinlets.

Minerals of the Pt–Fe and Pt–Pd Systems In addition to the aforementioned crystals of tetraferroplatinum, ores of the Maslovsky deposit contain numerous Pt alloys with Fe, with variable proportions of these metals: native Pt with a Fe admixture (Pt, Fe), compounds of the Pt_2Fe type, which occur in intimate aggregates with other PGM (Table 4.37).

It is known that the complete solid solution series (Pt,Fe)ss exist above $T>835$ °C, and three-ordered compounds, Pt_3Fe, PtFe, and $PtFe_3$, are formed in the subsolidus field in the system Pt–Fe. But some compositions of Pt–Fe minerals correspond to Pt_2Fe. Some explanations of this fact should be discussed (Evstigneeva 2010): (1) Pt_2Fe is another ordered phase in the system Pt–Fe; (2) the composition "Pt_2Fe" corresponds not to a homogeneous compound but to the fine "Pt_3Fe+PtFe" intergrowth; and (3) Pt_2Fe is the ferroan platinum stable under particular conditions T.L. Evstigneeva believes that the composition "Pt_2Fe" corresponds not to a homogeneous compound but to the fine "Pt_3Fe+PtFe" intergrowth. But we did not see such intergrowths in our samples, and we suppose the existing of the origin phase in Pt–Fe system.

Compounds of Pt and Pd also form a number of phases with variable proportions of these metals, ranging from native Pd with a Pt admixture (0.21, 25.6 wt %, 55.8 wt %, and 68.3 wt %). Practically all phases of this system contain admixtures of Fe (up to 1.25 wt %), Ni (0.3–0.6 wt %), and Cu (up to 5 wt %).

Intermetallic PGE *compounds with Te and Bi* are also widespread in the disseminated ores of the Maslovsky deposit (Table 4.38) and the Noril'sk 1 deposits (Distler et al. 1999), in which these compounds occur in the form of such minerals as michenerite, merenskyite, maslovite, and kotulskite. In the northern part of the Maslovsky deposit, minerals were found that contain the aforementioned elements together with Sn and Sb: $(Pd,Pt)_3SnTe$, (Pt,Pd) SnTe, and Pd_2 Bi (Sn,Sb).

Minerals in the *system Pt–Pd–Sn*. This system most often occurs in the form of euhedral crystals of minerals of the atokite–rustenburgite series (Table 4.38), which are often zonal. A new phase found in the northern part of the deposit has the formula $Pd_2(Sn,Sb)$. Another mineral identified there is palarstanide, which is a compound of Pd with Sn and As, $Pd_8(Sn,As)_3$.

Minerals in the *system Au–Ag* are predominant precious metal minerals in ores at the Maslovsky deposit. Numerous compounds of these elements compose a wealth of phases

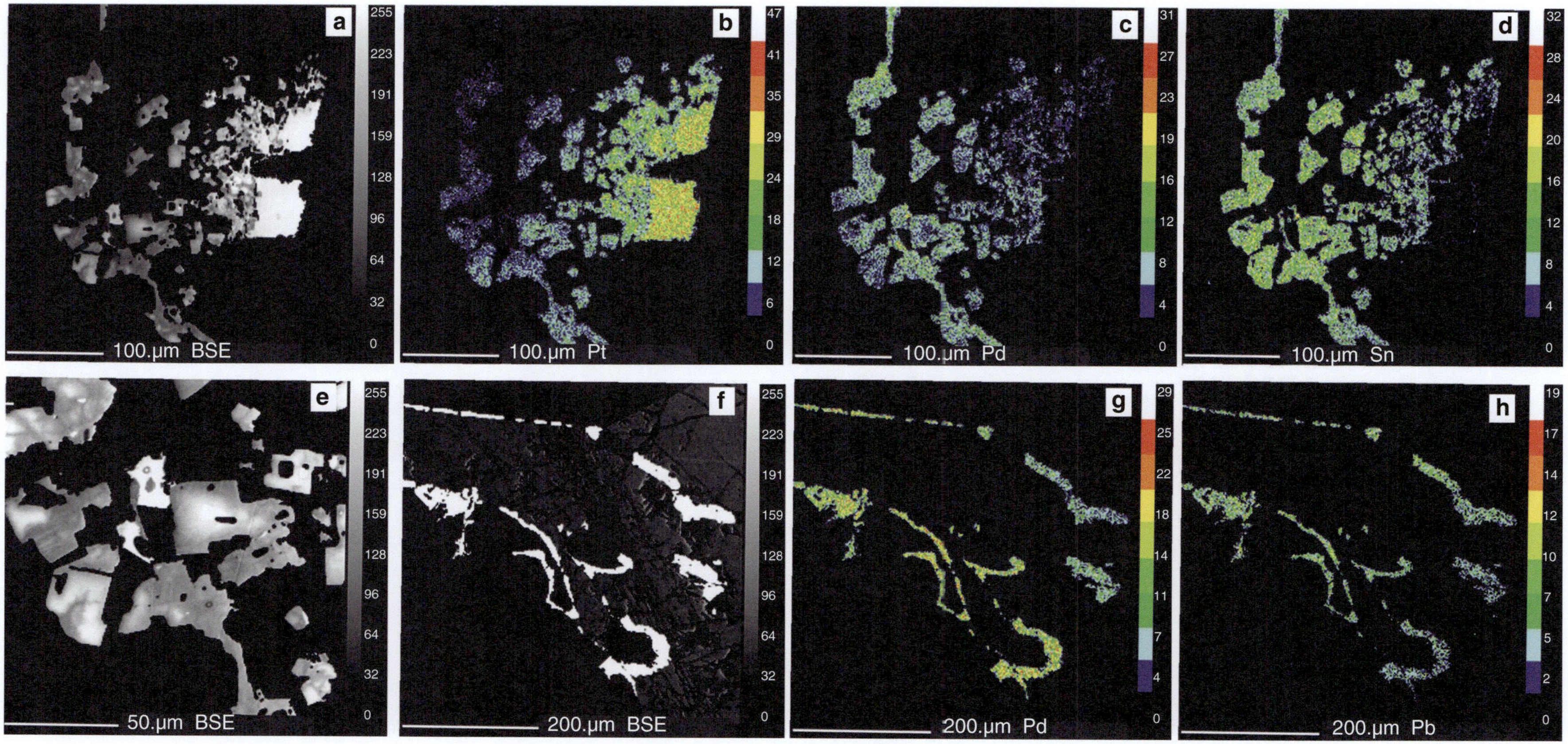

Fig. 4.63 PGM in sulfides and zvyagintsevite rims in disseminated ores of the northern part of the Maslovsky deposit
(**a**, **e**) BSE images of minerals in the system Pt–Pd–Sn; (**b**–**d**) images in characteristic X-ray radiation, (**b**) PtLa, (**c**) Pd La, (**d**) Sn La; (**f**–**h**) images of zvyagintsevite rim around chalcopyrite, (**f**) BSE image; (**g**–**h**) images in characteristic X-ray radiation, (**g**) PdLa, (**h**) PbMa (After Krivolutskaya et al. 2011)

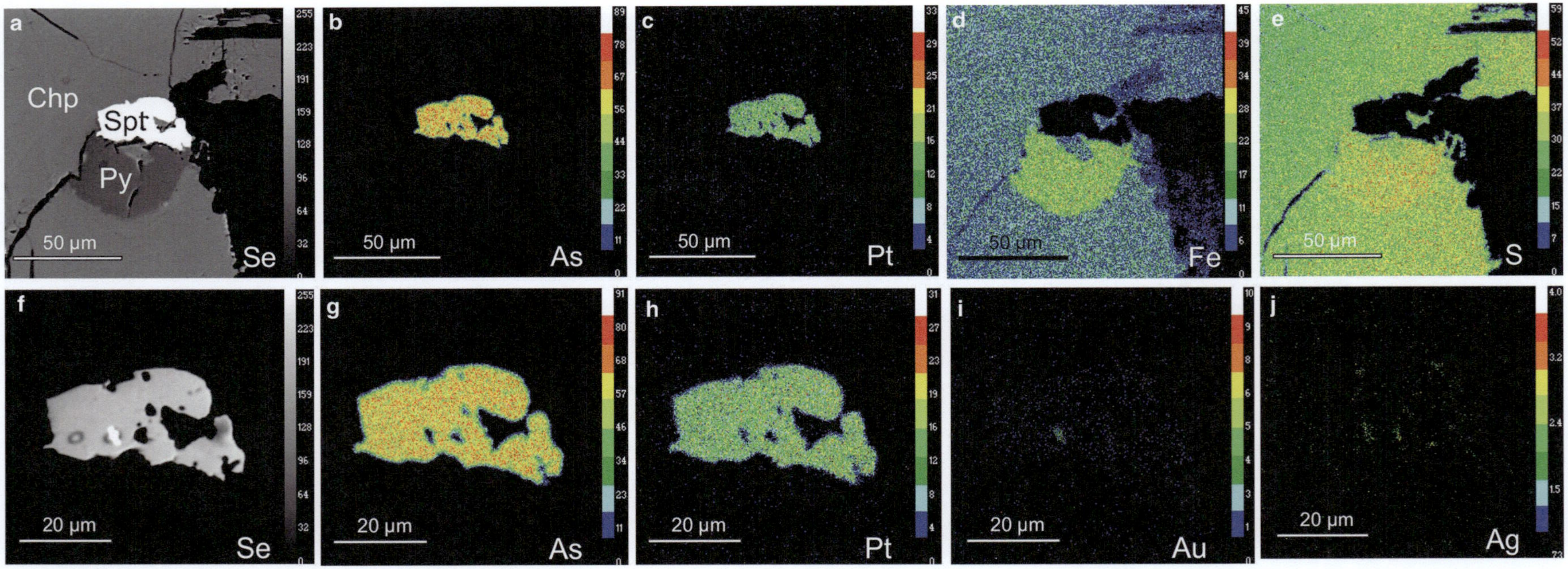

Fig. 4.64 Sperrylite grain with inclusion of native Au from disseminated ores of the Northern Maslovsky Intrusion
Images. (**a**) BSE and in characteristic X-rays radiation (**b**) As Lα, (**c**) Pt Lα, (**d**) Fe Kα, (**e**) S Kα; (**f–j**) in detail the same grain of sperrylite, (**f**) BSM, (**g**) As Lα, (**h**) Pt Lα, (**i**) Au Lα, (**j**) Ag Lα . Py, pyrite; Chp, chalcopyrite. Color bar shows the relative concentration of elements from low to high values (Here and on Figs 4.65–4.68 after Krivolutskaya et al. 2011)

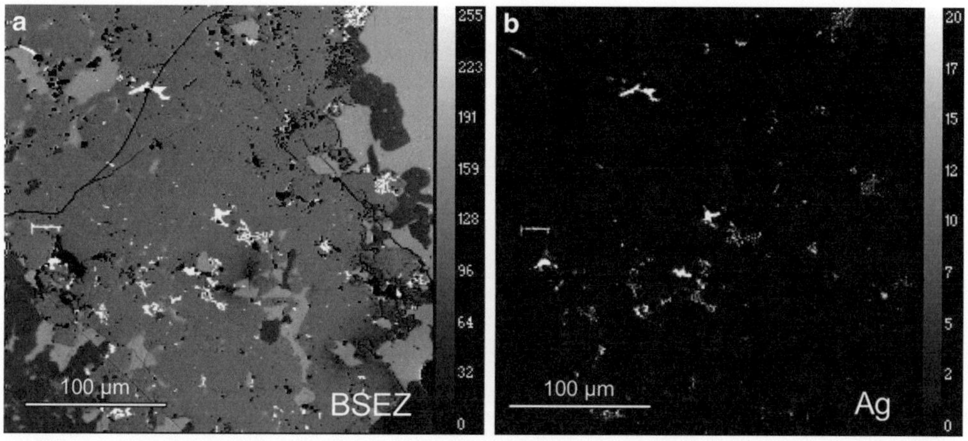

Fig. 4.65 Native Ag in bornite from disseminated ores of the Northern Maslovsky intrusion
Images: (**a**) BSM, (**b**) Ag L_{α}

(Table 4.39): native Au and Ag, kustelite, electrum, and a number of phases containing Pd. They most commonly occur as small single grains in chalcopyrite or, more rarely, in pyrrhotite and pentlandite, or among Pt and Pd minerals (e.g., in sperrylite; Krivolutskaya et al. 2012). The minerals are often zonal, with the cores richer in Cu (Fig. 4.67a–d). Minerals of this system were found in aggregates that compose veinlet-shaped accumulations in chalcopyrite, with the selvages enriched in Au and Ag and the cores enriched in Pd.

Elements of platinum group were established in the main ore-forming minerals as impurities as well. Figure 4.68 demonstrates their distribution in chalcopyrite, pyrrhotite, and pentlandite. These maps were carried out in Sudbury (analyst J. Petrus).

In recent years, laser ablation inductively coupled plasma mass spectroscopy (LA-ICP-MS) has been successfully used to map the distribution of elements in a variety of geological samples (e.g., Large et al. 2009; Ulrich et al. 2009; Woodhead et al. 2007; Woodhead et al. 2008). In this study, sample material was ablated by a Resonetics RESOlution M-50 laser ablation system and transported to a Thermo X Series II quadrupole ICP-MS where the masses of interest were analyzed. The RESOlution M-50 employs a 193-nm ArF excimer laser and a proprietary two-volume laser ablation cell (Müller et al. 2008). After the sample was loaded, air in the cell was evacuated and replaced with He. During the mapping sequence, the laser was operated with a repetition rate of 5 Hz, spot size of 10 µm, scan speed of 5 µm/s, and fluence of 10 J/cm^2. The vaporized material was entrained by 700 mL/min of He, combined with 2 mL/min of N_2, and 810 mL/min of Ar and delivered to the ICP-MS via approximately 2 m of nylon tubing. Under these conditions, a ThO^+/Th^+ ratio on NIST 612 (a synthetic glass trace element standard) of 0.5 % was typical. The map reported here consists of 106 rows of 1.250 mm in length, for a total area of 1.250 m m × 1.060 mm = 1.325 mm^2. The NIST 610 and Po725-T2 (a sulfide standard that is doped with approximately 40 ppm of PGEs and Au, Sylvester et al. 2005) standards were ablated before and after the map for calibration and quantification purposes. NIST 610 was used for S, Fe, Co, Ni, Cu, As, Se, Ag, Sb, and Re, while Po725-T2 was used for Ru, Rh, Pd, Os, Ir, Pt, and Au.

Data reduction and map construction were carried out using Iolite, a freely available add-on for Igor Pro© that was developed at the University of Melbourne (Hellstrom et al. 2008). The data was processed using the semiquantitative standardization method of Iolite's built in trace element data reduction scheme. In doing so, the concentrations of the analytes are computed assuming that the target minerals exhibit similar ablation characteristics to those of the standards used. The lack of an internal standard element present in all phases mapped is not ideal; however, owing to the 193 nm laser, ablation behavior between minerals does not generally deviate by more than 20 %.

The results of this study show that Ir, Ru, Os, and partially Rh are concentrated in pyrrhotite, while Pd and Ag (partially Pt and Rh) are present in pentlandite.

The Maslovsky deposit, whose northern part is a continuation of the Noril'sk 1 intrusion, is an important source of base and precious metals in Russia. It is thus important to study this deposit in both the applied and the academic aspect: understanding the distribution of precious metals in the ores provides a clue to the major relations and trends in the accumulation of Cu, Ni, and PGE in ultrabasite–basite melts. The pioneering results obtained on the geochemistry and mineralogy of ores from this deposit reveal important features of this deposits that make it different from other mineral deposits in the Noril'sk area.

First of all, this pertains to the distribution of valuable components in the ores. It was determined that practically all samples (of both ores and barren rocks) contain more Cu than Ni and more Pd than Pt. The latter is also typical of

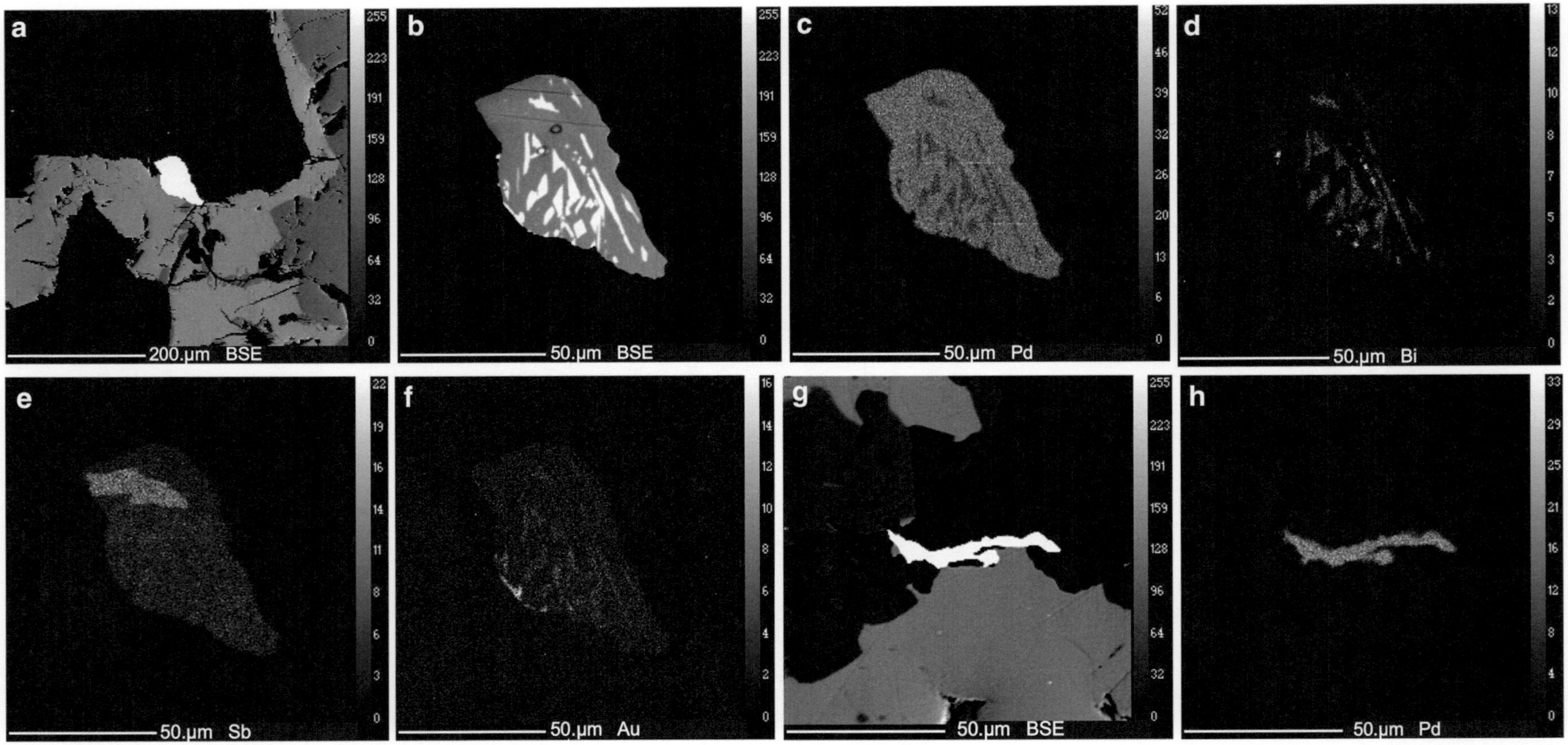

Fig. 4.66 Exsolution structures of solid solution in a grain of the Pd–Bi–Sb system and native Pd in disseminated ores from the northern part of the Maslovsky deposit
(**a**) Pd–Bi–Sb grain in the contact of ore and silicate minerals (overall view). (**b**–**f**) Exsolution texture of solid solution in a Pd–Bi–Sb grain: (**b**) BSE image. (**c**–**f**) Images in characteristic X-ray radiation: (**c**) Pd*L*a, (**d**) Bi*M*a, (**e**) Sb*L*a, (**f**) Au*L*a. (**g**–**h**) Pd grain: (**g**) BSM image, (**h**) image in X-ray radiation of Pd*L*a

Table 4.37 Minerals composition in the systems Pt–Fe and Pt–Pd, wt %

№	№ sample	Mineral	Pt	Pd	Rh	Os	Fe	Ni	Cu	Total
1.	OM-4/1010	Native Pd	0.06	99.8	0	0.03	0.04	0. 2	0.51	100.48
2.	OM-12 1121	Cu–Pd alloy	0.21	93.6	n/a	n/a	1.25	0.5	5.02	100.62
3.	OM-9/1123	Pt–Pd alloy	25.63	73.5	n/a	n/a	0.19	0.26	0.76	100.35
4.	OM-9/ 1123	Pt–Pd alloy	55.82	41.6	n/a	n/a	0.79	0.31	0.92	99.45
5.	OM-9/1123	Pt–Pd alloy	68.30	28.3	n/a	n/a	1.22	0.62	1.66	100.05
6.	OM-9/1123	Pt–Fe alloy	68.29	0.49	0.22	n/a	27.5	1.45	1.70	99.64
7.	OM-9/ 1123	Pt–Fe alloy	63.8	1.51	0.27	n/a	30.2	2.31	1.82	99.94
8.	OM-9/ 1123	Pt–Fe alloy	68.74	0.18	0.19	n/a	27.7	1.35	1.15	99.32
9.	OM-4/1010,7	Pt_2Fe	84.77	1.16	0.3	0.5	10.2	0.64	1.02	98.63
10.	OM-4/1010,7	Pt_2Fe	84.95	1.52	0.22	0.62	10.3	0.55	0.91	99.09
11.	OM-4/1010,7	Pt_2Fe	83.51	1.50	0.44	0.59	11	0.8	0.89	98.74
12.	OM-4/1010,7	Pt_2Fe	85.5	0.95	0.36	0.24	11.1	0.88	1.00	100.02
13.	OM-4/1010,7	Pt_2Fe	85.55	0.24	0.39	0.49	10.5	0.52	0.65	98.36
14.	OM-24/868,6	(Pt, Fe)	88.82	0.38	0.43	0.52	9.61	0.04	0.42	100.22
15.	OM-16/928.8	(Pt, Fe)	86.66	0.61	0.46	0.61	10.6	0.31	1.03	100.23
16.	OM-24/837,1	(Pt, Fe)	86.88	0.32	0.28	0.25	11.4	0.21	0.73	100.11
17.	OM-16/928.8	(Pt, Fe)	86.66	0.61	0.46	0.61	10.6	0.01	0.31	99.21
18.	OM-9/1123	PtFe*	75.81	0.31	n/a	n/a	15.9	3.86	3.86	99.76
19.	OM-9/1123	PtFe*	76.03	0.32	n/a	n/a	14.2	4.6	4.53	99.69

Notes: (1) OM-9 and OM-16, northern part of the Maslovsky deposit
(2) n/a, element was not analyzed; (3) * tetraferroplatinum
(9) $(Pt_{1,98}Pd_{0,05}Rh_{0,01}Os_{0,01})_{2,05}$ $(Fe_{0,83}Ni_{0,05}Cu_{0,07})_{0,95}$
(10) $(Pt_{1,98}Pd_{0,06}Rh_{0,01}Os_{0,01})_{2,06}$ $(Fe_{0,83}Ni_{0,04}Cu_{0,07})_{0,94}$
(11) $(Pt_{1,91}Pd_{0,06}Rh_{0,02}Os_{0,01})_{2,00}$ $(Fe_{0,88}Ni_{0,06}Cu_{0,06})_{1,00}$
(12) $(Pt_{1,91}Pd_{0,04}Rh_{0,02}Os_{0,01})_{1,98}$ $(Fe_{0,88}Ni_{0,07}Cu_{0,17})_{1,02}$
(13) $(Pt_{2,01}Pd_{0,01}Rh_{0,02}Os_{0,01})_{2,05}$ $(Fe_{0,86}Ni_{0,04}Cu_{0,05})_{0,95}$
After Krivolutskaya et al. (2011)

other ore-bearing intrusions in the Noril'sk district, but their predominant ores have Pd/Pt = 3, whereas ores at the Maslovsky deposit have this ratio as high as 5–6.

The first remarkable feature of the Maslovsky deposit is the occurrence of large PGE accumulations (mostly Pd intermetallics) in disseminated ores in the northern part of the deposit; these accumulations are often discernible in hand specimens with the unaided eye. As can be often seen in polished sections of the ores, precious metal minerals and their accumulation are sometimes as large as a few millimeters across, which makes these ores notably different from those at the Noril'sk 1 deposit, in which PGM are rarely larger than 50 μm and are usually <20 μm (they can be examined only under an electron microscope in concentrated separates from corresponding size fractions).

Another distinctive feature of the northern part of the Maslovsky deposit (compared to Noril'sk 1) is a greater number of PGM species in the disseminated ores, many of which remained unexamined so far. The comparison of the mineralogical features of these two deposits (Table 4.40) reveals their difference. While minerals of the atokite–rustenburgite group and Pt and Fe alloys predominate at Noril'sk 1, the ores of the Maslovsky deposit are dominated by zvyagintsevite. Moreover, the latter mineral is one of the most widely spread minerals in the vein ores of the former deposit. We determined this when studying the mineralogy of Vein 1 in picrite gabbro-dolerites at Noril'sk 1, which is exposed in the Medvezhy Creek open pit mine.

Ores at the Maslovsky deposit contain a variety of identified Pt–Fe compounds that has not been found in the ores of Noril'sk 1 as of yet. The identified native Pd and a number of Pt and Pd phases are the first finds at the Noril'sk group of deposits.

The disseminated ores of the southern part of the Maslovsky deposit differ from ores in the northern part of this deposit and Noril'sk 1. The sizes of PGM and the frequencies of their occurrence at the Southern Maslovsky deposit are much lower than at the other two deposits. PGM in the southern part of the deposit are dominated (90 %) by Au and Ag minerals and their compounds with Cu, Pd, Os, and other PGE.

Our data on the composition of the disseminated ores confirm our earlier conclusion (Krivolutskaya et al. 2012) that the Maslovsky deposit was formed under the effect of two discrete intrusive pulses, that gave two intrusive bodies differed not only in their geochemistry and the composition of their rock-forming minerals but also in the mineralogical composition of the ore mineralization. The northern part of the deposit is related to a separate

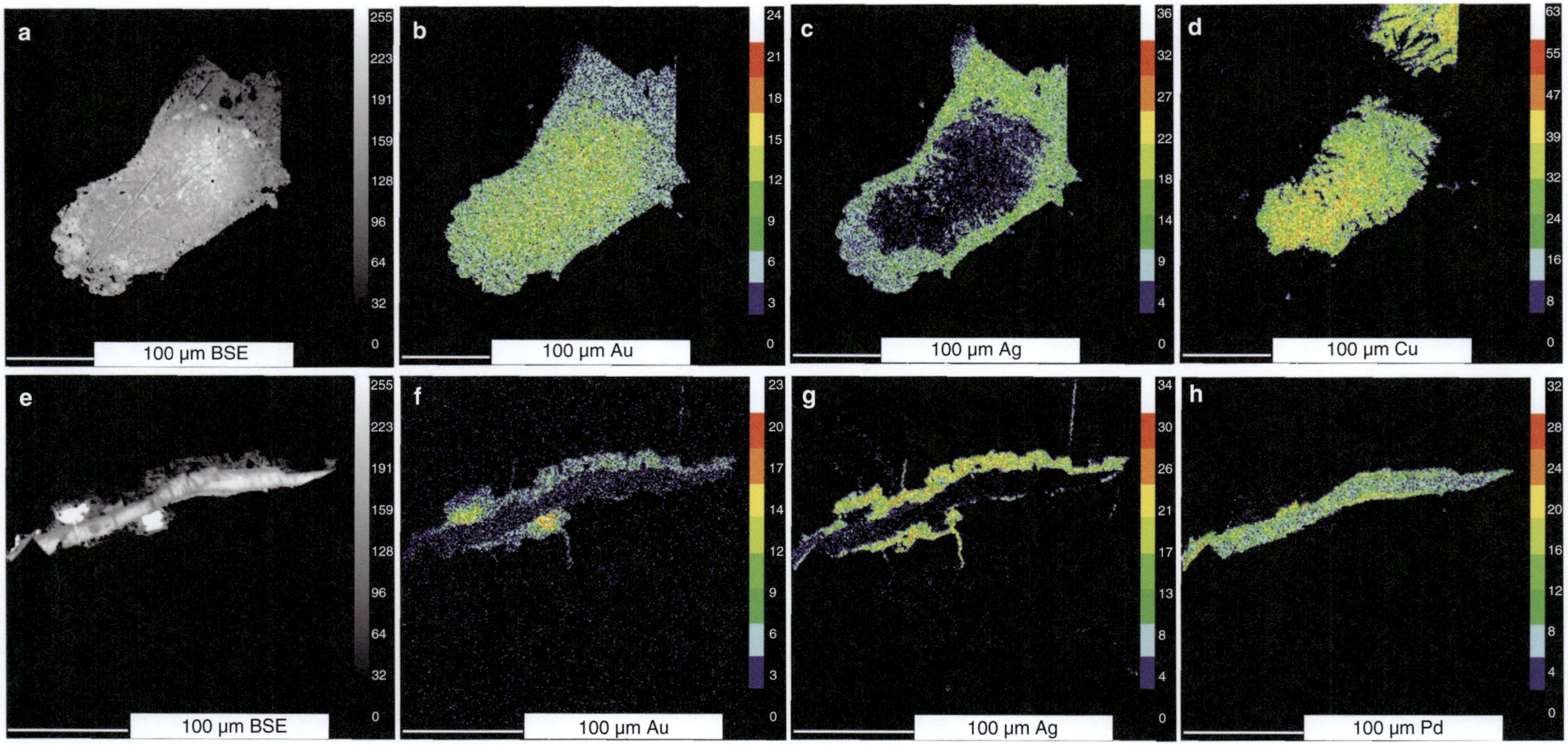

Fig. 4.67 Inner structure of Au–Ag and Pd alloys in disseminated ores from the southern part of the Maslovsky deposit
(**a–d**) zoned grain of Cu–Au–Ag alloy: (**a**) BSE image and (**b–d**) images in characteristic X-ray radiation, (**b**) Au*L*a, (**c**) Ag*L*a, (**d**) Cu*K*a, (**e–h**) veinlet composed of Au–Ag and Au–Pd alloys

Table 4.38 Mineral compositions in Pt–Pd–Sn, Pd–Te–Bi, and Pd–Pb systems, wt %

№	№ sample	Mineral	Pt	Pd	As	Sn	Sb	Pb	Fe	Ni	Cu	Te	Bi	Total
1.	OM-9/1123	Bi–moncheite	39.48				0.30					38.59	22.05	100.42
2.	OM-9/1123	Bi–kotulskite	16.23	27.13						0.00		24.40	33.37	101.13
3.	OM-9/1123	Pt–atokite	30.99	45.88		22.78	0.59		0.16	0.02	0.48			100.90
4.	OM-9/1132	Bi–kotulskite		38.86								23.94	36.66	99.46
5.	OM-9/1132	Paolovite	0.3	63.56		29.11	6.38							99.35
6.	OM-9/1123	Pt–atokite	20.56	54.56	0.03	24.42	0.22	–	0.07	0.13	0.03		0.04	100.06
7.	OM-5/1093	Pt–atokite	29.72	45.96	0.04	22.88	0.60	–	0.69	0.14	0.02		0.00	100.05
8.	OM-4/1010	Pt–atokite	29.41	46.24		22.81	0.47		0.18					99.11
9.	OM-1/1034	$Pd_5(Sn,Te\ As)_2$	0.01	68.63	0.75	18.47	12.23		0.12	0.05	0.01		0.05	100.32
10.	OM-4/1010	Pd–rustenburgite	52.31	27.57		18.89	1.15	0.32	0.24	0.21				100.69
11.	OM-24/941	Pd–rustenburgite	54.77	24.40		19.57	0.00	0.38	0.20	0.02				99.34
12.	OM-4/1023	Sperrylite	54.48	0.67	43.63				0.49	0.03	0.34	0.05	0.02	99.71
13.	OM-4/1025	Sperrylite	54.21	0.06	42.39				1.61	0.05	0.57	0.13	0.11	99.13
14.	OM-4/1025	Sperrylite	55.31		43.29				0.33	0.13	1.32	0.08	0.14	100.60
15.	OM-9/1123	Pt–atokite	31.00	45.96		22.88	0.00		0.36	0.02	0.18			100.40
16.	OM-4/1010	Pt–atokite	29.38	46.74		22.61	0.57		0.18					99.48
17.	OM-5/1093	Pt–atokite	27.72	47.98	0.04	22.43	0.60		0.69	0.14	0.92			100.52
19.	OM-24/855	$Pd_5(As,Sn)_2$	0.02	74.81	13.05	12.73	0.14		0.12			0.04	0.06	100.97
20.	OM-24/855	$Pd_5(As,Sn)_2$	0.24	74.72	16.98	7.41								99.35
21.	OM-24/833	$Pd_5(As,Sn)_2$	0.71	74.51	14.38	11.18	0.20							100.98
22.	OM-24/833	Zvyagintsevite	2.05	58.56	0.02	0.01	0.61	38.07						99.32
23.	Vein 1	Zvyagintsevite	0.05	60.25	0.01	0.01	0.76	38.10	0.43			0.13	0.14	99.88
24.	Vein 1	Zvyagintsevite	0.02	60.44	0.10	0.02	0.66	38.26	0.02			0.12	0.20	99.84
25.	Vein 1	Zvyagintsevite	0.01	60.89		0.04	0.09	39.25						100.28
26.	Vein 1	Zvyagintsevite	0.03	62.11				37.31	1.07					100.52
27.	Vein 1	Zvyagintsevite	0.00	61.27				36.71	0.27		0.86			99.11

1. $Pt_{1,00}(Te_{1,48}Bi_{0,52})_{2,00}$
2. $(Pd_{0,74}Pt_{0,24})_{0,98}(Te_{0,56}Bi_{0,46})_{1,02}$
3. $(Pd_{2,20}Pt_{0,79Cu0,01})_{3,00}(Sn_{0,99}Sb_{0,01})_{1,00}$
4. $Pd_{1,00}\ (Te_{0,51}Bi_{0,49})$
5. $Pd_{2,00}(Sn_{0,82}Sb_{0,18})_{1,00}$
6. $(Pd_{2,50}Pt_{0,49})_{2,99}\ Sn_{1,01}$
7. $(Pd_{2,20}Pt_{0,80}Fe_{0,01})_{3,01}(Sn_{0,98}Sb_{0,01})_{0,99}$
8. $(Pd_{2,20}Pt_{0,81})_{3,01}(Sn_{0,98})_{0,99}$
9. $Pd_{4,98}(Sn_{1,20}Te_{0,74}\ As_{0,08})_{2,02}$
10. $(Pt_{1,52}Pd_{1,47})_{2,99}(Sn_{0,94}Sb_{0,06}Pb_{0,01})_{1,01}$
11. $(Pt_{1,54}Pd_{1,45})_{2,99}(Sn_{0,94}Sb_{0,06}Pb_{0,01})_{1,01}$
12. $(Pt_{0,99}Pd_{0,01})_{1,00}\ As_{2,00}$
13. $(Pt_{0,99}\ Fe_{0,01})_{1,00}\ As_{2,00}$
14. $(Pt_{0,99\ Fe0,01})_{1,00}\ As_{2,00}$
15. $(Pd_{2,20}Pt_{0,81})_{3,01}Sn_{0,99}$
16. $(Pd_{2,23}Pt_{0,77})_{3,00}\ (Sn0_{,98}Sb_{0,02})_{1,00}$
17. $(Pd_{2,22}Pt_{0,77}Fe_{0,01})_{3,00}\ (Sn0_{,98}Sb_{0,02)1,00}$
18. $(Pd_{2,39}Pt_{0,54}Cu_{0,07})_{3,00}Sn_{1,00}$
19. $Pd_{4,99}\ (As_{1,24}Sn_{0,76}Sb_{0,01})_{2,01}$
20. $(Pd_{4,99}Pt_{0,01})_{5,00}\ (As_{1,56}Sn_{0,44})_{2,00}$
21. $(Pd_{4,95}Pt_{0,03})_{4,98}\ (As_{1,35}Sn_{0,6}Sb_{0,01})_{2,02}$
22. $(Pd_{2,94}Pt_{0,06})_{3,00}\ (Pb0_{,98}Sb_{0,02})_{1,00}$
23. $Pd_{3,00}\ (Pb_{0,97}Sb_{0,03})_{1,00}$
24. $Pd_{3,00}\ (Pb_{0,97}Sb_{0,03})_{1,00}$
25. $Pd_{3,00}\ Pb_{1,00}$
26. $Pd_{3,00}\ Pb_{1,00}$
27. $(Pd_{2,99}\ Cu_{0,01})_{3,00}\ Pb_{1,00}$

After Krivolutskaya et al. (2011)

Table 4.39 Minerals composition in the system Au–Ag, wt %

№	№ sample	Mineral	Ag	Au	Pd	Cu	Total
1.	OM-4/1003,5	(Au, Ag, Pd)	32.01	66.56	1.66	0.02	100.25
2.	OM-4/1005,2	Electrum	54.3	45.71	0.22	0.00	100.23
3.	OM-4/1005,2	(Au, Pd) Cu	1.04	66.68	7.58	25.21	100.51
4.	OM-24 868,5	Native Ag	98.15	1.12	0.00	0.01	99.28
5.	OM-24 941,4	Native Au	19.32	81.42	0.53	0.00	101.27
6.	OM-24 941,4	Native Au	18.7	80.79	0.6	0.00	100.09
7.	OM-24 /941,4	Native Au	19.32	81.42	0.53	0.00	101.27
8.	OM-24 941,4	Kustelite	34.01	65.56	0.23	0.02	99.82
9.	MS-31/527	Kustelite	70.96	25.14	2.86	0.00	101.02
10.	Vein 1	Kustelite	70.96	25.14	2.86	0.11	99.07
11.	Vein 1	(Au, Ag, Pd)	3.75	75.79	5.84	0.50	101.02

After Krivolutskaya et al. (2011)

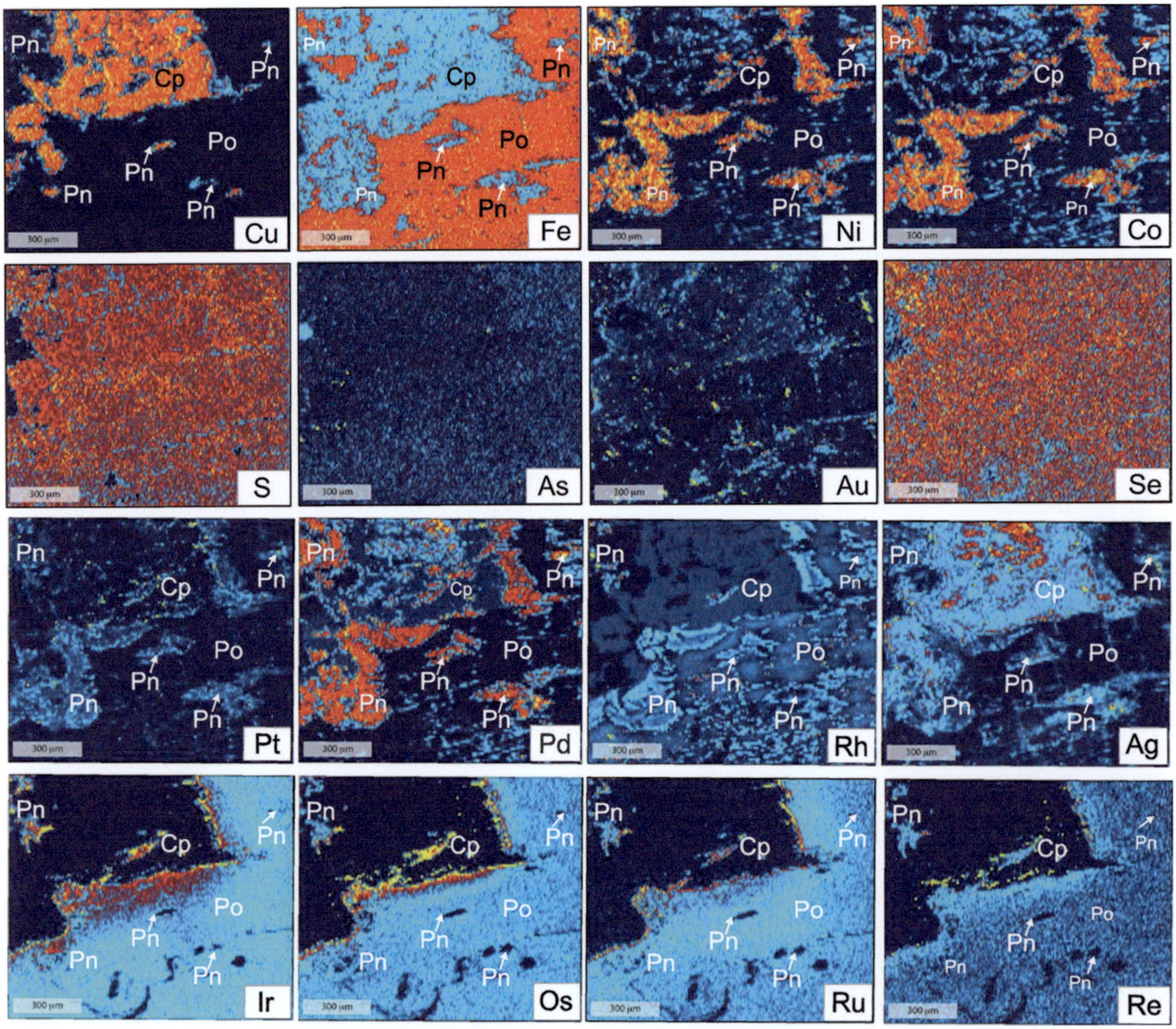

Fig. 4.68 Trace element distribution maps for the main sulfides found in the northern part of the Maslovsky deposit
Scale bar is 300 mm

intrusive body, whereas the northern part is likely a continuation of the Noril'sk 1 intrusion. The mineralogical compositions of the disseminated ores of Noril'sk 1 and the Northern Maslovsky deposit are generally similar, although the latter shows a number of unique features.

4.5.6.5 Sulfur Isotope Composition at the Maslovsky Deposit

We have analyzed sulfides (Table 4.41) from the lower part of the Southern Maslovsky intrusion (from the picritic and picritic-like gabbro-dolerites). This intrusion was selected for isotope study considering that its country rocks are basalts of the six lowermost formations of tuff–lava sequence in contrast to other massifs including Talnakh and Kharaelakh which are hosted in terrigenous–carbonate rocks. Thus by measuring sulfur isotope composition in this intrusion, one can test an influence of in situ sulfur contamination that has been supposed to be one of the main factors to achieve sulfur saturation and sulfide miscibility. The results of our studies are summarized in Table 4.41. The analyzed samples yielded an isotopically heavy sulfur composition with $\delta^{34}S$ >5‰ and 10.8‰ in two of the samples. These two samples represent picritic-like rocks with patches of plagioclase crystals and aggregates transitional downward to taxitic gabbro-dolerite within the vertical section of the body. The composition of the sulfide mineralization notably varies toward the lower contact of the intrusion and changes from pyrrhotite-dominated to chalcopyrite–pyrrhotite one. The change in sulfide species is likely related to the shift in sulfur isotope ratio toward heavier compositions.

Table 4.40 Main minerals of platinum group elements in ores of the Noril'sk ore junction

Northern Maslovsky	Southern Maslovsky	Noril'sk 1
Zvyagintsevit	Au–Ag allows	Pt–Fe allows
Pt–Fe allows	Pt–Fe allows Sperrylite	Atokite–Rustenburgite
Atokite–Rustenburgite Kotulskite		(Pd, Pt)$_5$ (Sb, Sn, As, Pb)$_2$
Moncheite		Kotulskite Sobolevskite, Sperrylite

Note: Table after Krivolutskaya et al. (2011)

Table 4.41 Isotope composition of sulfides from the Southern Maslovsky intrusion

№	N sample	$\delta^{34}S$, ‰
1.	OM-24/814	+5.1
2.	OM-24/814.5	+6.7
3.	OM-24/815	+5.2
4.	OM-24/816	+10,8
5.	OM-24/824	+5.0
6.	OM-24/830	+10.8
7.	OM-24-832.2	+5.0
8.	OM-24-843.6	+4.9
9.	OM-24-847	+6.2
10.	OM-24-848	+5.5
11.	OM-24-853.9	+6.0
12.	OM-24-859.3	+4.8

Note: Table after Krivolutskaya et al. (2012)

4.5.6.6 Comparison of the Noril'sk 1 and Maslovsky Intrusions

The Ni:Cu ratio is generally regarded as evidence of the parental magma composition: Ni-enriched ores are related with high-Mg rocks, whereas essentially Cu mineralization is located in basic intrusions. Basic–ultrabasic massifs of northern Transbaikalia demonstrate this tendency (Gongalsky and Krivolutskaya 1993). On the basis of rock geochemistry and composition of rock-forming minerals, we have suggested that the Maslovsky deposit consists of two intrusive bodies. The northern part of the deposit is a continuation of the Noril'sk 1 intrusion, and the southern part is a separate body. The weighted mean compositions (calculated on the basis of cross sections, cores OM–4, OM–24, and MS–31) of these massifs are very different (Table 4.42, Krivolutskaya et al. 2011), and it is interesting to compare the chemical and mineralogical composition of their mineralization.

The rocks of the Noril'sk 1 and the northern part of the Maslovsky intrusion have very similar compositions in terms of the SiO_2, TiO_2, FeO, and Na_2O contents (3.42). They essentially differ from the southern Maslovsky intrusion in

Table 4.42 Average mean composition of the intrusion of the Noril'sk Complex

Intrusion	Core no.	SiO_2	TiO_2	Al_2O_3	FeO	MnO	MgO	CaO	Na_2O
Northern Maslovsky	OM-4	46.71	0.70	11.82	12.66	0.24	16.68	8.91	1.75
Southern Maslovsky	OM-24	50.42	1.42	14.09	13.17	0.22	7.65	9.78	2.44
Noril'sk 1	MC-31	46.99	0.91	16.25	11.67	0.15	11.90	9.46	1.94

Core no.	K_2O	P_2O_5	Ni	Cu	Co	Zn	V	Cr	Ni:Cu
OM-4	0.42	0.09	1,173	1,135	107	153	248	n.a.	1.03
OM-24	0.64	0.17	185	184	53	70	216	n.a.	1.01
MC-31	0.52	0.15	311	410	59	75	181	3217	0.76

Note: Table after Krivolutskaya et al. (2011)

these oxides. But the MgO concentration varies in these three massifs from 16.68 in the northern Maslovsky and 11.90 in the Noril'sk 1 to 7.65 in the southern Maslovsky. The nickel content correlates with MgO contents in the rocks. Its behavior depends on the distribution between sulfide and silicate phases: the best correlation is typical of the southern Maslovsky intrusion ($R^2=0.71$), which contains the lowest contents of sulfides and Ni concentrates in olivine. The worst correlation ($R^2=0.38$) characterizes the Noril'sk 1 intrusion, where Ni accumulates in sulfides, in general. The behavior of Co is the same as Ni. It concentrates mostly in the olivine (174–206 ppm: Krivolutskaya et al. 2012) in these massifs, thus its contents correlate with MgO (R^2 in the range 0.61–0.79). In contrast, Cu accumulates in a sulfide melt and shows an independent behavior in the rocks. One can see on the Ni:Cu diagram that in the different intrusions, this ratio does not correlate with MgO in rocks and their weighted mean compositions (Table 4.42). Of course, this is a preliminary conclusion because it is based only on isolated drill holes.

Usually, the proportion of Ni exceeds that of Cu in the massive and disseminated ores of the Noril'sk region, as was shown for the Talnakh group deposits (Likhachev 1994). The same statement applies to the Noril'sk 1 and the northern Maslovsky deposits (Tables 4.42), where the Ni:Cu ratio reaches a value of 16.02. Disseminated ores of the southern Maslovsky deposit is characterized by lower values of this ratio, which varies from 0.78 up to 2.30. Thus, Cu dominates in most samples. This is the first unusual feature of the southern intrusion.

The second important peculiarity of the southern Maslovsky deposit is very low Pd/Pt ratio. The Noril'sk ores predominantly have Pd/Pt = 3 in general, and ores at the northern Maslovsky deposit have this ratio as high as 5–6, whereas most samples of the southern Maslovsky deposit are characterized by 0.62–1.62 (only one value is as high as 7, Table 4.36). This chemical composition of the ore is reflected in their mineralogical features. A comparison of the mineralogical peculiarities of these two parts of the deposit reveals the salient differences. The northern Maslovsky intrusion mostly contains Pd phases, whereas at the southern deposit, Au–Ag dominates (Table 4.40). Another distinctive feature of the northern part of the Maslovsky deposit is a greater number of PGM species in the disseminated ores, many of which remained unexamined so far (e.g., native Pd), and their large size.

Because the Ni:Cu and Pd/Pt ratios in ores at the northern Maslovsky and Noril'sk 1 deposits are similar, their ore mineralogy is comparable. But some differences exist. Whereas minerals of the atokite–rustenburgite group and alloys of Pt and Fe predominate at Noril'sk 1, the ores of the Maslovsky deposit are dominated by zvyagintsevite. Moreover, the latter mineral is one of the most widespread minerals in the vein ores of the Noril'sk deposit. This was determined when studying the mineralogy of Vein 1 in picritic gabbro-dolerites at Noril'sk 1, which is exposed in the Medvezhy Creek open-pit mine.

4.5.6.7 Conclusions

1. The Maslovsky deposit consists of two discrete intrusive bodies (Northern and Southern), which slightly differ in bulk rock geochemistry (Ta–Nb ratio) and are notably distinct in their trace element composition of their olivines.
2. The Northern Maslovsky intrusion closely resembles the Noril'sk 1 intrusion in terms of bulk rock geochemistry and mineral composition, while the Southern intrusion is a separate intrusive body.
3. The Southern Maslovsky intrusion was formed after early Nadezhdinsky time, and its geochemical features are very similar to those of the Talnakh and Kharaelakh Massifs. We thus suggest post-Nadezhdinsky origin for all intrusions of the Noril'sk Complex as a result of a separate magmatic pulse after the eruption of the Low Nadezhdinsky lavas.
4. The picrite units are determined to display principally different enrichment rates of trace elements in their rock-forming minerals, which are correlated with the ore potentials of these units. The best and most illustrative examples are HREE and Y in olivine from the picritic and, to a lesser degree, from the taxitic gabbro-dolerites, in which the Y concentrations may be as high as 3 ppm.
5. The assimilation of the host rocks took place only within thin marginal zones and cannot provide a sufficient amount of isotopically heavy sulfur to an amount sufficient to produce the sulfide ores.
6. The sulfide mineralization examined in the Northern Maslovsky intrusion is similar to the analogous mineralization at Noril'sk 1 in terms of the composition of rock-forming minerals and associations of sulfides and precious minerals.

4.6 Massif of the Mikchangdinsky Ore Junction

Geologists from NorilskGeology, Ltd. studied in detail new area to east from the city of Noril'sk. Several boreholes penetrated many basic–ultrabasic intrusion with sulfide mineralization. They located in Devonian sediments, in 1,000–1,200 m below the surface. It is suggested that this area represents a new ore junction. Here, we give new mineralogical and geochemical data from one massif named Mikchangdinsky.

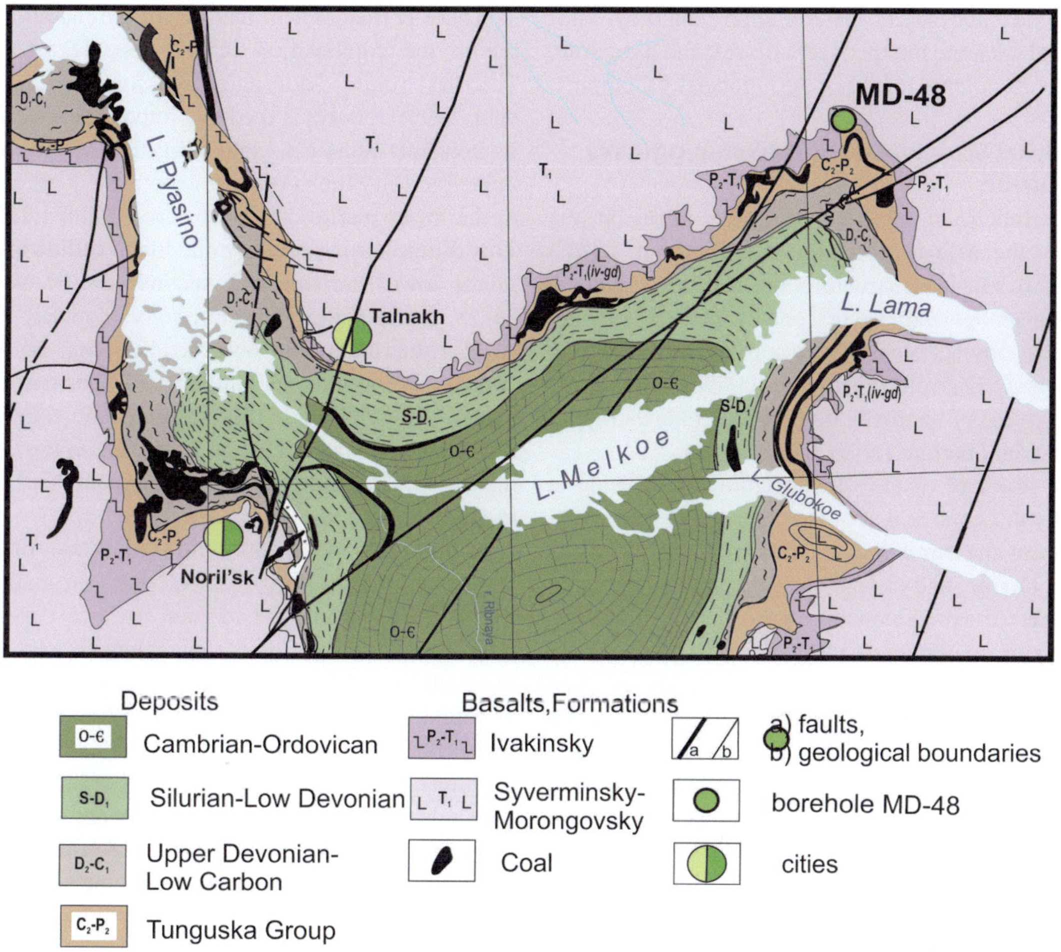

Fig. 4.69 Geological map of Mikchangda river basin Mikhailov (2003, unpublished)

4.6.1 Mikchangdinsky Massif

Discovered as a result of the geological prospecting works of NorilskGeology, Ltd., the Mikchangdinsky ultramafic–mafic intrusion (Figs. 4.2 and 4.69) is of interest because it may be a metalliferous massif. Therefore, understanding its internal structure, petrography, geochemistry, mineralogy and the conditions of its formation in relation to the Noril'sk Complex has practical importance. However, the most unusual properties of this massif are its rock-forming minerals, especially olivine, which differs from the olivines of many ultramafic–mafic massifs of this area. It is well known that the intrusions of the Noril'sk Complex that are related to the Early Triassic trap formation were formed in near-surface conditions. This history is evidenced by the structural and textural features of the rocks, the presence of glass in the endocontact zones, and the presence of glassy inclusions in the early liquidus minerals, olivine, and pyroxene. In rocks of the Mikchangdinsky massif, unique zoned olivines were found (Krivolutskaya et al. 2009a). For intrusive rocks, this phenomenon is quite rare globally because diffusion processes at subsolidus temperatures eliminate inhomogeneities in the minerals. Olivine with contrasting straight and reverse zoning is particularly typical of Martian meteorites (Connolly et al. 2006; Sarbadhikari et al. 2009). Terrestrial rocks have usually been noted to contain grain olivines with zone distribution of the main components (Fe and Mg) with an Fe gradient of 8–10 mol% (Kamenetsky et al. 2008). Therefore, discovering zoned olivine in igneous rocks of the Noril'sk area that exhibit a difference between the grain center and periphery greater than 20 mol% Fe is not only a unique phenomenon for the region but also for other ultramafic–mafic complexes around the world.

The Mikchangdinsky massif is located in the eastern part of the Noril'sk area (Fig. 4.2) in the Iken river valley (Mikchangda basin), in the northern periclinal zone of the Khantaysko–Rybninsky swell. This area is characterized by a dramatic reduction of the thickness of the volcanic rocks relative to the adjacent territories, particularly of the lower three formations. Nine tuff–lava pile formations have been identified in the studied area, i.e., the Ivakinsky, Syverminsky, Gudchikhinsky, Khakanchansky, Tuklonsky, Nadezhdinsky, Morongovsky, Mokulaevsky, and Kharayelakhsky formations. The Mikchangdinsky massif is located in Devonian sediments in the Zubovsky Formation, represented by dolomite, marl, and siltstone interbedded with salt and anhydrite, as well as gabbro-dolerite breaks of the Ogonersky Complex.

4.6.1.1 Internal Structure of the Mikchangdinsky Intrusion

The internal structure of the Mikchangdinsky intrusion was penetrated by the MD-48 borehole at a depth of 1,157–1,257 m (Fig. 4.70). In its structure, the following gabbro-dolerites were recognized: lower contact, picritic, olivine, olivine bearing, olivine, olivine free, upper picritic, olivine, and upper contact (Krivolutskaya et al. 2009b). The taxitic textures are typical of the rocks in the contact zones (upper 1,155–1,160 m and bottom 1,247–1,257 m).

Thus, the structure of the Mikchangdinsky intrusion is fundamentally the same as that of the other ore-bearing classic intrusions of the Noril'sk area. In almost all the rocks (olivine-free, olivine, and olivine-bearing gabbro-dolerites), poikilophytic texture predominates ("goroshchataya," Dodin et al. 1971) due to the presence of oikocrysts (rounded grains) of augite, containing numerous variously oriented laths of plagioclase (Fig. 4.70, photographic thin sections). Elongated laths of plagioclase (0.8–1 mm) are located either randomly as crystals inside augite or in parallel grain boundaries of clinopyroxene in the intergranular space. Crystals, usually located on the periphery of the oikocrysts of augite, vary from 0 to 20 vol.%. In picritic varieties, olivine is isometric or idiomorphic, 1.0–1.5 mm in size. It constitutes 60 % of the rock volume. Leists of plagioclase decrease up to short-prismatic crystals. There are small grains of orthopyroxene, magnetite, ilmenite, and apatite. In the interval from 1,204.7 to 1,217 m, there is more fine-grained structure than in the over- or underlying gabbro-dolerites. Lower taxitic gabbro-dolerites in the endocontact (1,247–1,257 m) are characterized by intense olivine and pyroxene replacement of secondary minerals (e.g., serpentine and amphibole) and by the presence of veinlets and disseminated sulfide minerals (up to 5 % of the rock volume).

The chemical composition of rocks from the Mikchangdinsky massif has been defined for the main types of rocks selected along the borehole. The results of these analyses are shown in Table 4.43. The distribution of the major rock-forming oxides corroborates the petrographic features of the rocks. There are two magnesium content maxima in this section: one at a depth of 1,169–1,170 m and another at 1,232–1,246 m, which correspond to the upper and lower picritic horizons (Fig. 4.70). The central part of the intrusion is characterized by an average MgO concentration of 7 wt % . The rocks generally have low-Ti (the concentration of TiO_2 does not exceed 1 wt %) and low potassium contents (K_2O=0.4 wt %). The concentration of Na_2O is significantly higher than that of potassium, reaching 2.33 wt %. It should also be noted that there is relatively high iron content in the rocks, varying from 12 to 15.5 wt % Fe_2O_3.

These petrochemical features of the massif are comparable to the composition of the upper Formations of the Siberian traps (Morongovsky–Mokulaevsky) (Lightfoot et al. 1993)). Nickel is the most important trace element, and its concentrations are highest in high-Mg rocks. Here, the upper picritic gabbro-dolerites contain a quarter of the nickel in the lower picrites (1,161.1–1,161.9 and 1,232–1,246 m). This difference is primarily due to the sulfide mineralization in the lower horizon and, secondarily, as will be shown below, to the lower concentrations of nickel in olivine. A similar distribution is observed for copper, while cobalt contents do not change similarly. Copper has a low-sulfide content, as well as highly insignificant concentrations of olivine, which has almost no influence on the Co content of the rocks. The behavior of trace elements will be discussed in the final section.

Olivine is found in almost all the horizons of the intrusion (except for a thin horizon of olivine-free gabbro-dolerites; Fig. 4.70). Morphology and grain size vary depending on the type of rock: in high-Mg varieties (picritic gabbro-dolerites), there are large subidiomorphic (1–2 mm on average 0.5 mm) or rounded grains (0.4–0.1 mm). In low-Mg rocks, olivine forms interstitial grains between plagioclase crystals (0.2–0.5 mm). In the intrusions of the Noril'sk region, olivines with a distinct zonal structure are very unusual, as revealed in the optical microscopy studies and confirmed by microprobe analyses (Fig. 4.71). Analyses of olivine were performed by N. Kononkova at the Vernadsky Institute of Geochemistry and Analytical Chemistry RAS (Cameca SX 100), and electron microprobe studies were performed by N. Krivolutskaya (JEOL JXA-8200 Superprobe) at the Max Planck Institute of Chemistry (Mainz, Germany) according to a specially developed technique (Sobolev et al. 2007) The determination accuracy of Fe mol % in olivine is 0.2 %.

The results of the measurements of the olivines from various horizons with the most contrasting composition are given in Table. 4.44. The most contrast in composition grains was detected in the olivine-bearing massive gabbro-dolerites with dolerite and poikilophytic structure (at depth 1,170–1,230 m, average content MgO=7.5 wt % (Fig. 4.69). The Fo in the olivines from different olivine grains of one sample varies from 12 to 20 mol% Fo. The degree of contrast between the central and peripherals zones of the grains can vary within a single sample. The maximum difference between the core and the rim established for one of the olivines is 20.1 mol% Fo. In picritic gabbro-dolerites (19–22 wt % MgO in rocks) from the upper and lower zones, the olivines are nearly uniform. The results of a detailed study of large olivine grains from one end to the other through the center (step is 100 mk) are shown in Fig. 4.71. The concentration of NiO directly correlates with the forsterite component content in the mineral (R^2=0.96), and MnO correlates with Fa (R^2=0.98). The behavior of calcium does not depend on

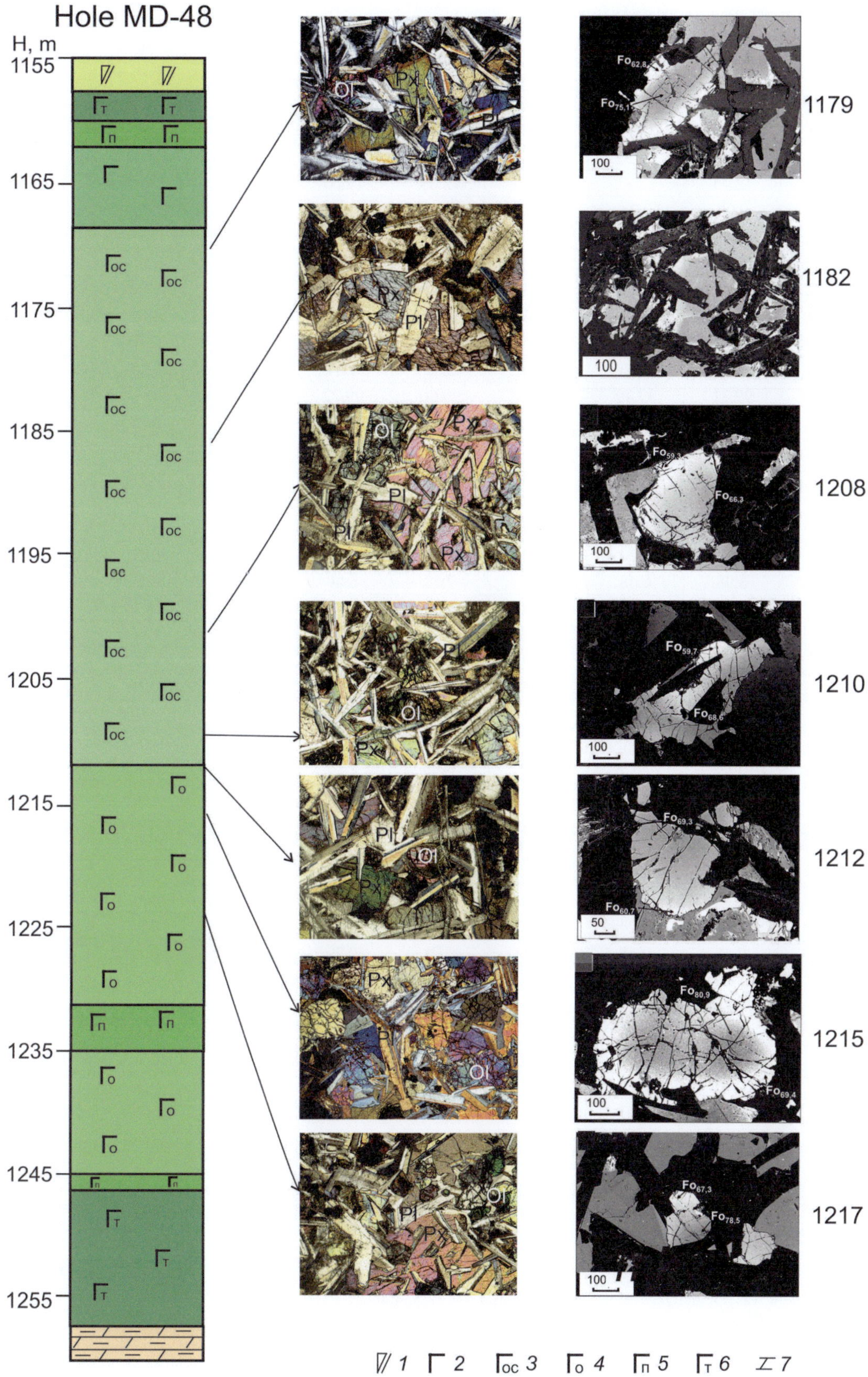

Fig. 4.70 Structure of the Mikchangdinsky massif (borehole MD-48) and microphotos of thin sections (the first column) and internal structure of olivine grains in back-scattered electrons (the second column)
(*1–7*) gabbro-dolerites: contactic (*1*), taxitic (*2*), olivine-free (*3*), olivine-bearing (*4*), olivine (*5*), picritic (*6*), sedimentary rocks (*7*). Lower boundary of thin sections is 4 mm, nicols crossing. Scale is given in microns (After Krivolutskaya et al. 2009b)

Table 4.43 Representative analyses of the rocks from the Mikchangdinsky intrusion (borehole MD-48)

№	1	2	3	4	5	6	7	8	9	10
Depth, m	1,257	1,161.1	1,161.9	1,171.4	1,179.5	1,208.5	1,212.5	1,217.2	1,232.8	1,246.8
SiO_2	47.68	42.29	44.45	48.66	48.83	47.79	47.38	46.45	40.14	40.24
TiO_2	1.03	0.54	0.84	1.16	1.16	1.03	0.94	0.95	0.47	0.37
Al_2O_3	15.95	10.90	12.39	15.02	14.66	15.85	15.56	13.43	7.65	7.44
Fe_2O_3	12.64	12.44	12.31	12.28	12.08	12.65	12.40	13.69	15.51	15.38
MnO	0.16	0.20	0.23	0.18	0.18	0.16	0.16	0.18	0.17	0.16
MgO	7.37	19.29	15.22	7.91	7.75	7.27	8.53	11.64	22.52	22.22
CaO	10.72	7.98	8.51	10.97	11.48	10.82	10.37	9.66	5.23	5.05
Na_2O	2.02	0.98	1.41	2.33	2.27	2.22	2.02	1.82	0.83	0.72
K_2O	0.37	0.31	0.47	0.45	0.40	0.38	0.35	0.35	0.19	0.20
P_2O_5	0.11	0.07	0.11	0.10	0.12	0.11	0.11	0.10	0.06	0.05
LOW	1.64	4.32	3.56	0.69	0.57	1.44	1.75	1.14	6.21	7.81
Total	99.69	99.31	99.49	99.75	99.50	99.72	99.57	99.41	98.97	99.63
Rb		6.83	9.33	10.11	8.54	8.89	8.94	8.87	4.72	
Ba		128	230	125	106	110	102	105	43	
Th		0.58	0.78	1.02	0.93	0.88	0.78	0.83	0.44	
U		0.22	0.30	0.37	0.36	0.33	0.31	0.30	0.16	
Nb		2.06	3.24	3.64	3.77	3.76	3.35	3.25	1.77	
Ta		0.13	0.20	0.23	0.26	0.23	0.20	0.20	0.11	
La		3.98	6.48	7.57	7.52	6.49	5.74	6.33	2.95	
Ce		8.73	14.15	16.66	16.85	14.49	12.80	14.24	6.54	
Pb		6.27	0.99	1.72	1.72	1.98	2.80	1.06	1.02	
Pr		1.18	1.90	2.28	2.31	2.00	1.73	1.96	0.91	
Nd		5.63	8.88	10.97	10.99	9.57	8.54	9.20	4.35	
Sr		217	208	208	214	251	232	210	123	
Sm		1.50	2.40	3.03	3.06	2.72	2.41	2.52	1.26	
Zr		50.4	58.9	73.9	80.6	73.9	69.9	66.2	36.1	
Hf		1.23	1.58	1.95	2.16	1.92	1.81	1.79	0.97	
Eu		0.54	0.81	1.02	1.01	0.93	0.80	0.90	0.42	
Ti		3,318	4,963	6,211	6,541	6,250	5,517	5,319	3,038	
Gd		1.97	2.96	3.88	3.77	3.42	2.84	3.23	1.60	
Tb		0.32	0.49	0.64	0.63	0.57	0.49	0.53	0.26	
Dy		2.13	3.24	4.28	4.34	3.81	3.33	3.66	1.80	
Ho		0.45	0.68	0.91	0.91	0.79	0.67	0.78	0.37	
Y		11.9	18.5	24.3	24.0	21.8	18.8	20.2	10.3	
Er		1.27	1.97	2.59	2.65	2.28	2.03	2.19	1.07	
Tm		0.18	0.29	0.37	0.37	0.33	0.29	0.31	0.15	
Yb		1.25	1.88	2.50	2.55	2.30	2.06	2.16	1.09	
Lu		0.19	0.27	0.37	0.37	0.34	0.30	0.32	0.16	
Ni		971	660	140	127	822	856	1,633	2,439	
Cu		570	579	111	127	1,508	1,274	2,480	2,919	
Zn		72	95	90	93	72	76	55	81	
Mn		1,636	1,806	1,434	1,374	1,286	1,211	1,380	1,387	
Sc		20.6	27.0	41.8	44.4	38.3	34.6	34.2	21.5	
Zr		50.4	58.9	73.9	80.6	73.9	69.9	66.2	36.1	
Co		108.6	84.8	50.8	47.5	83.6	84.9	94.5	163.3	

Note: Empty cell – element was not analyzed. Here and in Tables 4.44, 4.45, 4.46
After Krivolutskaya et al. (2009b)

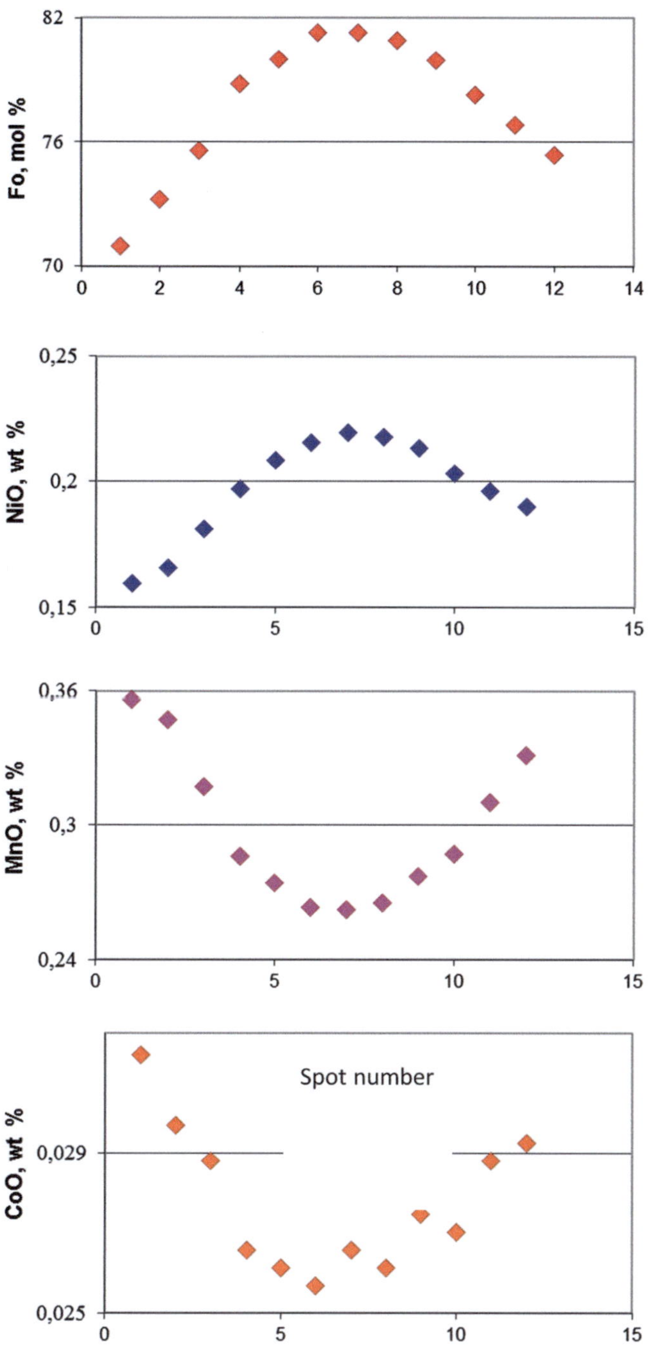

Fig. 4.71 Distribution major and rare elements in a zoned grain of olivine (sample MD-48/1193) in the Mikhangdinsky massif
After Krivolutskaya et al. (2009b)

the concentration of the main elements and is characterized by low $R^2 = 0.01$. However, the olivine grains may have not only a symmetrical zoning, as shown in Figs. 4.71 and 4.72 but also a more complicated shape, as in the intergrowths of olivine and plagioclase (Fig. 4.73).

The composition of olivine changes regularly in the intrusion and agrees well with the composition of the rocks (Fig. 4.74). Most magnesium grains are embedded in the lower and upper picritic gabbro-dolerites, where the forsterite component reaches 82.8 mol%. Most of the ferruginous material is found at the top of olivine-bearing gabbro-dolerites (51.0 mol% Fo).

The NiO content in the olivine from barren rocks varies from 0.05 wt % to 0.18 wt % with the increasing magnesium content of the mineral. The highest NiO content is observed in the rocks near the bottom of the massif, where there is sulfide dissemination. An opposite pattern can be observed in the correlation between NiO and Fe–Ni concentrations increase with increasing iron content in olivine (Fig. 4.75). As a result, on the chart of NiO–Fo, two opposing trends can be observed. Enrichment in nickel olivine from the lower horizons is likely the result of a long equilibrium with the sulfide melt, which proceeds by the redistribution of nickel between sulfide melt and mineral. A similar trend is observed for other mineralized intrusions of the Noril'sk region. It should be emphasized that in the upper horizon of picritic gabbro-dolerites, the reverse trend line is not observed. Despite the high-Mg content, they contain little NiO compared with similar olivines from olivine gabbro-dolerite. Calcium concentrations are highly variable in the studied olivines (from 0.08 to 0.42 wt %), and there is no correlation with the main elements. It is possible that this result is due to the redistribution of this element into the surrounding clinopyroxene. The manganese content in the olivine shows a direct correlation with the iron content

Plagioclase in the predominant mineral type is the earliest liquidus phase (alone or in its cotectic crystallization with olivine). According to its internal structure, similarity with olivine is observed: in picritic gabbro-dolerites, this similarity is represented by large table homogeneous crystals and varieties in composition, forming zonal lath-shaped crystals. Its composition also changes with the composition of olivine: in picritic gabbro-dolerites, the anorthite content is equal to 78–80 mol% An, while in olivine-bearing rocks of different varieties, the difference between the center and periphery of the grain is 22–24 and even 30 mol% An (e.g., Table. 4.45; № 28–29, 40–41, 36–37, in bold font in the table).

Pyroxene in rocks forms two species —monoclinic and orthorhombic. The first significantly dominates in the rocks and forms poikilitic grains, which often include crystals of plagioclase and/or olivine. Its composition changes regularly along the cross section of the intrusion (Fig. 4.74), with the most magnesium grains in picritic gabbro-dolerites, with their Mg# = 80. However, the Mg# pyroxene content in the rocks of the massif (in olivine and olivine-bearing gabbro-dolerites) is generally higher than the Fo content in olivine. For pyroxenes, as for olivines, there is typical contrasting zonation in which the rim of the grains differs from the center

Table 4.44 Representative analyses of olivines from the Mikchangdinsky intrusion (borehole MD-48), wt %

№	Depth, m	Fo, mol%	SiO_2	TiO_2	FeO	MnO	MgO	CaO	NiO	Cr_2O_3	Total
1.	1161.1	80.34	39.00	0.01	18.56	0.29	42.53	0.12	0.16	0.004	100.74
2.	1161.1	80.68	38.88	0.01	18.21	0.29	42.64	0.18	0.16	0.01	100.46
3.	1161.9	78.86	38.92	0.008	19.89	0.31	41.60	0.10	0.15	0.02	101.08
4.	1161.9	79.71	39.18	0.01	19.23	0.30	42.36	0.15	0.15	0.02	101.48
5.	1161.9	78.78	39.04	0.01	20.01	0.31	41.68	0.22	0.13	0.02	101.51
6.	1171.4	61.57	36.94	0.01	33.72	0.50	30.31	0.26	0.08	0.01	101.90
7.	1171.4	58.25	36.68	0.03	36.27	0.53	28.39	0.20	0.07	0.003	102.18
8.	1171.4	60.60	37.05	0.01	34.62	0.50	29.87	0.26	0.08	0.004	102.41
9.	1171.4	59.21	36.59	0.02	35.19	0.51	28.65	0.17	0.09	0.006	101.24
10.	1171.4	72.40	38.31	0.02	25.31	0.37	37.23	0.20	0.13	0.02	101.61
11.	1171.7	75.52	38.00	0.02	22.59	0.32	39.08	0.14	0.14	0.01	100.38
12.	1171.7	66.96	36.47	0.01	29.33	0.41	33.33	0.25	0.11	0.02	99.99
13.	1177	73.59	37.85	0.01	24.16	0.34	37.75	0.39	0.12	0.07	100.81
14.	1177	63.55	36.56	0.03	32.28	0.48	31.56	0.20	0.08	0.004	101.24
15.	1177	73.98	37.92	0.02	23.97	0.36	38.23	0.16	0.11	0.01	100.85
16.	1177	67.27	37.18	0.02	29.33	0.42	33.82	0.15	0.10	0.01	101.12
17.	1177	68.75	37.18	0.02	28.22	0.39	34.82	0.12	0.11	0.01	100.96
18.	1177	62.50	36.64	0.02	33.08	0.46	30.93	0.24	0.09	0.007	101.53
19.	1177	51.64	35.28	0.03	40.88	0.53	24.48	0.14	0.08	0.00	101.48
20.	1177	59.94	36.32	0.03	35.05	0.50	29.41	0.15	0.10	0.008	101.64
21.	1178.3	56.82	36.64	0.03	37.79	0.54	27.89	0.18	0.08	0.008	103.22
22.	1178.3	68.39	37.90	0.01	28.84	0.40	35.00	0.16	0.12	0.01	102.51
23.	1179.5	65.49	37.06	0.02	30.70	0.45	32.68	0.21	0.10	0.004	101.27
24.	1179.5	72.19	37.47	0.01	25.26	0.37	36.78	0.09	0.11	0.01	100.18
25.	1182.6	63.84	36.27	0.01	31.75	0.46	31.44	0.26	0.09	0.01	100.35
26.	1182.6	60.84	35.82	0.02	33.98	0.49	29.62	0.26	0.08	0.003	100.34
27.	1182.6	55.32	35.59	0.02	38.12	0.57	26.47	0.22	0.06	0.004	101.13
28.	1182.6	69.25	37.34	0.01	27.60	0.38	34.87	0.56	0.12	0.08	101.06
29.	1205.6	63.45	36.51	0.01	32.14	0.46	31.30	0.21	0.15	0.01	100.86
30.	1205.6	61.33	35.91	0.01	33.47	0.45	29.77	0.10	0.15	0.008	99.95
31.	1205.6	57.56	35.74	0.01	36.38	0.49	27.67	0.20	0.13	0.01	100.71
32.	1205.6	60.58	36.26	0.01	34.17	0.47	29.46	0.23	0.14	0.01	100.81
33.	1205.6	62.73	36.37	0.01	32.49	0.46	30.67	0.36	0.15	0.02	100.63
34.	1205.6	63.90	36.66	0.01	31.71	0.45	31.47	0.23	0.13	0.004	100.73
35.	1207.1	76.09	37.82	0.01	21.85	0.32	39.02	0.15	0.16	0.01	99.43
36.	1207.1	72.74	37.79	0.01	24.80	0.36	37.10	0.16	0.16	0.02	100.48
37.	1207.1	71.48	37.59	0.01	25.81	0.37	36.29	0.12	0.15	0.01	100.43
38.	1207.1	70.09	37.49	0.01	26.48	0.35	34.80	1.14	0.14	0.15	100.91
39.	1207.1	62.54	36.46	0.03	32.67	0.48	30.59	0.21	0.10	0.004	100.61
40.	1208.5	57.34	35.72	0.01	36.70	0.51	27.66	0.26	0.10	0.02	101.06
41.	1208.5	68.58	37.43	0.02	28.22	0.42	34.54	0.23	0.11	0.007	101.05
42.	1210.3	61.51	36.54	0.03	33.72	0.46	30.23	0.19	0.13	0.003	101.37
43.	1212.5	75.73	38.14	0.01	22.18	0.30	38.81	1.13	0.20	0.23	101.12
44.	1212.5	73.31	37.88	0.01	24.35	0.36	37.52	0.26	0.16	0.02	100.64
45.	1213.7	71.93	37.67	0.03	25.49	0.37	36.63	0.19	0.18	0.02	100.65
46.	1215	80.77	38.69	0.01	18.00	0.26	42.40	0.19	0.23	0.03	99.88
47.	1215	77.49	38.22	0.00	20.80	0.30	40.15	0.24	0.19	0.03	100.03
48.	1215	70.99	37.21	0.01	26.10	0.38	35.83	0.26	0.16	0.03	100.07

(continued)

Table 4.44 (continued)

№	Depth, m	Fo, mol%	SiO_2	TiO_2	FeO	MnO	MgO	CaO	NiO	Cr_2O_3	Total
49.	1217	67.01	37.29	0.03	29.38	0.42	33.48	0.15	0.19	0.004	100.96
50.	1217	68.01	37.48	0.01	28.55	0.41	34.05	0.25	0.20	0.01	100.97
51.	1217	71.76	37.85	0.01	25.61	0.37	36.51	0.23	0.21	0.01	100.81
131.	1217	67.52	37.48	0.01	29.15	0.42	33.99	0.19	0.20	0.009	101.47
132.	1217	65.89	37.31	0.02	30.28	0.43	32.81	0.23	0.21	0.007	101.31
133.	1232	74.18	37.63	0.01	23.45	0.33	37.78	0.66	0.15	0.01	100.13
134.	1232	76.14	38.11	0.02	22.03	0.32	39.42	0.21	0.16	0.007	100.33
135.	1232	74.43	37.72	0.03	23.42	0.35	38.24	0.15	0.15	0.00	100.15
136.	1232	72.91	37.67	0.03	24.71	0.36	37.31	0.17	0.16	0.001	100.48
137.	1232	72.58	37.52	0.03	24.95	0.37	37.04	0.18	0.16	0.00	100.34
152.	1232.8	78.12	38.15	0.01	20.17	0.29	40.38	0.31	0.22	0.03	99.64
153.	1232.8	76.63	37.80	0.01	21.14	0.28	38.89	1.21	0.21	0.11	99.78
161.	1233.1	77.84	38.87	0.01	20.55	0.30	40.49	0.25	0.19	0.02	100.71
163.	1236	76.86	38.75	0.01	21.34	0.31	39.76	0.30	0.22	0.01	100.72
164.	1236	77.27	38.68	0.01	20.98	0.30	40.01	0.23	0.22	0.01	100.47
165.	1236	77.04	39.00	0.02	21.30	0.31	40.10	0.20	0.20	0.005	101.14
166.	1236	75.31	38.33	0.03	22.63	0.33	38.71	0.16	0.24	0.003	100.44
167.	1236	75.47	38.66	0.02	22.57	0.33	38.95	0.15	0.25	0.001	100.94
168.	1236	77.83	38.77	0.03	20.56	0.30	40.49	0.12	0.20	0.01	100.50

Note: Analyses were carried out in GEOKHI RAS, analyst N. Kononkova

in 16 Mg# (Table 4.46, № 28–29, exemplary compositions of some grains singled bold). The main impurities in clinopyroxenes are titanium and chromium. The TiO_2 content reaches 1.68 wt % and that of Cr_2O_3—0.63 wt %. Orthopyroxene commonly occurs as rims around olivine, and its Mg# is often lower than the Mg# in clinopyroxene (Tab. 4.46, № 66–69). The maximum value is 62.72 Mg# from olivine gabbro-dolerites and in picritic rocks, Mg# = 76–77. Among the rare minerals, especially in those samples with the highest magnesium contents, chrommagnetite and ilmenite are present.

The crystallization conditions of the Mikchangdinsky intrusion were reproduced using COMAGMAT-3.65 (http://geo.web.ru/~ariskin/soft.html-id=comagmat.htm), based on the zoned olivine formed in the process of equilibrium crystallization . Calculations were performed for a series of samples taken along the section. The following results are those obtained for MD-48/1208. This rock consists of olivine (up to 10 wt %), plagioclase (35–40 wt %), clinopyroxene (25–30 wt %), a small amount of orthopyroxene (2–3 %), and titanomagnetite.

Considering the crystallization conditions of the initial melt, we calculated a pressure of 1 kbar (as the formation depth of the rocks did not exceed 1,200 m). The water content was increased by compatible elements; the concentration of K_2O in the rock was considered to be equal to 0.4 wt %. The oxygen fugacity was close to the buffer NNO-0.5 and was monitored by comparing the calculated data with the actual order of the crystallization minerals in the rock, particularly by the appearance of ilmenite and magnetite. According to the simulation results, the beginning of crystallization was associated with the appearance of liquidus plagioclase at 80.4 at $T = 1{,}197$ °C to which olivine almost immediately joined $Fo_{80.2}$ ($T = 1{,}184$ °C). After the temperature decreased to 1,135 °C, augite began to crystallize ($En_{43.4}\,Fs_{14.6}\,Wo_{41.0}$), followed by magnetite ($T = 1{,}083$ °C) and orthopyroxene ($T = 1{,}066$ °C). The measured data in this sample show mostly magnesium olivine (from 12 grains)—$Fo_{78.2}$ (lower than calculated), but there is a potential probability for the detection of minerals with higher contents of the forsterite component in this rock under detailed study. In addition, there is the possibility of partial reequilibration with the melt of this olivine when the temperature decreases. The observed zoned olivine from this sample (Fig. 4.70) occurs when the temperature gradient is 64 °C (1,152–1,088 °C, $Fo_{75.4–63.2}$). The composition of plagioclase surrounding the olivine grains is ($An_{70–72}$) very close to the calculated value ($An_{70.5}$) at the temperature of formation of the central olivine grain (1,152 °C).

In addition to the internal structure, the geochemistry and mineralogy of the Mikchangdinsky massif are compared with these parameters of the unique Talnakh and Noril'sk 1 ore-bearing intrusions. The studied section of the Mikchangdinsky massif has the greatest similarity with

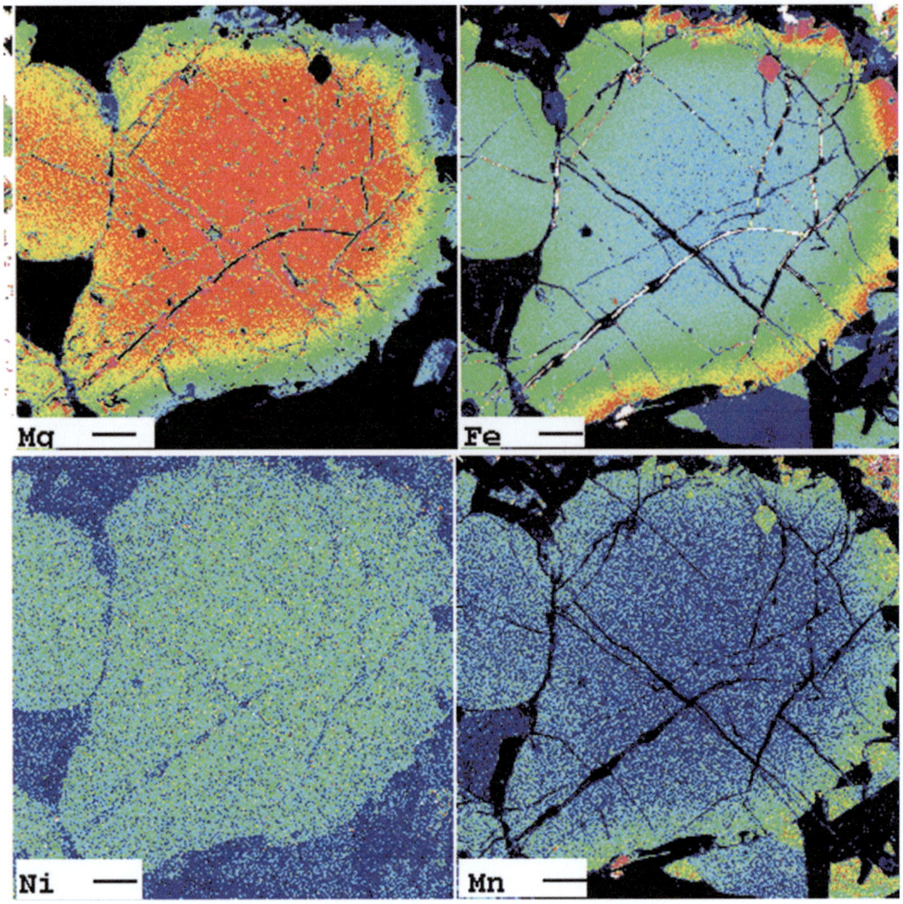

Fig. 4.72 Maps of MgO, FeO, NiO, MnO distributions in olivine grain 1 (sample MD-48/1193)
Olivine composition in the center of grain is $Fo_{83.2}$, rim is $Fo_{64.1}$, Scale is 100 microns. Here and on Fig. 4.73 analyst N. Kononkova, GEOKHI RAS

the section of the intrusion Noril'sk 1 in borehole G-22 (Fig. 4.28) discussed in the literature (Distler et al. 1999; Krivolutskaya and Rudakova 2009; Turovtsev 2002). In both intrusive bodies, there are high-Mg rocks in the upper and lower zones, represented by picritic gabbro- dolerites, while the central parts of these massifs consist of olivine-bearing and olivine gabbro-dolerites. The thicknesses of different horizons are similar in both sections. The composition of the rocks is also very similar in principal components and characterized by the features mentioned above (high iron content, reduced titanium and potassium). With respect to the Talnakh intrusion, the upper picritic gabbro-dolerites are a rare phenomenon; it is more typical to find rocks with elevated silica content in the upper parts of olivine-free massif gabbro-dolerites and gabbro-diorites (Krivolutskaya et al. 2001). However, the average mean weighted compositions of these intrusions are similar (Table 4.47). The author's calculation of the average weighted composition of the Mikchangdinsky massif based on borehole MD-48 shows that it is characterized by a higher content of magnesium than Talnakh, due mainly to the upper picrites. However, picritic gabbro-dolerite is present in its upper part, as in the Noril'sk 1. Naturally, calculations based on only one section cannot be representative because they reflect the differentiation of the ratio in only a specific section, while their thickness can vary significantly along the strike. Therefore, the most valuable compositions are obtained using data from multiple sections.

Such work was performed by M.B. Dneprovskaya for 29 boreholes for Talnakh (Dneprovskaya et al. 1987). Nevertheless, the approximate estimation obtained for the new individual massif (with limited factual material) helps to suggest preliminary conclusions about its mineralization. Considering the distribution range of rare elements normalized to the primitive mantle (Hofmann 1988; Fig. 4.76a), the topology of the spectra of the Mikchangdinsky rocks are notably close to Noril'sk 1 (Fig. 4.76b). Data

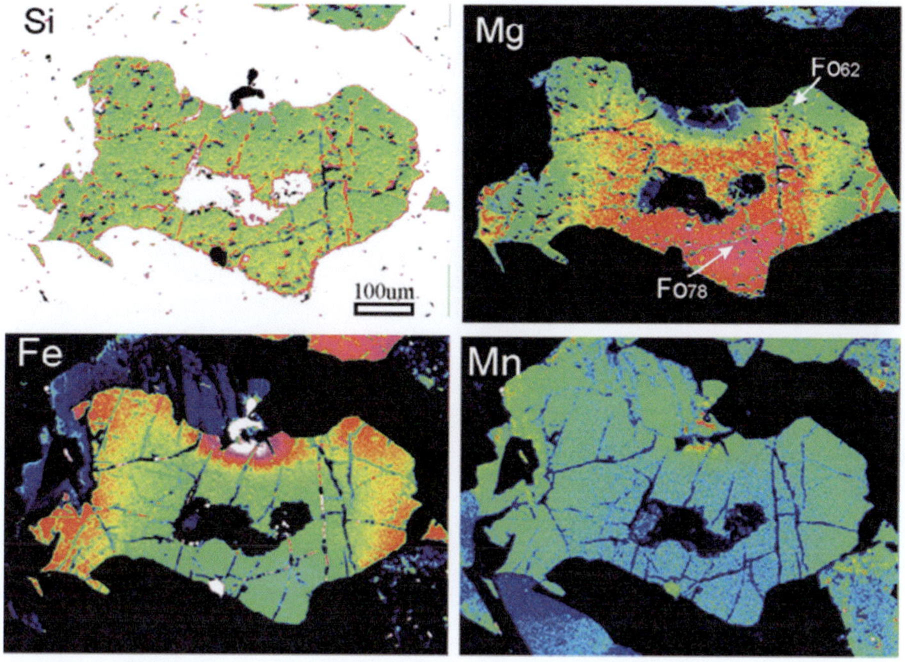

Fig. 4.73 Maps of SiO_2, MgO, FeO, and MnO distributions in olivine, grain 2 (sample MD-48O/1193)

regarding the contents of trace elements for the latter are taken from (Krivolutskaya and Rudakova 2009). They manifest the same characteristics: the presence of negative Ta–Nb, Ti and Pb, and Sr-positive anomalies and the same level of trace element accumulation, as well as the general inclination of the spectrum.

There was a significantly stronger lead anomaly in the rocks of Noril'sk 1 due to the presence of large amounts of sulfides in the rocks. Some differences were established only in the values of the anomalies of titanium and strontium, which may reflect some variation in the mineral composition of rocks (more titanomagnetite and plagioclase in some gabbro-dolerite varieties). In general, the spectra are very similar to the spectra of the continental crust.

The compositions of rock-forming minerals from the Mikchangdinsky massif are very close to the published ones from Talnakh and Noril'sk 1 (Krivolutskaya et al. 2001; Turovtsev 2002). The estimated crystallization temperatures for rock-forming minerals and the order of their appearances on the liquidus for Mikchangdinsky intrusion are comparable to those established for the crystallization of magma in the Talnakh intrusion (Krivolutskaya et al. 2001). However, the observed contrast zonation in olivine, pyroxene, and plagioclase in the Mikchangdinsky rocks indicates that the speed of cooling for rocks greatly exceeds the speed of cooling for other intrusions of this complex. We can assume that this was due to several factors. First, rapid cooling may have been a result of the structural position of the studied gabbro-dolerites of the Mikchangdinsky massif, specifically their location in the northern periclinal frame of the Khantaysko–Rybninsky swell. Here, magma rapidly intruded along a system of vertical fractures to the surface and cooled. We lack precise information on the depth of crystallization of the massif, but it could not exceed 1000 m. Second, an additional factor in the rapid cooling of the gabbros was the intrusion of melt in the unheated wall rocks because this block was not saturated with numerous intrusive bodies as was the case of the Kharaelakh or Noril'sk Trough, where many intrusions had been located previously. Third, the formation of the intrusion occurs within a large paleo-anticline structure, the Khantaysko–Rybninsky swell, where rock tuff–lava sequences are generally thin or absent.

Thus, the studied intrusion exposed by borehole MD-48 occurred close to the paleo-denudation surface, while the formation of massifs within the Noril'sk Trough had already occurred during (or immediately after) the formation of basalts. The above combination of geological factors led to fast cooling and the formation of unique olivine zoning in the ultramafic–mafic intrusion. The unzoned and very weakly zoned crystals of olivine in picritic gabbro-dolerites are explained by the presence of some intratelluric phases

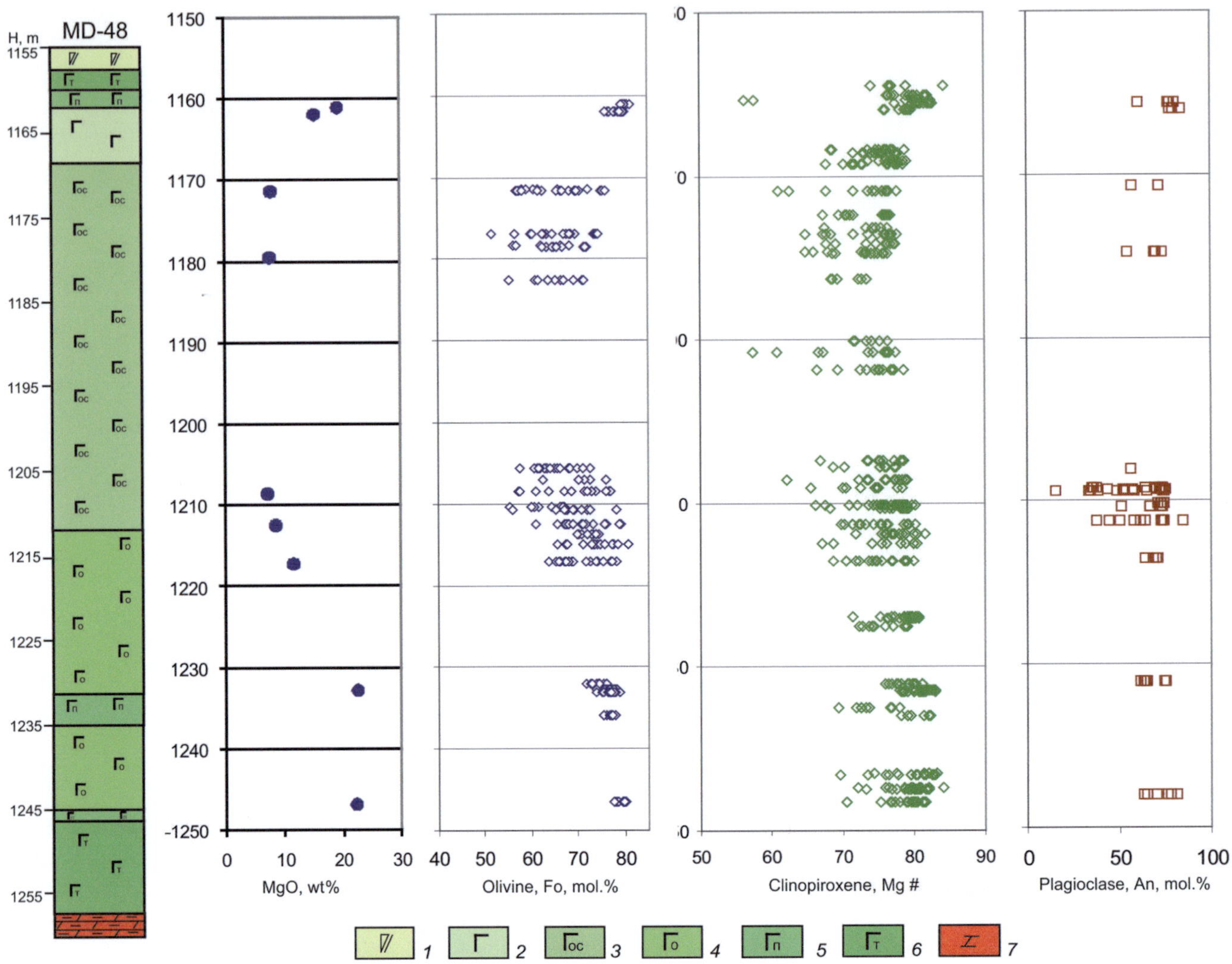

Fig. 4.74 Distribution of MgO in rocks and variations of olivine and pyroxene in section of the Mikchangdinsky massif (borehole MD-48) Legend: 1–6 see abbreviation, 7 – dolomites (After Krivolutskaya et al. 2009b)

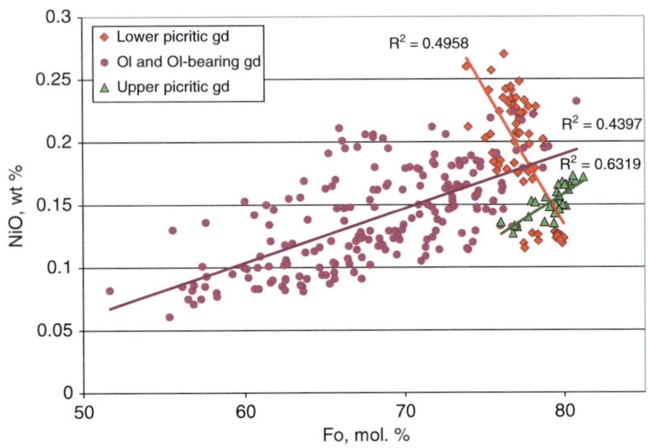

Fig. 4.75 NiO–Fo diagram for olivines from the Mikchangdinsky massif After Krivolutskaya et al. (2009b)

(olivine and possibly plagioclase) in the melt during its movement and the subsequent crystallization in the chamber. This enduring presence of the crystals in the magma led to reequilibration with melt and the formation of homogeneous grains.

Thus,

1. Mikchangdinsky massif in the eastern part of the Noril'sk ore district, according to the internal structure and petrogeochemical features of the rocks close to the ore-bearing intrusions, contains massive ores—Talnakh and Noril'sk 1.
2. The massif was formed from the parental magma, and it is almost identical in composition to the magma formed in the Talnakh intrusion. It crystallized at temperatures of 1,200–1,150 °C.

Table 4.45 Plagioclase composition from rocks of the Mikchangdinsky intrusion (borehole MD-48), wt %

№	№ sample	An. mol %	SiO_2	Na_2O	CaO	K_2O	FeO	Al_2O_3	Total
1.	1161_1a-1c-Pl-1	58.58	52.46	4.81	12.28	0.18	0.73	30.07	100.53
2.	1161_1a-1c-Pl-2	**74.98**	48.74	2.86	15.48	0.10	0.83	32.53	100.53
3.	1161_1a-1c-Pl-3	**76.45**	48.04	2.66	15.62	0.22	0.89	32.66	100.08
4.	1161_1a-1c-Pl-4	**79.32**	47.16	2.35	16.31	0.07	0.93	33.49	100.32
5.	1161_9-2k-Pl-1	**82.53**	34.12	1.53	13.04	0.11	1.02	23.19	73.00
6.	1161_9-2k-Pl-2	**78.53**	47.47	2.42	16.01	0.11	1.06	33.50	100.56
7.	1161_9-2k-Px-4	**76.63**	47.81	2.67	15.82	0.09	0.77	33.17	100.32
8.	1171_4b-2c-Pl-1	69.96	49.39	3.44	14.47	0.12	1.02	32.37	100.79
9.	1171_4b-2c-Pl-2	69.61	49.67	3.49	14.43	0.12	0.82	31.91	100.43
1.0	1171_4b-2-Pl-3	55.38	52.87	5.19	11.64	0.23	0.92	29.57	100.42
11.	1179_5a-2k-Pl-1	67.91	50.51	3.67	14.03	0.14	0.85	31.17	100.36
12.	1179_5a-3k-Pl-3	68.01	50.26	3.68	14.14	0.14	1.00	31.51	100.72
13.	1179_5a-3k-Pl-3	52.71	53.90	5.55	11.17	0.15	1.26	29.22	101.25
14.	1179-5a-3k-Pl-4	71.39	49.36	3.28	14.77	0.12	0.84	32.01	100.37
15.	11179-5a-5-Pl-4	47.04	55.35	6.34	10.18	0.23	1.10	28.06	101.26
16.	1206-pl-3k	54.10	53.95	5.32	11.32	0.26	0.97	28.85	100.67
17.	1208_4k-Pl-4a	70.64	49.71	3.35	14.54	0.12	0.91	32.03	100.66
18.	1208_4k-Pl-5k	33.64	59.41	7.66	7.01	0.59	0.58	26.00	101.25
19.	1208_4k-Pl-3	72.27	49.64	3.15	14.84	0.10	0.58	32.12	100.43
20.	1208_4k-Pl-2	62.53	52.02	4.29	12.93	0.18	0.75	30.33	100.49
21.	1208_5a-4k-Pl-1	71.23	49.68	3.30	14.77	0.12	0.65	32.07	100.58
22.	1208_4k-Pl-5a	68.08	50.22	3.64	14.03	0.16	0.84	31.33	100.22
23.	1208_4k-Pl-6	36.03	58.62	7.35	7.47	0.56	0.50	26.17	100.67
24.	1208-Pl-6a	71.90	49.67	3.21	14.84	0.12	0.56	31.95	100.34
25.	1208_7-Pl-4-1c	14.19	64.42	10.01	2.99	0.78	0.53	22.99	101.73
26.	1208_7-4-2	32.35	59.13	7.82	6.76	0.36	0.76	26.03	100.86
27.	1208-7-Pl-4-3k	62.43	50.84	4.27	12.82	0.18	1.02	30.85	99.99
28.	1208_7-Pl-5	**72.52**	48.79	3.14	14.99	0.10	0.77	32.57	100.36
29.	1208_7-Pl-5 k	**50.57**	53.87	5.83	10.77	0.26	0.95	28.72	100.40
30.	1208_7-Pl-6	56.10	52.51	5.18	11.97	0.21	0.99	30.06	100.92
31.	1208_7-Pl-6k	31.04	58.99	8.22	6.69	0.58	0.68	26.03	101.18
32.	1208_7-Pl-7c	**71.83**	49.21	3.20	14.76	0.11	0.77	32.31	100.35
33.	1208_7-Pl-7k	**72.62**	49.15	3.11	14.88	0.12	0.76	32.32	100.33
34.	1208_7-Pl-8c	35.85	57.95	7.55	7.63	0.47	0.76	26.60	100.94
35.	1208_7-Pl-8k	46.35	58.59	4.96	7.74	0.42	0.78	26.69	99.18
36.	1208_5-9m-Pl-1	**70.86**	49.69	3.34	14.68	0.14	0.87	31.94	100.65
37.	1208_5b-9m-Pl-	**40.92**	57.20	6.89	8.62	0.41	0.82	27.24	101.19
38.	1208_5a-9m-Pl-2	67.96	50.49	3.68	14.09	0.16	0.99	31.40	100.81
39.	1208_5a-9m-Pl-4	49.44	55.15	5.74	10.15	0.28	1.11	28.31	100.74
40.	1208_5a-9m-Pl-3	**70.72**	49.74	3.35	14.59	0.13	0.81	31.66	100.29
41.	1208_5a-9m-Pl-k	**54.63**	53.61	5.34	11.61	0.23	0.97	29.23	100.99
42.	1208_5a-5-Pl-1	71.00	49.80	3.33	14.75	0.12	0.83	32.01	100.84
43.	1208-5a-5-Pl-2	72.52	49.32	3.14	14.98	0.10	0.72	32.19	100.45
44.	1208-5a-5-Pl-3c	73.98	48.99	2.93	15.03	0.11	0.83	32.27	100.14
45.	48-1208-5a-Pl-4	72.81	49.15	3.11	15.02	0.09	0.91	32.22	100.50
46.	48-1210_7b_Pl-1	71.52	49.45	3.26	14.80	0.13	0.68	32.42	100.74
47.	48-1210_7b_Pl-2	72.01	49.28	3.20	14.85	0.12	0.82	32.55	100.82
48.	48-1210_7b_Pl-3	65.09	50.86	4.00	13.47	0.15	0.65	31.00	100.14
49.	48-1210_7b_Pl-4	49.13	55.19	5.82	10.15	0.32	0.78	28.49	100.75
50.	48-1210_7b_Pl-5	65.19	50.72	4.01	13.58	0.14	0.58	31.15	100.19

(continued)

Table 4.45 (continued)

№	№ sample	An. mol %	SiO_2	Na_2O	CaO	K_2O	FeO	Al_2O_3	Total
51.	48-1210_3_Px-3	71.59	49.57	3.26	14.86	0.13	0.76	32.17	100.74
52.	48-1210_3_Pl-1	70.91	49.47	3.33	14.67	0.11	0.91	31.85	100.35
53.	48-1210_3_Pl-2	69.94	49.57	3.43	14.41	0.15	1.02	31.96	100.53
54.	48-1210_3_Pl-3	72.70	48.87	3.08	14.81	0.12	0.73	32.09	99.71
55.	48-1210_3_Pl-4	72.71	49.06	3.11	14.97	0.13	0.80	32.16	100.24
56.	1212_5b-7c-c-Pl	72.11	49.09	3.19	14.89	0.10	0.73	32.38	100.38
57.	1212_5b-7-Pl-1k	**35.54**	58.36	7.58	7.55	0.53	0.83	26.30	101.14
58.	48-1212_5b-7-	47.75	54.85	6.15	10.15	0.33	0.94	28.40	100.80
59.	Pl-3	42.17	56.19	6.77	8.92	0.41	0.88	27.58	100.73
60.	1215_b_pl-1	71.96	48.98	3.18	14.73	0.13	0.71	32.45	100.17
61.	1215_b_pl-2	71.51	48.51	3.25	14.75	0.10	0.61	32.59	99.81
62.	1215_b_pl-3	72.84	48.29	3.08	14.92	0.12	0.79	32.54	99.74
63.	1215_b_pl-4	62.43	50.86	4.31	12.95	0.19	0.72	30.93	99.96
64.	1215_a_pl-1	71.97	49.30	3.19	14.78	0.11	0.87	31.97	100.21
65.	1215_a_pl-2c	**83.50**	46.64	1.84	16.86	0.07	0.40	33.95	99.76
66.	1215_a_pl-2k	**59.75**	52.43	4.65	12.48	0.20	0.71	29.80	100.28
67.	1215_a_pl-3k	72.06	48.49	3.18	14.83	0.10	0.60	31.74	98.95
68.	1215_a_pl-3kk	56.05	53.48	4.83	11.13	0.23	0.72	28.58	98.98
69.	1217_a_pl-1	67.21	50.65	3.75	13.87	0.13	0.66	31.25	100.31
70.	1217_a_pl-2	69.80	49.53	3.36	14.04	0.10	1.06	31.41	99.50
71.	1217_a_pl-3	68.31	50.43	3.62	14.09	0.14	0.72	31.29	100.28
72.	1217_a_px-5	97.96	51.89	0.21	18.33	0.02	9.05	3.38	82.88
73.	1217_a_pl-4	62.47	51.96	4.29	12.89	0.19	0.76	30.21	100.29
74.	1232_b_pl-1	73.70	48.81	3.00	15.16	0.13	0.67	32.60	100.36
75.	1232_b_pl-2	59.65	52.33	4.61	12.32	0.21	0.64	30.11	100.21
76.	1232_b_pl-3	61.39	51.54	4.48	12.88	0.20	0.72	30.40	100.22
77.	1232_b_pl-4	72.98	48.57	3.08	15.05	0.11	0.77	32.61	100.20
78.	1232_8g-5k-Pl-k	61.53	51.78	4.42	12.77	0.20	0.87	30.39	100.43
79.	1232_8g-5k-Pl-2	63.01	51.31	4.22	12.99	0.19	0.83	30.62	100.16
80.	1232_8g-5k-Pl-3	73.56	48.62	3.00	15.09	0.13	0.94	32.57	100.34
81.	1232_8g-5k-Pl-4	61.60	51.54	4.44	12.86	0.24	0.90	30.60	100.58
82.	1246_4-3k-Pl-1	76.14	47.87	2.71	15.64	0.08	1.04	33.27	100.62
83.	1246_4-3k-Pl-1c	63.00	50.96	4.21	12.96	0.14	0.92	30.84	100.03
84.	1246_4-3k-Pl-1k	67.75	49.78	3.69	14.00	0.11	0.79	31.77	100.13
85.	1246_4-3k-Pl-2	**74.16**	48.30	2.95	15.29	0.07	0.89	32.86	100.35
86.	1246_4-3k-Pl-2k	**79.91**	46.94	2.26	16.25	0.06	0.95	33.61	100.08
87.	1246_4-3k-Pl-3	61.16	51.97	4.48	12.76	0.14	0.96	30.51	100.82

Note: Here and in Table 4.45 analyses were carried out in MPI, analysts D. Kuzmin and N. Krivolutskaya. Bold numerals – data for one grain, center and rim

3. A distinctive feature of Mikchangdinsky massif is the contrasting zonation of rock-forming minerals. The olivine composition is most unusual, with a difference between the rim and center of 20 mol% Fo. Zoned olivine grains were found in all horizons, except for picritic gabbro-dolerites, where the difference between the core and rim was 1 mol% Fe on average.
4. The unusual mineral composition reflects the rapid cooling of rocks of the Mikchangdinsky massif, which resulted from its special geological and tectonic position. It is located within the Khantaysko–Rybninsky swell, close to paleosurface, and it lacked many comagmatic intrusive bodies at the time of the intrusion of the Mikchangdinsky magma.

4.6.1.2 Conclusions

The normal alkalinity ultrabasite–basite massifs can be subdivided into three groups based on the concentrations and ratios of the trace elements. The massifs with different mineralization (massive ores, rich disseminated ores, and poor disseminated ores) have similar distribution of rare elements.

Table 4.46 Pyroxene compositions from the rocks of the Mikchangdinsky intrusion (borehole MD-48), wt %

№	№ sample (depth, m)	Mg#	SiO_2	TiO_2	Al_2O_3	FeO	MnO	MgO	CaO	Na_2O	Cr_2O_3	NiO	Total
1.	1161_1a-1c-Px-1	80.98	51.26	1.67	3.27	6.85	0.19	16.36	20.32	0.38	0.11	0.03	100.43
2.	1161_9-2k-Px1	80.26	51.98	0.65	3.25	7.79	0.22	17.77	17.77	0.30	0.63	0.02	100.40
3.	48-1161_9-2k-Px2	78.62	51.57	0.80	3.10	8.54	0.24	17.62	17.68	0.24	0.37	0.01	100.18
4.	1161_9-2k-Px3	80.34	51.71	0.62	3.39	7.46	0.21	17.10	19.31	0.22	0.26	0.02	100.33
5.	1167_b_px-1	78.57	51.22	0.72	3.53	8.00	0.23	16.46	19.86	0.19	0.31	0.03	100.57
6.	1167_b_px-1k	70.37	51.02	0.95	2.83	12.05	0.33	16.05	17.49	0.20	0.01	0.05	100.99
7.	1167_b_px-2	78.72	51.77	0.63	3.07	7.96	0.22	16.51	19.89	0.19	0.27	0.04	100.56
8.	1167_b_px-2k	73.77	51.75	0.72	2.68	10.23	0.29	16.14	18.64	0.20	0.02	0.05	100.73
9.	1171-Px-2	60.14	50.66	0.86	2.25	17.35	0.47	14.68	14.04	0.16	0.01	0.001	100.49
10.	1179_5a-3-Px-2izm	62.88	50.24	1.20	2.67	14.28	0.37	13.57	17.93	0.20	0.00	0.01	100.48
11.	1179_5b-2k-Px	73.88	51.26	0.70	3.29	10.61	0.29	16.83	17.55	0.2	0.16	0.01	100.89
12.	1179_5b-Px-2	74.60	51.27	0.67	3.28	10.28	0.29	16.93	17.40	0.2	0.21	0.01	100.56
13.	48-1179_5b-Px-3	66.27	50.17	0.91	3.03	13.69	0.34	15.08	16.92	0.23	0.01	0.00	100.38
14.	1179_5a-3k-Px-6	67.04	50.05	1.06	3.37	12.84	0.32	14.65	18.07	0.27	0.03	0.01	100.69
15.	1179_5a-3k-Px-7	74.74	51.20	0.71	3.31	9.56	0.25	15.88	19.32	0.20	0.22	0.02	100.68
16.	1208-5a-4k-Px-1	76.46	52.04	0.65	3.21	9.48	0.30	17.27	17.79	0.18	0.19	0.01	101.14
17.	1208-5a-4k-Px-1 k	67.96	51.72	0.83	2.24	13.26	0.37	15.77	16.88	0.21	0.001	0.01	101.30
18.	1208_5a-4k-Px-2	78.61	51.31	0.75	3.75	8.02	0.21	16.54	19.59	0.16	0.36	0.03	100.73
19.	1208_5a-Px-3	79.12	50.88	0.79	4.24	7.56	0.20	16.06	19.93	0.19	0.55	0.01	100.43
20.	1208_4k-Px-4c	78.05	52.98	0.47	2.20	8.80	0.26	17.55	18.65	0.14	0.12	0.03	101.21
21.	1208_5a-4k-Px-4 k1	68.90	51.20	0.94	2.82	12.57	0.35	15.62	17.42	0.24	0.01	0.02	101.20
22.	1208_5v-7-Px-1 k	66.37	49.82	1.46	3.22	12.97	0.35	14.35	17.56	0.32	0.007	0.03	100.09
23.	1208_5b-7-Px-2 k	66.22	50.94	0.92	2.39	13.12	0.36	14.43	17.74	0.29	0.01	0.01	100.23
24.	1208_5a-9m-Px-1	71.31	51.42	0.87	3.02	11.32	0.29	15.78	18.08	0.21	0.04	0.03	101.07
25.	1208_5a-5m-Px-1	79.30	52.91	0.50	2.24	8.14	0.21	17.48	19.00	0.17	0.09	0.03	100.78
26.	1208-5a-5-Px2	77.47	53.01	0.44	2.06	9.17	0.27	17.70	17.81	0.15	0.16	0.02	100.79
27.	1210_7b_Px-1	77.07	51.57	0.69	3.31	8.76	0.24	16.52	19.15	0.19	0.13	0.008	100.58
28.	1210_7b_Px-2	**76.57**	51.88	0.65	2.95	9.03	0.25	16.56	19.20	0.19	0.14	0.02	100.88
29.	1210_7b_Px-2k	**60.31**	52.99	0.40	1.63	15.35	0.46	13.08	17.92	0.32	0.16	0.02	102.35
30.	1210_7b_Px-3	76.89	52.31	0.48	2.57	9.31	0.28	17.37	17.89	0.15	0.19	0.03	100.60
31.	1210_7b_Px-3k	70.08	50.95	1.02	2.64	11.98	0.31	15.73	17.80	0.24	0.001	0.02	100.70
32.	1210_3_Px-4	**79.43**	51.32	0.71	3.54	7.68	0.20	16.62	19.80	0.19	0.33	0.03	100.43
№	№ sample (depth, m)	Mg#	SiO_2	TiO_2	Al_2O_3	FeO	MnO	MgO	CaO	Na_2O	Cr_2O_3	NiO	Total
33.	1210_3_Px-4k	**67.74**	52.69	0.17	0.96	11.45	0.29	13.49	21.14	0.2	0.008	0.02	100.42
34.	1215_b_px-1	76.94	51.19	0.60	3.06	9.11	0.26	17.05	17.99	0.17	0.28	0.02	99.74
35	1215_b_px-1k	67.00	51.22	0.51	1.91	18.22	0.51	20.74	5.83	0.03	0.19	0.03	99.19
36.	1215_b_px-2	77.25	51.93	0.48	2.22	9.02	0.29	17.17	18.65	0.18	0.16	0.03	100.13
37.	1215_b_px-2k	64.91	51.92	0.22	0.97	12.71	0.35	13.19	20.47	0.23	0.10	0.01	100.18
39.	1215_a_px-1	74.06	50.10	0.71	3.70	10.47	0.27	16.77	17.18	0.19	0.23	0.02	99.66
40.	1215_a_px-2	71.61	51.11	0.86	2.91	11.82	0.33	16.72	16.51	0.19	0.08	0.01	100.55
41.	1225_b_ol-1	68.01	50.64	1.11	2.97	13.30	0.36	15.86	16.44	0.28	0.01	0.04	101.01

(continued)

Table 4.46 (continued)

№	№ sample (depth, m)	Mg#	SiO_2	TiO_2	Al_2O_3	FeO	MnO	MgO	CaO	Na_2O	Cr_2O_3	NiO	Total
42.	1217_a_px-1	75.41	52.30	0.70	2.63	10.36	0.31	17.82	16.84	0.18	0.07	0.02	101.23
43.	1217_a_px-2	81.04	51.81	0.61	3.44	7.08	0.18	16.97	19.99	0.12	0.39	0.05	100.64
44.	1217_a_px-3	81.61	51.59	0.60	3.74	6.68	0.18	16.63	20.42	0.12	0.68	0.04	100.68
45.	1217_a_px-3	81.64	51.69	0.62	3.77	6.79	0.18	16.94	20.15	0.18	0.67	0.04	101.04
46.	1217_a_px-4	79.14	52.67	0.60	3.10	8.52	0.25	18.13	18.06	0.15	0.32	0.05	101.86
47.	1217_a_px-5	76.96	52.04	0.71	3.01	9.18	0.26	17.20	18.53	0.18	0.13	0.03	101.27
48.	21232_b_px-1	80.89	51.42	0.71	2.96	7.20	0.20	17.10	19.59	0.21	0.20	0.03	99.64
49.	1232_b_px-2	82.08	52.09	0.57	3.33	6.91	0.19	17.75	19.22	0.14	0.53	0.03	100.77
50.	1232_b_px-3	76.86	53.61	0.87	1.68	14.68	0.35	27.36	2.39	0.00	0.20	0.04	101.20
51.	1232_b_px-4	77.41	53.73	0.82	1.68	14.44	0.35	27.76	2.23	0.04	0.30	0.05	101.41
52.	1232_b_px-5	77.11	53.74	0.79	1.76	14.59	0.33	27.56	2.32	0.02	0.18	0.05	101.35
53.	1232_8g-5k-Px-1	82.48	52.88	0.60	2.26	6.70	0.18	17.70	19.61	0.18	0.21	0.03	100.35
54.	1232_8g-5k-Px-2	80.25	52.60	1.14	2.26	7.84	0.21	17.87	18.54	0.21	0.08	0.03	100.79
55.	1232_8g-5k-Px-3	83.14	52.94	0.48	2.21	6.44	0.18	17.80	19.67	0.17	0.27	0.05	100.20
56.	1232_8g-5k-Px-4	80.69	52.43	1.23	2.11	7.58	0.21	17.76	19.08	0.27	0.03	0.05	100.75
57.	1232_8g-5k-Px-5	81.12	51.80	0.69	2.91	7.32	0.20	17.63	18.34	0.21	0.32	0.03	99.47
59.	1246_4-3k-Px-2	79.08	49.34	1.49	5.05	6.98	0.14	14.79	21.51	0.44	0.27	0.008	100.03
60.	1246_4-3k-Px-3	**81.68**	51.69	0.98	3.01	6.34	0.14	15.84	21.60	0.36	0.04	0.01	100.02
61.	1246_4-3k-Px-3 k	**70.68**	48.88	1.80	4.31	10.45	0.26	14.13	19.36	0.45	0.50	0.02	100.16
62.	1246_4-3k-Px-4	82.17	51.64	0.87	3.03	6.19	0.15	16.00	21.59	0.32	0.12	0.02	99.95
63.	1246_4-3k-Px-4 k	70.49	49.01	1.68	4.40	10.58	0.24	14.18	19.46	0.52	0.40	0.03	100.51
64.	1246_4-3k-Px-5	**79.14**	49.19	1.59	5.02	6.88	0.14	14.65	21.63	0.42	0.12	0.01	99.66
65.	1246_4-3k-Px-5 k	**68.93**	48.04	1.61	4.90	10.98	0.28	13.66	19.14	0.45	0.20	0.03	99.29
66.	1171_4b-2-Px-1	56.16	50.58	0.68	1.74	22.37	0.58	16.08	7.96	0.06	0.00	0.009	100.05
67.	1179-5a-Px-2 k	**53.30**	51.43	0.27	1.16	26.70	0.69	17.09	3.71	0.00	0.01	0.01	101.09
68.	1179_5a-3k-Px-2	**59.13**	51.52	0.81	1.91	23.31	0.60	18.91	4.33	0.24	0.02	0.008	101.64
69.	1208-5a-3k-Px-1	62.72	51.39	0.34	1.27	21.35	0.56	20.15	4.61	0.01	0.00	0.01	99.70

Table 4.47 Average mean composition of the Noril'sk Complex intrusions

Massif	Mikchangdinsky	Noril'sk 1	Talnakh		
Borehole	MD-48	G-22	KZ-1799	OUG-2	Σ29
Component	1	2	3	4	5
SiO_2	47.71	47.16	47.47	48.76	48.3
TiO_2	0.91	0.79	1.11	0.88	0.85
Al_2O_3	13.30	15.36	14.11	14.9	15.33
FeO	12.66	12.17	12.33	11.08	12.34
MnO	0.18	0.20	0.22	0.2	0.19
MgO	13.42	12.04	11.54	10.52	10
CaO	9.47	10.60	10.22	10.04	10.45
Na_2O	1.79	0.97	1.91	2.05	1.86
K_2O	0.45	0.39	0.56	0.65	0.58
P_2O_5	0.10	0.09	0.12	0.18	0.2
Cr_2O_3	–	0.23	0.13	0.14	0.1

Note: Data from: (1) Krivolutskaya et al. 2009a, b, (2) Krivolutskaya and Rudakova 2009, (3) Czamanske et al. 1994, (4) Krivolutskaya et al. 2001, (5) Dneprovskaya et al. 1987

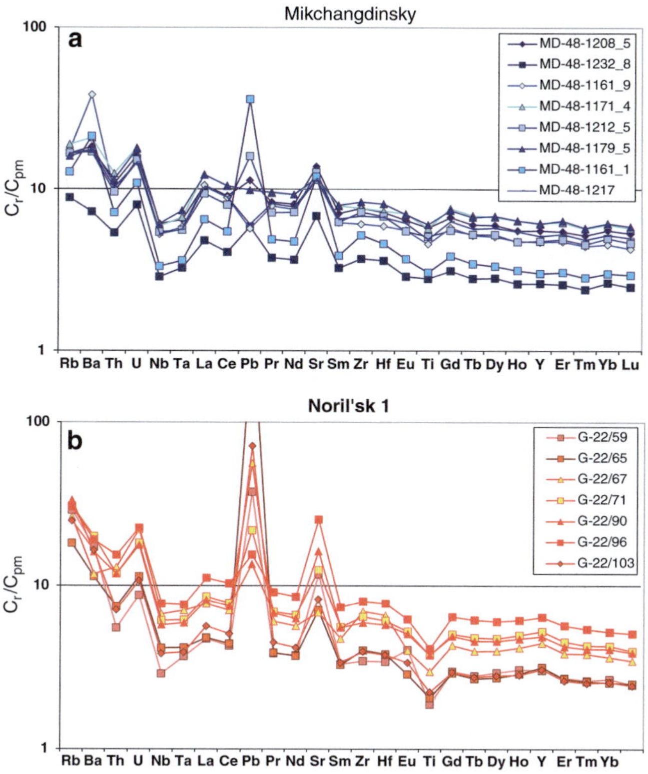

Fig. 4.76 Rare element patterns for rocks
(**a**) Mikchangdinsky massif, (**b**) Noril'sk 1 (After Krivolutskaya et al. 2009b)

The ore composition (Cu–Ni, PGE tenor) does not correlate with geochemistry or rocks. Massifs of the Noril'sk Complex are similar in terms of their internal structure.

References

Arndt NT, Czamanske GK, Walker RJ, Chauvel C, Fedorenko VA (2003) Geochemistry and origin of the intrusive hosts of the Noril'sk-Talnakh Cu-Ni-PGE sulfide deposits. Econ Geol 98:495–515

Barnes SJ, Kunilov VY (2000) Spinels and Mg ilmenites from the Noril'sk 1 and Talnakh intrusions and other Mafic rocks of the Siberian Flood Basalt Province. Econ Geol 95:1701–1717

Barnes S-J, Cox RA, Zientek ML (2006) Platinum-group element, gold, silver and base metal distribution in compositionally zoned sulfide droplets from the Medvezky Creek Mine, Noril'sk Russia. Contrib Mineral Petrol 152:187–200

Based Legend for geological map 1:50000 scale. Noril'sk seria (1993) Ed. V. Lyul'ko. Noril'sk (in Russian)

Begizov VD, Meshankina VI, Dubakina LS (1974) Palladoarsenide Pd2As – new natural arsenide of Pd from copper-nickel ores of the Oktyabr'skoe deposit. Vses Min Ob-vo. Ch. 103 1:104–107 (in Russian)

Budko IA, Kulagov EA (1968) New mineral talnakhite – cubic spice of chalcopyrite. Zap Vses Min Soc 97:63 (in Russian)

Cabri LJ (1988) Applications of proton and nuclear microprobe in ore deposit mineralogy and metallurgy. Nucl Instrum Methods Phys Res B30:459–465

Cabri LJ, Campbell JL, Laflamme JHG, Leigh RG, Maxwell JA, Scott JD (1985) Proton-microprobe analysis of trace elements in sulfides from some massive-sulfides deposits. Can Mineral 23:133–148

Chernova NA (1961) Differentiated trap intrusion in Zub Mount, Noril'sk region. Geol Ore Deposits 5:55–72 (in Russian)

Connolly HC, Zipfel J, Grossman JN et al (2006) The Meteoritical Bulletin, No.91. September. Meteoritics and Planetary Science. 41:1383–1418

Czamanske GK, Wooden JL, Zientek ML, Fedorenko VA, Zen'ko TE, Kent J, King BS, Knight RJ, Siems DF (1994) Geochemical and isotopic constraints on the petrogenesis of the Noril'sk-Talnakh ore-forming system. In: Lighfoot PC, Naldrett AJ (eds) Proceedings of the Sudbury-Noril'sk symposium, special publication 5, Geological Survey, Ontario, pp 313–342

Czamanske GK, Zen'ko TE, Fedorenko VA et al (1995) Petrographic and geochemical characterization of ore-bearing intrusions of the Noril'sk Type, Siberia; with discussion of their origin. Resour Geol Spec Issue 18:1–45

Distler VV (1994a) Platinum mineralization of the Noril'sk deposit, in geology and genesis of deposits of platinum metals, vol I. Nauka, Moscow, pp 3–341 (in Russian)

Distler VV (1994b) Platinum mineralization of the Noril'sk deposits. Proceedings of the Sudbury-Noril'sk symposium, Sudbury, Spec (5) 243–260

Distler VV, Dyuzhikov OA (1988) Formations of the Cu-Ni sulfide deposits. In: Genetic models for endogenic deposits. Nauka, Novosibirsk, pp 166–172 (in Russian)

Distler VV, Smirnov AV, Grokhovskaya TL et al (1979) Stratification, cryptic layering of differentiated trap intrusions, and conditions of the formation of sulfide mineralization. In: Conditions for the formation of magmatic ore deposits. Nauka, Moscow, pp 211–292 (in Russian)

Distler VV, Genkin AD, Grokhovskaya TL, Evstigneeva TL et al (1988) Petrology of the magmatic ore-forming process. Nauka, Moscow (in Russian)

Distler VV, Dyuzhikov OA, Kravtsov VF, Sluzhenikin SF, Turovtsev DM (1994) Low-sulfide platinum-formation of the Noril'sk District. In: Geology and genesis of deposits of platinum metals. Nauka, Moscow, pp 48–65 (in Russian)

Distler VV, Kulagov EA, Sluzhenikin SF, Laputina IP (1996) Quenched sulphide solutions in Noril'sk ores. Geol Ore Deposits 38:41–53

Distler VV, Sluzhenikin SF, Cabri LJ, Krivolutskaya NA, Turovtsev DM, Golovanova TA, Mokhov AV, Knauf VV, Oleshkevich OI (1999) Platinum ores of the Noril'sk Layered intrusions: magmatic and fluid concentration of noble metals. Geol Ore Deposits 41(3):214–237

Dneprovskaya MB, Frenkel MY, Yaroshevsky AA (1987) A quantitative model for layering in the Talnakh intrusion, Noril'sk Region. In: Simulating systems of ore mineralization). Nauka, Novosibirsk, pp 96–106 (in Russian)

Dobretsov NL, Kochkin YN, Krivenko AP, Kutolin VA (1971) Rock-forming pyroxenes. Nauka, Moscow (in Russian)

Dodin DA, Batuev BN (1971) Geology and petrology of the Talnakh differentiated intrusions and their metamorphic aureole. In: Petrology and ore resource potential of the Talnakh and Noril'sk differentiated intrusions. Nedra, Leningrad, pp 31–100 (in Russian)

Dodin DA, Batuev BN, Mitenkov GA et al (1971) Atlas of the rocks and ores of the Noril'sk copper–nickel deposits. Nedra, Leningrad (in Russian)

Dodin DA, Sluzhenikin SF, Bogomolov MA (2009) Ores and minerals of Noril'sk (unpublished, electron version, with authors' permission), Moscow (in Russian)

Dyuzhikov OA, Distler VV, Strunin BM et al (1988) Geology and ore potential of the Noril'sk ore district. Nauka, Moscow. Translated: Dyuzhikov OA, Distler VV, Strunin BM, Mkrtychyan AK, Sherman ML, Sluzhenikin SF, Lurye AM (1992) Geology and metallogeny of

sulfide deposits Noril'sk region USSR. Econ Geol Monogr 1, Spec vol 241, Ontario
Evstigneeva TL (2010) The "non-stoichiometry" of platinum group minerals. In: Eleventh international platinum symposium, Abstr., Ontario Geol. Surv., http://11ips.laurentian.caFilimonova
Evstigneeva TL, Genkin AD, Troneva NV, Filimonova AA, Tsepin AI (1973) Shadlunite – new sulfide of Cu, Fe, Pb, Mn and Cd from copper-nickel sulfide ores. Zap Vses Min Soc.Ch. 102:63–74 (in Russian)
Evstigneeva TL, Genkin AD, Kovalenker VA (1975) New mineral of Bi and Pd – sobolevskite and nomenclature of minerals from system PdBi — PdTe — PdSb. Zap Vses Min Ob-va Ch. 104:568–579 (in Russian)
Filimonova AA, Murav'eva IV, Evstigneeva TL (1974) Minerals from chalcopyrite group in Cu-Ni ores of the Noril'sk deposits. Geol Ore Deposits (5) 31–47 (in Russian)
Filimonova AA, Evstigneeva TL, Laputina IP (1980) Putoranite and Ni-putoranite – new minerals in chalcopyrite group. Zap Vses Min Soc Ch. 109. 3:335–341 (in Russian)
Genkin AD (1968) Minerals of platinum metals in ores of the Noril'sk 1 deposit. Nauka, Moscow (in Russian)
Genkin AD, Zvyagintsev OE (1962) Vysotskite, new sulfide of palladium and nickel. Zap Vses Min Soc 91:718–725 (in Russian)
Genkin AD, Murav'eva IV, Troneva NV (1966) Zvyagintsevite, natural intermetallic allow Pd, Pt, Pb and Sn. Geol Ore Deposits 8(3): 94–100 (in Russian)
Genkin AD, Evstigneeva TL, Troneva NV,Vyal'sov LP (1969) Polyarite – new mineral from copper-nickel sulfide ores. Zap Vses Min Soc 98:708–715 (in Russian)
Genkin AD, Evstigneeva TL, Vyal'sov LP, Laputina IP, Troneva NV (1970). Plumbopalladinite –Pd_3Pb_2- new mineral from copper-nickel sulfide ores. Geol Ore Deposits (5) 63–68 (in Russian)
Genkin AD, Evstigneeva TL, Vyal'sov LP, Laputina IP, Troneva NV. (1974) Paolovite – new mineral from copper-nickel sulfide ores. Geol Ore Deposits (1) 21–32 (in Russian)
Genkin AD, Evstigneeva TL, Troneva NV, Vyal'sov LP (1976) Mayakite- PdNiAs – new mineral from copper-nickel sulfide ores. Zap Vses Min Soc 105:698–703 (in Russian)
Genkin AD, Distler VV, Gladyshev GD et al (1981) Sulfide copper–nickel ores of the Noril'sk deposits. Nauka, Moscow, 281 p (in Russian)
Geology and ore deposits of the Noril`sk region (1994) In: Distler VV, Kunilov VE (eds) Guidebook of the VII international platinum symposium Moscow-Noril`sk. Moskovsky contact Press, Moscow
Ginzburg VL, Rogover GB (1960) Regularities of noble and nonferrous metals' distribution in main ore-forming and silicate minerals. Sovetskaya Geol 3:27–30 (in Russian)
Godlevsky MN (1959) Traps and ore-bearing intrusions of the Noril'sk District. Moscow, Gosgeoltekhizdat, 61 p, (in Russian)
Godlevsky MN, Shumskaya MI (1960) Chalcopyrite-millerite ores from Noril'sk 1 deposit. Geol Ore Deposits (2) 61–72 (in Russian)
Gongalsky BI, Krivolutskaya NA (1993) Chineysky layered pluton. Nauka, Novosibirsk, 187 p. (in Russian)
Grinenko LN (1967) Sulfur isotope composition of some Cu–Ni deposits of the Siberian platform. In: Petrology of trap basalts of the Siberian platform. Nedra, Moscow, pp 221–230 (in Russian)
Hellstrom J, Paton C, Woodhead J, Hergt J (2008) Iolite: software for spatially resolved LA-(quadrupole and MC)-ICP-MS analysis. In: Sylverster P (ed) Laser ablation ICP-MS in the earth sciences: current practices and outstanding issues. Mineral Assoc Can Short Course Ser. 40. 343–348
Hess, HH et al (1941) Pyroxenes of common Mafic Magmas. Am Mineral 26(9):515–535, (10):573–594
Hoernle KA, Tilton GR (1991) Sr–Nd–Pb isotope data for Fuerteventura (Canary Islands) basal complex and subaerial volcanics: applications to magma genesis and evolution. Schweiz Miner Petrogr Mitt 71:3–18
Hofmann AW (1988) Chemical differentiation of the earth: the relationship between mantle, continental crust and oceanic crust. Earth Planet Sci Lett 90:297–314
Jarosevich EJ, Nelen JA, Norberg JA (1980) Reference sample for electron microprobe. Geostand Newslett 4:43–47
Jochum KP, Dingwell DB, Rocholl A et al (2000) The preparation and preliminary characterisation of eight geological MPI-DING reference glasses for in-situ microanalysis. Geostand Newslett 24:87–133
Kamenetsky VS, Kamenetskaya MB, Sobolev AV et al (2008) Olivine in the Udachnaya-East kimberlite (Yakutia, Russia): morphology, compositional zoning and origin. J Petrol 49:823–839
Kamo SL, Czamanske GK, Amelin Y et al (2003) Rapid eruption of Siberian flood-volcanic rocks and evidence for coincidence with the Permian-Triassic boundary and mass extinction at 251 Ma. Earth Planet Sci Lett 214:75–91
Knauf VV (1996) On the metrologic provision of mineralogical studies. Zap Vses Mineral O-va 6:109–113 (in Russian)
Korovyakov IA, Nelyubin AE, Raikova ZA, Khortova LK (1963) The origin of Noril'sk trap intrusions containing copper–nickel ores. Gosgeotekhizdat, Moscow. Trudy Vses Nauch–Issled Inst Miner Syr'ya, 101p (in Russian)
Kovalenker VA, Genkin AD, Evstigneeva TL, Laputina IP (1974) Telargpalite – mew mineral of Pd, Ag, and Te from Cu-Ni ores of the Oktyabr'skoe deposit. Zap Min Soc 5:5 (in Russian)
Kovalenker VA, Laputina IP, Evstigneeva TL, Izoitko VM (1976) Talkusite, $Cu_{3-x}Tl_2Fe_{1+x}S_4$ – new sulfide of Tl from copper-nickel sulfide ores. Zap Vses Min Soc 105:202–206 (in Russian)
Kozyrev SM (2002) Mineralogy and behavior of platinum group minerals in the process of ore enrichment. In: Cabri LJ (ed) The geology, geochemistry, mineralogy and benefication of platinum-group elements
Krivolutskaya NA (2011) Formation of PGM–Cu–Ni deposits in the process of evolution of flood-basalt magmatism in the Noril'sk Region. Geol Ore Deposits 53(4):309–339
Krivolutskaya NA (2014) Evolution of trap magmatism and processes producing Pt-Cu-Ni mineralization in the Noril'sk area. KMK, Moscow, 320 p, (in Russian)
Krivolutskaya NA, Rudakova AV (2009) Structure and geochemical characteristics of trap rocks from the Noril'sk Trough, Northwestern Siberian Craton. Geochem Int 47(7):635–656
Krivolutskaya NA, Sobolev AV (2001) Magmatic inclusions in olivines from intrusions of the Noril'sk Region, Northwestern Siberian Platform: evidence for primary melts. Dokl Earth Sci 381(3):1047–1052
Krivolutskaya NA, Ariskin AA, Sluzhenikin SF, Turovtsev DM (2001) Geochemical thermometry of rocks of the Talnakh Intrusion: assessment of the melt composition and the crystallinity of the parental magma. Petrology 9(5):389–414
Krivolutskaya NA, Sobolev AV, Kuzmin DV, Svirskaya NM (2009a) Unique Zoned Olivines from an Ultrabasic—basic Massif in the Noril'sk District. Doklady Earth Sci 429A(9):1496–1500
Krivolutskaya NA, Mikhailov VN, Snisar SG, Gongalsky BI (2009b) Internal structure and composition of the Mikchangdinsky basic-ultrabasic massif in the Noril'sk area (Siberian trap provinces). Vestnik KRAUNTS № 14. C, pp 29–48 (in Russian)
Krivolutskaya NA, Gongalskiy BI, Yushin AA, Schlychkova TB, Kononkova NN, Tushentsova IN (2011) Mineralogical and geochemical characteristics of PGE-Cu-Ni ores of the Maslovsky deposit in the Noril'sk area, Russia. Can Mineral 49:1649–1674
Krivolutskaya NA, Sobolev AV, Snisar SG et al (2012) Mineralogy geochemistry and stratigraphy of the Maslovsky Pt–Cu–Ni sulfide deposit, Noril'sk Region, Russia: implications for relationship of ore-bearing intrusions and lavas. Miner Deposita 47:69–88
Kulagov EA, Evstigneeva TL, Yushko-Zakharova OE (1969) New nickel sulfide – godlevskite. Geol Ore Deposits 3:115–121 (in Russian)

Kursky AN, Vitozhents GC, Zdorova EP (1992) Methods for analysis for PGE. Rudy i Metally 3:27–46 (in Russian)

Large R, Danyushevsky L, Hollit C, Maslennikov V, Meffre S, Gilbert S, Bull S, Scott R, Emsbo P, Thomas H, Singh B, Foster J (2009) Gold and trace element zonation in pyrite using a laser imaging technique: implications for the timing of gold in orogenic Carlin-style sediment-hosted deposits. Econ Geol 104:635–668

Li C, Riplay EM, Naldrett AJ (2003) Compositional variations of olivine and sulfur isotopes in the Noril'sk and Talnakh intrusions. Econom Geol 98:69–86

Li CS, Ripley EM, Naldrett AJ (2009) A new genetic model for the giant Ni–Cu–PGE sulfide deposits associated with the Siberian flood basalts. Econ Geol 104:291–301

Likhachev AP (1965) The role of Leucocratic Gabbro in the origin of Noril'sk differentiated intrusions. Izv Akad Nauk SSSR Ser Geol (12):50–66 (in Russian)

Likhachev AP (1977) On the crystallization conditions of trap magmas in the Northwestern part of the Siberian Platform. Zap Vses Miner O–va 106:594–606 (in Russian)

Likhachev AP (1994) Ore-bearing intrusions of the Noril'sk Region. In: Proceeding of the Sudbury–Noril'sk symposium, Ontario Geol Surv Spec 5:185–201

Likhachev AP (1997) Trap magmatism and platinum–copper–nickel ore mineralization in the Noril'sk district. Otech Geol 10:8–19 (in Russian)

Likhachev AP (1996a) The Kharaelakh intrusion and its PGM–Cu–Ni ores. Rudy et Metals. (3):48–62 (in Russian)

Likhachev AP, (1996b) Emplacement dynamics of the Talnakh ore-bearing intrusions and related PGM–Cu–Ni Ores. Otechestvennay Geologiya (8):20–26 (in Russian)

Likhachev AP (2006) Platinum–copper–nickel and platinum deposits. Eslan, Moscow, 496 p, (in Russian)

Likhachev AP (2015) Composition and genesis of olivines from magmatic rocks in the Noril'sk area. Otechestvennay Geologiya 2:80–95 (in Russian)

Malich KN, Latypov RM (2011) Re–Os and S isotope constraints on timing and source heterogeneity of PGE–Cu–Ni sulfide ores: a case study at the Talnakh ore junction, Noril'sk province, Russia. Can Mineral 49:1653–1677

Malitch KN, Belousova EA, Griffin WL, Badanina IY, Pearson NJ, Presnyakov SL, Tuganova EV (2010) Magmatic evolution of the ultramafic–mafic Kharaelakh intrusion (Siberian Craton, Russia): insights from trace-element, U–Pb and Hf-isotope data on zircon. Contrib Miner Petrol 159:753–768

Malitch KN, Belousova EA, Griffin WL, Badanina (2013) Hafnium-neodymium constraints on source heterogeneity of the economic ultramafic-mafic Noril'sk-1 intrusion (Russia). Lithos. 03 164–167:36–46

Müller W, Shelley M, Miller P, Broude S (2008) Initial performance metrics of a new custom-designed ArF excimer LA-ICPMS system coupled to a two-volume laser-ablation cell. J Anal Atmos Spectrom 24:209–214

Murav'eva IV, Evstigneeva TL, Filimonova AA, Malov VS (1972) The first find of moikhukite in the Cu-Ni ores from the Oktyabr'skoe deposit, Noril'sk area. Geol Ore Deposits 3:27–29 (in Russian)

Naldrett AJ (1992) Model for the Ni–Cu–PGE ores of the Noril'sk Region and its application to other areas of flood basalts. Econ Geol 87:1945–1962

Naldrett AJ (2004) Magmatic sulphide deposits. Geology, geochemistry and exploration. Springer, Berlin/Heidelberg/New York, 727 p

Naldrett AJ (2009) Ore deposits related to flood basalts, Siberia. In: New developments in magmatic Cu–Ni and PGE deposits. Geological Publishing House, Beijing, pp 141–179

Naldrett AJ, Lightfoot PC (1999) Ni-Cu-PGE deposits of the Noril'sk region, Siberia: their formation in conduits for flood basalt volcanism. Geological association of Canada. Short Course Notes 13:195–250

Naldrett AJ, Lightfoot PC, Fedorenko VA et al (1992) Geology and geochemistry of intrusions and Flood Basalts of the Noril'sk region, USSR, with implication for origin of the Ni–Cu ores. Econ Geol 87:975–1004

Naldrett AJ, Fedorenko VA, Lightfoot PC, Kunilov VE, Gorbachev NS, Doherty W, Johan Z (1995) Ni-Cu-PGE deposits of the Noril'sk region: their formation in conduits for flood basalt volcanism. Trans Inst Mining Metall 104:B18–B36

Naldrett AJ, Fedorenko VA, Asif M, Lin S, Kunilov VE et al (1996) Controls on the composition of Ni-Cu sulfide deposits as illustrated by those at Noril'sk, Siberia. Econ Geol 91:751–773

Natorkhin IA, Arkhipova AI, Batuev BN (1977) Petrology of the Noril'sk type intrusions. Nedra, Leningrad, 236 p (in Russian)

Pavlov VE, Courtillot V, Bazhenov ML, Veselovsky RV (2007) Paleomagnetism of the Siberian traps: new data and a new overall 250 Ma pole for Siberia. Tectonophysics 443:72–92

Pavlov VE, Veselovskiy RV, Latyshev AV, Fluteau F, Fetisova AM (2011) Secular geomagnetic variations and volcanic pulses in the Permian-Triassic traps of the Norilsk and Maimecha-Kotui provinces. Izv Phys Solid Earth 47:402–417

Petrov YuM, Malich NS, Tuganova EV, Savushkin MP et al (2001) Legend to Igarka-Noril'sk serial Geological maps 1:200000 scale, vol. 1, 2. St-Petersburg (in Russian)

Rad'ko V A (1991) Model of dynamic differentiation of intrusive traps at the Northwestern Siberian trap. Geol Geophys 32(11):19–27

Razin LV, Borishanskaya SS (1970) Mineral occurrence forms of platinum metals and gold in the disseminated ores of the Noril'sk I deposit. Tr. TSNIGN, issue 87 (in Russian)

Rudakova AV, Krivolutskaya NA (2009) Structural and textural specific features of rocks of Trap Rock Association in the Norilsk Trough (NW of Siberian Platform). Moscow Univ Geol Bull 64(6):364–375

Ryabov VV (1984) About composition of upper horizons of the Noril'sk intrusions with rich chromite mineralization. In: Criteria of ore potential of magmatic rocks, Novosibirsk, pp 124–142 (in Russian)

Ryabov VV (1992) Olivines from Noril'sk intrusions as criteria of petrogeneses and ore producing process. Nauka, Novosibirsk (in Russian)

Ryabov VV, Zolotukhin VV (1977) Minerals of differentiated traps. Nauka, Novosibirsk, 392 p (in Russian)

Ryabov VV, Tsimbalist VG, Yakobi IY (1982) About PGE and chromium concentrations in the upper zones of the Noril'sk intrusions. Doklady Earth Sci 266(2):466–469 (in Russian)

Ryabov VV, Shevko AYa, Gora MP (2000) Igneous rocks of the Noril'sk district. V.1. Petrology of traps. Nonpareil Novosibirsk. V 1,2 (in Russian)

Ryabov VV, Shevko AY, Gora MP (2014) Trap magmatism and ore formation in the Siberian Noril'sk region.V 1, 2. Springer

Sarbadhikari AB, Day JMD, Lui Y et al (2009) Petrogenesis of olivine-phyric shergottite Larkman Nunatac 0613: implication for enriched components in martian basalts. Geochim Cosmochim Acta 73:2190–2214

Sluzhenikin SF, Distler VV (1998) Disseminated ores of the Talnakh ore field as a source of platinum metals. In: Dodin D (ed) Large and unique deposits of rare and noble metals, St Petersburg, pp 247–256 (in Russian)

Sluzhenikin SF, Mokhov AV (2015) Gold and silver in PGE–Cu–Ni and PGE ores of the Noril'sk deposits, Russia. Miner Deposita 50:465–492

Sluzhenikin SF, Distler VV, Dyuzhikov OA, Kravtsov VF, Kunilov VE, Laputina IP, Turovtsev DM (1994) Low-sulfide platinum mineralization in the Noril'sk differentiated intrusive bodies. Geol Ore Deposits 36:195–217

Sluzhenikin SF, Krivolutskaya NA, Rad'ko VA, Malich KN, Distler VV, Fedorenko VA (2014) Ultramafic-mafic intrusions, volcanic rocks and PGE-Cu-Ni sulfide deposits of the Noril'sk Province, Polar Siberia field trip guidebook. Yekaterinburg, 83 p

Smirnov MF (1966) Structure of Noril'sk Ni-bearing intrusions and their sulfide ores. Nedra, Moscow (in Russian)
Sobolev AV, Hofmann AW, Kuzmin DV et al (2007) Amount of recycled crust in sources of mantle-derived melts. Science 316:412–417
Sotnikov AA (1919) About exploration of the Noril'sk (Dudinsky) coal and copper deposit in term of creating of Northern Marine Route. Gubernsky typography, Tomsk 54 p (in Russian)
Spiridonov EM, Gritsenko YD (2009) Epigenetic low-grade metamorphism and Co-Ni-Sb-As mineralization in the Noril'sk ore area. Nauchny mir, Moscow, 214 p. (in Russian)
Stekhin AI (1994) Mineralogical and geochemical characteristics of the Cu-Ni ores of the Oktyabr'skoe and Talnakh deposits. In: Proceeding of the Sudbury-Noril'sk symposium, OGS special, vol 5, pp 217–230
Stoll B, Jochum KP, Herwig K, Amini M, Flanz M, Kreuzburg B, Kuzmin D, Willbold M, Enzweiler J (2008) An automated iridium-strip heater for LA-ICP-MS bulk analysis of geological samples. Geostand Geoanal Res 32:5–26
Sukhanova EN (1968) Zoning of orebodies, intrusions, and tectono-magmatic clusters and its applied implications. In: Geology and mineral resources of the Noril'sk mining district. NTO Tsvetmet, Noril'sk, pp 139–142 (in Russian)
Sylvester PJ, Cabri LJ, Tubrett MN, McMahon G, Laflamme JHG, Peregoedova A (2005) Synthesis and evaluation of a fused pyrrhotite standard reference material for platinum group element and gold analysis by laser ablation-ICPMS. 10th international platinum symposium, Oulu, August 7–11. Abstracts Volume, 16–20
Todt W, Cliff RA, Hanser A, Hofmann AW (1996) Evaluation of 202Pb-205Pb double spike for high-precision: earth processes: reading the isotope code. Geophys Monogr 95:429–436
Torgashin AS (1994) Geology of the massive and copper ores of the western part of the oktyabr'skoe deposit. In: Proceeding of the Sudbury-Noril'sk symposium, OGS special, vol 5, pp 231–242
Turovtsev DM (2002) Contact metamorphism of the Noril'sk intrusions. Nauchny Mir, Moscow, 293 p. (in Russian)
Ulrich T, Kamber BS, Jugo PJ, Tinkham DK (2009) Imaging element-distribution patterns in minerals by laser ablation – inductively coupled plasma – mass spectroscopy (LA-ICP-MS). Can Mineral 47:1001–1012
Urvantsev NN (1972) Some questions on formation of ore-bearing intrusion and ore. In: Cu-Ni ores of Talnakh ore junction. Nedra, Leningrad, pp 100–144 (in Russian)
Urvantsev NN (1974) Northern Siberian nickel-bearing area. Geol Gefiz. (3):3–11 (in Russian)
Ustinov VI, Grinenko VA (1965) Precision mass-spectroscopy method of determination of sulfur isotope. Nedra, Moscow (in Russian)
Woodhead JD, Hellstrom J, Hergt JM, Greig A, Mass R (2007) Isotopic and elemental imaging of geological materials by laser ablation inductively coupled plasma – mass spectroscopy. Geostd Geoanal Res 31:313–343
Woodhead JD, Hellstrom J, Paton C, Hergt J, Greig A, Maas R (2008) A guide to depth profiling and imaging application of LA-ICP-MS. In: Sylvester P (ed) Laser ablation ICP-MS in the earth sciences: current practices and outstanding issues. Mineral Assoc Can Short Course Ser 40, pp 135–145
Yakovleva ME (1947) Differentiated gabbro-diabase intrusion in Mount Chernaya. Doklady Acad Sci USSR (3):9–12 (in Russian)
Zen'ko TE, Czamanske GK (1994) Special and petrologic aspects of the Noril'sk and Talnakh ore junctions. In: Lightfoot PC, Naldrett AJ (eds) Proceedings of the Sudbury-Noril'sk symposium Ontario, 263–282
Zolotukhin VV (1997) Basaltic pegmatoids of the Noril'sk ore-bearing intrusions and the problem of its origin. Trudy IGG. Novosibirsk 834:90 p (in Russian)
Zolotukhin VV, Ryabov VV, Vasil'ev YR, Shatkov VA (1975) Petrology of the Talnakh ore-bearing differentiated trap intrusion. Nauka, Novosibirsk (in Russian)

5 Composition of the Parental Melts for the Intrusions

The compositions of the initial melts of the ore-bearing intrusions were studied based on the inclusions in the olivines and pyroxenes (major, rare and volatile components). These have compositions approximating those of the rocks. The magmas contained 0.5–0.7 % H_2O, up to 0.2 % Cl, and 300 ppm F. The magma composition of the Talnakh intrusion was estimated with the aid of KOMAGMAT software. It was demonstrated that the magma contained 10 % olivine and 7 % plagioclase crystals and that MgO content in melt was 8 wt %.

It has long been believed that the ores hosted in ultrabasite–basite intrusive bodies could have been produced by unusual picritic ore-bearing magmas, and this was reflected in the early models for the genesis of the Noril'sk deposits (Godlevsky 1959; Pertology and Ore 1978; Dyuzhikov et al. 1988). Indeed, such magmas are known to occur in nature and include the Fe-rich melts of the volcano El Laco in Chile (Frutos and Oyarzun 1975), magnetite lavas in Siberia (Ryabov and Pavlov 1991), ore-bearing melts in the Eastern Pacific in the area of the Juan de Fuca Ridge (Zhmodik 2002), etc. However, according to a hypothesis that became widely accepted over the past two decades, Cu–Ni deposits can be produced by melts of tholeiitic composition in an open system (Rad'ko 1991; Naldrett 1992, 2009; Naldrett et al. 1992, 1996; Li et al. 2009; Ariskin et al. 2009; 2013).

Differences in the estimated role of magmatism in the ore-forming processes largely stem from differences in the methodical approaches applied in evaluating the compositions of the parental melts of both the intrusions and the lavas. In the Noril'sk district, these have always been calculated as the weighted mean compositions of individual intrusive bodies (Dneprovskaya et al. 1987; Dneprovskaya and Dneprovsky 1988) or were thought to be represented by certain crystallization products of magmas, such as (1) picrite basalts and picrite–gabbro-dolerites (Dyuzhikov et al. 1988; Ryabov et al. 2000) or (2) gabbro-dolerites that form sills branching from ore-bearing massifs (Zolotukhin et al. 1986; Naldrett 1992). However, the former are cumulate rocks, and relationships between the latter and the intrusions were far from always reliably documented.

In light of this, we have recently attempted to utilize other techniques to obtain information on the parental melts of certain massifs in the Noril'sk ore district. These techniques involved (1) the direct evaluation of their composition by studying of the melts (naturally or experimentally quenched into glass) in olivines from picrites and picritic gabbro-dolerites in certain intrusions and (2) geochemical thermometry with the COMAGMAT-3.5 program package.

5.1 Melt Compositions Evaluated from Data on Melt Inclusions

Over the past two decades, the recovery of direct information on melt chemistries became possible through the study of melt inclusions in minerals due to advancements in modern high-resolution analytical techniques (Sobolev 1996, 1997; Inclusions in minerals…. 2005; Bodner and Student 2006; Ryabchikov et al. 2009 et ctr.). However, studies of PGE and Cu–Ni deposits worldwide still involve very few models that account for the genesis of these deposits on the basis of melt inclusion data (Bulgakova 1971; Vortsepnev 1978; Krivolutskaya and Sobolev 2001; Krivolutskaya et al. 2001b; Konnikov et al. 2005). Most of these studies were performed with data on crystalline and fluid inclusions (Ballhaus and Stumpfl 1986; Boudreau et al. 1986; Boudreau 1988). Publications devoted to the genesis of Siberian flood basalts with regard to melt inclusion data have dealt mostly with alkaline and subalkaline rocks in the province (Sobolev and Slutsky 1984; Ryabchikov et al. 2001, 2009; Sobolev et al. 2009a, b; Black et al. 2014).

5.1.1 Inclusions in Olivine

New data on the composition of magmatic inclusions in olivines from rocks related to the main stage of flood basalt

N.A. Krivolutskaya, *Siberian Traps and Pt-Cu-Ni Deposits in the Noril'sk Area*,
DOI 10.1007/978-3-319-17205-7_5

eruption are presented in this chapter. The composition of magmatic inclusions was studied based on olivines from the ore-bearing Talnakh, Kharaelakh, Northern Maslovsky, and Noril'sk 1 intrusions, as well as the barren Low Talnakh and Zelenogrivsky intrusions (Krivolutskaya and Sobolev 2001) and picrites of the Gudchikhinsky Formation (Sobolev et al. 2009a).

The following types of magmatic inclusions were detected in olivines (Krivolutskaya 2011; Fig. 5.1): (1) melt inclusions – glassy, partially crystallized (glass+gas+solid phase, including orthopyroxene, amphibole, ilmenite, and apatite) and completely crystallized (gas+solid phases) melt inclusions; (2) crystalline inclusions consisting of plagioclase, albite, pyroxene, Cr–spinel, and apatite; (3) fluid inclusions; and (4) composite inclusions (solid phase+melt or fluid). The crystallized melt inclusions were used to determine the initial melts; the glassy and fluid inclusions were used to determine the volatile components; and crystalline inclusions were used to estimate the temperature during magma crystallization. Despite general similarities, differences exist in the proportions of the aforementioned types of inclusions between particular intrusions. In the Talnakh, Kharaelakh, Low Talnakh, and Noril'sk 1 intrusions, the predominant role of melt in comparison with the other types of inclusions.

Melt inclusions are the most abundant (80 % of all inclusions). They differ in size and structure. Small inclusions (10–12 μm) are commonly glassy, whereas large inclusions (40–125 μm) are crystallized and are oval or close to negative crystals in shape. Orthopyroxene, ilmenite, pyrrhotite, chalcopyrite, amphibole, and phlogopite occur as daughter minerals. The crystallized inclusions are coarse or fine grained, often with a very fine-grained rim (5 μm) around a coarsely crystalline core (40–50 μm). To determine the compositions of the melt in the inclusions, they were heated to the melting temperature of the last daughter mineral (1,200–1,250 °C). The samples were held at $T>1{,}200$ °C for no longer than 5 min. In most runs, partial homogenization (melt+fluid) was achieved. Subsequently, the samples were quenched and polished. The concentrations of major elements in the melts and host olivines were determined by a microprobe. The homogeneity of the melts obtained in experiments was verified through the comparison of measurements at different points. The measured compositions of homogenized inclusions were recalculated to equilibrium with the host mineral using the PETROLOG-2 program (Danyushevsky et al. 1999). According to the calculations, the oxygen fugacity (= −7.4 to −7.5) was close to the NNO buffer (±0.5) at *T*hom = 1,200–1,250 °C. The data obtained (Table 3.10) show that the inclusions in the olivines from picrites of the Gudchikhinsky and Tuklonsky formations are characterized by a narrow range of MgO contents (8–10 wt %). The inclusions from intrusive rocks reveal a wider range (3–11 wt % MgO), indicating melt fractionation (Table 5.1): the Al_2O_3 and SiO_2 contents are correlated (Fig. 5.2a). In addition, the melt inclusions are enriched in titanium (Fig. 5.2b) and alkali metals. Thus, intratelluric olivine crystallized from an evolved melt. The Cl contents in the inclusions are low and are correlated with the K contents. Such a correlation is most typical of the Noril'sk 1 intrusion (Fig. 5.3). The Cl concentrations in the ore-bearing and barren intrusions are similar. As was shown by Krivolutskaya and Sobolev (2001), the distribution patterns of trace elements normalized to the composition of the primitive mantle are divided into two types: (1) close to that of the host rocks but differing due to a slight Eu minimum (Zelenogrivsky and Low Talnakh intrusions and the picrites of the Gudchikhinsky Formation) and (2) obviously distinct from the host rocks due to higher concentrations of all trace elements (Talnakh, Kharaelakh, and Noril'sk 1 intrusions). This situation is illustrated by Fig. 5.4, which shows the results of a picritic gabbro-dolerite from the Noril'sk 1 intrusion (a sample taken from Borehole G-22 at a depth of 64 m) as an example. The trace element contents of the rock are much lower than the inclusions hosted in the olivine (average of 21 inclusions, unpublished data of the author). Thus, the high concentrations of trace elements are characteristic of the melts forming ore-bearing intrusions. The degree of enrichment in trace elements depends on the Mg# of the host mineral, which reflects the progress of the fractionation of the initial magma.

Crystalline inclusions are typical of olivine. Small grains (1–3 to 10–12 μm) are distinguished by fused faces, whereas larger inclusions (>15 μm) are euhedral. Cr–spinel (largely, Cr-magnetite) inclusions in olivine make it possible to determine the oxygen fugacity of the melt (Table 5.2). The Cr–spinels from the intrusions and picrobasalts are markedly different in composition. The Cr–spinel inclusions in the olivine from picritic gabbro-dolerite of the Low Talnakh intrusion are enriched in V_2O_3. The corresponding data points occupy a special field in the $V_2O_3 - Fe^{2+}/Fe^{3+}$/diagram (Fig. 5.5). The ferrous/ferric ion ratio is calculated from the stoichiometry of the spinel. The compositional fields of the Cr–spinels incorporated into the olivine of picrites of the Gudchikhinsky and Tuklonsky formations overlap and are separated from the Cr–spinel fields of the intrusive rocks. The variation in composition of minerals from the Tuklonsky Formation is much wider. The Cr–spinels from picritic gabbro-dolerite of the Talnakh, Imangdinsky, and Noril'sk 1 intrusions are enriched in chromium, Cr# = Cr/(Al + Cr) = 0.77–0.81, whereas the Cr# of the spinels from the Lower Talnakh intrusion is 0.46–0.57. Picrites of the Tuklonsky and Gudchikhinsky formations are characterized by transitional Cr# of spinels (0.65, on average). The spinels from the Talnakh intrusion are distinguished by extremely high Ti contents (6.77 wt %, on average) and are classified as Ti–chromite. Plagioclase is the most abundant mineral occurring as inclusions in the olivine. The fused grains of this mineral are up to 40 μm in size and are especially frequent in the olivine from intrusions of the Talnakh ore

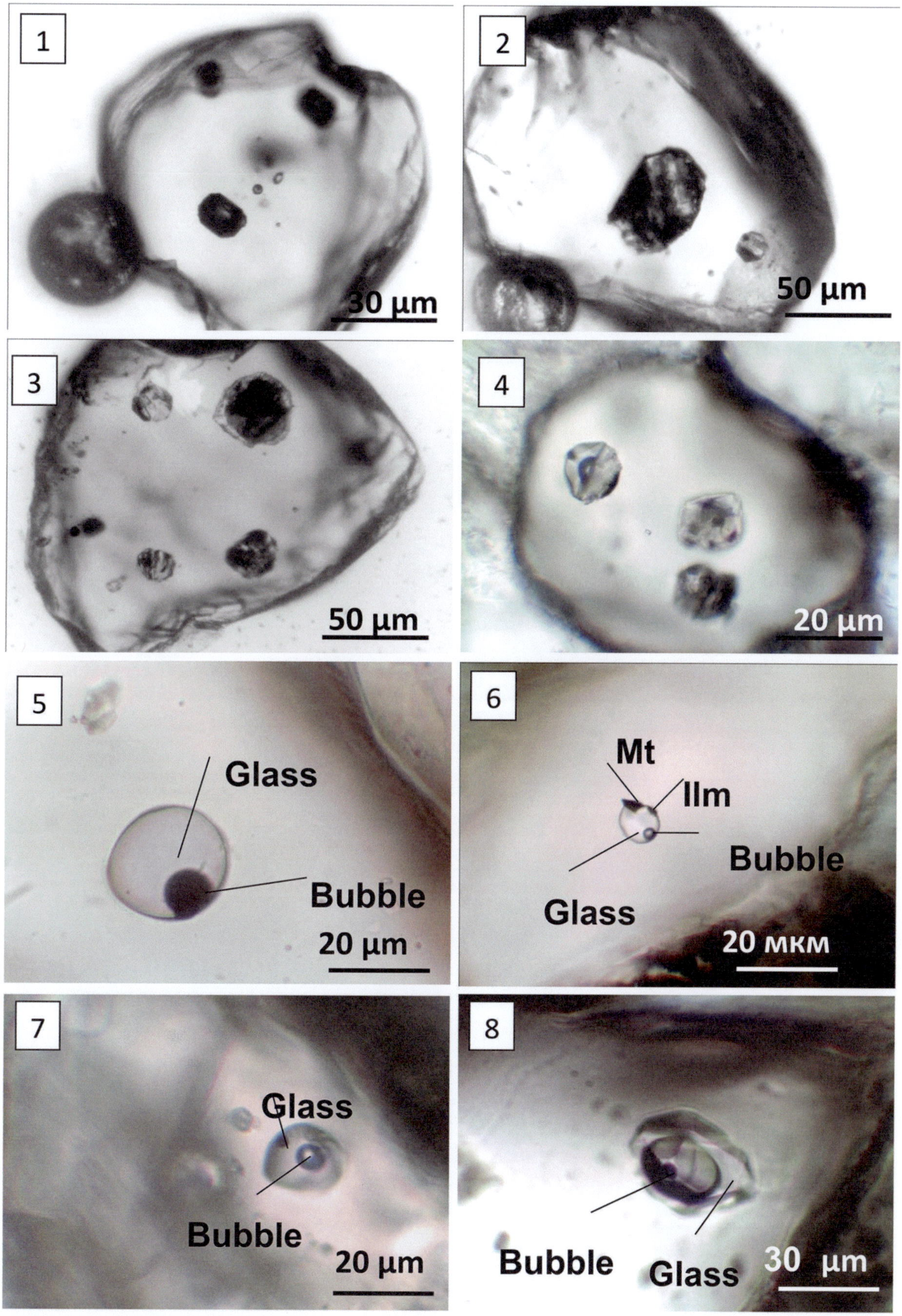

Fig. 5.1 Photomicrographs of magmatic inclusions in olivines
(1–4) general view of olivine grains with inclusions. (5–14) Melt inclusions: (5–8) glassy, (9–11) partly crystallized, (12–14) completely crystallized. (15, 16) Composite inclusions: (15) Cr–spinel + fluid, (16) Cr–spinel + glass. (17–20) Fluid inclusions. (21–24) Crystalline inclusions: (21) plagioclase, (22) Cr–spinel, (23) plagioclase, (24) apatite. *Opx* orthopyroxene, *Pl* plagioclase, *Ap* apatite, *Cr-Mt* chrome magnetite, *Ilm* ilmenite. After Krivolutskaya (2011)

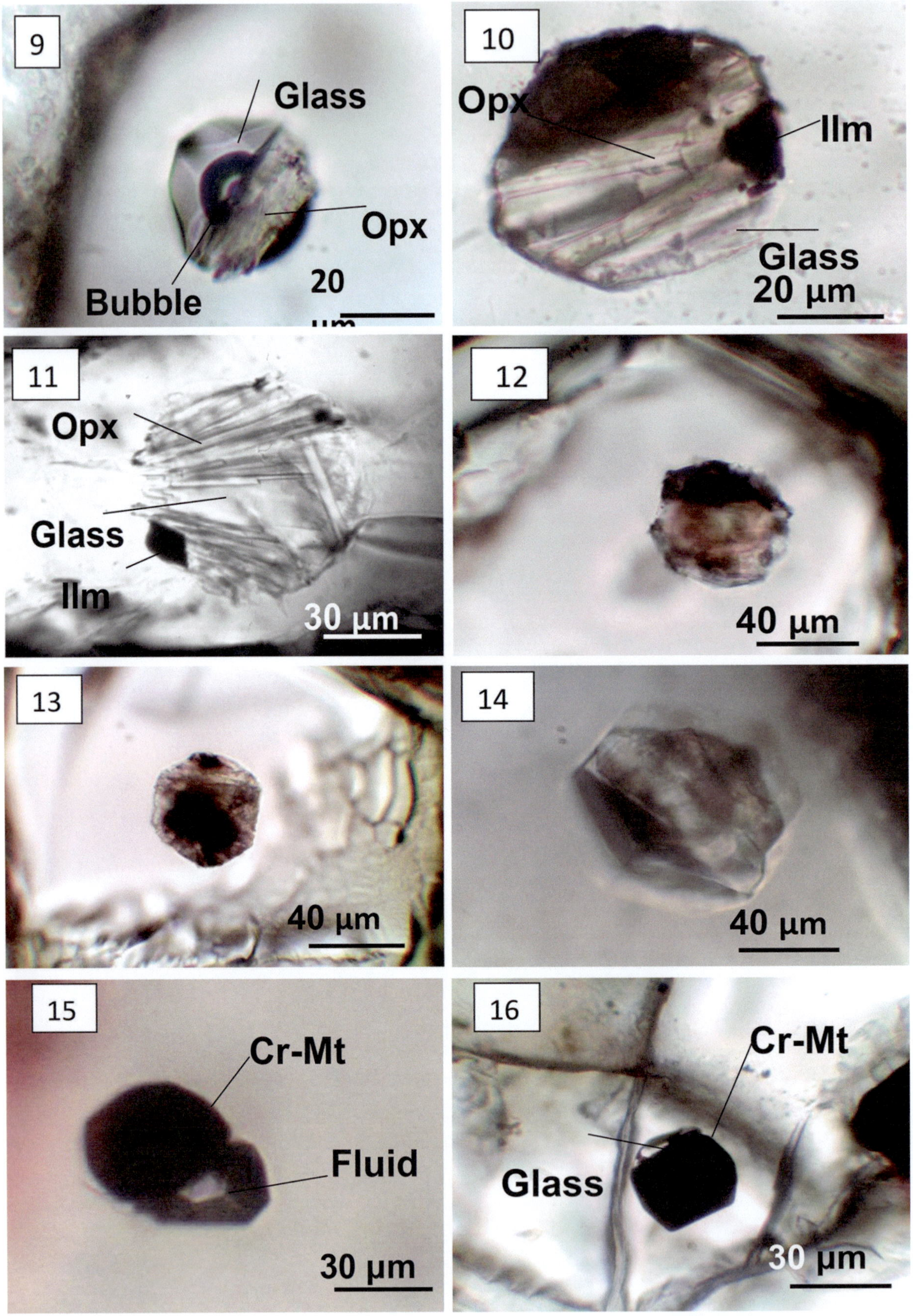

Fig. 5.1 (continued)

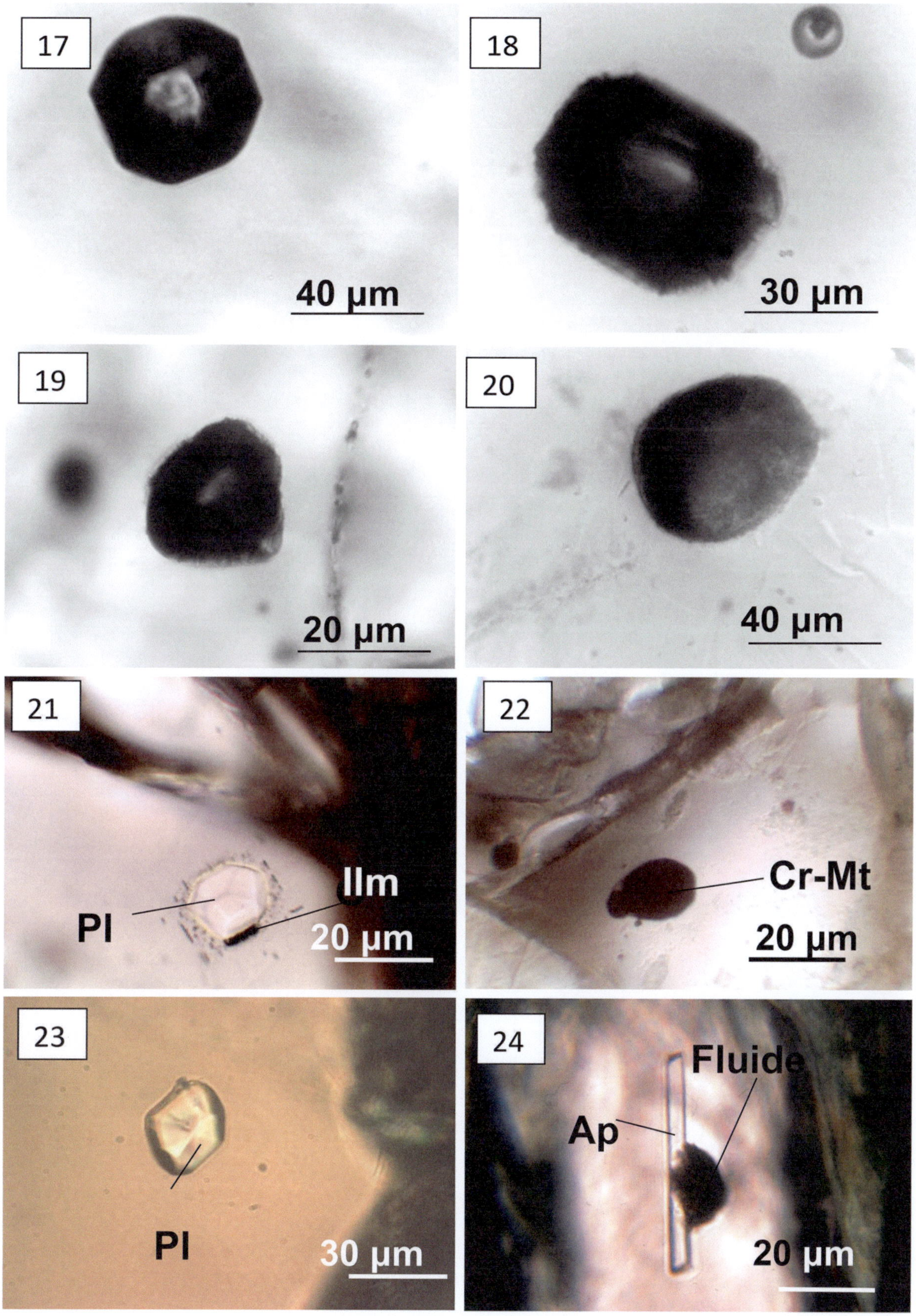

Fig. 5.1 (continued)

Table 5.1 Composition of heated melt inclusions from olivine, wt %

No	No sample	SiO_2	TiO_2	Al_2O_3	FeO	MnO	MgO
1.	4,051/18	51.86	1.36	13.04	11.04	0.33	5.12
2.	3,018/19.8	52.38	2.54	12.01	10.43	0.19	6.50
3.	228–15a	54.86	0.96	13.40	8.30	0.15	5.34
4.	228–18	54.68	2.12	13.34	11.19	0.17	5.59
5.	OM–5/1,220	50.93	1.90	10.39	14.37	0.28	5.77
6.	TG31/820	53.07	1.14	13.28	11.35	0.14	4.52
7.	TG31/820	53.48	0.82	13.30	11.93	0.21	7.56
8.	F-233/343	49.83	0.56	17.74	10.80	0.24	7.06
9.	F-233/343	58.73	0.18	14.59	8.24	0.27	5.91
10.	F-233/343	50.83	1.07	12.94	12.16	0.24	8.37
11.	F-233/343	50.07	1.04	12.57	12.40	0.18	8.39
12.	F-233/343	51.44	1.01	13.49	10.70	0.28	7.46
13.	F-233/343	53.95	1.27	15.09	8.88	0.21	5.90
14.	F-233/343	51.98	1.24	16.93	8.83	0.28	6.00
15.	F-233/350	49.41	0.82	13.98	14.68	0.32	8.02
16.	F-233/350	49.66	0.92	14.61	14.19	0.30	7.06
17.	F-233/350	50.39	0.92	15.15	14.08	0.28	5.85
18.	CY-315-5	49.02	0.25	18.15	13.53	0.17	7.70
19.	CY-315-4	48.91	0.02	18.53	12.93	0.16	7.41
20.	KhS-51/130	54.06	1.31	12.89	10.61	0.17	8.77
21.	KhS-51/130	54.79	1.43	12.92	11.34	0.22	7.84
22.	KhS-51/130	53.79	1.40	10.15	9.64	0.16	9.80
23.	KhS-51/130	54.23	1.29	12.39	10.44	0.14	9.07
24.	KhS-51/130	52.36	1.54	13.18	10.28	0.13	9.00
25.	KhS-51/130	53.07	1.75	13.26	7.68	0.12	9.50
26.	KhS-51/130	51.21	1.65	13.45	11.97	0.17	8.79
27.	KhS-51/130	52.65	1.55	13.13	9.91	0.14	8.98

	CaO	Na_2O	K_2O	Cl	S	P_2O_5	Сумма	Fo, mol %
1.	10.09	4.75	0.93	0.01	0.08	0.47	99.07	73.24
2.	10.45	3.06	1.17	0.02	0.04	0.29	99.08	78.65
3.	11.18	2.67	1.22	0.09	0.03	0.54	98.7	80.06
4.	7.24	2.94	1.24	0.1	0.08	0.41	99.1	79
5.	9.37	3.12	0.33	0.61	0.06	1.66	98.81	80.08
6.	9.23	3.25	1.33	0.02	0.12	0.19	97.6	78.77
7.	7.98	2.85	1.4	0.22	0.11	0.02	99.88	78.41
8.	9.49	1.64	1.51	0.02	0.01	0.08	98.97	78.13
9.	8.09	2.1	0.77	0.25	0.11	0.32	99.56	78.35

Table 5.1 (continued)

	CaO	Na_2O	K_2O	Cl	S	P_2O_5	Сумма	Fo, mol %
10.	9.71	2.51	0.9	0.01	0.03	0.11	98.88	79.58
11.	10.98	2.32	0.75	0.08	0.01	0.12	98.91	79.49
12.	11.22	2.5	0.74	0.02	0.04	0.14	99.03	79.84
13.	9.27	3.34	1.09	0.01	0.03	0.15	99.2	79.3
14.	11.07	1.79	0.68	0.02	0.01	0.33	99.16	78.42
15.	8.86	1.63	0.66	0.01	0.11	0.13	98.69	72.35
16.	9.01	1.69	0.67	0.01	0.03	0.16	98.45	71, 89
17.	9.52	1.73	0.69	0.02	0.01	0.12	98.75	68.98
18.	8.44	0.91	0.2	0.03	0.01	0.03	98.44	75.77
19.	10.9	1.35	0.07	0.07	0.03	0.06	100.6	76.76
20.	8.83	2.24	0.54	0.01	0.05	0.15	99.62	81.96
21.	6.7	3.46	1.09	0.02	0.14	0.22	100.2	80.05
22.	11.38	2.41	0.64	0.02	0.06	0.17	99.59	82.61
23.	9.18	2.33	0.54	0.1	0.06	0.18	99.87	81.69
24.	9.63	2.5	0.55	0.12	0.04	0.15	99.38	81.77
25.	10.19	2.45	0.61	0.11	0.03	0.16	98.81	82.42
26.	9.95	2.21	0.58	0.1	0.05	0.15	100.2	80.37
27.	9.36	2.46	0.59	0.15	0.05	0.16	98.99	82.09

Note: № sample is borehole number/depth, m (F-233/350) and field number (CY-315, 228–18); inclusions from olivines of the intrusions: 1–2 Talnakh, 3–4 Noril'sk 1, 5 Maslovsky, 6,7 Low Talnakh, 8–17 Zelenogrivsky, 18–19 Tuklonsky Formation, 20–27 Gudchikhinsky Formation. Here and in Tables 5.2 and 5.3, analyses were carried in GEOKHI RAS, analyst N. Kononkova; After Krivolutskaya (2011)

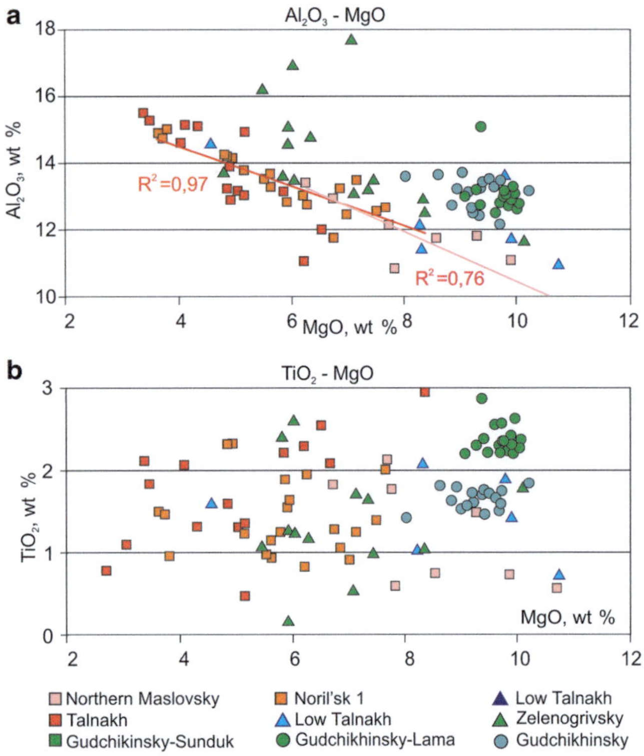

Fig. 5.2 Compositions of melt inclusions in olivines
(**a**) AlO_2O_3–MgO, (**b**) TiO_2–MgO

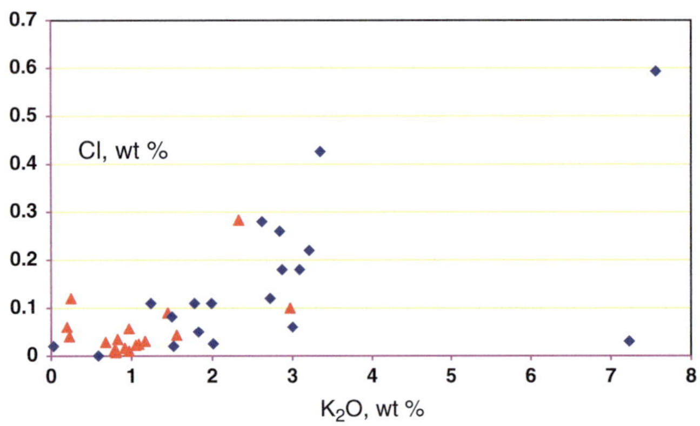

Fig. 5.3 Cl–K_2O diagram for melt inclusions in olivines
Red marks are ore-bearing intrusions, Noril'sk 1, Talnakh; *blue marks* are barren intrusions, Low Talnakh and Zelenogrivsky

cluster and much less frequent in the Zelenogrivsky intrusion. The plagioclase compositions of the inclusions are given in Table 5.3.

5.1.2 Inclusions in Pyroxene

Pyroxenes from various units of the gabbro-dolerites often contain melt inclusions up to 30–40 μm across. However, these inclusions are usually spatially constrained to cleavage planes; thus, it is hard to obtain their homogenized glasses.

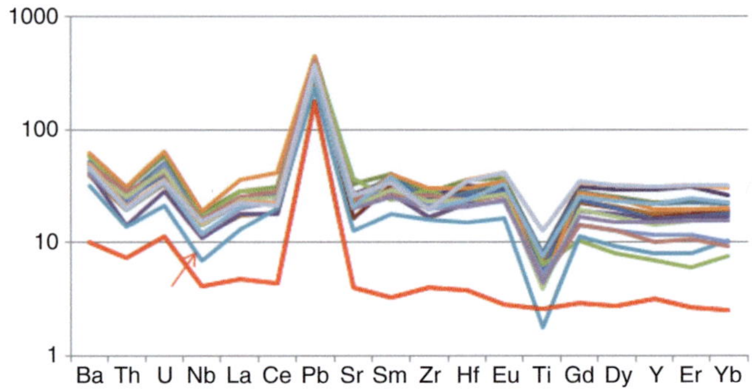

Fig. 5.4 Rare element patterns for inclusions in olivines and host-rock Sample G-22/64, Noril'sk 1 intrusion, picritic gabbro-dolerites

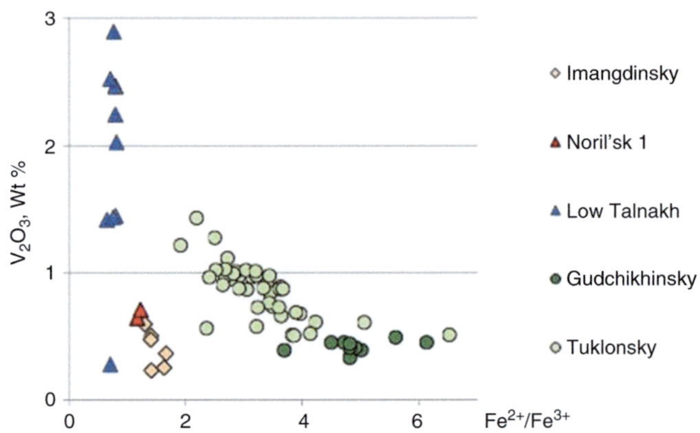

Fig. 5.5 Spinel compositions from magmatic rocks

We were lucky to find unique glassy inclusions in pyroxene from the upper part of the Southern Maslovsky intrusion. In this intrusion, pyroxene is an early liquidus phase; hence, data on these inclusions are of particular interest. The inclusions are occasionally as large as 70 μm in diameter, but most range from 30 to 40 μm (Fig. 5.6). Along with glass and a bubble, the inclusions often contain titanomagnetite as a daughter phase or as captured crystals and grains. The inclusions completely homogenize at temperatures of 1,170–1,190 °C, depending on the Mg# of their host pyroxene. The inclusions were determined to contain volatile components, including 0.5–0.9 wt % H_2O and up to 0.2 wt % Cl. They contain (as melt inclusion in olivine) high rare earth element concentrations (Fig. 5.7).

5.2 Fluids in Magmas

One of the necessary prerequisites for the accumulation of uniquely high concentrations of metals in products of hyperbasite–basite magmas is often thought to be an unusual fluid regime. Thus, fluids with extremely different compositions were suggested. Some researchers believe that the parental

Table 5.2 Composition of spinel from olivine, wt %

	1	2	3	4	5	6	7	8
Component	IM-29	IM-30	IM-31	IM-1	IM-1_4	IM-1_5	G-22/64-4	G-22/64-5
SiO_2	0.38	0.41	0.37	0.29	0.44	0.34	0.47	1.18
TiO_2	2.33	3.69	3.65	2.52	1.67	3.13	1.28	1.42
Al_2O_3	9.34	8.36	7.79	9.14	10.39	8.56	5.36	5.57
Cr_2O_3	30.27	29.76	26.92	27.48	34.20	32.81	31.52	31.30
V_2O_3	0.50	0.23	0.47	0.60	0.36	0.25	0.64	0.71
FeO_{tot}	50.05	49.82	53.61	52.48	46.36	48.23	51.62	50.81
MnO	0.40	0.40	0.42	0.43	0.44	0.42	0.45	0.42
MgO	3.53	4.45	2.88	3.32	3.71	3.57	3.03	3.97
NiO	0.15	0.15	0.16	0.19	0.15	0.15	0.20	0.19
ZnO	0.20	0.16	0.23	0.24	0.23	0.20	0.25	0.25
Summa	97.15	97.52	96.52	96.68	97.99	97.66	94.83	95.86
FeO	29.51	29.32	31.20	29.65	29.00	30.07	27.97	27.97
Fe_2O_3	22.83	22.78	24.91	25.37	19.29	20.18	26.28	25.38
Summa 1	99.43	99.71	98.99	99.22	99.88	99.68	97.45	98.36
Fo, wt %	76.0	76.0	75.5	76.2	77.1	77.4	79.3	76.4
Fe^{2+}/Fe^{3+}sp	1.44	1.43	1.39	1.30	1.67	1.66	1.18	1.22
Fe^{2+}/Fe^{3+}m	4.52	4.49	4.34	3.96	5.51	5.44	3.50	3.67

	9	10	11	12	13	14	15	16	17	18
Component	4051/18-1	4051/18-2	4051/18-3	4051/18-4	4051/18-5	4051/18-6	4051/18-7	TG-31/827-1	TG-31/827-2	TG-31/827-3
SiO_2	0.32	0.58	0.55	2.86	0.55	0.50	0.49	0.50	0.50	0.49
TiO_2	4.51	7.93	5.88	5.24	22.30	6.17	7.23	3.73	2.54	0.53
Al_2O_3	8.72	4.40	6.50	5.02	2.60	6.36	4.61	6.75	7.94	4.16
Cr_2O_3	45.48	15.78	21.41	16.63	10.64	20.57	16.77	9.08	9.46	9.17
V_2O_3	0.24	0.83	0.74	0.77	0.51	0.83	0.70	2.46	2.47	1.42
FeO_{tot}	26.79	64.77	59.44	65.14	58.87	59.88	64.37	71.58	70.88	75.89
MnO	0.20	0.51	0.44	0.33	0.42	0.40	0.31	0.34	0.36	0.27
MgO	14.07	2.80	3.03	2.44	3.27	3.23	2.74	2.94	3.01	1.53
NiO	0.33	0.13	0.07	0.10	0.00	0.06	0.22	0.00	0.01	0.09
ZnO	0.09	0.17	0.30	0.24	0.05	0.07	0.07	0.10	0.09	0.06
Summa	100.74	97.91	98.37	98.75	99.19	98.07	97.51	97.49	97.25	93.60
FeO	16.24	35.48	33.68	37.09	47.75	33.71	34.94	32.19	31.08	29.94
Fe_2O_3	11.72	32.55	28.63	31.18	12.35	29.09	32.71	43.78	44.23	51.07
Summa 1	101.92	101.17	101.24	101.88	100.43	100.99	100.78	101.88	101.68	98.72
Fo, mol %	76.1	73.6	73.6	70.7	73.2	74.4	74.1	77.8	77.7	78.3
Fe^{2+}/Fe^{3+}sp	1.54	1.21	1.31	1.32	4.30	1.29	1.19	0.82	0.78	0.65
Fe^{2+}/Fe^{3+}m	4.95	3.61	3.99	4.05	18.95	3.92	3.52	2.16	2.03	1.61

Note: After Krivolutskaya (2011)

(continued)

Table 5.2 (continued)

Component	19 TG –31/827-4	20 TG –31/827-5	21 F-233/350	22 F-233/350	23 F-233/350	24 530-12-9	25 530-12-11	26 530-12-18
SiO_2	0.50	0.56	0.42	0.49	0.70	0.39	0.33	0.41
TiO_2	2.05	4.12	1.02	7.14	4.31	1.84	1.86	2.43
Al_2O_3	9.64	5.61	3.55	3.88	5.61	14.38	14.12	16.04
Cr_2O_3	12.18	8.79	25.18	26.03	17.76	42.73	42.54	35.59
V_2O_3	1.45	2.25	0.71	0.81	0.49	0.51	0.50	0.95
FeO_{tot}	67.47	72.24	64.41	51.74	66.00	32.09	31.84	36.78
MnO	0.31	0.34	0.48	0.52	0.44	0.32	0.31	0.30
MgO	3.27	2.98	2.22	5.32	2.99	6.82	6.64	6.26
NiO	0.15	0.21	0.26	0.31	0.11	0.06	0.06	0.06
ZnO	0.02	0.12	0.07	0.00	0.14	0.19	0.20	0.20
Summa	97.02	97.21	98.31	96.23	98.53	99.36	98.42	99.04
FeO	30.38	32.06	30.04	30.01	32.67	25.45	25.32	27.08
Fe_2O_3	41.21	44.65	38.20	24.15	37.04	7.37	7.25	10.77
Summa 1	101.15	101.68	102.14	98.65	102.24	100.07	99.13	100.10
Fo	77.4	78.4	78.83	78.62	78.38	76.78	76.55	75.21
Fe^{2+}/Fe^{3+}sp	0.82	0.80	0.90	1.43	1.01	3.84	3.88	2.79
Fe^{2+}/Fe^{3+}m	2.17	2.09	2.46	4.48	2.85	16.34	16.60	10.79

Note: Calculation FeO, Fe_2O_3 were carried based spinel stoichiometry; *FeOtot* total Fe; Summa and Summa I – summa before and after calculations correspondingly; Fe^{2+}/Fe^{3+} sp Fe^{2+}/Fe^{3+} m ratios two- and tree-valence iron in spinel and melt according (Maurel and Maurel 1982); *Fo* forsterite component in host olivine (mol %); numbers analyses from massifs: 1–6 Imangdinsky, 7,8 Noril'sk 1, 9–15 Talnakh, 16–20 Low Talnakh, 21–23 Zelenogrivsky, 24–26 picrites from Nadezhdinsky Formation

Table 5.3 Composition of plagioclase from olivine in picritic gabbro-dolerites, wt %

No	No sample	SiO_2	Al_2O_3	FeO	CaO	Na_2O	K_2O	Сумма	An, mol %	Fo, mol %
1.	4,051/18-1	48.28	32.40	1.63	16.17	2.03	0.10	100.61	81.51	73.69
2.	4,051/18-2	47.29	28.76	5.01	13.32	2.26	0.17	96.81	76.53	74.52
3.	4,051/18-3	49.39	32.08	1.43	15.82	2.37	0.11	101.20	78.69	74.61
4.	4,051/18-4	50.17	31.34	1.32	14.74	3.73	0.27	101.57	68.62	71.99
5.	4,051/18-5	47.74	32.46	1.70	15.39	2.34	0.19	99.82	78.45	69.13
6.	4,051/18-6	47.88	33.13	1.97	15.24	2.27	0.20	100.69	78.82	69.13
7.	4,051/18-7	47.37	33.11	1.71	15.70	2.20	0.14	100.21	79.81	68.92
8.	4,051/18-8	47.60	32.62	1.86	14.93	2.55	0.20	99.76	76.44	69.49
9.	4,051/18-9	48.13	32.13	1.25	15.49	2.43	0.15	99.58	77.88	74.07
10.	4,051/18-0	48.70	31.09	1.24	15.90	2.36	0.25	99.54	78.85	74.68
11.	3,018	48.43	31.72	1.30	16.59	2.01	0.06	100.10	82.04	74.64
12.	TG-31/820	48.1	34.08	1.55	17.11	1.91	0.12	102.87	83.21	78.35
13.	TG-31/820a	49.74	32.11	1.44	14.81	2.97	0.37	101.44	73.40	78.36
14.	TG-31/820б	47.67	34.45	1.23	17.31	1.83	0.1	102.59	83.96	78.91
15.	TG-31/829	46.29	32.64	2.3	16.75	1.68	0.1	99.76	84.65	78.60
16.	TG-31/839	48.51	32.56	1.5	15.76	2.44	0.05	100.82	78.14	78.23
17.	TG-31/839a	47.4	32.42	0.57	15.81	2.19	0.03	98.42	79.98	78.23
18.	TG-31/827в	47.59	33.46	1.6	16.62	1.69	0.03	100.99	84.48	78.05
19.	TG-31/827a	47.22	33.39	1.73	15.84	2.29	0.16	100.63	79.28	77.91
20.	TG 31/827c	47.19	32.93	1.72	15.89	2.15	0.18	100.06	80.32	77.91
21.	TG-31/829c	47.54	32.64	1.1	17.18	1.64	0.08	100.18	85.29	77.70
22.	F-233/343	48.46	32.37	1.52	16.04	2.24	0.11	100.74	79.82	79.17
23.	F-233/343a	48.65	32.40	1.34	16.05	2.28	0.11	100.82	79.58	79.05
24.	F-233/343c	47.46	31.62	1.54	15.96	2.08	0.10	98.76	80.97	78.68

Note: *An, mol %* anorthite in plagioclase, mol %; *Fo*-forsterite (mol %) in host olivine, intrusions: 1–11 Talnakh, 12–21 Low Talnakh, 22–24 Zelenogrivsky

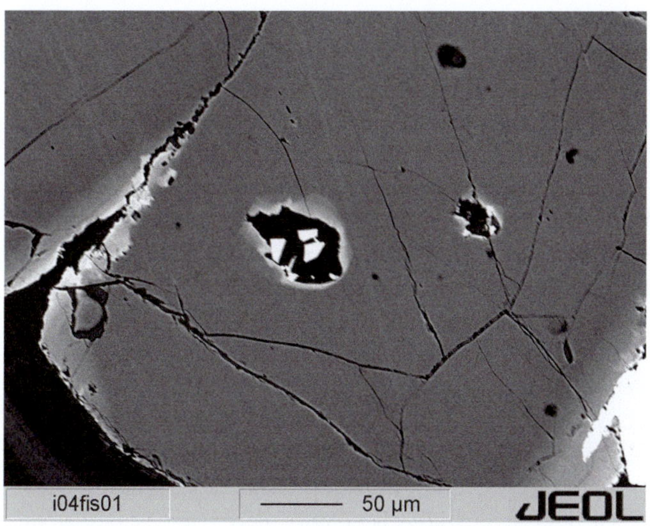

Fig. 5.6 Melt inclusion (*black*) in pyroxene from gabbro-dolerites of Southern Maslovsky intrusion (*white* crystals – magnetite) Sample OM-24/666

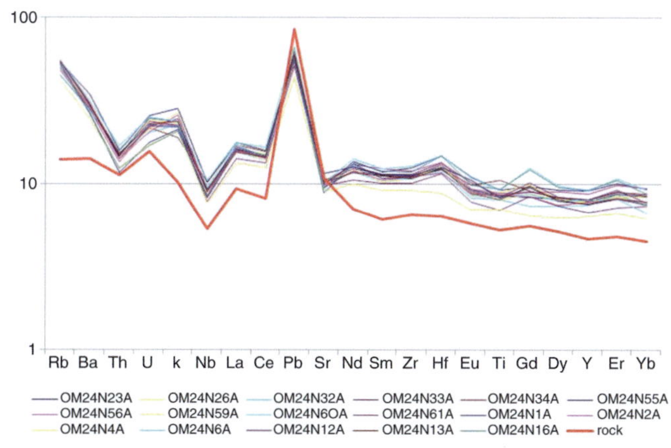

Fig. 5.7 Rare element patterns for inclusions in pyroxene and host-rock Sample OM-24/666, the Southern Maslovsky intrusion

melts were enriched in water and chlorine (Distler et al. 1999), whereas others argued that the melts should have been rich in methane, hydrogen sulfide, etc. (Aplonov 1995). The hypothesis that the melts were enriched in water is underlain by data from the contact aureoles of the ore-bearing intrusions (Turovtsev 2002), which are believed to be formed under the effects of fluids that segregated from the crystallizing melts and Cl by virtue of the fact that the ores contain Pt–Pd compounds with As, Sb, Te, and Se, which can be transported in the form of chloride complexes even at relatively low temperatures (Boudreau et al. 1986; Boudreau 1988). Another perspective is based on chromatographic data obtained from whole-rock samples. According to this hypothesis, the ore-forming processes involved mostly reduced gases: H_2, CH_4, CO, N_2, etc. (Neruchev and Prasolov 1995). Data from fluid inclusions in minerals from trap rocks (Bulgakova 1971; Ryabchikov et al. 2001, 2009; Sobolev et al. 2009a) indicate that the fluids were dominated by H_2O and CO_2.

Some researchers believe that the contact aureoles of the intrusions are composed only of metamorphic rocks. If so, these rocks could have formed under the influence of a continuously active heat source, which could be visualized only as a long-lived magmatic system. Such a system could be generated by either melt flowing in only one direction through a channel (Rad'ko 1991; Naldrett 1992) or oscillating melt flow (Likhachev 1996). In this instance, the melts should not contain anomalously high concentrations of fluids.

Our data provide support for the latter hypothesis (Krivolutskaya et al. 2001b, 2004). The fluid inclusions in olivine from the picrite–gabbro-dolerites compose approximately 1–3 % of the total volume of all of the aforementioned inclusions and are usually no larger than a few microns in diameter, only occasionally reaching 40 μm (Fig. 5.1). Morphologically, the small and large inclusions are similar: they are shaped either as negative crystals or are spherical.

First, the inclusions were cooled to −165 to −190 °C in a cooling and heating stage at the Institute of Geology of Ore Deposits, Petrography, Mineralogy, and Geochemistry, Russian Academy of Sciences (analyst T.L. Krylova). No visible change occurred due to the extremely low density of the fluid. The direct determination of the gas composition in fluid inclusions with Raman spectroscopy (analyst: J. Debussie, Nancy, France) detected only low H_2O and CO_2 concentrations and confirmed the absence of other gases (hydrogen, nitrogen, and methane). Crystalline phases—magnesite and organic carbon—were identified on the inner surfaces of large inclusions at high magnification (Krivolutskaya et al. 2001b).

For the precise determination of the H_2O content in the melts, glassy inclusions in olivine were used to avoid loss of water by heating of crystallized inclusions. As is shown by Raman spectroscopy, the round or oval glassy inclusions are filled with up to 80–85 vol. % glass and contain an empty bubble (10–15 vol. %). Close to the intrusive contact, in the lower part of the taxitic unit, the number and size of glassy inclusions reaching 35–40 μm in diameter increase. The H_2O content in glass reaches 1.2 wt % (ion microprobe; analyst: S.G. Simakin, Institute of Microelectronics and Information Science, RAS) and exhibits a weak correlation with potassium. The concentration levels of volatile components (H_2O, F, and Cl) in the ore-bearing and barren intrusions are equal (Krivolutskaya and Sobolev 2001). The extremely low H_2O contents (0.03–0.25 wt %) were observed in picrites of the Gudchikhinsky Formation (Sobolev et al. 2009a). Thus, the crystallizing melt was not enriched in water, chlorine, and reduced gases.

To specify additional evidence for the enrichment of the initial melt in fluids—wide contact aureoles around intrusions—the composition of the water participating in their formation was estimated. The stable O, H, and C isotopes in gabbro-dolerite intrusions and country rocks were studied for this purpose (Pokrovsky et al. 2005). The central parts of intrusive bodies are characterized by the oxygen isotopic composition typical of magmatic water ($\delta^{18}O = 5.5‰$), whereas in the contact zone, the $\delta^{18}O$ values increase to 18‰. These data obtained for the Noril'sk 1, Talnakh, Zelenogrivsky, and Lower Talnakh intrusions do not confirm hydrothermal reworking of country rocks under the effect of magmatic fluids released from crystallizing magma. The enrichment of the contact zone in the heavy oxygen isotope indicates that these zones have been reworked by meteoric rather than magmatic water. The involvement of meteoric water in postmagmatic alteration is quite natural because the emplacement of enormous magmatic masses, such as at the Siberian Platform, must initiate fluid flow in the country rocks.

5.3 Composition and Temperature of Melts Estimated with the KOMAGMAT-3.5 Program

The phase composition of the initial melt and its temperature were estimated for the Talnakh intrusion using the KOMAGMAT-3.5 program (Ariskin and Barmina 2000) on the assumption that a melt with suspended crystals was emplaced into a magma chamber. The formation of the internal structure of this intrusion as a result of magma fractionation was shown in terms of this model; these data were obtained with A. Ariskin (Krivolutskaya et al. 2001a). Previously, a discrepancy in the composition of the chilled margin and the weighted average composition of the intrusion (7–8 and 10–12 wt % MgO, respectively) was adduced as the main argument against this model. For the calculations, data for the Talnakh intrusion (structure, geochemistry of rocks and minerals) were taken from Chap. 4.

The fact that the weighted mean composition of the Noril'sk-type intrusions is notably more magnesian than that of the typical "common flood basalt" (Table 5.4) led to the conclusion that a picritic basalt parental magma existed and was derived from a deep-seated source (Godlevsky 1959). These concepts were further developed by researchers who assumed that the flood basalt parental magma has a composition intermediate between typical picrite and flood basalt (Magnesium basic…1984). It was also assumed that this magma was primary, i.e., it was derived by the melting of a mantle source at relatively high pressures and temperatures.

Table 5.4 Average mean composition of the Talnakh intrusion (wt %) according to different estimations

Component	1	2	3
SiO_2	47.47	48.30	48.76
TiO_2	1.11	0.85	0.88
Al_2O_3	14.11	15.33	14.90
FeO	12.33	12.34	11.08
MnO	0.22	0.19	0.20
MgO	11.54	10.00	10.52
CaO	10.22	10.45	10.04
Na_2O	1.91	1.86	2.05
K_2O	0.56	0.58	0.65
P_2O_5	0.12	0.20	0.18
Cr_2O_3	0.13	0.10	0.14

Note: 1—based on data from KZ-1799 and KZ-1713 (Czamanske et al. 1994). 2—based on 29 boreholes (Dneprovskaya et al. 1987); *3* borehole OUG-2; Fe recalculated as FeO. Tables 5.4, 5.5 and 5.6 – After Krivolutskaya et al. (2001a)

The analysis of the fractionation schemes of such melts demonstrates that the likely origin of known "intrusive flood basalt types" features a single source that contained 10–12 wt % MgO and had a composition close to the weighted mean composition of the Noril'sk I intrusion (Zolotukhin and Laguta 1985; Zolotukhin and Vasil'ev 1986). This finding implies an important role of the processes of pre-chamber differentiation, which may have occurred in magmatic columns or intermediate chambers (Likhachev 1965, 1978; Oleinikov 1979), mainly via the gravitational separation of crystals and melt. The modern interpretation of these processes is based on finer isotopic and geochemical differences (such as Mg#, TiO_2, REE patterns, Th/Ta ratios, etc.) between flood basalt magmas, which can be interpreted as indications of the composition of the mantle source, the depth (pressure) and grade of its partial melting, and the extent of the crustal contamination of the magma (Naldrett et al. 1992; Arndt et al. 1993; Lightfoot et al. 1993; Hawkesworth et al. 1995). However, regardless of the picritic or picrite-like composition of the parental magmas, the concept of the deep-seated evolution of magnesian melts implies that these processes are responsible for the main differences between flood basalt magmas, and the rocks crystallizing from them (including those of differentiated intrusions) inherit petrochemical and geochemical indications of this diversity. At the same time, it is sometimes thought that the average compositions of solid flood basalts correspond to the partial melts of basaltic composition, whose source may have been mantle pyroxenites (Kutolin 1972). Within the guidelines of this hypothesis, insignificant or nonexistent fractionation of the parental flood basalt magma in deep-seated chambers is postulated, and the diversity of intrusive rocks is considered to be the result of contamination and in-chamber differentiation (Feoktistov 1978). Hence, the problems of the Mg# of the parental magmas and the role of the magmas' differentiation at depth lie at the heart of discussions about the genesis and ore potential of ultramafic–mafic intrusions in the Siberian Platform.

We believe that these problems can be resolved by determining the thermodynamic and dynamic parameters of magma intrusions for each individual intrusive body. The list of parameters should include the temperature and composition of the parental melt, its crystallization degree during intrusion, and the proportion and composition of mineral phases suspended in the melt. The amassment of such data would enable one to more reliably trace the possible genetic relationships between the magmas of discrete flood basalt intrusions and to gain better insight into the spatial distribution of the initial temperature, composition, and phase parameters. These problems were first attacked by a computer simulation of the inner structures of flood basalt intrusions in the Siberian Platform on the basis of the convection–accumulation mechanism of in-chamber differentiation (Frenkel' et al. 1985, 1988). The examples of the Kuz'movka (on the Podkamennaya Tunguska River), V-304, and Vavukan (in the upper reaches of the Vilyui River) sills were utilized to demonstrate that the parental magma contained no more than 8 vol. % intratelluric crystals during its intrusion, with the solid phase consisting of excess *Ol* or the *Ol*+*Pl* assemblage. The application of these approaches to the Talnakh intrusion made it possible to explain the principal features of its inner structure as the result of chamber replenishment by a magma batch with higher proportion of crystals: 15–20 % intratelluric phases with olivine and plagioclase simultaneously present at the liquidus (Dneprovskaya et al. 1987). Further developing this approach, we present the first estimates of the intrusion parameters of the Talnakh magma, which were obtained by the solution of the inverse geochemical problem by the computer simulation of melt–crystal equilibria in basaltic systems (Ariskin and Barmina 2000).

Based on a review of the literature, the only method previously utilized to evaluate the composition of parental flood basalt magmas was the calculation of the weighted mean compositions of the respective intrusions. Data of these types are fairly extensive and were used to distinguish associated types of intrusions and to develop schemes for their petrochemical classification (Vilensky and Oleinikov 1970). However, this approach fails to resolve several problems; in particular, why the composition of the chill zones does not correspond to the average composition of the respective intrusions. This situation is clearly illustrated by the examples of the Noril'sk I and Talnakh intrusions: the average compositions of both massifs are significantly more magnesian than the compositions of the inner-contact gabbro-dolerites (Krivolutskaya et al. 2001a, b). The main cause of these inconsistencies is the fact that the weighted mean composition of an intrusion can be accurately calculated only if full information of the volumetric percentages of all rock varieties is available. No correct assessment can be obtained on the basis of average chemical compositions or data from a few vertical sections (even if they are thoroughly studied) because this technique yields compatible estimates only for nearly ideal sheet-shaped bodies. Additionally, using inner-contact rocks to model the parental melt is also problematic because of possible changes during the interaction with the wall rocks. Furthermore, one of the most important identification criteria of inner-contact rocks, i.e., their fine-grained or subaphyric texture, is also questionable (see, e.g., Hoover 1989). This criterion implies that the magma was homogeneous overheated melt. Several lines of evidence obtained over recent years indicate that melts injected into magma chambers contain variable amounts of crystalline phases (Frenkel' et al. 1988; Marsh 1989; Chalokwu et al. 1993). This information provides the basis for another approach to the problem of intrusive magma and allows for different reconstructions of the parental melt compositions and the mechanisms of their in-chamber differentiation (Ariskin 1999). The technique of geochemical thermometry was developed in the 1980s as a means to assess the initial parameters of state based on information "recorded" in the whole-rock compositions of volcanic and intrusive mafic rocks (Frenkel' et al. 1987). This approach provides full phase and chemical interpretations of igneous rocks, including estimates of the original temperatures and composition of entrapped (intercumulus) melts (Barmina et al. 1988). In certain cases, geochemical geothermometry also makes it possible to assay the phase composition of the intruded magma and the original composition of the intratelluric crystalline phases (Barmina et al. 1989; Chalokwu et al. 1993).

We applied these approaches to the taxitic and picritic gabbro-dolerites of the Talnakh intrusion, whose inner structure is relatively simple and which shows typical features of differentiated ultramafic–mafic massifs of the Noril'sk type. As noted by earlier researchers, the taxitic and picritic units should have played a determining role in the genesis of the sulfide mineralization (Petrology and Ore Potential … 1978). We will now consider certain fundamental terms and physicochemical principles important for the analysis of geochemical thermometric data (Frenkel' et al. 1987; Ariskin and Barmina 2000).

5.3.1 Specification of the Petrological Terminology

To unambiguously interpret the results presented below, we will first specify the terms of magma and magmatic melt. One can find a broad diversity of definitions of these terms, with emphasis on the aggregate state and rock-forming role of magmas in geological processes. For example, Kuznetsov (1990) stressed that magmas are physically homogeneous (silicate melt + dissolved volatile components) or, more often, heterogeneous (melt + crystals) systems, "whose distinctive feature is fluidity, a property manifesting itself at >25 % liquid in the mixture." Current discussions of the aggregate state of magmas are focused on the permitted proportions of crystals and melt that characterize the unequal distribution of crystalline material from the full solidification boundary (the walls of the magma conduit or chamber) toward the interior of the reservoir. The analysis of this situation leads to the recognition of two principal areas (Sinton et al. 1992; Jaupart and Tait 1995). The zone developing near the solidus surface has a low concentration of melt, which fills in the space between crystals. Touching by their edges and faces, the crystals compose a rigid continuous framework, the so-called rigidus. In accordance with some of its rheological properties, this mixture behaves similarly to a solid but does not lose its plasticity. With a decrease in the percentage of crystalline material, the rigidus zone grades into crystalline mush and, furthermore, to a magmatic suspension with <25 % solid phases (Marsh 1989). The rheological boundary between the rigidus and mush is thought to correspond to a sharp (by a few dozen orders of magnitude) change in the dynamic viscosity at a concentration of crystalline material of approximately 50–60 vol. % (Bergantz 1990; Sharapov et al. 1997). This value is referred to as critical crystallinity. The amounts of the complementary residual liquid define the critical melt fraction (see, e.g., Renner et al. 2000).

Hence, we will use the term *magma* to mean heterogeneous silicate systems with a concentration of individual crystals and/or their aggregates below the critical crystallinity. The liquid constituent of magma is magmatic melt. When overheated (i.e., contains no suspended solid phases), magma consists only of silicate liquid. In compliance with these definitions, the bulk chemical composition of any magma can be expressed through the compositions and proportions of suspended solid phases ($1 \leq j \leq M$) and melt (l). The bulk composition (melt + gas + solid phase) of the magma that formed intrusions was more magnesian than the initial magma of the volcanic rocks. Both magma as a whole and magmatic melts, the principal constituents of magma, can be regarded as rock-forming systems, even if magmas are highly crystalline. We would like to stress this point because of the recent tendency to apply the term *magmatic melt* to compositions derived from the study of microscopic fragments of vitreous mesostasis or melt inclusions in minerals. In the latter instance, it seems to be reasonable to use, following A.V. Sobolev's proposal of the term *differential melts*, which is intended to highlight the bulk-composition heterogeneity of the final melt (a feature detectable when magmatic inclusions are examined in individual crystals (Sobolev 1997)). Natural melts (which serve as a transport medium, deliver crystalline fractionation products to the surface, and crystallize to form magmatic rocks) were suggested to be termed *integral melts*. They are thought to be produced by the mixing of differential liquids during the pre-chamber stage. Within the framework of this terminology, we will focus on integral magmatic melts.

5.3.2 Justification of Geochemical Thermometry

One of the basic assumptions is as follows: During the initial moment of the origin of rocks of a massif, the compositions of the minerals and melt are related by the equations of thermodynamic equilibrium, which involve, as one of their parameters, the temperature of the rock's origin. This temperature corresponds to the initial conditions of the existence of the melt–crystal assemblage and affects, as an equilibrium factor, the distributions of major and trace elements between the phases. Another fundamental assumption is that the whole diversity of rocks in an examined massif can provide samples whose chemical variations were caused only by changes in the initial phase proportions. These samples represent rock groups that were produced by melt at the same temperature and minerals of the same composition. An important feature of these rocks is that their equilibrium melting may yield series of melts that are characterized by distinct evolutionary trajectories in temperature–composition plots, but their melting curves inevitably have points that correspond to a temperature at which the compositions of all liquids are identical, i.e., these lines should intersect either at a point or within a compact area. It is easy to realize that this composition should correspond to the composition of the entrapped liquid (the same for all rocks), and the deduced temperature is the formation temperature of these rocks. This statement can be proved by contradiction: if no such intersection occurs, this would mean that no mass balance condition holds for the chosen rock compositions, or the initial compositions of the minerals and melt are not related by thermodynamic equilibrium laws. Thus, to determine the initial formation parameters of a given intrusive rock, one should (1) select a few additional samples whose entrapped melt composition is supposedly similar, (2) conduct a physical experiment on or numerical simulation of the equilibrium melting of this group of rocks, (3) determine or compute the compositions of the resultant melts, and (4) construct the evolutionary trends for the model liquids in a temperature–composition plot and determine the point (or area) of their intersection.

5.3.2.1 Application of the Method

In the application of this method to intrusive mafic rocks, the constancy of the entrapped or residual magmatic liquid allows for two variants of their genetic and geological interpretation. First, this may have resulted from the local (on a scale of a few centimeters to a few meters or, at most, dozens of meters for large layered complexes) heterogeneity of the rocks and, as the limiting case, microscopic-scale cyclic layering. Second, this could have been caused by the weak fractionation of the magma, which implies the existence of similar formation conditions for rocks throughout a given geologic body. Ideally, it is most desirable to have a few groups of genetically interrelated rocks distributed over the vertical section of the intrusion or throughout its volume. This would make it possible to refine the character of melt evolution during the in-chamber differentiation and the possible compositional heterogeneity of the magmatic liquid in different parts of the intrusion. From the perspective of estimating the composition of the original (during its intrusion) magmatic melt, the most informative data can be provided by inner-contact and near-contact rocks, for which it is quite realistic to suggest the compositional homogeneity of the entrapped liquid. The geochemical thermometry results for the near-contact *Ol–Pl* cumulates of the Skaergaard intrusion, which were sampled within 10 m of the contact and are referred to as the Side Marginal Group (Ariskin 1999), are an illustrative example of the efficiency of this approach. It is difficult to determine whether an analogous research can be conducted on the rocks of the Talnakh intrusion because this approach requires systematic information on the chemistry of the inner-contact gabbro-dolerites with compositions that vary as much as possible. However, numerous analyses of the taxitic and picritic gabbro-dolerites from the lower parts of the massif are available from the literature. Provided the Talnakh magma was not overheated and entrained a certain amounts of intratelluric olivine (in addition to, possibly, plagioclase) into the chamber, the compositional variations of rocks from the lower part of the massif can be regarded as resulting from the redistribution and sorting of crystals without significant fractionation of silicates in the main volume of the magmatic melt (Dneprovskaya et al. 1987). When used in this case, the geochemical thermometry method allows one to constrain the formation parameters of the taxitic and picritic gabbro-dolerites, whose temperature, mineral chemistry, and melt composition should be similar to those of the parental intrusive magma. The correctness of this assumption and the estimates obtained may be tested by the consistency of the calculated and observed compositions of the rock-forming minerals.

The approach proposed above can be realized with the COMAGMAT-3.5 computer program (Ariskin 1999; Ariskin and Barmina 1999). The program was developed to simulate the crystallization of basalts of low and moderate alkalinity within predetermined ranges of pressure and oxygen fugacity. Its application to geochemical thermometry is justified by the fact that equilibrium crystallization and melting processes are reversible, and the calculation of a melting trajectory can be replaced by the calculation of the equilibrium crystallization of the respective melt. When relatively few compositions are involved, the intersection points of the calculated evolutionary paths can be found by the visual analysis of a series of *T–X* diagrams. However, the analytical errors and uncertainties involved in the phase equilibrium model cause certain errors in the determined intersection of the calculated crystallization trajectories and, hence, this intersection occurs in a *T–X* plot as an area with closely spaced evolutionary lines (see below). When using the COMAGMAT program in association with tholeiite-like systems, the errors of the temperature estimates are on the order of the employed thermometers, i.e., 10–15 °C, with the major component concentration accuracy of 0.5–1 wt % (Ariskin and Barmina 2000).

5.3.3 Simulation of Phase Equilibria

To perform thermometric calculations, we used the chemical compositions of rocks recovered from Hole OUG-2 (Table 5.4) and Holes KZ-1713 and KZ-1799 farther north (Czamanske et al. 1995). The use of additional data was due to the need to enhance the representativeness of the petrological material and the intention to rely on data from relatively closely spaced sections. Their spatial proximity let us suggest that there were only insignificant differences between the compositions of the melts entrapped by the rocks of the lower units, in which no significant lateral geochemical heterogeneity was detected (Dneprovskaya and Dneprovsky 1988). Thus, we used only the compositions of picritic ($n = 13$) and taxitic ($n = 6$) gabbro-dolerites for the calculations: 11 samples from Hole KZ-1799, seven from Hole KZ-1713, and five from Hole OUG-2.

5.3.3.1 Selection of Starting Compositions

Recent versions of the COMAGMAT program enable one to simulate melt–crystal equilibria that involve five silicates (*Ol*, *Pl*, *Aug*, *Pig*, and *Opx*) and two oxides (Ti–*Mt* and *Ilm*) but no crystallizing sulfides. An attempt to use this model to calculate phase equilibria in rocks with significant amounts of sulfides may result in changes in the configurations of the fields of certain minerals and shifts in the model lines for the melt evolution. Therefore, the original chemical compositions should be recalculated to determine the amounts of FeO in the silicate matrix based on the overall sulfur and base-metal concentrations in each individual sample. The recalculation technique implies that Cu and Ni are contained only in chalcopyrite and pentlandite, respectively, and it is assumed that the proportion of Ni accommodated in olivine is insignificant and does not principally affect the Fe balance between the sulfide and silicate constituents of the rock. The calculation succession involves (1) the determination of the bulk amounts (in wt %) of chalcopyrite and pentlandite in the rocks based on data on their actual compositions and the concentrations of Cu and Ni in the rock, (2) the calculation of the amounts of S contained in these minerals, (3) the calculation of the excess S (based on the data of a chemical analysis) that should be assigned to pyrrhotite, (4) the calculation of the FeO concentration in the silicate matrix as a difference between the whole-rock and "sulfide" FeOtot concentrations, and (5) the normalization of the composition calculated to 100 anhydrous residue. The amount of iron in the sulfides was calculated using the following mineral compositions: chalcopyrite was calculated as stoichiometric $CuFeS_2$ (i.e., Cu = 34.64, Fe = 30.42, and S = 34.94 wt %), and the averaged microprobe analyses of other minerals (obtained from disseminated ores from Hole OUG-2) were used. In accordance with these data, we assumed the following average compositions (wt %) of pentlandite: Ni = 32, Fe = 34, and S = 34 for the picritic gabbro-dolerites and Ni = 35, Fe = 31, and S = 34 for taxitic gabbro-dolerites. The average pyrrhotite compositions (wt %) are Fe = 61 and S = 39 for picrites and Fe = 60 and S = 40 for taxites. All sampled compositional data on the silicate constituents of the rocks used in the calculations of the phase equilibria are available from the authors (Krivolutskaya et al. 2001a).

Additionally, we conducted a series of calculations for three compositions that represented the weighted mean assays for rocks of the Talnakh intrusion (Table 5.4). They were based on the data of M.B. Dneprovskaya from 29 holes (Dneprovskaya et al. 1987), the data of T.E. Zen'ko and V.A. Fedorenko et al. on Holes KZ-1799 and KZ-1713 (Czamanske et al. 1995), and our results on rocks from Hole OUG-2. All of the cited compositions characterize weighted mean estimates for the silicate constituents of the Talnakh rocks, which were calculated by the same techniques as those used for individual samples. These initial compositions are interesting because they potentially contain analogous information on the intrusion temperature and on the composition of the intratelluric phases and magmatic melt, which is "recorded" in the bulk compositions of rocks from lower units. When an ideal sheet-shaped body is considered, these compositions should not be principally different and should, in fact, represent the composition of the parental magma. In reality, it is impossible to correctly evaluate the weight proportions of all rock varieties that compose both continuous units and small bodies, lenses, etc. Therefore, we consider these weighted mean compositions to be model systems, which, perhaps, do not always correspond to the parental magma but are nevertheless useful in checking the correctness of the geochemical thermometric results obtained for the actual igneous rocks.

5.3.3.2 Calculation Conditions

An important aspect of geochemical thermometry is specifying the pressure and redox conditions of the parental magma intrusion. While studying the outer contact zones of the

Noril'sk-I intrusion, we determined that the host limestone contains monticellite rocks, and this allowed us to determine that the complex had been intruded under conditions of the gehlenite–monticellite depth facies (Godlevsky 1959). This estimate is consistent with data on the lithostatic load created by rocks overlying intrusions of the Noril'sk type (500–1,000 bar; Magnesium basic…1984). In compliance with this information, we calculated the crystallization equilibria using a pressure of 0.5 kbar. The estimates for the oxygen fugacity are less definite. These values are usually based on the whole-rock Fe_2O_3/FeO proportion, which should depend, similar to melts, on the temperature and composition. The calculations that were conducted for the rocks of magnesian intrusions in the northwestern Siberian Platform make use of the simplest relations of Fudali (1965): 1,200 °C yielded log values from −7.2 to −8.2 (Magnesium basic…1984). This interval overlaps with the range of buffer equilibria from approximately NNO+0.5 to QFM. In modeling the crystallization trajectories, we assumed that the conditions were more reduced, QFM – 0.5, because it is known that the Fe_2O_3/FeO ratios of rocks are usually disturbed by surface oxidation and the respective Fe_2O_3/FeO values become overestimated compared with the actual values of the magmatic melts (Carmichael and Ghiorso 1990).

In the geochemical thermometry calculations, it is necessary to specify the water concentration of the parental melts, which are the full-melting products of certain rocks, in our case, taxitic and picritic gabbro-dolerites. No such melts exist in nature, but their compositions should be used as the starting point in the calculations of the equilibrium crystallization trajectories (see the section devoted to the fundamentals of geochemical thermometry techniques). Hence, the initial water concentration of the melts of the taxitic and picritic gabbro-dolerites was taken to be equal to 0.1 wt %, a value that ensured melt undersaturation with respect to water at the assumed pressure and 0–70 % crystallinity (Al'meev and Ariskin 1996). This condition implies that the Talnakh rocks contain no primary magmatic amphibole and is consistent with the insignificant concentrations of water-bearing phases, e.g., biotite (Ryabov and Zolotukhin 1977). Phase equilibria were calculated for each sample successively, as the crystallinity of the melt progressively increased, with a step of 1 mol % crystallinity. The maximum crystallinity of the model systems did not exceed 80 % (20 % captured liquid).

5.3.4 Results of Geochemical Thermometry

5.3.4.1 Crystallization Order

For 19 of the 23 calculated trajectories, the crystallization of excess *Ol* was determined for the temperature interval of 1,250–1,540 °C. The second crystallizing phase was *Pl* (its minimum melting temperature is 1255°C), and the third was clinopyroxene (*Aug*). In four cases, cotectic *Ol*+*Pl* crystallization was identified at temperatures of ~1,200–1,235 °C, and the next mineral to crystallize was augite.

The initial crystallization temperature of augite never exceeded 1,189 °C. The character of the textural relationships of *Cpx* with *Ol* and *Pl* observed in the taxitic and picritic gabbro-dolerites suggests the absence of *Cpx* from the primary liquidus mineral assemblage (see above), and, hence, the calculated value of ~1,190 °C can be regarded as the lower limit for the probable temperature of the parental magma during emplacement. The anomalously high liquidus temperatures of high-Mg picritic gabbro-dolerites (T>1,450 °C) testify to the cumulative nature of these rocks, which were produced by the accumulation of olivine crystals. This follows from the calculated composition of the high-temperature olivine (91–92 % *Fo*), which is much more magnesian than the most primitive *Ol* (Fo_{82}) encountered in the rocks of the lower units. It is interesting to analyze the distribution structure of the liquidus temperatures of olivine). Note that the absence of calculated temperatures in the range of ~1,300–1,450 °C causes the apparently bimodal character of the histogram. We believe that this is a direct implication of the contrasting character of MgO distribution (and also the distribution of *Ol*) over the section of taxitic and picritic gabbro-dolerite units. This fact suggests that the parental magma parameters are constrained to temperatures below ~1,300 °C. In this connection, it is expedient to return to the initial crystallization of plagioclase. The highest calculated melting temperatures of this mineral are approximately 1,250 °C, but these evaluations pertain to the rocks with small amounts of excess *Ol* and, perhaps, *Pl*. Apparently, the actual crystallization temperatures of plagioclase could not exceed the liquidus values for the subcotectic rocks, when *Ol* and *Pl* started to crystallize concurrently. According to our calculations, these temperatures were 1,200–1,235 °C. Hence, provided the parental magmatic melt was saturated with respect to both *Ol* and *Pl* but undersaturated with respect to *Aug*, the probable temperature range of the parental magmas should be constrained to 1,190–1,235 °C. The considerations and evaluations presented above were based on the crystallization succession of the melts of the taxitic and picritic gabbro-dolerites and a textural interpretation of these rocks, which implies that clinopyroxene was absent from and plagioclase was present in the primary liquidus assemblage. Genetic models of this type inevitably involve uncertainties. The correctness of the assays could be justified by an intersection of (or closer spacing between) the calculated evolutionary lines for the melt near 1,200 °C.

5.3.4.2 Estimated Composition of the Parental Melt

Figure 5.8 shows the compositional trends of melts, calculated for the samples as functions of temperature over the interval of 1,120–1,350 °C for Al_2O_3 and FeO. The 17 lines in the plot represent, with a high degree of confidence, a

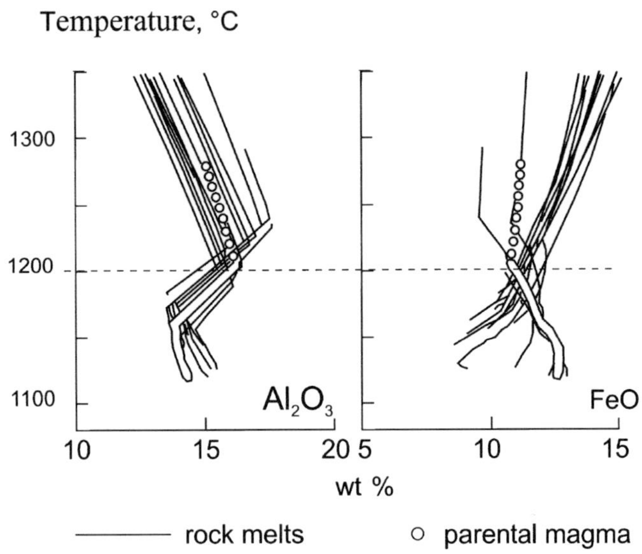

Fig. 5.8 Result of modeling crystallization Talnakh rocks on diagram T–Al_2O_3, FeO
After Krivolutskaya et al. (2001a)

single set of cotectics, in spite of the significant differences between the initial compositions. The other six trajectories demonstrate a wider scatter in the FeO and SiO_2 concentrations of the melts (the causes of this phenomenon remain unclear); thus, they were excluded from further consideration. The plot also shows the evolution trajectory of the melt during the equilibrium crystallization of the weighted mean composition of the Talnakh intrusion (based on our data from Hole OUG-2; Table 5.4). These data demonstrate that the convergence and intersections of the model lines can be reliably identified based on the concentrations of Al_2O_3 and FeO, Fig. 5.8 (and, to a lesser degree, SiO_2 and TiO_2) and are restricted to the temperature interval of 1,180–1,220 °C. These temperature and compositional intervals also include the trajectory for the weighted mean composition. Based on the results of the geochemical thermometry, we estimate the temperature of the parental Talnakh magma was 1,200 ± 20 °C (with an error slightly smaller than those of the geothermometers used in the COMAGMAT program). We estimated an average composition for this magmatic melt, which was calculated with the use of the 17 compositions, each of which is represented by an individual model line at 1,200 °C. Apparently, this composition can be grouped with typical tholeiitic ferrobasalt that is somewhat enriched in K. It contains approximately 8 wt % MgO and, thus, is notably more magnesian than the average composition of the "normal" flood basalt on the Siberian Platform but still does not attain the values of 10–12 wt % MgO that should determine, according to Zolotukhin et al., the picrite-basaltic character of the parental melt of the Siberian flood basalts (Magnesium basic…1984). Indeed the concentration 12 wt % MgO is a characteristic of a primary magma (with crystals), not a melt.

5.3.4.3 Chemistry of Minerals in the Original Assemblage

In the section devoted to the method of geochemical thermometry, we noted that the method is based on the assumption of the equilibrium of component distribution between minerals and melt in the original assemblage of each rock. Consequently, small differences between the estimated compositions of the intercumulus melts should produce similar calculated estimates for the original compositions of the minerals at the determined formation temperature of the rock. Because all of the 17 trajectories at 1,200 °C portrayed in Fig. 5.8 suggest an *Ol* + *Pl* assemblage, one can determine the compositions of these liquidus phases and assume that they represent the original characteristics of the Talnakh magma. The composition of olivine and plagioclase in each of the samples evaluate and the uncertainty intervals for the simulation result at 1,200 °C. The calculations for the average values at this temperature yield 81.3 ± 1.1 mol % *Fo* in *Ol* and 80.5 ± 2.4 mol % *An* in *Pl*. These values are in good agreement with microprobe analyses of the minerals, which demonstrate that the maximum Mg# of *Ol* in the taxitic gabbro-dolerites corresponds to $Fo_{82–83}$, and the highest temperature plagioclase contains no more than 83 % *An*.

5.3.4.4 Phase Composition of the Parental Magma for the Rocks of Lower Units

A noteworthy feature of geochemical thermometry is its ability to assay the phase proportion of the parental magma and the proportions of cumulus grains and melt in the original cumulates, from which the rock eventually crystallized. These proportions can be found by tackling the problem of thermodynamic equilibrium for each sample at the determined temperature of rock formation. The phase parameters of the parental magma can be estimated in an analogous manner, i.e., from the constructed trajectories of equilibrium crystallization for weighted mean compositions. Table 5.5 presents the full phase and chemical characteristics of the melt–crystal associations, which represent, at 1,200 °C, the aforementioned models for the composition of the Talnakh intrusion (Table 5.4). As is seen from these data, the *Ol* + *Pl* (with the predominance of *Ol*) cotectic assemblage is identified in all instances, with the overall percentage of intratelluric phenocrysts in the magma during its intrusion amounting to 10.8–14.0 wt %. The composition of the liquid constituent of the parental magma is also consistent with the estimate obtained from the contact gabbro-dolerites (Table 5.6). Figure 5.9 presents the model distribution of cumulative *Ol* + *Pl* over the generalized section for the lower part of Holes KZ-1799 and OUG-2. The thicknesses of the picritic and taxitic gabbro-dolerite units in this section are given as percentages of the overall thickness of the intrusive rocks penetrated by each of the holes. The upper pair of plots illustrates the relationships between the calculated amounts of original *Ol* crystals and the actual MgO and Ni concentrations of the rocks. Apparently, our geochemical thermometric data point to a gradual increase in the amounts of cumulus crystals from a few percent near the lower contact of the taxitic gabbro-dolerites to 50–60 % in the upper part of the picrite unit. Unfortunately, the strong correlation between the chemical and modal parameters cannot be regarded as a direct validation of the cumulus hypothesis because the *Ol* contents were calculated from the chemical compositions by the solution of the inverse problem. An independent proof could be provided by data on Ni, but the presence of sulfides obscures a correlation between this element and the modelled amounts of olivine. Analogous distributions are shown for the modelled plagioclase and alumina (Fig. 5.9). The results indicate that the amounts of *Pl* crystals (which are supposedly intratelluric) entrapped in the lower front vary from 10 to 30 % in the taxites and are much lower, 2–5 %, in the picrites. These values correlate not only with Al_2O_3 but also with the distribution of Sr, the concentrations of which were not taken into account in the geochemical thermometric simulations. This correspondence can be treated as an independent criterion pointing to the presence of early plagioclase crystals in the parental Talnakh magma. It is also interesting to compare the evolution of the amounts of complementary magmatic melts, which were calculated for each rock based on thermometric data and the distribution of incompatible elements. These relationships for P, Y, Sm, La, and Th are demonstrated in the lower plots of Fig. 5.9. For both holes, there appears to be a concurrent decrease in the amounts of melt from 70 to 85 % in the upper portion of the taxites to 45–50 % in the picrites. This decrease is strongly correlated with the distribution of P_2O_5, Y, Sm, La, and Th concentrations, which also decrease by factors of 2–2.5 toward the upper part of the picrate unit. Data on incompatible elements are independent in our approach, and this also testifies to an important role of the proportions of the parental melt and crystals in the formation of the chemical composition of the lower contact units.

The model simulations presented above pertain to the most complicated (in the structural sense) zone of the Talnakh intrusion, i.e., the lower units of the near-contact, taxitic, and picritic gabbro-dolerites. The genesis of these gabbro-dolerites has been discussed for years, and they have always attracted much attention because of their ore-generating role. This discussion results from several petrological and geological observations that are usually put forth as arguments

Table 5.5 Chemical and phase characteristics of parental magma for Talnakh intrusion based on modeling KOMAGMAT package

Component	1	2	3
Composition of primary melt, wt % (T=1,200°C)			
SiO_2	48.75	48.98	50.08
TiO_2	1.30	0.95	1.00
Al_2O_3	15.42	15.74	15.88
FeO	12.00	12.35	10.94
MnO	0.22	0.19	0.20
MgO	7.94	7.91	7.81
CaO	11.43	11.01	10.90
Na_2O	2.16	1.99	2.25
K_2O	0.65	0.64	0.74
P_2O_5	0.14	0.22	0.21
Phase composition of parental magma, wt %			
Melt	86.0	89.2	88.5
Ol	11.3 ($Fo_{80.3}$)	7.0 ($Fo_{79.8}$)	8.6 ($Fo_{81.8}$)
Pl	2.7 ($An_{78.1}$)	3.8 ($An_{80.1}$)	2.9 ($An_{78.5}$)
Ol+Pl	14.0	10.8	11.5

Note: 1–3 different average mean compositions (Table 5.4); *Ol*+*Pl* common part of intratelluric crystals at the moment of intrusion (*Ol* olivine, *Pl* plagioclase). After Krivolutskaya et al. (2001a)

Table 5.6 Composition of parental melt, contact gabbro-dolerites of the Talnakh intrusion, and typical volcanic rocks of the Siberian platform

Component, wt %	Talnakh intrusion: Model melt T=1,200°C (n=17)	Talnakh intrusion: Contactic gabbro-dolerites, Average (n=2)	Talnakh intrusion: Contactic gabbro-dolerites, KЗ-1,799/1,341.9	Average composition of trap rocks: Mokulaevsky Formation, mk_1	Average composition of trap rocks: Mokulaevsky Formation, mk_2	Average composition of trap rocks: Flood basalts, Putorana (n=300)
	1	2	3	4	5	6
SiO_2	49.44 (0.71)	45.30	48.57	49.70	49.10	49.85
TiO_2	1.20 (0.22)	1.60	1.21	1.00	1.14	1.30
Al_2O_3	15.44 (0.46)	14.32	15.35	15.80	16.30	15.53
FeO	11.43 (0.66)	13.16	12.63	11.26	11.26	12.04
MnO	0.22 (0.02)	0.20	0.22	0.17	0.19	0.19
MgO	8.06 (0.19)	7.99	7.24	7.31	7.22	7.30
CaO	11.53 (0.71)	11.04	10.40	11.20	11.20	11.43
Na_2O	1.82 (0.26)	2.02	1.93	1.80	2.05	1.93
K_2O	0.69 (0.22)	0.86	0.90	0.33	0.28	0.29
P_2O_5	0.19 (0.04)	0.12	0.15	0.09	0.11	0.12

Note: 1—standard deviation (1σ), 2—(Dodin and Batuev, 1971). 3–5—data from (Geology and ore deposits of the Noril'sk region 1994): *mk1* glomeroporphyritic basalts of the low Mokulaevsky Formation; *mk2* aphiric basalts of the upper Mokulaevsky; 6—data from Nesterenko et al. (1991). Fe was calculated in FeO; all compositions were normalized to 100 wt %

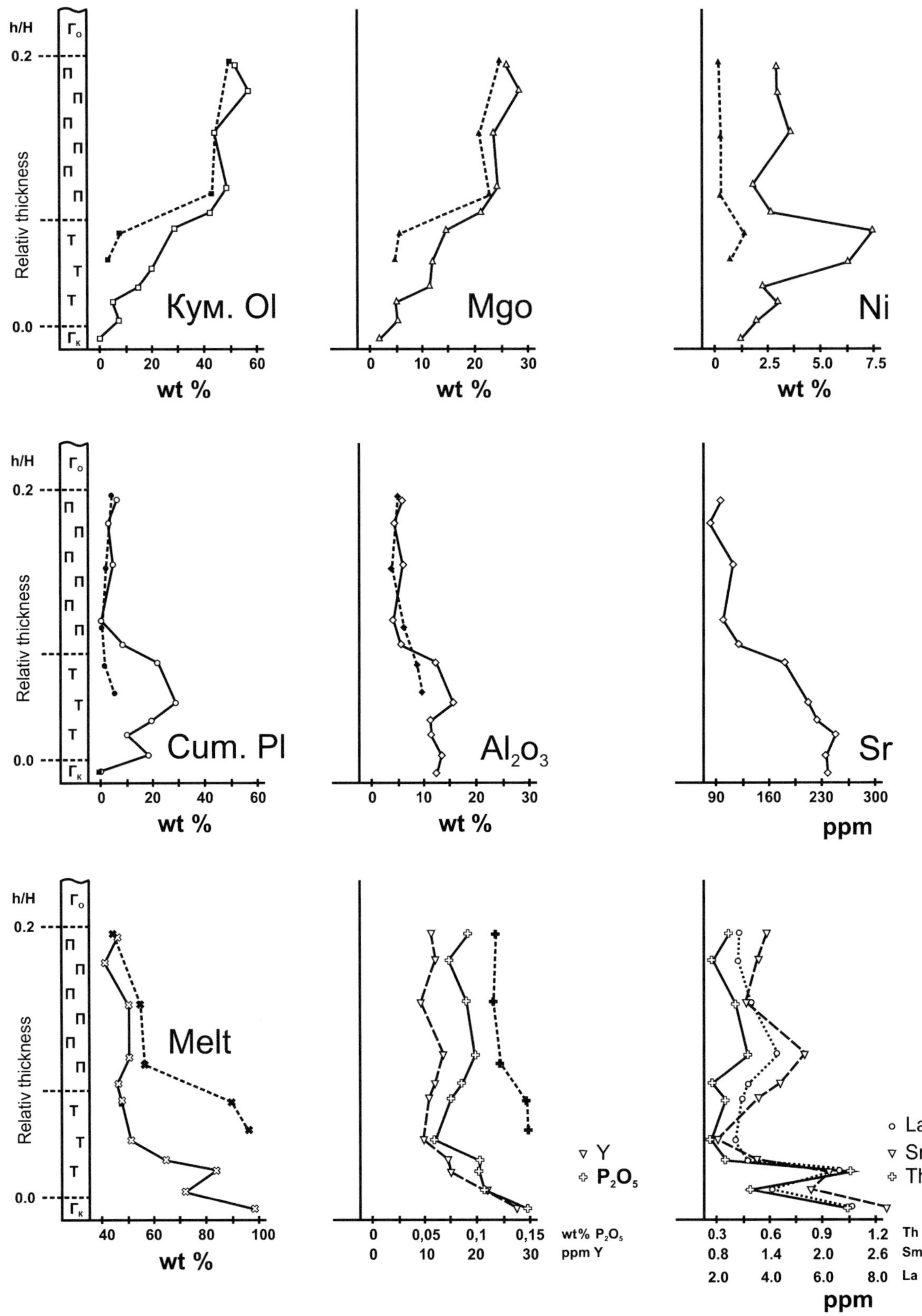

Fig. 5.9 Model distribution of the amounts of cumulus phases and melt compared with the concentrations of some major and trace elements Concentrations of elements are given in accordance with chemical analyses of the picritic and taxitic gabbro-dolerites from Holes OUG-2 (Tables 4.6 and 4.7) and KZ-1799 (Czamanske et al. 1995). Open symbols represent data on Hole KZ-1799; solid symbols are data on Hole OUG-2. After Krivolutskaya et al. 2001a

against hypotheses of the origin of layered intrusions by means of crystallization differentiation. These petrological and geological features are as follows: (1) inconsistencies between the compositions of contact gabbro-dolerites and the weighted mean composition of the intrusion; (2) the "suspended" position of picritic gabbro-dolerites in the magmatic succession; (3) traces of "magma flow" in the picritic units; (4) sharp boundaries between the picritic gabbro-dolerites and overlying olivine gabbro-dolerites; (5) tapering of the picritic units toward the peripheral parts and weak correlations between their thicknesses and that of the intrusion as a whole; (6) the restriction of taxitic gabbro-dolerites to the bottom of the intrusion; and (7) the presence of "wandering" leucogabbro bodies in the magmatic sequence and the non-systematic increase in the thicknesses of taxitic gabbroids toward the peripheral parts of the intrusive body. To interpret these and several other facts, a great variety of hypotheses have been proposed, which include (1) fluid–magmatic liquid immiscibility associated with the origin of picritic units from an individual picritic magma, which was segregated from the bulk of the parental magmatic melt (Ryabov et al. 2000); (2) the relatively early development of the taxitic gabbro-dolerites and leucogabbro due to the still earlier intrusion of *Pl*-enriched magmatic mush into the chamber (Likhachev 1965, 1978); (3) the later origin of the taxitic units due to magmatic and postmagmatic leucocratization and sulfurization of the rocks of the Main Layered Series (Zolotukhin 1997); (4) the origin of the taxitic gabbro-dolerites due to the influence of transmagmatic fluids on the gabbro-diorites (Zotov 1979, 1989); among others.

Our reconstructions of the formation conditions of the taxitic and picritic gabbro-dolerites and the chemical and phase characteristics of the Talnakh magma provide a consistent solution of most of the problems discussed above based on the concept of a heterogeneous parental magmatic system with allowance for the sorting and accumulation of intratelluric material. The main results of the conducted analysis include the good agreement (within the accuracy of the method) between the composition of the contact gabbro-dolerites and the calculated composition of the parental melt (Tables 5.6). This situation testifies to the realistic character of our models because the origin of the composition of the contact zones has not yet been adequately interpreted. Conceivably, insignificant amounts of intratelluric phenocrysts contained in the rocks can be accounted for by the known effect of pressing of crystals away from contact zones as the magma flows and fills intrusive chambers (Naslund and McBirney 1996; Sharapov et al. 1997). From the perspective of regional magmatism, the parental magmatic melt has a typical tholeiitic composition, which corresponds to ferrobasalt that is slightly enriched in K_2O. The comparison between the major-component composition of the parental melts and that of the volcanic flood basalts of the Noril'sk district reveals the closeness of the former to the late-stage volcanics of the Siberian Platform, particularly to the aphyric and glomerophyric basalts of the Mokulaevsky Formation (mk_1 and mk_2, respectively; and Morongovsky Formation as well; Geology and Ore… 1994). Because we compare only the composition of the melt (but not the magma as a whole), it seems to be more appropriate to compare it with aphyric varieties, although they are less magnesian than the glomerophyric basalts. It is also interesting to note that the composition of the parental melt is close to the average compositions of basalts on the Putorana Plateau, which account for more than 90 vol. % of the plateau basalts of the Siberian Platform (Nesterenko et al. 1991; Table 5.6). Another important result concerns the original *Ol* + *Pl* cotectic nature of the Talnakh magma. This idea is not new: the occurrence of early *Pl* and *Ol* in the Talnakh magma was suggested by Korovyakov et al. (1963), Oleinikov (1979), Zolotukhin et al. (1975, 1997), Godlevsky and Likhachev (1984), and other researchers. This problem was considered in much detail by Likhachev (1965, 1977, 1982, 1997), who has also arrived at the conclusion that the intruded magma was saturated with respect to *Ol* and *Pl*. The gravitationally induced segregation of minerals occurred when the magma ascended to the surface, resulting in a chamber that was initially filled with plagioclase-rich mush. This mush was then pushed away and forced to the peripheral portions of the massif by the main volume of melt, which was enriched in olivine. Within the realm of this hypothesis, the taxitic gabbro-dolerites and leucogabbro were produced by early batches of the magma; this hypothesis explains the position of these rocks near the bottom and their occurrence as "wandering" bodies among other gabbro-dolerites.

Our geochemical thermometric results do not contradict this cotectic interpretation but demonstrate that the average Talnakh magma contained approximately 3 % intratelluric *Pl* (Table 5.5). This estimate should somehow agree with the amount of leucogabbroids, which are unevenly distributed throughout the volume of the Talnakh intrusion. However, the method used in our research cannot explain the genesis of the early phases of olivine and, particularly, plagioclase, for which the compositions An_{90} and more calcic were reported. It is quite probable that this scatter of values is explained by the fact that some crystals are intratelluric phenocrysts and, thus, xenogenic with respect to the melt that entrained them, as was suggested by Likhachev (1997). The results of our computer simulation of phase equilibria suggest that the picritic and taxitic gabbro-dolerites could be produced by the same melt. This phenomenon explains several geologic observations. First, these gabbro-dolerites are traces of melt flow in the picrite unit and the gradual tapering toward the peripheral parts of the intrusion and the chamber's narrower places, which can be explained as resulting from the accumulation of denser intratelluric *Ol* phenocrysts in deeper zones of the intrusion. The uneven distribution of these crystals caused the internal heterogeneity of the picritic unit, and the unit's sharp boundary with the overlying olivine gabbro-dolerites can be naturally interpreted as a consequence of the practically full settling of the intratelluric phase. In this connection, it is pertinent to recall that the sharp character of the upper boundary of the picritic gabbro-dolerite unit was first obtained by the direct modeling of in-chamber differentiation of the Talnakh magma in compliance with the convection–accumulation mechanism (Dneprovskaya et al. 1987). Finally, one of the most disputable problems is the spatial distribution of taxitic and picritic gabbro-dolerites.

The fact that the *Ol*-enriched and denser picritic rocks rest on less dense taxites caused the researchers of the latter to refer to it as "suspended" in the sense of the hydrodynamic instability of the system (Zolotukhin 1997). It should be mentioned that the occurrence of a zone of MgO enrichment upward from the lower contact (the so-called S-shaped profile) resulted from an increase in the accumulation grade of settling olivine crystals associated with the concurrent displacement of the crystallization front in the opposite direction (Marsh 1989; Jaupart and Tait 1995). A similar situation can be observed in several differentiated mafic massifs. Data on the modeling of the dynamics of this process indicate that upward-directed olivine accumulation results from a decrease in the displacement velocity of this front with increasing distance from the contact (due to a decrease in the heat flow from the intrusive chamber; Frenkel' et al. 1988). This conclusion is fully compatible with our geochemical thermometric data, which indicate an increase in the percentage of cumulative *Ol* and *Pl* upward from the lower contact (Fig. 5.9). Simultaneously, a quantitative interpretation was provided for the well-known distribution profile of incompatible components (P, Ti, K, Na, Y, La, Sm, and Th) in the cross section of the Talnakh intrusion with a small minimum in the upper part of the taxitic gabbro-dolerite unit. According to our calculation results, this minimum corresponds to the smallest fraction of the parental magmatic liquid. Relationships of this type seem to be apparent, but they have never been considered in other genetic models.

5.4 Conclusions

1. Composition of melt for ore-bearing intrusion was close to tholeiitic basalt with elevated MgO content (8 wt %), H_2O = 0.5–0.7 %, and Cl up to 0.2 %. There is no difference in melt composition between ore-bearing and barren (Low Talnakh-type) intrusions.
2. We used a geochemical thermometry technique and the COMAGMAT computer model to estimate the phase and chemical parameters of the parental magma of the Talnakh intrusion and the melt–crystal mixtures that gave rise to the taxitic and picritic gabbro-dolerites in the lower zone of the massif. These results demonstrate the cotectic (*Ol* + *Pl*) nature of the Talnakh magma, which was intruded at a temperature of approximately 1,200 °C and contained 10–15 % intratelluric phenocrysts, predominantly *Ol* (7–11 %). The composition of the parental magmatic melt (the liquid constituent of the magma) corresponded to tholeiitic ferrobasalt with somewhat elevated concentrations of MgO (~8 wt %) and K_2O (~0.7 wt %) relative to the "normal" flood basalts of the Siberian Platform. The composition of the parental melt was most similar to that of rocks produced during the late volcanic stage.
3. The composition of the parental magmatic melt was determined to correspond to the composition of the liquid that gave rise to the intercumulus materials in the rocks of the taxite and picrite units. The compositional variations of the rocks from the lower part of the Talnakh intrusion can be explained as the result of variations in the proportions of *Ol* and *Pl* intratelluric crystals and the intercumulus liquid. In light of these data, quantitative interpretations can be made for the known profile of the distribution of incompatible elements.

References

Al'meev RR, Ariskin AA (1996) Computer simulation of melt–mineral equilibria in a water-bearing basaltic system. Geochem Int 7:624–636

Aplonov VS (1995) Fluid regime and platimun potencial problems of basic differentiated intrusions. Platimun of Russia. M. Geoinformmark, pp 102–106 (in Russian)

Ariskin AA (1999) Phase equilibria modeling in igneous petrology: use of COMAGMAT model for simulating fractionation of ferro-basaltic magmas and the genesis of high-alumina basalt. J Volcanol Geotherm Res 90:115–162

Ariskin AA, Barmina GS (1999) An empirical model for the calculation of spinel–melt equilibrium in Mafic igneous systems at atmospheric pressure: II. Fe–Ti oxides. Contrib Mineral Petrol 134:251–263

Ariskin AA, Barmina GS (2000) Simulation of phase equilibria at basalt magma crystallization. Nauka, Moscow (in Russian)

Ariskin AA, Konnikov EG, Danyushevskii LV et al (2009) The Dovyren intrusive complex: problems of petrology and Ni sulfide mineralization. Geochem Int 47(5):425–453

Ariskin AA, Danyushevsky LV, Bychkov KA et al (2013) Modeling solubility of Fe-Ni sulfides in basaltic magmas: the effect of nickel. Econ Geol 11:1983–2003

Arndt NT, Czamanske GK, Wooden JL, Fedorenko VA (1993) Mantle and crustal contributions to continental flood volcanism. Tectonophysics 223:39–52

Ballhaus CG, Stumpfl EF (1986) Sulfide and platinum mineralization in the Merensky reef: evidence from hydrous silicates and fluid inclusions. Contrib Mineral Petrol 94:193–204

Barmina GS, Ariskin AA, Koptev-Dvornikov EV, Frenkel' MY (1988) Estimation of the compositions of primary cumulative minerals in differentiated traps. Geokhimiya 8:1108–1119

Barmina GS, Ariskin AA, Frenkel' MY (1989) Petrochemical types and crystallization conditions of Plagiodolerites in the Kronotsky Peninsula, Eastern Kamchatka. Geokhimiya 2:192–206

Bergantz GW (1990) Melt fraction diagrams: the link between chemical and transport models. Rev Mineral 24:239–257

Black BA, Hauri EH, Elkins-Tanton LT, Brown SM (2014) Sulfur isotopic evidence for sources in Siberian traps. Earth Planet Sci Lett 394:58–69

Bodner RJ, Student JJ (2006) Melt inclusions in plutonic rocks: petrography and microthermometry. In: Melt inclusions in plutonic rocks. Mineralogical Association of Canada, Ottawa, pp 1–26

Boudreau AE (1988) Investigations of the stillwater complex 4: the role of volatiles in the petrogenesis of the J.M. Reef, Minneapolis Adit section. Can Mineral 26:193–208

Boudreau AE, Mathez EA, McCallum IS (1986) Halogen geochemistry of the stillwater and bushweld complex: evidence for transport of the platinum-group elements by Cl-rich fluids. J Petrol 27:967–986

Bulgakova EN (1971) Physicochemical conditions of formation of the Noril'sk differentiated trap intrusions. In: Traps of the Siberian platform and their metallogeny, Irkutsk, pp 36–37 (in Russian)

Carmichael ISE, Ghiorso MS (1990) The effect of oxygen fugacity on the Redox state of natural liquids and their crystallizing phases. In: Nicholls J, Russell JK (eds) Modern methods of igneous petrology: understanding magmatic processes. Rev Miner 24:191–212

Chalokwu CI, Grant NK, Ariskin AA, Barmina GS (1993) Simulation of primary phase relations and mineral compositions in the Partridge River intrusion, Duluth Complex. Implications for the Parent Magma Composition, Minnesota. Contrib Mineral Petrol 114:539–549

Czamanske GK, Wooden JL, Zientek ML, Fedorenko VA, Zen'ko TE, Kent J, King BS, Knight RJ, Siems DF (1994) Geochemical and isotopic constraints on the petrogenesis of the Noril'sk-Talnakh ore-forming system. In: Lighfoot PC, Naldrett AJ (eds) Proceedings of the Sudbury-Noril'sk symposium, special publication 5, Geological Survey, Ontario, pp 313–342

Czamanske GK, Zen'ko TE, Fedorenko VA et al (1995) Petrographic and geochemical characterization of ore-bearing intrusions of the Noril'sk type, Siberia; with discussion of their origin. Res Geol Spec Iss 18:1–45

Danyushevsky LV, Della-Pasqua FN, Sokolov S (1999) Re-equilibration of melt inclusions trapped by magnesian olivine phenocrysts from subduction-related magmas: petrological implications. Contrib Mineral Petrol 138(1):68–83

Distler VV, Sluzhenikin SF, Cabri LJ, Krivolutskaya NA, Turovtsev DM, Golovanova TA, Mokhov AV, Knauf VV, OleshkeYich OI (1999) Platinum ores of the Noril'sk Layered intrusions: magmatic and fluid concentration of noble metals. Geol Ore Deposits 41(3):214–237

Dneprovskaya MB, Dneprovsky NV (1988) Quantitative description of geochemical regularities in the structures of geological objects using random functions, with the Talnakh intrusion as an example. Geokhimiya 1:128–137

Dneprovskaya MB, Frenkel' MY, Yaroshevsky AA (1987) A quantitative model for layering in the Talnakh intrusion, Noril'sk region. In: Simulating systems of ore mineralization. Nauka, Novosibirsk, pp 96–106 (in Russian)

Dodin DA, Batuev BN (1971) Geology and petrology of the Talnakh differentiated intrusions and their metamorphic aureole. In: Petrology and ore resource potential of the Talnakh and Noril'sk differentiated intrusions. Nedra, Leningrad, pp 31–100 (in Russian)

Dodin DA, Batuev BN, Mitenkov GA, Izoitko VM (1971) Atlas of rocks and ores of the Noril'sk copper–nickel deposits. Nedra, Leningrad (in Russian)

Dyuzhikov OA, Distler VV, Strunin BM et al (1988) Geology and ore potential of the Noril'sk region. Nedra, Moscow

Feoktistov GD (1978) Petrology and formation conditions of trap sills. Nauka, Novosibirsk (in Russian)

Frenkel' MYa, Yaroshevskii AA, Koptev-Dvornikov EV et al (1985) Crystallization mechanism of layering in layered intrusions. Zap Vses Mineral O–va Part CXIV, (3):257–274 (in Russian)

Frenkel' MY, Ariskin AA, Barmina GS et al (1987) Geochemical thermometry of igneous rocks: principles and examples of application. Geokhimiya 11:1546–1562

Frenkel' MY, Yaroshevsky AA et al (1988) Dynamics of the chamber differentiation of basic magmas. Nauka, Moscow (in Russian)

Frutos J, Oyarzun JM (1975) Tectonic and geochemical evidence concerning the genesis of El Laco magnetite lava flow deposit. Econ Geol 70(5):988–990

Fudali RF (1965) Oxygen fugacities of basaltic and andesitic magmas. Geochim Cosmochim Acta 29:1063–1075

Geology and Ore Deposits of the Noril'sk Region (1994) Proceedings of the VII international platinum symposium, with a special session of IGCP project Moscow – Noril'sk

Godlevsky MN (1959) Traps and ore-bearing intrusions of the Noril'sk region. Gosgeoltekhizdat, Moscow, 61 p. (in Russian)

Godlevsky MN, Likhachev AP (1984) Types and distinctive features of ore-bearing formations of copper-nickel deposits. In: Geology and metallogeny of copper deposits. Springer, New York, pp 124–134 (in Russian)

Hawkesworth CJ, Lightfoot PC, Fedorenko VA et al (1995) Magma differentiation and mineralization in the Siberian continental flood basalts. Lithos 34:61–81

Hoover JD (1989) The chilled marginal Gabbro and other contact rocks of the Skaergaard intrusion. J Petrol 30:441–476

Inclusions in minerals and processes in the Earth's mantle (2005). Max-Plank-Institute, Mainz, 103 p

Jaupart C, Tait S (1995) Dynamics of differentiation in magma reservoirs. J Geophys Res 100:17615–17636

Jaupart C, Tait S (2005) Dynamics of differentiation in magma reservoirs. J Geophys Res 100:17615–17636

Konnikov EG, Kovyazin SV, Nekrasov AN, Simakin SG (2005) Interaction of magmatic fluids and mantle magmas with lower crustal rocks: evidence from inclusions in the minerals of intrusions. Geochem Int 43(10):939–958

Korovyakov IA, Nelyubin AE, Raikova ZA, Khortova LK (1963) The origin of Noril'sk trap intrusions containing copper–nickel ores. Gosgeotekhizdat, Moscow. Tr. Vses. Nauch.–Issled. Inst. Miner. Syr'ya, Novaya ser., 1963, issue 9 (in Russian)

Krivolutskaya NA (2011) Formation of PGM–Cu–Ni deposits in the process of evolution of flood-basalt Magmatism in the Noril'sk region. Geol Ore Depos 53(4):309–332

Krivolutskaya NA, Sobolev AV (2001) Magmatic inclusions in olivines from intrusions of the Noril'sk region, Northwestern Siberian platform: evidence for primary melts. Dokl Earth Sci 381(3):1047–1052

Krivolutskaya NA, Ariskin AA, Sluzhenikin SF, Turovtsev DM (2001a) Geochemical thermometry of rocks of the Talnakh intrusion: assessment of the melt composition and the crystallinity of the parental magma. Petrology 9(5):389–414

Krivolutskaya NA, Sobolev AV, Sluzhenikin SF et al (2001b) Melt inclusions in olivines of the Talnakh type intrusions. In: Proceedings of the X international conference on thermobarogeochemistry. VNIISIMS, Aleksandrov, pp 141–157 (in Russian)

Krivolutskaya NA, Sobolev AV, Sluzhenikin SF, Pokrovsky BG (2004) Olivine-hosted magmatic inclusions from the Noril'sk intrusions: application to origine of Pt-Cu-Ni deposits (Russia). metallogeny of the Pacific northwest: tectonics, magmatism and metallogeny of active continental margins. In: Khanchuk AI, Gonevchuk GA, Mitrokhin AN , Simanenko LF, Cook NJ, Seltmann R (ed) Dalnauka, Vladivostok, pp 296–299

Kutolin VA (1972) Problems of the petrochemistry and petrology of basalts. Nauka, Moscow (in Russian)

Kuznetsov YA (1990) Problems of origin and analysis of associations of magmatic bodies, in selected works, vol III. Nauka, Novosibirsk (in Russian)

Li CS, Ripley EM, Naldrett AJ (2009) A new genetic model for the giant Ni–Cu-PGE sulfide deposits associated with the Siberian flood basalts. Econ Geol 104:291–301

Lightfoot PC, Hawkesworth CJ, Hergt J et al (1993) Remobilisation of the continental lithosphere by a mantle plume: major- and trace-element, Sr-, Nd- and Pb-isotope evidence from Picritic and Tholeiitic lavas of the Noril'sk District, Siberian Trap, Russia. Contrib Mineral Petrol 114:171–188

Likhachev AP (1965) The role of Leucocratic Gabbro in the origin of Noril'sk differentiated intrusions. Izv Akad Nauk SSSR Ser Geol 12:50–66 (in Russian)

Likhachev AP (1977) On the crystallization conditions of trap magmas in the Northwestern part of the Siberian platform. Zap Vses Mineral O–va 5:594–606 (in Russian)

Likhachev AP (1978) On the formation conditions of ore-bearing and non-bearing Mafic–Ultramafic magmas. Dokl Akad Nauk SSSR 238(2):447–450 (in Russian)

Likhachev AP (1982) Formation conditions of copper–nickel deposits. Sov Geol 6:31–46 (in Russian)

Likhachev AP (1996) Emplacement dynamics of the Talnakh ore-bearing intrusions and related PGM–Cu–Ni ores. Otech Geol 8:20–26 (in Russian)

Likhachev AP (1997) Trap magmatism and platinum–copper–nickel ore mineralization in the Noril'sk District. Otech Geol 10:8–19 (in Russian)

Magnesian Basic Rocks from the Western Siberian Platform and the Problem of Nickel-Bearing Potential (1984) Sobolev VS (ed). Nauka, Novosibirsk (in Russian)

Marsh BD (1989) Magma chambers. Annu Rev Earth Planet Sci 17:439–474

Maurel C, Maurel P (1982) Etude experimentale de l'equilibre Fe2+––Fe3+ dans les spinelles chromiferes et les liquides silicates basiques coexistants a 1 atm. C R Acad Sci Paris. 285:209–215

Naldrett AJ (1992) A model for the Ni–Cu–PGE ores of the Noril'sk region and its application to other areas of flood basalt. Econ Geol 87:1945–1962

Naldrett AJ (2009) Ore deposits related to flood basalts, Siberia, in new developments in magmatic Cu-Ni and PGE deposits. Geological Publishing House, Beijing, pp 141–179

Naldrett AJ, Lightfoot PC, Fedorenko VA et al (1992) Geology and geochemistry of intrusions and flood basalts of the Noril'sk region, USSR, with applications for the origin of the Ni–Cu ores. Econ Geol 87:975–1004

Naldrett AJ, Fedorenko VA, Asif M et al (1996) Controls on the composition of Ni–Cu sulfide deposits as illustrated by Those at Noril'sk, Siberia. Econ Geol 91:751–773

Naslund HR, McBirney AR (1996) Mechanisms of formation of igneous layering. In: Cawthorn RG (ed) Layered intrusions. Elsevier, Amsterdam, pp 1–43

Neruchev SS, Prasolov EM (1995) Fluid-geochemical model for platinum deposits related to flood basalt magmatism. Platimun of Russia. M. Geoinformmark, pp 94–101 (in Russian)

Nesterenko GV, Tikhonenkov PI, Romashova TV (1991) Basalts of the Putorana plateau. Geokhimiya 10:1419–1425

Oleinikov BV (1979) Geochemistry and ore genesis in platform basic rocks. Nauka, Novosibirsk (in Russian)

Petrology and Ore Potential of the Traps of the Northern Siberian Platform (1978) Zolotukhin VV, Vilensky AM (eds). Nauka, Novosibirsk (in Russian)

Pokrovsky BG, Sluzhenikin SF, Krivolutskaya NA (2005) Interaction conditions of Noril'sk trap intrusions with their host rocks: isotopic (O, H, and C) evidence. Petrology 13(1):49–72

Rad'ko VA (1991) Model of dynamic differentiation of intrusive traps at the Northwestern Siberian trap. Geol Geofiz 32(11):19–27 (in Russian)

Renner J, Evans B, Hirth G (2000) On the rheologically critical melt fraction. Earth Planet Sci Lett 181:585–594

Ryabov VV, Pavlov AL (1991) Magnetite lava in traps of the Siberian platform. Dokl Earth Sci 319:1193–1197

Ryabchikov ID, Solovova IP, Ntaflos T et al (2001) Subalkaline picrobasalts and plateau basalts from the Putorana plateau (Siberian Continental Flood Basalt Province): II. Melt inclusion chemistry, composition of "Primary" magmas and *P–T* regime at the base of the Superplume. Geochem Int 39(5):432–446

Ryabchikov ID, Kogarko LN, Solovova IP et al (2009) Physicochemical conditions of magma formation at the base of the Siberian Plume: insight from the investigation of melt inclusions in the Meymechites and Alkali Picrites of the Maimecha–Kotui Province. Petrology 17(3):287–299

Ryabov VV, Zolotukhin VV (1977) Minerals of differentiated traps. Nauka, Novosibirsk (in Russian)

Ryabov VV, Shevko AY, Gora MP (2000) Igneous rocks of the Noril'sk region, vol 1, 2. Nonparel, Novosibirsk (in Russian)

Sharapov VN, Cherepanov AN, Popov VN, Lobov AG (1997) Dynamics of cooling of Mafic melt filling a funnel-shaped intrusive reservoir. Petrology 5(1):10–22

Sinton J, Langmuir C, Bender J, Detric R (1992) What is a magma chamber? Ridge Events 3(1):46–48

Sobolev AV (1996) Melt inclusions in minerals as a source of principle petrological information. Petrology 4(3):209–220

Sobolev AV (1997) The problems of origin and evolution of mantle magmas. Doctoral (Geol.–Min.) dissertation, Vernadsky Inst. Geokhim. Analit. Khim., Akad. Nauk SSSR, Moscow, 225 p (in Russian)

Sobolev AV, Slutsky AB (1984) Composition and crystallization conditions of the initial melt of Siberian Meimechites in connection to the general problem of Ultramafic magmas. Geol Geophys 25(12):97–110 (in Russian)

Sobolev AV, Krivolutskaya NA, Kuz'min DV (2009a) Petrology of the parental melts and mantle sources of Siberian trap magmatism. Petrology 17(3):253–286

Sobolev AV, Sobolev SV, Kuz'min DV et al (2009b) Siberian Meimechites: origin and relationship between traps and kimberlites. Geol Geophys 50:999–1033

Turovtsev DM (2002) Contact metamorphism of the Noril'sk intrusions. Nauchny Mir, Moscow, 293 p (in Russian)

Vilensky AM, Oleinikov BV (1970) The principal factors of diversity and classification problems of traps of the Siberian platform. In: Geology and petrology of intrusive traps of the Siberian platform. Nauka, Moscow, pp 5–25 (in Russian)

Vortsepnev VV (1978) Temperature, pressure, and geochemical conditions of formation of the Talnakh copper–nickel deposit. Cand. Sc. (Geol.–Min.) dissertation. Moscow State University, Moscow (in Russian)

Zhmodik AS (2002) Composition and thermodynamic conditions of development of magmatic ore mineralization in basalts of the cleft segment of the Juan de Fuca Ridge Cand. Sci (Geol–Mineral) Novosibirsk: Inst Geol Geophys, 25 p (in Russian)

Zolotukhin VV (1997) Mafic pegmatoids of the Noril'sk ore-bearing intrusions and the problem of ore mineralization of Noril'sk type, Tr Ob"edin Inst Geol Geofiz Mineral. Sib, Novosibirsk. Otd. Ross. Akad. Nauk, issue 834 (in Russian)

Zolotukhin VV, Laguta ON (1985) On the fractionation of Magnesian Mafic melts and trap diversity of Siberian platform. Dokl Akad Nauk SSSR 280(4):967–972 (in Russian)

Zolotukhin VV, Vasil'ev Y (1986) In: Vasilenko BV (ed) Problems of platform magmatism: examples from the Siberian platform. Nauka, Novosibirsk, (in Russia

Zolotukhin VV, Ryabov VV, Vasil'ev YR, Shatkov VA (1975) Petrology of the Talnakh ore-bearing differentiated trap intrusion). Nauka, Novosibirsk (in Russian)

Zolotukhin VV, Vilensky AM, Dyuzhikov OA (1986) Basalts of the Siberian platform. Nauka, Novosibirsk (in Russian)

Zotov IA (1979) The genesis of trap intrusions and metamorphic rocks of Talnakh. Nauka, Moscow (in Russian)

Zotov IA (1989) Transmagmatic fluids in magmatism and ore mineralization. Nauka, Moscow (in Russian)

6 Relationship Between the Lavas and the Ore-Bearing Massifs

The geologic relationships between the basalts and ore-bearing intrusions are described based on the example of the Southern Maslovsky massif, which intruded basalts of the Nadezhdinsky Formation. Therefore, these rocks were formed in post-Nadezhdinsky time. The ore-bearing intrusions are similar to the volcanic rocks of the Morongovsky Formation in terms of their rare elements and isotopic compositions (Sr, ε_{Nd}), but they contain higher MgO concentrations (10–12 wt % on average) and a heavier S isotope composition (up to 18‰ compared with 6–7‰ in the basalts).

The possibility of comagmatic relationships between intrusive and volcanic rocks has long been a matter of discussion and is not yet settled. Some researchers continue to distinguish volcano-plutonic associations (Rad'ko 1991; Dyuzhikov et al. 1992; Naldrett 2009; Li et al. 2009; Lightfoot and Zotov 2014), whereas others believe that massifs in the Noril'sk Complex were produced from derivatives of other magmas (Likhachev 1994, 2006; Latypov 2002, 2007; Malitch et al. 2010, 2014; Krivolutskaya et al. 2012a).

The principal candidates for comagmatic rocks of the ore-bearing intrusions were first thought to be the picrites of the Gudchikhinsky Formation because all of the rocks are highly magnesian. However, later data on the distribution of trace elements in the rocks demonstrated the differences between the Gudchikhinsky picrites and rocks of the Nori'lsk Complex. Gudchikhinsky rocks are the most primitive melts in the Noril'sk area. They show almost no evidence of their crustal contamination (Sobolev et al. 2009), which was suggested for other volcanites. These rocks have no negative Ta–Nb anomalies, which are typical of all of the trap rocks, and exhibit negative, but not positive, Pb anomalies. Therefore, the Gudchikhinsky melts are principally different from the melts that gave rise to the ore-bearing massifs: their trace element patterns are not as enriched, including that these rocks are depleted in HREE and show no negative Ta–Nb anomalies. Volcanic rocks of the Gudchikhinsky Formation and ore-bearing picritic gabbro-dolerites strongly differ in isotope composition, especially ε_{Nd} and $^{87}Sr/^{86}Sr$ (+5.5; 0,703 and +1; 0,706 consequently). Another remarkable feature is the extremely high Ni concentration in the olivine, which is almost twice as high in olivine as the ore-bearing units and has elevated Ca concentrations (0.40 and 0.22 wt % NiO; 0.30 and 0.12 wt % CaO for Fo_{82}, respectively). This result indicates that the picrite basalts and picrite gabbro-dolerites in intrusions of the Noril'sk Complex could not be derivatives of a single parental melt, as was previously hypothesized.

In their models of the ore-forming processes, A. Naldrett and P. Lightfoot (Naldrett and Lightfoot 1994; Naldrett et al. 1992; Lightfoot et al. 1994; Keays and Lightfoot 2007) attached much importance to the rocks in the Tuklonsky Formation as possible crystallization products of the parental magma of the ore-bearing intrusions. This conclusion is based on the elevated Mg# of the Tuklonsky rocks and their geochemical similarities with the gabbro-dolerites of the intrusions of the Noril'sk Complex. However, the vertical section of the Tuklonsky Formation was so far examined only in the eastern portion of the area, within the Sunduk paleovolcanic structure (in the vicinity of Lake Glubokoe, Fig. 6.1—section 1 F (Lightfoot et al. 1994)). The rocks were added as an important constituent to the vertical section constructed based on data from the Kharaelakh depression (Boreholes SG-9 and SG-32) (Lightfoot et al. 1990, 1993; Brügman et al. 1993) and assumed to be the reference for the area. This combination resulted in the "occurrence" of a thick (240 m) unit of magnesian rocks. However, the Tuklonsky high-Mg lavas (from 8–9 to 16 wt % MgO) were found only within a small (10 by 3 km) paleovolcanic structure at Mount Sunduk (Lightfoot et al. 1994). The possible analogues of these rocks were thought (Fedorenko et al. 1996) to be picrite basalts exposed in the Mikchangdinsky flow. However, we proved (Krivolutskaya et al. 2012b) that these rocks are cumulates of the Nadezhdinsky Formation.

Therefore, the vertical sections of the Tuklonsky rocks in the Noril'sk Trough and elsewhere in the area are dominated

N.A. Krivolutskaya, *Siberian Traps and Pt-Cu-Ni Deposits in the Noril'sk Area*,
DOI 10.1007/978-3-319-17205-7_6

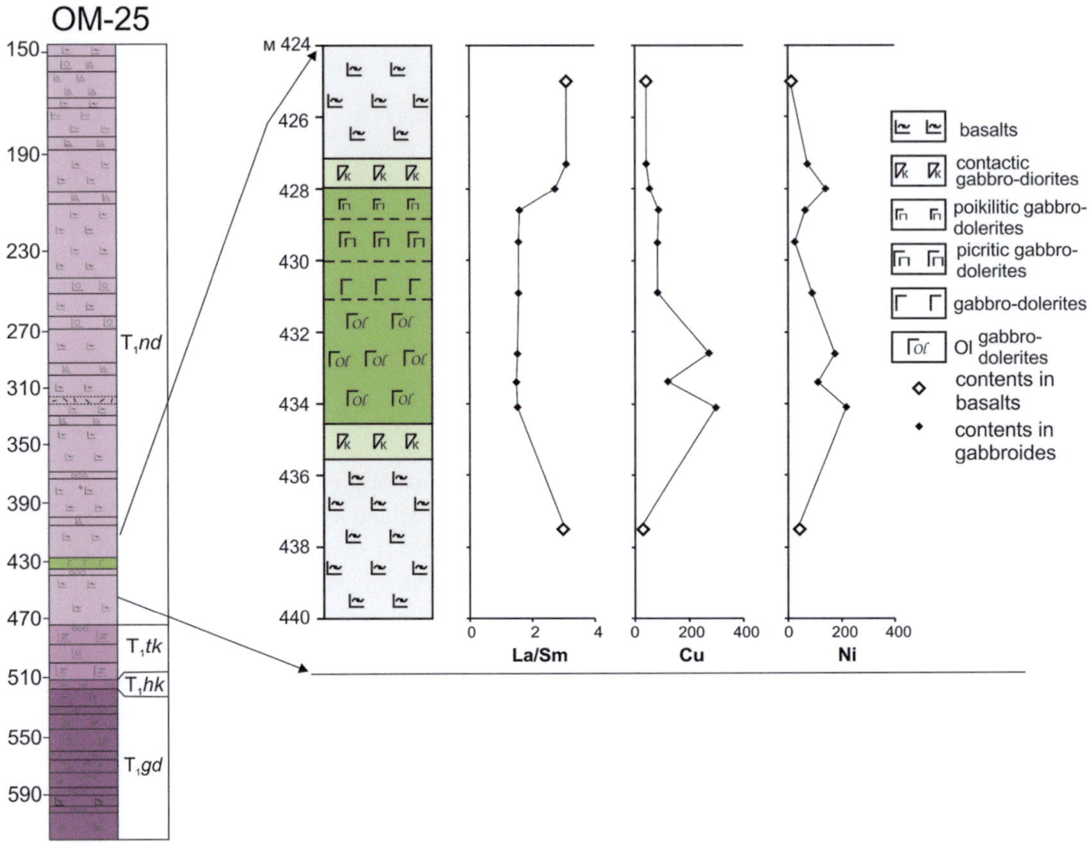

Fig. 6.1 Position of the Maslovsky sill inside tuff-lavas sequence After Krivolutskaya and Rudakova (2009)

by tholeiitic basalts (containing close to 7 wt % MgO), and locally occurring, thin bodies of picritic varieties are rare. The weighted mean composition of the rocks of the formation exposed over a length of >100 km is generally similar to the composition of the overlying tholeiite basalts. Comparing the geochemistry of the Tuklonsky rocks and gabbro-dolerites of the Noril'sk Complex, it is apparent that, because they have similar distributions of trace elements, the former rocks are depleted in U and enriched in Eu. Therefore, they have different isotopic compositions: for example, Tuklonsky basalts and picrobasalts have low $\varepsilon_{Nd}=-4, -5$, while in the intrusions, it varies approximately 0. Moreover, the olivine compositions in the hypothetical intrusive and volcanic varieties are notably different; the NiO concentrations are 0.11 and 0.21 wt % in the picrite basalts and picritic gabbro-dolerites, respectively, and the CaO concentrations are 0.21 and 0.12 wt % for Fo_{78} in these rocks. It follows that there is no justification to suggest that the Tuklonsky magma could be parental for the ore-bearing intrusions, as was hypothesized earlier based only on the elevated Mg# of the rocks.

In the models of A. Naldrett and his followers (Lightfoot et al. 1993; Brügman et al. 1993), the rocks composing the lower portion of the Nadezhdinsky Formation should be genetically related to the Tuklonsky basalts, and their depletion in base metals was predetermined by the crystallization of sulfides from the parental Tuklonsky melt contaminated with the host rocks. If the suggested mechanism had operated, the fields of ore-bearing intrusions should not have included any rocks of the Tuklonsky Formation because massifs of the Noril'sk Complex and the basalts of the Nadezhdinsky Formation are the interaction products of the Tuklonsky magma and terrigenous-carbonate sediments. Consequently, the picritic gabbro-dolerites of the ore-bearing massifs should have been identical to picritic basalts of the Nadezhdinsky Formation. However, their mineralogy and geochemistry are remarkably in contrast to the hypothetical intrusive comagmatic rocks, which were not saturated with sulfur (Likhachev 2006); the examined vertical section also includes (along with intrusions of the Noril'sk Complex) rocks from all of the aforementioned formations, which casts doubt onto the plausibility of the mechanism suggested for the origin of the ores.

Later this model was transformed into a two-stage model (Li et al. 2009). It was suggested that "early sulfide segregation took place in a deep staging chamber due to contamination with granitic crustal materials in the lower parts of the upper crust", as suggested previously by other researchers (Naldrett et al. 1992; Naldrett and Lightfoot 1994; Arndt et al. 2003). The magma then rose to form the weakly mineralized intrusions and erupted to the surface to form the Nd_{1-2} lavas, leaving a sulfide liquid with relatively low tenors of Ni, Cu, and PGE in the staging chamber. The PGE-poor sulfide liquid in the chamber was then upgraded in chalcophile elements (PGE, Ni, and Cu) by the Morongovsky magma forming a PGE-rich sulfide liquid. The PGE-rich sulfide liquid remained in the staging chamber while the Morongovsky magma erupted to form the Morongovsky lavas. New, S-unsaturated magma from the mantle continued to enter the chamber and progressively dissolved the PGE-rich sulfide liquid in the chamber to form a PGE-enriched magma. The PGE-enriched magma then rose to the upper parts of the upper crust where it reacted with anhydrite-bearing evaporite country rocks and became sulfide saturated, thereby producing immiscible sulfide liquid with high PGE concentrations as well as high $\delta^{34}S$ values. The sulfide liquid became lodged in the hydraulic traps of the plumbing system at Kharaelakh.

This hypothesis explains the differences between the sulfur isotopes, PGE, Cu, and Ni tenors in weakly mineralized and ore-bearing intrusions, i.e., Low Talnakh and Kharaelakh. However, it does not account for the large difference between the Re-Os compositions of the sulfides from these intrusions.

6.1 Geologic Relationships Between the Lavas and the Intrusions of the Noril'sk Complex

The possible comagmatic relationships between the ore-bearing intrusions and lavas were examined within the Noril'sk Trough, where intrusive rocks with disseminated or massive ores (at the Noril'sk 1, Maslovsky, Chernogorsky, and Noril'sk 2 deposits) are hosted in the uppermost units of the volcanic rocks, i.e., in the middle portion of the tuff–lava unit. In other regional folded structures, these massifs are hosted mostly in the Devonian sedimentary rocks underlying the volcanics (the Kharaelakh and Talnakh intrusions in the Kharaelakh Trough and the Vologochansky intrusion in the Vologochansky Trough), and hence, it is more difficult to correlate them. An important fact is that the Noril'sk Trough hosts rocks of all formations used in the models of various researchers (Gudchikhinsky, Tuklonsky, Nadezhdinsky, and Morongovsky); thus, it is possible to obtain insight into their role in the genesis of the ores.

We selected the geologic sections penetrated by boreholes OM-6 and OM-25 as references. The latter borehole penetrated a sill of the Noril'sk Complex, which is hosted in volcanic rock, is a tongue of the Maslovsky intrusion, and contains high-grade stringer-disseminated Cu–Ni ore mineralization. Despite its relatively insignificant thickness (12 m), this sill is clearly differentiated from the olivine gabbro-dolerite to leucogabbro. It was attributed to the Noril'sk Complex based on detailed data about the composition of its rocks, which were proven to be identical to rocks in the Noril'sk 1 Massif (Krivolutskaya and Rudakova 2009).

The volcanics hosting the sill were subdivided into eight suites, and the tongue appeared to cut through the rocks of the lower portion of the Nadezhdinsky Formation; these rocks were distinguished by the high La/Sm ratios and low Cu and Ni concentrations (Lightfoot et al. 1990; Fig. 6.1). To determine whether the sill is affiliated with the Noril'sk Complex, we compared its geochemistry with that of the Noril'sk 1 Massif (Tables 6.1 and 6.2). It follows that the Maslovsky intrusion and, hence, other comparable ore-bearing intrusions were emplaced after the eruption of the early flows of the Nadezhdinsky Formation, i.e., the ore-bearing intrusions cut across volcanic rocks that are thought to be comagmatic with the picrites of the Gudchikhinsky Formation and the basalts of the Tuklonsky and Nadezhdinsky formations.

This result suggests the hypothesis that the rocks can be comagmatic with rocks of the Morongovsky Formation (Li et al. 2009), whose geochemistry is the closest to that of the rocks of the Noril'sk Intrusive Complex.

However, we have found a barren dyke in Cape Kamenny in Lake Lama whose geochemistry (MgO concentration of 8.92 wt %, TiO_2 concentration of 0.81 wt %, and trace

Table 6.1 Concentrations of rare elements in intrusive rocks of the Noril'sk Trough (Noril'sk 1 intrusion), ppm

No	1	2	3	4	5	6	7	8	9
	Sample No								
Element	59	65.2	67.9	71.8	78	90	96.8	103.8	425
Rb	12.6	10.4	8.43	15.1	18.7	24.8	22.4	11.6	23.3
Ba	103	68.6	71.0	120	104	97.1	114	99.3	309
Th	0.44	0.60	1.05	0.95	1.08	0.95	1.24	0.57	4.16
U	0.17	0.23	0.44	0.36	0.41	0.35	0.45	0.21	1.03
Nb	1.79	2.56	4.13	3.77	4.30	3.55	4.78	2.37	10.5
Ta	0.13	0.25	0.25	0.22	0.95	0.21	0.27	0.14	0.68
La	2.89	2.94	4.74	5.25	5.74	5.00	6.82	3.46	20.9
Ce	6.86	7.06	11.3	12.4	13.5	11.9	16.5	8.11	43.5
Pr	0.93	0.94	1.46	1.67	1.74	1.64	2.20	1.08	5.19
Sr	213	128	124	226	257	295	460	150	301
Nd	4.44	4.40	6.72	7.84	8.11	7.46	10.15	4.96	20.7
Sm	1.28	1.27	1.83	2.16	2.30	2.14	2.86	1.31	4.35
Zr	33.5	39.0	67.4	62.5	65.8	57.5	77.6	38.3	150
Hf	0.92	1.02	1.76	1.63	1.72	1.54	2.08	1.01	3.54
Eu	0.59	0.42	0.58	0.76	0.82	0.74	0.91	0.49	1.16
Ti	2,930	2,555	3,630	4,660	4,565	4,730	5,390	2,830	6,035
Gd	1.53	1.49	2.20	2.58	2.66	2.51	3.30	1.51	4.30
Tb	0.26	0.25	0.37	0.45	0.46	0.43	0.58	0.26	0.69
Dy	1.87	1.74	2.54	3.02	3.15	2.91	3.83	1.79	4.37
Y	11.9	12.4	17.5	20.6	20.8	19.1	25.1	12.0	23.9
Ho	0.43	0.41	0.59	0.70	0.72	0.67	0.87	0.40	0.92
Er	1.13	1.12	1.60	1.89	1.90	1.76	2.36	1.09	2.56
Tm	0.17	0.17	0.24	0.27	0.29	0.26	0.35	0.16	0.37
Yb	1.11	1.05	1.50	1.76	1.84	1.69	2.15	1.06	2.31
Lu	0.16	0.16	0.22	0.25	0.27	0.25	0.32	0.16	0.35
Cu	7,940	2,180	2,590	132	277	81	91.4	643	43.7
Ni	7,487	2,743	2,875	556	295	210	164	1,239	14.1
Co	176	143	143	77.6	60.9	51.1	50.3	36.2	37.7

No	10	11	12	13	14	15	16	17	18
	Sample No								
Element	427.3	428	428.6	429.5	430.9	432.6	433.4	434.1	437.5
Rb	48.5	37.1	4.30	27.3	3.51	6.81	3.85	2.57	63.5
Ba	459	355	91.1	196	74.7	163	100	117	495
Th	3.29	2.23	0.94	0.91	0.82	1.07	1.02	1.00	3.70
U	0.88	0.71	0.36	0.36	0.31	0.41	0.39	0.39	0.91
Nb	8.52	6.37	3.80	4.95	5.39	4.18	4.15	5.33	9.17
Ta	0.53	0.41	0.23	2.72	0.36	0.28	0.26	0.37	0.57
La	16.4	12.0	6.30	5.77	5.62	6.66	6.57	6.44	17.5
Ce	34.1	25.1	14.4	13.1	12.9	15.3	15.2	14.9	36.7
Pr	4.10	3.10	1.97	1.79	1.76	2.12	2.11	2.07	4.42
Sr	405	397	308	310	271	242	258	265	257
Nd	16.08	12.4	9.12	8.44	8.08	9.88	9.69	9.69	17.5
Sm	3.41	2.82	2.49	2.34	2.30	2.77	2.79	2.69	3.81
Zr	117	95.6	80.1	74.2	72.0	87.6	86.1	85.3	133
Hf	2.84	2.21	1.95	1.87	1.71	2.20	2.09	2.08	3.19
Eu	1.00	0.94	1.05	0.78	0.88	1.00	1.13	0.95	0.95
Ti	4,908	4,460	5,766	5,588	5,448	6,328	5,989	6,607	5,297
Gd	3.37	2.92	2.96	2.84	2.69	3.30	3.30	3.28	3.97
Tb	0.55	0.48	0.50	0.48	0.47	0.56	0.57	0.56	0.63

(continued)

Table 6.1 (continued)

No	10	11	12	13	14	15	16	17	18
	Sample No								
Element	427.3	428	428.6	429.5	430.9	432.6	433.4	434.1	437.5
Dy	3.50	3.08	3.29	3.16	3.06	3.76	3.70	3.62	4.01
Y	18.7	17.1	18.7	17.4	17.4	21.3	20.9	21.1	21.6
Ho	0.74	0.65	0.72	0.69	0.64	0.81	0.82	0.78	0.81
Er	2.00	1.79	2.00	1.88	1.80	2.26	2.25	2.19	2.31
Tm	0.29	0.27	0.31	0.29	0.27	0.34	0.34	0.33	0.35
Yb	1.87	1.67	1.96	1.78	1.75	2.19	2.13	2.08	2.18
Lu	0.28	0.26	0.29	0.27	0.26	0.33	0.32	0.32	0.34
Cu	41.7	53.0	90.4	83.1	82.1	273	123	299	29.1
Ni	72.4	139	63.7	26.1	88.4	173	112	215	43.5
Co	42.0	39.8	41.7	29.6	35.3	56.8	51.2	60.2	37.2

Note: Sample No is the depth in borehole G-22. After Krivolutskaya and Rudakova (2009)

Table 6.2 Average mean composition of the Noril'sk Complex intrusions and supposed comagmatic formations, wt %

Intrusion, formation	N	SiO_2	TiO_2	Al_2O_3	FeO	MnO	MgO	CaO	Na_2O	K_2O	P_2O_5	Cr_2O_3
Noril'sk 1 borehole MS-31	25	47.75	0.89	15.93	11.46	0.14	11.67	9.28	1.91	0.5	0.15	0.32
Noril'sk 1 borehole G-22	14	47.16	0.79	15.36	12.17	0.2	12.04	10.6	0.97	0.39	0.09	0.23
Maslovsky borehole OM-4	21	47.36	0.89	15.27	11.98	0.26	12.25	10.28	1.17	0.43	0.1	0.02
mk	9	50.3	1.3	15.99	11.81	0.2	6.93	12.08	1.08	0.16	0.12	0.02
mr	12	49.72	1.19	15.9	11.98	0.21	7.29	11.93	1.26	0.37	0.14	0.02
nd	10	51.86	1.11	15.75	10.94	0.18	6.89	11.38	1.03	0.72	0.13	0.02
tk	4	50.75	1	15.78	9.34	0.21	6.74	14.61	1.04	0.38	0.1	0.04
gd_2*	5	46.76	1.06	7.72	14.13	0.19	22.41	6.24	0.74	0.48	0.1	0.17
mk*	8	49.34	1.22	16.65	11.68	0.19	6.92	11.57	2.03	0.3	0.1	–
mr*	6	50.01	1.11	16.49	11.31	0.18	6.99	11.43	1.97	0.38	0.11	–
nd*	2	50.97	1.02	16.69	10.43	0.18	6.64	11.11	1.95	0.85	0.15	–
tk*	9	50.18	0.89	15.88	10	0.17	9.09	11.24	2.12	0.34	0.08	–
gd_2*	5	48.64	1.63	10	13.59	0.2	16.69	7.61	0.98	0.53	0.12	–

Note: 1—element was not determined; 2—all compositions were recalculated in 100 %; 3—formations (*mk* Mokulaevsky, *mr* Morongovsky, *nd* Nadezhdinsky, *tk* Tuklonsky (author's data)); 4—gd* Gudchikhinsky (after) (Lightfoot et al. 1993); 5—Noril'sk 1, hole G-22 (average mean composition of rocks based on borehole G-22). N – analyses number. After Krivolutskaya et al. (2012a)

element patterns) is closely similar to that of the rocks in the Noril'sk Complex and cuts across the basalts of the Morongovsky Formation (Fig. 6.2); several similar dykes at the contact of the Nadezhdinsky Formation are mapped below (Fig. 6.3).

It is thus highly probable that intrusions of the Noril'sk Complex were emplaced in post-Morongovsky time, but the absence of disseminated sulfides provides no conclusive arguments in support of this hypothesis.

Fig. 6.2 Dyke of the Noril'sk Complex in basalts of the Morongovsky formation

6.2 Comparison of the Geochemistry of the Lavas and Intrusive Rocks of the Noril'sk Complex

The aforementioned data on the inner structure and composition of tuff–lava and intrusive rocks recovered from boreholes in the northern and eastern portions of the Noril'sk depression were demonstrated (see above) to show that the ore-bearing massifs and rocks of the Gudchikhinsky, Tuklonsky, and Nadezhdinsky formations are not comagmatic. This unambiguously follows from the fact that the sill of the Maslovsky intrusion cuts the volcanic rocks of these formations and from the differences between the trace element and mineralogical compositions of the rocks.

Another possible candidate for a volcanic analogue of the Noril'sk intrusions is the Morongovsky Formation (Rad'ko 1991; Arndt et al. 2003). The ore-bearing massifs are localized below their basalts and are similar to them in several geochemical parameters, such as the concentrations of trace elements and their distribution (Lightfoot et al. 1993; Brügman et al. 1993). At the same time, the concentrations of the major oxides, primarily MgO, in these rocks are principally different from those in rocks of the Noril'sk Complex (Fig. 6.4). Throughout its entire field, the Morongovsky Formation comprises exclusively tholeiitic basalts (6–7 wt % MgO), and no rocks with elevated Mg# have been found either in this formation or in the overlying formations. Conversely, practically all rocks of the ore-bearing massifs contain olivine. It is hard to visualize the complete settling and removal of olivine and sulfides from a melt volume of 50,000 km^3 in intrusive chambers that make up less than 10 % by volume. Indeed, in addition to bent magma feeders where these phases could accumulate, there could be numerous magmatic conduits of other geometries that did not facilitate this process. These could be vertical channels and fissures, whose presence at that time follows from the occurrence of Morongovsky-age central-type volcanoes that were mapped in the valley of the Mikchangda River (geological survey conducted by Noril'skgeologia and the author) and documented by several researchers (Petrology and Ore Potential 1978). Volcanic edifices in the fields of the Morongovsky Formation should then have contained magnesian rocks with elevated contents of sulfides. Neither these rocks nor sulfides have been found anywhere in the rocks of the formation, which are exposed over hundreds of kilometers along their strikes.

Finally, the geochemical parameters of the Morongovsky and all overlying formations such as ε_{Nd} (Fedorenko et al. 1996; Lightfoot et al. 1993; Wooden et al. 1993) and, particularly, S isotopic composition (Grinenko 1985; Ripley et al. 2003, 2010: $\delta^{34}S = 1–5‰$ in the basalts and 18‰ in the intrusive rocks) are remarkably different from those in the ore-bearing intrusions. The ore-bearing intrusions could have conceivably been emplaced even after all of the tuff–lava rocks were formed because some zircons from the Kharaelakh Massif were dated at 220 Ma (Malitch et al. 2010), which is generally consistent with the geological data.

6.3 Conclusions

The data presented above show that the ore-bearing intrusions are not directly genetically related to the lavas but were produced during a separate pulse of magmatic activity in post-Nadezhdinsky time. It is important to emphasize that if there was an open system (with a long flowing melt along the chamber), we could not observe such quenched rocks in endocontacts intrusions that have been found in the Maslovsky and Mikchangdinsky intrusions – with the presence of glass and the contrast-zoned rock-forming minerals.

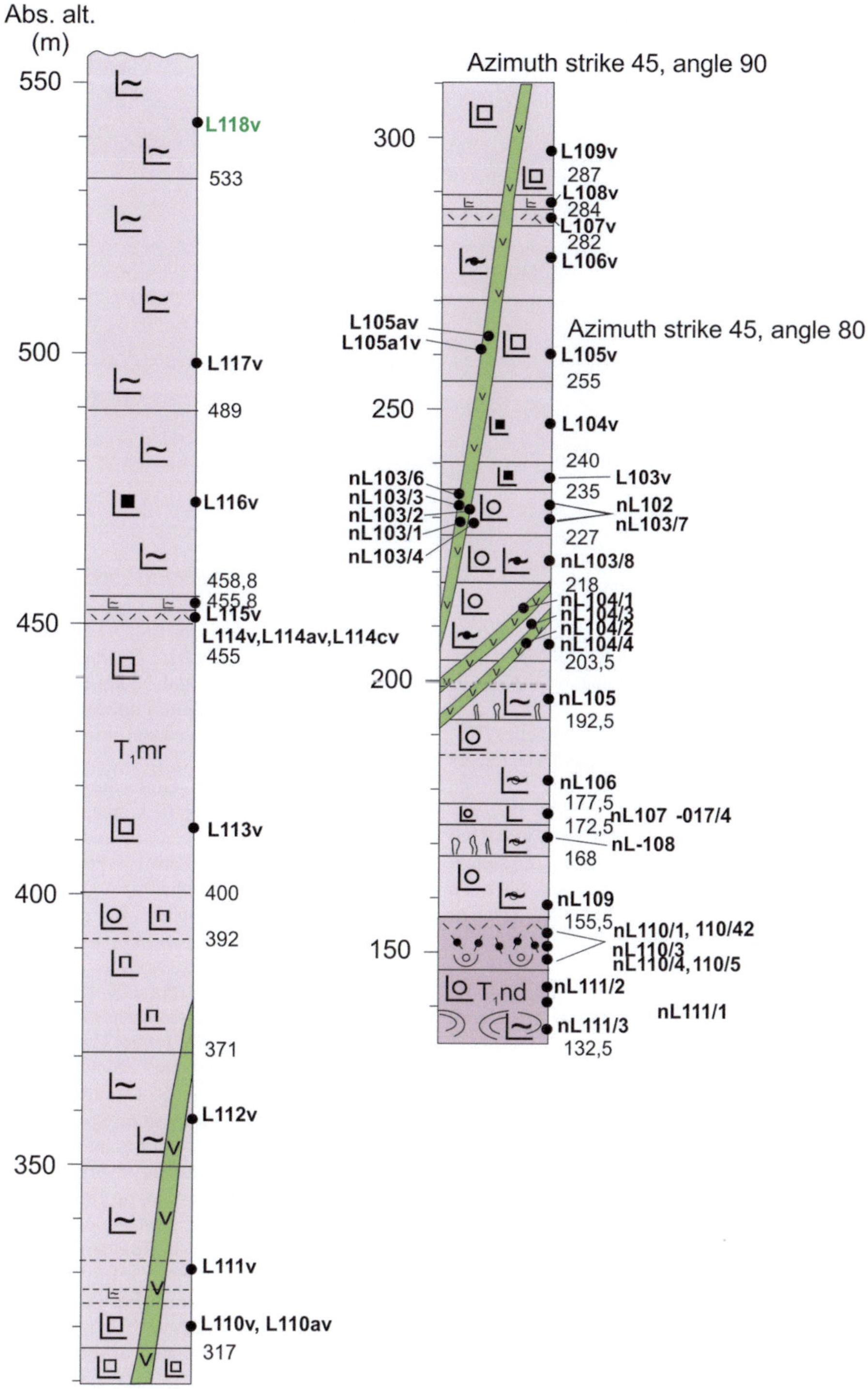

Fig. 6.3 Structure of volcanic rocks of the Morongovsky formation with dykes of the Noril'sk Complex (Section was studied with A. Rudakova) L118v – sample number and its location

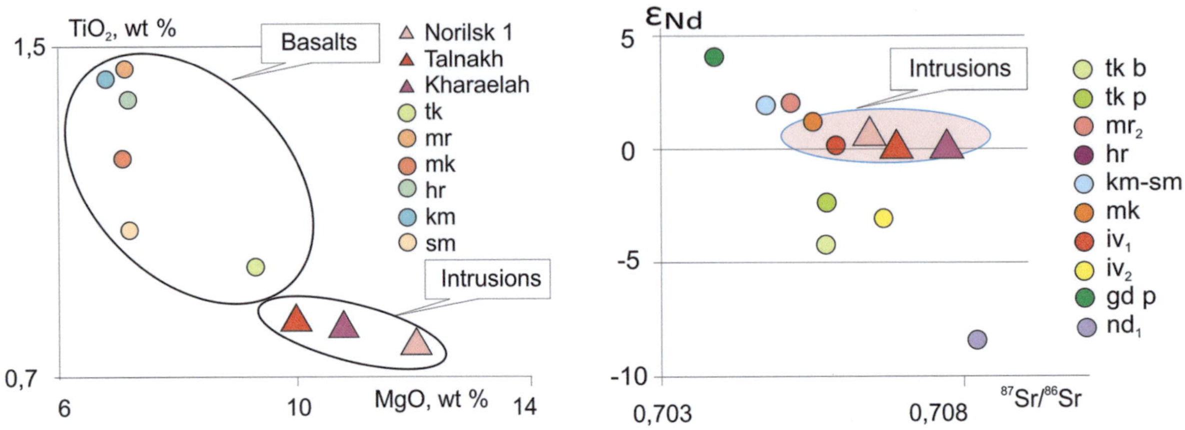

Fig. 6.4 TiO_2–MgO and ε_{Nd}–^{86}Sr/8^7Sr for magmatic rocks of the Noril'sk area

References

Arndt NT, Czamamanske GK, Walker RJ et al (2003) Geochemistry and origin of the intrusive hosts of the Noril'sk-Talnakh Cu-Ni-PGE sulfide deposits. Econ Geol 98:495–515

Brügman GE, Naldrett AJ, Asif M, Lightfoot PC, Gorbachev NS, Fedorenko VA (1993) Siderophile and chalcophile metals as tracers of the evolution of the Siberian Trap in the Noril'sk region, Russia. Geochim Cosmochim Acta 57:2001–2018

Dyuzhikov OA, Distler VV, Strunin BM et al (1988) Geology and ore potential of the Noril'sk Ore District. Nauka, Moscow. Translated: Dyuzhikov OA, Distler VV, Strunin BM, Mkrtychyan AK, Sherman ML, Sluzhenikin SF, Lurye AM (1992) Geology and metallogeny of sulfide deposits Noril'sk region USSR. Econ Geol Monogr 1. Ontario, Spec vol, 241 p

Fedorenko VA, Lightfoot PC, Naldrett AJ et al (1996) Petrogenesis of the flood-basalt sequence at Noril'sk, North Central Siberia. Int Geol Rev 38:99–135

Grinenko LN (1985) Sources of sulfur of the nickeliferous and barren gabbrodolerite intrusions of the northwest Siberian platform. Int Geol Rev 27:695–708

Keays RP, Lightfoot PC (2007) Siderophile and chalcophile metal variations in Tertiary picrites and basalts from West Greenland with implications for the sulphide history of continental flood basalt magmas. Mineral Deposita 42:319–336

Krivolutskaya NA, Rudakova AV (2009) Structure and geochemical characteristics of Trap rocks from the Noril'sk Trough, Northwestern Siberian craton. Geochem Int 47:635–656

Krivolutskaya NA, Sobolev AV, Snisar SG, Gongalskiy BI et al (2012a) Mineralogy, geochemistry and stratigraphy of the Maslovsky Pt–Cu–Ni sulfide deposit, Noril'sk Region, Russia. Mineral Deposita 47:69–88

Krivolutskaya NA, Sobolev AV, Mikhailov VN, Plechova AA, Kostitsyn YA, Roschina IA, Fekiacova Z (2012b) Parental melt of the Nadezhdinsky formation: geochemistry, petrology and connection with Cu-Ni deposits (Noril'sk area, Russia). Chem Geol 302:87–105

Latypov RM (2002) Phase equilibria constraints on relations of ore-bearing intrusions with flood basalts in the Noril'sk region, Russia. Contrib Mineral Petrol 143:438–449

Latypov RM (2007) Noril'sk- and Lower Talnakh-type intrusions are not conduits for overlying flood basalts: insights from residual gabbroic sequence of intrusions. Appl Earth Sci 116:215–225

Li CS, Ripley EM, Naldrett AJ (2009) A new genetic model for the giant Ni–Cu–PGE sulfide deposits associated with the Siberian flood basalts. Econ Geol 104:291–301

Lightfoot PC, Zotov IA (2014) Geological relationships between intrusions, country rocks and Ni-Cu-PGE sulfides of the Kharaelakh intrusion, Noril'sk region: implication for the role of sulfide differentiation and metasomatism in their genesis. Northwest Geol 47(1):1–34

Lightfoot PC, Naldrett AJ, Gorbachev NS (1990) Geochemistry of the Siberian trap of the Noril'sk area, USSR, with amplication for the relative contributions of crust and mantle to flood basalt magmatism. Contrib Mineral Petrol 104:631–644

Lightfoot PC, Hawkesworth CJ, Hergt J et al (1993) Remobilisation of the continental lithosphere by a mantle plume: major-, trace-element, and Sr-, Nd-, and Pb-isotopic evidence from Picritic and Tholeiitic lavas of the Noril'sk District, Siberian Trap, Russia. Contrib Mineral Petrol 114:171–188

Lightfoot PC, Naldrett AJ, Gorbachev NS, Fedorenko VA, Howkesworth CJ, Hergt J, Doherty W (1994) Chemostratigraphy of Siberian Trap Lavas, Noril'sk District: implication for the source of flood basalt magmas and their associated Ni-Cu mineralization. In: Proceeding of the Sudbury – Noril'sk symposium, Special vol 5. Ontario, pp 283–312

Likhachev AP (1994) Ore-bearing intrusions of the Noril'sk region. In: Proceedings of the Sudbury-Noril'sk symposium, Special vol 5. Geological Survey, Ontario, pp 185–201

Likhachev AP (2006) Platinum–copper–nickel and platinum deposits. Eslan, Moscow, 496 p (in Russian)

Malitch KN, Belousova EA, Griffin WL, Badanina IY, Pearson NJ, Presnyakov SL, Tuganova EV (2010) Magmatic evolution of the ultramafic–mafic Kharaelakh intrusion (Siberian Craton, Russia): insights from trace-element, U–Pb and Hf-isotope data on zircon. Contrib Miner Petrol 159:753–768

Malitch KN, Latypov RM, Badanina IY, Sluzhenikin SF (2014) Insights into ore genesis of Ni-Cu-PGE sulfide deposits of the Noril'sk Province (Russia): evidence from copper and sulfur isotopes. Lithos 2014(05):014

Naldrett AJ (2009) Ore deposits related to flood basalts, Siberia. In: New developments in magmatic Cu–Ni and PGE deposits. Geological Publishing House, Beijing, pp 141–179

Naldrett AJ, Lightfoot PC (1994) The Ni-Cu-PGE ores of the Noril'sk region, Siberia: a model for giant magmatic sulfide deposits associated with flood basalts. In: Proceeding in symposium on giant ore deposits, Queen's University, May 1992. Society of Economic Geology Special Publication, vol 2, pp 81–123

Naldrett AJ, Lightfoot PC, Fedorenko VA et al (1992) Geology and geochemistry of intrusions and flood basalts of the Noril'sk region, USSR, with applications for the origin of the Ni-Cu ores. Econ Geol 87:975–1004

Petrology and Ore Potential of the Traps of the Northern Siberian Platform (1978) Zolotukhin VV, Vilenskii AM (eds) Nauka, Novosibirsk

Rad'ko VA (1991) Model of dynamic differentiation of intrusive traps at the Northwestern Siberian Trap. Geol Geophys 32(11):19–27

Ripley EM, Lightfoot PC, Li C et al (2003) Sulfur isotopic studies of continental flood basalts in the Noril'sk region: implications for the association between lavas and ore-bearing intrusions. Geochim Cosmochim Acta 67:2805–2817

Ripley EM, Li C, Craig H et al (2010) Micro-scale S isotope studies of the Kharaelakh intrusion, Noril'sk region, Siberia: constraints on the genesis of coexisting anhydrite and sulfide minerals. Geochim Cosmochim Acta 74:634–644

Sobolev AV, Krivolutskaya NA, Kuzmin DV (2009) Petrology of the parental melts and mantle sources of Siberian trap magmatism. Petrology 17:253–286

Wooden JL, Czamanske GK, Fedorenko VA et al (1993) Isotopic and trace-element constraints on mantle and crustal contributions to Siberian continental flood basalts, Norilsk area, Siberia. Geochim Cosmochim Acta 57:3677–3704

7 Basic–Ultrabasic Intrusions in the Kola-Karelia Area

Geochemical features of basic–ultrabasic intrusions, including massifs with Pt–Cu–Ni mineralization, have been studied in the Kola–Karelia area. Major and rare elements were determined in several samples from famous Ni deposits as Monchepluton, Fedorova–Pansky Tundras, and Burakovsky as compared to intrusions of the South Kovdor area, perspective on founding sulfide mineralization. All intrusions of the Kovdor area were related to drusite (corona) complex. Their geochemical study demonstrated their difference in distribution of rare elements. According to these data, they were subdivided into three groups. The first one is close to Monchepluton rocks with rich Ni mineralization. The geochemical features of these massifs are very similar to the features of the Noril'sk Intrusive Complex. The patterns of these massifs are close to the crust; they have negative Ta–Nb and positive Pb anomalies. Their age is 2,410 ± 10 Ma showing their later formation as compared with layered intrusions.

The Kola Peninsula is the oldest region in Russia where Ni deposits were explored. It contains large deposits (Monchepluton) of Ni and so as weekly mineralized intrusions including massifs of South Kovdor area. Their geochemical features were studied and compared with the features of massifs of the Noril'sk Intrusive Complex.

Deciphering the history of the formation and evolution of the continental crust within the Baltic Shield requires detailed petrogeochemical studies of the ultrabasic–basic massifs, which are indicators of diastrophic epochs (Polkanov 1939; Sudovikov 1937). In this context, the intrusions that are widely developed within the Belomorian Mobile Belt (BMB) and united into the so-called drusite complex are of special interest (Early Precambrian… 2005). The characteristic features of their rocks, first described by E. Fedorov, are the coronal rims of orthopyroxene, clinopyroxene, amphibole, and garnet surrounding olivine, which were termed coronal or drusite textures (Fedorov 1886). It was suggested that they were formed by slow equilibrium magma crystallization at extreme depth owing to the interactions between crystallizing minerals and magmatic melt. The decisive role of metamorphism in the formation of drusite textures was first noted by Machkovtsev (1927).

Later, the drusite textures were subdivided by A. A. Strona (1929) into reaction magmatic (orthopyroxene) and celyphytic (garnet and amphibole), which were formed under metamorphic conditions. During the subsequent high-grade metamorphism, the rocks often lost these distinctive features and were transformed into metagabbro, garnet gabbro, and garnet–biotite-bearing plagioamphibolites. Initially, the drusite complex combined peridotite with gabbro rocks (Fedorov 1886), which were interpreted as being differentiation products of single primary magma (Mashkovtsev 1927). Later, the drusites were subdivided into two series the peridotite–gabbro and gabbro–anorthosite bodies, which are produced by the injection of magma into a mobile framework during the formation of small rootless bodies (Igneous Rocks 1988). A more detailed classification of the magmatic bodies was developed by Shurkin et al. (1962), who distinguished between pre-orogenic, early orogenic, and late orogenic groups of ultrabasic–basic rocks.

Interest in this complex was not only provoked by the specific textural–structural features of its rocks but also by their ore potential. Most researchers (Amelin and Semenov 1990; Stepanov 1981; Malov and Sharkov 1978; Sharkov et al. 1997; Sharkov and Smolkin 1997) compare the rocks with the layered intrusions that widely developed within the Kola–Lapland–Karelian province. Initially, this analogy was based on the similarity between the rock types and the evolution of their chemical and mineral composition as well as the discovery of copper–nickel mineralization in some drusite massifs (Yudom–Navolok, Kovdozero, and others) (Stepanov 1981). This viewpoint was later supported by Sm–Nd and U–Pb isotope data, which showed that the drusite complex and the layered intrusions were formed at the same Sumian stage, which began during the Paleoproterozoic era within the Baltic Shield (Balashov et al. 1993; Bogdanova and Bibikova 1993; Amelin et al. 1995; Amelin and Semenov 1996).

N.A. Krivolutskaya, *Siberian Traps and Pt-Cu-Ni Deposits in the Noril'sk Area*,
DOI 10.1007/978-3-319-17205-7_7

The drusite massifs were assumed to mark the onset of the extension of the Belomorian allochthon and its early rifting stage (Stepanov 1981; Precambrian Magmatic… 1985; Smolkin 1992) and are syngenetic with layered plutons consolidated under stable and rigid cratons. The differences in their morphology were explained by the differences in the tectonic conditions during the emplacement of these intrusions in various blocks. There is an opinion that the drusite massifs were derived from the same magma generation areas as the layered intrusions, but unlike the "localized" magmatism of the latter, the former represent an example of dispersed mafic–ultramafic magmatism (Sharkov et al. 1994). In both cases, magmas produced a relatively siliceous and high-magnesian (boninite-like) series of the rocks.

However, the problems of the origin of the drusite complex and its metallogenic association remain unresolved. Its heterogeneity was characterized by A. Stepanov (1981), who distinguished different age complexes of gabbro–anorthosites (which partly correspond to the early orogenic drusites of A. Shurkin), lherzolite–gabbronorites (early orogenic–late orogenic drusites), and coronite gabbro (partly late orogenic drusites) based on their geologic relationships. The development and improvements in isotope measurement technique have revealed that the former two complexes are close in absolute age (2,440 ± 10 Ma (Efimov and Kaulina 1997; Slabunov et al. 2001), whereas the third complex is significantly younger (2,115 ± 25; Stepanova et al. 2003). It has also been determined that the layered plutons of the Karelia–Kola region are not simultaneous complexes but are comprised of older (2,507–2,470 Ma for the Monchepluton and Fedorova–Pana Tundras intrusion) and younger massifs (2,441–2,437 Ma for the Umbarechka–Imandra Complex and the intrusions of eastern and northern Karelia (Smolkin et al. 2005).

The difficulties in the interpretation of the origin and role of the drusite massifs in the geological evolution of the region are related to insufficient geochemical knowledge of the numerous ultrabasic–basic bodies that are united in this complex. Detailed geological and geochemical studies, including the isotopic characteristics of the rocks, have only spanned an insignificant part of a giant family of this type of intrusions, which number several thousand magmatic bodies. Among them are the well-exposed massifs of the western Belomorian region: Vorony I., Yudom–Navolok, and Pezhostrov as well as the Tolstik, Romanovskaya, Kovda, and Zhemchuzhnyi Peninsulas and other reference objects. The contents of major, trace, rare earth, and ore elements in the rocks as well as the Sm–Nd isotope data were used to interpret the origin and evolution of the parental magmas of these massifs (Sharkov et al. 1994; Lobach-Zhuchenko et al. 1998). However, this information is insufficient to characterize the entire complex of ultrabasic–basic rocks, which reaches from the western Belomorian region to the boundary of Finland. This explains the contradictory views of the nature and ore prospects of these objects, and these circumstances have provoked comprehensive petrogeochemical (including isotope) research of the practically unstudied massifs in the central part of the Belomorian Belt (South Kovdor region) that is ascribed to the drusite complex. The major rock types of several layered plutons of the Kola Peninsula and Karelia have also been studied for comparison, and the results have served as a basis for the separation of the intrusive bodies into several groups.

7.1 Brief Geology

The Belomorian Mobile Belt is bounded by the deep-seated Lapland fault in the north and by a fault system extending from the Pliajarvi Lake to the Vetreny Belt Range in the southwest (Fig. 7.1). It is predominantly made up of tonalite gneisses, amphibolites, and aluminous gneisses, which are intersected by charnockites and basic–ultrabasic massifs of different composition and age.

For a long time, the belt was considered to be part of the Saamian Mobile Belt (Kratts et al. 1980), but at present, the BMB is considered to be an independent metamorphic belt that evolved through several Archean basement stages under granulite (early stage) and amphibolite facies (Early Precambrian… 2005). It is comprised of Archean and Paleoproterozoic rock complexes, including greenstone belts.

According to modern concepts of its geological structure, the Belomorian Belt shows lateral zoning (Bibikova et al. 1993; Slabunov and Bibikova 2001; Miller et al. 2002) caused by the presence of several tectonic nappes. Two of them, the Kovdozero and Orijarvi nappes, as well as a fragment of the Eisk greenstone belt, are located in part of the study area characterized by a block mosaic pattern. Most of the study area is occupied by Archean rocks (AR_2 basement complex) that are characterized by biotite plagiogneisses and migmatized tonalite–trondhjemite–granodiorites. They are overlain by the Belomorian metamorphic complex (AR_2bl), which is characterized by garnet–biotite, kyanite–garnet–biotite, and amphibole–biotite gneisses intercalated with amphibolites that contain lenses and thin stratal bodies of ultrabasic rocks.

Objects and methods. The geochemical study was based on 66 samples collected from 19 drusite massifs of the South Kovdor area (Fig. 7.2). The frequency of sampling varied depending on the degree of exposure of the intrusive bodies; the 403.0-m massifs, Sorkajoki and Poioiva (Tables 7.1 and 7.2), were studied in the most detail. To compare their geochemical properties, we took samples from the following ore-bearing layered intrusions of the Kola region: Monchepluton (Mt. Travyanaya, Mt. Nittis, and Mt. Sopcha), the Fedorova and Pansky Tundras, and Umbarechensky as well as gabbro–anorthosites of the Monchetundra Massif

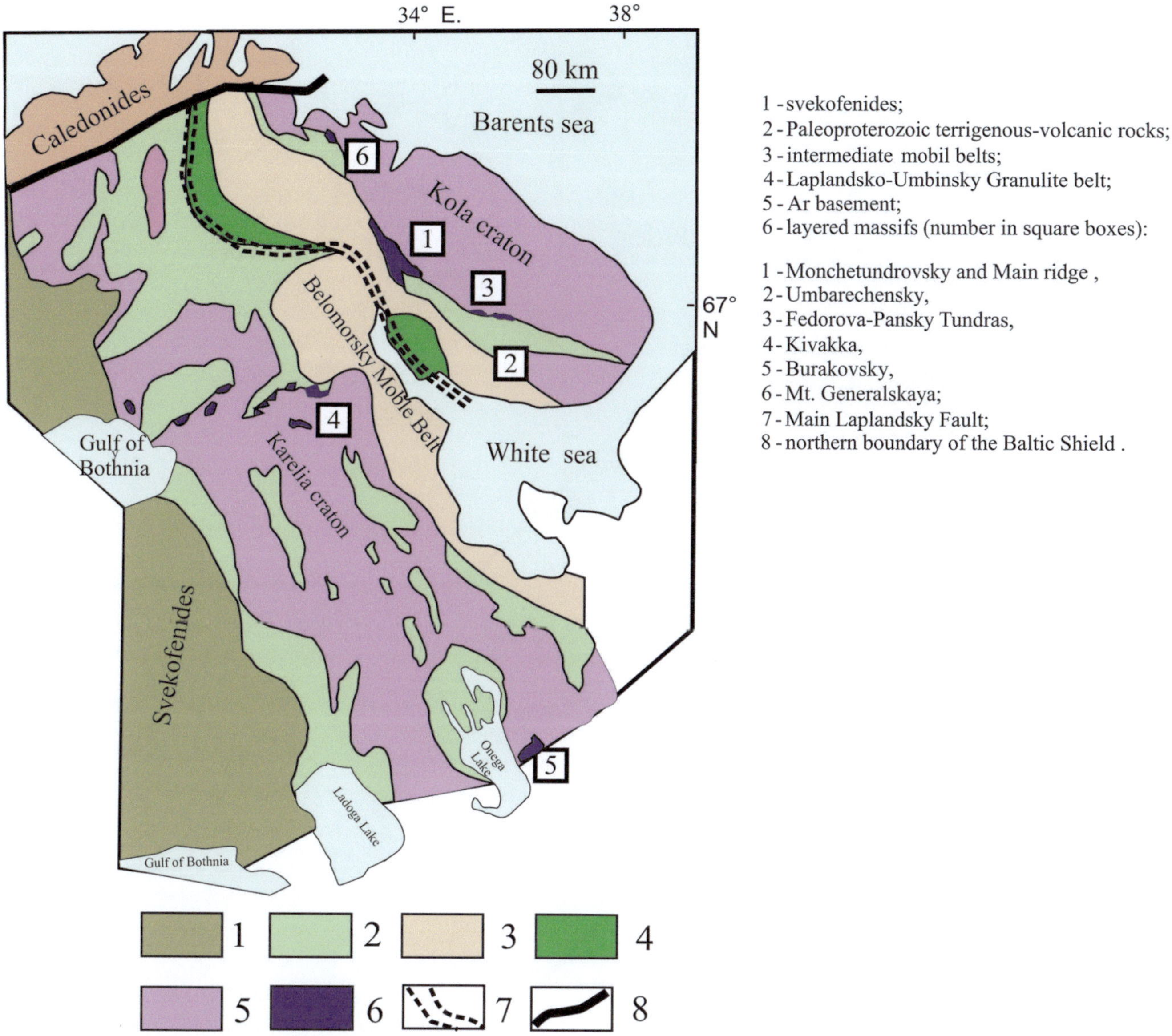

Fig. 7.1 Schematic map of the Karelia–Kola region
After Sharkov et al. (1997)

(Main Ridge). In addition, we also used representative samples from different zones of such large Karelian plutons as Burakovsky and Kivakka that were kindly provided by G. S. Nikolaev and Ya. V. Bychkova (Vernadsky Institute of Geochemistry and Analytical Chemistry, Russian Academy of Sciences).

The following analyses were performed. (1) The major oxide contents in all samples were determined by X-ray fluorescence analysis (XRF performed with an AXIOS Advanced PANalytical spectrometer at the Vernadsky Institute of Geochemistry and Analytical Chemistry, Russian Academy of Sciences, Central Laboratory of Matter Analysis by I. A. Roshchina and T. V. Romashova). (2) Trace element abundance was determined following the standard regime using inductively coupled plasma mass spectrometry (ICPMS with an ELAN 6100 DRC and ELAN 6100 DRC Software Kit, May 2000, PerkinElmer SCIEX Instruments, at the Institute of Mineralogy, Geochemistry, and Crystal Chemistry of Rare Elements by D. Z. Zhuravlev). The detection limits (DL) of the elements varied from 1–5 ppb for heavy- and medium-weight elements (uranium, thorium, rare earth elements, and others) to 20–50 ppb for light elements (beryllium and others). The measurement accuracy accounted for 3–10 rel. % of the element contents of more than 20–50 DL. (3) The composition of the rock-forming minerals in the drusite complex was studied using an X-ray microprobe (EPMA performed with a SX100 Cameca at the Vernadsky Institute of Geochemistry and Analytical Chemistry, Russian Academy of Sciences by N. N. Kononkova).

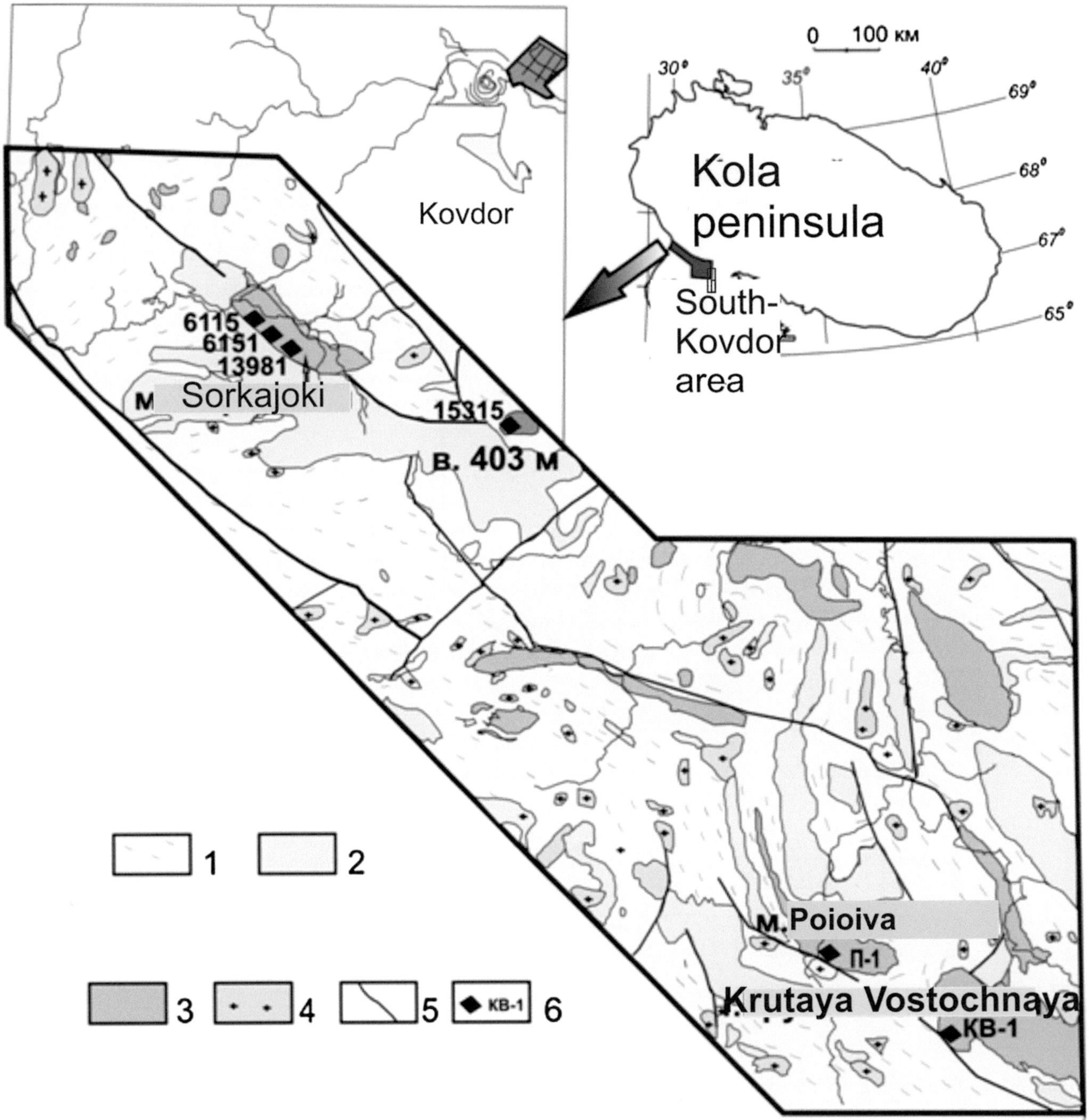

Fig. 7.2 Schematic geological map of the southern Kovdor area
(*1*, *2*) Basement complex (Middle Archean): (*1*) biotite plagiogneisses and (*2*) amphibolites, (*3*) ultrabasite–basite massifs (Early Proterozoic–Sumian), (*4*) plagioclase–microcline and microcline granites (Early Proterozoic–Karelian), (*5*) faults, and (*6*) sampling sites and sample numbers. (After Krivolutskaya et al. 2010a)

Minerals and oxides physicochemically similar to the analyzed material were used as standards for the microprobe analysis. The lower limit of the analyzed concentrations is 0.02 wt % for elements from Mg to Zn and 0.06 wt % for F and Na. The beam size was 2 μm, and the operating conditions were an accelerating voltage of 15 kV and a beam current of 40 nA.

7.2 Petrography of the Drusite Complex Massifs

Several hundred drusite complex massifs are distributed over the entire study area, and their sizes typically range from 0.2 to 0.5 km, occasionally 1.5–3 km, across with lengths of 5–7 km. They are usually clustered as chains con-

Table 7.1 Chemical composition of rocks from massifs in the South Kovdor area, wt %

No.	No. and massif name	Sample No	SiO_2	TiO_2	Al_2O_3	FeO	MnO	CaO	MgO	Na_2O	K_2O	P_2O_5	Cr_2O_3	LOW	Total	Rock
								Type I								
1.	1. Sorkajoki	13,941	44.28	0.28	5.30	11.68	0.14	3.62	32.18	0.77	0.20	0.04	0.68	0.10	99.28	Pl lherzolite
2.		13,961	49.32	0.24	2.92	12.32	0.19	8.57	24.75	0.47	0.11	0.03	0.45	0.01	99.39	Pl lherzolite
3.		13,971	48.09	0.24	3.16	12.26	0.18	4.12	29.86	0.54	0.16	0.03	0.60	0.66	99.91	Pl lherzolite
4.		13,981	45.18	0.28	4.26	12.42	1.16	3.78	31.18	0.70	0.23	0.05	0.56	0.01	99.93	Pl lherzolite
5.		13,991	44.92	0.30	3.82	14.15	0.20	3.58	30.14	0.66	0.28	0.05	0.37	0.32	98.80	Pl lherzolite
6.		14,001	47.58	0.55	5.98	12.37	0.18	5.20	23.96	1.11	0.53	0.09	0.32	1.15	99.03	Ol gabbronorite
7.		17,641	42.56	0.21	3.72	13.75	0.17	3.22	33.87	0.58	0.12	0.03	0.75	0.01	99.00	Pl lherzolite
8.		17,651	47.94	0.40	4.94	13.17	0.18	5.75	25.22	0.85	0.40	0.05	0.39	0.01	99.31	Ol gabbronorite
9.		17,661	49.08	0.36	6.21	11.68	0.17	6.48	23.02	1.02	0.30	0.05	0.34	0.51	99.22	Ol gabbronorite
10.		20,181	43.57	0.19	2.86	11.12	0.13	3.18	37.08	0.48	0.13	0.04	0.75	0.01	99.54	Pl lherzolite
11.		20,141	46.17	0.06	0.52	14.65	0.24	5.63	31.73	0.06	0.01	0.01	0.71	0.01	99.81	lherzolite
12.		20,391	44.12	0.36	4.00	12.53	0.20	7.88	26.40	0.10	0.01	0.04	0.17	3.11	98.93	lherzolite
13.	2. Poioiva	17,781	49.58	0.48	7.65	11.41	0.15	7.06	21.45	1.21	0.32	0.06	0.40	0.01	99.78	Ol gabbronorite
14.		17,791	46.58	0.33	6.37	12.36	0.15	5.91	24.66	0.97	0.18	0.04	0.52	1.59	99.67	Ol gabbronorite
15.		17,901	48.88	0.30	6.36	11.42	0.17	8.24	22.14	0.80	0.14	0.03	0.38	0.01	98.87	Ol gabbronorite
16.		17,911	49.35	0.31	4.84	12.09	0.18	7.41	23.44	0.64	0.15	0.04	0.45	0.01	98.91	Pl lherzolite
17.		20,911	47.06	0.40	4.52	14.33	0.20	4.64	26.43	0.74	0.10	0.08	0.57	0.07	99.14	Pl lherzolite
18.		22,711	49.65	0.49	10.55	11.11	0.15	8.63	16.52	1.58	0.36	0.06	0.20	0.10	99.40	gabbronorite
19.	3. Mt. Rumimuristo	22,521	49.77	0.31	4.71	12.47	0.19	4.57	25.92	0.73	0.25	0.05	0.45	0.01	99.43	Pl lherzolite
20.	4. Kamenka River	20,671	48.94	0.37	4.45	12.32	0.18	5.56	25.88	0.63	0.21	0.04	0.55	0.01	99.15	Pl lherzolite
								Type II								
21.	5. Mt. Perchatka	16,091	50.06	0.52	8.56	11.37	0.16	7.22	19.46	1.43	0.50	0.09	0.23	0.10	99.70	Ol gabbronorite
22.	6. Massif at 463 m in elevation	17,731	47.60	0.37	7.87	10.71	0.07	9.20	19.92	0.73	0.02	0.04	0.32	2.60	99.46	Ol gabbronorite
23.		17,741	52.46	0.43	5.72	9.69	0.11	9.11	20.17	0.71	0.03	0.05	0.47	0.35	99.30	Ol gabbronorite

(continued)

Table 7.1 (continued)

No.	No. and massif name	Sample No	SiO_2	TiO_2	Al_2O_3	FeO	MnO	CaO	MgO	Na_2O	K_2O	P_2O_5	Cr_2O_3	LOW	Total	Rock
24.		17,751	48.15	0.38	4.51	13.12	0.18	6.68	24.44	0.71	0.18	0.07	0.53	0.74	99.68	Pl lherzolite
25.		17,761	49.05	0.36	4.94	12.43	0.18	7.18	24.33	0.70	0.15	0.04	0.46	0.01	99.82	Pl lherzolite
26.	7. Massif at 403 m in elevation	10,051	51.92	1.85	11.58	15.30	0.20	9.34	5.05	2.32	0.78	0.19	0.02	0.64	99.18	Gr plagioamphibolite
27.		10,061	52.88	2.00	13.54	13.45	0.19	8.88	4.08	3.25	0.53	0.22	0.01	0.42	99.45	Gr metagabbro
28.		10,071	46.32	2.46	15.02	18.12	0.30	9.42	3.87	2.72	0.44	0.31	0.01	0.70	99.68	Gr plagioamphibolite
29.		11,511	52.41	1.07	11.91	13.25	0.21	10.74	6.46	2.12	0.25	0.10	0.02	0.43	98.97	Metagabbro
30.		11,512	52.53	1.18	12.63	13.73	0.21	10.55	5.87	1.99	0.28	0.12	0.01	0.44	99.54	Metagabbro
31.		11,513	51.54	2.52	10.85	17.22	0.24	9.43	5.21	1.92	0.55	0.32	0.01	0.24	100.05	Metagabbro
32.		11,514	52.64	0.91	12.15	12.79	0.19	10.77	6.82	1.97	0.26	0.09	0.02	0.51	99.12	Metagabbro c Gr
33.		11,531	52.12	0.84	13.03	11.80	0.20	11.26	6.88	1.95	0.28	0.07	0.02	0.70	99.14	Amf. gabbronorite
34.		11,571	47.85	3.94	17.05	15.85	0.13	7.92	1.55	3.39	0.62	0.06	0.01	0.96	99.32	Plagioamphibolite
35.		22,231	52.73	0.89	14.12	11.63	0.20	10.94	6.85	1.95	0.17	0.06	0.01	0.24	99.79	Gabbro
37.		22,232	50.22	1.43	13.62	16.21	0.20	9.87	4.22	2.34	0.38	0.15	0.01	0.32	98.96	Plagioamphibolite
38.		22,234	53.27	1.59	13.15	15.42	0.20	9.53	2.92	2.61	0.41	0.18	0.01	0.30	99.58	Metagabbro
39.		22,421	52.95	2.49	12.18	15.76	0.22	8.54	2.72	2.57	0.86	0.32	0.01	0.19	98.81	Metagabbro
40.		22,441	52.74	1.99	11.36	14.86	0.21	9.71	5.02	2.35	0.69	0.20	0.02	0.01	99.16	Gabbro
41.	8. Massifs at 391.2 m in elevation	13,081	51.02	0.59	9.22	11.26	0.15	6.82	17.56	1.76	0.59	0.07	0.16	0.17	99.38	Ol gabbronorite
42.		16,191	49.65	0.38	5.30	12.15	0.19	6.76	22.42	0.90	0.32	0.07	0.35	0.45	98.94	Ol metagabbronorite
43.		16,201	51.61	0.70	6.16	11.14	0.18	7.60	19.57	1.04	0.13	0.09	0.19	0.77	99.19	Ol gabbronorite
44.	9. Mt. Grob Tundra	11,891	40.75	2.25	17.34	18.43	0.14	12.74	5.88	0.88	0.18	0.18	0.01	0.20	98.97	Gr amphibolite
45.		11,901	35.56	1.27	16.56	18.04	0.16	12.82	9.87	0.62	0.12	0.11	0.02	3.75	98.90	Plagioamphibolite
46.		15,211	50.04	1.01	17.85	10.37	0.21	6.45	7.88	3.46	0.28	0.14	0.02	1.15	98.85	Gr amphibolite
47.		18,841	52.76	0.55	11.62	9.69	0.18	9.70	12.85	1.59	0.30	0.08	0.08	0.55	99.94	Metagabbro
48.		18,842	53.58	0.62	12.50	10.83	0.18	10.07	9.53	1.77	0.48	0.06	0.07	0.01	99.70	Gabbro
49.	10 Mt. Kontiovara	12,461	45.24	0.23	4.96	9.51	0.11	5.48	29.16	0.03	н/о	0.02	0.31	4.70	99.76	Lherzolite

No.	No. and massif name	Sample No	SiO_2	TiO_2	Al_2O_3	FeO	MnO	CaO	MgO	Na_2O	K_2O	P_2O_5	Cr_2O_3	LOW	Total	Rocks
		Type III														
50.	11. Mt. Ragutchane	13,311	50.14	0.75	14.09	11.83	0.23	13.05	5.96	2.00	0.40	0.06	0.03	0.57	99.11	Metasomatite with Gr
51.		13,312	49.68	0.70	14.94	10.14	0.26	17.48	3.96	0.77	0.04	0.06	0.02	0.74	98.79	Gr plagioamphibolite
52.	12. Mt. Krutaya Vostochnaya	20,771	48.94	0.38	4.45	12.32	0.18	5.56	25.88	0.63	0.21	0.04	0.55	0.01	99.13	Ol–Pl вебстерит
53.	13. Western Poioiva	18,661	51.32	0.42	10.72	9.86	0.15	8.52	15.97	1.71	0.34	0.06	0.21	0.10	99.38	Metagabbro
54.	14. Lais-Tundra	15,141	54.75	1.78	10.83	12.45	0.17	9.24	4.92	3.78	0.50	0.20	0.01	0.38	99.01	Gabbro
55.		18,401	52.62	0.51	15.26	8.38	0.14	10.08	7.18	2.38	0.60	0.07	0.02	1.42	98.66	Plagioamphibolite
57.	15. Nora River	12,424	52.98	0.79	12.31	11.81	0.20	10.68	6.52	2.16	0.64	0.08	0.01	0.93	99.12	Plagioamphibolite
58.	16. Mt. Nora	12,881	49.80	0.44	9.52	11.37	0.16	11.30	14.56	0.95	0.05	0.07	0.11	0.93	99.26	Gabbronorite
59.		12,891	51.76	1.92	10.24	17.68	0.29	9.58	5.58	1.17	1.00	0.23	0.02	0.45	99.92	G plagioamphibolite
60.		12,901	50.18	0.45	9.05	12.10	0.17	11.40	14.33	0.81	0.01	0.02	0.12	0.60	99.23	Gabbronorite
61.		10,841	53.00	0.79	9.82	10.95	0.21	10.30	11.40	0.77	0.78	0.15	0.08	0.87	99.11	Amphibolite
62.	17. Mt. Levoiva	10,381	54.98	0.90	14.16	10.24	0.20	9.93	4.63	2.71	0.35	0.07	0.02	1.06	99.25	Gr плагиоамфиболит
63.		14,742	49.78	1.21	13.38	15.33	0.24	10.11	6.62	1.58	0.36	0.16	0.03	0.69	99.49	Gr amphibolite
64.		10,481	50.07	1.23	13.13	15.56	0.20	10.09	6.57	0.79	0.29	0.16	0.05	0.73	98.87	Gr amphibolite
65.	18. Mt. Ristivara	10,691	54.38	0.67	11.23	11.05	0.18	9.36	8.03	1.71	0.95	0.08	0.04	1.77	99.44	Gr amphibolite
66.	19. Massif at 408.8 m in elevation	16,981	52.50	2.07	10.62	17.52	0.24	8.60	3.45	2.56	1.12	0.23	0.01	0.15	99.06	Gr plagioamphibolite

Note: *Gr* garnet, *Ol* olivine, *Pl* plagioclase. Here and in Table 7.2, analyses were done in GEOHKI RAS, analysts I. Rostshina and T. Romashova

Tables 7.1, 7.2, 7.3, 7.4, 7.5, 7.6, and 7.7 – After Krivolutskaya et al. (2010a)

Table 7.2 Chemical composition of rocks from layered plutons in the Kola–Karelia area, wt %

No.	Pluton name	Sample No	SiO_2	TiO_2	Al_2O_3	FeO	MnO	CaO	MgO	Na_2O	K_2O	P_2O_5	Cr_2O_3	LOW	Total	Rock
1.	*Monchepluton*															
	Mt. Travyanaya	1007/98	50.23	0.20	6.64	10.49	0.14	5.11	24.86	0.77	0.07	0.03	0.53	0.44	99.51	Ol gabbronorite
	Dunite block	1063/98	43.65	0.16	3.97	12.39	0.18	3.32	34.82	0.55	0.05	0.03	0.24	0.01	99.36	Peridotite
		1054/98	40.52	0.09	1.90	12.46	0.15	1.57	39.48	0.09	0.02	0.02	0.65	2.89	99.83	Dunite
	Mt. Sopcha	1011/98	54.32	0.15	3.28	8.72	0.15	2.82	29.32	0.30	0.01	0.02	0.70	0.06	99.85	Pyroxenite
		1017/98	44.52	0.05	1.15	12.11	0.12	0.99	37.98	0.01	0.01	0.01	1.55	0.66	99.16	Peridotite
	Mt. Nyud	1025/98	53.47	0.03	17.54	6.51	0.14	10.98	8.32	2.23	0.01	0.01	0.01	0.03	99.29	Gabbro–anorthosite
2.	*Umbarechensky*	02002/8.3	55.32	0.37	5.58	11.05	0.21	4.32	20.92	1.08	0.43	0.06	0.25	0.01	99.36	Orthopyroxenite
		03002/14.9	54.05	0.24	14.94	5.69	0.11	10.82	9.58	2.38	0.59	0.04	0.02	0.86	99.85	Metagabbronorite
		45	56.58	0.47	7.03	10.08	0.19	4.92	18.02	1.32	0.58	0.07	0.22	0.01	99.83	Ol gabbronorite
3.	*Fedorova–Pansky Tundras*	52.7/BGF428	52.53	0.12	18.63	6.11	0.11	9.77	9.06	2.03	0.09	0.01	0.01	0.34	99.16	Pyroxenite
		260.7/BGF428	52.48	0.11	18.52	6.14	0.11	9.47	10.06	2.04	0.02	0.02	0.01	0.89	99.29	Pyroxenite
		53/BGF428	53.68	0.12	15.28	5.16	0.11	11.55	11.58	1.74	0.08	0.01	0.01	0.37	99.61	metagabbronorite
		211/BGF425	54.25	0.26	2.54	11.69	0.21	2.53	26.38	0.25	0.07	0.02	0.51	0.34	99.49	Harzburgite
		154.6/BGF428	56.58	0.05	21.72	3.32	0.05	6.90	3.09	5.76	0.41	0.01	0.01	1.1	99.33	leucogabbro
	Fedorovy Tundras	243.4/BGF425	56.28	0.05	20.62	3.79	0.06	6.05	4.42	5.82	0.43	0.01	0.01	1.44	98.82	Leucogabbro
		154.75/BGF428	53.77	0.16	4.62	9.85	0.19	3.83	26.38	0.44	н/о	0.01	0.35	0.01	99.87	Harzburgite
		243.65/BGF425	58.46	0.10	12.02	5.03	0.12	10.32	10.22	1.62	0.06	0.01	0.01	1.25	99.69	Leucogabbro
4.	*Kivakka*	кв-970	41.36	0.09	2.64	13.65	0.22	1.97	36.62	0.42	0.05	0.03	0.07	1.77	99.05	Harzburgite
		кв-409	52.92	0.22	4.64	10.58	0.18	3.58	26.17	0.51	0.03	0.03	0.61	0.32	99.01	Bronzitite
		кв-422	51.42	0.21	17.45	6.06	0.11	8.82	12.84	1.72	0.17	0.03	0.09	0.93	98.99	Bronzitite
		кв-441	51.66	0.24	16.32	6.74	0.14	11.75	8.91	2.33	0.17	0.04	0.01	0.5	99.61	Norite
		кв-475	42.2	0.09	2.96	13.80	0.22	1.92	37.05	0.38	н/о	0.02	0.06	1.14	99.22	Harzburgite
5.	*Burakovsky*	20–958.8	39.92	0.02	0.26	9.44	0.15	0.37	47.5	0.10	0.27	0.08	0.10	2.20	98.89	Dunite
		200–396	51.76	0.21	1.34	7.70	0.19	14.04	20.75	0.39	0.12	0.02	0.10	3.71	99.79	Pyroxenite
		200–108	52.91	0.19	16.46	5.56	0.12	10.31	10.41	1.99	0.32	0.02	0.03	1.00	99.84	Ol gabbronorite
		45–89.2	49.85	1.14	16.19	13.55	0.14	8.96	4.91	3.50	0.49	0.11	0.03	0.01	98.80	Mt. gabbronorite

Table 7.3 Rare elements' concentrations in rocks of the drusite complex, ppm

No.	1	2	3	4	5	6	7	8	9	10	11	12
Sample No	10,051	11,511	11,531	11,571	22,232	22,441	13,961	13,971	13,981	17,641	17,661	20,141
Rb	26	13	15	26	14	25	5	7	9	6	11	3
Ba	225	94	94	88	126	253	84	123	89	59	109	19
Th	2.48	1.11	0.80	0.29	1.35	3.01	0.46	0.51	0.68	0.44	0.92	0.15
U	0.61	0.26	0.21	0.39	0.32	0.70	0.15	0.15	0.18	0.13	0.23	0.06
Nb	10.2	4.9	3.9	2.0	7.4	10.9	1.2	1.0	1.3	3.4	1.6	0.6
Ta	0.65	0.34	0.30	0.20	0.48	0.73	0.07	0.07	0.11	0.07	0.13	0.05
La	11.4	5.56	6.00	1.14	7.18	12.4	2.59	3.03	4.04	2.47	4.80	0.73
Ce	30.1	14.6	12.9	2.84	14.2	32.3	5.76	6.22	8.45	5.43	10.8	1.55
Pr	3.97	3.97	1.82	0.44	2.36	4.39	0.74	0.76	1.05	0.67	1.38	0.18
Nd	17.5	8.92	8.15	2.31	10.8	20.6	3.17	3.22	4.17	2.70	5.72	0.71
Sr	175	136	133	155	134	183	77.4	118	81.6	67.4	113	9.12
Sm	4.42	2.66	2.27	0.92	3.27	5.45	0.77	0.75	0.88	0.60	1.27	0.16
Zr	152	78	65	36.8	102	154	20.2	19	26	17	37	10
Hf	3.65	1.95	1.58	1.00	2.63	4.02	0.54	0.49	0.66	0.44	0.96	0.22
Eu	1.33	0.98	0.81	0.68	1.26	1.67	0.22	0.22	0.26	0.22	0.38	0.04
Ti	11,120	6,402	5,726	24,341	9,081	11,651	1,551	1,256	1,374	1,032	1,955	289
Gd	4.74	3.35	2.80	1.28	4.30	6.12	0.90	0.71	0.90	0.61	1.30	0.17
Tb	0.80	0.60	0.50	0.24	0.79	1.01	0.15	0.12	0.14	0.10	0.21	0.03
Dy	5.02	3.97	3.23	1.59	5.28	6.31	0.92	0.76	0.85	0.61	1.28	0.15
Ho	1.06	0.87	0.70	0.34	1.18	1.33	0.19	0.15	0.18	0.13	0.27	0.04
Y	32.1	26.6	22.0	10.6	35.7	39.9	5.96	4.72	5.56	3.82	8.38	1.19
Er	3.04	2.49	2.01	0.99	3.38	3.77	0.56	0.43	0.51	0.38	0.75	0.10
Tm	0.46	0.38	0.30	0.15	0.51	0.56	0.08	0.07	0.08	0.05	0.11	0.02
Yb	2.87	2.35	1.93	0.96	3.21	3.54	0.53	0.43	0.48	0.36	0.70	0.09
Lu	0.45	0.36	0.29	0.15	0.49	0.53	0.08	0.06	0.07	0.05	0.11	0.02
Be	1.23	0.75	0.68	0.76	0.94	1.26	0.47	0.40	0.48	0.50	0.67	0.32
V	324	289	275	1,282	298	322	173	106	94.3	81.4	137	68.7
Cr	116	121	122	19.1	45.8	125	3,381	3,305	3,204	4,053	2,464	3,965
Mn	1,672	1,611	1,524	1,130	1,566	1,646	1,687	1,505	1,308	1,376	1,329	1,873
Co	50	52	50	31	47	49	112	98	126	136	97	131
Ni	28	32	33	13	19	67	545	634	1,601	1,663	481	1,005
Cu	86	79	92	174	85	163	14	38	24	20	29	5
Zn	144	106	88	81	124	144	81	77	76	84	70	52
Ga	21.8	17.0	15.7	24.9	20.6	21.0	5.0	4.7	5.5	5.0	7.8	1.5
Mo	1.4	1.6	0.9	1.3	1.3	1.3	2.0	1.0	1.5	1.1	1.3	0.9
Cs	0.59	0.44	0.53	1.30	0.35	0.48	0.23	0.43	0.23	0.19	0.38	0.35
Hf	3.65	1.95	1.58	1.00	2.63	4.02	0.54	0.49	0.66	0.44	0.96	0.22

(continued)

Table 7.3 (continued)

No.	13	14	15	16	17	18	19	20	21	22	23	24
Sample No	16,091	17,781	17,901	20,671	20,771	20,773	17,741	17,751	12,881	10,841	13,311	13,312
Rb	17	10	6	7	17	8	1	6	3	12	25	9
Ba	192	128	75	76	142	94	67	183	12	140	61	19
Th	1.28	0.99	0.54	0.63	2.13	1.59	0.53	0.73	0.09	1.11	0.23	0.22
U	0.32	0.21	0.12	0.13	0.47	0.43	0.18	0.16	0.05	0.53	0.17	0.09
Nb	2.52	1.90	1.09	1.33	5.57	8.75	1.85	1.56	0.93	3.04	2.23	2.56
Ta	0.17	0.12	0.07	0.09	0.34	0.53	0.13	0.11	0.07	0.24	0.16	0.15
La	8.0	5.8	8.0	3.4	7.2	8.4	3.9	4.4	0.9	5.5	3.0	2.6
Ce	17.1	12.8	7.16	7.68	21.8	18.4	9.83	9.54	2.64	13.9	8.02	6.85
Pr	2.14	1.62	0.94	0.98	3.40	2.98	1.34	1.21	0.44	1.79	1.26	1.09
Nd	8.70	6.85	4.04	4.19	16.8	14.5	5.80	5.17	2.45	7.60	6.41	5.64
Sr	181	116	97.3	69.5	125	478	54.1	135	66.1	97.5	119	104
Sm	1.90	1.51	0.97	0.97	4.90	4.20	1.39	1.20	0.93	2.05	2.04	1.92
Zr	57	43	24	24	166	117	42	31	22	61	52	46
Hf	1.49	1.10	0.67	0.67	4.16	2.95	1.08	0.90	0.65	1.73	1.45	1.32
Eu	0.54	0.48	0.30	0.31	1.63	1.35	0.42	0.36	0.40	0.73	0.69	0.72
Ti	2,923	2,522	1,799	1,739	14,981	9,654	2,468	2,025	2,537	4,148	5,685	5,184
Gd	1.84	1.55	1.10	1.04	6.02	5.34	1.50	1.27	1.34	2.20	2.76	2.68
Tb	0.29	0.26	0.18	0.17	1.06	0.94	0.24	0.21	0.26	0.38	0.48	0.49
Dy	1.80	1.59	1.12	1.06	6.82	6.20	1.53	1.33	1.72	2.28	3.10	3.14
Ho	0.38	0.34	0.25	0.23	1.47	1.37	0.32	0.29	0.38	0.48	0.69	0.70
Y	11.3	10.3	7.19	6.79	45.7	40.6	9.74	8.22	11.6	14.3	21.4	21.7
Er	1.04	0.96	0.69	0.65	4.21	3.80	0.90	0.80	1.15	1.44	2.00	2.06
Tm	0.15	0.15	0.10	0.10	0.62	0.57	0.13	0.12	0.18	0.22	0.32	0.31
Yb	0.95	0.91	0.65	0.63	3.93	3.61	0.84	0.75	1.19	1.37	2.05	1.97
Lu	0.14	0.14	0.10	0.10	0.59	0.54	0.13	0.12	0.19	0.21	0.31	0.30
Be	0.64	0.68	0.66	0.60	1.15	0.84	0.56	0.43	0.36	0.97	1.55	2.31
V	161	162	147	125	460	326	185	146	187	177	304	266
Cr	1,852	2,830	2,719	3,409	46.2	44.9	3,195	3,395	1,253	953	266	261
Mn	1,283	1,352	1,419	1,456	1,613	1,543	1,132	1,546	1,309	1,653	2,163	2,246
Co	88	92	92	114	52	40	68	110	67	66	60	50
Ni	395	425	289	484	29	36	212	433	109	166	54	48
Cu	47	42	46	24	43	109	8	63	22	42	80	70
Zn	83	79	68	73	149	75	59	103	64	115	104	71
Ga	10.8	9.1	7.3	6.1	22.3	17.8	8.6	7.3	10.8	12.6	17.1	18.2
Mo	1.33	1.18	1.20	1.72	1.17	1.27	0.93	1.25	0.71	2.00	1.33	1.15
Cs	0.58	0.19	0.10	0.12	0.10	0.11	0.06	0.18	0.49	0.51	1.07	2.19
Hf	1.49	1.10	0.67	0.67	4.16	2.95	1.08	0.90	0.65	1.73	1.45	1.32

Element	11,891	15,211	10,381	14,742	10,691	12,424	12,461	16,981	13,081	16,201	18,401
	25	26	27	28	29	30	31	32	33	34	35
Rb	3	6	6	8	43	10	1	19	21	6	20
Ba	26	42	64	57	188	79	7	403	188	207	219
Th	0.59	0.73	0.40	0.84	3.11	0.23	0.12	4.11	1.41	1.61	1.22
U	0.15	0.22	0.17	1.71	0.77	0.10	0.05	0.53	0.41	0.43	0.26
Nb	5.60	6.90	2.30	5.30	3.40	6.20	1.10	8.20	2.82	3.45	2.87
Ta	0.40	0.42	0.15	0.44	0.31	0.13	0.09	0.50	0.23	0.26	0.17
La	6.50	6.20	4.70	9.60	9.60	4.60	0.90	7.60	8.45	9.70	7.90
Ce	18.0	16.0	8.00	26.0	20.0	11.0	3.00	24.0	18.8	20.9	18.2
Pr	2.77	2.29	1.59	3.90	2.40	1.55	0.38	3.83	2.35	2.70	2.17
Nd	14.2	11.3	7.6	18.5	9.70	7.40	1.90	18.6	9.60	11.4	8.99
Sr	372	170	212	135	99.0	112	51.0	176	151	124	288
Sm	4.56	3.51	2.26	4.99	2.32	2.27	0.59	5.41	2.10	2.56	2.02
Zr	113	112	60	84	80	58	16	172	66	86	57
Hf	3.19	2.99	1.60	2.44	1.97	1.55	0.43	4.63	1.71	2.14	1.45
Eu	1.52	1.18	0.89	1.40	0.71	0.79	0.19	1.72	0.64	0.67	0.77
Ti	13,132	7,928	5,661	7,439	4,148	5,632	1,790	12,108	3,284	3,889	3,301
Gd	5.80	4.41	2.98	5.11	2.57	3.02	0.86	6.27	2.14	2.57	2.21
Tb	1.03	0.77	0.53	0.82	0.43	0.54	0.15	1.05	0.35	0.41	0.35
Dy	6.47	4.99	3.50	5.13	2.75	3.51	0.97	6.65	2.12	2.49	2.13
Ho	1.31	1.12	0.77	1.10	0.59	0.75	0.21	1.38	0.43	0.52	0.45
Y	37.0	32.1	24.0	33.5	18.9	22.6	6.50	40.2	13.0	15.7	13.9
Er	3.60	3.31	2.27	3.31	1.67	2.15	0.60	3.90	1.24	1.47	1.23
Tm	0.51	0.51	0.35	0.52	0.25	0.32	0.09	0.58	0.18	0.21	0.18
Yb	3.13	3.23	3.13	3.40	1.57	2.04	0.56	3.60	1.14	1.32	1.20
Lu	0.46	0.48	0.35	0.54	0.24	0.31	0.09	0.53	0.17	0.20	0.18
Be	1.10	2.81	0.66	1.00	0.92	0.57	0.40	1.12	0.75	0.76	0.65
V	590	462	300	260	207	307	124	292	163	187	176
Cr	27	179	136	145	404	113	2,161	94	1,699	1,620	100
Mn	1,101	1,247	1,532	2,002	1,420	1,637	1,049	1,861	1,279	1,296	1,035
Co	67	71	42	42	54	52	97	49	81	80	41
Ni	16	30	62	18	52	27	1,055	100	404	396	34
Cu	12	23	15	38	69	43	37	50	45	40	56
Zn	152	176	31	121	87	120	67	144	77	102	68
Ga	43.9	50.0	16.5	21.6	15.7	16.5	6.2	22.8	12.4	11.4	17.0
Mo	0.76	1.21	1.01	0.95	0.99	1.67	1.41	1.56	1.09	1.43	1.05
Cs	0.14	0.35	0.38	0.41	0.54	0.36	0.27	0.23	0.69	0.28	0.77
Hf	3.19	2.99	1.60	2.44	1.97	1.55	0.43	4.63	1.71	2.14	1.45

Note: Massifs: 1–6—Massif at 403 m; 7–12—Sorkajoki; 13—Mt. Perchatka; 14–15—Poioiva; 16—Kamenka River; 17–18—Mt. Kruaya; 19–20—massif at 463.9 m; 21–22—Mt. Nora; 23–24—Mt. Ragutchane; 25–26—Grob Tundra; 27–28—Mt. Levoiva; 29—Mt. Ristivara; 30—Nora River; 31—Mt. Kontiovara; 32—massif at 408.8 m; 33–34—massif at 391.2 m; 35—Lais-Tundra

Table 7.4 Rare elements' concentrations in rocks of the layered plutons, ppm

N	1	2	3	4	5	6	7	8	№п/п	9	10	11	12	13	14
N sample	1007/98	11	11	1882/164.8	16	20/736.8	36281	10		10	11/224.5	02002/8.3	45	BGF/52.7	BGF/260.7
Rb	3	3	1	3	1	12	2	1	Rb	1	4	17	20	3	1
Ba	50	39	51	56	5	43	11	100	Ba	5	79	170	198	55	49
Th	0.24	0.20	0.13	0.32	0.03	0.21	0.10	0.11	Th	0.05	0.16	1.63	1.86	0.09	0.10
U	0.06	0.05	0.04	0.08	0.04	0.09	0.04	0.04	U	0.02	0.06	0.42	0.48	0.03	0.03
Nb	0.8	0.6	0.5	0.6	0.2	1.2	0.4	0.5	Nb	0.5	0.6	2.2	2.3	0.4	0.5
Ta	0.06	0.07	0.04	0.04	0.03	0.06	0.03	0.12	Ta	0.11	0.04	0.35	0.18	0.06	0.06
La	2.1	1.8	1.0	2.3	0.2	1.2	0.6	1.0	La	0.4	1.9	7.0	8.0	1.2	1.3
Ce	5	4	2	5	0	3	1	2	Ce	1	4	15	17	2	3
Pr	0.59	0.50	0.27	0.60	0.06	0.42	0.15	0.27	Pr	0.11	0.61	1.80	2.04	0.30	0.39
Nd	2.5	2.2	1.1	2.4	0.2	1.9	0.7	1.2	Nd	0.4	2.8	7.1	8.0	1.2	1.7
Sr	102	64	49	288	9	29	14	59	Sr	9	296	123	129	280	190
Sm	0.62	0.52	0.26	0.51	0.05	0.42	0.16	0.44	Sm	0.09	0.80	1.39	1.62	0.30	0.46
Zr	16	14	10	14	2	11	6	9	Zr	3	14	43	48	6	10
Hf	0.43	0.38	0.24	0.37	0.05	0.32	0.16	0.25	Hf	0.07	0.40	1.20	1.34	0.16	0.29
Eu	0.23	0.18	0.11	0.24	0.02	0.09	0.06	0.11	Eu	0.03	0.49	0.34	0.36	0.26	0.26
Ti	1290	889	478	821	136	802	671	803	Ti	305	1712	2068	2301	735	887
Gd	0.71	0.53	0.27	0.51	0.05	0.42	0.20	0.29	Gd	0.08	1.04	1.31	1.52	0.33	0.52
Tb	0.11	0.09	0.04	0.09	0.01	0.07	0.03	0.05	Tb	0.02	0.18	0.20	0.24	0.06	0.09
Dy	0.75	0.56	0.25	0.55	0.06	0.42	0.24	0.37	Dy	0.10	1.21	1.25	1.40	0.36	0.56
Ho	0.16	0.12	0.06	0.12	0.01	0.09	0.06	0.08	Ho	0.02	0.27	0.27	0.30	0.08	0.13
Y	4.9	3.7	1.8	3.8	0.5	3.0	1.7	2.3	Y	0.8	8.1	7.8	8.9	2.4	3.8
Er	0.52	0.38	0.16	0.36	0.04	0.27	0.18	0.25	Er	0.08	0.80	0.78	0.85	0.26	0.39
Tm	0.08	0.05	0.03	0.05	0.01	0.05	0.03	0.05	Tm	0.01	0.12	0.12	0.13	0.04	0.06
Yb	0.52	0.38	0.21	0.35	0.05	0.32	0.21	0.35	Yb	0.10	0.79	0.78	0.84	0.29	0.38
Lu	0.08	0.06	0.03	0.06	0.01	0.05	0.04	0.04	Lu	0.02	0.12	0.12	0.13	0.04	0.06
Be	0.32	0.45	0.39	0.50	0.37	0.50	0.28	0.40	Be	0.36	0.52	0.62	0.64	0.41	0.47
V	121	70	51	73	30	87	82	102	V	62	158	154	128	66	85
Cr	3712	1769	3457	806	15216	4816	4734	3681	Cr	6137	15	2047	1714	114	152
Mn	1403	1253	1236	726	822	1465	1456	1314	Mn	1308	1212	1623	1303	812	1004
Co	123	145	153	42	137	92	99	84	Co	155	50	78	62	47	73
Ni	1728	2189	1869	87	3174	613	505	290	Ni	2392	38	89	80	149	275
Cu	449	273	12	10	4	70	20	34	Cu	300	102	41	24	201	64
Zn	70	62	72	43	47	79	56	58	Zn	71	61	98	65	39	50
Ga	7.5	4.4	2.6	12.9	1.7	4.1	3.0	3.9	Ga	2.1	17.3	7.7	8.1	14.4	11.2
Mo	1.04	0.82	1.68	0.78	0.81	0.86	0.87	2.10	Mo	1.38	1.08	1.28	0.85	1.04	0.70
Cs	0.07	0.07	0.07	0.11	0.04	0.59	0.10	0.06	Cs	0.05	0.16	0.49	0.54	0.21	0.08
Hf	0.43	0.38	0.24	0.37	0.05	0.32	0.16	0.25	Hf	0.07	0.40	1.20	1.34	0.16	0.29
Ta	0.06	0.07	0.04	0.04	0.03	0.06	0.03	0.12	Ta	0.11	0.04	0.35	0.18	0.06	0.06

Note: 1–7 = Monchepluton: 1-Travyanaya, 2-9-Mt. Moncha, 10–11 = Mt. Sopcha, 12 =Main Ridge; 13–14 = Umbarechensky, 15–18 = Pansky Tundra,

15	16	17	№	18	19	20	21	22	23	24	25	26	27
BGF/211	BGF/154	FG243.65	N sample	FG211.8	KV-409	KV-422	KV-441	KV-475	KV-970	20–959.8	200–396	200–108	45–89.2
4	5	21	Rb	4	1	3	9	5	1	20	2	4	6
64	29	182	Ba	57	40	42	77	112	24	6	11	82	163
0.06	0.19	0.06	Th	0.06	0.08	0.19	0.28	0.38	0.12	0.11	0.20	0.48	0.65
0.03	0.05	0.06	U	0.06	0.03	0.05	0.07	0.09	0.04	0.04	0.05	0.10	0.18
1.0	0.5	0.7	Nb	0.6	0.4	0.6	1.1	0.9	0.6	0.3	0.4	0.8	1.1
0.12	0.06	0.07	Ta	0.02	0.03	0.06	0.07	0.06	0.03	0.09	0.04	0.11	0.07
1.4	1.5	2.9	La	1.1	0.8	2.0	2.7	3.6	1.0	0.3	1.3	2.6	3.3
3	3	4	Ce	2	2	4	6	8	2	0	4	5	7
0.40	0.43	0.36	Pr	0.31	0.23	0.49	0.72	1.05	0.27	0.06	0.58	0.67	0.85
1.9	1.8	1.2	Nd	1.4	1.0	2.0	3.0	4.5	1.1	0.2	3.0	2.8	3.6
273	26	830	Sr	267	50	66	275	338	42	8	37	370	435
0.48	0.41	0.36	Sm	0.38	0.31	0.44	0.66	1.06	0.23	0.05	0.90	0.66	0.87
9	15	11	Zr	6	6	14	17	26	8	7	10	15	22
0.27	0.43	0.27	Hf	0.19	0.17	0.38	0.48	0.69	0.20	0.17	0.32	0.44	0.65
0.25	0.14	0.85	Eu	0.22	0.09	0.16	0.31	0.44	0.07	0.01	0.27	0.40	0.47
742	1265	369	Ti	618	431	1096	1190	1662	492	112	1234	1279	7071
0.54	0.53	0.53	Gd	0.44	0.24	0.49	0.67	1.20	0.26	0.05	1.17	0.78	0.98
0.09	0.09	0.04	Tb	0.08	0.04	0.08	0.11	0.19	0.04	0.01	0.18	0.12	0.16
0.57	0.65	0.54	Dy	0.48	0.29	0.54	0.70	1.24	0.27	0.05	1.08	0.80	0.99
0.12	0.16	0.23	Ho	0.10	0.06	0.13	0.15	0.26	0.06	0.01	0.22	0.16	0.21
3.6	4.8	0.8	Y	2.9	1.8	3.9	4.6	7.9	1.9	0.5	6.3	4.6	5.9
0.36	0.54	0.17	Er	0.30	0.19	0.39	0.45	0.76	0.20	0.04	0.58	0.45	0.55
0.05	0.10	0.01	Tm	0.04	0.03	0.06	0.07	0.12	0.03	0.01	0.08	0.06	0.08
0.34	0.70	0.07	Yb	0.28	0.23	0.44	0.47	0.74	0.21	0.06	0.49	0.40	0.50
0.05	0.12	0.01	Lu	0.04	0.03	0.07	0.08	0.11	0.03	0.01	0.07	0.06	0.07
0.40	0.39	0.51	Be	0.37	0.32	0.35	0.46	0.48	0.28	0.51	0.43	0.44	0.54
92	95	26	V	77	40	117	82	136	39	11	142	114	1173
75	2713	81	Cr	113	726	3382	1097	21	597	998	1389	227	43
753	1508	467	Mn	811	1317	1504	858	1054	1421	1044	1521	941	1093
37	98	32	Co	33	134	80	39	43	147	161	91	51	79
92	1542	97	Ni	74	1750	255	93	38	1939	3481	579	92	22
19	228	577	Cu	10	30	25	26	93	16	7	330	19	266
35	100	52	Zn	35	72	69	46	50	68	59	49	42	88
13.0	5.3	25.8	Ga	10.8	3.1	5.9	12.7	15.2	2.7	1.0	3.1	14.7	23.5
1.56	0.90	12.40	Mo	1.68	0.89	0.67	0.75	0.65	1.49	1.04	0.67	1.01	0.81
0.25	0.35	1.63	Cs	0.26	0.09	0.13	0.22	0.21	0.07	0.37	0.08	0.09	0.15
0.27	0.43	0.27	Hf	0.19	0.17	0.38	0.48	0.69	0.20	0.17	0.32	0.44	0.65
0.12	0.06	0.07	Ta	0.02	0.03	0.06	0.07	0.06	0.03	0.09	0.04	0.11	0.07

19–20 = Fedorova Tubdra, 20–23 = Kivakka, 24–27 = Burakovsky

Table 7.5 Olivine composition from drusite massifs, wt %

No.	Sample No	Massif	Fo	SiO_2	MgO	CaO	MnO	FeO	NiO	Total	MgOr
1.	20,671	Kamenka	79.71	38.96	42.35	0.11	0.3	19.21	0.28	101.2	25.88
2.	20,671		79.54	38.11	42.24	0.02	0.25	19.37	0.31	100.3	25.88
3.	20,671		78.94	38.22	41.92	0.01	0.3	19.94	0.28	100.7	25.88
4.	20,671		79.59	38.7	42.22	0.01	0.28	19.3	0.24	100.7	25.88
5.	20,671		80.12	41.58	40.64	0.04	0.22	17.97	0.24	100.7	25.88
6.	20,671		79.03	38.37	41.88	0.01	0.24	19.81	0.27	100.6	25.88
7.	20,671		78.70	38.42	41.72	0.01	0.3	20.13	0.23	100.8	25.88
8.	20,671		78.79	38.63	41.63	0.02	0.26	19.98	0.27	100.8	25.88
9.	20,671		79.01	38.44	42.23	0.03	0.26	20	0.25	101.2	25.88
10.	17,761	Massif at 463.9 m	77.52	38.76	40.85	0.04	0.25	21.11	0.29	101.3	24.33
11.	17,761		77.93	38.35	40.76	0.02	0.22	20.57	0.32	100.3	24.33
12.	17,761		78.45	38.8	41.44	0.03	0.32	20.29	0.26	101.2	24.33
13.	17,761		78.39	38.69	41.1	0.02	0.26	20.2	0.33	100.6	24.33
14.	16,091	Mt. Perchatka	72.58	37.7	37.28	0.02	0.28	25.11	0.4	100.8	19.46
15.	16,091		73.03	37.56	37.53	0.04	0.32	24.7	0.44	100.6	19.46
16.	16,091		73.08	37.67	37.84	0.02	0.13	24.85	0.37	100.9	19.46
17.	16,091		72.95	37.89	37.85	0.02	0.29	25.01	0.44	101.5	19.46
18.	16,091		73.24	38.04	37.83	0.03	0.33	24.64	0.4	101.3	19.46
19.	16,091		74.18	37.63	37.77	0.03	0.27	23.43	0.39	99.53	19.46
20.	16,091		74.17	37.28	37.52	0.04	0.29	23.29	0.38	98.81	19.46
21.	13,971	Sorkajoki	81.43	38.85	43.27	0	0.32	17.59	0.3	100.3	29.86
22.	13,971		81.25	38.72	42.65	0.02	0.31	17.54	0.3	99.54	29.86
23.	13,971		81.84	38.41	43.64	0.02	0.23	17.26	0.27	99.83	29.86
24.	13,971		81.85	38.62	43.5	0.03	0.29	17.19	0.3	99.93	29.86
25.	13,971		81.81	38.46	43.48	0.03	0.25	17.23	0.3	99.76	29.86
26.	13,971		81.96	38.85	43.63	0.01	0.23	17.12	0.28	100.1	29.86
27.	13,971		82.07	38.98	43.84	0	0.33	17.07	0.31	100.5	29.86
28.	13,971		82.15	38.76	43.64	0.02	0.27	16.9	0.29	99.88	29.86
29.	13,971		81.66	38.61	43.54	0.02	0.22	17.43	0.33	100.1	29.86
30.	13,971		81.86	38.94	43.57	0	0.19	17.21	0.32	100.2	29.86
31.	17,641		84.00	38.45	45.28	0.01	0.25	15.37	0.29	99.65	33.87
32.	17,641		83.67	38.27	44.81	0	0.24	15.59	0.32	99.23	33.87
33.	17,641		83.91	37.94	45.27	0.02	0.14	15.47	0.32	99.17	33.87
34.	17,641		83.73	39.2	44.64	0.02	0.26	15.46	0.36	99.96	33.87
35.	17,641		83.66	39.07	44.8	0.01	0.26	15.6	0.33	100.1	33.87
36.	17,641		84.14	39	45.32	0.11	0.32	15.23	0.33	100.3	33.87
37.	17,641		84.11	39.1	45.19	0.08	0.29	15.22	0.37	100.3	33.87
38.	17,641		84.36	38.75	45.62	0.06	0.2	15.07	0.3	100	33.87
39.	17,641		84.17	38.86	45.35	0.03	0.26	15.2	0.31	100	33.87
40.	17,641		83.66	39.32	45.1	0.02	0.27	15.7	0.29	100.7	33.87
41.	17,641		83.57	39.22	44.58	0.11	0.26	15.62	0.31	100.1	33.87
42.	17,641		83.65	39.03	44.77	0.05	0.29	15.6	0.33	100.1	33.87
43.	17,641		83.46	39.2	44.44	0.02	0.25	15.7	0.3	99.91	33.87
44.	17,641		83.32	39.17	43.98	0.04	0.24	15.69	0.31	99.44	33.87
45.	17,641		83.80	39.15	44.76	0.14	0.23	15.42	0.31	100	33.87
46.	17,641		83.47	38.49	44.76	0.18	0.16	15.8	0.31	99.71	33.87
47.	20,141	Sorkajoki	81.04	37.86	42.7	0.02	0.28	17.81	0.31	98.97	31.73
48.	20,141		81.23	38.12	42.74	0.02	0.24	17.6	0.29	99.02	31.73
49.	20,141		81.19	38.12	43.02	0	0.29	17.76	0.26	99.45	31.73
50.	20,141		81.15	38.49	42.74	0.01	0.31	17.7	0.25	99.49	31.73
51.	20,141		81.28	38.53	42.64	0.01	0.31	17.51	0.3	99.31	31.73
52.	20,911		77.85	37.3	40.18	0	0.27	20.38	0.27	98.4	26.43

(continued)

Table 7.5 (continued)

No.	Sample No	Massif	Fo	SiO_2	MgO	CaO	MnO	FeO	NiO	Total	MgOr
53.	20,911		77.60	37.43	39.73	0.01	0.27	20.44	0.28	98.16	26.43
54.	20,911		77.42	37.84	39.7	0.02	0.26	20.64	0.29	98.76	26.43
55.	22,521	Mt. Rumumistro	77.29	38.1	40.35	0.03	0.23	21.13	0.32	100.2	25.92
56.	22,521		77.38	38.8	40.47	0.02	0.28	21.09	0.34	101	25.92
57.	22,521		77.24	37.92	40.1	0.03	0.23	21.06	0.28	99.61	25.92
58.	22,521		77.23	37.84	39.94	0.06	0.27	20.99	0.28	99.37	25.92
59.	18,661	Poioiva	69.84	37.46	35.07	0.04	0.39	26.99	0.34	100.3	15.97
60.	18,661		69.62	37.38	34.9	0	0.32	27.15	0.31	100.1	15.97
61.	ОП-18	Anisimov	73.81	37.63	38.41	–	0.17	24.29	0.43	100.9	
62.	Л-704	Pezhostrov	75.43	37.9	38.83	–	0.34	22.55	0.28	101.5	
63.	ОГН-705		73.56	37.51	37.5	–	0.21	24.03	0.58	99.83	
64.	L-B1	Voroniy	80.09	36.67	42.3	–	0.23	18.74	0.65	98.59	
65.	L-B7		72.24	37.07	36.93	–	0.21	25.29	0.59	100.1	20.99
66.	L-П-16	Shang	79.73	37.87	44.68	–	0.38	20.25	0.38	100.1	
67.	ОН-П30	Yudom	77.95	38.08	41.06	–	0.25	20.7	0.5	100.6	
68.	L-П37	Navolok	76.21	37.61	39.45	–	0.28	21.95	0.57	99.86	

Note: No 1–60, After Krivolutskaya et al. (2010a); No 61–68, after Sharkov et al. (2004); MgOr–MgO content in rocks; "–" is under D.L.

Table 7.6 Composition of chrome–spinels and ilmenite from drusite massifs, wt %

No.	Sample No	MgO	Al_2O_3	TiO_2	V_2O_3	Cr_2O_3	FeO	MnO	ZnO	Summa	Massif	Cr#
1.	20,671	5.23	22.99	0.63	0.91	38.36	32.2	0.43	0.19	100.93	Kamenka	0.63
8.	20,671	5.12	24.93	1.02	0.58	35.23	32.52	0.38	0.44	100.22	Kamenka	0.59
2.	17,161	3.07	16.23	1.64	0.44	36.64	40.92	0.4	0.02	99.35	Massif at 463.1 m	0.69
3	17,761	6.1	25.96	0.28	0.4	37.64	29.87	0.42	0.09	100.77	Massif at 463.1 m	0.59
4.	17,761	6.96	30.16	0.17	0.31	33.56	29.15	0.39	0.2	100.9	Massif at 463.1 m	0.53
5.	12,461	1.27	0.03	52.3	0	0.04	45.51	1.16	0.12	100.43	Mt. Kontiovara	
6.	12,461	1.16	0.01	52.58	0.29	0.04	45.4	1.16	0.04	100.68	Mt. Kontiovara	
7.	12,461	1.3	0.02	52.28	0.17	0.04	45.63	1.15	0.04	100.58	Mt. Kontiovara	
11.	13,971	3.5	12.17	1.05	0.43	37.77	41.59	0.46	0.42	97.38	Sorkajoki	0.76
12.	13,971	7.09	13.71	0.63	0.3	46.8	29.62	0.43	0.01	98.59	Sorkajoki	0.77
13.	17,641	3.88	10.23	7.23	0.72	35.21	37.71	0.5	0.04	95.52	Sorkajoki	0.77
14.	17,641	4.34	10.8	11.27	0.41	33.72	36.85	0.57	0.03	97.99	Sorkajoki	0.76
15.	16,091	3.22	25.56	1.54	0.66	33.52	33.72	0.31	0.99	99.53	Perchatka	0.57

Note: Cr# = $Cr_2O_3/(Cr_2O_3 + Al_2O_3)$

Table 7.7 Garnet compositions from drusite massifs, wt %

No.	Sample No	MgO	Al_2O_3	CaO	FeO	MnO	SiO_2	Total	Massif
1.	22,711	8.72	21.6	5.42	22.8	1.33	38.67	98.54	Poioiva
2.	15,211	7.48	21.9	8.01	24.62	0.3	39.3	101.6	Grob Tundra
3.	15,211	6.75	21.43	7.87	25.5	0.32	38.99	100.86	Grob Tundra
4.	15,211	6.96	21.68	8.7	24.01	0.39	39.27	101.01	Grob Tundra
5.	10,061	2.01	20.82	10.53	28.38	1.44	38.05	101.23	Massif at 403 m
6.	10,061	1.76	20.96	10.54	28.28	1.47	38.06	101.07	Massif at 403 m
7.	22,232	1.84	21.25	10.38	29.46	1.3	37.93	102.16	Massif at 403 m
8.	22,232	1.91	21.23	10.88	28.62	0.86	38.23	101.73	Massif at 403 m
9.	20,771	1.67	21.03	9.92	29.34	1.92	37.71	101.59	Mt. Krutaya
10.	20,771	1.75	21.3	10.03	29.23	1.94	38.4	102.65	Mt. Krutaya
11.	20,773	1.88	21.3	9.42	28.75	1.99	38.17	101.51	Mt. Krutaya
12.	20,773	1.74	21.34	9.78	28.35	2.54	37.98	101.73	Mt. Krutaya
13.	20,773	2.17	21.38	9.79	28.55	1.16	37.42	100.47	Mt. Krutaya

sisting of several bodies whose position is controlled by the general fold structure of the area; they are restricted to the bedding planes of the host gneiss members. Larger intrusions typically occur as individual bodies. Morphologically, small massifs usually form rootless bodies of lenticular and phacolithic shape, while large massifs form subvertical stratal intrusions. Together with the host rocks, they were metamorphosed under the amphibolite facies, which caused changes in their morphology and mineral composition. Most of the intense alterations occurred where the massifs made contact with the host rocks, which is where ultrabasic rocks were often transformed into amphibolites and crystalline schists.

In terms of the structural–textural features of the rocks and the degree of their metamorphic transformations, the studied massifs of the drusite complex of the South Kovdor area can be subdivided into the following types: (1) Large (a few kilometers long), very weakly altered, basic–ultrabasic intrusions (Sorkajoki, Poioiva, Mt. Perchatka, Mt. Rumimuristo, and the Kamenka River) having an extended platelike or isometric shape. (2) Medium and, more rarely, larger intrusions (a few hundred meters to kilometers) mainly composed of gabbroids with subordinate amounts of ultrabasic rocks, which are variably altered with the preserved relicts of primary magmatic textures (the massifs at 403 m in elevation, 463.9 m, Mt. Grob Tundra, 391 m, Mt. Krutaya Vostochnaya, the upper reaches of the Nora River and Mt. Kontiovara). (3) Small and medium (a few hundred meters) composed of strongly metamorphosed ultrabasic–basic bodies transformed into garnet amphibolites and plagioamphibolites (the massifs of Mt. Lais-Tundra, Mt. Ristivara, Mt. Nora, Mt. Ragutchane, Mt. Krutaya, 408 m in elevation, the middle reaches of the Nora River, the Mt. Levoiva, and the western flank of the Poioiva Massif).

Drusite textures were identified in all three types of massifs.

The typomorphic massif of the *first type* is the large (up to 7 km long and 1.5 km thick) sheeted, steeply dipping *Sorkajoki Massif* that extends in the NW direction and cuts across the fold structures of the Archean gneisses. The massif is mainly composed of fresh fine- to medium-grained ultrabasic rocks with subordinate basic rocks with massive structures. The predominant rocks are lherzolites and the less common plagioclase websterites, troctolites, olivine gabbronorites, and harzburgites. Olivine (15–75 %) occurs as large, euhedral crystals up to 1.5 mm in size and small, rounded poikilitic inclusions (0.2–0.5 mm) in pyroxene. More rarely, it occurs in plagioclase. The composition of the olivine (Table 7.5) varies within $Fo_{81.1–84.3}$, and the clinopyroxene content in the rocks (En_{49} Fs_5 Wo_{46}) varies widely, reaching up to 75 vol.%. Orthopyroxene ($En_{81–85}$ $Fs_{13–17}$ $Wo_{2–5}$) is observed as colorless, elongated crystals up to 1 mm in size (5–55 %). The rocks of this massif contain the most Cr–spinel (Cr# = 0.77; Table 7.6).

Polysynthetically twinned grains of plagioclase ($An_{57–59}$) fill the interstices between mafic minerals as well as form poikilitic inclusions in the pyroxenes. Its highest content in the rocks was no more than 20 %. In some samples, grains of olivines and, partly, orthopyroxenes, which occur at the contact zone with the plagioclase, are often fringed by one- or two-layer (<0.02 mm) amphibole–pyroxene rims, which form coronal or drusite textures. The rims can occupy up to 55 vol.% of the rock. All samples contain evenly disseminated chromite (1–3 %) with a Cr_2O_3 content of up to 46.8 wt %.

In terms of the composition of the derivatives, the close analogue of the Sorkajoki Massif is the *Poioiva Massif*, which is dominated by olivine gabbronorites and gabbronorites. The rocks consist of rounded grains of olivine $Fo_{77–78}$ and its relicts (5–10 %), large xenomorphic grains of clinopyroxene ($En_{53–60}$ $Fs9$ $Wo_{31–38}$), and smaller orthopyroxene crystals. Plagioclase is observed as large laths ($An_{56–55}$) with ingrowths of pyroxenes and olivines, or it forms irregularly shaped grains that fill the interstices between grains of mafic minerals. Its content in the rocks varies from 20 to 40 %. At the contact point with plagioclase olivine, it is fringed by the rims described above. The secondary minerals are chlorite, biotite, and pelite materials. The rocks have gabbro, gabbroophitic, hypidiomorphic, poikilitic (in places), and coronal textures. The *Mt. Rumimuristo Massif*, which is characterized by a higher metamorphic grade, is also ascribed to this type. The weakly altered rocks of this massif contain olivine, whose composition is very close to the composition of the olivine from the Poioiva Massif ($Fo_{77–78}$). However, practically, the entire plagioclase is newly formed, which follows from its fairly acid composition $An_{34–39}$. The intense metamorphic overprint is confirmed by the discovery of high-Mg amphibole (Mg# = 81–82). The *Mt. Perchatka* Massif is primarily made up of gabbronorites with subordinate lherzolites. Among the intrusions of this type, this massif contains the olivine with the most Fo and narrow variations of the forsterite component ($Fo_{72.5–74}$). This is also typical of clinopyroxene, which contains the least amount of magnesium (Mg# = 78.4–79.6) among the clinopyroxenes compared to other massifs. Its composition varies within the following limits: $En_{62–59}$ $Fs_{10–06}$ $Wo_{31–28}$. The magnesian number of orthopyroxene is lower than that of clinopyroxene and varies from 76.4 to 78.9 ($En_{76–78}$ $Fs_{18–24}$ Wo_2).

The *second type* of massifs includes the medium-sized (hundreds of meters to a few kilometers), strongly altered bodies with relicts of primary rocks and variable proportions of magmatic and metamorphosed rocks. The most characteristic massif of this type is the *Massif at 403.0-m* elevation, which is mainly composed of metagabbroids and plagioamphibolites. The metagabbroids differ from the plagioamphibolites by the presence of relicts with clinopyroxene inclusions in amphibole (up to 20–30 %) with relatively high Fe (Mg# 70), relicts of gabbro, ophitic textures, and a spotted structure. The amphibolites were presumably formed after the gabbroids during regional metamorphism. Also among the intrusive bodies of this type is the *Massif of 463.9 m*. The unaltered rocks of this massif (lherzolite, olivine gabbro) contain olivine $Fo_{78.5-77}$ as well as orthopyroxenes (En_{80-82} Fs_{22-16} Wo_2) and clinopyroxenes (En_{55-60} Fs_{10} Wo_{37-33}). Olivine associates with Cr–spinellids containing 33.6–37 wt % Cr_2O_3 and elevated Zn contents (up to 0.2 wt %). The rocks have a nematoblastic fine- to medium-grained texture and a lenticular–gneissose foliated structure. In some samples, rocks are transformed into amphibolites. Amphibole (Mg# 81.5) accounts for up to 90 vol.%, and additionally, there is chlorite, carbonate, and rutile. Olivine gabbro and metagabbro in the 391.2-m-high massif are also ascribed to this type.

Plagioamphibolites consist mainly of amphibole, which is represented by green hornblende, more rarely the almost colorless, greenish mineral of the actinolite–tremolite series (35–75 %, Mg#=30–40–57) and sericitized, pelitic plagioclase (25–50 %) with occasional polysynthetic twins. These rocks typically contain up to 15 % Fe–garnet (almandine), which often forms pseudomorphs and contains inclusions of other minerals, such as quartz (1–5 %). Accessory minerals are represented by titanite, apatite, zircon, and oxide minerals (Cr–spinellids with Cr_2O_3 contents of 33.6–37.6 wt %; Table 7.6). Amphibole is replaced by red–brown biotite, epidote, and chlorite. The rocks have a nematogranoblastic, sometimes cataclastic, and poikiloblastic texture and a banded–spotted, lenticular–banded to vaguely banded structure.

One more massif in this group is the *Mt. Grob Tundra Massif* represented by fine-grained gabbroids that are often transformed into amphibole metagabbro and amphibolites that are often with garnet. The gabbros and metagabbros consist of large euhedral plagioclase (25–50 %) with polysynthetic twins and poikilitic inclusions of pyroxene (up to 55 %) that are almost completely replaced by fine-grained actinolite–tremolite aggregate (Mg#=55.5–82). Amphibole is replaced by biotite (2–3 %), epidote (5–7 %), and chlorite (10–15 %). The rock has a fine-, medium-, to coarse-grained, poikilitic, and gabbro texture.

Plagioamphibolites differ from amphibolites due to the presence of large (up to 10 mm) grains and porphyroblasts of garnet (FeO up to 21.7 wt %; Table 7.7), blue–green hornblende, and a variable amount of plagioclase (5–50 %). Their peculiar feature is a nematogranoblastic, porphyroblastic, and poikiloporphyroblastic texture and a massive, sometimes gneissose, structure.

The third type is represented by numerous small intrusive bodies mainly consisting of metamorphic rocks, amphibolites, and plagioamphibolites, often with garnet and quartz and relicts of primary gabbroids. The intrusions of this type are similar in mineral composition and structural–textural features. The characteristic *Mt. Lais-Tundra Massif* is represented by plagioamphibolites and garnet amphibolites with a gneissose, thin-banded texture. The rocks consist of 45–50 % amphibole (Mg# 40.5–45), which is partly actinolitized or chloritized and replaced by epidote–chlorite aggregate, and 45–50 % of a very acidic (An_{15-13}), strongly sericitized plagioclase and quartz. Plagioclase often forms diablastic intergrowths with quartz, which shows simultaneous extinction. The accessory minerals are apatite and rutile. *The Massif of the middle reaches of the Nora River* has a very permanent rock composition. It is made up of inequigranular medium- to fine-grained plagioamphibolites with gneissose structure. Amphibole occupies up to 60–70 vol.% and has a more magnesian composition compared to that from the Lais-Tundra Massif (Mg# 48.5–54.0). Strongly sericitized plagioclase accounts for 35 vol.%, and quartz accounts for up to 3 %.

7.3 Brief Geology of the Layered Plutons

The drusite massifs were compared to several well-studied layered plutons, which are considered to be reference objects of the peridotite–pyroxenite–gabbronorite formation (Fig. 7.1). The list of the rock samples from the layered intrusions is given in Table 7.2. Brief information on the structure and composition of these intrusions from the literature is reported below with indications of the sampling localities (Figs. 7.3a–d and 7.6e).

Monchepluton represents a typical layered ultrabasic–basic intrusion that is approximately 1,000 m thick and contains deposits and occurrences of Cu–Ni, sulfide, chromite, and PGE ores of magmatic genesis. The massif has an arch shape and consists of two chambers (Fig. 7.3a). One, which is 7 km long, is oriented in the NE direction and expressed topographically by the Nittis, Kumuzh'ya, and Travyanaya mountains. From the bottom upward, it is made up of a thick basal zone of quartz-bearing norites and gabbronorites from 10 to 100 m, harzburgites (100–200 m), a zone of alternating harzburgites and orthopyroxenites (250–400 m), and orthopyroxenites (300–700 m) with lenses of nodular chromites (Mt. Kumuzh'ya). The structure of the northeastern chamber is complicated by the presence of the Dunitovyi block, which contains the Sopcha Lake chromite deposit. The total thickness of the rock section increases from the south northward from 200–300 m to 800–1,000 m. The second chamber is 9 km long and extends in the eastern direction via the summits of the Sopchuaivench and Poazuaivench mountains (Smolkin et al. 2004). It is composed (from the bottom upward) of quartz-bearing gabbronorites and norites, melanocratic norites with lenses, and intercalations of harzburgites and norites, meso- and leucocratic norites, and gabbronorites with beds of sulfide-bearing dunite–harzburgites that are 1–5 m thick in the upper section (Mt. Sopcha, ore layer 330). The total thickness of the rock section of the eastern chamber varies from 300–400 to 600–800 m thick. Samples were taken from the section of both chambers: Mt. Nittis, Mt. Travyanaya, Mt. Nyud, and Mt. Sopcha (Fig. 7.3a).

The Monchetundra Massif is part of one of the largest gabbro–anorthosite plutons in the Baltic Shield, the Main Range

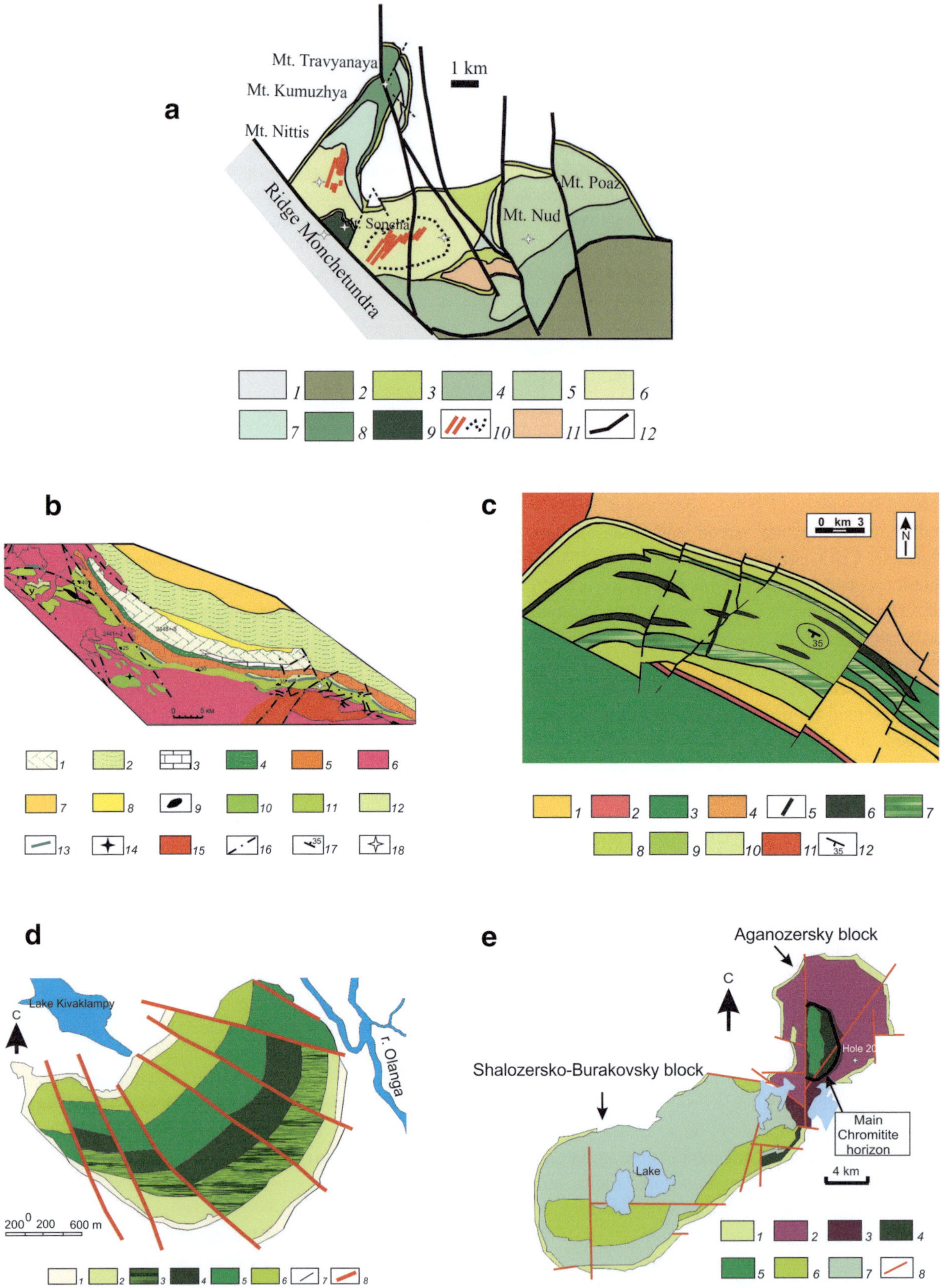

Fig. 7.3 Schematic geological maps of the layered intrusions of the Karelia–Kola region.
(**a**) Scheme of the geological structure of Monchepluton. (*1*) Gabbronorites and anorthosites of the Monchetundra Massif, (*2*) metavolcanics; (*3*) gabbronorites, gabbros, and anorthosites of the foothills of Vurechuaivench; (*4*) melanorites; (*5*) norites; (After Smolkin et al. 2004)

Massif (440 km^2 in area). The submeredional fault splits the pluton into two tectonic blocks, Chuna–Volch'I Tundras and Moncha Tundra. A 5-km vertical section of the Main Range pluton is subdivided into three zones (Sharkov 2006): the lowermost, 500–600-m-thick gabbronorite zone (sampled for comparison); the up to 2-km-thick middle gabbronorite–anorthosite zone, and the no less than 2.5-km-thick uppermost zone, which consists of coarse-grained massive and taxitic gabbro–anorthosites. The pluton differs in its presence of coronal textures at the contact between the olivine and plagioclase and its relatively weakly expressed hidden layering.

The Umbarechensky Massif is part of the Umbarechensky–Imandra Complex (Smolkin et al. 2004). It has a sheet-like shape with variable thickness from 0.5 to 1.5 km thick, and its outcrops extend for approximately 70 km (Fig. 7.3b). The main part of the massif is located in the southern limb of the synclinal structure between the felsic volcanics of the Seidorechka Formation and the Archean Basement, while its western part is completely localized among the rocks of the Archean complex. The pluton is subdivided into six zones consisting of diverse gabbroids that are dominated by mesocratic gabbronorites containing micropegmatite intergrowths of quartz with acid plagioclase. The lowermost part of the massif contains chromite interlayers (samples were taken from the gabbronorites and olivine-bearing varieties), while the upper part is represented by mineralized gabbro with abundantly disseminated Ti–magnetite.

The Fedorova–Pana pluton (80 km^2 in area) extends in the northwestern direction for a distance of more than 25 km and dips to the southwest at 30–35° (Latypov and Chistyakova 2000). The thickness of the section is approximately 4 km. At the present-day erosion level, this pluton is faulted into several blocks (Fig. 7.3c), the largest of which (from west to east) are the Fedorova, Western Pana, and Eastern Pana tundras. The pluton predominantly consists of gabbronorites, leucogabbro, and anorthosites as well as pyroxenites and harzburgites. Several horizons of low-sulfide PGE mineralization were identified in the upper part of the section. We sampled the section of the Western Pana Tundra that is subdivided into the following zones (from the bottom upward): (1) marginal zone (50–60 m, taxitic gabbronorites), (2) norite (40–50 m), and (3) gabbronorite (approximately 4000 m).

The Kivakka intrusion is ascribed to the Olanga group of the layered peridotite–gabbronorite intrusions of North Karelia. The host rocks are represented by upper Archean migmatized biotite and amphibole and granite gneisses. The intrusion has an inverted cone morphology whose axis dips at 40° to the northwest. The exposed part of the massif is represented by the uppermost part of the cone, which consists of layered series from the marginal chill zone to the roof. The massif is subdivided into the lower gabbronorite contact zone (approximately 100 m), the layered series (1,700 m), and the uppermost-contact zone (up to 50 m) (Bychkova et al. 2007). In terms of cumulus assemblage, the layered series consists of the following zones that were sampled for study: olivinites (dunites), norites, gabbronorites, and gabbronorites with pigeonite (Fig. 7.3d).

The Burakovsky pluton is situated in East Karelia. The host rocks are the Baltic Shield's oldest rocks, which are represented by tonalite granite gneisses with small fragments of metakomatiites. The pluton has an irregular oval shape that is curved in map view. Its length is 50 km with a width of 13–17 km, and it consists of two large blocks, Aganozero and Shalozero–Burakovsky, which touch at their upper sections (Chistyakov et al. 2000, 2002; Nikolaev and Ariskin 2005). They are separated by a fault and have autonomous internal structures. The funnel shape of the Aganozero block is extended in the submeridional direction. It is made up of ultrabasic rocks that are replaced from the bottom upward by basic rocks that form a small, synclinal structure in its center.

Fig. 7.3 (continued) (*6*) orthopyroxenites; (*7*) intercalation of harzburgites, olivine pyroxenites, and orthopyroxenites; (*8*) harzburgites and rocks of the near-bottom series of the Nittis–Kumuzh'ya–Travyanaya (NKT) mounts; (*9*) plagiodunites and chromitites of the Dunitovyi block; (*10*) sulfide rocks of the NKT and Mt. Sopcha ore fields; (*11*) felsic volcanics of Mt. Aivarench; and (*12*) faults. (**b**) Geological scheme of the Umbarechka–Imandra intrusive complex (After Smolkin et al. 2004). Supracrustal rocks: (*1–5*) Lower Proterozoic, (*1*) volcanoterrigenous rocks of the Selenoozerskaya Formation and (2) tholeiitic basalts of the Mitrijarvi Formation, (*3*) terrigenous-carbonate rocks of the Umba Formation, (*4*) high-Mg basalts of the Polisarka Formation, (*5*) felsic rocks, (*6*) Archean gneisses. Intrusive rocks: (*7*) nepheline syenites of the Khibiny tundras; (*8*) syenites of the Soustov Massif; (*9*) ultrabasic intrusions; (*10–13*) rocks of the Imandra lopolith: (*10*) melanorites of the lower layer, (*11*) gabbronorites of the Main and (*12*) ferrogabbro-diorites of the Roof zone, (*13*) marker gabbro–anorthosite horizons and overlaying ore gabbro; (*14*) Bol'shaya Varaka chromite deposit; (*15*) Archean granites; (*16*) faults; (*17*) bedding of layering. *U* designates the Umbarechka Massif of the Imandra lopolith. (**c**) Geological scheme of the Paleoproterozoic Western Pana Tundra intrusion (After Latypov and Chistyakova 2000). (*1–3*) Lower Proterozoic volcanogenic–sedimentary rocks of the Strel'na Group of the Imandra–Varzuga zone: (*1*) metaandesites of the Seidorechka Formation, (*2*) quartzites of the Seidorechka Formation, (*3*) metabasalts of the Kuksha Formation, (*4*) alkaline granites of the Belye Tundras, (*5*) dikes of gabbro-dolerites and quartz dolerites, (*6–10*) intrusive rocks of the Western Pana Tundra, (*6*) magnetite gabbros with inverted pigeonite, (*7*) layered horizons, (*8* and *9*) trachytoid and massive rocks, (*10*) rocks of the norite and marginal zones, (*11*) Archean gneisses and granitoids of the Keivy geoblock, and (*12*) faults. (**d**) Geological scheme of the Kivakka Massif (After Bychkova et al. 2007). Lower- and upper-contact zones; (*2*) olivinite zone; (*3–4*) norite zone, (*3*) subzone of bronzite–norite intercalations, (*4*) norite subzones; (*5*) gabbronorite zone; (*6*) pigeonite gabbronorite zone; (*7*) geological boundaries; and (*8*) faults. (**e**) Geological scheme of the Burakovsky intrusion (After Chistyakov et al. 2000). (*1*) Marginal series; (*2–7*) layered series, (*2*) dunite, (*3*) reridotite, (*4*) pyroxenite, (*5*) gabbronorite, (*6*) pigeonite gabbronorite, (*7*) magnetite gabbro-diorite; and (*8*) faults. Sampling localities are shown by *boxes*

The Shalozero–Burakovsky block extends in the northeastern direction and has a lopolith-like shape; it mainly consists of basic rocks, while ultrabasic rocks occur at the deeper levels. The layered series of both blocks have similar structures. From the bottom upward, they are subdivided into five zones: ultrabasic, pyroxenite, gabbronorite, pigeonite gabbronorite, and magnetite gabbronorite–diorite. The latter zone only developed in the Shalozero–Burakovsky block. Samples were collected from boreholes 20, 200 (Aganozero block) and 45 (Shalozero–Burakovsky block), which are shown in Fig. 7.3e.

7.4 Petro- and Geochemical Features of the Rocks of the Drusite Massifs and the Reference Samples of the Layered Plutons

7.4.1 Major Elements

The rocks from all of the studied massifs show continuous and wide variations in composition. Most of the rocks have normal alkalinity, except for a few subalkaline varieties (Tables 7.1 and 7.2). In the Harker binary diagrams (Fig. 7.4), the contents of the major rock-forming oxides show very significant variations (wt %): MgO, 1.55–47.52; SiO_2, 35.56–58.46; TiO_2, 0.02–3.94; FeO, 3.32–18.43; K_2O, 0.01–1.12; Na_2O, 0.01–5.82; Al_2O_3, 0.52–21.72, and P_2O_5, 0.01–0.32. The maximum compositional variability was found in strongly differentiated, layered intrusions consisting of ultrabasic and basic rocks.

The widest variations in Mg were identified in the rocks of the Burakovsky Massif (47.5–4.91 wt % MgO). In the Monchepluton, the rocks with the highest Mg were sampled from the Dunitovyi block (MgO up to 39.48 wt %), while the gabbroids of Mt. Nyud contain only 8.32 wt %. Significant variations in MgO are also noted in the Fedorova Tundra Massifs, where concentrations of this element decrease from 26.38 to 9.59 wt % from the ultrabasic rocks to the gabbroids, and in the Pana Tundra Massif (MgO = 26.38–3.32 wt %). The MgO content in the Kivakka Massif varies within the same range (37.05–8.91 wt %), and the rocks of the Umbarechka Massif have high contents of SiO_2 and MgO (55–56 and 18–20 wt %, respectively). The norites of the Main Range Massif were only analyzed for trace elements.

In terms of the major element composition, the rocks of the drusite complex show strong variations, but their data points delineate individual fields. The end members of this set are two massifs, Sorkajoki (high Mg) and 403.0 m in height (low Mg), but they are represented by few samples. The compositions of the other massifs are similar to the compositions of the former or latter intrusions.

The Sorkajoki Massif is made up of Cr-rich rocks with the highest Mg concentrations (MgO = 23.02–37.08 wt %), which form distinct trends in almost all of the diagrams, including SiO_2–MgO ($R^2 = 0.6$, Fig. 7.4a). The rocks of the 403.0-m-high massif are sharply different with their low MgO contents (1.55–6.88 wt %) and elevated FeO (up to 18.1 wt %) (Fig. 7.4). The contrast in the composition of these two massifs is emphasized by a sharp difference in TiO_2, P_2O_5, and total alkalis ($K_2O + Na_2O$); the elevated contents of which are typical of the Massif as high as 403.0 m. The comparison of the drusite massifs with the reference layered intrusions testifies that, in terms of the differentiation index and major oxide composition, the Monchegorsk pluton is similar to the massifs of the Sorkajoki River, Poioiva, Mt. Perchatka, Mt. Lais-Tundra, the middle reaches of the Nora River, and Mt. Rumimuristo. They are not only characterized by elevated MgO contents but also by low contents of TiO_2 (<0.5 wt %) (Fig. 7.4b) as well as by similar phosphorus behavior in the studied samples.

The second group represented by high-Fe and high-Ti rocks comprises the Massifs of 403 m in height, Mt. Kontiovara, Kamenka R., Mt. Ragutchane, some of the rocks of Mt. Levoiva, and the dikes of Mt. Grob Tundra. They are characterized by low Mg and elevated P contents. Thus, the massifs are clustered into two groups: high-Mg massifs with elevated Cr content (Sorkajoki group) and relatively low-Mg massifs enriched with Fe, Ti, and P (403 m in height massif type, Fig. 7.6b). The 391.2-m and 463.9-m massifs and those of the upper and middle reaches of the Nora River, Mt. Krutaya (Fig. 7.7), and the western flank of the Poioiva River occupy an intermediate position.

7.4.2 Behavior of the Trace Elements in the Rocks

The trace element distribution patterns in the rocks of the layered plutons and drusite massifs were compared using primitive, mantle-normalized spidergrams (normalized by Hofmann 1988) and can be subdivided into three groups (Figs. 7.5, 7.6, and 7.7; Tables 7.3 and 7.4). The first group includes all of the rocks of the layered plutons and is distinguished by two main characteristics: a negative Ta–Nb anomaly and sharp enrichment by light elements. This group also displays positive Sr and a weakly expressed negative Ti anomaly. The HREE distribution is flat with a slight upward concavity at the end of the pattern, which is regarded as the signature of initial boninitic magmas. Such patterns are typical of the Umbarechensky and Kivakka massifs (Fig. 7.5a), especially of the Monchepluton and Monchetundra massifs (Fig. 7.5b), as well as the Burakovsky and Fedorova–Pana plutons (Fig. 7.5c). The abundances of the trace elements in the rocks of the latter massif vary within an order of magnitude with their lowest values being found in the dunites and olivine norites.

The same types of spidergrams are typical of most of the rocks of the drusite complex: peridotites of the Sorkajoki

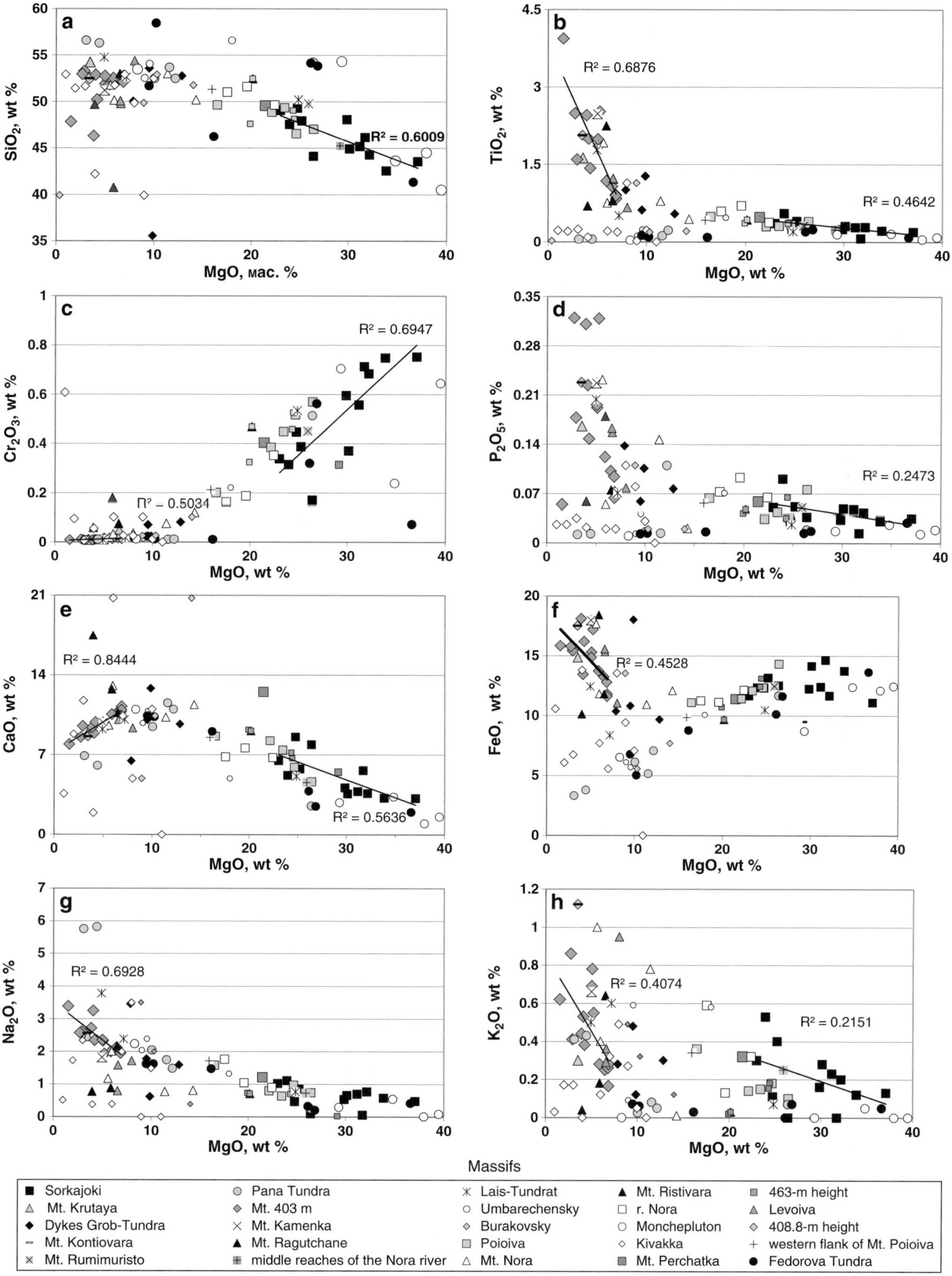

Fig. 7.4 Harker diagrams for the rocks of the drusite complex and layered plutons

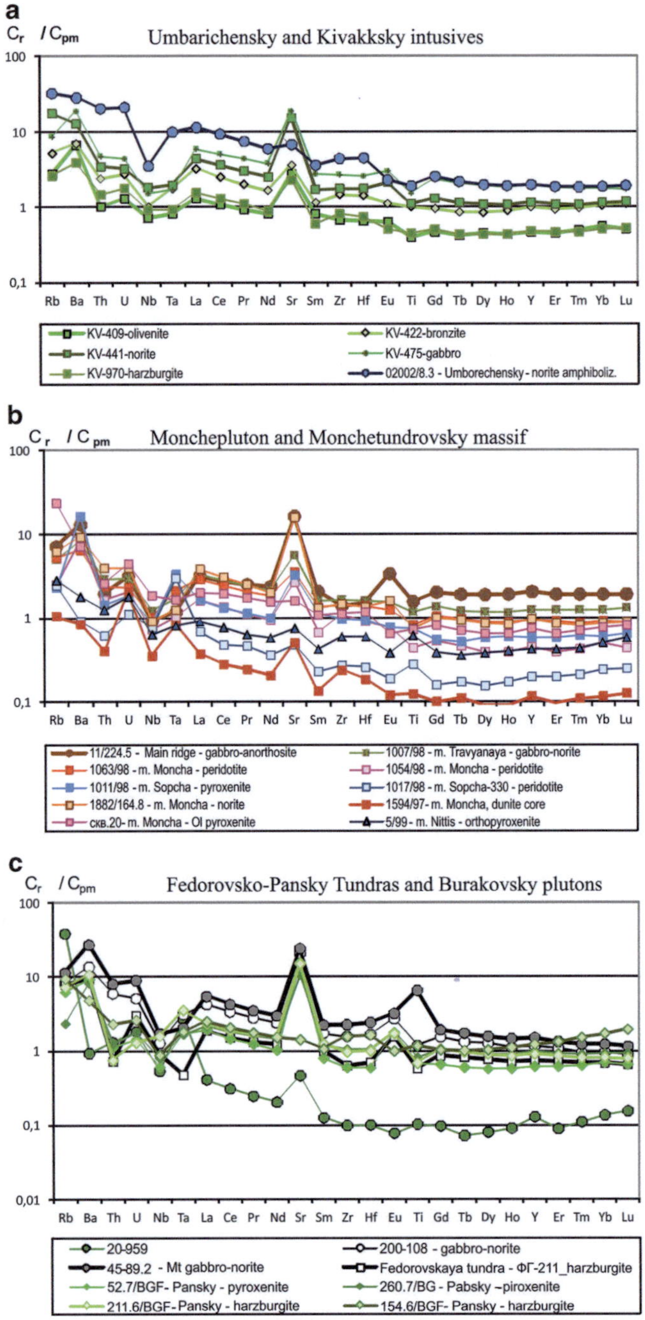

Fig. 7.5 Rare element patterns for massifs Poioiva, Kamenka River, Mt. Perchatka, 463.9-m high, Nora River, Lais-Tundra, Mt. Ristivara

Massif (Fig. 7.6a). The second group of patterns was only identified in several drusite massifs (403.0 m in height and others, Fig. 7.6b). They are characterized by an insignificant slope on the left-hand side, which indicates weak enrichment by incompatible elements, and the spectra sometimes exhibit negative Sr and Nb anomalies. Such spidergrams are typical of the metagabbros and gabbronorites of the 403.0-m-high massif. The gabbronorites of the Poioiva intrusion, Mt. Perchatka, and the Kamenka River, 463.9 m in height; the

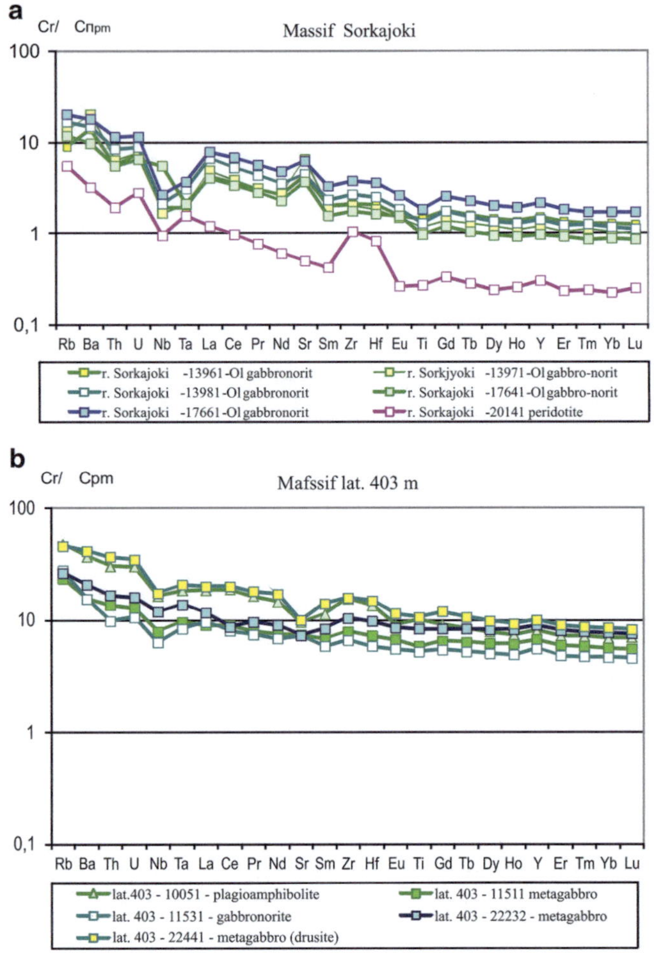

Fig. 7.6 Rare element patterns for massifs Sorkajoki, 403 m high

upper reaches of the Nora River (Fig. 7.7); and the amphibolites of Mt. Lais-Tundra and Mt. Ristivara. The spidergrams of the third group are similar to the patterns described above but show no enrichment, being sometimes depleted of the light trace elements. These are mainly metagabbros and amphibolites of Mt. Levoiva, the middle reaches of the Nora River, Mt. Nora, Mt. Ragutchange, Mt. Grob Tundra, and the rocks of Mt. Kontiovara (Fig. 7.8). Rock groups distinguished on the basis of geochemistry differ from those characterized by petrographic features, which mainly reflect the degree of alteration of the primary rocks. This leads us to conclude that differences in trace element patterns could not be caused by metamorphic processes but reflect the primary characteristics of the parental magmas of these massifs (see later).

However, they somewhat differ from the reference layered plutons in their more fractionated patterns, which is determined by two ratios (Fig. 7.9), $(La/Sm)_n$ (2.5 on average, Fig. 7.9) and $(Gd/Lu)_n$. The latter ratio varies from 1.1 to 1.9, indicating the absence of an upward trend from Tm to Lu (as in the Monchepluton). In turn, the Th/Nb ratio, the measure of a negative Ta–Nb anomaly, is variably expressed in these

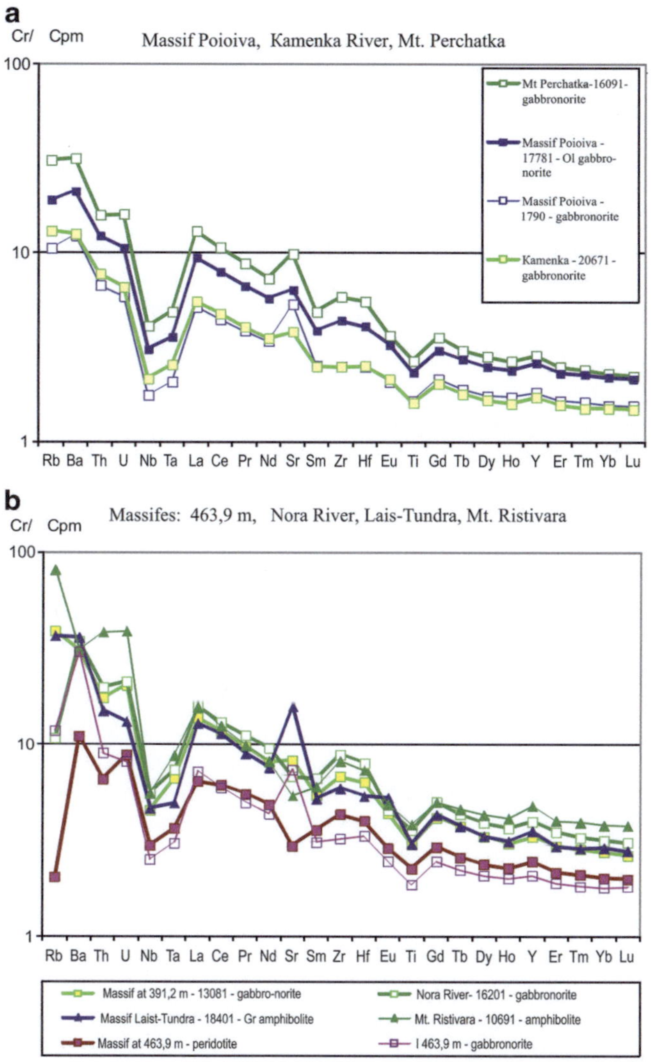

Fig. 7.7 Rare element patterns for massifs Poioiva, Kamenka River, Mt. Perchatka, 463.9-m high, Nora River, Lais-Tundra, Mt. Ristivara

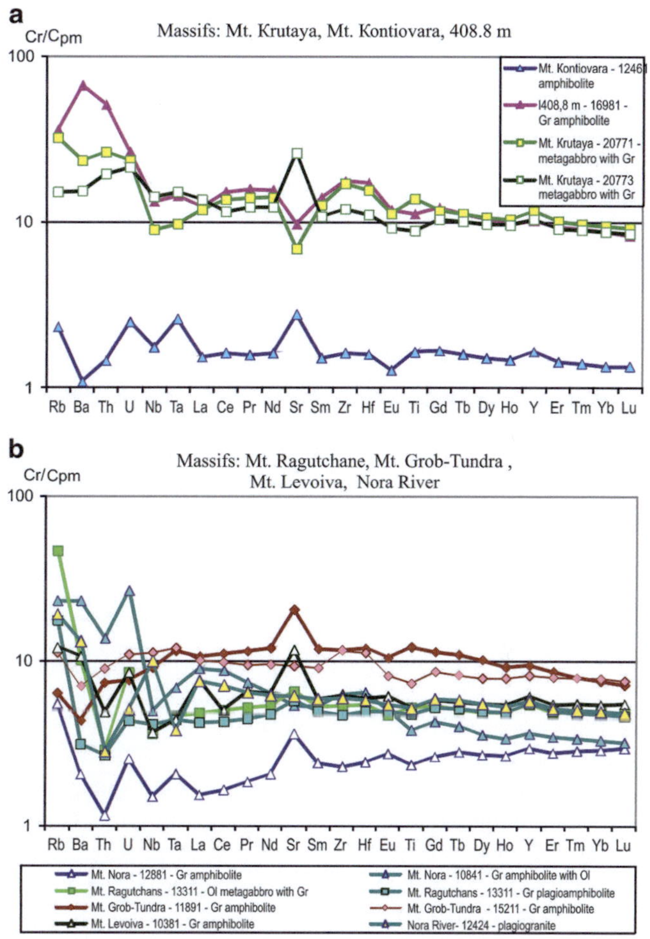

Fig. 7.8 Rare element patterns for massifs Mt. Krutaya, Mt. Kontiovara, 408.8-m Mt. Ragutchane, Mt. Grob Tundra, Mt. Levoiva, Nora River

massifs. The deepest anomaly is observed in the patterns of most of the rocks of the Sorkajoki Massif, the upper reaches of the Nora River, the Kamenka River, 469.8 m in height, Mt. Perchatka, and the Poioiva Massif, where (Th/Nb) varies within 3–4 (Fig. 7.9a). The maximum value of this ratio was found in the Mt. Ristivara Massif, which exceeds the highest values of the layered plutons and those of the Umbarechensky Massif in particular; $(Th/Nb)_n = 5.7–6.4$. The presence of this anomaly and its depth, together with the light element enrichment, indicates a significant role of crustal contamination of the parental magmas that gave birth to these massifs.

7.4.3 Behavior of the Rare Earth Elements

For a more detailed comparison of the studied rocks, the REE distribution patterns were considered separately (Fig. 7.10). In terms of the REE distribution patterns, the rocks of the drusite complex are subdivided into two groups. The first, fairly uniform group is comprised of the rocks of the massifs of Sorkajoki, Poioiva, Mt. Perchatka, the Kamenka River, 463.9 m in height, the upper reaches of the Nora River, Mt. Lais-Tundra, and Mt. Ristivara (Fig. 7.10a). Their patterns are almost parallel with each other and show significant LREE enrichment; the normalized La contents vary from 4.02 to 15.67. The right-hand sides of the patterns are subparallel to the horizontal axis, and this group includes gabbronorites, olivine gabbronorites, and amphibolites. The rocks of this group show no Eu anomaly.

The second group of REE patterns is less homogeneous (Fig. 7.10b). They have similar or parallel HREE patterns [the normalized Lu contents vary from 4.63 (Mt. Ragutchange Massif, olivine metagabbro) to 9.24 (Mt. Krutaya, metagabbro)] but show wide variations in LREE contents. Some rocks are rich in these elements relative to the primitive mantle (leucogabbro from the 403-m-high massif, plagioamphibolite from Mt. Levoiva, metagabbro of Mt. Krutaya), whereas

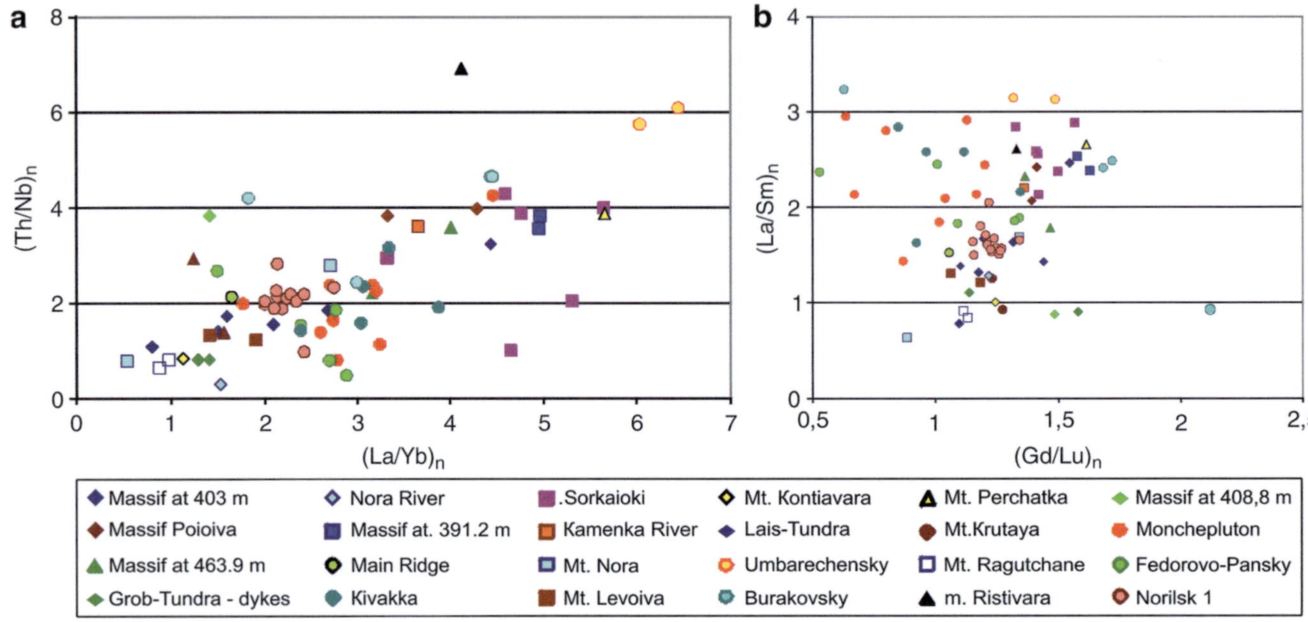

Fig. 7.9 Diagrams (**a**) $(Th/Nb)_n$–$(La/Yb)_n$ and (**b**) (La/Sm)n–Gd/Lu)n for rocks from the drusite complex

the other rocks are not. Among the latter are the metagabbro and garnet amphibolites of the massifs of Mt. Grob Tundra and at 403 m in height. The rocks of the Mt. Ragutchane Massif (olivine metagabbro and garnet amphibolites) have somewhat depleted LREE patterns. Two patterns fall beyond the typical range of this group. The first is that of the amphibolite from Mt. Kontiovara, which has an REE pattern similar to the others, but the REE contents are an order of magnitude lower (approximately 1 compared to 10 for the normalized contents of the other rock varieties).

Their analysis revealed the strong difference between the reference rocks of the layered plutons. It should first be noted that the rocks of the Monchepluton have lower REE contents compared to the rocks of the other intrusions (Fig. 7.11a). Their lowest contents [depleted relative to the primitive mantle (Crock/CPM < 1)] are typical of the ultrabasic rocks and dunites of the Dunitovyi block and the peridotites of Mt. Sopcha. The normalized patterns of these rocks show a concave upward trend as a consequence of strong MREE depletion. The LREE content is somewhat higher than HREE, and it should be noted that this feature, to a varying extent, is typical of other rocks: the orthopyroxenites of Mt. Nittis and, partly, the gabbronorites of Mt. Travyanaya. At the same time, some peridotites and dunites show patterns of REE enrichment with PM-normalized contents close to or even higher than one. A negative Eu anomaly was only found in one sample. A small positive Eu anomaly, at the expense of a significant amount of plagioclase, is present in the gabbroid patterns (norites and leucogabbro).

Only the dunites of the Burakovsky pluton (Fig. 7.11b), whose REE contents are even lower than in similar rocks from the Dunitovyi block, are similar to the Monchepluton. The study sample is also depleted of MREE (below 0.01 PM). The ultrabasic rocks of the Kivakka Massif have higher REE contents compared to similar rocks of the Monchepluton which have lower contents relative to the remaining rock varieties from the other massifs. The PM-normalized REE patterns in the rocks of the Kivakka, Burakovsky, and Fedorova–Pana Tundra plutons are mostly above that of the primitive mantle and show significant LREE/HREE enrichment. Their prominent feature is the presence of a well-expressed, positive Eu anomaly. This group mainly incorporates gabbroids (norites and gabbronorites) as well as pyroxenites. The Umbarechensky pluton differs slightly from the aforementioned massifs. In general, its trace element patterns are similar in shape, but its rocks (gabbronorites and amphibolized norites) are significantly enriched by LREE (La > 10) and have a negative Eu anomaly.

Taking into account the multiple occurrences of ultrabasic–basic magmatic bodies in the Karelia–Kola–Lapland region as well as the absence of stratified deposits, it was necessary to determine the place of the drusite complex and its role in the general evolution of the Earth's crust in the region. Since drusites are mainly distributed within the

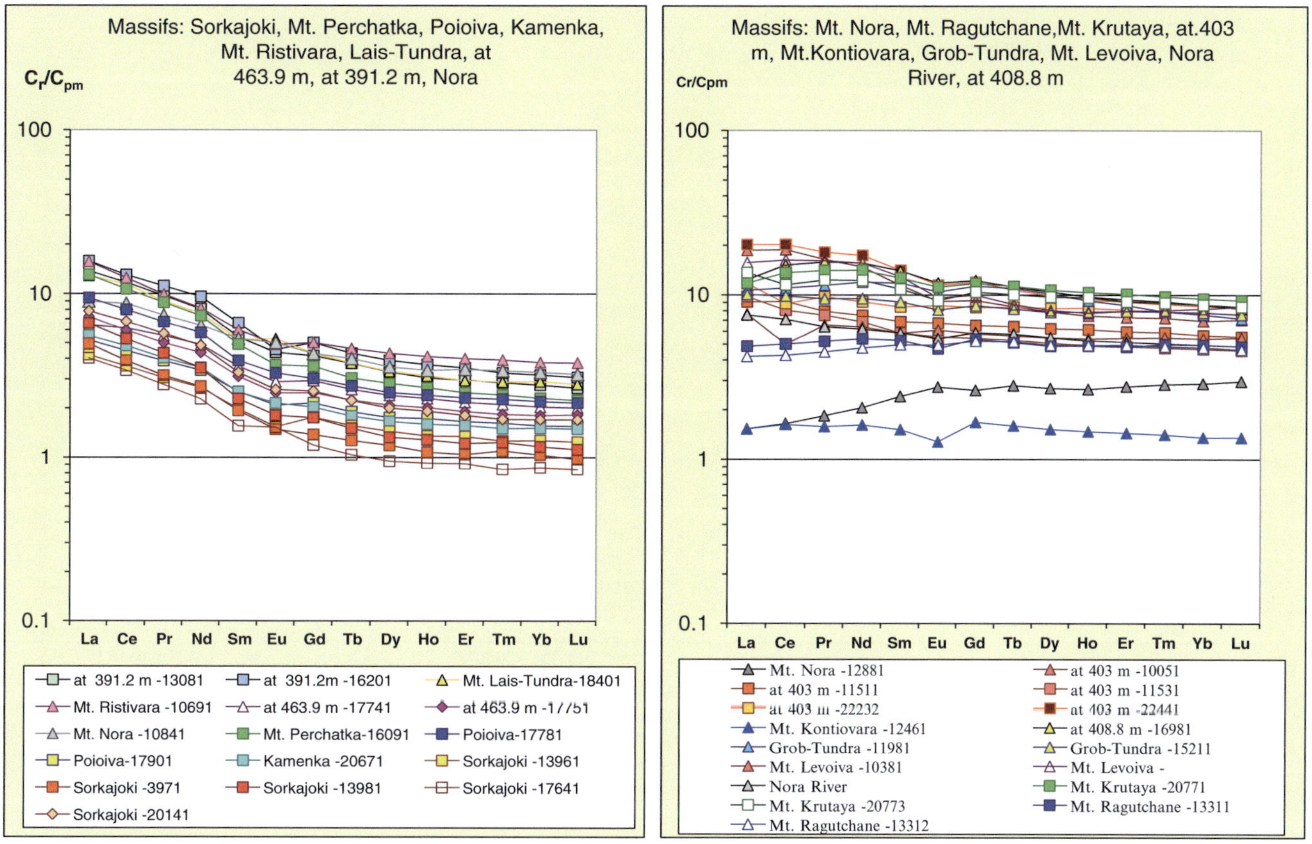

Fig. 7.10 REE patterns for massifs of the drusite complex

Belomorian Mobile Belt, we compared them with compositionally similar massifs from adjacent territories (the Karelian and Kola Peninsula).

The discovery of basic and ultrabasic rocks with coronal textures (Fedorov 1886) and their subsequent integration into the drusite complex were of extreme significance to the understanding of the geological evolution of the region.

In the early stages of the study of these magmatic complexes (1904–1960s), attention was primarily focused on the geological methods, which allowed for the establishment of the spatiotemporal relationships between the rocks. Since the works of A. Shurkin et al. (1962), the problem of the integration of gabbro–anorthosites and lherzolite massifs into a single complex was a matter of hot debate. It was believed that gabbro–anorthosite drusites were Archean in age while peridotites were Proterozoic, but both of these types share a common origin. In spite of intense isotope studies, the problem of their genesis, in particular the composition of parental melts, remains controversial. It was suggested that they were derived either from boninite-like (Bogdanova and Bibikova 1993) or picrite melts (Lobach-Zhuchenko et al. 1998; Kratts et al. 1980) or formed via the subsequent intrusion of magmas of different compositions (Early Precambrian… 2005). These contradictions are mainly related to uncertainty in the genetic models of the coronal textures, on the basis of which they were ascribed to a single complex. The mechanisms of the formation of drusite textures at different stages of rock evolution (early and late magmatic and metamorphic stages) have been discussed in the literature for a long time.

One geologist believed that they represented the reaction rims between the mafic minerals and melt while others regarded them as products of retrograde metamorphism. A considerable contribution to this problem was the determination of the physicochemical parameters of the formation of drusite structures. In particular, L. V. Larikova (2000, 2002) showed that they are formed at certain temperatures (660–690 °C) and pressures (6–8 kbar). Corona growth is controlled by the mechanism of prograde metamorphism, which

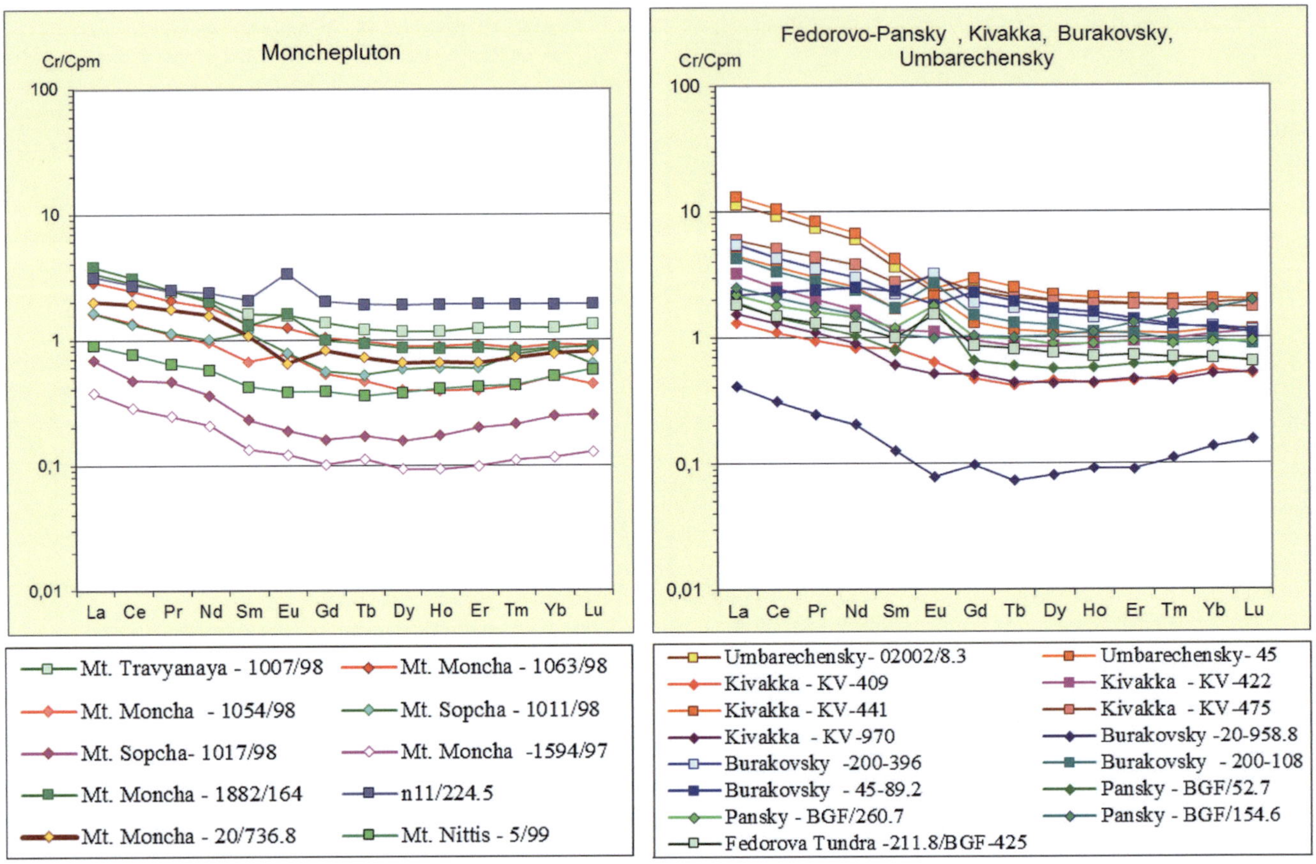

Fig. 7.11 REE patterns for layered intrusions of the Kola–Karelia area

was experimentally confirmed (Larikova and Zaraisky 2002). This mechanism results in the formation of different types of drusite structures in both olivine-bearing and olivine-free rocks. With increasing water pressure and subsequent retrograde metamorphism, the structures are decomposed, and the rocks are transformed into garnet amphibolites. The formation of coronal textures at the contact between the olivine and plagioclase was considered by A. Tomilenko and S. Kovyazin (Tomilenko and Kovyazin 2008) to be the example of the Korosten anorthosite pluton. The authors distinguished several generations of rock-forming minerals that compose the rocks: olivine, orthopyroxene I, clinopyroxene I, and plagioclase I. The structure of the coronas around the olivine in these rocks differs from those in the ultrabasic–basic rocks of the Belomorian complexes. The inner zone is typically composed of orthopyroxene II, while the outer zone consists of symplectite intergrowths of plagioclase II, orthopyroxene III, clinopyroxene II, and orthoclase. However, the study of fluid–crystalline inclusions in olivine fissures led the authors to conclude that the drusite textures were formed via the reaction between olivine and orthopyroxene with the participation of CO_2-rich postmagmatic fluids under subsolidus conditions at $T = 980$–860 °C and $P > 5$ kbar.

Petrographic studies have shown that coronal textures are not only observable in the small rootless bodies but also in the large intrusions from the central part of the Belomorian block. Moreover, drusite textures were also found in the gabbro–anorthosites of the Main Range (Slabunov et al. 2001). Nonetheless, in spite of the similar structural–textural features of the rocks in the central part of the Belomorian block, our study showed that the massifs integrated in the drusite complex differ in their geochemical composition. In this respect, they are subdivided into three groups, which differ from the types distinguished on the basis of petrography: (1) high-Mg group (MgO > 20 wt %) with elevated Cr contents, light trace element-enriched patterns, and a distinctly expressed negative Ta–Nb anomaly (reference: Sorkajoki Massif). The massifs of this group are petrogeochemically similar to the layered plutons; (2) low-Mg (MgO < 10 wt %) or high-Fe group with elevated contents of Ti and P, subhorizontal trace element patterns, slight LREE enrichment, and an extremely weak Ta–Nb anomaly (403.0 m high massif); (3) moderate-MgO group (10–20 wt %) with flat trace element distribution patterns, occasionally depleted of light trace elements, and having no Ta–Nb anomaly (Mt. Grob Tundra dikes). The observed geochemical differences do not depend on the degree of metamorphic transformation. This is confirmed by the fact that different rocks, such as the gabbroids and plagioamphibolites of the 403-m-high massif, have similar trace element distribution patterns (Fig. 7.9). At the same time, the massifs united in a single type on the basis of structural and textural features have different petrogeochemical characteristics and are ascribed to different petrogeochemical groups.

The massifs of the first group are mainly represented by large peridotite bodies and strongly approximated layered plutons that were emplaced under a thick lithosphere. Their trace element patterns are very close to the average composition of the mature continental crust. The massifs were formed from tholeiitic (highly magnesian?) magmas under syntectonic conditions at the peak of the evolution of the paleoriftogenic structure, which is confirmed by their dike-like morphology. This group presumably also includes the small rootless bodies of the drusite complex, which are close to the layered plutons in terms of age and composition and are located among strongly deformed Archean folded rocks.

The second group of intrusions is independently significant and differs sharply from the first group in terms of its petrogeochemical features: high iron contents, alkalis, and phosphorus. Their analogues occur in the northern part of the Belomorian Belt and within the Kolvitsa granulite belt in the form of massifs of the clinopyroxenite–wehrlite formation. In the area of the Main Range, they cut across the gabbro–anorthosites, i.e., they are younger with respect to the gabbro–anorthosites and layered plutons. The third group of massifs is mainly represented by gabbros, metagabbros, and plagioamphibolites with gabbroid relics as well as by garnet amphibolites. Geochemically, they differ sharply from the two other groups in their trace element patterns, being comparable to the volcanics of the Purnach and Kuksha formations. These rocks are situated at the base of the rift zone, and some of the plagioamphibolites might not represent the transformation products of the ultrabasic–basic massifs. Instead, they could be metamorphosed volcanics of the tholeiitic series, which are widely developed in the Paleoproterozoic paleorift structures, such as the Pana–Varzuga and Vetreny belts.

Thus, the established differences between the massifs suggest different phase and chemical compositions of the corresponding parental magmas as well as different conditions during their crystallization. Owing to regional metamorphism within the Belomorian block, the geochemically and genetically different rocks acquired a specific structural appearance.

The massifs of the first group were emplaced into relatively mobile zones of the Belomorian Belt. The compositions of their parental magmas were significantly modified by crustal contamination, which is reflected by the presence of a negative Ta–Nb anomaly and LREE enrichment. *The massifs of the second group* were formed from magmas that were significantly less contaminated by crustal material. In terms of composition, they are close to the rocks that occasionally occur in other regions of the Belomorian block. Judging from the elevated Ti, P, and alkalis contents, their parental melts were of subalkaline affinity, which can be related to increased magma generation depth and metasomatic protolith reworking.

The massifs of the third group are made up of gabbroids and strongly metamorphosed rocks of different geneses. In addition to the intrusive rocks, there are bodies that strongly approximate Sumian tholeiitic basalts at the base of the Imandra–Varzuga section (the Purnach and Kuksha formations). These volcanics occupy large volumes starting in the Early Paleoproterozoic volcanism at the Kola Peninsula and lesser volumes in the Karelian craton. These results cast some doubts on the affiliation of the magmatic rocks with a single drusite complex only on the basis of their structural features and require correlation with and allowance for geochemical data.

7.5 Isotope Characteristics of the Drusite Massifs

The drusite massifs of the first group have been suggested to have been produced by dispersed mafic–ultramafic magmatism in extension zones, as opposed to "focused" magmatism in rigid blocks, which gave rise to the formation of layered plutons in similar magma generation areas (Sharkov et al. 2004). To understand the specification of the different drusite massifs, we have studied their isotope composition (Krivolutskaya et al. 2010b).

The Sm–Nd and U–Pb data generally support the hypothesis that drusite massifs in western Belomorie were formed between 2,460 and 2,434 Ma (Shurkin et al. 1962; Efimov and Kaulina 1997; Alexejev et al. 2000) and can be classified with a single Early Proterozoic complex. However, despite extensive petrological–geochemical (and isotopic) data on the basite–ultrabasite massifs in the Kola Peninsula and Karelia, many aspects of their genesis remain uncertain because of the limited number of examined intrusive bodies. For example, our data on ultrabasite–basite massifs with drusite textures (Krivolutskaya et al. 2010a) in the central Belomorian Belt show that the composition of these massifs is extremely heterogeneous, so combining them into a single complex only based on the similarities in the fabrics of their rocks can hardly be warranted.

It was important to trace the possible genetic links between the intrusions in the southern Kovdor area with known drusite intrusions and layered plutons to gain insight into the geological evolutionary history of the area and the genesis of its mineral deposits. Our publication is centered on these problems. Because the most interesting massifs in the area that were examined by the authors belong to the first group (lherzolite–gabbronorite intrusions according to A. Stepanov (1981) and their geochemistry is closest to that of layered plutons accompanied by Cu-Ni and PGE ore mineralization, our isotopic studies were focused on rocks composing the largest of these massifs: Sorkajoki, Poioiva, Mount Krutaya, and Massif of height 403 m (Fig. 7.2).

7.6 Materials and Methods

The problems formulated above were approached simultaneously using the Sm–Nd and U–Pb isotopic systems of rock-forming minerals and zircons from the rocks, which enabled us to date the crystallization of the massifs most accurately and reliably.

The Sm–Nd studies were conducted on five representative samples of the least altered rocks from the ultrabasite–basite massifs in the Kovdor area: Sorkajoki (samples 6,115, 13,981, and 6,151—plagioclase lherzolites), Poioiva (sample 17,781, olivine gabbronorite), and elevation at 403 m (sample 11,531, amphibolized gabbronorite). The geochemistry of these rocks has been described in detail earlier. Our samples had a mass of 1–2 kg, and the rock-forming minerals (olivine, clinopyroxene, orthopyroxene, and plagioclase) were separated for further isotope analysis through conventional techniques (bromoform application and electromagnetic separation) and hand selection of monomineralic separates under a binocular magnifier, which yielded approximately 95 % pure final mineral separates.

The Nd isotopic composition and the Sm and Nd concentrations in the mineral and whole-rock samples were determined by isotopic dilution at the Center for Isotopic Studies of the Karpinskii All-Russia Research Institute of Geology (CIS VSEGEI) in St. Petersburg. The analyses were carried out with 50–300-mg samples of monomineralic fractions (analyst B. V. Belyatsky), which were preliminarily washed in nitric acid to remove surface contamination before being pulverized in an agate mortar to approximately 200-mesh size. The sample was then spiked with a ^{150}Nd–^{149}Sm mixed isotopic tracer and dissolved in an $HF+HNO_3+HClO_4$ mixture at 120 °C for 5 days in an autoclave. The further separation of elements for mass spectrometric analysis was conducted by ion exchange and chromatographic techniques as described in (Richard et al. 1976) with minor modifications (Amelin and Semenov 1996).

The laboratory blanks were 0.005 ng for Sm and 0.01 ng for Nd and did not require any additional isotopic ratio corrections in the calculations of the initial isotopic composition of the sample. The concentrations and isotopic composition of Sm and Nd were measured on a Triton Ti (ThermoElectron) solid-state multicollector mass spectrometer. In preparation for the analysis of each set of samples (18–20 samples), the 100-ng JNdi-1 internationally certified standard was analyzed. The average error of the analyses (2σ) was 0.005 % for the ^{143}Nd/^{144}Nd ratio and 0.03 % for the ^{149}Sm/^{147}Sm ratio. The analytical errors in the concentrations of the elements were evaluated at 0.5–1 %. The relative errors at low concentrations (a few hundred ppm) were higher and were evaluated separately based on the current precision capability of replicate analyses; these errors never exceeded 10 %. The analytical errors in the isotopic composition of Nd were no higher than 0.005 % based on the multiple replicate analyses of the BCR-2 standard sample (outer reproducibility). The error in the Nd isotopic composition of the individual samples did not exceed the values specified for each of the samples in the table (internal reproducibility). The ^{143}Nd/^{144}Nd ratio during our experiments was 0.512106 ± 5 (2σ, six analyses) for the JNdi-1 standard and 0.512648 ± 4 for the BCR-2 standard.

In order to date the rocks by the U–Pb zircon method, we selected 10-kg hand specimens of the rocks from the Mt. Krutaya (sample KV-1) and Poioiva (sample P-1) massifs. Zircons were separated from the rocks using bromoform and magnetic separation; the zircon yield was 5–10 mg. The local isotopic analysis was carried out on 50 grains larger than 100 μm.

Suitable areas for the analysis were selected under an optical and CamScan MX2500S electron scanning microscope equipped with a CLI/QUA2 (Benthan, United Kingdom) cathodoluminescence setup. The U/Pb isotopic ratios were measured on a SHRIMP-II (ASIPty, Australia) at the Center for Isotopic Studies of the Karpinskii All-Russia Research Institute of Geology (CIS VSEGEI) in St. Petersburg by the method described in (Williams 1998). The intensity of the current of the beam of negatively charged oxygen ions was 4 nA, and the craters burned off at the analytical spots were 15–18 μm in diameter. The primary processing of the raw data was conducted with the SQUID 2.0 software (Ludwig 2006). The U/Pb ratios were normalized to 0.0668 of the TEMORA zircon standard value (dated at 416.75 Ma (Black and Kamo 2003).

The precision of the analytical procedure was evaluated based on the replicate analyses of the 91500 zircon standard. The proportion of the number of analyses conducted on the standards and our zircons in each analytical session was 1:2. The errors of the individual analyses (isotopic ratios and ages) are reported in the table at the 1σ level, and the errors of the calculated concordant ages and concordia intercepts are reported at the 2σ level. The concordia plots were constructed with the ISOPLOT/EX 3.11 computer program (Ludwig 2005).

REE and trace elements in zircons were analyzed on a Cameca IMS-4 F ion microprobe at the Institute of Microelectronics and Informatics, Russian Academy of Sciences, in Yaroslavl (analyst S. G. Simakin). The local analyses were carried out on the same samples that were dated by U–Pb methods (SHRIMP-II) with the ion beam focused in the vicinity of the earlier analytical spots of the U–Pb analyses or immediately within the craters. The measurement technique corresponded to that described in (Smirnov et al 1995; Nosova et al. 2002). When reaching the surface of the sample, the primary beam of oxygen ions had an energy of 14.5 kV and was focused at a 20-μm diameter spot.

The analytical cycle was comprised of five repeated measurements, and the counting time depended on the intensity of the signal and was determined automatically by controlling the counting statistics but did not exceed 30 s. The absolute concentrations of each element were calculated from the measured intensities of positive single-atom secondary ions of elements normalized to the intensities of secondary silicon 30Si + ions using relative sensitivity coefficients, $Ci = Ii/ \times Ki \times$. The calibration curves were constructed based on known certified standard samples. The SiO_2 concentrations at each spot were independently determined by an electron microprobe. The concentrations of $^{153}Eu+$, $^{174}Yb+$, $^{158}Gd+$, $^{167}Er+$, $^{138}Ba+$, $^{139}La+$, $^{140}Ce+$, $^{141}Pr+$, and $^{149}Sm+$ were calculated following the methods described in (Bottazzi et al. 1994; Belousova et al. 2002).

7.6.1 Sm/Nd Isotopic Ratios of Whole-Rock Samples

The results obtained on the Sm–Nd isotopic system and the isochron ages of the basic–ultrabasic rocks from the Poioiva and Sorkajoki massifs and the massif in 403-m elevation are summarized in Table 7.8 and graphically represented in Fig. 7.12. As seen in Table 7.8, the ranges in the concentrations of Nd (2.75–3.01 ppm) and Sm (0.622–0.666 ppm) in the plagioclase lherzolites of the Sorkajoki Massif are very insignificant and can be accounted for by fluctuations in mineralogical composition. However, these variations significantly differ in the Poioiva Massif (11.7 and 8.7 ppm for Nd and 2.39 and 2.54 ppm for Sm).

The differences in the concentrations of these elements in the major minerals are much lower; the Nd and Sm concentrations in the olivine are 0.405–0.854 and 0.127–0.200 ppm for *I*si and 30 $CSiO_2$, respectively, and the analogous values are 0.335–0.649 and 0.122–0.235 ppm for orthopyroxene, 3.47–6.31 and 1.08–2.14 ppm for clinopyroxene, and 5.07–8.71 and 0.74–1.4 ppm for the plagioclase. These differences turned out to be high enough to bring about a significant spread in both the Sm/Nd ratio and the Nd isotopic composition of the minerals.

Based on these data, we calculated regression lines and constructed Sm–Nd isochron diagrams. The ages evaluated for the Poioiva Massif and the massif at 403-m elevation coincide with the following errors: $T = 2507 \pm 91$ Ma, MSWD = 2.9 (sample 17,781), and T = 2,518 ± 66 Ma, MSWD = 0.28 (sample 11,531). For the Nd isotopic composition of the rocks, $\varepsilon_{Nd}(T) = +1.05 \pm 0.70$ and $+0.70 \pm 0.37$.

We calculated the isochrons for each of our three samples of plagioclase lherzolites from the Sorkajoki Massif and plotted them with specified values for age and Nd initial isotopic composition: $T = 2{,}494 \pm 37$ Ma and ε_{Nd} $(T) = +1.01 \pm 0.25$ (sample 6,115); 2,493 ± 130 Ma and ε_{Nd} $(T) = +0.37 \pm 1.03$ (sample 6,151); and 2,472 ± 37 Ma and ε_{Nd} $(T) = +1.25 \pm 0.29$ (sample 13,981). The consistency of these values within the errors enabled us to calculate a common isochron for the three samples based on 15 points. Thus, the calculated age is 2,485 ± 51 Ma, ε_{Nd} $(T) = +0.88 \pm 0.39$, and MSWD = 5.

Our data on the Sm–Nd isotopic systematics of rock-forming minerals and whole-rock samples of lherzolites from the three massifs provide grounds to suggest that the massifs were roughly simultaneously emplaced and the rocks were

Table 7.8 Sm–Nd isotope composition of rocks and minerals from massifs in the South Kovdor area

Sample No		[Sm]	[Nd]	$^{147}Sm/^{144}Nd$	$^{143}Nd/^{144}Nd$	±2σ	ε_{Nd} (T)	$\varepsilon_{Nd}(2.41)$
Sorkajoki Massif								
6,115	wr	0.66	2.74	0.14	0.51	0.000008	1.06	0.57
6,115	cpx	1.65	5.14	0.19	0.51	0.000004	0.72	0.69
6,115	opx	0.12	0.33	0.22	0.51	0.000009	1.19	1.42
6,115	pl	0.84	5.35	0.09	0.51	0.000006	0.89	−0.11
6,115	ol	0.14	0.43	0.20	0.51	0.000007	1.07	1.17
6,151	wr	0.63	3.00	0.12	0.51	0.000005	−0.28	−0.96
6,151	cpx	1.62	5.35	0.18	0.51	0.000003	0.39	0.26
6,151	opx	0.22	0.57	0.23	0.51	0.000007	0.77	1.12
6,151	pl	0.79	5.25	0.09	0.51	0.000008	0.88	−0.15
6,151	ol	0.14	0.53	0.16	0.51	0.000005	−0.06	−0.36
13,981	wr	0.62	2.83	0.13	0.51	0.000004	1.45	0.82
13,981	cpx	1.22	3.94	0.18	0.51	0.000008	1.03	0.95
13,981	opx	0.23	0.64	0.21	0.51	0.000011	1.16	1.38
13,981	pl	0.74	5.07	0.08	0.51	0.000006	1.13	0.07
13,981	ol	0.20	0.85	0.14	0.51	0.000005	1.85	1.31
Poioiva Massif								
11,531	wr	2.54	8.70	0.17	0.51	0.000004	0.58	0.31
11,531	cpx	2.14	6.30	0.20	0.51	0.000002	0.79	0.90
11,531	pl	1.07	5.16	0.12	0.51	0.000002	0.66	−0.25
17,781	wr	2.39	11.6	0.12	0.51	0.000004	1.54	0.60
17,781	cpx	1.07	3.47	0.18	0.51	0.000004	1.30	1.19
17,781	opx	0.16	0.40	0.24	0.51	0.000010	0.82	1.39
17,781	pl	1.40	8.70	0.09	0.51	0.000007	0.61	−0.67
17,781	ol	0.12	0.40	0.19	0.51	0.000004	1.02	0.94

Note: Concentrations of elements are given in ppm, error of Nd isotope composition ($^{143}Nd/^{144}Nd$) is done at 95 % level of significance (2δ). *wr* rocks, *pl* plagioclase, *cpx* clinopyroxene, *ol* olivine, *opx* orthopyroxene. Samples 6115, 6151 и 13981 are plagioclase lertholites from Sorkajoki Massif; samples 11531 (metagabbro) and 17781 (Ol gabbronorite) were taken from Poioiva Massif. Primary Nd isotope composition (ε_{Nd}) were calculated based on (1) isochron age of sample (T) and 2.41 Ga (Data from U–Pb in zircons) (ε_{Nd} = +2.41 Ga). Analyst B. V. Belyatsky (VNII Ocean geology, St. Petersburg, Russia). Tables 7.8, 7.9, 7.10, and 7.11 – After Krivolutskaya et al. (2010b)

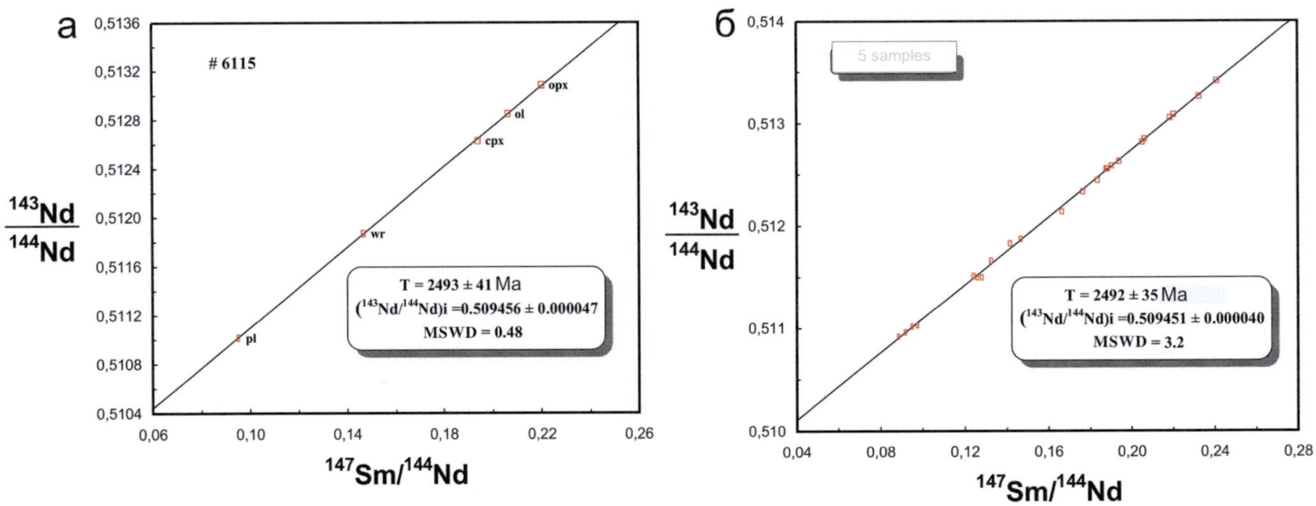

Fig. 7.12 Sm–Nd isochron diagrams for drusite massifs

derivatives of a common enriched mantle source. The relatively large errors in the Sm–Nd age values of the massifs called for more accurate dating of their crystallization age by the U–Pb zircon method. For this purpose, we have collected additional rock samples from the Poioiva and Mount Krutaya massifs but failed to find a sufficient amount of rock material at the Sorkajoki and 403-m elevation massifs.

7.6.2 U/Pb Isotopic Ratios of Zircons

Morphology of zircons from rocks of the Poioiva and Mount Krutaya massifs. The morphologies of the zircon grains in our samples suggest that each fraction contains various genetic types of the mineral (Fig. 7.13). The most typical grains are long, prismatic, euhedral crystals with shiny, well-pronounced faces with the minimum number of crystallographic forms: one prism [100] and dipyramids [101], [202], or [211]. These zircons were interpreted as magmatic (Figs. 7.13a, c), and crystals of this type are either pinkish (sometimes with crimson shades) or colorless. Additionally, the rocks contain turbid zircons with rounded edges, which were thought to be xenogenic (Fig. 7.13b) and differ significantly in morphology from the magmatic zircons depending on the source. Some of these grains may contain ancient cores, but the intensity of their color and fracturing did not allow us to unambiguously identify these cores. The magmatic zircons can either be similar to the intrusive zircons or be significantly different from them (Fig. 7.13d). Even if the captured xenogenic zircons were magmatic, their morphology and crystallographic forms show certain distinctive features, such as the character of their surface, caused by their interaction with the high-temperature melt (partial dissolution in it) and the "smoothing" of crystal elements and the preservation of major crystal features (in contrast to sedimentary zircons, which are usually strongly abraded).

The most obvious differences between these zircon populations are in their inner structures as seen in cathodoluminescence (Fig. 7.14). Our zircon grains exhibit pronounced concentric growth zoning (Fig. 7.14a, c, d) and, in certain instances, cores in older zircons (growth seeds or captured, more ancient zircons, Fig. 7.14a, d). In contrast, the young zircons are homogeneous, display no pronounced growth zoning, and show dark cathodoluminescence (Fig. 7.14b, e, f). Only occasional cross sections perpendicular to the elongation of some grains exhibit growth zoning (Fig. 7.14d) with all of the other sections either yielding peculiar cathodoluminescent images (e.g., Fig. 7.14f) or showing sectorial zoning, which is quite typical.

The morphological varieties of the zircons that were distinguished were studied by SIMS to elucidate their compositional characteristics (Table 7.9). The C1 chondrite-normalized trace element patterns of the zircons (Fig. 7.15) from rocks from the massifs in the southern Kovdor area are generally analogous to the patterns of zircons from layered intrusions; they are enriched in HREE, have higher U concentrations than those of Th, and show positive Yb–Nb, Ce, and Hf anomalies and negative P–Ti–Sr and Ba–La anomalies. In spite of the topological similarities between these patterns, the patterns of the young and old zircons display significant differences, and these differences are practically the same for both of our samples from the Poioiva Massif and the massif of Mount Krutaya. These differences are most conspicuously pronounced in the LREE region at, first of all, La, Pr, and Nd, whose concentrations are much higher (by one order of magnitude) in the young magmatic zircons than in the old ones. Young zircons from the latter massif contain lower concentrations of HREE. Another noteworthy feature is the relatively low Y/Nb ratios and more clearly pronounced Ce anomalies in zircons from the Poioiva Massif. It is also worth noting that two grains of the zircon had practically equal normalized U and Th concentrations. The old and young zircons also have different U/Th ratios as was determined in the course of our isotopic studies and will be demonstrated below.

As revealed by comparing the trace element patterns of the zircons, the aforementioned geochemical features of zircons from ultrabasic–basic massifs in the southern Kovdor area are given in a series of diagrams (Krivolutskaya et al. 2010b). The variations in the LREE and MREE compositions of the zircons as their LREE/MREE [(La/Gd)n and (Pr/Gd)n] ratios correlate with their MREE/HREE [(Gd/Yb)n] ratios. This diagram clearly demonstrates the differences between the young and old zircons, which are pronounced by the steeper patterns in the latter due to their higher (La/Gd)*n* and (Pr/Gd)n ratios. To characterize the general situation regarding the compositional variations of zircons in ultrabasite–basite rocks from various geodynamic environments, the diagrams also display the composition of zircons from ophiolites, oceanic crustal rocks, Early Archean gneisses, and continental gabbroids (Belousova et al. 2002; Hoskin and Ireland 2000; Grimes et al. 2007; Whitehouse and Kamber 2003). As expected, the composition points of our zircons fall within the gabbroid field. In the diagrams showing the correlations between the Eu and Ce anomalies (Eu/Eu*–Ce/Ce*) and variations in the Hf concentration (in ppm) of the zircons in correlation with their Th/U ratios the situation is more ambiguous. The scatter of the data

Fig. 7.13 Transmitted light micrographs of examined zircon grains
(**a**, **b**) The massifs of Mount Krutaya Vostochnaya (sample KV-1): inherited magmatic zircons are fractured grains with an intense color, pronounced growth zoning, and partly preserved crystal faces (Fig. 7.14a, marked with an *arrow*); typical magmatic zircons are colorless prismatic crystals with pronounced crystal faces and transparent fragments of these grains (Fig. 7.14b, marked with *arrows*). (**c**, **d**) Poioiva Massif (sample P-1): newly formed magmatic zircons are indicated by *arrows* (Fig. 7.14c); inherited magmatic and metamorphic zircons are extensively fractured grains with rounded faces, more intensely colored (Fig. 7.14d)

points of the zircons from the southern Kovdor area is much greater, and all of those points lie far above (in terms of Eu/Eu*) the field of the continental gabbroids.

The samples from the Poioiva and Mount Krutaya massifs contain two varieties of zircons of different morphology, inner structure, and composition. These zircons are deemed to have different ages. To test this hypothesis and date these zircon populations, we determined their U/Pb ratios.

U–Pb isotopic analysis of zircons. Tables 7.10 and 7.11 show the SIMS SHRIMP-II analyses of the zircons, and their U–Pb concordia plots are shown in Fig. 7.16. As was anticipated, zircons in each of the samples fall into at least two age populations (Fig. 7.16a). This is pronounced in the diagrams by the occurrence of two or more groups of data points that roughly correspond to 2,400 and 2,700 Ma. These two zircon age populations have distinct geochemical characteristics. One of the distinctive features of magmatic zircons from basic rocks is their elevated Th concentrations (their Th/U ratios are much greater than one; from 1.3 to 2.6 for zircons from sample KV-1 and from 1.2 to 4.0 for sample P-1), hence the absence of cathodoluminescence (dark grains in cathodoluminescence images) and the absence of growth zoning or poorly pronounced growth zoning (Fig. 7.14b, e). The well-pronounced zoning of the old zircons is due to alternating pale and dark cathodoluminescence stripes from systematic variations in the concentrations of trace elements (U and REE first of all) (Fig. 7.14a and d and occasional grains in Fig. 7.14a and f). For example, the Th/U ratios of the old zircons vary from 0.2 to 0.6 for sample KV-1 and from 0.4 to 1.8 for sample P-1.

For the Mount Krutaya Massif (sample KV-1), three data points of the two analyzed old zircon grains are approxi-

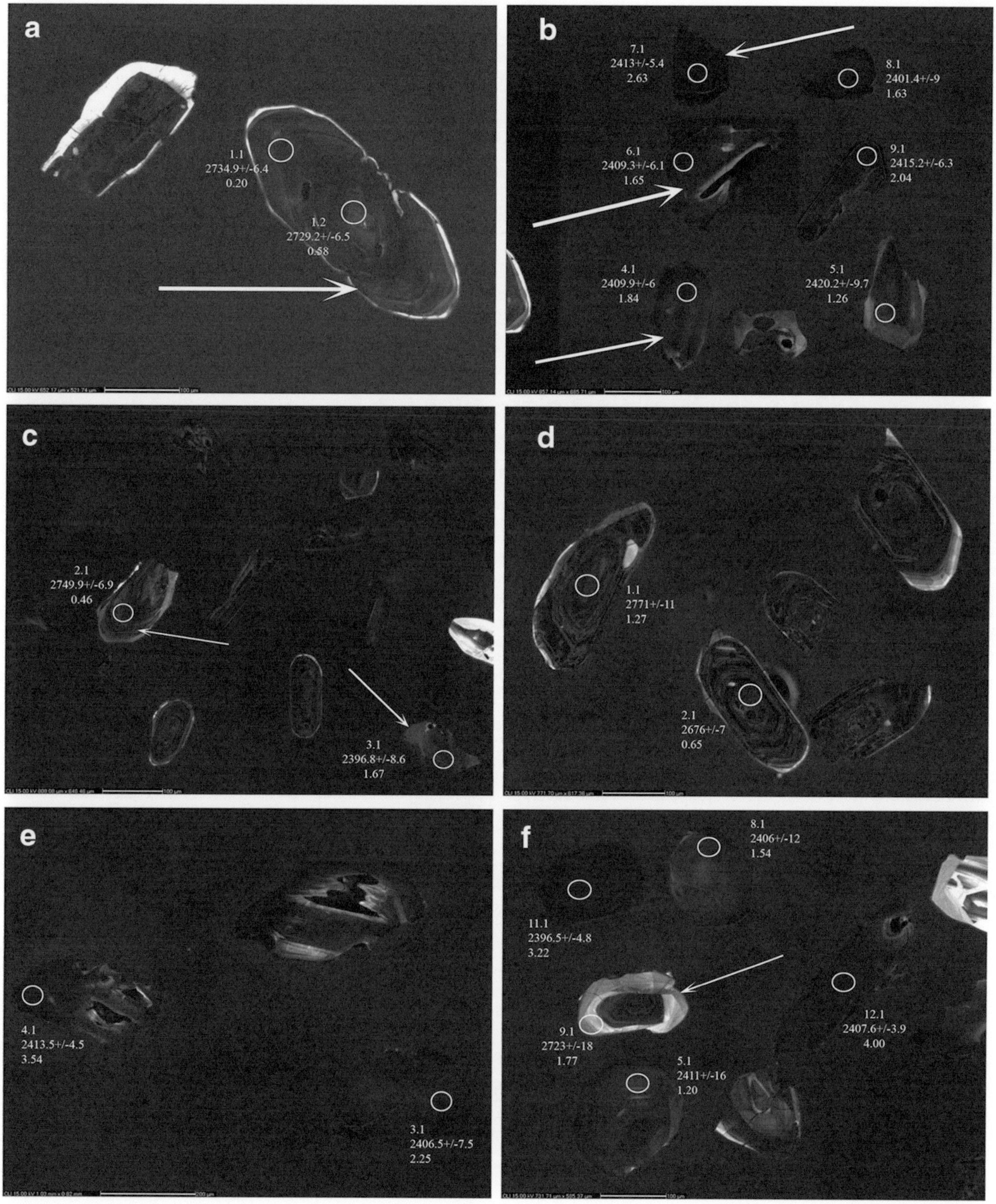

Fig. 7.14 Cathodoluminescence images of zircon grains with analytical spots and their numbers. The numbers correspond to those in Table 7.2, $^{206}Pb/^{238}U$ age values, and Th/U ratios
(**a–c**) Zircons from rocks of the massif of Mount Krutaya Vostochnaya (sample KV-1); (**d–f**) zircons from rocks of the Poioiva Massif (sample P-1). *Arrows point* to differences in the structure of young (homogeneous) and old zircons

Table 7.9 Rare elements' constants in zircons from drusite massifs in the South Kovdor area, ppm

	2,734	2,729	2,749	2,420	2,409	2,415	2,751	2,776	2,771	2,676	2,408	2,406	2,411	2,414
Massif	Mt. Krutaya (Kv-1) gabbronorite						Poioiva Massif (P-1) Pl lherzolite							
Element	N1.1	N1.2	N2.1	N5.1	N4.1	N9.1	N7.1	N6.1	N1.1	N2.1	N12.1	N8.1	N5.1	N4.1
P	343.0	270.6	202.7	345.9	446.3	524.7	280.2	465.0	415.1	773.2	860.5	311.3	310.3	755.7
Ti	11.85	22.64	53.38	23.81	19.85	25.75	8.61	8.64	26.46	64.74	15.73	11.29	12.74	22.64
Sr	0.41	1.05	1.22	0.34	0.54	18.75	7.15	0.48	1.41	27.13	0.620	0.42	0.35	0.83
Y	208.9	705.4	385.0	613.6	1,958	2,774	657.9	3,041	1,833	1,167	4616	945.2	649.8	3,386
Nb	37.17	32.24	13.92	27.51	26.26	27.52	23.73	13.91	9.37	20.11	17.35	11.04	9.45	24.85
Ba	1.23	2.35	2.80	1.11	1.16	17.88	6.22	0.82	2.98	13.88	1.25	0.76	1.12	2.18
La	1.35	3.99	6.07	0.18	0.21	1.33	0.87	0.17	10.25	46.76	0.529	0.13	0.08	0.84
Ce	23.01	72.66	88.46	8.85	22.25	33.03	46.57	48.86	137.4	290.4	87.24	16.72	14.24	203.1
Pr	1.17	2.81	6.57	0.30	0.55	1.08	0.85	1.00	10.10	30.70	1.94	0.30	0.13	1.01
Nd	7.06	13.75	57.76	2.77	7.59	9.16	6.51	16.24	70.28	151.4	36.30	3.43	1.65	12.81
Sm	2.28	6.76	28.62	2.79	9.95	10.25	2.14	24.66	20.41	21.92	50.46	4.23	2.40	20.54
Eu	0.57	1.37	8.13	0.72	2.28	2.62	0.60	5.73	5.00	4.21	11.69	1.12	0.55	3.71
Gd	4.74	21.19	31.30	12.92	46.74	57.05	10.70	107.1	61.96	43.29	191.5	18.57	10.74	94.15
Dy	13.70	57.52	34.09	46.98	164.4	215.2	46.60	291.5	167.7	90.09	490.8	71.72	45.92	309.5
Er	36.96	101.6	61.27	102.4	318.4	452.3	119.5	465.6	307.0	209.0	656.0	160.3	109.6	531.6
Yb	101.8	193.0	147.9	186.5	509.3	738.7	272.5	686.0	542.1	411.5	815.2	270.2	190.4	793.9
Lu	21.03	32.65	29.29	31.68	81.27	115.4	45.59	105.6	94.85	69.59	114.1	44.00	30.98	111.6
Hf	11,407	11,391	9,984	6,306	6,495	6,671	10,611	6,828	7,578	8,458	6,984	6,398	7,300	6,906
Th	70.78	442.9	126.9	221.6	1,185	2,304	211.4	135.4	354.7	497.4	5,474	524.4	268.1	12,385
U	464.4	747.1	456.9	269.2	897.6	1,676	622.5	212.3	355.3	1,236	1,892	441.4	299.1	4,925
(Sm/La)n	2.70	2.71	7.54	23.68	75.36	12.28	3.93	223.0	3.19	0.75	152.7	49.81	43.39	39.18
(Lu/Gd)n	35.34	12.27	7.45	19.52	13.84	16.10	33.91	7.85	12.19	12.80	4.74	18.86	22.96	9.43
Ce/Ce*	4.41	5.25	3.39	9.00	15.78	6.65	13.06	28.02	3.27	1.86	20.85	19.76	31.45	53.40
Eu/Eu*	0.53	0.35	0.83	0.37	0.32	0.33	0.38	0.34	0.43	0.42	0.36	0.39	0.33	0.25
T°(Ti in)	756	818	913	823	805	831	728	728	834	937	782	752	763	818

Note: Element contents (ppm) were determined by SIMS (Cameca IMS–4 F. IMI RAS. Yaroslavl. Russia. analyst S. Simakin) at the same points where U–Pb data were obtained; $^{207}Pb/^{206}Pb$ age is done for the same zircon grain. Anomalies were calculated: $Ce/Ce^* = Ce/\sqrt{(La \times Pr)}$ и $Eu/Eu^* = Eu/\sqrt{(Sm \times Gd)}$. T° (Ti in)—T crystallization of zircon according to Ferry and Watson (2007): $T°C = 5080/(6.01 - \lg(Ti)) - 273.1$. Upward are sample numbers. N1.1—number of point in zircon

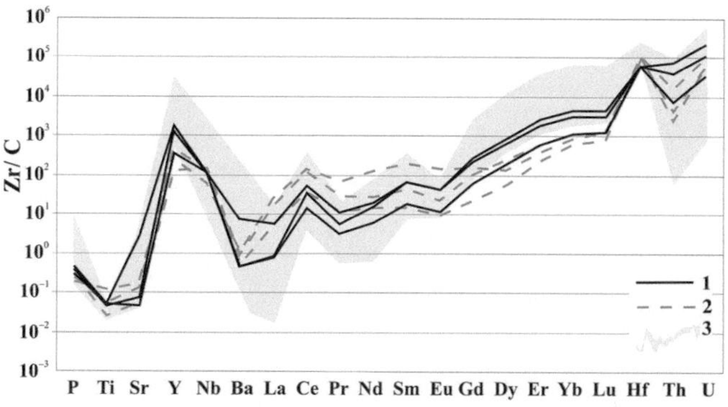

Fig. 7.15 C1 chondrite-normalized trace element patterns of zircons from (**a**) the massif of Mount Krutaya Vostochnaya (sample KV-1) (*1*) Inherited zircon dated at 2,733 Ma, (*2*) magmatic zircons dated at 2,408 Ma (Table 7.9), and (*3*) composition of zircons from basic–ultrabasic layered massifs. (Belousova et al. 2002)

mated by a linear trend that intercepts the concordia at points corresponding to 2,733 ± 9.0 and −384 ± 420 Ma (Fig. 7.16b). If the U–Pb system of grain 2.1 was disturbed and the grain partly lost its Pb near the surface during "modern" times, then the recalculated trend with a lower intercept at the zero point ("modern" time) would yield a well-defined upper intercept with the concordia corresponding to 2,737.3 ± 7.5 Ma at MSWD = 2.5. Two concordant points from the first grain date the process of crystallization at 2,731 ± 11 Ma (at concordance MSWD = 0.35 and a concordance probability = 0.56). It is thus reasonable to conclude that the magma of the Mount Krutaya Massif captured and entrained zircons from rocks with an age of 2,731–2,737 Ma, and these rocks either hosted the massif or occurred beneath its crystallization chamber.

The young zircons of this sample define a compact group in the concordia plot that is slightly spread along a trend line emanating from the origin of the coordinates. This allowed us to calculate this linear trend and its intercept point with the concordia corresponding to an age of 2,410.2 ± 5.1 Ma at MSWD = 0.9 (Fig. 7.16c). The calculation of the concordant age of this group yields a zircon crystallization age of 2,407.8 ± 8.0 Ma (concordance MSWD = 7.5 and concordance probability = 0.006). The high MSWD value and, hence, the low concordance probability are caused by a slight shift in the data points relative to the concordia (small losses of radiogenic Pb) and their spread, as seen from the position of the weighted mean point in Fig. 7.16. It is thus reasonable to assume that the crystallization age of the zircons in this massif is 2,407.8 ± 8.0 Ma.

For the Poioiva Massif (sample P-1) (Fig. 7.16), five of the six analyses of the old zircon grains yield a compact and nearly concordant group, and one grain shows traces of significant radiogenic Pb losses (approximately 26 % discordance). The discordia constructed for the whole set of six analyses intercepts the concordia at points corresponding to 2,756 ± 33 and 502 ± 310 Ma, MSWD = 2.6. The significant errors of the age values are explained by the location of two reversely discordant points, which also have the greatest analytical errors due to the lowest Pb and U concentrations found in these grains. The concordant age, based on the five most concordant points, is 2,754 ± 13 Ma (concordance MSWD = 0.31 and concordance probability = 0.58). It is thus reasonable to assume that zircon from the host rocks has a crystallization age of 2,754 ± 13 Ma.

The U–Pb isotopic data points of the young magmatic zircons compose a practically concordant group in the concordia plot (the discordance is no higher than 6 %). Nevertheless, in calculating the concordant age based on all six analyses, we obtained the concordance probability of 0 at MSWD = 17 and an age of 2,406.1 ± 14 Ma. If only analyses with discordance of no higher than 5 % are taken into consideration (four points), then an age value of 2,407 ± 8.3 at MSWD = 10.8 and a probability of 0.001 are obtained. The best estimate of the crystallization age of the magmatic zircons is likely to come from the data obtained by calculating the intercept of the discordia for the whole set of young zircons (six points and the origin of the coordinates, i.e., "zero age"), which is 2,406.5 ± 4.6 Ma (MSWD = 1.4). Our data on the U–Pb systematics of zircons from sample P-1 led us to conclude that these zircons crystallized, along with the massif itself, at 2,406.5 ± 4.6 Ma, and the melt was emplaced into Archean crystalline rocks at 2,754 ± 13 Ma.

The geological evolution of the Kola–Lapland– Karelia province occurred over a time period of 2.5–2.4 Ga when numerous ultrabasite–basite massifs of various morphologies and compositions were formed and are now localized in both mobile belts and rigid cratons. These are small rootless bodies and large layered intrusions, and the similarities in their composition and age provide grounds to believe that this time span was marked by the occurrence of a microcontinent (Smolkin et al. 2005). The crust of the Belomorian Mobile Belt was then affected by a super plume that was located beneath the eastern part of the Baltic Shield and was responsible for the development of rift belts and coeval magmatism; the latter produced numerous bodies of ultrabasic and basic rocks (Balaganskii 2002).

Table 7.10 Results of local U–Pb isotope analysis of zircon from massif Mt. Krutaya (Kv-1)

Point number	% $^{206}Pb_c$	[U]	[Th]	$^{232}Th/^{238}U$	[^{206}Pb*]	(1) $^{206}Pb/^{238}U$ age		(2) $^{206}Pb/^{238}U$ age		(3) $^{206}Pb/^{238}U$ age		(1) $^{207}Pb/^{206}Pb$ age	
1.1	0.05	243	46	0.20	110	2,714	±60	2,707	±83	2,714	±61	2,735	±6
1.2	0.07	364	203	0.58	170	2,790	±62	2,818	±91	2,793	±66	2,729	±6
2.1	0.07	272	122	0.46	109	2,464	±57	2,381	±68	2,465	±59	2,750	±7
3.1	0.04	268	434	1.67	102	2,358	±55	2,349	±68	2,357	±68	2,397	±9
4.1	–	569	1,015	1.84	219	2,387	±54	2,382	±68	2,384	±69	2,410	±6
5.1	–	199	243	1.26	75	2,357	±55	2,342	±68	2,354	±64	2,420	±10
6.1	0.00	586	934	1.65	223	2,361	±54	2,349	±66	2,358	±66	2,409	±6
7.1	0.01	1,327	3,375	2.63	492	2,313	±53	2,290	±64	2,322	±76	2,413	±5
8.1	0.02	438	691	1.63	164	2,331	±54	2,314	±66	2,333	±66	2,401	±9
9.1	0.01	984	1,940	2.04	376	2,372	±54	2,361	±67	2,363	±71	2,415	±6

		Isotope ratios						
Point number	D, %	(1) $^{207}Pb^*/^{206}Pb^*$	±%	(1) $^{207}Pb^*/^{235}U$	±%	(1) $^{206}Pb^*/^{238}U$	±%	*r*
1.1	1	0.18	0.39	13.66	2.7	0.52	2.7	0.99
1.2	−2	0.18	0.39	14.08	2.8	0.54	2.7	0.99
2.1	12	0.19	0.42	12.26	2.8	0.46	2.8	0.98
3.1	2	0.15	0.5	9.41	2.8	0.44	2.8	0.98
4.1	1	0.15	0.35	9.62	2.7	0.44	2.7	0.99
5.1	3	0.15	0.57	9.54	2.9	0.44	2.8	0.98
6.1	2	0.15	0.36	9.50	2.7	0.44	2.7	0.99
7.1	4	0.15	0.32	9.29	2.7	0.43	2.7	0.99
8.1	3	0.15	0.53	9.31	2.8	0.43	2.8	0.98
9.1	2	0.15	0.37	9.58	2.7	0.44	2.7	0.99

Note: Here and in Table 7.11, the value corresponds to the error-sigma; Pbc and Pb *—ordinary and radiogenic lead. The error in the calibration standard in the measurement session was 0.91 %. Age as determined by the appropriate isotope ratios is given in million years. Element concentrations are given in ppm. (1)—the usual Pb corrected from the measured ion current ^{204}Pb; (2)—the usual Pb corrected assuming concordance $^{206}Pb/^{238}U$ and $^{207}Pb/^{235}U$ ages; (3)—the usual Pb corrected assuming concordance $^{206}Pb/^{238}U$ and $^{208}Pb/^{232}Th$ ages. D = 100 (1- ($^{206}Pb/^{238}U$ age)/($^{207}Pb/^{206}Pb$ age)—the coefficient of discordance, *r*—correlation coefficient of error axes $^{207}Pb/^{235}U$ – $^{206}Pb/^{238}U$. Here and in Table 7.11. Analyst B. V. Belyatsky (VNIIOkeangeologia)

However, the wide application of isotopic techniques in recent geological studies of volcanic–plutonic complexes in the Karelia–Kola area with the aim of dating these complexes (first of all, by high-precision U–Pb zircon techniques) made it possible to elucidate many details of the evolutionary history of Paleoproterozoic magmatism in this territory (Lobach-Zhuchenko et al. 1998; Bibikova et al. 2004).

This pertains, first of all, to the origin of large layered plutons hosted in rigid crustal blocks, which are now considered to be classic in the analysis of magmatism of this age in the territory. These bodies were classified into two groups whose emplacement was separated by a time span of approximately 20 Ma (Smolkin et al. 2004). The older group was emplaced at 2,507–2,470 Ma (Mount General'skaya, Monchepluton, Fedorova–Pansky Tundras, and Umbaozero massifs). The younger group has an age of 2,441–2,437 Ma and includes layered intrusions in eastern and northern Karelia and Finnish Lapland (the Umbarechensky–Imandra Complex, Olanga group of intrusions, Burakovsky plutons, and massifs in northern Finland), which were placed within a relatively narrow time span of 2,441–2,437 Ma (Smolkin et al. 2004, 2005). The older group was formed before the development of the ancient rift system, whereas the massifs of the younger group were placed during the active functioning of the rift.

It was thought that massifs of the drusite complex were formed simultaneously with massifs of the younger group. However, more detailed studies (Slabunov et al. 2005) indicate that the latter can also be subdivided into two age groups: the older (Alexejev et al. 2000) gabbro–anorthosites of the Kotozero and Pezhostrov massifs and the Kolvitsa Massif (2,450 Ma) and lherzolites–gabbronorites of the Kovdozero Massif (2,440 ± 10 Ma) (Efimov and Kaulina 1997) and Shobozero Massif (2,435 ± 5 Ma) (Slabunov et al. 2005) and the younger coronite gabbro (dikes and small

Table 7.11 Data of local U–Pb analyses (SHRIMP-II) for zircons from Poioiva Massif (P-1)

Point number	% $^{206}Pb_c$	[U]	[Th]	$^{232}Th/^{238}U$	[^{206}Pb*]	(1) $^{206}Pb/^{238}U$ age		(2) $^{206}Pb/^{238}U$ age		(3) $^{206}Pb/^{238}U$ age		(1) $^{207}Pb/^{206}Pb$ age	
2.1	0.32	612	384	0.65	205	2,117	±49	2,005	±53	2,141	±52	2,676	±7
11.1	0.02	562	1,752	3.22	203	2,263	±51	2,233	±61	2,286	±83	2,397	±5
3.1	0.02	531	1,159	2.25	192	2,264	±52	2,233	±62	2,259	±71	2,407	±8
12.1	–	1,190	4,603	3,99	438	2,297	±52	2,271	±62	2,322	±98	2,408	±4
8.1	0.02	281	417	1.54	103	2,301	±54	2,276	±65	2,299	±65	2,406	±12
5.1	0.02	154	178	1,20	57.6	2,335	±56	2,317	±68	2,332	±65	2,411	±16
4.1	–	2,672	9,146	3.54	1,010	2,349	±53	2,333	±65	2,356	±91	2,414	±5
7.1	0.23	300	102	0.35	134	2,689	±61	2,668	±83	2,691	±63	2,751	±10
6.1	0.08	81	51	0.66	36.1	2,694	±66	2,663	±89	2,692	±71	2,776	±19
1.1	0.85	156	192	1.27	70.6	2,706	±61	2,690	±83	2,728	±71	2,771	±11
10.1	0.01	54	40	0.77	25.2	2,810	±66	2,839	±99	2,803	±72	2,746	±13
9.1	0.43	32	55	1.77	15.5	2,860	±73	2,931	±120	2,837	±91	2,723	±18

		Isotope ratios						
Point number	D, %	(1) $^{207}Pb^*/^{206}Pb^*$	±%	(1) $^{207}Pb^*/^{235}U$	±%	(1) $^{206}Pb^*/^{238}U$	±%	*r*
2.1	26	0.18	0.42	9.78	2.7	0.38	2.7	0.98
11.1	6	0.15	0.28	8.96	2.7	0.42	2.7	0.99
3.1	6	0.15	0.44	9.02	2.8	0.42	2.7	0.98
12.1	5	0.15	0.23	9.18	2.7	0.42	2.7	0.99
8.1	5	0.15	0.72	9.19	2.9	0.42	2.8	0.96
5.1	3	0.15	0.95	9.38	3.0	0.43	2.9	0.94
4.1	3	0.15	0.27	9.46	2.7	0.44	2.7	0.99
7.1	2	0.19	0.63	13.63	2.8	0.51	2.8	0.97
6.1	3	0.19	1.1	13.88	3.2	0.51	3.0	0.93
1.1	2	0.19	0.68	13.91	2.8	0.52	2.7	0.97
10.1	−2	0.19	0.77	14.35	3.0	0.54	2.9	0.96
9.1	−5	0.18	1.1	14.45	3.3	0.55	3.2	0.94

intrusions of garnet gabbro, 2,110 Ma (Mitrofanov et al. 1995). It has been demonstrated for the Zhemchuzhnyi Massif that the lherzolites are related to the gabbro–anorthosites and that the drusite rocks were formed and metamorphosed within a narrow age interval, 2,356 ± 4 Ma for the magmatic zircons and 2,354 ± 2 Ma for the metamorphic zircons (Balaganskii et al. 1997).

Hence, the setting and genesis of drusite massifs in extension zones were recently constrained fairly unambiguously; they were separated by major collision events and were formed within a narrow time span early in the rifting process. Judging from published data, this time span lasted 35–40 Ma. Some researchers believe that the drusite complex was formed within a much more narrow time span of 4–1 Ma (Balaganskii et al. 1997; Smolkin et al. 2005; Belousova et al. 2002). Discrepancies between these estimates are likely explained by differences in the approach to grouping the rocks into the drusite complex. If all rocks with drusite textures are considered, the total time span of their origin should have been as long as tens of millions of years, but if the massifs are differentiated according to their composition, this time span should be reduced to a few million years.

Thus, the magmatic zircons in the gabbro–anorthosite complex provide evidence that they crystallized at 2,440–2,450 Ma; the most widely spread lherzolites–gabbronorites crystallized at 2,435 ± 5 Ma, and the coronate gabbro crystallized at 2,115 ± 7 Ma (Alexejev et al. 2000; Slabunov et al. 2005; Mitrofanov et al. 1995; Stepanova et al. 2003).

The very first geochronological data obtained for massifs in the South Kovdor area are generally consistent with our results. Our data on the Sm–Nd isotopic systematics of rock-forming minerals and whole-rock samples of lherzolites and olivine gabbronorites from two massifs, Sorkajoki [2,485 ± 15 Ma at MSWD = 5.0 and $\varepsilon_{Nd}(T) = +0.88 \pm 0.39$] and Poioiva [2,507 ± 51 Ma at MSWD = 2.9 and $\varepsilon_{Nd}(T) = +1.05 \pm 0.70$], suggest that they crystallized simultaneously and their rocks were derivatives from an undepleted mantle source. At the same time, the initial Nd isotopic com-

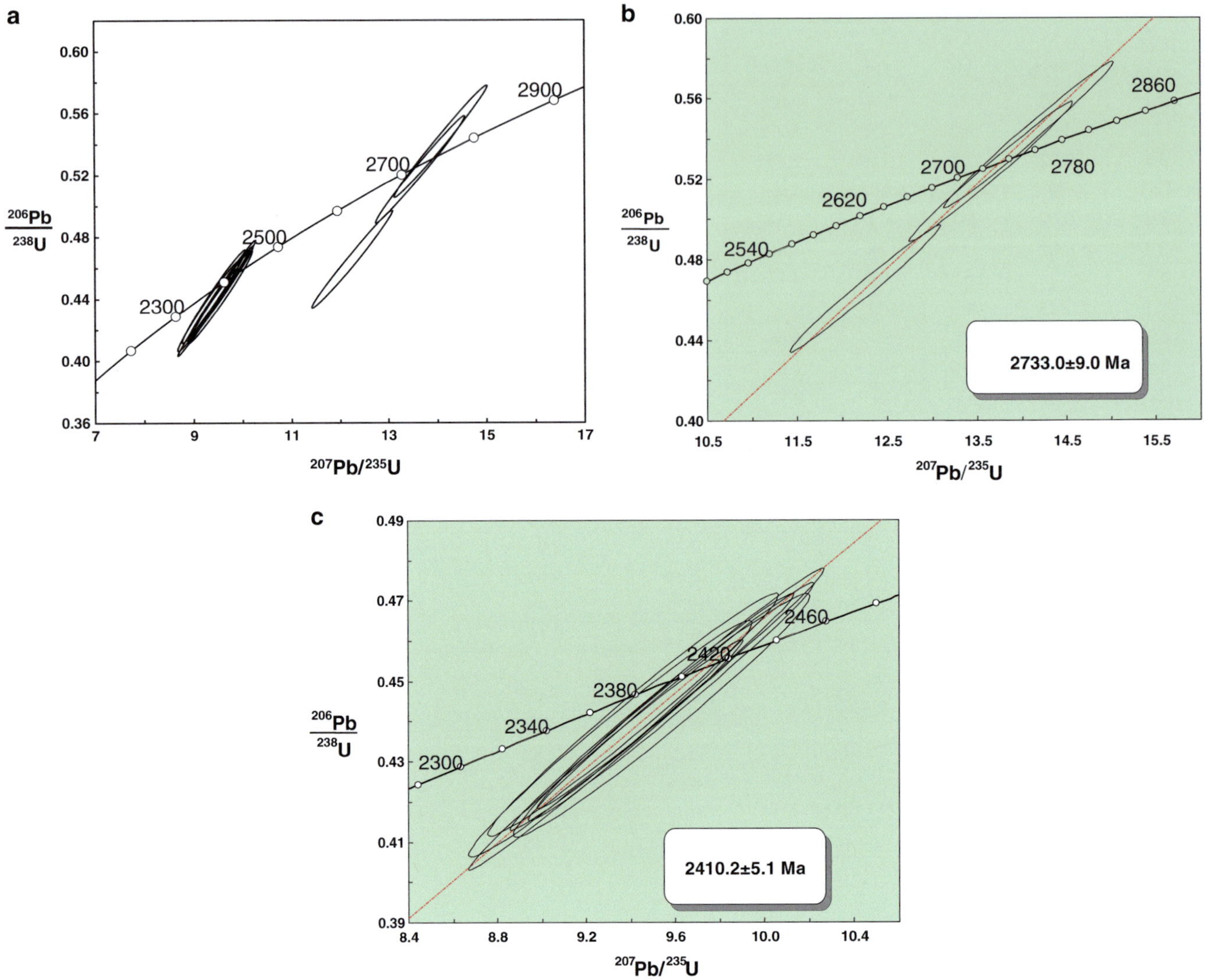

Fig. 7.16 U–Pb concordia diagrams for examined zircons from the massif of Mount Krutaya Vostochnaya
(**a**) The diagram presents all measurements; point numbers are the same as in Table 7.2. Error ellipses correspond to 95 % significance level (2σ). (**b**) Discordia based on three measured analytical spots at two inherited grains (Table 7.10, analytical spots 1.1, 1.2, and 2.1). (**c**) Discordia based on seven analyses and passing through the origin of coordinates ("zero age"); the *gray* ellipse corresponds to the calculated weighted concordant age value (2,407.8 ± 8.0 Ma) of this set of analyses. Sample KV-1

position of the intrusions is notably different from that of the drusite massifs and layered plutons (except for a single sample from the Monchepluton, whose distinctive feature is an even more enriched mantle source (ε_{Nd} from −2 to −4) and which makes it possible to relate them to a hypothetical Paleoproterozoic plume) (Sharkov 2006). The massifs examined in the southern Kovdor province had a source consisting of undepleted mantle material whose isotopic composition was close to that of the Archean komatiite basalts of the Karelia–Kola area (Puchtel et al. 2001; Vrevsky et al. 2003; Matrenichev and Vrevsky 2004; Early Precambrian… 2005; Smolkin 1992).

The U–Pb zircon ages of the southern Kovdor massifs are younger, 2,406.5 ± 4.6 Ma for the Poioiva Massif and 2,407.8 ± 8.0 Ma for the massif of Mount Krutaya, and they were emplaced into a block of crystalline Archean rocks (2,754–2,730 Ma). The thermal metamorphism of the rocks of these massifs after their crystallization is "documented" in the U–Pb systematics of the zircons as the insignificant discordance of the magmatic zircons (up to 6 % of newly formed zircons), which is more significant for the Late Proterozoic zircons (up to 26 %).

Finally, although the independently obtained results using two different geochronologic techniques on rocks of the Poioiva Massif in the South Kovdor area show a certain discrepancy between the age values (2,507 ± 51 Ma by Sm/Nd and 2,406.5 ± 4.6 Ma by U/Pb), we believe that this discrepancy is not a crucial problem for constraining the age and conditions of the origin of the massif and the analogous intrusive bodies in the Kovdor area.

The older age value obtained by the Sm/Nd method is explained by the significant analytical errors involved in determining the Sm/Nd ratios of the minerals and whole-rock samples of the ultrabasite massifs, in which the Sm and Nd concentrations are too low to obtain reliable dates. With regard to the much higher precision of the U–Pb zircon method applied to these rocks, it is natural to assume that these dates are more reliable. Because of this, we recalculated ε_{Nd} to $T=2.41$ Ga for the Sorkajoki Massif and obtained ε_{Nd} from −0.96 to +0.82 ± 0.50 with the analogous values for the Poioiva Massif being +0.60 ± 0.50. In conclusion, it should be mentioned that our newly obtained data on the inner structure, mineralogy, petrologic–geochemical specifics (Krivolutskaya et al. 2010a), and isotopic composition of the rocks from certain ultrabasic–basic massifs in the southern Kovdor area (Sorkajoki, Poioiva, Mount Krutaya Vostochnaya, and others) led us to combine these bodies into a single intrusive complex, which has never before been recognized in this area. In spite of the occurrence of drusite textures in the rocks of the massifs, this complex differs from typical drusites in the morphology of its intrusive bodies, their geochemical specifics, their younger age, and the undepleted source of the parental magmas.

7.7 Conclusions

1. The intrusions of the drusite complex in the central part of the Belomorian Mobile Belt are compositionally heterogeneous. Their rocks are subdivided into three geochemical groups: (a) high-magnesian and high-Cr peridotites (lherzolites–wehrlites), olivine gabbronorites, and gabbronorites, which are similar to the reference rocks from the layered intrusions of the peridotite–pyroxenite–gabbronorite formation of the Kola and Karelian regions; (b) low-Mg, high-Fe gabbros, metagabbro, and garnet metagabbros with elevated contents of phosphorus and alkalis; and (c) moderate-Mg, olivine-free, and olivine-bearing gabbros and amphibolites.
2. The application of two geochronologic techniques (Sm–Nd on whole-rock samples and minerals and U–Pb on zircons) allowed us to obtain the first age data of large ultrabasic–basic intrusions in the central part of the Belomorian Belt. The massifs were provisionally ascribed to the drusite complex based on the occurrence of corona textures in the rocks.
3. The four examined massifs (Sorkajoki, Poioiva, Massif at 403-m elevation, and Mount Krutaya) were placed simultaneously in the Paleoproterozoic and were roughly coeval with numerous magmatic bodies in the Karelia–Kola territory, i.e., layered plutons hosted in rigid cratons and drusite massifs in mobile belts.
4. The Sm–Nd crystallization ages of the massifs are as follows: 2,507 ± 51 Ma at MSWD = 2.9 and $\varepsilon_{Nd}=+1.05\pm0.70$ for the Poioiva Massif, 2,485 ± 51 Ma at MSWD = 5.0 and $\varepsilon_{Nd}=+0.88\pm0.39$ for the Sorkajoki Massif, and 2,518 ± 66 Ma at MSWD = 0.28 and $\varepsilon_{Nd}=+0.7\pm0.37$ for the massif at height 403 m. The genesis of these intrusions was characterized by their relationship to an undepleted mantle source (ε_{Nd} from +0.25 to +1.37), whereas the parental magmas of other ultrabasic–basic intrusions of this age in the area were derived from an even more strongly enriched source ($\varepsilon_{Nd}=2$ to −4), which suggests that these intrusions could be related to a Paleoproterozoic plume.
5. The crystallization of magmatic zircon, a process occurring late in the evolutionary histories of the two massifs, was roughly synchronous: 2,406.5 ± 4.6 Ma for the Poioiva Massif and 2,407.8 ± 8.0 for the massif of Mount Krutaya. Thus, these massifs are younger than the magmatic bodies of the drusite complex with which they were previously classed. The massifs were emplaced into Archean crystalline rocks (dated at 2,754–2,730 Ma).
6. Earlier data on the geochemistry of rocks in the central part of the Belomorian Mobile Belt and on their isotopic composition indicate that the intrusions described herein are not typical drusite massifs (according to the conventional understanding) but compose an individual intrusive complex in the Karelia–Kola area.
7. Our results demonstrate that the occurrence of corona texture in rocks of ultrabasite–basite massifs is insufficient for the classification of these massifs with the drusite complex. Relationships with this complex must be established with regard to the morphology of the intrusions and the geochemistry and isotopic characteristics of their rocks.
8. Geochemical features of the massifs in the South Kovdor area (the first type) are very similar with the features of the ore-bearing intrusions in the Noril'sk area. Their patterns and isotope characteristics have crustal signatures (Ta–Nb-negative and Pb-positive anomalies, ε_{Nd} varies around 0).

References

Alexejev N, Zinger T, Belyatsky B, Balagansky B (2000) Age of crystallization and metamorphism of the Pestovoy Gabbro_Anorthosites, northern Karelia, Russia. In: Proceedings of the 5th SVEKALAPKO workshop, Lammi, Finland, p 3

Amelin Yu V, Semenov VS (1990) On the age and source of Lower Proterozoic PGE-Bearing layered intrusions of Karelia. In:

Proceedings of the All_Union conference on isotopic dating of endogenous ore formations, Kiev, pp 40–42 (in Russian)

Amelin Yu V, Semenov VS (1996) Nd and Sr isotope geochemistry of Mafic layered intrusions in the Eastern Baltic Shield: implication for sources and contamination of Paleoproterozoic Continental Mafic magmas. Contrib Mineral Petrol 124:255–272

Amelin Yu V, Heaman LM, Semenov VS (1995) U–Pb geochronology of layered intrusions in the Eastern Baltic Shield: implication for timing and duration of Paleoproterozoic Continental rifting. Precambrian Res 75:31–46

Balaganskii VV (2002) Extended abstract of doctoral dissertation in geology and mineralogy. St. Petersburg, 42 p (in Russian)

Balaganskii VV, Kudryashov NM, Balashov Yu A et al (1997) The age of the Zhemchuzhnyi Drusite Massif, North-western White Sea Area: U–Pb isotopic data and geological implications. Geochem Int 35:127–136

Balashov Yu A, Bayanova TB, Mitrofanov FP (1993) Isotope data on the age and genesis of layered basic–ultrabasic intrusions in the Kola Peninsula and Northern Karelia, Northeastern Baltic Shield. Precambrian Res 64(1–4):197–205

Belousova EA, Griffin WL, O'Reilly SY, Fisher NI (2002) Igneous Zircon: trace element composition as an indicator of source rock type. Contrib Mineral Petrol 143:602–622

Bibikova EV, Skiold T, Bogdanova SV (1993) Geochronology of the Belomorides: interpretation of multistage geological history. Geochem Int 10:1393–1411

Bibikova EV, Bogdanova SV, Glebovitskii VA et al (2004) Evolution of the Belomorian Belt: NORDSIM U–Pb Zircon dating of the Chupa Paragneisses, magmatism, and metamorphic stages. Petrology 12:195–210

Black LP, Kamo SL (2003) TEMORA 1: a New Zircon standard for U–Pb geochronology. Chem Geol 200:155–170

Bogdanova SV, Bibikova EV (1993) The "Saamian" of the Belomorian Mobile Belt: new geochronological constraints. Precambrian Res 64:131–152

Bottazzi P, Ottolini L, Vannucci R, Zanetti A (1994) An accurate procedure for the quantification of rare earth elements in silicates. In: SIMS IX proceedings. Wiley, Chichester, pp 927–930

Bychkova Ya V, Koptev_Dvornikov EV, Kononkova NN, Kameneva EE (2007) Composition of rock_forming minerals in the Kivakka Layered Massif, Northern Karelia, and systematic variations in the chemistries of minerals in the rhythmic layering subzone. Geochem Int 45:131–151

Chistyakov AV, Bogatikov OA, Grokhovskaya TL et al (2000) The Burakov layered pluton, Southern Karelia: petrological and isotope geochemical evidence for the juxtaposition of two intrusions. Dokl Earth Sci 372:698–704

Chistyakov AV, Sharkov EV, Grokhovskaya TL et al (2002) Petrology of the Europe largest Burakovka early Paleoproterozoic layered pluton (Southern Karelia, Russia). Russ J Earth Sci 4(1):44

Early Precambrian of the Baltic Shield (2005) Glebovitskii VA (ed). St. Petersburg, Nauka (in Russian)

Efimov AA, Kaulina TB (1997) Geological features and U–Pb dating (First data) of the Southeastern part of the Kovdozero Basic–Ultrabasic Massif (Puakhta Block). In: Proceedings of the conference on the Belomorian Mobile Belt, Geology, geodynamics, geochronology, KarNTs RAN, Petrozavodsk, Russia, p 31 (in Russian)

Fedorov ES (1886) On a new group of igneous rocks. Izv. Mosk. Sel'khoz. Inst. Book 1, pp 168–187 (in Russian)

Ferry JM, Watson EB (2007) New thermodynamic models and revised calibrations for the Ti in Zircon and Zr in_Rutile thermometers. Contrib Mineral Petrol 154:429–437

Grimes CB, John BE, Kelemen PB et al (2007) Trace element chemistry of Zircons from oceanic crust: a method for distinguishing detrital Zircon provenance. Geology 35:643–646

Hofmann AW (1988) Chemical differentiation of the earth: the relationship between mantle, continental crust, and oceanic crust. Earth planet. Sci Lett 90:297–314

Hoskin PWO, Ireland TR (2000) Rare earth element chemistry of Zircon and its use as a provenance indicator. Geology 28:627–630

Igneous Rocks (1988) Ultrabasic rocks, vol 5. Nauka, Moscow (in Russian)

Kratts KO, Khil'tova Ya V, Vrevskii AB (1980) Stages and types of the evolution of the Precambrian crust of ancient shields. Nauka, Leningrad (in Russian)

Krivolutskaya NA, Smolkin VF, Svirskaya NM, Mamontov VP, Fanygin AS et al (2010a) Geochemical features of the Drusite Massifs, the Central Part of the Belomorian Mobile Belt: I. Distribution of major and trace elements in the rocks. Geochem Int 48:465–491

Krivolutskaya NA, Belyatsky BV, Smolkin VF, Mamontov VP, Fanygin AS, Svirskaya NM (2010b) Geochemical features of the Drusite Massifs, the Central Part of the Belomorian Mobile Belt: II. Study Sm-Nd isotope system in rocks and U-Pb system in zircons. Geochem Int 11:1132–1153

Larikova TL (2000) Genesis of Drusitic (Corona) textures around olivine and orthopyroxene during metamorphism of Gabbroids in Northern Belomorie, Karelia. Petrology 8:384–401

Larikova TL (2002) Extended Abstract of Candidate Dissertation in Geology and Mineralogy. IGEM, Moscow, 16 p (in Russian)

Larikova TL, Zaraisky GP (2002) Experimental modeling of corona textures. Exp Geosci 10(1) (in Russian)

Latypov RM, Chistyakova Yu S (2000) Mechanism of differentiation of the Western Pana Tundra intrusion. Kol'sk.Nauchn. Ts. RAN. Apatity, 248 p (in Russian)

Li CS, Ripley EM, Naldrett AJ (2009) A new genetic model for the giant Ni–Cu-PGE sulfide deposits associated with the Siberian flood basalts. Econ Geol 104:291–301

Lobach-Zhuchenko SB, Arestova NA, Chekulaev VP et al (1998) Geochemistry and petrology of 2.40–2.45 Ga magmatic rocks in the North-Western Belomorian Belt, Fennoscandian Shield, Russia. Precambrian Res 92:223–250

Ludwig KR (2005) User's manual for isoplot/Ex, Version 3.11. A geochronological toolkit for Microsoft Excel. Geochronology Center Special Publication, Berkeley

Ludwig KR (2006) SQUID 2.0. A user's manual. Geochronology Center Special Publication, Berkeley

Malov ND, Sharkov EV (1978) Composition of parental melts and conditions of crystallization of the early Precambrian Intrusions of the Belomorian Drusite complex. Geokhimiya 7:1032–1039

Mashkovtsev S (1927) On the problem of the Belomorian Drusites. Tr Leningr Ob-va Estestvoisp 8(4):43–78 (in Russian)

Matrenichev VA, Vrevsky AB (2004) Isotopic–geochemical model for the upper mantle evolution of the Baltic Shield. Geochem Int 42:86–91

Miller Yu V, Myskova TA, Mil'kevich RI (2002) Supracrustal rocks in the tectonic windows of the marginal part of the Karelian Craton (Northwestern Belomorian Province). Geotectonics 36:11–23

Mitrofanov FP, Balagansky BB, Balashov Yu A (1995) U–Pb age of Gabbro–Anorthosite Massifs in the Lapland Granulite Belt. Nor Geol Unders Spec Publ 15:67–83

Nikolaev GS, Ariskin AA (2005) Burakovo–Aganozero layered Massif in the Trans-Onega area: II. Structure of the marginal series and the estimation of the parental magma composition by geochemical thermometry techniques. Geochem Int 43:646–665

Nosova AA, Sazonova LV, Narkisova VV, Simakin SG (2002) Minor elements in Clinopyroxene from Paleozoic volcanics of the Tagil Island Arc in the Central Urals. Geochem Int 40:219–235

Polkanov AA (1939) Pre-Quaternary geology of the Kola Peninsula and Karelia or the easternmost part of the Fennoscandian Shield. In: Proceedings if the 17th session international geological. congress, 2, pp 27–58

Precambrian Magmatic Formations of the Northeastern Baltic Shield (1985) Leningrad, Nauka (in Russian)

Puchtel LS, Arndt NT, Hofmann AW, Haas KM (1998) Petrology of Mafic Lavas within the Onega Plateau, Central Karelia: evidence for 2.0 Ga plume_related continental crustal growth in the Baltic Shield. Contrib Mineral Petrol 130:143–153

Puchtel IS, Brьgmann GE, Hofman AW et al (2001) Os isotope systematics of Komatiitic Basalts from the Vetreny Belt, Baltic Shield: evidence for a chondritic source of the 2.45 Ga plume. Contrib Mineral Petrol 140:588–599

Richard P, Shimizu N, Allegre CJ (1976) ^{16}3Nd/^{146}Nd, a natural tracer. An application to oceanic basalts. Earth planet. Sci Lett 31:65–74

Sharkov EV (2006) Formation of Layered Intrusions and Related Mineralization. Nauchnyi Mir, Moscow, 367 p (in Russian)

Sharkov EV, Smolkin VF (1997) The Early Proterozoic Pechenga–Varzuga Belt: a case of Precambrian. Precambrian Res 82:133–151

Sharkov EV, Lyakhovich Vl V, Ledneva GV (1994) Petrology of the Paleoproterozoic Drusite complex of the Belomorian region by the example of the Pezhostrov Massif, Northern Karelia. Petrology 2(5):511–531

Sharkov EV, Smolkin VF, Krasivskaya IS (1997) Early Proterozoic igneous Province of Siliceous high-Mg Boninite_like rocks in the Eastern Baltic Shield. Petrology 5:448–465

Sharkov EV, Krasivskaya IS, Chistyakov AV (2004) Dispersed Mafic–ultramafic intrusive magmatism in Early Paleoproterozoic mobile zones of the Baltic Shield: an example of the Belomorian Drusite (Coronite) complex. Petrology 12:561–582

Shurkin KA, Gorelov NV, Sal'e ME (1962) Belomorian complex of Northern Karelia and the Southwestern Kola Peninsula. Akad. Nauk SSSR, Moscow, 187 p (in Russian)

Slabunov AI, Bibikova EV (2001) The Meso- and Neo-Archaean of the Karelian and Belomorian Provinces, Baltic Shield, Geology, isotope geochemistry and geodynamic reconstructions. In: Extended abstracts of 4th international Archaean symposium, Perth, Australia, pp 359–361

Slabunov AI, Volodichev OI, Balaganskii VV et al (2005) Belomorian mobile belt: general features of geological structure and evolution. In: Belomorian mobile belt and its analogues: geology, geochronology, geodynamics, and metallogeny, Petrozavodsk, pp 6–12 (in Russian)

Smirnov VK, Sobolev AV, Batanova VG et al (1995) Quantitative SIMS analysis of melt inclusions and host minerals for trace elements and H_2O. EOS Trans Spring Meet Suppl Am Geol Union 76:270

Smol'kin VF, Sharkov EV, Chistyakov AV, Krasivskaya IS (2005) Comparative analysis of the intrusions of the Belomorian Drusite Complex, layered intrusions of Karelia and Kola Peninsula, and Comagmatic Paleoproterozoic Volcanics. In: Belomorian mobile belts and its analogues: geology, geodynamics, metallogeny, Petrozavodsk, pp 282–285 (in Russian)

Smolkin VF (1992) Early Precambrian Komatiitic and picritic magmatism of the Baltic Shield. Nauka, St. Petersburg, 162 p, (in Russian)

Smolkin VF, Tessalina SP (2007) Genetic problems of the Paleoproterozoic layered intrusions on the basis of isotopic data. In: Proceedings of the all_Russian conference on geodynamics, magmatism, sedimentogenesis, and metallogeny of Northwestern Russia, Petrozavodsk, Izd. KarNTs RAN, Petrozavodsk, pp 379–383 (in Russian)

Smolkin VF, Fedotov Zh A, Neradovskii Yu N et al (2004) Layered intrusions of the Monchegorsk ore district: petrology, isotopy, and deep structure. Apatity 1:2 (in Russian)

Smolkin VF, Sharkov EV, Chistyakov AV, Krasivskaya IS (2005) Comparative analysis of the intrusions of the Belomorian Drusite Complex. Layered intrusions of Karelia and Kola Peninsula and Comagmatic Paleoproterozoic volcanic rocks. In: Belomorian mobile belt and its analogues: geology, geochronology, geodynamics, and metallogeny, Petrozavodsk, pp 282–285 (in Russian)

Stepanov VS (1981) Precambrian basic magmatism of the Western Belomorian region. Nauka, Leningrad, p 169 (in Russian)

Stepanova AV, Larionov AN, Bibikova EV (2003) Early Proteorozoic (2.1 Ga) Fe-Tholeiitic magmatism of the Belomorian Province, Baltic Shield: geochemistry and geochronology. Dokl Earth Sci 390:607–610

Strona AA (1929) On some drusites of the Kandalaksha Bay of the White Sea. Mater Obshch Prikl Geol 127:55

Sudovikov NG (1937) A brief review of Pre-Quaternary geology of Karelia. In: Tr. 17th session MGK **2**, pp 15–25; Early Precambrian of the Baltic Shield (2005) Glebovitsky VA (ed). Nauka, St. Petersburg (in Russian)

Tomilenko AA, Kovyazin SV (2008) Development of corona textures around olivine in anorthosites of the Korosten' Pluton, Ukrainian Shield: mineralogy, geochemistry, and fluid inclusions. Petrology 16(1):87–103

Vrevsky AB, Matrenichev VA, Ruzh'eva MS (2003) Petrology of Komatiites from the Baltic Shield and isotope geochemical evolution of their mantle sources. Petrology 11:532–561

Whitehouse MJ, Kamber BS (2003) A rare earth element study of complex Zircons from Early Archaean Amitsoq Gneisses, Godthabsfjord, South_West Greenland. Precambrian Res 126:363–377

Williams IS (1998) U_Th-Pb geochronology by ion microprobe. In: McKibben MA, Shanks, WC III, Ridley WI (eds) Applications of microanalytical techniques to understanding mineralizing processes. Rev Econ Geol 7:1–35

8 Assimilation of the Host Rocks by Basic Magma

The contact zones around the intrusions were studied to estimate the extent of assimilation of the host rocks by the magmas. Assimilation by the Maslovsky intrusion, which intruded the Nadezhdinsky basalt, is absent (central zone of the upper contact) or occurs in a narrow zone (1 m). The Kharaelakh intrusion, which intruded Devonian sediments, did not assimilate anhydrite, based on the isotope distribution of the rare elements. The anhydrite's isotopic composition (Sr, Nd, Pb) does not permit it to be regarded as a contaminant in the sulfide ore formation.

When solving the problems regarding the formation of sulfide ores at deposits of various genetic types, one of the key factors is the isotopic composition of sulfur. Such an approach implies the definition of either the mantle or the crustal source of sulfur. This problem becomes especially topical when considering the igneous copper–nickel deposits localized in ultrabasite–basite complexes. The mantle origin of the melts that formed these deposits and the ores proper is usually recognized by most geologists and is based mostly on the data of isotopic analysis for sulfur in ore-forming sulfides. The isotopic composition of sulfur usually changes within narrow limits, from −2 to +2 %, and is close to the value of troilite from the Canyon Diablo iron meteorite (a standard for mantle rocks). This represents a postulate in geochemistry.

As was mentioned above, the most extraordinary feature of the deposits of the Noril'sk group is the disproportionality between the thicknesses of the intrusions and related sulfide ores, with the ores sometimes comprising as much as 15 vol.% of the intrusions (Likhachev 1996a). Given the low sulfur solubility (0.2 wt %) in basaltic melts at shallow depths at which the intrusions of the Noril'sk Complex were emplaced and crystallized, it is difficult to imagine how all sulfur of the sulfides could come solely from the parental magma of a given intrusive body because the entire volume of the sulfides requires almost two orders of magnitude more sulfur than can be dissolved in the melt.

To resolve this inconsistency, a hypothesis was advanced that the sulfur was borrowed from an external source, i.e., that the parental magma extracted sulfur from the host rocks (Godlevsky and Grinenko 1963) either in situ or en route to the chamber in which the magma crystallized. This hypothesis is largely based on the composition of the rocks hosting the ore-bearing massifs, more specifically, on the occurrence of thick anhydrite beds in the Silurian–Devonian sedimentary rocks that most commonly host the intrusions. A principally important argument in support of the paramount importance of the assimilation of the host rocks by the parental melt is the heavy sulfur isotopic composition of the Noril'sk ores (Grinenko 1967, 1984; Gorbachev and Grinenko 1973; Kovalenker et al. 1974). It has been demonstrated (Grinenko 1985) that the larger the deposit, the heavier the sulfur isotopic composition (up to $\delta^{34}S = +18‰$), which appears to provide support to the hypothesis of assimilation. However, the material balance calculations and the analysis of the geological settings of several intrusions in the area led Grinenko (1984) to conclude that anhydrite was not involved in the assimilation process and that the heavy sulfur isotopic composition of the intrusions resulted from the accumulation of hydrocarbons in intermediate chambers, with these hydrocarbons borrowed from oil and gas pools hosted in the Vendian–Cambrian sedimentary rocks.

However, if the process of assimilation of the host rocks by magma occurred, we must observe the appropriate petrographical and geochemical features. The processes of the interactions between the melts and their host rocks are traditionally analyzed with reference to granitic magmas, with basaltic melts being considered in this aspect much more rarely. In various models for the genesis of Cu–Ni and Fe–Ti–V ores related to ultrabasite–basite complexes, much importance has long been attached to the assimilation of sedimentary rocks by basaltic magmas (Konnikov 1986; Konnikov et al. 2005; Naldrett 1992, 2005, 2009; Arndt et al. 2003). The assimilation of sulfur from host rocks is one of the key points of the new model of the genesis of the Noril'sk deposits, which was

N.A. Krivolutskaya, *Siberian Traps and Pt-Cu-Ni Deposits in the Noril'sk Area*,
DOI 10.1007/978-3-319-17205-7_8

suggested by Rad'ko (1991) and further developed by A. Naldrett and his colleagues (Naldrett 1992, 2005; Arndt 2005; Li et al. 2009a). This model was formulated as an alternative to the hypothesis that the Noril'sk ores were generated in a closed system (Godlevsky 1959; Dyuzhikov et al. 1988; Distler et al. 1988). According to this hypothesis, the intrusions are regarded as feeders for the lavas erupted at the surface. As an argument in support of this hypothesis, its authors quoted the fact that the intrusive rocks contain anhydrite (Li et al. 2009b). Nevertheless, the magmatic nature of the anhydrite was not proved.

In view of this, it is petrologically interesting to evaluate how much host rocks cannot actually be assimilated by mafic melts. Elucidation of this issue is also important for the theory of magmatic ore-forming processes.

The unique Pt–Cu–Ni ore deposits in the Noril'sk district, which are giant sulfide bodies (are 2 km wide, 4 km long, and up to 50 m thick and are related to relatively thin, 150–200 m, gabbro-dolerite intrusions), provide a unique possibility of studying these processes. The seemingly straightforward solution of this problem is that sulfur can be borrowed from the terrigenous–sedimentary rocks that host the ore-bearing massifs and that contain much anhydrite-bearing rocks (up to individual beds of pure anhydrite). This mineral was then viewed as a potentially important source of sulfur for the origin of the ores at the level of modern intrusive chambers. This suggestion is based on data of L.N. Grinenko, who established positive correlations between the sulfur contents of the intrusions and their sulfur isotopic composition (Grinenko 1985), which provided a basis for the hypothesis that the sulfides were formed in situ.

In spite of the seemingly obvious nature of this hypothesis, which is underlain only by the sulfur isotopic composition, the possible assimilation of the host rocks by magmas in the Noril'sk district has never been critically analyzed from any other viewpoints. Had the assimilation process occurred, it would have affected the petrography and geochemistry of the rocks, particularly in contact zones of the intrusions, and the average composition of the intrusions, which are hosted in rocks of various compositions. Indeed, in rare cases, we noted the changes in rock compositions of intrusive bodies. For example, in the 1 m endocontact zone of the Kharaelakh intrusion, quartz notites appear. Similar rocks can be observed in other intrusions.

Our study was focused on the inner structure and composition of contact zones between the intrusions and their host rocks and on elucidating the extent of the contamination during the crystallization of the melts in the modern intrusive chambers. However, the situation when the massifs are hosted in anhydrite-bearing sedimentary rocks is obviously of the greatest interest. An example of this setting is offered by the Kharaelakh intrusion and its uniquely large ore resources. This example was used to analyze other variations in the composition of the rocks from the outer to the inner contacts and farther inward of the intrusion. This publication is based on the results of this study and their discussion.

We performed detailed studies of the western part of the Kharaelakh and the central portion of the Talnakh intrusions and examined the host and intrusive rocks of an apophysis of the Kharaelakh Massif, which was penetrated by Borehole ZF-12 (which was drilled by Noril'skGeology Ltd. when this portion of the deposit was explored), contains rich sulfide ores, and is hosted in Devonian rocks. To monitor the lateral geochemical variability of the intrusive rocks, we compared our data with data on rocks recovered by a borehole (borehole OUG-2) drilled through the central part of the Talnakh intrusion. The samples selected when the core from the borehole was documented were analyzed for major components and trace elements and for ratios of radiogenic isotopes. The intrusive rocks were analyzed by XRF on an AXIOS Advanced (PANalytical) spectrometer at the Vernadsky Institute (analysts I.A. Roshchina and T.V. Romashova). The characteristic radiation of elements was excited by an X-ray tube with a Rh anode (tube loading up to 4 kW). For the concentration ranges typical of the rock samples, the relative errors (two standard errors calculated from the reproducibility of standard reference samples) were as follows (%): 1.2 for SiO_2; 3.5 for Al_2O_3; 6.2 FeO; 8.0 for Na_2O, MgO, P_2O_5, K_2O, CaO, and TiO_2; 14.0 for Cr_2O_3; and 17.0 for MnO. The rocks were analyzed for trace elements by ICP-MS at the Institute of Experimental Mineralogy, Russian Academy of Sciences, IEM (analyst V.K. Karandashev), in standard mode. The sensitivity of the spectrometer throughout the entire mass scale was calibrated against standard reference solutions that contained all elements for which our samples were analyzed. To control the quality of the analyses and account for the drift in the sensitivity of the spectrometer, the analyses of our samples were alternated with those of a monitor, which was a BCR-2 standard reference sample of basalt (GeoRem database at http://georem.mpch-mainz.gwdg.de/). The detection limits of elements were from 1–5 ppb for elements with heavy and intermediate masses (U, Th, REE, and others) to 20–50 ppb for light elements (Be and others). The analyses were accurate to 3–10 % for elemental concentrations >20–50 times their detection limits.

The chemical composition of sedimentary rocks was studied at the Institute of Mineralogy, Geochemistry, and Crystal Chemistry of Rare Elements (analyst B.I. Volkov). The Sr and Sm–Nd isotopic systems of the rocks were studied at the Vernadsky Institute (analyst A.A. Plechova and Yu. A. Kostitsyn) and at the Center for Isotopic Studies of the

Karpinsky All-Russia Research Institute of Geology (at which data on the Pb isotopic systems were also obtained by the analyst B.V. Belyatsky).

The Talnakh intrusion dips at gentle angles (4°–7°) northeastward and is generally conformable with the host rocks. Its northern and southern parts are hosted in tuff–lava units and sedimentary rocks, respectively. In the context of our research, the most interesting setting of the intrusion with massive ores was that in anhydrite-bearing sedimentary rocks, as is typical of the Kharaelakh intrusion.

8.1 Inner Structure of the Western Part of the Kharaelakh Intrusion

According to the data of Noril'skGeology Ltd., the southwestern part of the intrusion (Figs. 8.1 and 8.2) splits into relatively thin branches, whose composition is often remarkably different: some of them consist only of highly magnesian rocks (so-called picritic gabbro-dolerites), whereas others are made up of leucogabbro. This portion of the intrusion contains practically no clearly differentiated bodies; nevertheless, vein sulfide ores were found there. The thickness of intrusive rocks in this part of the ore field exceeds 100 m, and they trend for 1.5 km. These rocks are subconformable with the stratification of the host Devonian sedimentary rocks. The intrusive bodies penetrated by the lower part of borehole ZF-12 (Figs. 8.2 and 8.3) are hosted in Devonian anhydrite–carbonate terrigenous rocks, which are usually thought to be the main sulfur source for the sulfide ores. This body is interesting because of the striking inconsistency between the volumes of its silicate and sulfide constituents: the thickness of the gabbro-dolerite is 50 m, and the thickness of the related massive ore is 11 m, i.e., 20 % of the vertical section is composed of sulfides. Moreover, the body is weakly differentiated, which is atypical of the central portions of the ore-bearing intrusions, whose rocks vary from ultramafic to mafic (from below upward). The rocks of this body cannot be subdivided in such a series.

The intrusive body is dominated by medium-to-coarse-grained olivine gabbro-dolerites, which commonly have a poikilophitic texture, with plagioclase laths cemented by clinopyroxene (Fig. 8.4b). The rocks sometimes have typical doleritic textures (Fig. 8.4c, d). The contact varieties found in the upper and lower contacts of the massif are finer grained and are often porphyritic (Fig. 8.4a). The groundmass has a doleritic or hypidiomorphic granular texture.

The major rock-forming minerals are plagioclase, clinopyroxene, and olivine; the minor minerals are titanomagnetite, Cr-magnetite, and sulfides. The olivine (or, sometimes, olivine-bearing) gabbro-dolerites at lower levels sometimes contain biotite. The rare minerals are apatite and zircon. The rocks show evidence of their secondary alterations. This pertains, first of all, to plagioclase, which is significantly (sometimes completely) replaced by saussurite; the olivine is replaced by serpentine and bowlingite. The only mineral that is practically unaltered in all rock varieties is clinopyroxene.

Table 8.1 summarizes data on the chemistries of silicate minerals in the intrusive rocks penetrated by the lower part of borehole ZF-12. Clinopyroxene in the predominant rock variety (olivine-bearing gabbro-dolerites) is more magnesian than in the contact gabbro-dolerites. The Mg# of clinopyroxene varies from 77.2 to 83.5 (Table 8.1, nos. 1–7), and such

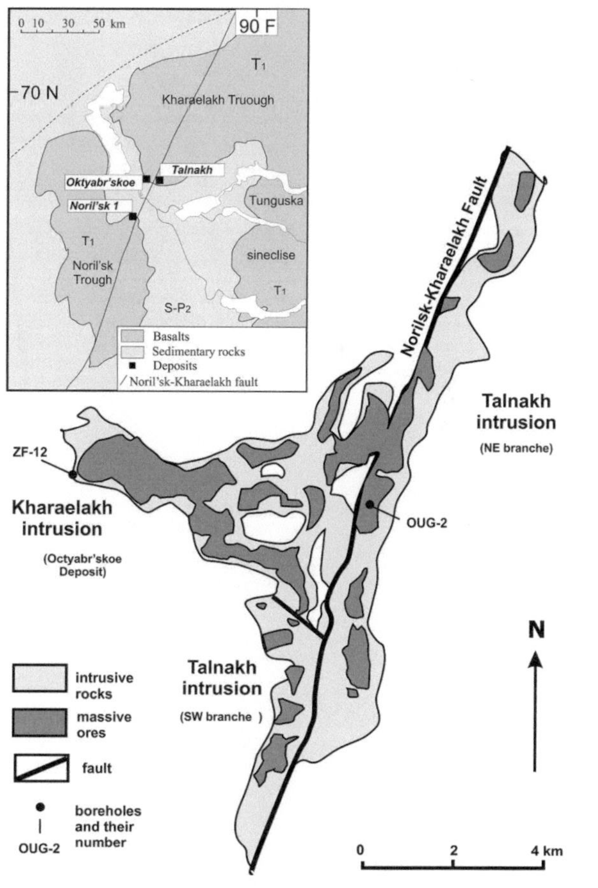

Fig. 8.1 Projections of the Oktyabr'skoe and Talnakh deposits onto a horizontal plane
Here and Figs. 8.2, 8.3, 8.4, 8.5, 8.6, 8.7, 8.8 after Krivolutskaya et al. 2014

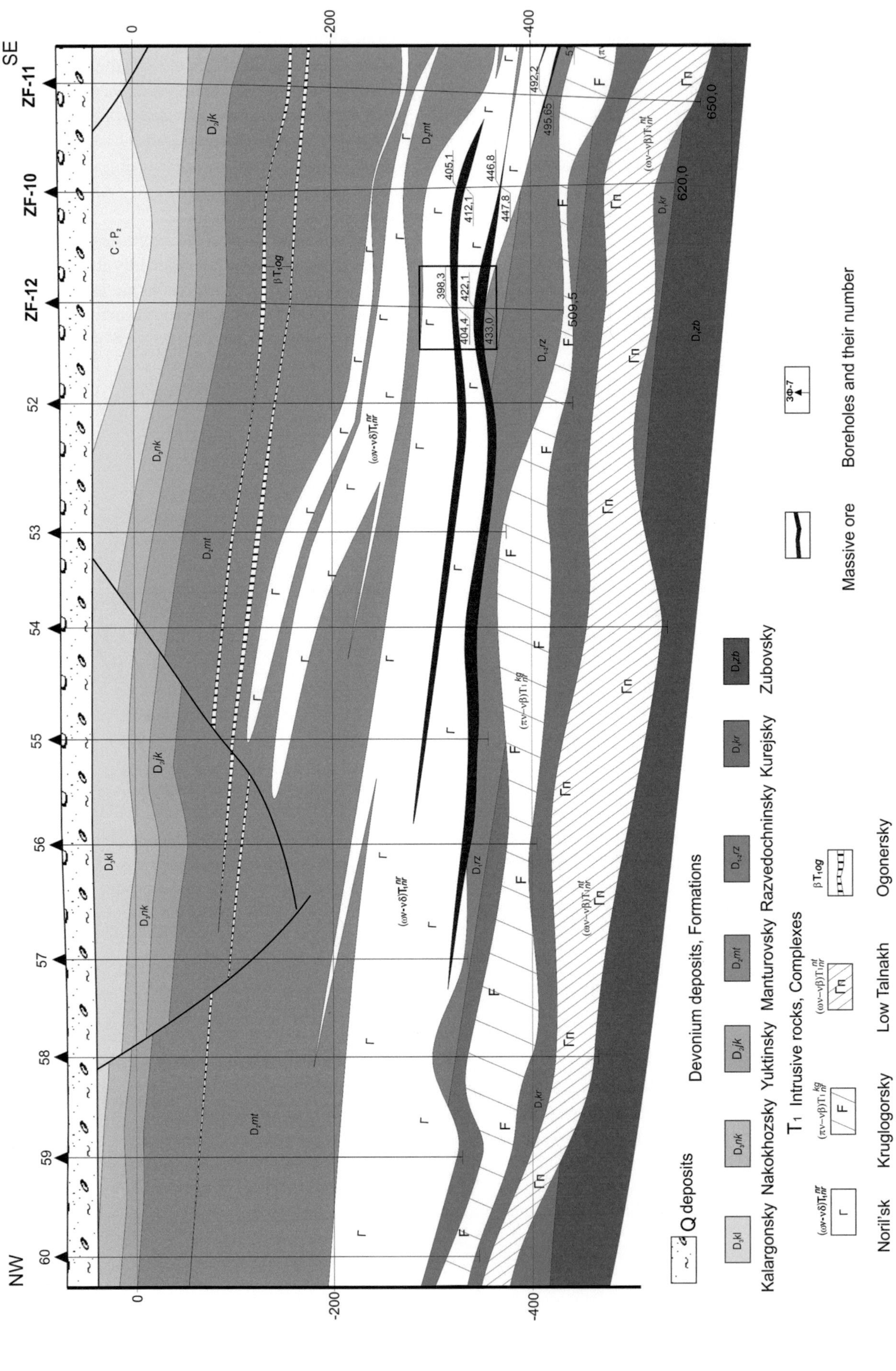

Fig. 8.2 Vertical geological section through the western part of the Kharaelakh intrusion (prepared by I.A. Matveev based on data from Noril'skGeology Ltd

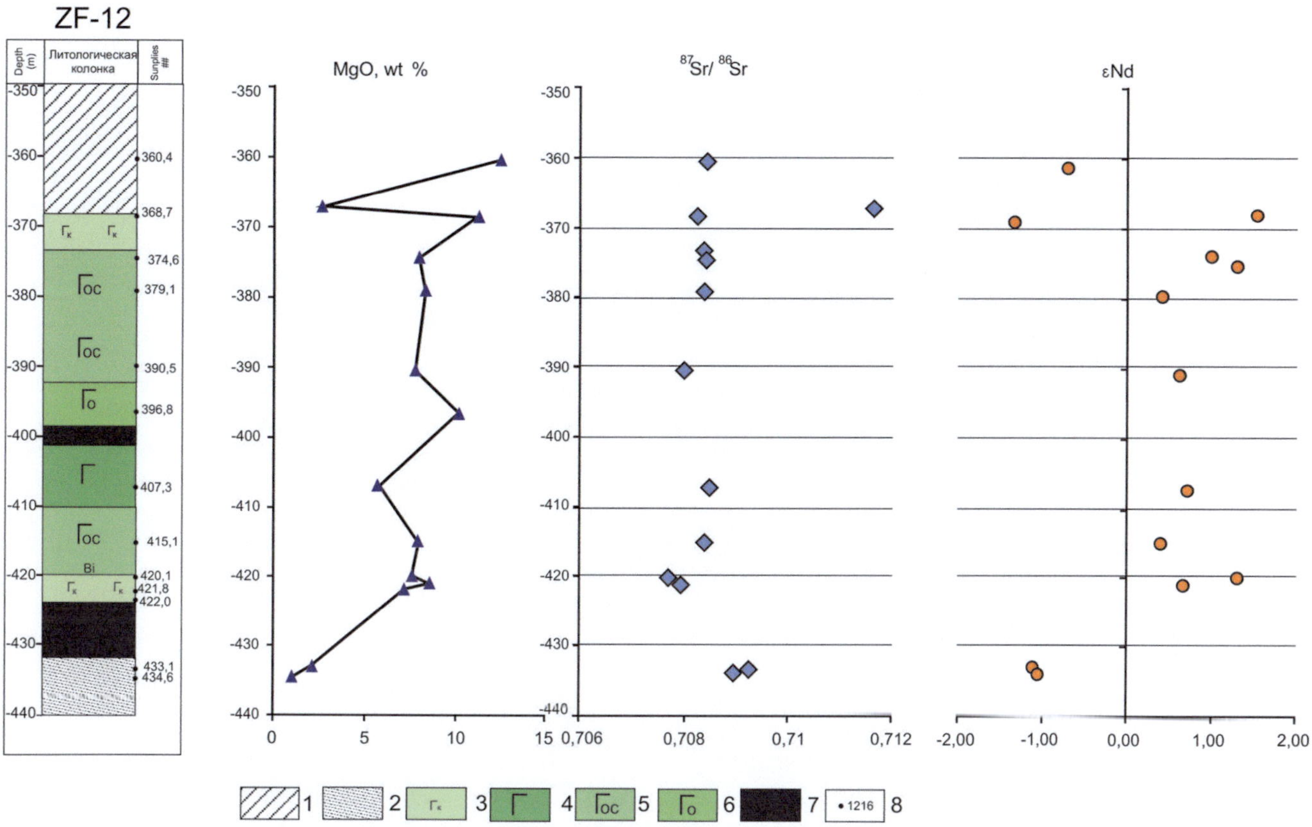

Fig. 8.3 Inner structure of an apophyse of the Kharaelakh intrusion (based on materials recovered by borehole ZF-12) and variations in the MgO concentration, $^{87}Sr/^{86}Sr$ isotopic ratio, and ε_{Nd} value
(1, 2) Devonian rocks: (1) anhydrite-bearing carbonate rocks, (2) predominantly terrigenous rocks. (3–5) gabbro-dolerites (3) contact, (4) olivine-free, (5) olivine-bearing (including biotite-bearing)), (6) olivine, (7) massive sulfide ores, (8) sampling sites and numbers

variations can quite often be detected within a single grain, with Mg# decreasing from the grain core to its peripheries. In the contact gabbro-dolerites, clinopyroxene of analogous composition (Mg#=81.2) was found in phenocrysts (Fig. 8.4a; Table 8.1, no. 18). This mineral typically contains a large amount of Al_2O_3 and TiO_2 (3.81 wt % and 0.5 wt %, respectively). The groundmass clinopyroxene of these rocks has little variation in composition within the individual grains and in the sample as a whole (Mg#=70.2–75.5). The situation with the plagioclase is analogous, and this mineral is much more calcic in the olivine-bearing gabbro-dolerites (An=82–83 mol % on average, although rare grains can be more sodic; see Table 8.1, nos. 8–13), whereas plagioclase in the contact rocks is more sodic (An=40–60 mol %; Table 8.1, nos. 24–30). Olivine in the olivine-bearing gabbro-dolerites varies in composition from 67.3 to 73.4 mol % forsterite. The olivine is poor in Ni and relatively rich in Ca (compared to olivine in the picritic gabbro-dolerites).

The host rocks recovered by the borehole from the upper and lower contact zones of the intrusion and belonging to the Devonian Manturovsky Formation have different compositions. The upper-contact varieties are terrigenous-carbonate rocks containing up to 25 vol.% sulfate and up to 10 % carbonate material (sample ZF-12/360, Table 8.2). Anhydrite was found in the rocks in the form of (0.5–1.0 mm) veinlets and single grains in the groundmass (Figs. 8.4e and 8.5). Serpentine hornfels is rare among the upper outer-contact rocks (sample ZF-12/368.7, Fig. 8.4f) but is predominant in the lower outer-contact zone and replaces mudstone and sandstone containing minor amounts of carbonate material (Table 8.2, Fig. 8.4g, h).

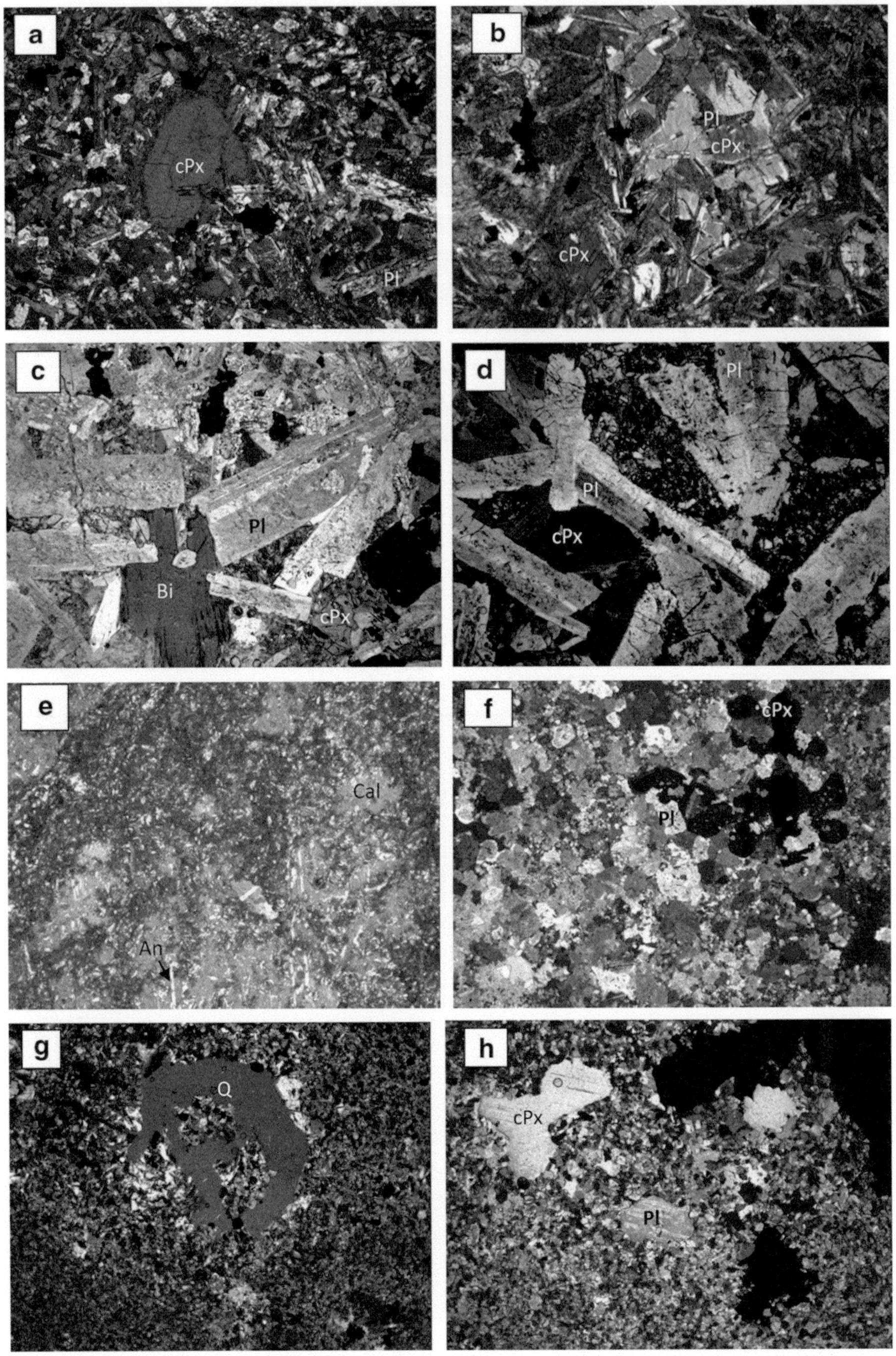

Fig. 8.4 Rock micrographs (see *left-hand* column in Fig. 8.3 for rock position)
(**a**) Porphyritic contact gabbro-dolerite, clinopyroxene phenocrysts in a microdoleritic groundmass (sample ZF-12/421.8); (**b**) poikilophitic olivine-bearing gabbro-dolerite (sample ZF-12/390.5); (**c**) coarse crystalline biotite-bearing gabbro-dolerite of dolerite structure (sample ZF-12/420.1); (**d**) coarse crystalline gabbro-dolerite of dolerite structure (sample ZF-12/407.3); (**e**) microcrystalline anhydrite–carbonate rock (sample ZF-12/360.4); (**f**) hornfelsic pyroxene–plagioclase rock (sample ZF-12/368.7); (**g**) porphyritic pyroxene–plagioclase hornfels replacing silty sandstone (quartz grain at the center; sample ZF-12/433.1); (**h**) inequigranular hornfelsized silty sandstone (sample ZF-12/434.6). The *lower* side of the micrographs is 0.8 mm. Minerals: *Pl*—plagioclase, *Cpx*—clinopyroxene, *Bt*—biotite, *Qz*—quartz, *Cal*—calcite, *An*—anhydrite

Table 8.1 Composition of rock-forming minerals from Kharaelakh intrusion (borehole ZF-12), wt %

No	No обр.	SiO_2	Na_2O	Al_2O_3	CaO	K_2O	Cr_2O_3	TiO_2	FeO	MnO	MgO	NiO	Total	Mg#	An mol. %	Fo mol. %
1.	390.5	53.85	0.28	1.27	21.65	0.00	0.02	0.19	8.09	0.38	15.36		101.10	77.20		
2.	390.5	51.80	0.34	3.09	21.24	0.00	0.04	0.57	7.17	0.16	15.91		100.32	79.82		
3.	390.5	52.45	0.28	2.83	19.99	0.00	0.43	0.40	6.52	0.17	17.28		100.36	82.53		
4.	390.5	52.11	0.31	3.54	21.75	0.00	0.41	0.46	6.21	0.14	15.92		100.85	82.05		
5.	390.5	51.77	0.31	3.70	21.89	0.00	0.65	0.45	5.67	0.12	16.07		100.63	83.48		
6.	390.5	51.23	0.32	3.92	21.04	0.00	0.49	0.62	6.94	0.16	15.67		100.39	80.10		
7.	390.5	51.82	0.30	3.55	21.86	0.00	0.56	0.43	6.60	0.27	15.19		100.59	80.40		
8.	390.5	49.43	1.75	31.35	16.00	0.07			0.68		0.12		99.40		83.48	
9.	390.5	48.87	1.49	32.20	16.94	0.03			0.53		0.10		100.16		86.25	
10.	390.5	52.68	3.62	30.73	11.84	0.27			0.70		0.09		99.93		64.44	
11.	390.5	48.97	1.20	32.83	17.42	0.04			0.59		0.11		101.16		88.95	
12.	390.5	48.24	1.66	31.86	16.60	0.05			0.63		0.07		99.10		84.72	
13.	390.5	48.23	1.66	31.87	16.44	0.06			0.71		0.08		99.05		84.53	
14.	390.5	38.20		0.01	0.25		0.00	0.01	24.54	0.47	38.14	0.10	101.62			73.48
15.	390.5	38.20		0.01	0.25		0.00	0.01	24.98	0.50	37.83	0.09	101.77			72.97
16.	390.5	37.26		0.01	0.16		0.00	0.03	28.08	0.49	34.33	0.10	100.36			68.55
17.	390.5	37.11		0.01	0.17		0.00	0.02	28.93	0.51	33.37	0.09	100.12			67.28
18.	421.8	52.21	0.23	3.81	20.27	0.00	1.07	0.50	6.51	0.18	15.81		100.59	81.24		
19.	421.8	51.94	0.25	1.47	19.48	0.02	0.06	0.53	10.85	0.32	14.50		99.42	70.44		
20.	421.8	51.23	0.79	4.06	18.74	0.12	0.03	0.56	9.01	0.13	15.61		100.28	75.54		
21.	421.8	52.03	0.50	1.76	16.35	0.17	0.28	0.34	11.91	0.28	15.70		99.32	70.15		
22.	421.8	52.13	0.38	1.74	20.80	0.03	0.23	0.47	9.54	0.37	13.38		99.07	71.43		
23.	421.8	52.55	0.30	1.04	20.24	0.00	0.09	0.36	10.65	0.38	14.27		99.88	70.49		
24.	421.8	50.96	3.69	28.22	11.86	0.65	0.07	0.03	0.97	0.05	0.06		96.56		64.01	
25.	421.8	47.42	2.25	30.45	15.70	0.11	0.00	0.02	0.94	0.03	0.05		96.97		79.43	
26.	421.8	49.15	2.43	30.20	11.74	1.77	0.00	0.04	0.88	0.08	0.10		96.39		72.78	
27.	421.8	51.06	3.88	22.57	2.86	4.59	1.68		5.43	0.30	0.78		93.12		28.97	
28.	421.8	57.81	5.98	24.34	7.66	0.68		0.02	0.52		0.04		97.03		41.48	
29.	421.8	58.76	6.04	24.53	7.32	0.70	0.02	0.06	0.44	0.00	0.03		97.90		40.14	
30.	421.8	52.65	3.94	27.90	12.48	0.30	0.02	0.06	0.96	0.04	0.09		98.44		63.67	

Note: (1) № sample is depth for borehole ZF-12 (Fig. 8.3). 390.5, *Ol*-bearing gabbro-dolerites; 421.8, contact gabbro-dolerites. (2) Mg# = (MgO/(MgO + FeO)*100)) for pyroxene, *An* anorthite component in plagioclase (mol %), *Fo* forsterite component in olivine (mol %), (3) Empty boxes mean element was not analyzed; 40 analyses were carried out at the Vernadsky Institute by analyst N. Kononkova. Tables 8.1, 8.2, 8.3, 8.4, 8.5, and 8.6 – After Krivolutskaya et al. (2014)

Table 8.2 Composition of host rocks for western apophyse of the Kharaelakh intrusion (borehole ZF-12), wt %

No	1	2	3	4	5	6
Depth, m	360.4	367.2	368.2	373.2	433.1	434.6
SiO_2	29.7	57.39	22.96	43.04	62.66	67.66
TiO_2	0.51	0.83	0.03	0.89	1.26	1.00
Al_2O_3	8.29	20.13	6.86	16.26	16.55	12.08
Fe_2O_3	1.16	1.84	1.65	3.20	0.44	0.30
FeO	3.97	4.80	1.49	6.14	0.61	0.54
MnO	0.15	0.02	0.08	0.14	0.03	0.02
MgO	12.49	2.64	11.28	10.11	2.11	0.99
CaO	21.93	1.39	26.42	13.34	3.44	4.97
Na_2O	0.20	3.51	0.11	1.82	6.18	3.52
K_2O	0.09	2.78	0.01	0.51	0.38	0.09
P_2O_5	0.01	0.15	<0.01	0.10	0.09	0.33
SO_3	11.43	1.38	14.14	1.35	2.90	4.63
H_2O^+	7.72	2.49	7.23	2.32	2.62	3.15
H_2O^-	0.42	0.16	0.47	0.14	0.12	0.12
CO_2	1.43	0.05	6.44	0.32	0.24	0.17
Total	99.50	99.56	99.46	99.68	99.63	99.57
Low	10.70	3.20	14.33	3.48	3.06	3.40

Note: Analyses were carried out at Institute of Mineralogy, Crystal Chemistry, and Geochemistry of Rare Elements by analyst B.I. Volkov

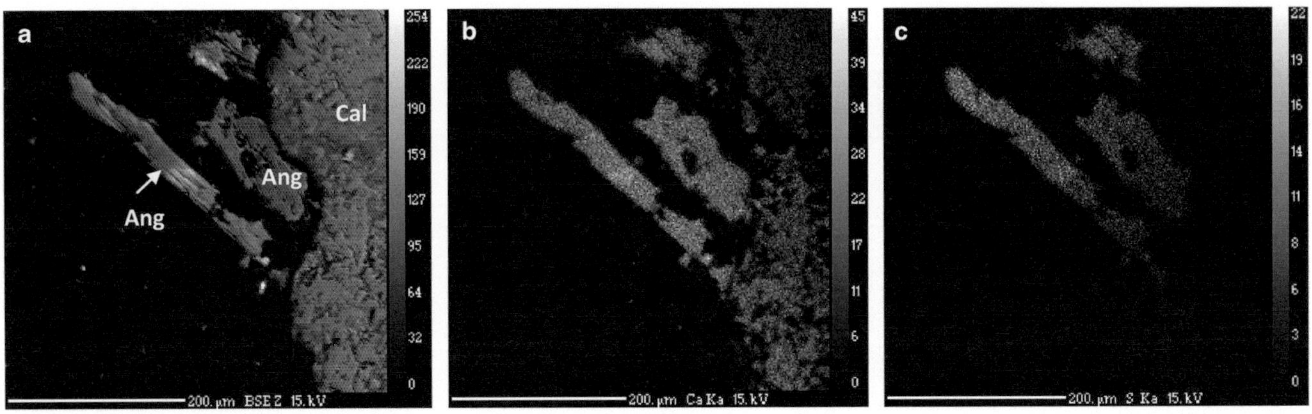

Fig. 8.5 Anhydrite (Ang) and calcite (Cal) grains in a carbonate–anhydrite rock
(**a**) BSE image; (**b**, **c**) images in characteristic X-ray radiation: (**b**) Ca *Kα*, (c) S *Kα*

8.1.1 Chemical Variability of the Rocks

The composition of the intrusive rocks only insignificantly varies in the vertical section (Table 8.3) in both major oxides and trace elements. Figure 8.3 displays the variations in the concentration of MgO, one of the most informative major oxides in both sedimentary and intrusive rocks, which have contrasting concentrations of this component. The MgO concentration of the intrusive rocks varies from 5.9 to 9.3 wt %, with the lowest concentrations typical of sulfide-enriched rocks (Table 8.3). Note that the MgO concentration in the central portions of the Talnakh intrusion reaches up to 22.4 wt % in picritic gabbro-dolerites (Krivolutskaya et al. 2001), but these rocks are absent from these magmatic body. The CaO content in the gabbroids also varies insignificantly, from 10.0 to 13.6 wt %, whereas the host rocks contain up to 21.8 wt % MgO (Tables 8.2 and 8.3). Thus, the contact zones are not enriched in CaO, as could be expected with regard for assimilation, but the concentration of this component drastically increases with the transition from the sedimentary to intrusive rocks.

The concentrations of certain elements normalized to the primitive mantle are shown as normalized patterns in Fig. 8.6 (Table 8.4). The patterns of the intrusive rocks (Fig. 8.6a) are practically identical and only insignificantly differ in the concentrations of the elements, which reflects the weak fractionation of the melt and the corresponding crystallization of rocks having various Mg#. The patterns of the surrounding rocks from both the lower- and the upper outer-contact zones

Table 8.3 Composition of the rocks for Kharaelakh intrusion (borehole ZF-12), wt %

Oxide\depth, m	374.6	379.1	390.5	396.8	407.3	415.1	420.0	421.1	422.0
SiO_2	49.47	47.60	19.97	39.12	48.40	47.60	46.41	50.00	45.71
TiO_2	1.12	1.03	0.08	0.50	1.61	1.02	1.21	0.84	1.05
Al_2O_3	14.54	14.52	5.93	8.39	14.73	15.43	14.94	15.31	13.52
Fe_2O_3	11.74	12.01	50.94	15.68	14.30	11.02	13.25	10.47	18.05
MnO	0.17	0.16	0.15	0.13	0.23	0.17	0.18	0.17	0.24
MgO	7.93	8.32	7.75	10.17	5.88	7.94	7.61	8.59	7.16
CaO	11.14	12.08	3.89	20.19	10.01	12.67	12.43	11.33	10.18
Na_2O	2.82	2.36	0.12	0.06	2.71	2.68	2.23	1.85	2.34
K_2O	0.68	0.95	0.19	0.02	1.75	0.99	0.88	1.22	0.33
P_2O_5	0.12	0.13	0.07	0.16	0.19	0.11	0.12	0.10	0.14
S_{tot}	0.27	0.85	10.91	5.58	0.20	0.36	0.74	0.11	1.28

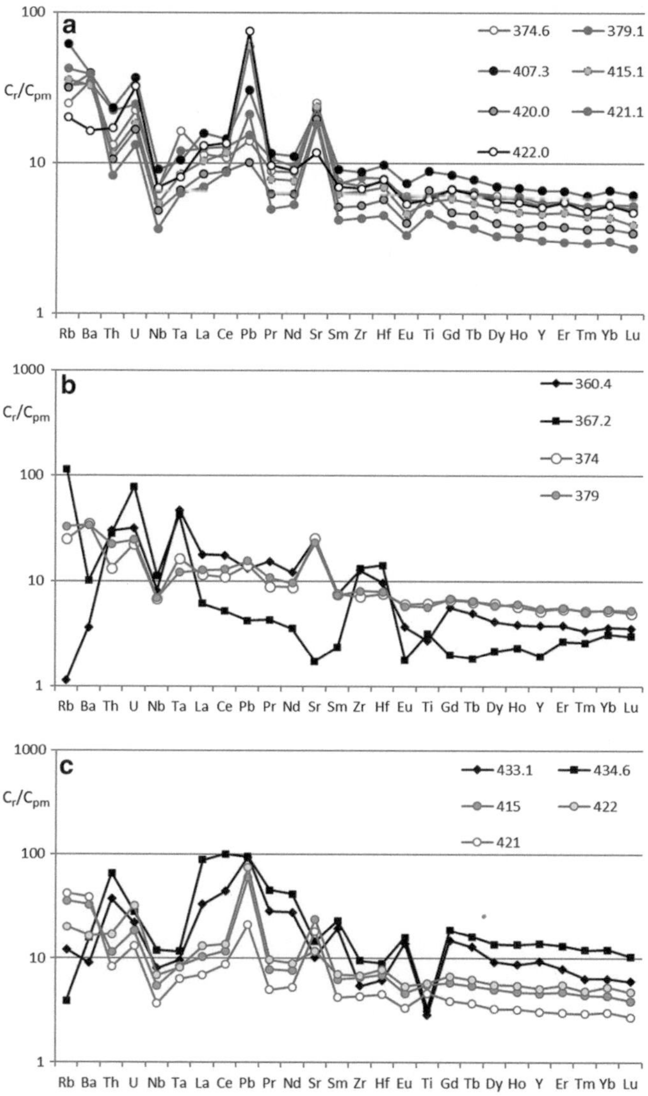

Fig. 8.6 Primitive mantle-normalized trace element patterns of rocks recovered by borehole ZF-12.
Intrusive rocks; (**a**) rocks from the central part of massif; (**b**) rocks from the upper outer-contact zone; (**c**) rocks from the lower outer-contact zone

are notably different from the patterns of the intrusive rocks: these are strongly enriched in U and Pb and depleted in Sr and Ti, as can be clearly seen in the normalized patterns of the rocks, which show the corresponding negative anomalies. It is important to emphasize that no patterns of intermediate types that should have corresponded to the suggested mixing of compositionally different components (sedimentary rocks and basaltic melt) have ever been detected in rocks of either of the upper parts of the intrusion.

This result also follows from the contrasting variations in certain characteristic elemental ratios that are typical of major rock types (sedimentary and magmatic), for example, the La/Sm ratio.

8.1.2 Distribution of Radiogenic Isotopes in the Rocks

Figure 8.3 and Table 8.5 illustrate major trends in the behavior of Sr, Nd, and Pb isotopes in the vertical section of the intrusive body. The $^{87}Sr/^{86}Sr$ ratio of the gabbro-dolerites varies from 0.70769 to 0.70847 (Table 8.5) and drastically changes with the transition from the intrusive to host rocks, first and foremost, in the roof of the massif, and reaches a value of 0.71163. The lower outer-contact rocks have lower values of this ratio: 0.70922. Similar to the major components, this ratio drastically changes, and intermediate values are absent. The situation with Sr is analogous to that with ε_{Nd} (Fig. 8.3), which varies relatively little around zero in the central part and decreases to negative values in the host rocks.

Data on the Pb isotopic composition were not obtained for all of our samples in which the Sr and Sm–Nd isotopic systems were studied. Nevertheless, these data also confirm the results obtained by other isotopic methods for most of the vertical section. Particularly interesting rocks occur in the lower outer-contact zone because these rocks have elevated Pb isotopic ratios: $^{206}Pb/^{204}Pb = 20.240$ and $^{208}Pb/^{206}Pb = 40.907$ (Table 8.5). We found radiogenic Pb in sedimentary anhydrite form the Noril'sk district (Krivolutskaya 2011).

Table 8.4 Concentration of rare elements in rocks from borehole ZF-12, ppm

Depth, m	360.4	367.2	368.7	373.2	374.6	379.1	390.5	396.8	407.3	415.1	420.0	421.1	422.0	433.1	434.6
Element	1	2	3	4	5	6	7	8	9	10	11	12	13	14	15
Li	65.0	62.9	72.4	90.2	45.5	32.3	66.6	20.8	20.9	34.6	20.3	23.2	36.3	14.9	10.1
Be	0.75	1.6	0.85	0.71	0.46	0.75	0.03	2.6	0.66	0.46	0.48	0.33	0.93	2.9	1.7
Sc	8.4	18.7	2.5	28.9	32.1	32.9	7.8	8.9	54.7	40.7	35.2	34.4	33.3	5.5	17.9
V	55.0	152	27.7	224	267	249	22.0	186	398	225	223	173	255	54.1	91.2
Cr	42.1	89.6	33.1	340	330	305	412	35.1	94.8	435	390	504	431	120	175
Co	20.4	22.7	8.6	50.5	55.1	65.8	691	156	55.5	51.8	48.5	41.4	93.0	4.4	3.0
Ni	24.7	38.7	39.0	110	135	397	12,886	2,446	77.3	262	401	204	1,868	115	78.9
Cu	733	13.3	87.8	106	135	804	15,805	6,437	196	2,036	470	357	2,572	2,040	222
Zn	130	16.5	67.3	98.4	85.9	88.5	194	82.7	134	179	94.8	118	143	81.6	106
Ga	11.4	20.5	4.6	17.2	14.7	14.6	4.1	12.7	19.8	14.8	11.4	10.8	16.4	17.9	10.0
Rb	0.62	61.8	0.42	12.1	13.3	17.6	3.4	1.0	33.4	19.2	17.0	22.8	10.8	6.5	2.1
Sr	454	31.5	212	430	461	418	66.9	118	420	434	358	332	214	186	265
Y	15.0	7.7	18.5	21.4	20.7	21.6	1.6	32.5	26.2	18.3	15.4	12.2	20.3	37.0	55.1
Zr	123	129	9.4	148.1	69.3	78.5	6.6	126	85.1	63.8	51.1	42.0	66.3	53.1	92.6
Nb	5.06	6.88	2.97	5.88	4.18	4.27	0.64	7.31	5.62	3.35	3.01	2.28	4.28	4.95	7.42
Mo	0.64	1.71	0.63	1.17	1.90	1.92	0.58	0.86	1.04	3.24	1.45	1.42	1.55	2.43	5.08
Sn*	2.04	1.15	0.45	1.01	0.86	1.47	2.69	4.09	0.96	1.04	0.67	0.97	1.97	0.91	2.78
Cs	0.03	0.45	0.04	0.57	1.23	0.30	0.44	0.16	0.21	0.32	0.24	0.34	0.45	0.09	0.06
Ba	22.1	61.0	4.6	208	212	206	51.3	3.1	237	200	240	237	99.4	54.8	95.7
La	10.9	3.8	5.8	6.8	7.0	7.8	0.67	20.7	9.7	6.4	5.2	4.3	8.0	20.5	54.5
Ce	28.0	8.3	17.0	18.0	17.5	20.6	1.7	57.4	23.2	18.9	14.3	14.0	21.9	70.9	162
Pr	3.70	1.04	2.51	2.52	2.14	2.60	0.20	6.02	2.81	1.90	1.52	1.21	2.35	6.89	11.05
Nd	14.38	4.24	10.47	10.96	10.44	11.40	0.89	24.13	13.25	9.12	7.44	6.32	10.71	33.28	49.74
Sm	2.94	0.92	2.38	3.04	2.91	2.84	0.23	5.02	3.51	2.44	1.98	1.63	2.70	7.48	8.97
Eu	0.54	0.26	0.26	0.73	0.88	0.84	0.07	0.97	1.07	0.68	0.59	0.49	0.79	2.03	2.33
Gd	2.90	1.02	2.83	3.30	3.43	3.48	0.28	5.75	4.32	2.99	2.42	2.01	3.46	7.64	9.62
Tb	0.46	0.17	0.47	0.58	0.60	0.61	0.05	0.98	0.73	0.51	0.43	0.35	0.58	1.22	1.53

Depth, m	360.4	367.2	368.7	373.2	374.6	379.1	390.5	396.8	407.3	415.1	420.0	421.1	422.0	433.1	434.6
Element	1	2	3	4	5	6	7	8	9	10	11	12	13	14	15
Dy	2.62	1.39	3.08	3.70	3.92	3.71	0.29	6.26	4.52	3.21	2.55	2.10	3.58	5.94	8.75
Ho	0.55	0.33	0.64	0.82	0.82	0.86	0.06	1.34	0.98	0.68	0.54	0.46	0.78	1.25	1.93
Er	1.58	1.12	1.81	2.23	2.29	2.34	0.17	3.81	2.75	1.98	1.58	1.26	2.32	3.32	5.51
Tm	0.22	0.17	0.26	0.31	0.34	0.33	0.03	0.53	0.39	0.29	0.24	0.19	0.31	0.41	0.78
Yb	1.51	1.30	1.77	2.23	2.19	2.23	0.17	3.42	2.76	1.82	1.53	1.26	2.21	2.66	5.06
Lu	0.23	0.19	0.25	0.32	0.32	0.34	0.03	0.49	0.40	0.25	0.22	0.18	0.30	0.38	0.66
Hf	2.58	3.79	0.38	3.47	2.03	2.12	0.21	2.64	2.51	1.86	1.55	1.21	2.10	1.65	2.39
Ta	1.65	1.49	0.38	1.14	0.58	0.42	0.04	0.66	0.37	0.30	0.23	0.22	0.29	0.34	0.41
W	0.21	0.86	0.09	0.26	0.27	0.37	0.07	0.26	0.48	0.58	0.24	0.23	0.55	0.65	1.08
Pb	2.30	0.74	14.13	5.63	2.47	2.70	12.19	2.97	5.38	10.54	1.77	3.71	3.20	16.48	1.68
Th	2.44	2.28	1.16	1.61	1.08	1.82	0.08	7.48	1.88	0.94	0.86	0.68	1.39	3.05	5.36
U	0.64	1.60	0.15	0.46	0.45	0.50	0.07	0.57	0.75	0.37	0.34	0.27	0.66	0.45	0.58

Note: Depth is place of sample in borehole ZF-12; analyses were carried out at IEM by analyst V. Karandashev

Table 8.5 Ratios of radiogenic isotopes in rocks of the Kharaelakh intrusion (borehole ZF-12)

Depth, m	Sm, ppm	Nd, ppm	$^{147}Sm/^{144}Nd$	$^{143}Nd/^{144}Nd$	±2sigma, abs	Rb, ppm	Sr, ppm	$^{87}Rb/^{86}Sr$	$^{87}Sr/^{86}Sr$	±2sigma, abs	$(^{87}Sr/^{86}Sr)t$	ε Nd	$^{206}Pb/^{204}Pb$	$^{207}Pb/^{204}Pb$	$^{208}Pb/^{204}Pb$
360.4*	3.402	16.50	0.12461	0.512484	0.000003	2.45	477	0.0148	0.708486	0.000009	0.7084334	–0.70	18,556	15,636	38,663
367.2	2.12	10.29	0.12433	0.512597	0.000015	87.7	42.7	5.951	0.732883	0.000008	0.711633513	1.53			
368.2*	2.476	10.50	0.14247	0.512481	0.000005	2.55	221	0.0333	0.708355	0.000014	0.7082357	−1.33	18,322	15,551	38,053
373.2*	3.126	11.88	0.15898	0.512627	0.000003	18.54	475	0.1130	0.708751	0.000009	0.7083478	1.00	18,516	15,574	38,283
374.6	2.67	9.85	0.16389	0.512651	0.000009	17.18	517	0.0961	0.708747	0.000015	0.70840413	1.31			
379.1	3.13	12.11	0.15606	0.512593	0.000010	21.2	463	0.1323	0.708852	0.000011	0.70837939	0.42			
390.5*	0.273	1.057	0.15579	0.512603	0.000014	5.65	83	0.1977	0.708685	0.000018	0.7079787	0.63	18,197	15,539	37,926
407.3	3.60	13.59	0.16015	0.512615	0.000015	45.9	485	0.2738	0.709450	0.000007	0.70847217	0.72			
415.1	2.54	9.93	0.15482	0.512590	0.000008	25.4	465	0.1581	0.708932	0.000010	0.70836744	0.41			
420.1*	2.972	11.33	0.15858	0.512643	0.000006	23.08	372	0.1798	0.708333	0.000006	0.7076905	1.31	18,745	15,583	38,389
421.1	2.32	9.30	0.15087	0.512597	0.000006	27.9	371	0.2174	0.708704	0.000008	0.70792784	0.68			
433.1	8.91	41.5	0.12988	0.512472	0.000008	11.62	277	0.1213	0.709657	0.000010	0.70922432	−1.10			
434.6*	10.50	56.20	0.11296	0.512446	0.000003	3.78	370	0.0296	0.709046	0.000014	0.7089408	−1.05	20,240	15,721	40,907

Note: Analyses were carried out in GEOKHI RAS by analysts A. Plechova* and at VSEGEI by B. Belyatsky, using (2 sigma); $t=251$ Ma

8.2 Upper Zone of the Talnakh Intrusion

A strongly differentiated intrusive body 155 m in total thickness was studied in the central part of the Talnakh, using materials of borehole OUG-2 (Table 8.6, Fig. 8.7). The petrography and mineralogy of the rocks are described elsewhere (Chap. 4, this study; Dodin and Batuev 1971; Likhachev 1996b; Krivolutskaya et al. 2001; Turovtsev 2002; Sluzhenikin et al. 2014). All three of the aforementioned series occur in the vertical section of the body: (1) upper gabbro, (2) main layered, and (3) lower gabbro. The bottom portion of the intrusion contains thick (up to 40 m bodies of massive

Table 8.6 Composition of rocks from the Talnakh intrusion, Borehole OUG-2

No sample	1,085	1,097	1,116	1,130	1,151	1,173	1,181	1,186	1,191	1,200
No	1	2	3	4	5	6	7	8	9	10
SiO_2	50.25	49.48	48.24	48.11	49.96	49.48	46.57	49.78	49.56	46.69
TiO_2	1.37	1.98	0.63	1.47	1.26	0.88	0.81	0.89	1.42	0.72
Al_2O_3	14.66	14.13	16.28	13.45	14.16	16.71	17.32	19.42	13.35	17.32
FeO	11.7	14.04	9.82	13.96	12.12	9.14	9.8	8.10	12.34	9.56
MnO	0.24	0.26	0.18	0.24	0.21	0.19	0.15	0.17	0.39	0.16
MgO	6.4	5.3	11.73	6.28	7.29	7.27	8.87	6.39	6.73	10.05
CaO	10.82	10.03	10.97	10.1	12.19	13.20	10.98	12.42	11.62	10.84
Na_2O	1.83	1.00	1.53	2.32	2.22	2.50	1.98	2.45	2.84	1.81
K_2O	1.25	0.81	0.35	0.83	0.50	0.54	0.49	0.53	1.06	0.46
P_2O_5	0.11	0.12	0.07	0.25	0.13	0.10	0.14	0.09	0.17	0.15
Cr_2O_3	0.04	0.01	0.08	0.00	0.00	0.08	0.07	0.07	0.05	0.09
Low	1.67	1.78	0.87	0.9	0.13	0.19	2.14	0.18	0.77	2.19
Total	100.34	98.95	100.76	97.91	100.18	100.28	99.32	100.49	100.31	100.04
S	0.28	0.52	0.91	0.2	0.08	0.06	0.07	0.04	0.69	0.05
Rb	40.8	l.d.	12.6	23.4	17.9	16.0	14.7	13.2	37.9	3.4
Ba	505	n.a.	181	219	147	169	154	185	217	130
Th	1.25	n.a.	1.03	1.36	1.58	0.40	0.82	1.11	0.82	0.69
U	0.56	n.a.	0.28	0.58	0.46	0.21	0.37	0.42	0.35	0.35
Nb	4.70	n.a.	2.38	4.94	4.22	2.36	2.78	3.74	4.90	2.85
Ta	0.33	n.a.	0.17	0.35	0.26	0.15	0.21	0.22	0.30	0.22
La	8.91	n.a.	3.8	9.87	7.5	5.4	5.41	7.1	8.1	4.87
Ce	20.74	n.a.	8.5	22.71	16.9	12.3	12.73	16.3	18.8	11.94
Pb	35.21	n.a.	35	86	80	44	80	359	2,410	44
Pr	2.80	n.a.	1.17	3.05	2.32	1.66	1.74	2.19	2.45	1.60
Nd	12.7	n.a.	5.54	13.9	11.19	7.87	7.9	10.04	11.73	7.3
Sr	411	n.a.	233	250	242	262	268	275	288	230
Sm	3.48	n.a.	1.53	3.74	3.11	2.19	2.12	2.67	3.19	1.95
Zr	92.5	n.a.	40	99	74	67	61	59	70	59
Hf	2.41	n.a.	1.13	2.71	1.97	1.67	1.70	1.67	1.81	1.61
Eu	1.18	n.a.	0.58	1.41	1.02	0.79	0.80	0.97	1.09	0.75
Ti	7,850	n.a.	3,869	8,832	7,610	5,249	4,807	5,472	8,608	4,562
Gd	4.10	n.a.	2.04	4.48	3.77	2.68	2.53	3.14	4.00	2.29
Tb	0.70	n.a.	0.34	0.76	0.65	0.43	0.44	0.52	0.69	0.39
Dy	4.56	n.a.	2.22	5.01	4.19	2.77	2.85	3.34	4.27	2.61
Ho	0.98	n.a.	0.47	1.07	0.87	0.56	0.62	0.70	0.90	0.55
Y	28.3	n.a.	11.2	30.1	21.7	14.2	17.0	17.3	22.6	14.5
Er	2.76	n.a.	1.28	3.02	2.44	1.57	1.73	1.91	2.45	1.56
Tm	0.41	n.a.	0.19	0.46	0.33	0.22	0.26	0.27	0.35	0.24
Yb	2.60	n.a.	1.33	2.85	2.36	1.53	1.63	1.85	2.52	1.52
Lu	0.40	n.a.	0.20	0.44	0.36	0.23	0.25	0.28	0.37	0.23
Ni	92.9	n.a.	1,603	56	136	275	254	415	148	287
Cu	59.0	n.a.	3072	158	399	716	134	1,970	4,047	91
Zn	70.5	n.a.	151	186	168	230	61	143	559	68
Mn	1,530	n.a.	1,436	1,674	1,794	1,500	1,017	1,687	3,323	1,096
Co	42.9	n.a.	105	49	59	83	47	61	117	52
V	301	n.a.	170	345	332	241	182	222	355	178

(continued)

Table 8.6 (continued)

No sample	1,203	1,211	1,216	1,221	1,222	1,225	1,227	1,230	1,231	1,234	1,259
No	11	12	13	14	15	16	17	18	19	20	21
SiO_2	48.15	40.73	36.16	32.97	48.02	38.09	44.98	47.34	18.4	22.42	56.63
TiO_2	0.90	0.54	0.82	0.36	2.23	0.29	0.59	0.54	0.17	0.27	0.57
Al_2O_3	16.25	8.03	9.32	6.97	12.00	12.76	16.43	14.01	6.65	6.28	13.53
FeO	10.77	17.55	18.15	21.55	17.71	16.27	10.97	10.75	33.35	32.11	16.08
MnO	0.26	0.28	0.29	0.23	0.65	0.25	0.14	0.15	0.13	0.07	0.25
MgO	9.48	22.57	15.33	17.89	5.72	12.69	7.62	6.64	0.95	1.28	1.97
CaO	11.55	6.09	6.63	4.17	9.68	7.18	12.71	9.22	5.08	4.98	4.71
Na_2O	1.98	0.85	0	0	2.52	0	0.16	1.94	0	0.39	4.08
K_2O	0.30	0.23	0.24	0.32	0.89	0.5	0.3	0.79	0.15	0.35	1.53
P_2O_5	0.09	0.07	0.10	0.04	0.21	0.02	0.02	0.14	0.00	0.04	0.04
Cr_2O_3	0.09	0.92	0.81	0.09	0.07	0.14	0.08	0.03	0.10	0.07	0.00
LOW	0.51	4.32	4.38	6.32	1.42	3.8	2.13	2.01	9.74	7.91	2.02
Total	100.3	102.1	92.24	90.91	101.1	91.99	96.14	93.56	74.74	76.19	101.4
S	0.49	4.26	0.52	6.56	0.03	4.96	2.68	4.85	29.95	28.65	1.01
Rb	8.9	9.7	9.6	12.7	38.5			6.4			35.0
Ba	159	97	121	84	441			170			1,911
Th	0.73	0.62	1.02	0.32	2.31			4.82			10.76
U	0.32	0.20	0.47	0.16	1.20			0.52			2.09
Nb	3.33	1.67	3.42	1.54	7.29			2.20			10.99
Ta	0.23	0.10	0.27	0.10	0.54			0.16			0.83
La	6.0	3.3	170.8	2.75	11.7			20.09			14.3
Ce	13.7	7.5	308	6.33	28.1			31.83			27.6
Pb	380	1,874	359	2,410	263			380			32
Pr	1.84	0.98	31.93	0.81	3.99			5.97			2.91
Nd	8.69	4.56	96.2	3.6	19.61			22.0			10.63
Sr	239	133	134	93	257			275			299
Sm	2.29	1.22	11.88	0.95	5.53			3.01			2.01
Zr	70	28	75	27	414			46			164
Hf	2.01	0.81	2.11	0.72	9.66			1.32			4.64
Eu	0.89	0.49	2.63	0.36	1.86			1.13			0.62
Ti	5,431	3,382	4,868	2,198	13,552			2,735			3,540
Gd	2.79	1.47	6.92	1.14	7.17			2.38			1.71
Tb	0.47	0.25	0.90	0.19	1.24			0.39			0.26
Dy	3.02	1.56	4.71	1.23	8.29			2.38			1.57
Ho	0.62	0.32	0.91	0.26	1.79			0.48			0.32
Y	15.4	8.1	30.8	7.7	43.7			16.1			8.2
Er	1.71	0.96	2.45	0.74	5.07			1.33			0.95
Tm	0.24	0.13	0.34	0.11	0.71			0.20			0.16
Yb	1.79	0.96	2.00	0.74	5.03			1.29			1.25
Lu	0.27	0.16	0.29	0.11	0.74			0.19			0.23
Ni	556	4,930	6,450	8,706	81			7,097			2,579
Cu	3,084	17,236	7,774	10,598	686			12,086			9,236
Zn	1,209	654	267	128	358			98			96
Mn	2,148	2,186	1,851	1,510	5480			1,018			1,877
Co	96	250	241	303	82			208			195
V	220	203	201	87	610			196			118

Note: Empty box means element was not analyzed; oxides are given in wt % and elements in ppm, l.d. - under det. limited, n.a. - not analysed

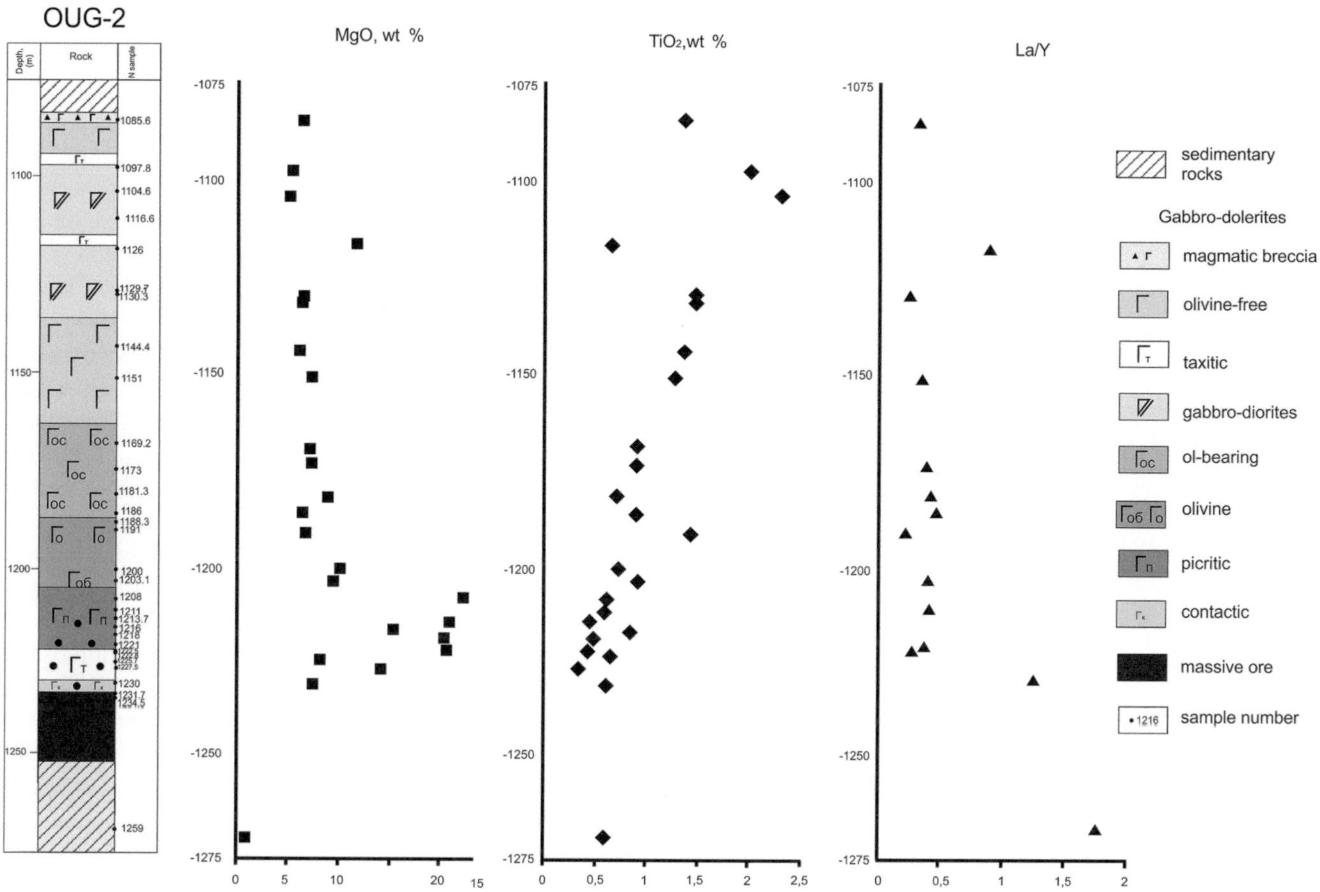

Fig. 8.7 Inner structure of the central portion of the Talnakh intrusion (based on materials recovered by Borehole OUG-2) and the distributions of the MgO and TiO_2 concentrations and the La/Y ratio in the vertical section of the intrusion

pyrrhotite–chalcopyrite ore of reticulate structure (large pyrrhotite segregations are surrounded by thin chalcopyrite rims), which are spatially restricted to the contact of the massif with underlying carboniferous terrigenous rocks of the Middle Carboniferous–Early Permian Tunguska Group.

This section is noted for the presence of a thick (up to 30 m) zone of gabbro-diorite composing the uppermost one-third of the massif. It underlies gabbro-dolerite and hybrid metasomatic rocks (2–3 m thick). The latter contain quartz, potassic feldspar, sodic plagioclase, biotite, and hornblende, along with fragments of quartzite and carboniferous rocks. This zone is quite often thought to be related to the overlying rocks. However, all geochemical parameters indicate that the gabbro-diorite was produced by the crystallization differentiation of the parental melt, which is obvious from the normalized patterns of trace elements: the topologies of all of these patterns are similar and close to those of the intrusive rocks (Fig. 8.8). Elevated concentrations of all trace elements are typical of the inner-contact rocks that occur at the boundary with the massive ores and are more magnesian (up to 5 wt % MgO).

The host rocks of the Tunguska Group notably differ from the inner-contact rocks: e.g., sample OUG-2/1259, whose rock is depleted in all REE and enriched in LILE (Table 8.6). As seen from Fig. 8.8, neither this body nor that described above contains any rocks whose normalized elemental composition patterns would be intermediate between those of the sedimentary and magmatic rocks, and this also pertains to the hybrid metasomatic rocks in the upper zone (sample OUG-2/1085).

As was demonstrated above with reference to the Talnakh intrusion, assimilation processes practically did not operate in the magmatic chambers. In considering the interactions of the melt with the surrounding rocks, the most interesting information can be derived from the behavior of trace elements, which are more susceptible than major components to compositional variations.

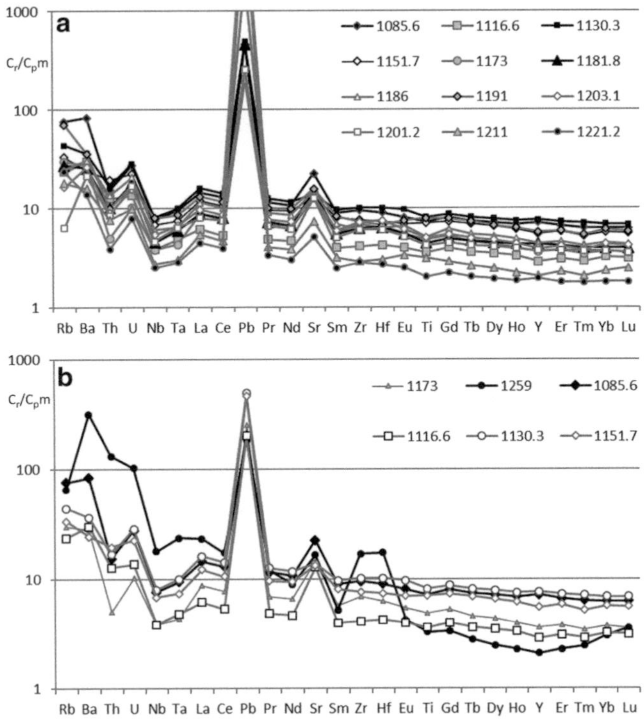

Fig. 8.8 Primitive mantle-normalized trace element patterns of rocks recovered by borehole OUG-2 from the central part of the Talnakh Massif
(**a**) Trace element patterns of rocks from the central portion of the vertical section; (**b**) patterns of inner-contact rocks of the intrusion and host rocks (*black*). The numbers of patterns correspond to depth (in m) down the hole

8.3 Contact Zones of the Maslovsky Intrusion

This behavior was demonstrated (Krivolutskaya and Rudakova 2009; Krivolutskaya et al. 2012) for the Maslovsky intrusion accompanied by high-grade ore mineralization. This characterization of the compositional variations was performed using the La/Sm ratio (borehole OM-25), which can vary very significantly and drastically with the transition from the outer-contact rocks (tholeiites of the lower portion of the Nadezhdinsky Formation; Fig. 8.9c, Table 8.7), which have higher La/Sm ratios (up to 8) than any other tholeiites in the tuff–lava unit of the Noril'sk district, to the inner-contact and inner parts of the intrusive body (which consist of gabbroids with the usual La/Sm ratios of 1–2). One can see the distribution of Cu and the change in the La/Sm ratio (the most contrasting elements and ratio) in basalts and gabbro. Rocks enriched by La/Sm rocks (and depleted in Cu) were found only in narrow contact zones (approximately 1 m).

In the central part of the Southern Maslovsky intrusion (OM-24), one can observe the rapid change of these parameters, demonstrating the absence of contamination (Fig. 8.9b).

Analogous conclusions can be drawn from the sulfur isotopic composition of sulfides at the Maslovsky deposit (Table 8.8). While ores at the Southern Maslovsky deposit have $\delta^{34}S = 10.8‰$ (Krivolutskaya et al. 2012), that of the host basalts of the Nadezhdinsky formation is no greater than 5‰ (Ripley et al. 2003). The composition of the anhydrite itself is in conflict with the hypothesis of its assimilation. This conflict is most clearly pronounced in the concentrations of radiogenic isotopes in the mineral, first, in the $^{206}Pb/^{204}Pb = 24$ (Krivolutskaya 2011) (Fig. 8.10, Table 8.9).

The idea that the massive ores could be formed by the assimilation of anhydrite-bearing rocks by the basite melt was formulated, first, with reference to the Oktyabr'skoe deposit, whose main orebody is hosted in Devonian evaporite-bearing rocks. However, the highest $\delta^{34}S$ values (up to +18‰) were obtained by L.N. Grinenko (1985) for sulfide ores of the Noril'sk 1 deposit, which are localized in the Middle Carboniferous–Early Permian terrigenous–sedimentary rocks of the Tunguska Group. L.N. Grinenko has also conducted material balance calculations with the aim of confirming that the melt assimilated carbonate–sulfate rocks and explaining the heavy sulfur isotopic composition of the sulfides. These calculations have not, however, confirmed that such a process could occur. This lack of confirmation led L. Grinenko to suggest that the melt could be enriched in hydrocarbons, which are abundant in the Vendian sedimentary rocks in East Siberia that host the deposit and contain oil and gas pools (Grinenko 1984, 1985). However, the mechanism of this enrichment was obscure because, as was demonstrated for oil pools in this area (Polyansky and Reverdatto 2006), hydrocarbons should migrate away from where intrusions were emplaced, and the composition of these hydrocarbons should thereby be modified. Moreover, the hydrocarbons themselves are very poor in sulfur (4–5 % on average; Yashchenko 2004; Polishchuk and Yashchenko 2005) and are dominated by methane and other gases, and the gases only rarely contain more than 10 % sulfur. Obviously, the uniquely large sulfide reserves should have required vast amounts of sulfur (and, hence, hydrocarbons); however, no such amounts of hydrocarbons occur in the area. Therefore, the source of the sulfur remains uncertain.

Nevertheless, the idea that the magmas assimilated anhydrite was explored by foreign geologists, starting in 1990. This hypothesis was promoted by several scientists (Lightfoot et al. 1993; Arndt et al. 2003; Li et al. 2003; Keays and Lightfoot 2007, 2010). As indisputable proof of the assimilation of anhydrite (from the host rocks) by the magma, the fact is presented that anhydrite was found as a rock-forming mineral in the intrusive rocks from Talnakh (Li et al. 2009b), and it was hypothesized that this mineral crystallized from the melt. Note that one of the published photos of anhydrite among silicates shows a strongly altered rock with secondary sulfides: instead of chalcopyrite and pyrrhotite, which are the major ore-forming sulfides, the rock contains typical pyrite (or, perhaps,

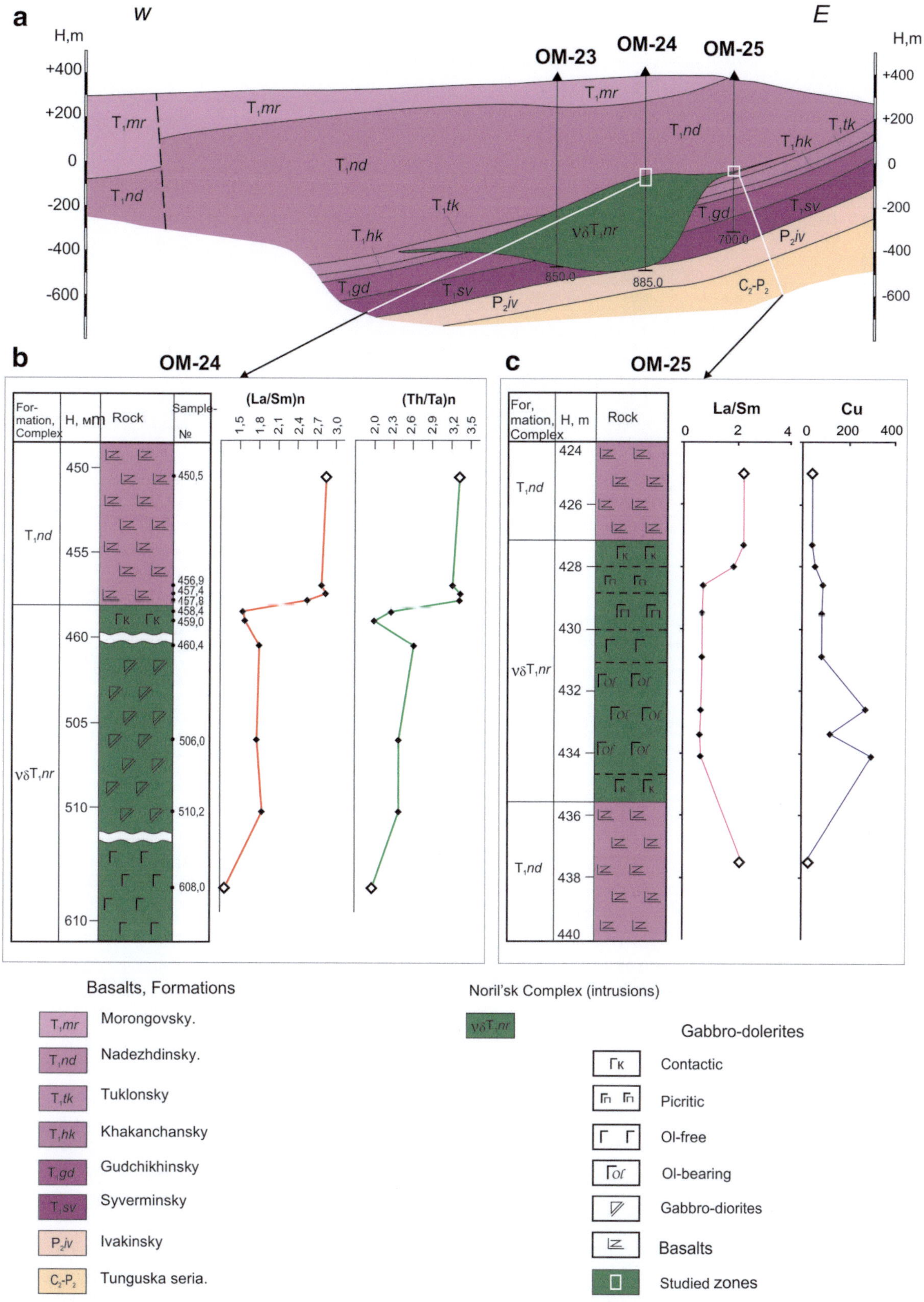

Fig. 8.9 (**a**) Vertical section through the Southern Maslovsky intrusion (based on materials from NorilskGeology Ltd., modified after Krivolutskaya et al. 2014)
(**b**) Variations in primitive mantle-normalized (Hofmann 1988) elemental ratios of the outer- and inner-contact zones of the intrusion (based on materials recovered by borehole OM-24); (**c**) variations in the La/Sm ratio and Cu concentration in rocks of the sill of the South Maslovsky intrusion and host basalts (Krivolutskaya and Rudakova 2009)

Table 8.7 Composition of rocks from upper-contact zone of the Southern Maslovsky intrusion (borehole OM-24)

No	1	2	3	4	5	6	7	8	9	10
No sample	450.5	456.9	457.4	457.8	458.4	459	460.4	506.2	510.2	608
SiO_2	52.53	52.53	52.04	51.37	50.50	50.54	52.80	55.25	56.42	45.83
TiO_2	1.03	1.03	1.04	1.03	1.10	1.08	0.93	1.75	1.64	2.73
Al_2O_3	14.78	14.78	14.72	14.41	15.12	14.96	14.27	13.07	12.91	12.60
Fe_2O_3	10.35	10.35	10.44	11.02	11.10	10.69	10.55	13.79	14.19	19.37
MnO	0.15	0.15	0.17	0.17	0.15	0.17	0.15	0.29	0.26	0.27
MgO	6.06	6.06	5.71	6.16	6.25	6.47	5.33	3.01	2.45	5.50
CaO	8.47	8.47	9.71	9.84	10.28	10.94	9.33	5.06	5.20	10.21
Na2O	2.58	2.58	2.53	2.33	2.19	1.93	2.46	3.92	3.31	2.19
K_2O	1.66	1.66	1.85	1.52	0.90	1.11	1.31	1.55	1.34	0.33
P_2O_5	0.13	0.13	0.13	0.13	0.14	0.12	0.13	0.35	0.48	0.12
LOW	2.06	2.06	1.47	1.82	2.06	1.81	1.98	2.80	1.95	0.83
Total	99.80	99.80	99.81	99.80	99.79	99.81	99.23	98.47	100.14	99.96
Rb	24.52	45.73	24.51	1.69	3.02	3.10	13.23	43.66	33.87	6.55
Ba	328	295	220	8	89	82	183	465	392	186
Th	3.55	3.47	3.55	3.15	1.23	1.15	1.56	3.68	5.07	1.19
U	0.95	0.86	1.05	0.85	0.46	0.42	0.48	1.19	1.74	0.41
Nb	8.26	8.38	8.11	7.74	3.87	4.00	4.61	15.29	16.65	4.29
Ta	0.54	0.55	0.54	0.48	0.27	0.25	0.29	0.93	0.98	0.26
La	17.7	16.7	19.2	13.4	7.3	7.0	9.0	25.0	29.9	7.7
Ce	38.1	36.5	42.2	29.8	17.5	16.7	20.6	54.6	67.0	17.4
Pr	4.47	4.29	5.02	3.60	2.34	2.20	2.86	7.38	9.00	2.46
Sr	269	103	206	52	196	209	243	168	287	458
Nd	18.2	16.8	19.7	14.8	10.9	10.1	13.6	34.2	41.5	11.8
Sm	3.92	3.76	4.27	3.29	2.98	2.79	3.83	8.75	10.60	3.42
Zr	129	127	126	115	86	83	109	300	336	96
Hf	3.28	3.27	3.30	2.93	2.29	2.15	2.94	7.19	8.47	2.53
Eu	1.12	1.03	1.21	1.01	1.09	0.93	1.35	2.53	2.99	1.22
Ti	5,518	5,267	5,347	4,740	5,881	5,524	5,254	11,434	10,189	16,469
Gd	4.03	3.65	4.23	3.38	3.63	3.37	4.87	10.11	12.29	4.27
Tb	0.65	0.61	0.69	0.55	0.63	0.57	0.81	1.67	2.04	0.71
Dy	4.08	3.86	4.28	3.57	4.05	3.65	5.54	11.23	13.38	4.78
Y	20.8	19.3	21.1	17.8	20.6	19.3	31.1	62.0	76.9	27.0
Ho	0.85	0.82	0.91	0.73	0.87	0.79	1.16	2.29	2.78	0.99
Er	2.34	2.22	2.46	2.01	2.40	2.20	3.41	6.85	8.17	2.91
Tm	0.36	0.35	0.37	0.30	0.35	0.34	0.50	1.00	1.17	0.42
Yb	2.24	2.22	2.25	1.92	2.25	2.14	3.42	6.87	8.09	2.83
Lu	0.35	0.33	0.34	0.28	0.34	0.33	0.51	1.01	1.19	0.42
Cu	41.2	52.4	58.1	16.0	83.2	129.8	90.9	9.4	18.4	170.1
Ni	31.3	68.0	31.0	50.8	225.5	220.6	15.2	8.0	13.2	45.2
Co	41.9	44.6	41.9	28.7	52.2	62.7	57.6	32.3	30.4	65.2

Note: (1) № sample corresponds to the depth (m) in borehole OM-24, and (2) oxides are given in wt % and elements in ppm. Tables 8.7 and 8.8 (After Krivolutskaya et al. 2012)

marcasite) rims; this provokes doubts in a magmatic genesis of the anhydrite.

At the same time, anhydrite is indeed widespread at the deposits, not only as a typically mineral of the host sedimentary rocks but also as younger hydrothermal veinlets and metasomatic segregations redeposited in the course of secondary hydrothermal processes both in the outer-contact zone of the intrusion and in its inner portions. Anhydrite segregations analogous to those documented by C. Li were found in the Vologochansky Massif, which is hosted by the same rocks that host the Kharaelakh intrusion (to which the Oktyabr'skoe deposit is related) but contains no massive ore. Hence, anhydrite presence in magmatic rocks cannot be regarded as an argument in support of the assimilation hypothesis and, particularly, as an indication that the massif should contain sulfides.

Table 8.8 Isotope composition of the disseminated ores from the Maslovsky deposit

No	Sample	$\delta^{34}S$, ‰
1.	OM-24/832.2.	+5.0.
2.	OM-24/843.6.	+4.9.
3.	OM-24/847.	+6.2.
4.	OM-24/848.	+5.5.
5.	OM-24/850.0.	+5.1.
6.	OM-24/853.9.	+6.0.
7.	OM-24/859.3.	+4.8.
8.	OM-24/862.0.	+6.7.
9.	OM-24/863.2.	+5.2.
10.	OM-24/863.4.	+5.0.
11.	OM-24/870.0.	+10.5.
12.	OM-24/870.4.	+10.8.

Table 8.9 Isotope composition of anhydrites from Devonian deposits

No sample	Ang-1	Ang-2	KS-56/1754
Sm, ppm	0.192	0.654	0.762
Nd, ppm	1.038	3.150	3.482
$^{147}Sm/^{144}Nd$	0.11184	0.12540	0.13222
$^{143}Nd/^{144}Nd$	0.512543	0.512573	0.512525
2σ	0.000008	0.000005	0.000004
Rb, ppm	0.495	9.749	1.104
Sr, ppm	3,848	288.8	1,294
$^{87}Rb/^{86}Sr$	0.00037	0.09768	0.00247
2σ	4.78	0.59	2.00
$^{87}Sr/^{86}Sr$	0.708637	0.708949	0.708926
2σ	0.000010	0.000012	0.000017
$^{206}Pb/^{204}Pb$	20.817	23.310	24.528
2σ	0.012	0.014	0.011
$^{207}Pb/^{204}Pb$	15.635	15.769	15.829
2σ	0.012	0.014	0.010
$^{208}Pb/^{204}Pb$	38.220	39.062	40.478
2σ	0.015	0.015	0.012
$(^{87}Sr/^{86}Sr)_t$	0.70864	0.70860	0.70892
ε_{Nd}	0.87	1.02	–0.14

Note: $(^{87}Sr/^{86}Sr)t$ and $(^{143}Nd/^{144}Nd)t$ – Sr and Nd ratios for 251 Ma
Analyses were carried out in VSEGEI, Saint Petersburg, by analyst B. Belyatsky. After Krivolutskaya (2011)

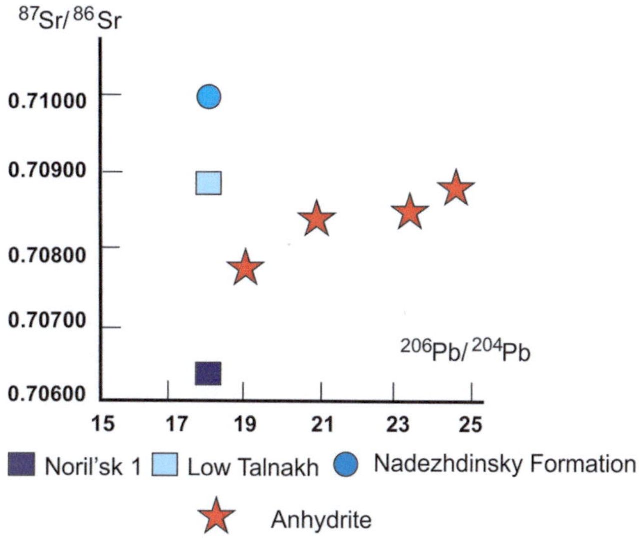

Fig. 8.10 Diagram $^{87}Sr/^{86}Sr$ versus $^{206}Pb/^{204}Pb$ for rocks of the Noril'sk district and anhydrite, after the data presented in Table 8.5
Intrusions: (1) Noril'sk 1, (2) Low Talnakh, (3) basalt of the Nadezhdinsky Formation, (4) anhydrite (After Krivolutskaya 2011)

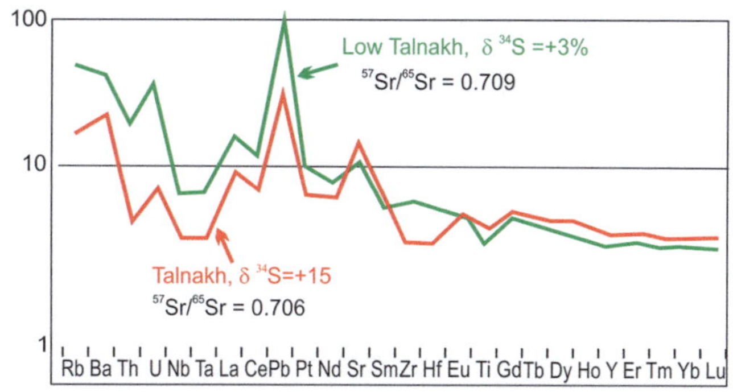

Fig. 8.11 Trace element contents normalized to primitive mantle (PM) in rocks of the Talnakh (1) and Lower Talnakh (2) intrusions

A serious argument that contradicts the important role of the assimilation of host rocks by basaltic magmas in the origin of the ore is the absence of a correlation between S and Sr isotopes in intrusive rocks (Fig. 8.11). Thus, for the essentially contaminated Low Talnakh intrusion, high $^{87}Sr/^{86}Sr$ ratios are typical (0.709), which has a good correlation with the enriched patterns. However, S isotopes in sulfides from this intrusion are similar to the mantle data (usually $+1-+3\delta^{34}S$, sometimes up to +6). In contrast, the Talnakh intrusion (it is believed to be strongly contaminated by host rocks due to the presence of huge massive sulfide) has a low $^{87}Sr/^{86}Sr$ of 0.705–0.706 and a high $\delta^{34}S$ of +15‰. As an additional factor of ore enrichment by heavy sulfur isotope, one can regard fractionation of sulfur isotopes in the process of sulfide melt crystallization and appearance of different minerals which was established by A. Likhachev and V. Strizhov in experiments (1977).

Had the magmas actively assimilated the rocks, the composition of the intrusions hosted in the rocks of contrastingly different composition (carbonate–sulfide, terrigenous, and basalt) would have also significantly differed. However, our calculations of the weighted mean compositions of the intrusions hosted in various terrigenous–sedimentary and volcanic rocks indicate that these compositions are practically identical (Krivolutskaya 2011) (Table 8.10). This observation is obvious from the normalized patterns of trace elements (Table 8.11), which are much more stable parameters of intrusions than the major components. The concentrations

Table 8.10 Average mean composition of the Noril'sk Complex intrusions, wt %

Massif	Mikchangdinsky	Noril'sk 1	Talnakh	Talnakh	Talnakh
Borehole	MD-48 (Tunguska Group)	G-22 (Tunguska Group)	KZ-1799 (Ivakinsky Formation)	ОUГ-2	29 boreholes (terrigenous-carbonate rocks)
SiO_2	47.71	47.16	47.47	48.76	48.3
TiO_2	0.91	0.79	1.11	0.88	0.85
Al_2O_3	13.30	15.36	14.11	14.9	15.33
FeO	12.66	12.17	12.33	11.08	12.34
MnO	0.18	0.20	0.22	0.2	0.19
MgO	13.42	12.04	11.54	10.52	10
CaO	9.47	10.60	10.22	10.04	10.45
Na_2O	1.79	0.97	1.91	2.05	1.86
K_2O	0.45	0.39	0.56	0.65	0.58
P_2O_5	0.10	0.09	0.12	0.18	0.2
Cr_2O_3	–	0.23	0.13	0.14	0.1

Note: The weighted average chemical compositions were calculated from the following sources: (1) author's data –MD-48, G-22, OUG-2 (2) Cza manske and Zen'ko (1994) – KZ-1799; Dneprovskaya et al. (1987) – 29 boreholes. After Krivolutskaya (2011)

Table 8.11 Average mean compositions of rare elements in rocks of the Noril'sk Complex intrusions and low crust, ppm

Massif	Kharaelakh	Talnakh	Noril'sk 1	Mikchangdinsky	Low
Borehole	TG-21	OUG-2	G-22	MD-48	Crust
No	1	2	3	4	5
Rb	23	20	18	9	11
Ba	134	222	100	111	259
Th	1.29	1.16	0.93	0.84	1.2
U	0.50	0.47	0.35	0.31	0.2
Nb	4.32	3.77	3.65	3.39	5
La	6.37	7.29	4.97	6.42	8
Ce	14.6	16.5	11.9	14.3	20
Pr	1.87	2.28	1.57	1.96	2.4
Nd	8.39	10.22	7.29	9.35	11
Sr	224	262	259	212	348
Sm	2.12	2.66	2.05	2.61	2.8
Zr	64	77	59	69	68
Hf	1.65	2.11	1.55	1.83	1.9
Eu	0.79	0.99	0.71	0.88	1.1
Ti	4,065	6,404	4,220	5,686	4,916
Gd	2.38	3.64	2.40	3.27	3.1
Tb	0.41	0.60	0.41	0.54	0.48
Dy	2.72	3.85	2.81	3.69	3.1
Ho	0.62	0.82	0.64	0.77	0.68
Y	18	23	19	21	16
Er	1.66	2.29	1.72	2.23	1.9
Tm	0.25	0.34	0.26	0.32	0.24
Yb	1.58	2.15	1.63	2.19	1.5
Lu	0.24	0.33	0.24	0.32	0.25

Note: Average mean composition for different intrusions: 1–2, unpublished data of author; 3, Krivolutskaya and Rudakova 2009; 4, Krivolutskaya et al. 2009; 5, Rudnick and Gao, 2003

of the latter components strongly depend on the inner structure of the intrusions, which was controlled not by the composition of the parental magma but by its crystallization conditions, which are reflected in the compositional heterogeneity of the magmatic body, not only in its vertical section but also in the lateral distribution of the crystallization products (Krivolutskaya 2011).

Thus, our data on the contact zones of the Kharaelakh and Talnakh intrusions (and on the previously studied contact zone of the Maslovsky intrusion) indicate that assimilation processes could operate only very locally or did not operate at all at the level of modern intrusive chambers. This conclusion is supported by the sulfur isotopic composition of the sulfides, by the concentrations of radiogenic Pb and Sr isotopes in the anhydrite, and by the similarities between the compositions of various ore-bearing massifs hosted in rocks of different compositions: anhydrite-bearing, terrigenous, and volcanic. These facts testify that the assimilation of host rocks by the melts could not play any appreciable role in the generation of the unique Pt–Cu–Ni ores in the Noril'sk district.

In this case, a question arises: what causes the heavy isotopic composition of sulfur in the sulfide ores of the Noril'sk deposits? To explain this peculiarity, we thoroughly examined rare derivatives of mantle magmas, i.e., picrite basalts of the Gudchikhinsky Formation. Among these rocks, we found rocks insignificantly enriched in the crustal component and rocks containing absolutely no Ta–Nb and Pb anomalies, which are typical for crustal rocks (Chap. 4). On the basis of thorough examination of the geochemical peculiarities (major, rare, and radiogenic elements), an analog of these volcanic rocks was found to be the Dyumtaleysky massif, southwestern Taimyr.

The revealed isotopic composition of sulfur is quite atypical for the mentioned igneous rocks and ores. In contrast to the expected low values of $\delta^{34}S$, enrichment in the heavy sulfur isotope was found. The following values of $\delta^{34}S$ were defined earlier for rocks of the Gudchikhinsky Formation (Ripley et al. 2003): from −4 to +8.7; note that the latter value was characteristic for a picrite flow in the section of Mt. Sunduk and defined as a product of the most primitive magma crystallization (Fig. 2.9, sample no. SY-50).

The Dyumtaleysky massif is a clearly layered lopolith-like intrusive body, whose upper part is composed of gabbroids of an increased titanomagnetite content, while the lower part is of peridotite composition (wehrlites, lherzolites). The REE distribution is shown in Fig. 8.12, where one can clearly see the similarity between distribution curves for gabbroids and picrites of the Gudchikhinsky Formation. The differences are caused only by the presence of ore mineralization in the intrusive body: an increased titanomagnetite content yields a positive titanium anomaly and that of sulfides yields a positive lead anomaly. The thickness of rocks at the studied LP-1 borehole is 600 m; sulfide ores form several horizons of diseminated mineralization and thin veins at depths of 1,390, 1,490, and 1,600 m. The composition of ores does not differ from that typical for ordinary Cu–Ni deposits: the major ore minerals are chalcopyrite, pyrrhotine, and pentlandite; among the auxiliary minerals are cubanite, bornite, and millerite; among the rare minerals is michenerite. The isotopic composition of sulfur from the Dyumtaleysky massif sulfides (Krivolutskaya 2014) indicates that these ores are characterized by a heavy isotopic

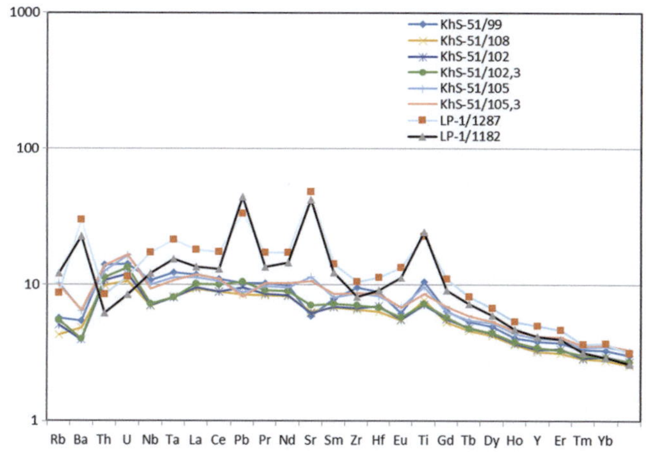

Fig. 8.12 Trace element contents normalized to primitive mantle in rocks of the Dyumtaleysky intrusion and Gudchikhinsky picrites

composition of sulfur ($\delta^{34}S$ + 12.8‰) as the standard ore-bearing intrusions of the Noril'sk I, Talnakh, and Kharaelakh (from +3 to +16 ‰ (Grinenko 1985)) despite the fact that the magmas that shaped this massif have principally different geochemical characteristics and origins.

This unique phenomenon can be interpreted as the presence of initially heavy sulfur in mantle rocks (Likhachev 2006). This point can explain the anomalous isotopic composition of sulfur from Noril'sk deposit ores, in comparison to other Cu–Ni deposits of the world; therefore, we must reject the possible participation of anhydrite that essentially differs from the intrusive rocks in the composition of radiogenic isotopes. Note that almost all Cu–Ni and Pt deposits were formed in the Proterozoic, while the Noril'sk deposits were formed in the Early Triassic. This age difference is of fundamental importance because the composition of the magma source (in particular, the mantle source) that shaped these ores must have been essentially different. By the beginning of the Phanerozoic, a thick continental crust had formed on the Earth, and this envelope was increasingly involved into subduction; therefore, this envelope had an increased influence on the mantle substrate composition, in comparison to the Precambrian. It is very likely that sulfides suffered biogenic reworking on the Earth's surface and then sunk by subduction; in the Siberian region, accuditions are not clearly understood yet: its composition might be partially (essentially) altered at high pressures, temperatures, valence changes, etc. This thought is verified by the data on the isotopic composition of sulfur from the ancient deposits of the Siberian craton. S. Kryazhev et al. (2003) demonstrated that a heavy sulfur isotopic composition of sulfides (up to $\delta^{34}S$ = 20‰) was typical of certain Siberian deposits (in the Yenisei Range) already in the Proterozoic, when anhydrite-bearing rocks were yet to be deposited. The rocks enriched in the heavy sulfur isotope were recently proved to be first formed on the Earth in the Early Proterozoic (Farquhar et al. 2010).

Very often as evidence of a broad process of assimilation, magmas surrounding rocks provide data on halos of metamorphic and metasomatic rocks around intrusions. But our results are in contradiction with this opinion.

Thus, our data obtained on the oxygen isotopic composition of rocks from five hyperbasite–basite massifs (Pokrovsky et al. 2005), three of which are accompanied by economic Pt–Cu–Ni ore mineralization (Noril'sk 1, Talnakh, and Kharaelakh) and two contain only poor mineralization (Low Talnakh and Zelenaya Griva), as well as analogous data on the hornfels and metasomatites hosting these intrusions, suggest that the oxygen isotopic composition of the magmas before their emplacement to where the differentiated intrusions are located now or before their surface eruptions was analogous to the oxygen isotopic composition of common mantle melts ($\delta^{18}O = 5.5 \pm 0.5‰$). The broad variations in the $\delta^{18}O$ (from −0.8 to 13.7‰) in the intrusive rocks and basalts were caused mostly by hydrothermal–metasomatic processes at a subordinate role of crustal assimilation during the magmatic stage. The hydrogen isotopic composition of the trap rocks (δ D from −94 to −137‰) suggests that the main source of the hydrothermal fluids was atmospheric waters.

8.4 Conclusions

The study of contact zones of the Kharaelekh, Talnakh, and Southern Maslovsky intrusions demonstrates absence or local (<1 m) appearance of the assimilation processes. So anhydrite was not the source of sulfur for Noril'sk ores. It was suggested that the source of magmas in Noril'sk region was enriched in heavy sulfur isotope. It was found in mantle magmas for the first time (Dyumtaleysky intrusion).

References

Arndt NT (2005) The conduits of magmatic ore deposits. Mineral Assoc Can Short Course 35:161–182

Arndt NT, Czamanske GK, Walker RJ et al (2003) Geochemistry and origin of the intrusive hosts of the Noril'sk–Talnakh Cu–Ni–PGE sulfide deposits. Econ Geol 98:495–515

Distler VV, Grokhovskaya TL, Evstigneeva TL et al (1988) Petrology of the magmatic sulfide ore formation. Nauka, Moscow, p 1988 (in Russian)

Dneprovskaya MB, Frenkel' MY, Yaroshevsky AA (1987) Quantitative model of the layering formation of the Talnakh intrusion. In: Development of the models of the ore-forming systems. Nauka, Novosibirsk, pp 96–106 (in Russian)

Dodin DA, Batuev BN (1971) Geology and petrology of differentiated intrusions and their metamorphic aureole. In: Petrology and ore potential of the Talnakh and Noril'sk differentiated intrusions. Nedra, Leningrad, pp 31–100 (in Russian)

Dyuzhikov OA, Distler VV, Strunin BM et al (1988) Geology and ore potential of the Noril'sk region. Nedra, Moscow (in Russian)

Farquhar J, Canfield DE, Odoro H (2010) Connection between sulfur cycle evolution, sulfur isotopes, sediments, and base metal sulfide deposits. Econ Geol 105:509–533

Godlevsky MN (1959) Trap basalts and ore-bearing intrusions of the Noril'sk region. Gosgeoltekhizdat, Moscow, p 61 (in Russian)

Godlevsky MN, Grinenko LN (1963) Some sulfur isotope data on the Noril'lsk deposits. Sov Geol 1:27–39 (in Russian)

Gorbachev NS, Grinenko LN (1973) Sulfur isotope ratios in sulfides of the Oktyabr'skoe sulfide deposit, Noril'sk region, and problems of their genesis. Geokhimiya 8:1127–1136 (in Russian)

Grinenko LN (1967) Sulfur isotope composition of some Cu–Ni deposits of the Siberian platform. In: Petrology of trap basalts of the Siberian platform. Nedra, Moscow, pp 221–230 (in Russian)

Grinenko LN (1984) H_2S-bearing gas reservoirs as sulfur source during sulfurization of magmas of economically ore intrusions of the Noril'sk region. Dokl Akad Nauk SSSR 278(3):730–732 (in Russian)

Grinenko LN (1985) Sources of sulfur of the nickeliferous and barren gabbrodolerite intrusions of the northwest Siberian Platform. Int Geol Rev 27:695–708

Hofmann AW (1988) Chemical differentiation of the earth: the relationship between mantle, continental crust and oceanic crust. Earth Planet Sci Lett 90:297–314

Keays RR, Lightfoot PC (2007) Siderophile and chalcophile metal variations in Tertiary picrites and basalts from West Greenland with implications for the sulphide history of continental flood basalt magmas. Miner Deposita 42:319–336

Keays RR, Lightfoot PC (2010) Crustal sulfur is required to form magmatic Ni–Cu sulfide deposits: evidence from chalcophile element signatures of Siberian and Deccan trap basalts. Miner Deposita 45:241–257

Konnikov EG (1986) Differentiated ultramafic-mafic complexes of the Transbaikal Region. Nauka, Novosibirsk (in Russian)

Konnikov EG, Kovyazin SV, Nekrasov AN et al (2005) Interaction of magmatic fluids and mantle magmas with lower crustal rocks: evidence from inclusions in the minerals of intrusions. Geochem Int 43(10):939–958

Kovalenker VA, Gladyshev GD, Nosik LP (1974) Sulfur isotope composition of sulfides from the deposits of the Talnakh ore cluster in relation with their Se potential. Izv Akad Nauk SSSR Ser Geol 2:10–35 (in Russian)

Krivolutskaya NA (2011) Formation of PGM – Cu_Ni deposits in the process of evolution of flood-basalt magmatism in the Noril'sk region. Geol Ore Depos 53(4):309–339

Krivolutskaya NA (2014) Mantle origin of heavy isotopes of sulfur in ores of the Noril'sk deposits. Dokl Earth Sci 454(Part 1):76–78

Krivolutskaya NA, Rudakova AV (2009) Structure and geochemical characteristics of trap rocks from the Noril'sk Trough, northwestern Siberian Craton. Geochem Int 7(7):635–656

Krivolutskaya NA, Mikhailov VN, Snisar SG, Gongalsky BI (2009) Internal structure and composition of the Mikchangdinsky basic-ultrabasic massif in the Noril'sk area (Siberian trap provinces). Vestn KRAUNTS 14:29–48 (in Russian)

Krivolutskaya NA, Ariskin AA, Sluzhenikin SF et al (2001) Geochemical thermometry of rocks of the Talnakh intrusion: assessment of the melt composition and the crystallinity of the parental magma. Petrology 9(5):389–414

Krivolutskaya NA, Sobolev AV, Snisar SG et al (2012) Mineralogy, geochemistry and stratigraphy of the Maslovsky Pt-Cu-Ni sulfide deposit, Noril'sk region, Russia: implications for relationship of ore-bearing intrusions and lavas. Miner Deposita 47:69–88

Krivolutskaya NA, Plechova AA, Kostitsyn Yu A, Belyatsky BV, Roshchina IA, Svirskaya NM, Kononkova NN (2014) Geochemical aspects of the assimilation of host rocks by basalts during the formation of Norilsk Cu–Ni ores. Petrology 22(2):128–150

Kryazhev SG, Glukhov AP, Rusinova OV et al (2003) Isotopic geochemical regime of the formation of the Sovetskoe gold–quartz deposit. In: Applied geochemistry. IMGRE, Moscow. issue 4, pp 154–164 (in Russian)

Li C, Ripley EM, Naldrett AJ (2003) Compositional variations of olivine and sulfur isotopes in the Noril'sk and Talnakh intrusions. Econ Geol 98:69–86

Li CS, Ripley EM, Naldrett AJ (2009a) A new genetic model for the giant Ni–Cu-PGE sulfide deposits associated with the Siberian flood basalts. Econ Geol 104:291–301

Li C, Naldrett AJ, Shmitt AK et al (2009b) Magmatic anhydrite – sulfide assemblages in plumbing system of the Siberian traps. Geology 37:259–262

Lightfoot PC, Howkesworth CJ, Hergt J et al (1993) Remobilisation of the continental lithosphere by a mantle plume: major-, trace element, and Sr-, Nd-, and Pb isotopic evidence from picritic and tholeiitic lavas of the Noril'sk district, Siberian trap, Russia. Contrib Mineral Petrol 114:171–188

Likhachev AP (1996a) Kharaelakh intrusion and its Pt–Cu–Ni ores. Rudy Met 3:48–62 (in Russian)

Likhachev AP (1996b) Dynamics of the emplacement of the Talnakh ore-bearing intrusion and related Pt–Cu–Ni deposits. Otechestvennaya Geol 8:20–26 (in Russian)

Likhachev AP (2006) Platinum–copper–nickel and platinum deposits. Eslan, Moscow (in Russian)

Likhachev AP, Strizhov VP (1977) Fractionation of sulfur isotopes. Dokl Akad Nauk SSSR 236(1):223–226 (in Russian)

Naldrett AJ (1992) A model for the Ni–Cu–PGE ores of the Noril'sk region and its application to other areas of flood basalts. Econ Geol 87:1945–1962

Naldrett AJ (2005) A history of our understanding of magmatic Ni–Cu sulfide deposits. Can Mineral 43:2069–2098

Naldrett AJ (2009) Ore deposits related to flood basalts, Siberia. In: New developments in magmatic Cu–Ni and PGE deposits. Geological Publishing House, Beijing, pp 141–179

Pokrovsky BG, Sluzhenikin SF, Krivolutskaya NA (2005) Interaction conditions of Noril'sk trap intrusions with their host rocks: isotopic (O, H, and C) evidence. Petrology 13(1):49–72

Polishchuk YM, Yashchenko IG (2005) Study of relation of oils properties with geothermal characteristics of oil provinces. Vestn SVNTs DVO RAN 3:26–34 (in Russian)

Polyansky OP, Reverdatto VV (2006) Role of fluid in heat and mass transfer during evolution of sedimentary basins with trap magmatism. In: Fluids and geodynamics. Nauka, Moscow, pp 219–243 (in Russian)

Rad'ko VA (1991) Model of geodynamic differentiation of the intrusive traps of the northwestern Siberian platform. Geol Geofiz 11:19–27 (in Russian)

Ripley EM, Lightfoot PC, Li C et al (2003) Sulfur isotopic studies of continental flood basalts in the Noril'sk region: implications for the association between lavas and ore-bearing intrusion. Geochim Cosmochim 67:2805–2817

Rudnick RL, Gao S (2003) Composition of the continental crust. Treatise on geochemistry, vol 3. Elsivier-Pergamon, Oxford, pp 1–64

Sluzhenikin SF, Krivolutskaya NA, Rad'ko VA, Malitch KN, Distler VV, Fedorenko VA (2014) Ultramafic-mafic intrusions, volcanic rocks and PGE-Cu-Ni sulfide deposits of the Noril'sk Province, Polar Siberia. Field Trip Guidebook, Yekaterinburg, 83 p

Turovtsev DM (2002) Contact metamorphism of the Noril'sk intrusions. Nauchny Mir, Moscow, p 325 (in Russian)

Yashchenko IG (2004) Cyclic changes of sulfur and paraffin contents depending on the age of oil-bearing rocks. Izv Tomsk Politekh Univ 307(3):54–59 (in Russian)

9 Resume: Interpretation of Results - Possible Magma Sources and Ore-Forming Processes

The very first studies of the Noril'sk deposits were launched by N. N. Urvantsev in 1919, and these deposits have now been studied for almost one century. During these years, the unique ores of the Talnakh and Oktyabr'skoe deposits were discovered. Their resources, composition, and mineralogical diversity have no equivalents anywhere on the planet. The uniqueness of the Noril'sk ore-bearing massifs within the extensive class of magmatic Pt–Cu–Ni deposits (the setting of the Noril'sk massifs in the world's largest Siberian flood basalt province, their Early Triassic age, and their combination of sulfide and low sulfide types of PGE mineralization within the same massifs) continues to attract the keen interest, in both purely theoretical and applied aspects, of several researchers. The principal relations and trends revealed in the origin of the Noril'sk deposits largely facilitate exploration for analogous ores not only in the Noril'sk district itself but also throughout the Siberian Platform and elsewhere worldwide.

Several aspects of the genesis of the Noril'sk deposits are, however, still uncertain or not fully understood. First of all, the mechanism responsible for the enrichment of metals and the ensuing origin of the rich ore mineralization with regard to the small volumes of the silicate melts compared to the volumes of the sulfide orebodies is unknown. Several researchers attempted to explain this phenomenon from various standpoints. Some geologists (Godlevsky 1959; Likhachev 1965, 1996, 2006; Dyuzhikov et al. 1988) believed that the unusual inner structures of the deposits can be accounted for by their generation by means of unusual ore-bearing magmas. Conversely, some other scientists (Rad'ko 1991; Naldrett 1992) were prone to think that the deposits were produced by usual tholeiitic melts in the course of their long-lasting ascending flow to the surface. It was thus assumed that the magmatic systems were either open or closed during the generation of the ores. Almost all genetic schemes attach much importance to the assimilation of host rocks by the melts. This pertains, first of all, to anhydrite in the host rocks as a source of sulfur. Much attention was also paid to the regime of fluids as a predominant or subordinate agent of the enrichment of valuable components in the ores.

The author's long-term studies of the massifs provided extensive data on the geology, geochemistry, and mineralogy of magmatic rocks in the Noril'sk district. In certain instances, these data allowed to settle some disputable problems (such as the composition of the melts and volatiles in them and as the possible extent of assimilation) or constrain the boundary conditions of certain hypotheses (Krivolutskaya 2014). The results are briefly summarized below.

The detailed studies of the inner structure and geology of the volcanic rocks in various tectonic structures of the district made it possible to distinguish two major episodes (rifting and trap magmatism itself) in the origin of the rocks of the tuff–lava sequence, which comprise four cycle formations: the Ivakinsky–Syverminsky, Gudchikhinsky, Nadezhdinsky–Tuklonsky, and Morongovsky–Samoedsky. The basalts of the stratigraphically lowermost formations (Ivakinsky, Syverminsky, Gudchikhinsky, and Nadezhdinsky) were found only within the Noril'sk–Igarka ancient rift structure (Fig. 3.43) and give insight into the rift evolutionary episode of the area. The composition of these formations is principally different from those of the other volcanic products, including those in the Tuklonsky and Morongovsky–Samoedsky Formations, whose rocks were found far away from the Noril'sk district (at the Putorana Plateau and Tunguska Syneclise); these rocks affiliate with the trap association itself (Fig. 3.43). The rocks generated during rifting typically have elevated Ti concentrations and elevated Gd/Yb ratios, which suggests that garnet occurred in the source material from which the parental magmas were derived (this source was closely similar to the source of OIB). The exception is the basalts of the Nadezhdinsky Formation, which are poor in Ti but are enriched in LILE. Analogous rocks were found in rift zones in the West Siberian Platform (Al'mukhamedov et al. 2004; Saunders et al. 2005; Reichow et al. 2005).

N.A. Krivolutskaya, *Siberian Traps and Pt-Cu-Ni Deposits in the Noril'sk Area*,
DOI 10.1007/978-3-319-17205-7_9

The pioneering and principally new result in studying of volcanism in the Noril'sk district was the discovery of two simultaneously operating magmatic centers (in the western and eastern parts of the territory), which characterize different evolutionary episodes of the volcanic rocks. The western center pertained to the latest evolution of the rift system, whereas the eastern one marked the onset of trap magmatism itself. These conclusions were derived from data on the inner structure and geochemistry of rocks of the Nadezhdinsky and Tuklonsky Formations, which occur, respectively, in the western and eastern portions of the territory and alternate in the vertical section of its central part (Figs. 3.23 and 3.24) because of the simultaneous generation of tuff (whose composition corresponds to that of the Nadezhdinsky rocks) and basalts of the Tuklonsky Formation. Magmatic activity within the rift was likely more intense, because the younger Nadezhdinsky lavas extend eastward and overly the Tuklonskaya basalts. The latter are not as widely spread as the younger products of typical trap magmatism (the Morongovsky and overlying formations). It follows that the elevated thickness of volcanic rocks in the northwestern Siberian Platform is explained not by the activity of a single powerful magmatic center (as was hypothesized previously) but by a combination of products of rift and platform evolutionary episodes in this territory.

Although detailed data were obtained on the structure and chemistry of the tuff–lava sequence in the Noril'sk district and the evolution of the volcanic processes was reproduced, the spatiotemporal evolution of these processes can be realistically visualized only based on the reproduced evolutionary history of magmatism throughout the whole Siberian Platform. This is a subject for further studies.

Intrusive rocks widespread in the Noril'sk district are classified (based on their distribution of trace elements and on isotopic data) into three types, which correspond to the Gudchikhinsky (Dymtaleysky type of intrusions), Nadezhdinsky (Low Talnakh type), and Morongovsky (Noril'sk type) Formations.

Proceeding from the geology and relations between the basalts and intrusions, it was established that the massifs of the Noril'sk Complex were emplaced in post-Nadezhdinsky time, i.e., they were generated during the trap episode of the magmatic evolution. A comparison of the major component and trace element composition of the ore-bearing massifs and lavas of this episode reveals their significant differences: in the MgO concentrations (10–12 wt % in the intrusions and 6–7 wt % in the volcanic rocks), in the TiO_2 concentrations (Fig. 6.4) and La/Yb ratio of the rocks of the Noril'sk Complex, and in the isotopic composition (first of all, the sulfur isotopic composition—$\delta^{34}S$ from +1 to −7 and up to 18‰ in the basalts and intrusions, respectively). In the absence of geological data on relations between the intrusions and lavas and with regard for their geochemical differences, it was concluded that the ore-bearing intrusions and volcanic rocks are not comagmatic. As was suggested by most researchers, the massifs of the Noril'sk Complex were emplaced during an individual pulse of magmatic activity in the post-Nadezhdinsky time period. Many fewer arguments were collected in support of the hypothesis that the ore-bearing magmas were emplaced in post-Morongovsky time period and, conceivably, even after the accumulation of the whole volcanic pile (Malitch et al. 2010; Ivanov et al. 2013), and this issue awaits its further thorough studying.

Differences in the estimated nature of the parental magma that produced the intrusions with rich ore mineralization largely stem from differences in the approaches to these estimations: the composition of the magmas either was estimated as the weighted mean of the composition of the massifs or was assumed to be analogous to the composition of the magnesian intrusions, or else was taken up to be close to the composition of peripheral sills. We were the first to widely apply a new approach to estimating the composition of the melts that produced the magmatic rocks: this composition was evaluated from data on melt inclusions in the early liquidus phases (olivine and pyroxene). The results of this study allowed us to directly evaluate the concentrations of major, trace, and volatile components in the magmas. For example, it was demonstrated that the ore-bearing massifs were produced by magnesian melts (up to 8 wt % MgO), which contained intratelluric olivine and plagioclase phenocrysts. The parental magma had crustal geochemical signatures: their normalized trace element patterns show pronounced negative Ta–Nb and positive Pb anomalies. The fluid was aqueous solution (the H_2O concentration in the parental melt was 0.5–0.7 wt %) poor in CO_2 and sometimes containing elevated Cl concentrations (up to 0.2 wt %). The basalts and intrusions mostly have lower $Mg\# = MgO/(MgO + FeO) \times 100 = 37–38$ than those of typical mantle rocks, which have Mg# >50. It was calculated that the Mg# was 55–57 for picritic basalt of the Tuklonsky Formation, 59–63 for the Nadezhdinsky Formation, and 56–57 for the picritic gabbro–dolerites (Krivolutskaya et al. 2012). These rocks contain low-magnesian olivine (<*Fo84*). This testifies that the parental melts of the rocks were not in equilibrium with a peridotite source but were cumulates of tholeiitic magmas whose characteristics were close to those of crustal rocks ($\varepsilon_{Nd} = 0 \pm 1.5$; $^{87}Sr/^{86}Sr = 0.706 \pm 0.1$). The products of typical mantle magmas (picrites of the Gudchikhinsky Formation, Ayan picrites, and meymechites, which occur outside the Noril'sk district) are restricted only to the ancient rift structure in the northern Siberian Platform and are not directly related to the traps.

The characteristics of most rocks of the trap association listed above are traditionally explained by the crustal contamination of the mantle melts, which could reportedly take place at certain depths or immediately where the melts crystallized and could play a decisive role in the origin of the sulfide ores (Lightfoot et al. 1993; Naldrett 1992; Li et al. 2009a, b). However, data on the distribution of major and trace elements in the contact zones of the intrusions with their host rocks and on the isotopic composition of the rocks (their ε_{Nd}, $^{87}Sr/^{86}Sr$, $\delta^{34}S$) are at variance with any significant in situ contamination. The heavy sulfur isotopic composition of the sulfide ores (up to 18‰ according to Grinenko 1985), which is atypical of magmatic deposits, was (and still is) explained (Li et al. 2009a, b) by the assimilation of sulfate-bearing rocks by the melts, with such rocks (with thick anhydrite beds) widespread in the Noril'sk district. Our pioneering data on the sulfur radiogenic isotopes in the anhydrite are in conflict with the hypothesis that this mineral could serve as a sulfur source for the Noril'sk ores. The fact that the average composition of the intrusions is independent on the stratigraphic setting of these intrusions, which can be hosted by limestone, sandstone, and/or basalt, provides further support for the idea that no assimilation took place at the depths of the chambers in deep zonez of the crust which the melts crystallized. No evidence of magmatic assimilation of the host rocks, a process that is believed to play a decisive role in the genesis of rocks of the trap association (Arndt et al. 2003), has ever been published in the literature.

L. Grinenko (1984) suggested that the ore-bearing magma could be enriched in sulfur, which could be borrowed from the Vendian–Cambrian hydrocarbon deposits of the Siberian Platform. However, as early as 1973, L. Perchuk stressed that ascending magmas should affect the wall rocks rather than assimilate them. This was convincingly proved by O. P. Polyansky and V. V. Reverdatto (2006) and illustrated by the example of trap sills emplaced into carbonaceous rocks: the thermal effect of the emplaced magma forces the hydrocarbons from the crystallization chamber, and the melt does not assimilate them. Moreover, the hydrocarbon deposits are poor in sulfur (contain <10 % H_2S), so that the total volume of sulfur was not comparable to that contained in the Noril'sk deposits.

The reason for the heavy sulfur isotopic composition of ores in the Noril'sk district is still uncertain. Data obtained over the past decade on the sulfur isotopic composition of basalts and ores from some intrusions in the Taimyr Peninsula likely provide a clue to this problem. The highest $\delta^{34}S$ values in rocks of all of the formations were detected in the primitive Gudchikhinsky picrites ($\delta^{34}S=+8.7$‰; Ripley et al. 2003) which are geochemically similar to rocks from the Dyumtaleysky massif ($\delta^{34}S=12.2$‰; Krivolutskaya and Gongalsky 2013, 2014; Krivolutskaya 2014). These data support that the mantle source of magmas in the Noril'sk district was enriched in the heavy sulfur isotope. A. P. Likhachev (1978) came up with this idea but he had no clear evidences of this. So this enrichment could occur, for example, if sulfides that were biogenically recycled at the Earth's surface were then brought to the mantle by subduction. However, it is hard to believe that crustal material involved in global circulation (up to the Earth's core) can somehow be preserved and remain unmodified during a time span of 400 Ma. But it is the age difference that could be responsible for the unusual composition of the Noril'sk ores, because the mantle source in the Early Mesozoic was remarkably different from that in the Proterozoic one, when practically all Cu–Ni deposits worldwide were produced.

Another possible interpretation of these data can be based on the fact that rocks enriched in the heavy sulfur isotope were generated in the Earth starting in the Early Proterozoic (Farquhar et al. 2010). Such deposits are also known in the southern Siberian Platform, where the Proterozoic ores have $\delta^{34}S=20$‰ (Kryazhev et al. 2003). Evidence supporting their biogenic genesis is still insufficient. Theoretically, such sulfides could be remobilized by the parental melts and brought to where they were later deposited in the Noril'sk district.

Also, it cannot be ruled out that the sulfur isotopic composition could be modified during the crystallization of the sulfide orebodies. The fractionation of sulfur isotopes in the sulfide melt was proved experimentally during the onset of liquid immiscibility and decomposition into Cu-rich and pyrrhotite constituents (Kovalenker et al. 1974; Likhachev and Strizhov 1977). It is also reasonable to hypothesize that such processes could play an important role in the vast volumes of sulfide melts (1–2 km^3) in the Noril'sk district and were largely responsible for the shift in the sulfur isotopic composition of the sulfide system. An argument in support of the hypothesis that this process can operate in nature is provided by a direct correlation between the concentration of the heavy sulfur isotope and the volume of the deposit (Grinenko 1985).

These results presented above allowed the author to suggest her own interpretation of the genesis of the Noril'sk deposits. Mineralogical and geochemical data on magmatic rocks in the Noril'sk district and the limited operation of assimilation processes testify that the parental melts of the tuff–lava sequence of the trap episode of volcanism (from the Tulonsky Formation to Samoedsky) and the intrusions of the Noril'sk Complex were originated from lower crust rather than mantle (Krivolutskaya 2011, 2014).

The evaluation of the composition of the lower crust is a challenging problem by itself, and this problem can be solved by analyzing (1) the rare exposures of lower crustal rocks, (2) xenoliths entrained by magmas of various composition, and (3) geophysical data. It was thought (Rudnick and Fountain 1995; Bogatikov et al. 2010) that the lower crust is made up mostly of mafic and ultramafic rocks. However, with regard to its heterogeneity, the average composition of the crust should be close to andesite (Rudnick and Gao 2003;

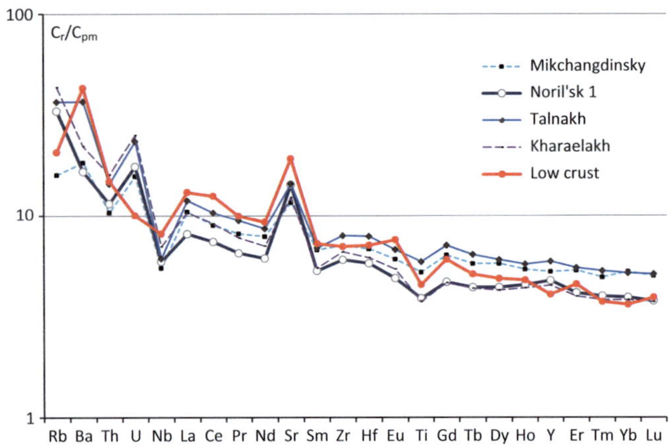

Fig. 9.1 Rare element patterns for ore-bearing intrusions (author's data) and low crust (Rudnick and Gao 2003)

Table 8.10). The former composition reported by these authors is close to that of the "average trap" and the model melt of the Talnakh intrusion. For comparison with the ore-bearing intrusions, the concentrations of trace elements are listed according to the latter composition, which was recommended as a reference one. As can be seen in Fig. 9.1, the concentrations of trace elements in rocks of the Noril'sk Complex are closely similar to those in the lower crust. The trace element patterns are even more "crustal" than the reference pattern of the lower crust, because the slope of their right-hand parts is more gentle and they have much higher U concentrations. The isotopic parameters of the ore-bearing intrusions ($^{87}Sr/^{86}Sr = 0.705–0.706$ and $\varepsilon_{Nd} = -2$) are also lower crustal. It follows that, according to their composition, lower crustal rocks can be a probable candidate for the source material of the Noril'sk magmas, but this does not eliminate the question whether this melting process could occur. The key issue of this problem is the temperature in the lower crust, which is usually evaluated at 600–800 °C.

It was lately demonstrated that the seismic parameters of rocks that are usually utilized to interpret the inner structure and composition of deep Earth's zone are more strongly dependent on temperature than composition (Kuskov et al. 2014). The geophysical data of Kola Superdeep Borehole have been extrapolated to the middle and lower crust (Gorbatsevich et al. 2013). Taking into account a surface heat flux 2.2 °C/100 m, these authors have got the temperatures of 710–790 °C in the middle crust, and one can suppose 1,120–1,200 °C in the lower crust. Thus, even decompressional melting alone is able to generate melts typical of ore-bearing intrusions in various areas.

Of course, these are other factors facilitating this process. It was recently discovered, with the application of modern geophysical techniques, that the structure of the Moho (M) discontinuity beneath the continents is complicated. The discontinuity was proved to be thinly layered (the thicknesses of individual layers are a few kilometers), and the transition from the crust to mantle corresponds to a change not in the rock compositions but in their rheology, as follows from the results obtained by precritical reflection techniques and deep seismic sounding (Pavlenkova 2006). Moreover, gravimetric data testifies that isostatic crustal equilibrium occurs at the M discontinuity but not in the bottom part of the asthenosphere. This provides support for the hypothesis that rheologically weakened layers can occur, and their material can likely flow. The observed rock characteristics are explained by most researchers by elevated fluid contents (Lobkovsky et al. 2004). The role of fluid in crustal and mantle processes was emphasized by L. Perchuk (1973), who believed that the principal driving force of melting processes was fluid.

A combination of several factors (high temperature, a nearly molten state of the material, high fluid contents in the rocks, and stress buildup in the lithosphere) could result in laterally extensive (as large as hundreds of kilometers, as follows from seismic data) horizontal basalt magma chambers in weakened zones in the bottom portion of the continental crust. A possibility of formation of large volume of crust melt was shown in publication (Sobolev et al. 2011). Details of the melting mechanism can, however, be elucidated only based on specialized studies.

The melting of the lower crust can explain more logically than assimilation the following geological facts:

1. The occurrence of homogeneous lava flows over vast territories (e.g., the Nadayansky plagiophyric tholeiite flow in the lower part of the Mokulaevsky Formation covers an area of 48,000 km^2; Starosel'tsev 1989). The possible scale of naturally occurring assimilation processes has still not been estimated, nor has the time needed for the homogenization of all parameters of the melt, newly produced by magma interaction with its host rocks or by the mingling of different melts. It is still uncertain as to whether such a process could occur, considering the high accumulation rate of the whole tuff–lava sequence of the Siberian Platform.
2. The absence of sulfides from basalts of all of the formations. If a single mechanism had operated during the crustal contamination of the mantle magmas, a certain portion of the sulfide droplets would have been brought to the surface, and hence, the basalt should had hosted mineralized zones or sulfide-rich lavas, if not mineral deposits. However, the volcanic rocks are practically completely devoid of sulfide minerals or contain only very low concentrations of these minerals.
3. Deposits of different volume are hosted in certain structures. Again, if a universal mechanism of melt contamination with host rocks had operated, deposits or at least occurrences of Cu–Ni ores should have been found else-

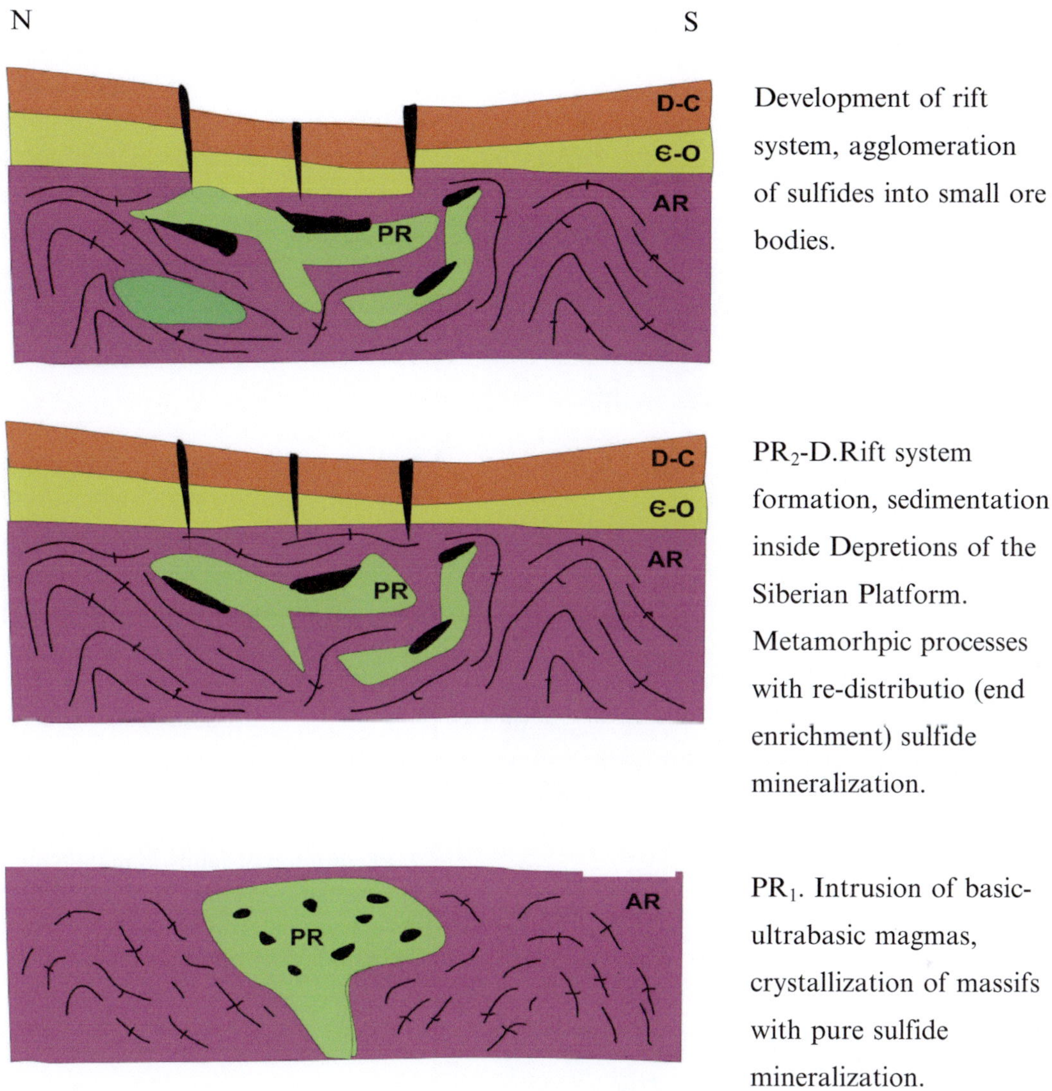

Fig. 9.2 Schema of magmatism evolution for the first etap

where in the Siberian Platform. Numerous dolerite sills in the southern part of the latter locally account for up to 50 % of the total volume of the magmatic rocks and are mostly exposed at the surface, which very much simplifies searches for mineral deposits in this area compared to the northern part of the platform, but no sulfide ores have been found the as of yet.

The occurrence of ore mineralization only in certain crustal zones is controlled by the unusual structures of the latter. The origin of uniquely large deposits can thereby be interpreted as resulting from the melting of a source material already enriched in sulfides. The metals were most likely enriched during a number of episodes (Figs. 9.2 and 9.3), including (1) their initial synthesis in Proterozoic time and (2) mobilization and entrainment with melts in Early Mesozoic time period. This leads to the idea that the Noril'sk deposits were produced over a long time period.

The disseminated sulfides started to crystallize already during the growth of the continental crust, when the Proterozoic ultramafic–mafic magmas were emplaced and gave rise to large intrusive bodies, which typically hosted, however, merely small orebodies of Cu–Ni ores worldwide. Such bodies are also known in the Siberian Platform and are related to the Chineysky, Luktursky, Dovyrensky, Chaisky, and other massifs. This process is the most probable in long-lived rift structures, which were formed in the Paleoproterozoic in various areas, including the northern Siberian craton, as was mentioned by several researchers (Dyuzhikov et al. 1988; Gorbunov et al. 2011; Dodin et al. 2011).

The long-lasting (from the Proterozoic to Mesozoic) residence of the sulfide in lower crustal environments led to the enrichment of the metals in the course of metamorphic processes at the crust–mantle interface. The melting of vast volumes of material (likely under the effect of a plume or induced by other mechanisms) during the origin of trap asso-

<table>
<tr><td>T_1mr-sm</td><td>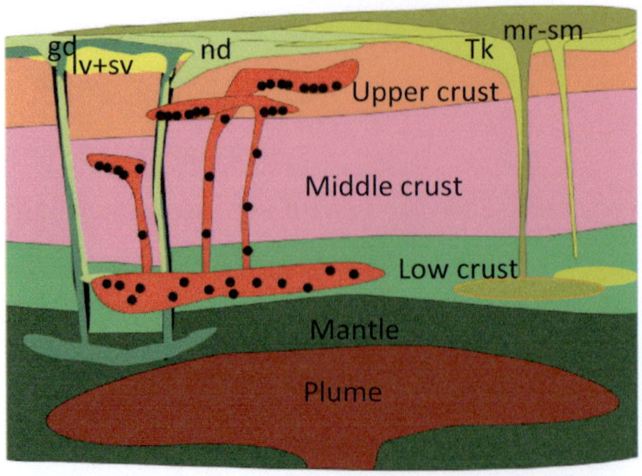
</td><td>4. Melting of Lower crust containing sulfides, saturation of magmas by sulfide (inside rift zones). Transport of sulfide melt (in form of small drops) to the surface. Location of sulfide-bearing magmas predominatingly in terrigenouse rocks (depth 1-3 km). Settling of sulfides and their crystallization.</td></tr>
<tr><td>T_1nd+tk</td><td>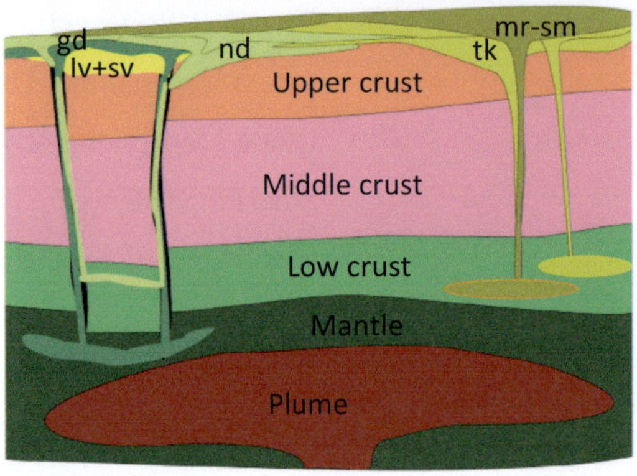
</td><td>3. Appearing of diffused rifting zones inside Siberian Platforme along with rifting zonez in the Northern part of the Platform. Simultaneous developing processes on West and East of the Noril'sk region: explosion of the Low Nadezhdinsky tuffs and effusion of the Tuklonsky basalts.</td></tr>
<tr><td>T_1gd</td><td>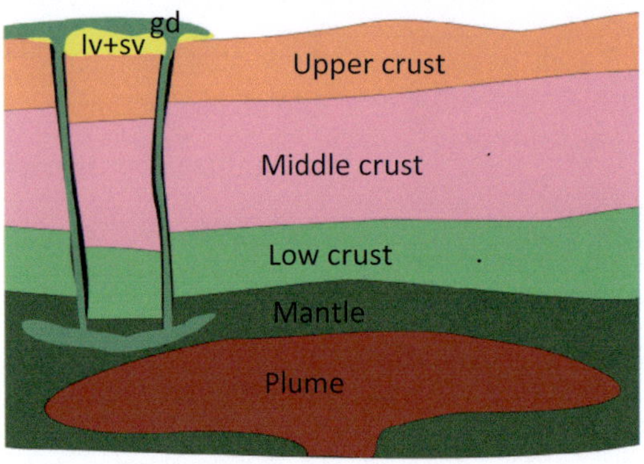
</td><td>2. Penetration of faults in mantle due to increase of exertions. Forming of magmatic chambers and effusion of the Gudchikhinsky picrite lavas.</td></tr>
<tr><td>T_1iv+sv</td><td>1.
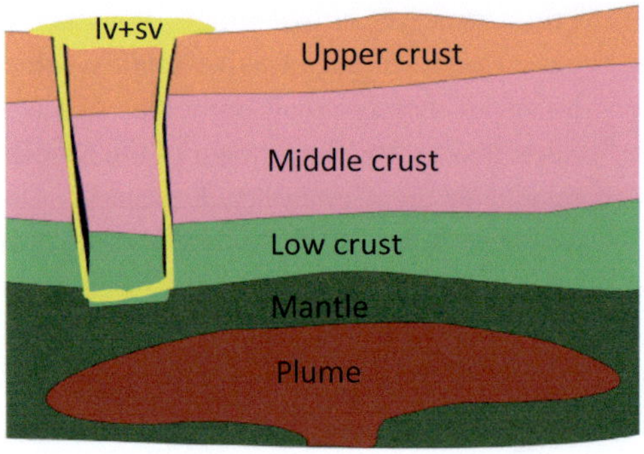
</td><td>1.Activation of ancient paleorift zonez (Noril'sk-Igarka and Yenisey-Khatangsky). Effusion of subalcaline lavas of the Ivakinsky and Syverminsky Formations.</td></tr>
</table>

Fig. 9.3 Schema of magmatism evolution for the second etap

ciations in the Early Mesozoic and the emplacement of melts into upper crustal levels aided the remobilization of the sulfides and their transport to the surface.

The absence of correlations between geochemical characteristics of the melts and ores can be explained by the magmas that played a predominantly transport role. The possibility of the transfer of great volumes (up to 10 %) of sulfide melts in the form of small droplets was experimentally demonstrated (Likhachev 2006).

The unique PGE enrichment of sulfides in the Noril'sk ores can be explained by the long-lasting residence of the sulfide droplets in the melt when they were transferred from the source to the crystallization chamber (for a distance of more than 40 km). In this situation, the mechanism of the enrichment of sulfide globules in PGE (Naldrett 1992) should have operated even more efficiently than at the movement of the melt only within the platform cover (i.e., for a distance of 1–2 km). Conceivably, this mechanism accounts for the higher PGE concentrations of the Noril'sk ores than at other Cu–Ni deposits worldwide that are related to Proterozoic massifs in the basements of platforms. It cannot be ruled out that the source material was originally enriched in PGE and base metals, which could be mobilized from, for example, black shales in the source.

The crystallization of the ore-bearing magmas began during their ascent to the surface and terminated in chambers that were hosted mostly in Devonian sedimentary rocks as the most lithologically favorable for melt emplacement. The whole volume of the sulfide was deposited where the intrusions crystallized.

Our analysis of the composition of ore-bearing intrusions in various parts of Russia (northern Transbaikalia, Karelia and Kola Peninsula, and the Eastern Sayan Range) reveals amazing compositional similarities of the ore-bearing rocks: regardless of their age, all of them are geochemically similar (including their isotopic composition) and show geochemical parameters typical of the lower crust. The only exception is the sulfur isotopic composition of sulfides in the orebodies, which is usually remarkably different from the sulfur isotopic composition of Noril'sk deposits and which usually only insignificantly varies about zero ($\delta^{34}S = 0 \pm 2‰$).

It should be mentioned that the scheme suggested above for the origin of the Noril'sk deposits is consistent with the methodological fundamentals of the origin of ore deposits suggested by L. Ovchinnikov (1988), who repeatedly attracted attention to the accompanying character of ore-forming processes and their random occurrence in nature. Instead of a comprehensive analysis of the situation, the interrelated system of geological factors is often substituted for interpretations of a few selected processes and phenomena (for instance, sulfur solubility, assimilation, etc.) that are thought to play a decisive role.

L. Ovchinnikov (1988, p. 242) wrote that "the origin of an ore deposit is predated by a long-lasting and complicated evolutionary prehistory." "Genes of the would-be ore were produced during the earliest geological evolution of the Earth and its crust." The probability of the accumulation of economic concentrations of metals in the Earth's crust is low because "this always requires additional "efforts" of nature to concentrate usually scattered metals… All…logical errors in studying ore-forming processes… stem, first and foremost, from ignoring the probabilistic nature of laws in geology and from that no theory can, in principle, be developed if this probability is ignored." These ideas were, in a sense, confirmed by the negative results of long-lasting searches for any exploration guides for massifs with rich Cu–Ni ores in the Noril'sk district.

9.1 General Conclusions

1. The volcanic rocks of the Noril'sk districts were formed during two episodes (rifting and trap per se) and in four stages.
2. The ore-bearing intrusions have no comagmatic analogues among the rocks of the tuff–lava sequence and were formed by an independent pulse of magmatic activity during the trap episode of the regional evolutionary history.
3. The intrusions crystallized from tholeiite melts of elevated Mg#.
4. The fluid regime of trap magmatism was not anomalous: the fluid was H_2O-CO_2; the parental melts of the ore-bearing and barren massifs had similar and low concentrations of volatile components (H_2O, CO_2, and F).
5. Assimilation processes did not play any appreciable role in producing the ore deposits.
6. The parental melts of the magmatic rocks of the trap association were derived from lower crustal rocks.

References

Al'mukhamedov AI, Medvedev AY, Zolotukhin VV (2004) Evolutions of Permo-Triassic basalts of the Siberian platform in time and space. Petrology 12(4):339–353

Arndt NT, Czamanske GK, Walker RJ, Chauvel C, Fedorenko VA (2003) Geochemistry and origin of the intrusive hosts of the Noril'sk-Talnakh Cu-Ni-PGE sulfide deposits. Econ Geol 98:495–515

Bogatikov OA, Kovalenko VI, Sharkov EV (2010) Magmatism, tectonic, geodynamic of the earth. Nauka, Moscow (in Russian)

Dodin DA, Zoloev KK, Koroteev VA, Chernyushev NM (2011) Platinum of Russia: the state and problems. Platin Russia 7:12–51 (in Russian)

Dyuzhikov OA, Distler VV, Strunin BM et al (1988) Geology and ore potential of the Noril'sk ore district, Nauka, Moscow. Translated: Dyuzhikov OA, Distler VV, Strunin BM, Mkrtychyan AK, ShermanML, Sluzhenikin SF, Lurye AM (1992) Geology and metallogeny of sulfide deposits Noril'sk region USSR. Econ Geol Monogr 1, Ontario, Spec vol 24

Farquhar J, Wu N, Canfield DE, Odoro H (2010) Connections between sulfur cycle evolution, sulfur isotopes, sediments, and base metal sulfide deposits. Econ Geol 105:509–533

Godlevsky MN (1959) Traps and ore-bearing intrusions of the Noril'sk district. Gosgeoltekhizdat, Moscow, 61p (in Russian)

Gongalsky BI, Krivolutskaya NA (2014) Geochemical peculiarities and mineralization of Mafic magmas: lack of correlation. Dokl Earth Sci 459(1):470–473

Gorbatsevich FF, Vetrin VP, Trishina OM, Kovalevsky MV (2013). Model of properties variations and conditions for rocks of upper, middle and lower continental crust in Kola-Norway block, Kola peninsula. Physico-chemical and petrochemical studies in Earth Sciences Moscow, pp 82–85 (in Russian)

Gorbunov GI, Dyuzhikov OA, Sharkov EV (2011) Conditions of formations of big and unique Pt-Cu-Ni deposits. Platin Russ 7:64–73 (in Russian)

Grinenko LN (1984) H_2S-bearing gas reservoirs as sulfur source during sulfurization of magmas of economically ore intrusions of the Noril'sk region. Dokl Akad Nauk SSSR 278(3):730–732 (in Russian)

Grinenko LN (1985) Sources of sulfur of the nickeliferous and barren gabbrodolerite intrusions of the northwest Siberian platform. Int Geol Rev 27:695–708

Ivanov AV, He H, Yan L, Ryabov VV, Shevko AY, Stanislav V, Palesskii SV, Irina V, Nikolaeva IV (2013) Siberian Traps large igneous province: evidence for two flood basalt pulses around the Permo-Triassic boundary and in the Middle Triassic, and contemporaneous granitic magmatism. Earth Sci Rev 122:58–76

Kovalenker VA, Gladyshev GD, Nosik LP (1974) Sulfur isotope composition of sulfides from the deposits of the Talnakh ore cluster in relation with their Se potential. Izv Akad Nauk SSSR Ser Geol 2:10–35 (in Russian)

Krivolutskaya NA (2011) Formation of PGM-Cu-Ni deposits in the process of evolution of flood-basalt magmatism in the Noril'sk region. Geol Ore Depos 53(4):309–339

Krivolutskaya NA (2014) Evolution of trap magmatism and processes producing Pt-Cu-Ni mineralization in the Noril'sk area. KMK, Moscow, 320 p (in Russian)

Krivolutskaya NA, Gongalskiy BI (2013) Dyumtaleysky massif, Taimyr, Russia: a magmatic derivative from a complex mantle source. Mineral deposit research for a high-tech word. In: 12th biennial SGA meeting, proceedings, Sweden, Uppsala, vol 3. Ore deposits associated with mafic and ultramafic rocks, pp 1036–1039

Krivolutskaya NA, Sobolev AV, Snisar SG, Gongalskiy BI, Hauff B, Kuzmin DV, Tushentsova IN, Svirskaya NM, Kononkova NN, Schlychkova TB (2012) Mineralogy, geochemistry and stratigraphy of the Maslovsky Pt-Cu-Ni sulfide deposit, Noril'sk region, Russia: implications for relationship of ore-bearing intrusions and lavas. Mineral Deposita 47:69–88

Kryazhev SG, Glukhov AP, Rusinova OV et al (2003) Isotopic geochemical regime of the formation of the Sovetskoe Gold–Quartz deposit. In: Applied geochemistry. IMGRE, Moscow, issue 4, pp 154–164 (in Russian)

Kuskov OL, Kronrod VA, Prokof'ev AA, Pavlenkova NI et al (2014) Petrological–geophysical models of the internal structure of the lithospheric mantle of the Siberian Craton. Petrology 22:17–44

Li CS, Ripley EM, Naldrett AJ (2009a) A new genetic model for the giant Ni–Cu-PGE sulfide deposits associated with the Siberian flood basalts. Econ Geol 104:291–301

Li C, Naldrett AJ, Shmitt AK et al (2009b) Magmatic anhydrite – sulfide assemblages in plumbing system of the Siberian traps. Geology 37:259–262

Lightfoot PC, Howkesworth CJ, Hergt J et al (1993) Remobilisation of the continental lithosphere by a mantle plume: major-, trace element, and Sr-, Nd-, and Pb isotopic evidence from picritic and tholeiitic lavas of the Noril'sk district, Siberian trap, Russia. Contrib Mineral Petrol 114:171–188

Likhachev AP (1965) The role of Leucocratic Gabbro in the origin of Noril'sk differentiated intrusions. Izv Akad Nauk SSSR Ser Geol 12:50–66 (in Russian)

Likhachev AP (1978) On the formation conditions of ore-bearing and non-bearing Mafic–Ultramafic magmas. Dokl Akad Nauk SSSR 238:447–450 (in Russian)

Likhachev AP (1996) Emplacement dynamics of the Talnakh orebearing intrusions and related PGM–Cu–Ni ores. Otechestvennaya Geol 8:20–26 (in Russian)

Likhachev AP (2006) Platinum–copper–nickel and platinum deposits. Eslan, Moscow, 496 p, (in Russian)

Likhachev AP, Strizhov VP (1977) Fractionation of sulfur isotopes. Dokl Akad Nauk SSSR 236(1):223–226 (in Russian)

Lobkovsky LI, Nikishin AM, Khain VE (2004) Modern problems of geotectonic and geodynamics. M., Nauchny Mir, 611 p (in Russian)

Malitch KN, Belousova EA, Griffin WL, Badanina IY, Pearson NJ, Presnyakov SL, Tuganova EV (2010) Magmatic evolution of the ultramafic–mafic Kharaelakh intrusion (Siberian Craton, Russia): insights from trace-element, U–Pb and Hf-isotope data on zircon. Contrib Mineral Petrol 159:753–768

Naldrett AJ (1992) A model for the Ni–Cu–PGE ores of the Noril'sk region and its application to other areas of flood basalts. Econ Geol 87:1945–1962

Ovchinnikov LN (1988) Origin of ore deposits. Nedra, Moscow, 355 p, (in Russian)

Pavlenkova NI (2006) Fluid regime of upper covers of the earth according geophysical data. In: Fluids and geodynamics. Nauka, Moscow, pp 201–218 (in Russian)

Perchuk LL (1973) Thermodynamic regime of deep-seated petrogenesis. Nauka, Moscow, 318 p, (in Russian)

Polyansky OP, Reverdatto VV (2006) Role of fluid in transport of heat inside sedimentary basins with trap magmatism. In: Fluids and geodynamics. Nauka, Moscow, pp 219–243 (in Russian)

Rad'ko VA (1991) Model of geodynamic differentiation of the intrusive traps of the northwestern Siberian platform. Geol Geofiz 11:19–27

Reichow MK, Saunders AD, White RV, Al'mukhamedov AI, Medvedev AY (2005) Geochemistry and petrogenesis of basalts from the West Siberian Basin: an extension of Permo-Triassic Siberian traps. Russiaю Lithos 79:425–452

Ripley EM, Lightfoot PC, Li C et al (2003) Sulfur isotopic studies of continental flood basalts in the Noril'sk region: implications for the association between lavas and ore-bearing intrusion. Geochim Cosmochim Acta 67:2805–2817

Rudnick RL, Fountain DM (1995) Nature and composition of the continental crust: a low crustal perspective. Rev Geophys 33(3):267–309

Rudnick RL, Gao S. (2003) Composition of the continental crust. Treatise on geochemistry, vol 3. Elsivier-Pergamon, Oxford pp 1–64

Saunders AD, England RW, Reichow MK, White RV (2005) A mantle plume origin for the Siberian traps: uplift and extension in the West Siberian Basin, Russia. Lithos 79:407–424

Sobolev SV, Sobolev AV, Kuzmin DV, Krivolutskaya NA, Petrunin AG, Arndt NT, Radko VA, Vasilev YR (2011) Linking mantle plumes, large igneous provinces and environmental catastrophes. Nature 477:312–316

Starosel'tsev VS (1989) Tectonic of lavas' plateau. Nedra, Moscow, 258 p (in Russian)

Appendix 10

N.A. Krivolutskaya, *Siberian Traps and Pt-Cu-Ni Deposits in the Noril'sk Area*,
DOI 10.1007/978-3-319-17205-7_10

Table 10.1 Rock compositions of the Maslovsky deposit

No	1	2	3	4	5	6	7	8
No sample	OM-4/854.8	OM-4/881.7	OM-4/907.6	OM-4/1,007.7	OM-4/1,008.2	OM-4/1,010.8	OM-4/1,013.2	OM-4/1,015
SiO_2	47.95	49.06	51.30	42.62	41.97	42.19	40.91	42.31
TiO_2	1.29	1.13	0.36	0.41	0.40	0.38	0.38	0.31
Al_2O_3	13.52	15.12	16.44	6.57	5.96	6.54	6.69	5.55
FeO	11.51	11.90	7.46	15.62	15.67	15.49	16.69	16.11
MnO	0.76	0.20	0.14	0.22	0.21	0.24	0.22	0.25
MgO	7.11	7.11	8.49	29.59	31.00	30.59	29.60	31.26
CaO	11.72	11.50	12.53	4.92	4.18	4.38	4.35	3.78
Na_2O	2.64	2.90	2.64	0.42	0.48	0.46	0.48	0.41
K_2O	0.64	0.54	0.62	0.22	0.28	0.25	0.22	0.20
P_2O_5	0.15	0.11	0.06	0.06	0.07	0.05	0.05	0.04
Total	97.30	99.59	100.07	100.64	100.22	100.55	99.60	100.22
Rb	20.6	19.0	20.8	6.8	9.4	7.5	6.2	8.2
Ba	261	105	123	82	62	103	96	87
Th	1.12	0.85	1.36	0.40	0.43	0.27	0.39	0.18
U	0.62	0.50	0.57	0.12	0.16	0.11	0.14	0.07
Nb	4.99	3.29	3.02	1.52	1.44	1.40	1.72	0.82
Ta	0.32	0.23	0.19	0.12	0.11	0.11	0.14	0.07
La	8.60	5.90	6.90	2.80	2.27	2.38	2.77	2.05
Ce	19.09	14.11	15.63	6.39	5.46	5.02	5.81	4.30
Pr	2.59	1.98	2.09	0.86	0.76	0.67	0.80	0.55
Nd	12.20	9.54	9.70	4.01	3.70	3.07	3.63	2.57
Sr	297	304	243	107	83	91	83	86
Sm	3.34	2.75	2.62	1.11	1.06	0.84	1.00	0.69
Zr	94.6	92.8	89.0	26.9	25.0	22.2	34.0	14.8
Hf	2.38	2.41	2.28	0.72	0.67	0.61	0.89	0.41
Eu	1.24	1.09	0.82	0.42	0.31	0.39	0.40	0.33
Ti	7,481	6,407	2,003	2,515	2,375	2,237	2,516	1,678
Gd	3.99	3.40	3.21	1.36	1.28	1.04	1.19	0.87
Tb	0.65	0.58	0.53	0.23	0.21	0.17	0.20	0.15
Dy	4.40	3.82	3.68	1.52	1.45	1.18	1.34	1.01
Ho	0.90	0.78	0.77	0.32	0.30	0.24	0.30	0.22
Y	22.71	19.96	19.24	8.51	7.96	6.53	7.89	5.98
Er	2.57	2.17	2.11	0.93	0.89	0.73	0.87	0.66
Tm	0.35	0.30	0.29	0.13	0.12	0.11	0.13	0.10
Yb	2.37	2.07	2.08	0.91	0.86	0.76	0.87	0.72
Lu	0.36	0.31	0.31	0.14	0.13	0.11	0.14	0.12

No	9	10	11	12	13	14	15
No sample	OM-4/1,016	OM-4/1,020.2	OM-4/1,022	OM-4/1,026.5	OM-4/1,029.3	OM-4/1,030.5	OM-4/1,031.4
SiO_2	41.57	42.44	41.79	41.91	38.75	43.25	43.35
TiO_2	0.28	0.43	0.29	0.29	2.83	0.46	0.46
Al_2O_3	5.28	5.85	5.15	5.07	9.42	5.91	6.67
FeO	16.78	15.96	16.28	16.28	19.37	14.38	14.96
MnO	0.25	0.25	0.23	0.23	0.10	0.22	0.23
MgO	30.75	30.42	31.15	31.80	5.44	31.28	28.25
CaO	3.95	3.81	4.10	3.58	10.89	4.51	5.14
Na_2O	0.48	0.59	0.33	0.29	2.16	0.44	0.46
K_2O	0.24	0.22	0.16	0.21	0.16	0.24	0.21
P_2O_5	0.05	0.06	0.04	0.05	0.03	0.06	0.05
Total	99.63	100.01	99.55	99.71	89.18	100.77	99.79
Rb	5.5	5.6	6.5	7.1	2.5	9.2	7.0
Ba	86	89	40	48	134	60	87
Th	0.33	0.27	0.33	0.29	1.69	0.40	0.41
U	0.10	0.09	0.09	0.12	0.61	0.14	0.16
Nb	1.18	1.75	0.96	1.04	14.14	1.65	1.60
Ta	0.09	0.26	0.08	0.08	1.11	0.12	0.12
La	2.47	3.03	1.68	1.99	4.45	2.25	3.10
Ce	5.49	6.47	4.37	4.60	12.42	5.46	7.08
Pr	0.72	0.86	0.63	0.63	2.09	0.77	0.97
Nd	3.29	3.87	2.95	2.87	11.11	3.70	4.52
Sr	97	98	85	90	371	86	109
Sm	0.85	1.08	0.83	0.78	3.69	1.10	1.27
Zr	24.4	28.9	26.3	27.1	203.5	32.5	32.2
Hf	0.61	0.66	0.62	0.67	5.18	0.87	0.82
Eu	0.38	0.40	0.26	0.30	0.86	0.34	0.44
Ti	1,853	2,281	1,786	1,715	17,064	2,682	2,664
Gd	1.02	1.25	1.03	0.98	4.61	1.41	1.53
Tb	0.17	0.20	0.17	0.17	0.80	0.23	0.26
Dy	1.13	1.30	1.19	1.15	5.41	1.62	1.74
Ho	0.24	0.28	0.26	0.25	1.12	0.35	0.37
Y	6.68	7.85	6.87	6.79	30.41	9.32	9.85
Er	0.72	0.81	0.74	0.73	3.21	1.02	1.08
Tm	0.10	0.11	0.11	0.11	0.45	0.14	0.15
Yb	0.74	0.80	0.75	0.76	3.01	0.99	1.03
Lu	0.11	0.12	0.12	0.12	0.43	0.15	0.16

(continued)

Table 10.1 (continued)

No	16	17	18	19	20	21	21
No sample	OM-4/1,032.3	OM-4/1,032.7	OM-4/1,037.4	OM-4/1,042.4	OM-4/1,046.2	OM-4/1,047.5	OM-4/1,050.4
SiO_2	43.80	43.30	43.08	42.74	43.79	44.01	44.45
TiO_2	0.44	0.36	0.45	0.45	0.45	0.52	0.50
Al_2O_3	7.43	5.75	6.30	7.32	7.28	8.23	9.59
FeO	14.46	15.55	15.89	16.05	14.90	14.86	13.75
MnO	0.22	0.21	0.22	0.23	0.23	0.23	0.21
MgO	27.30	29.62	28.77	26.48	26.54	24.99	23.65
CaO	5.34	4.80	4.70	5.28	5.47	5.67	6.71
Na_2O	0.57	0.40	0.41	0.70	0.61	0.65	0.85
K_2O	0.23	0.19	0.16	0.20	0.23	0.32	0.24
P_2O_5	0.04	0.05	0.04	0.07	0.05	0.06	0.06
Total	99.84	100.24	100.03	99.52	99.55	99.55	100.03
Rb	6.1	7.0	5.7	5.6	6.5	11.7	8.0
Ba	72	73	90	92	126	127	103
Th	0.26	0.38	0.42	0.39	0.44	0.49	0.42
U	0.09	0.13	0.14	0.14	0.14	0.17	0.17
Nb	1.24	1.72	2.44	1.49	1.76	2.03	1.82
Ta	0.11	0.12	0.18	0.11	0.14	0.16	0.15
La	2.30	2.30	3.33	3.11	3.65	3.94	3.25
Ce	5.17	5.53	7.66	6.93	8.08	8.76	7.30
Pr	0.72	0.77	1.03	0.93	1.09	1.16	0.98
Nd	3.48	3.71	4.81	4.35	5.15	5.41	4.63
Sr	95	100	90	101	118	122	142
Sm	1.00	1.04	1.30	1.17	1.45	1.52	1.28
Zr	25.1	36.7	34.0	30.2	32.4	38.4	36.4
Hf	0.65	0.94	0.93	0.81	0.82	1.01	0.93
Eu	0.39	0.36	0.45	0.47	0.57	0.55	0.50
Ti	2,093	2,813	3,760	2,669	2,703	3,195	3,010
Gd	1.29	1.33	1.61	1.49	1.80	1.83	1.60
Tb	0.22	0.22	0.27	0.24	0.30	0.31	0.26
Dy	1.45	1.53	1.80	1.64	2.02	2.07	1.79
Ho	0.31	0.32	0.37	0.35	0.42	0.43	0.37
Y	8.60	8.81	10.10	9.41	11.65	11.87	10.27
Er	0.95	0.97	1.08	1.02	1.23	1.25	1.10
Tm	0.14	0.14	0.15	0.15	0.18	0.18	0.16
Yb	0.91	0.97	1.06	0.96	1.21	1.24	1.08
Lu	0.14	0.15	0.16	0.15	0.18	0.19	0.16

No	22	23	24	25	26	27	28
No sample	OM-4/1,054.7	OM-4/1,062	OM-4/1,095	OM-4/1,097.2	OM-24/506.2	OM-24/512.2	OM-24/529.5
SiO_2	45.01	48.26	47.70	49.58	56.60	57.45	54.02
TiO_2	0.65	1.78	1.14	1.11	1.79	1.67	1.93
Al_2O_3	9.63	14.40	15.89	14.26	13.39	13.14	13.07
FeO	13.73	13.75	11.55	11.34	14.13	14.45	17.84
MnO	0.23	0.24	0.25	0.69	0.30	0.26	0.30
MgO	22.13	6.32	7.17	6.72	3.09	2.50	3.14
CaO	6.81	11.00	11.25	10.55	5.18	5.29	5.86
Na_2O	1.00	2.57	2.13	2.06	4.01	3.37	2.80
K_2O	0.38	0.41	0.58	0.51	1.59	1.36	0.54
P_2O_5	0.07	0.24	0.14	0.15	0.36	0.49	0.46
Total	99.65	98.97	97.85	97.03	100.44	99.99	99.96
Rb	12.5	10.4	17.5	17.8	43.7	33.9	12.8
Ba	155	146	214	245	465	392	254
Th	0.52	1.43	0.87	1.10	3.68	5.07	5.29
U	0.20	0.65	0.37	0.54	1.19	1.74	1.85
Nb	2.25	6.03	4.07	4.02	15.29	16.65	17.68
Ta	0.16	0.40	0.25	0.25	0.93	0.98	1.06
La	4.09	10.29	6.98	4.97	25.04	29.89	29.19
Ce	9.06	23.98	16.28	12.70	54.60	67.03	65.07
Pr	1.22	3.38	2.24	1.92	7.38	9.00	8.81
Nd	5.74	16.31	10.78	9.70	34.23	41.46	40.67
Sr	158	215	247	220	168	287	276
Sm	1.63	4.68	2.97	2.85	8.75	10.60	10.61
Zr	45.5	129.4	68.8	78.6	299.8	336.3	347.9
Hf	1.24	3.29	1.74	2.01	7.19	8.47	8.86
Eu	0.62	1.63	1.22	1.14	2.53	2.99	2.87
Ti	4,008	10,237	6,497	6,352	11,434	10,189	11,934
Gd	2.01	5.68	3.63	3.71	10.11	12.29	12.37
Tb	0.34	0.98	0.61	0.62	1.67	2.04	2.05
Dy	2.27	6.40	4.08	4.10	11.23	13.38	13.67
Ho	0.48	1.37	0.85	0.84	2.29	2.78	2.84
Y	12.84	34.77	21.09	21.84	62.02	76.93	77.16
Er	1.39	3.73	2.28	2.37	6.85	8.17	8.47
Tm	0.20	0.52	0.31	0.33	1.00	1.17	1.22
Yb	1.39	3.69	2.23	2.30	6.87	8.09	8.49
Lu	0.21	0.53	0.32	0.33	1.01	1.19	1.26

(continued)

Table 10.1 (continued)

No	29	30	31	32	33	34	35
No sample	OM-24/537.8	OM-24/544.8	OM-24/565.9	OM-24/575.5	OM-24/584.5	OM-24/597.4	OM-24/608
SiO_2	46.35	54.73	53.66	49.85	50.65	48.58	46.29
TiO_2	2.05	1.81	2.29	2.74	2.70	2.37	2.75
Al_2O_3	13.41	13.00	13.21	11.87	12.81	12.48	12.72
FeO	18.36	16.84	16.19	18.92	17.27	17.10	19.57
MnO	0.18	0.33	0.28	0.27	0.29	0.26	0.27
MgO	6.27	1.98	3.61	4.60	4.41	5.86	5.55
CaO	11.16	5.86	6.05	8.06	8.52	10.23	10.31
Na_2O	2.14	3.22	3.68	2.83	2.70	2.58	2.21
K_2O	0.42	1.55	1.24	0.66	0.58	0.55	0.33
P_2O_5	0.07	0.55	0.16	0.23	0.08	0.14	0.12
Total	100.41	99.88	100.39	100.04	100.04	100.14	100.14
Rb	9.3	39.6	32.0	18.3	13.0	13.2	6.6
Ba	117	550	354	192	185	183	186
Th	0.99	5.81	2.82	2.15	1.97	1.56	1.19
U	0.29	2.03	0.91	0.74	0.58	0.48	0.41
Nb	3.33	19.37	10.53	7.54	7.33	4.61	4.29
Ta	0.21	1.15	0.64	0.46	0.45	0.29	0.26
La	6.92	35.95	19.18	12.12	13.01	9.01	7.75
Ce	15.06	78.75	42.31	27.59	27.54	20.61	17.41
Pr	2.06	10.46	5.71	3.84	3.83	2.86	2.46
Nd	9.72	47.74	26.74	18.06	17.95	13.57	11.77
Sr	200	168	268	233	251	243	458
Sm	2.72	12.17	7.05	5.01	5.02	3.83	3.42
Zr	71.6	395.3	201.6	137.7	151.1	108.7	95.8
Hf	2.00	9.72	5.12	3.63	3.98	2.94	2.53
Eu	0.94	3.43	2.28	1.62	1.62	1.35	1.22
Ti	12,590	11,183	14,391	17,246	16,672	12,254	16,469
Gd	3.46	14.21	8.52	6.17	6.02	4.87	4.27
Tb	0.59	2.32	1.38	1.02	1.01	0.81	0.71
Dy	3.87	15.55	9.28	6.86	6.80	5.54	4.78
Ho	0.82	3.18	1.93	1.41	1.39	1.16	0.99
Y	22.02	87.95	52.50	38.59	38.57	31.13	27.04
Er	2.39	9.44	5.67	4.17	4.15	3.41	2.91
Tm	0.35	1.35	0.80	0.61	0.61	0.50	0.42
Yb	2.45	9.21	5.58	4.10	4.22	3.42	2.83
Lu	0.35	1.38	0.81	0.61	0.62	0.51	0.42

No	36	37	38	39	40	41	42
No sample	OM-24/616.5	OM-24/647.2	OM-24/650.8	OM-24/655.8	OM-24/676.7	OM-24/688.2	OM-24/696.7
SiO_2	45.40	50.78	51.59	51.27	50.49	51.18	51.11
TiO_2	2.79	1.13	0.96	1.08	1.03	1.14	1.10
Al_2O_3	11.97	14.67	14.10	13.76	15.66	14.39	15.14
FeO	20.51	11.30	10.30	11.00	10.44	10.21	10.43
MnO	0.26	0.21	0.17	0.20	0.20	0.20	0.17
MgO	5.97	6.89	7.23	7.17	7.04	7.20	7.09
CaO	10.95	12.31	11.73	12.31	12.85	13.15	12.26
Na_2O	2.14	2.43	2.91	2.76	2.18	2.17	2.32
K_2O	0.35	0.38	0.91	0.59	0.35	0.38	0.44
P_2O_5	0.11	0.12	0.10	0.12	0.09	0.10	0.13
Total	100.46	100.21	100.00	100.27	100.35	100.13	100.20
Rb	6.8	8.6	23.9	17.8	7.0	8.1	9.7
Ba	143	123	149	144	112	131	141
Th	1.05	1.06	1.02	1.02	1.06	1.15	1.32
U	0.33	0.39	0.36	0.36	0.37	0.41	0.46
Nb	3.61	3.71	3.57	3.57	3.75	3.91	4.39
Ta	0.23	0.23	0.21	0.21	0.22	0.23	0.27
La	6.66	7.14	6.81	6.60	7.01	7.24	8.12
Ce	15.34	16.22	15.46	15.37	16.08	16.94	18.62
Pr	2.16	2.24	2.15	2.14	2.19	2.34	2.57
Nd	10.52	10.69	10.29	10.32	10.45	10.99	12.16
Sr	311	232	223	241	241	237	244
Sm	3.06	2.97	2.93	2.96	2.91	3.10	3.37
Zr	80.7	80.4	75.6	73.8	76.2	78.9	92.9
Hf	2.22	2.18	2.02	2.00	2.03	2.13	2.46
Eu	1.02	1.05	1.03	1.05	1.03	1.07	1.13
Ti	15,932	6,951	6,798	6,480	6,684	7,010	6,895
Gd	3.91	3.76	3.68	3.67	3.52	3.83	4.13
Tb	0.66	0.63	0.63	0.62	0.60	0.64	0.69
Dy	4.48	4.24	4.29	4.15	4.01	4.24	4.61
Ho	0.93	0.88	0.88	0.84	0.83	0.88	0.96
Y	24.62	23.28	23.56	22.65	22.09	23.43	25.55
Er	2.69	2.55	2.58	2.49	2.41	2.59	2.83
Tm	0.39	0.37	0.37	0.35	0.35	0.36	0.41
Yb	2.62	2.52	2.47	2.43	2.36	2.53	2.68
Lu	0.39	0.37	0.37	0.36	0.34	0.37	0.40

(continued)

Table 10.1 (continued)

No	43	44	45	46	47	48	49
No sample	OM-24/705.4	OM-24/717.9	OM-24/733	OM-24/758.6	OM-24/769.8	OM-24/789.9	OM-24/803.3
SiO_2	50.31	50.50	50.43	50.05	49.39	48.76	48.37
TiO_2	1.15	1.08	0.94	1.04	0.85	0.75	0.80
Al_2O_3	15.04	14.45	15.96	15.85	18.80	18.36	16.91
FeO	10.73	10.73	9.55	9.85	9.42	9.15	10.10
MnO	0.16	0.18	0.18	0.17	0.13	0.14	0.17
MgO	7.21	7.49	7.33	7.77	7.59	9.53	11.20
CaO	13.10	13.25	13.00	12.98	12.00	11.53	10.77
Na_2O	2.20	2.12	2.23	2.15	2.16	2.07	1.95
K_2O	0.35	0.29	0.38	0.31	0.30	0.30	0.44
P_2O_5	0.13	0.12	0.09	0.10	0.08	0.09	0.10
Total	100.38	100.23	100.09	100.28	100.74	100.70	100.83
Rb	7.8	6.7	7.1	6.3	7.3	7.6	11.1
Ba	118	100	135	106	96	86	99
Th	1.04	0.91	0.89	1.01	0.81	0.78	0.74
U	0.35	0.32	0.32	0.33	0.27	0.27	0.27
Nb	3.64	3.62	3.22	3.27	2.86	2.65	3.07
Ta	0.22	0.22	0.20	0.20	0.17	0.16	0.18
La	6.93	6.58	6.15	6.16	5.84	5.65	5.35
Ce	15.79	15.22	14.09	14.07	13.41	12.56	12.38
Pr	2.20	2.13	1.93	1.94	1.79	1.68	1.70
Nd	10.53	10.27	9.23	9.31	8.39	7.84	8.01
Sr	234	213	248	212	238	237	220
Sm	2.95	2.90	2.55	2.65	2.29	2.09	2.15
Zr	77.0	69.1	70.8	71.7	56.3	57.3	56.7
Hf	2.08	1.85	1.81	1.90	1.49	1.52	1.51
Eu	1.05	1.02	0.96	0.93	0.81	0.79	0.78
Ti	6,802	6,867	6,049	6,197	5,207	4,552	4,840
Gd	3.67	3.60	3.10	3.30	2.68	2.48	2.59
Tb	0.62	0.60	0.53	0.55	0.46	0.42	0.44
Dy	4.21	4.08	3.54	3.70	2.96	2.84	2.89
Ho	0.86	0.84	0.74	0.76	0.62	0.58	0.59
Y	23.02	22.28	19.76	20.38	16.35	15.41	15.91
Er	2.51	2.41	2.16	2.25	1.79	1.71	1.76
Tm	0.36	0.35	0.31	0.32	0.25	0.25	0.25
Yb	2.42	2.35	2.10	2.15	1.73	1.65	1.77
Lu	0.36	0.34	0.31	0.32	0.25	0.24	0.26

Note: (1) No analyses: 1–25 – Northern Maslovsky. 26–49 – Southern Maslovsky. Here and in Tables 10.2, 10.3 after Krivolutskaya et al. (2012)

Table 10.2 Olivine composition from intrusive rocks of the Maslovsky and Noril'sk 1 deposits

No	1	2	3	4	5	6	7	8	9	10	11	12
No sample	OM-4/1,006	OM-4/1,006	OM-4/1,006	OM-4/1,006	OM-4/1,006	OM-4/1,006	OM-4/1,010.8	OM-4/1,010.8	OM-4/1,010.8	OM-4/1,010.8	OM-4/1,010.8	OM-4/1,010.8
Fo,mol.%	80.64	79.98	80.34	80.75	79.96	80.89	79.75	79.85	79.98	79.91	79.83	80.01
SiO_2	39.06	39.24	39.15	39.47	39.35	39.43	39.16	39.08	39.23	39.12	39.21	39.34
TiO_2	0.02	0.03	0.03	0.02	0.03	0.02	0.03	0.03	0.03	0.03	0.02	0.03
Al_2O_3	0.02	0.01	0.01	0.01	0.01	0.03	0.00	0.01	0.01	0.01	0.02	0.01
FeO	18.26	18.85	18.57	18.17	18.88	18.09	19.01	18.90	18.82	18.96	19.00	18.94
MnO	0.28	0.29	0.28	0.28	0.29	0.27	0.29	0.29	0.29	0.29	0.29	0.29
MgO	42.65	42.22	42.58	42.76	42.25	42.96	41.99	42.02	42.16	42.29	42.17	42.51
CaO	0.13	0.10	0.10	0.11	0.14	0.13	0.11	0.12	0.10	0.09	0.16	0.11
NiO	0.21	0.26	0.23	0.22	0.24	0.22	0.24	0.24	0.24	0.23	0.25	0.28
CoO	0.02	0.02	0.02	0.02	0.02	0.02	0.02	0.02	0.02	0.02	0.02	0.02
Cr_2O_3	0.01	0.01	0.01	0.01	0.01	0.02	0.01	0.01	0.01	0.01	0.01	0.01
Total	100.70	101.06	101.01	101.11	101.25	101.22	100.91	100.76	100.95	101.07	101.18	101.56
Sr	0.00	0.00	0.00	0.00	0.00	0.00	0.00	0.00	0.00	0.00	0.01	0.00
Y	0.71	1.19	0.77	0.54	1.64	0.42	1.57	1.28	1.03	0.73	1.45	0.73
Dy	0.07	0.08	0.07	0.04	0.15	0.03	0.13	0.09	0.09	0.06	0.13	0.05
Ho	0.02	0.03	0.02	0.01	0.05	0.01	0.05	0.04	0.03	0.02	0.05	0.02
Er	0.10	0.21	0.12	0.08	0.27	0.07	0.27	0.19	0.15	0.12	0.23	0.11
Tm	0.02	0.05	0.03	0.02	0.05	0.01	0.05	0.05	0.03	0.03	0.05	0.03
Yb	0.19	0.45	0.28	0.13	0.49	0.13	0.52	0.54	0.32	0.29	0.44	0.29
Lu	0.04	0.08	0.05	0.03	0.09	0.03	0.10	0.09	0.06	0.05	0.08	0.05
Ni	1,751	2,111	1,811	1,771	2,001	1,808	1,940	1,996	1,888	1,871	1,961	2,217
Ti	159	210	182	145	208	123	209	215	217	199	151	198
Cu	0.83	0.79	0.77	0.85	0.85	0.78	0.91	0.81	0.78	0.74	1.04	0.83
Zn	112	120	116	114	125	114	125	122	121	125	127	124
Mn	2,200	2,298	2,227	2,202	2,282	2,150	2,273	2,275	2,245	2,267	2,287	2,292
Sc	9.11	9.42	8.29	7.81	9.01	8.27	9.60	8.71	8.13	7.01	9.43	7.93
Al	146.53	86.90	99.56	79.73	79.88	80.50	80.07	79.73	80.21	80.90	79.64	79.89
Zr	0.175	0.25	0.25	0.17	0.19	0.10	0.19	0.35	0.24	0.18	0.12	0.32
Co	182	172	182	181	178	181	181	179	184	185	182	173
Ge	0.78	0.89	0.81	0.66	0.79	0.85	0.85	0.78	0.80	0.84	0.77	0.73
V	9.40	7.58	7.84	7.28	8.36	8.53	9.74	8.36	9.23	8.86	12.31	8.26
Li	5.63	4.47	4.63	3.69	4.76	3.74	5.02	4.40	5.08	4.14	4.90	4.01

(continued)

Table 10.2 (continued)

No	13	14	15	16	17	18	19	20	21	22	23
No обр.	OM-4/1,010.8	OM-4/1,015	OM-4/1,015	OM-4/1,015	OM-4/1,015	OM-4/1,015	OM-4/1,015	OM-4/1,015	OM-4/1,015	OM-4/1,015	OM-4/1,015
Fo,mol.%	79.83	79.42	79.50	79.56	80.19	79.63	79.89	79.79	79.51	79.82	79.95
SiO_2	39.15	39.12	39.07	39.00	39.27	39.25	39.32	39.29	39.17	39.16	39.21
TiO_2	0.02	0.03	0.03	0.03	0.02	0.03	0.02	0.03	0.03	0.03	0.02
Al_2O_3	0.01	0.01	0.01	0.01	0.01	0.01	0.02	0.01	0.01	0.01	0.01
FeO	18.94	19.32	19.23	19.14	18.65	19.18	18.96	19.03	19.18	19.02	18.90
MnO	0.29	0.28	0.29	0.29	0.28	0.28	0.28	0.28	0.28	0.28	0.28
MgO	42.04	41.83	41.82	41.78	42.33	42.07	42.24	42.14	41.75	42.20	42.29
CaO	0.14	0.11	0.11	0.11	0.13	0.11	0.14	0.11	0.17	0.11	0.15
NiO	0.23	0.28	0.26	0.25	0.24	0.27	0.25	0.23	0.26	0.26	0.24
CoO	0.02	0.02	0.02	0.02	0.02	0.02	0.02	0.02	0.02	0.02	0.02
Cr_2O_3	0.01	0.01	0.01	0.01	0.01	0.01	0.01	0.01	0.01	0.01	0.01
Total	100.88	101.04	100.88	100.67	100.99	101.26	101.31	101.18	100.91	101.15	101.16
Sr	0.00	0.004	0.00	0.00	0.005	0.00	0.005	0.002	0.005	0.004	0.003
Y	1.33	1.86	1.63	1.49	0.65	1.56	1.60	1.45	1.43	1.04	0.82
Dy	0.12	0.16	0.13	0.12	0.06	0.12	0.14	0.11	0.12	0.08	0.07
Ho	0.04	0.06	0.04	0.04	0.02	0.05	0.05	0.04	0.04	0.03	0.03
Er	0.22	0.31	0.25	0.23	0.12	0.26	0.24	0.22	0.23	0.16	0.13
Tm	0.04	0.07	0.05	0.05	0.02	0.07	0.04	0.05	0.05	0.04	0.03
Yb	0.48	0.66	0.57	0.55	0.21	0.58	0.45	0.50	0.46	0.33	0.29
Lu	0.08	0.12	0.10	0.10	0.04	0.10	0.08	0.09	0.09	0.07	0.05
Ni	1,868	2,173	2,022	1,893	1,874	2,148	1,976	1,886	2,010	2,146	2,001
Ti	160	220	220	205	173	200	175	217	176	216	170
Cu	1.03	0.82	0.74	0.75	0.84	0.72	0.95	0.78	0.83	0.79	0.69
Zn	123	134	139	126	125	130	127	125	131	128	123
Mn	2,285	2,238	2,252	2,233	2,224	2,267	2,250	2,248	2,276	2,246	2,231
Sc	9.39	9.68	8.83	8.07	7.84	8.99	9.70	8.73	8.74	7.81	6.84
Al	79.98	79.63	79.74	79.84	80.17	79.74	79.57	79.64	107.17	96.72	100.38
Zr	0.13	0.18	0.25	0.21	0.15	0.19	0.13	0.20	0.08	0.22	0.13
Co	185	177	180	175	181	177	181	183	186	181	182
Ge	0.87	0.84	0.89	0.86	0.81	0.85	0.80	0.82	0.81	0.81	0.82
V	10.79	15.54	14.09	11.81	10.29	13.54	15.31	14.67	15.52	12.72	6.44
Li	4.19	4.46	5.43	4.26	2.90	2.74	4.64	4.22	3.99	4.55	3.45

No	24	25	26	27	28	29	30	31	32	33	34	35
No sample	OM-4/1,020.2	OM-4/1,020.2	OM-4/1,020.2	OM-4/1,020.2	OM-4/1,020.2	OM-4/1,020.2	OM-4/1,026.5	OM-4/1,026.5	OM-4/1,026.5	OM-4/1,026.5	OM-4/1,026.5	OM-4/1,026.5
Fo, mol.%	79.72	79.60	79.20	79.38	79.15	79.55	79.66	79.97	79.52	79.58	79.89	79.57
SiO_2	39.19	39.32	39.20	38.93	39.15	39.11	39.23	39.19	39.09	39.19	39.12	39.04
TiO_2	0.02	0.03	0.03	0.03	0.02	0.03	0.03	0.02	0.03	0.03	0.03	0.03
Al_2O_3	0.02	0.01	0.01	0.01	0.01	0.009	0.01	0.02	0.01	0.01	0.01	0.01
FeO	19.06	19.26	19.52	19.32	19.58	19.18	19.15	18.88	19.19	19.20	18.90	19.17
MnO	0.29	0.29	0.29	0.29	0.29	0.29	0.28	0.28	0.28	0.28	0.28	0.29
MgO	42.03	42.13	41.68	41.72	41.68	41.84	42.06	42.27	41.78	41.98	42.13	41.87
CaO	0.14	0.13	0.13	0.12	0.15	0.14	0.15	0.12	0.14	0.12	0.14	0.12
NiO	0.25	0.25	0.27	0.25	0.25	0.26	0.24	0.24	0.24	0.24	0.24	0.24
CoO	0.02	0.02	0.02	0.02	0.02	0.02	0.02	0.02	0.02	0.02	0.02	0.02
Cr_2O_3	0.02	0.01	0.01	0.01	0.01	0.01	0.01	0.01	0.01	0.01	0.01	0.01
Total	101.07	101.49	101.19	100.72	101.20	100.92	101.21	101.08	100.82	101.12	100.91	100.84
Sr	0.007	0.005	0.005	0.002	0.004	0.005	0.000	0.000	0.006	0.007	0.18	0.006
Y	1.30	1.60	1.81	1.54	1.56	1.48	1.98	0.40	1.87	2.07	0.64	2.10
Dy	0.11	0.14	0.17	0.10	0.14	0.14	0.19	0.02	0.17	0.16	0.04	0.16
Ho	0.04	0.05	0.06	0.04	0.05	0.05	0.06	0.01	0.06	0.06	0.02	0.06
Er	0.22	0.27	0.31	0.22	0.24	0.24	0.30	0.06	0.28	0.31	0.10	0.30
Tm	0.04	0.05	0.06	0.06	0.05	0.05	0.06	0.01	0.06	0.06	0.02	0.06
Yb	0.43	0.51	0.64	0.57	0.53	0.47	0.48	0.16	0.55	0.60	0.22	0.62
Lu	0.07	0.09	0.12	0.10	0.09	0.08	0.08	0.03	0.10	0.10	0.04	0.10
Ni	1,974	1,978	2,099	1,965	1,925	2,051	1,874	1,902	1,913	1,945	1,899	2,020
Ti	124	223	212	224	153	192	207	172	207	240	213	225
Cu	1.12	1.24	1.12	0.91	1.16	1.06	0.88	0.71	1.00	0.73	0.78	0.89
Zn	143	149	143	142	144	141	161	155	158	156	153	158
Mn	2,258	2,282	2,289	2,254	2,239	2,271	2,192	2,162	2,196	2,208	2,168	2,238
Sc	9.77	9.55	10.41	9.26	9.93	8.68	10.29	8.11	11.12	10.09	7.52	10.03
Al	138.23	85.02	84.16	81.36	103.12	84.69	89.85	103.20	99.35	86.02	140.50	82.30
Zr	0.15	0.16	0.15	0.22	0.09	0.12	0.17	0.10	0.14	0.21	0.24	0.18
Co	194	193	181	190	189	188	192	190	195	193	192	190
Ge	0.90	0.87	0.74	0.87	0.85	0.92	0.84	0.87	0.87	0.88	0.80	0.80
V	15.70	18.24	16.26	15.19	17.06	15.48	21.52	13.86	20.07	19.67	13.43	17.78
Li	4.20	5.75	5.51	4.65	4.13	4.39	4.71	3.34	5.41	4.59	2.50	4.45

(continued)

Table 10.2 (continued)

No	36	37	38	39	40	41	42	43	44	45	46
No sample	OM-4/1,026.5	OM-4/1,026.5	OM-4/1,026.5	OM-4/1,026.5	OM-4/1,026.5	OM-4/1,026.5	OM-4/1,030.6	OM-4/1,030.6	OM-4/1,030.6	OM-4/1,030.6	OM-4/1,030.6
Fo, mol.%	79.60	79.81	79.54	79.76	79.75	79.73	79.07	79.25	79.26	79.03	79.37
SiO_2	39.19	39.33	39.29	39.26	39.21	39.25	39.18	39.21	39.14	39.30	39.15
TiO_2	0.03	0.03	0.03	0.03	0.03	0.03	0.02	0.03	0.03	0.03	0.03
Al_2O_3	0.01	0.01	0.01	0.01	0.02	0.01	0.01	0.02	0.01	0.01	0.01
FeO	19.14	19.00	19.26	19.06	19.05	19.11	19.70	19.54	19.54	19.77	19.48
MnO	0.28	0.28	0.29	0.28	0.28	0.29	0.30	0.30	0.30	0.30	0.29
MgO	41.91	42.12	42.00	42.14	42.07	42.15	41.74	41.86	41.88	41.79	42.04
CaO	0.12	0.12	0.13	0.15	0.16	0.15	0.14	0.12	0.10	0.12	0.11
NiO	0.24	0.24	0.25	0.24	0.24	0.23	0.22	0.22	0.22	0.22	0.23
CoO	0.02	0.02	0.02	0.02	0.02	0.02	0.02	0.02	0.02	0.02	0.02
Cr_2O_3	0.01	0.01	0.01	0.02	0.01	0.01	0.01	0.01	0.01	0.01	0.01
Total	101.00	101.21	101.32	101.26	101.13	101.28	101.37	101.37	101.28	101.58	101.39
Sr	0.007	0.001	0.005	0.005	0.004	0.011	0.006	0.002	0.014	0.001	0.000
Y	1.77	1.13	1.65	1.56	1.79	1.57	1.52	1.29	0.92	1.34	0.80
Dy	0.15	0.09	0.14	0.15	0.17	0.15	0.14	0.11	0.08	0.12	0.06
Ho	0.06	0.03	0.05	0.04	0.06	0.04	0.05	0.04	0.03	0.04	0.02
Er	0.25	0.17	0.26	0.24	0.25	0.24	0.23	0.20	0.15	0.22	0.14
Tm	0.05	0.03	0.06	0.05	0.05	0.05	0.05	0.04	0.03	0.05	0.02
Yb	0.51	0.35	0.58	0.46	0.55	0.46	0.51	0.43	0.36	0.50	0.30
Lu	0.09	0.06	0.10	0.07	0.10	0.08	0.09	0.07	0.06	0.09	0.06
Ni	1,96	1,959	1,975	1,886	2,037	1,892	1,782	1,829	1,804	1,787	1,798
Ti	201	205	218	209	183	213	196	183	214	217	187
Cu	0.89	0.89	0.89	0.91	0.89	0.89	0.76	0.77	0.59	0.72	0.55
Zn	154	154	155	147	158	151	132	130	134	136	136
Mn	2,180	2,228	2,229	2,187	2,248	2,242	2,329	2,300	2,315	2,317	2,322
Sc	9.18	9.25	10.45	9.41	9.38	9.18	8.52	8.54	7.47	9.16	7.17
Al	95.21	89.34	85.88	93.79	94.34	88.81	101.52	109.88	81.26	93.35	105.56
Zr	0.18	0.22	0.16	0.15	0.12	0.15	0.12	0.14	0.19	0.16	0.14
Co	195	195	189	191	184	193	183	183	184	183	183
Ge	0.87	0.90	0.88	0.78	0.87	0.83	0.82	0.93	0.77	0.85	0.71
V	17.72	17.24	17.24	17.42	19.81	19.62	16.54	16.98	14.06	17.67	11.79
Li	3.95	3.96	4.19	5.03	3.42	4.77	6.28	5.28	5.09	6.62	4.34

No	47	48	49	50	51	52	53	54	55	56
No sample	OM-4/1,030.6	OM-4/1,030.6	OM-4/1,030.6	OM-4/1,035.1	OM-4/1,035.1	OM-4/1,035.1	OM-4/1,035.1	OM-4/1,035.1	OM-4/1,035.1	OM-4/1,035.1
Fo,mol.%	79.08	78.89	79.12	78.95	77.98	79.24	78.29	79.25	78.48	78.25
SiO_2	39.13	39.16	39.11	39.01	39.19	38.75	39.06	39.23	39.24	39.10
TiO_2	0.03	0.03	0.03	0.03	0.03	0.03	0.03	0.03	0.03	0.03
Al_2O_3	0.01	0.01	0.02	0.00	0.01	0.01	0.01	0.01	0.01	0.01
FeO	19.68	19.83	19.65	19.78	20.65	19.33	20.34	19.47	20.17	20.36
MnO	0.29	0.29	0.29	0.29	0.30	0.29	0.30	0.29	0.29	0.30
MgO	41.72	41.57	41.77	41.62	41.02	41.49	41.14	41.72	41.26	41.08
CaO	0.13	0.13	0.14	0.12	0.12	0.11	0.135	0.144	0.148	0.138
NiO	0.23	0.24	0.22	0.22	0.26	0.23	0.240	0.223	0.224	0.266
CoO	0.02	0.02	0.02	0.02	0.02	0.02	0.02	0.02	0.02	0.02
Cr_2O_3	0.01	0.00	0.01	0.01	0.007	0.01	0.00	0.01	0.00	0.00
Total	101.28	101.32	101.28	101.15	101.64	100.36	101.30	101.16	101.43	101.34
Sr	0.01	0.001	0.004	0.00	0.00	0.004	0.006	0.00	0.002	0.000
Y	1.12	1.44	1.30	0.92	1.52	0.43	1.44	0.80	1.52	1.53
Dy	0.10	0.11	0.12	0.08	0.13	0.03	0.13	0.06	0.15	0.13
Ho	0.04	0.04	0.04	0.03	0.05	0.01	0.04	0.02	0.04	0.05
Er	0.19	0.27	0.22	0.15	0.25	0.08	0.26	0.12	0.25	0.26
Tm	0.04	0.05	0.04	0.04	0.06	0.01	0.05	0.02	0.04	0.05
Yb	0.37	0.51	0.50	0.33	0.54	0.17	0.48	0.23	0.43	0.58
Lu	0.07	0.09	0.09	0.06	0.11	0.04	0.09	0.04	0.08	0.10
Ni	1,878	1,929	1,800	1,815	2,022	1,808	1,853	1,738	1,794	2,087
Ti	184	191	182	185	219	209	187	196	194	199
Cu	0.80	0.57	0.79	0.73	0.67	0.66	0.76	0.65	0.46	0.67
Zn	136	132	134	133	143	126	131	124	138	141
Mn	2,398	2,286	2,393	2,342	2,362	2,275	2,339	2,245	2,338	2,353
Sc	7.53	7.97	8.75	7.90	9.04	6.64	7.67	7.19	9.66	9.22
Al	99.44	89.28	103.56	89.38	79.81	84.58	73.54	86.24	109.89	84.64
Zr	0.10	0.14	0.12	0.17	0.13	0.14	0.11	0.18	0.14	0.14
Co	191	188	185	185	179	182	179	178	189	180
Ge	0.90	0.77	0.72	0.75	0.85	0.81	0.81	0.80	0.89	0.82
V	14.45	14.46	19.97	19.00	28.06	13.40	27.01	14.13	33.46	26.91
Li	5.20	3.25	6.10	5.85	4.98	4.83	4.85	5.66	5.76	5.15

(continued)

Table 10.2 (continued)

No	57	58	59	60	61	62	63	64	65	66	67
No sample	OM-4/1,035.1	OM-4/1,035.1	OM-4/1,035.1	OM-4/1,035.1	OM-4/1,039.4	OM-4/1,039.4	OM-4/1,039.4	OM-4/1,039.4	OM-4/1,039.4	OM-4/1,039.4	OM-4/1,046.2
Fo, mol.%	78.63	78.80	78.19	79.43	78.36	79.12	78.05	78.12	77.94	79.05	77.65
SiO_2	39.29	39.19	38.87	39.41	38.98	39.17	38.79	39.06	39.13	39.32	39.05
TiO_2	0.02	0.03	0.03	0.03	0.02	0.03	0.01	0.02	0.03	0.02	0.02
Al_2O_3	0.01	0.01	0.00	0.01	0.01	0.01	0.01	0.02	0.01	0.02	0.01
FeO	20.11	19.92	20.40	19.40	20.25	19.58	20.52	20.53	20.65	19.79	20.92
MnO	0.30	0.30	0.30	0.29	0.30	0.30	0.31	0.31	0.31	0.30	0.31
MgO	41.52	41.53	41.03	42.03	41.13	41.61	40.93	41.11	40.92	41.89	40.77
CaO	0.16	0.13	0.12	0.12	0.16	0.13	0.18	0.19	0.15	0.15	0.16
NiO	0.22	0.23	0.23	0.22	0.23	0.22	0.23	0.23	0.24	0.23	0.22
CoO	0.02	0.02	0.02	0.02	0.02	0.02	0.02	0.02	0.02	0.02	0.02
Cr_2O_3	0.008	0.01	0.009	0.01	0.004	0.007	0.000	0.006	0.002	0.009	0.004
Total	101.69	101.38	101.04	101.58	101.13	101.10	101.04	101.52	101.48	101.77	101.52
Sr	0.001	0.000	0.000	0.005	0.007	0.006	0.006	0.004	0.004	0.002	0.005
Y	1.43	0.59	1.40	0.60	0.99	0.72	1.23	1.12	1.35	0.79	1.44
Dy	0.14	0.04	0.13	0.05	0.10	0.07	0.10	0.10	0.12	0.07	0.15
Ho	0.05	0.02	0.04	0.02	0.03	0.02	0.04	0.04	0.04	0.02	0.05
Er	0.23	0.10	0.24	0.11	0.18	0.11	0.19	0.15	0.24	0.11	0.22
Tm	0.04	0.02	0.05	0.02	0.03	0.02	0.04	0.03	0.05	0.02	0.05
Yb	0.49	0.27	0.56	0.22	0.33	0.24	0.36	0.33	0.45	0.27	0.49
Lu	0.09	0.05	0.10	0.04	0.05	0.04	0.06	0.06	0.08	0.05	0.09
Ni	1,872	1,816	1,975	1,770	1,853	1,855	1,763	1,842	1,945	1,858	1,772
Ti	167	206	226	198	153	201	137	108	183	136	143
Cu	0.88	0.67	0.74	0.66	0.88	0.80	0.89	1.12	0.88	1.02	1.06
Zn	144	135	141	127	135	134	134	139	145	139	149
Mn	2,372	2,336	2,353	2,292	2,422	2,309	2,313	2,379	2,439	2,361	2,480
Sc	9.59	7.28	8.99	7.88	8.94	7.76	8.47	9.48	8.60	8.82	9.06
Al	115.91	98.88	75.47	106.22	105.35	98.92	97.85	148.18	92.81	118.32	108.45
Zr	0.10	0.14	0.17	0.12	0.12	0.18	0.09	0.07	0.09	0.10	0.08
Co	185	185	185	183	190	187	181	178	181	208	214
Ge	0.79	0.76	0.81	0.83	0.82	0.82	0.79	0.86	0.80	0.83	0.85
V	32.31	20.64	25.17	16.55	23.11	17.80	26.29	30.86	24.45	20.05	28.47
Li	4.81	5.71	5.70	5.45	4.86	6.37	5.03	6.00	4.97	4.88	6.01

No	68	69	70	71	72	73	74	75	76	77	78
No sample	OM-4/1,046.2	OM-4/1,046.2	OM-4/1,046.2	OM-4/1,046.2	OM-4/1,046.2	OM-4/1,050.4	OM-4/1,050.4	OM-4/1,050.4	OM-4/1,050.4	OM-4/1,050.4	OM-4/1,050.4
Fo, мол.%	77.64	77.75	77.36	77.46	80.42	76.79	78.37	78.21	77.63	76.96	76.73
SiO_2	38.99	39.02	38.99	39.09	39.34	38.94	39.03	39.03	39.12	39.03	38.87
TiO_2	0.028	0.029	0.034	0.025	0.023	0.021	0.032	0.016	0.025	0.025	0.023
Al_2O_3	0.016	0.013	0.009	0.011	0.022	0.026	0.023	0.026	0.023	0.018	0.025
FeO	20.94	20.83	21.15	21.00	18.49	21.56	20.28	20.39	20.90	21.43	21.67
MnO	0.31	0.31	0.31	0.32	0.27	0.32	0.30	0.30	0.31	0.32	0.32
MgO	40.78	40.83	40.52	40.49	42.60	40.00	41.21	41.03	40.69	40.15	40.07
CaO	0.17	0.15	0.13	0.24	0.13	0.19	0.16	0.21	0.15	0.17	0.19
NiO	0.23	0.25	0.30	0.28	0.24	0.27	0.24	0.26	0.25	0.23	0.25
CoO	0.02	0.02	0.02	0.02	0.02	0.02	0.02	0.02	0.02	0.02	0.02
Cr_2O_3	0.005	0.002	0.00	0.002	0.02	0.00	0.005	0.01	0.003	0.001	0.00
Total	101.50	101.47	101.49	101.50	101.18	101.36	101.33	101.31	101.51	101.42	101.46
Sr	0.003	0.004	0.001	0.006	0.002	0.003	0.004	0.001	0.006	0.015	0.002
Y	1.29	1.12	1.76	0.85	0.36	1.30	1.32	0.65	1.01	1.55	1.38
Dy	0.13	0.11	0.16	0.08	0.03	0.12	0.12	0.06	0.07	0.14	0.12
Ho	0.04	0.03	0.06	0.02	0.01	0.04	0.04	0.02	0.03	0.04	0.04
Er	0.20	0.19	0.28	0.14	0.05	0.22	0.23	0.09	0.17	0.28	0.20
Tm	0.04	0.04	0.06	0.03	0.01	0.04	0.04	0.01	0.03	0.05	0.04
Yb	0.46	0.38	0.70	0.35	0.09	0.42	0.45	0.19	0.36	0.53	0.48
Lu	0.08	0.08	0.11	0.06	0.01	0.08	0.08	0.04	0.05	0.10	0.08
Ni	1,803	1,943	2,369	2,251	1,913	2,180	1,930	2,059	1,992	1,862	2,034
Ti	141	146	208	120	133	130	175	99	152	150	150
Cu	0.97	0.86	0.72	1.23	0.65	1.17	0.69	1.09	0.68	0.97	1.03
Zn	150	144	153	152	121	158	140	146	150	157	159
Mn	2,483	2,432	2,483	2,498	2,144	2,557	2,339	2,361	2,459	2,521	2,561
Sc	9.07	8.75	8.81	9.59	7.61	10.77	9.49	10.16	9.51	9.95	9.63
Al	116.74	98.15	84.65	120.52	144.31	144.23	105.50	177.70	120.85	131.78	128.26
Zr	0.09	0.10	0.12	0.05	0.09	0.06	0.11	0.07	0.09	0.08	0.08
Co	211	209	207	212	203	206	214	209	212	217	217
Ge	0.79	0.79	0.71	0.72	0.76	0.79	0.79	0.81	0.77	0.92	0.89
V	29.09	23.13	27.52	37.37	10.87	41.48	31.82	23.49	24.77	33.26	32.78
Li	5.68	5.22	6.12	6.21	5.04	5.28	7.11	5.15	5.44	5.25	5.49

(continued)

Table 10.2 (continued)

No	79	80	81	82	83	84	85	86	87	88	89	90
No sample	OM-24/807	OM-24/807	OM-24/807	OM-24/807	OM-24/807	OM-24/807	OM-24/807	OM-24/807	OM-24/807	OM-24/807	OM-24/807	OM-24/807
Fo, mol.%	75.82	78.15	78.00	71.17	73.54	74.29	73.29	74.94	71.92	76.33	73.46	71.47
SiO_2	38.02	38.48	38.37	37.78	38.11	38.11	37.89	38.13	37.92	38.31	37.78	37.73
TiO_2	0.01	0.01	0.01	0.02	0.01	0.01	0.01	0.01	0.01	0.01	0.01	0.01
Al_2O_3	0.02	0.03	0.03	0.01	0.02	0.02	0.02	0.03	0.02	0.02	0.02	0.02
FeO	22.32	20.27	20.38	26.21	24.27	23.58	24.45	23.00	25.63	21.78	24.12	25.88
MnO	0.33	0.30	0.30	0.40	0.36	0.35	0.37	0.34	0.38	0.33	0.36	0.39
MgO	39.24	40.67	40.51	36.29	37.83	38.22	37.64	38.57	36.82	39.39	37.46	36.36
CaO	0.28	0.30	0.31	0.18	0.28	0.22	0.22	0.29	0.28	0.30	0.25	0.23
NiO	0.18	0.19	0.19	0.16	0.16	0.17	0.17	0.18	0.15	0.18	0.17	0.16
CoO	0.02	0.02	0.02	0.02	0.02	0.03	0.02	0.02	0.02	0.02	0.03	0.02
Cr_2O_3	0.03	0.03	0.03	0.01	0.02	0.01	0.02	0.03	0.01	0.02	0.02	0.01
Total	100.47	100.33	100.17	101.11	101.12	100.75	100.85	100.62	101.28	100.39	100.25	100.85
Sr	0.003	0.002	0.002	0.004	0.000	0.002	0.011	0.008	0.000	0.005	0.032	0.005
Y	0.35	0.23	0.19	1.01	0.36	0.66	0.56	0.28	0.38	0.26	0.40	0.88
Dy	0.03	0.01	0.01	0.09	0.04	0.06	0.05	0.03	0.02	0.02	0.04	0.09
Ho	0.01	0.00	0.00	0.03	0.01	0.02	0.01	0.01	0.01	0.01	0.01	0.03
Er	0.05	0.03	0.02	0.14	0.04	0.11	0.08	0.04	0.05	0.03	0.06	0.13
Tm	0.008	0.008	0.004	0.03	0.01	0.02	0.01	0.00	0.01	0.005	0.01	0.02
Yb	0.09	0.07	0.06	0.33	0.12	0.21	0.17	0.08	0.10	0.06	0.11	0.24
Lu	0.01	0.01	0.01	0.06	0.02	0.03	0.03	0.01	0.02	0.01	0.02	0.04
Ni	1,431	1,514	1,491	1,256	1,335	1,330	1,330	1,412	1,239	1,427	1,330	1,282
Ti	47	39	42	71	41	46	37	61	66	56	62	74
Cu	0.79	1.21	1.05	0.69	0.77	0.79	0.83	0.72	0.62	0.93	0.84	0.70
Zn	173	151	146	210	185	167	168	174	204	161	179	193
Mn	2,604	2,406	2,351	3,114	2,858	2,753	2,894	2,667	2,870	2,521	2,680	3,015
Sc	9.19	9.25	9.04	8.81	9.41	9.14	8.93	9.13	9.71	8.94	8.74	9.21
Al	151.19	168.59	168.02	122.80	149.51	134.13	151.88	159.56	148.58	158.58	153.39	134.32
Zr	0.04	0.03	0.03	0.08	0.03	0.05	0.08	0.04	0.03	0.03	0.03	0.04
Co	182	185	182	191	190	189	190	199	204	198	197	206
Ge	0.75	0.83	0.86	0.82	0.83	0.77	0.80	0.85	0.70	0.79	0.82	0.75
V	15.49	14.77	14.57	15.79	17.83	19.63	12.33	15.67	17.25	14.67	18.02	16.39
Li	6.20	4.26	4.37	7.37	6.03	6.33	6.36	5.74	6.15	5.22	5.94	6.89

No	91	92	93	94	95	96	97	98	99	100	101	102
No sample	OM-24/807	OM-24/807	OM-24/807	OM-24/811	OM-24/811	OM-24/811	OM-24/311	OM-24/811	OM-24/811	OM-24/811	OM-24/811	OM-24/811
Fo, мол.%	77.85	76.22	75.89	80.22	79.08	78.35	78.00	77.80	77.90	79.56	79.43	78.38
SiO_2	38.70	38.44	38.45	38.82	38.56	38.63	38.51	38.70	38.48	38.91	38.95	38.59
TiO_2	0.01	0.01	0.00	0.01	0.01	0.01	0.01	0.01	0.01	0.01	0.01	0.01
Al_2O_3	0.03	0.02	0.02	0.03	0.03	0.02	0.02	0.03	0.02	0.02	0.02	0.03
FeO	20.60	22.04	22.24	18.53	19.53	20.19	20.46	20.71	20.52	19.15	19.24	20.10
MnO	0.31	0.33	0.33	0.28	0.30	0.31	0.32	0.32	0.32	0.29	0.30	0.31
MgO	40.61	39.61	39.27	42.16	41.42	40.98	40.70	40.72	40.56	41.80	41.67	40.88
CaO	0.29	0.30	0.29	0.27	0.24	0.24	0.25	0.23	0.24	0.24	0.25	0.22
NiO	0.19	0.19	0.18	0.20	0.20	0.19	0.18	0.18	0.19	0.20	0.20	0.19
CoO	0.02	0.02	0.02	0.02	0.02	0.02	0.02	0.02	0.02	0.02	0.02	0.02
Cr_2O_3	0.02	0.03	0.03	0.03	0.02	0.02	0.02	0.01	0.02	0.03	0.03	0.03
Сумма	100.81	101.00	100.87	100.39	100.36	100.66	100.53	100.96	100.41	100.71	100.72	100.42
Sr	0.002	0.004	0.007	0.008	0.008	0.01	0.008	0.004	0.02	0.004	0.005	0.07
Y	0.22	0.24	0.33	0.28	0.32	0.69	0.40	0.63	0.52	0.41	0.54	0.61
Dy	0.01	0.01	0.02	0.02	0.03	0.07	0.04	0.06	0.05	0.04	0.05	0.05
Ho	0.00	0.009	0.01	0.01	0.01	0.02	0.01	0.02	0.01	0.01	0.02	0.02
Er	0.04	0.03	0.05	0.05	0.04	0.09	0.06	0.09	0.07	0.06	0.07	0.10
Tm	0.00	0.008	0.009	0.009	0.01	0.02	0.01	0.02	0.01	0.01	0.01	0.01
Yb	0.06	0.08	0.09	0.09	0.10	0.25	0.13	0.18	0.16	0.14	0.16	0.20
Lu	0.01	0.01	0.02	0.02	0.02	0.04	0.02	0.04	0.03	0.02	0.03	0.03
Ni	1,520	1,487	1,428	1,617	1,542	1,502	1,493	1,452	1,456	1,590	1,600	1,535
Ti	51	54	58	67	63	74	62	89	64	65	71	66
Cu	1.37	0.93	0.87	0.80	0.83	0.73	0.88	0.84	0.91	0.93	0.88	1.14
Zn	156	161	170	120	128	138	146	140	136	125	126	117
Mn	2,405	2,509	2,670	2,159	2,269	2,456	2,469	2,486	2,499	2,292	2,324	2,442
Sc	9.24	8.80	9.36	8.98	9.78	10.23	10.53	10.05	10.51	9.12	9.76	10.13
Al	174.77	167.94	165.91	162.03	176.84	165.22	161.21	150.36	151.07	150.47	149.93	178.10
Zr	0.02	0.03	0.03	0.04	0.03	0.06	0.03	0.06	0.05	0.05	0.06	0.05
Co	198	197	200	181	181	184	184	185	183	183	184	183
Ge	0.79	0.82	0.75	0.81	0.84	0.90	0.84	0.88	0.67	0.86	0.81	0.89
V	14.03	13.93	15.04	10.65	15.37	11.85	12.00	11.20	11.48	9.79	9.91	10.99
Li	4.26	5.12	5.84	4.53	4.70	5.56	4.53	5.65	5.15	5.07	5.47	6.20

(continued)

Table 10.2 (continued)

No	103	104	105	106	107	108	109	110	111	112	113	114
No sample	OM-24/811	OM-24/811	OM-24/811	OM-24/811	OM-24/811	OM-24/814.5	OM-24/814.5	OM-24/814.5	OM-24/814.5	OM-24/814.5	OM-24/814.5	OM-24/814.5
Fo, мол.%	78.28	75.68	78.69	78.43	78.19	79.13	79.25	80.00	79.02	80.54	79.33	79.33
SiO_2	38.63	38.35	38.43	38.45	38.14	38.73	38.90	38.84	38.65	38.92	38.92	38.93
TiO_2	0.01	0.01	0.01	0.01	0.01	0.01	0.01	0.02	0.01	0.01	0.02	0.01
Al_2O_3	0.03	0.02	0.02	0.02	0.03	0.03	0.03	0.03	0.02	0.02	0.02	0.03
FeO	20.23	22.53	19.80	20.06	20.20	19.40	19.37	18.65	19.45	18.17	19.30	19.24
MnO	0.31	0.34	0.31	0.31	0.32	0.30	0.30	0.29	0.30	0.28	0.30	0.30
MgO	40.91	39.33	40.99	40.91	40.63	41.26	41.50	41.85	41.08	42.18	41.54	41.42
CaO	0.27	0.25	0.24	0.22	0.27	0.25	0.26	0.19	0.23	0.26	0.17	0.26
NiO	0.19	0.19	0.19	0.20	0.19	0.21	0.21	0.21	0.20	0.22	0.21	0.21
CoO	0.02	0.02	0.02	0.02	0.02	0.02	0.02	0.02	0.02	0.02	0.02	0.02
Cr_2O_3	0.03	0.02	0.02	0.02	0.02	0.02	0.03	0.02	0.02	0.03	0.02	0.03
Сумма	100.66	101.09	100.07	100.25	99.86	100.27	100.65	100.16	100.03	100.14	100.54	100.47
Sr	0.00	0.00	0.03	0.09	0.001	0.01	0.008	0.02	0.006	0.01	0.01	0.003
Y	0.41	0.43	0.43	0.59	0.37	0.44	0.50	0.67	0.49	0.26	0.95	0.40
Dy	0.03	0.05	0.04	0.06	0.02	0.04	0.05	0.06	0.04	0.03	0.09	0.03
Ho	0.01	0.01	0.01	0.01	0.01	0.01	0.01	0.01	0.01	0.01	0.03	0.01
Er	0.06	0.06	0.06	0.09	0.05	0.07	0.08	0.10	0.07	0.04	0.16	0.06
Tm	0.01	0.01	0.01	0.02	0.01	0.01	0.01	0.02	0.02	0.01	0.03	0.01
Yb	0.14	0.10	0.12	0.20	0.12	0.16	0.17	0.21	0.16	0.11	0.30	0.11
Lu	0.02	0.02	0.02	0.04	0.02	0.03	0.03	0.04	0.04	0.01	0.06	0.02
Ni	1,516	1,444	1,543	1,502	1,531	1,636	1,683	1,703	1,583	1,766	1,680	1,708
Ti	57	51	55	68	49	77	83	147	95	74	170	59
Cu	0.82	0.78	0.85	0.89	0.88	1.15	1.41	0.82	0.98	1.07	0.72	1.19
Zn	138	164	130	131	133	133	134	119	137	123	132	139
Mn	2,449	2,645	2,430	2,401	2,513	2,332	2,318	2,258	2,376	2,224	2,399	2,362
Sc	10.95	9.44	9.56	9.47	10.16	10.57	11.26	9.17	10.68	8.73	9.93	9.80
Al	181.11	163.90	151.80	172.54	179.44	182.93	197.10	131.37	173.43	168.03	124.46	174.88
Zr	0.05	0.04	0.04	0.05	0.04	0.05	0.06	0.13	0.06	0.05	0.14	0.04
Co	185	186	182	178	180	181	183	182	182	183	186	184
Ge	0.95	0.72	0.81	0.80	0.80	0.73	0.88	0.81	0.84	0.73	0.88	0.85
V	12.72	13.68	11.13	10.52	11.43	13.28	15.02	10.64	12.47	11.54	10.16	12.95
Li	5.23	5.97	5.68	5.28	4.56	5.35	4.82	6.93	4.84	3.59	7.15	4.44

No	115	116	117	118	119	120	121	122	123	124	125	126	128	129
No sample	OM-24/814.5	Nor –1/120	Nor –1/120	Nor –1/120	Nor –1/120	Nor –1/120	Nor –1/120	Nor –1/120	Nor –1/120	Nor –1/120	Nor –1/120	Nor –1/120	Nor –1/120	Nor –1/120
Fo, мол.%	79.17	79.80	79.19	79.41	79.56	79.29	81.00	79.47	79.23	78.93	80.41	80.30	79.51	79.45
SiO_2	38.86	38.67	38.61	38.47	38.44	38.58	38.88	38.44	38.60	38.30	38.60	38.77	37.87	38.14
TiO_2	0.02	0.01	0.02	0.02	0.03	0.02	0.01	0.02	0.02	0.02	0.02	0.02	0.02	0.03
Al_2O_3	0.03	0.01	0.01	0.00	0.01	0.01	0.02	0.02	0.00	0.01	0.02	0.00	0.02	0.01
FeO	19.4	18.8	19.37	19.13	18.96	19.23	17.81	19.09	19.32	19.52	18.26	18.41	19.00	19.04
MnO	0.30	0.29	0.29	0.29	0.29	0.30	0.27	0.29	0.29	0.30	0.28	0.29	0.29	0.29
MgO	41.40	41.74	41.34	41.37	41.40	41.29	42.59	41.44	41.33	41.01	42.03	42.10	41.35	41.29
CaO	0.21	0.18	0.11	0.10	0.11	0.11	0.11	0.12	0.14	0.11	0.13	0.13	0.12	0.11
NiO	0.21	0.22	0.25	0.28	0.27	0.26	0.23	0.23	0.27	0.23	0.23	0.23	0.26	0.28
CoO	0.02	0.02	0.02	0.02	0.02	0.02	0.02	0.02	0.02	0.02	0.02	0.02	0.02	0.02
Cr_2O_3	0.02	0.01	0.009	0.01	0.01	0.01	0.03	0.01	0.01	0.01	0.02	0.01	0.01	0.01
Total	100.52	100.03	100.06	99.74	99.56	99.86	100.01	99.72	100.04	99.56	99.64	100.02	99.00	99.26
Sr	0.001	0.01	0.01	0.00	0.00	0.00	0.00	0.00	0.01	0.00	0.00	0.00	0.005	0.65
Y	0.63	1.07	1.54	1.85	1.71	1.97	0.62	1.88	1.82	2.32	1.30	1.55	1.83	1.99
Dy	0.06	0.09	0.13	0.17	0.15	0.17	0.06	0.16	0.18	0.18	0.10	0.16	0.15	0.18
Ho	0.02	0.03	0.05	0.06	0.06	0.06	0.02	0.05	0.06	0.07	0.04	0.05	0.06	0.06
Er	0.08	0.16	0.25	0.30	0.27	0.32	0.09	0.31	0.27	0.40	0.19	0.24	0.30	0.33
Tm	0.02	0.03	0.06	0.06	0.05	0.07	0.01	0.06	0.06	0.08	0.04	0.05	0.06	0.07
Yb	0.19	0.37	0.52	0.63	0.51	0.65	0.18	0.60	0.62	0.76	0.37	0.46	0.64	0.67
Lu	0.04	0.06	0.11	0.12	0.11	0.13	0.03	0.11	0.11	0.14	0.07	0.07	0.12	0.12
Ni	1,696	1,820	2,048	2,298	2,171	2,083	1,850	1,901	2,200	1,820	1,843	1,894	2,094	2,211
Ti	122	77	208	200	190	192	127	173	182	187	165	175	180	225
Cu	1.10	1.10	0.97	0.90	1.06	1.06	0.76	0.94	1.19	1.03	1.26	0.98	0.92	1.83
Zn	134	132	140	133	133	132	120	130	140	137	126	125	126	130
Mn	2,379	2,294	2,293	2,308	2,279	2,315	2,135	2,273	2,306	2,302	2,214	2,230	2,251	2,297
Sc	10.52	10.30	8.83	9.90	9.76	11.65	8.07	10.13	9.83	11.20	12.12	10.91	10.75	10.40
Al	158.56	98.29	83.06	81.91	93.16	100.91	106.33	96.25	101.74	97.53	139.51	110.82	99.21	92.96
Zr	0.07	0.10	0.22	0.49	0.22	0.23	0.19	0.28	0.19	0.27	0.34	0.21	0.24	0.26
Co	186	196	196	185	189	184	195	194	194	192	194	196	180	178
Ge	0.95	0.81	0.77	0.82	0.78	0.86	0.86	0.85	0.83	0.84	0.98	0.65	0.81	0.85
V	13.00	8.06	6.55	5.78	6.30	6.06	6.26	5.95	6.34	6.30	8.66	7.74	7.32	6.95
Li	5.81	2.04	2.14	1.80	2.16	2.31	1.92	2.05	1.00	2.70	4.54	2.75	4.25	6.35

Note: No analyses: 1–130 – Northern Maslovsky, 131–191 – Southern Maslovsky, 192–213 – Noril'sk 1

Table 10.3 Pyroxene composition from intrusive rocks of the Maslovsky deposit (wt %)

No	No sample	MgO#	SiO_2	TiO_2	Al_2O_3	FeO	MnO	MgO	CaO	Na_2O	Cr_2O_3	Total
1	OM-4/835.3	78.73	51.95	0.71	2.95	8.06	0.23	16.73	19.10	0.23	0.59	100.56
2	OM-4/836.1	83.62	52.54	0.39	2.91	6.00	0.16	17.18	20.37	0.18	0.74	100.48
3	OM-4/836.1	82.74	51.12	0.51	3.72	6.17	0.14	16.58	20.41	0.22	0.88	99.78
4	OM-4/836.1	82.48	51.09	0.47	3.39	6.27	0.16	16.54	20.48	0.21	0.70	99.33
5	OM-4/836.1	83.39	51.60	0.39	2.89	6.07	0.15	17.10	20.30	0.21	0.67	99.41
6	OM-4/836.1	76.08	51.22	0.77	3.01	8.99	0.23	16.03	19.33	0.23	0.22	100.05
7	OM-4/836.1	78.92	50.78	0.52	3.03	7.50	0.21	15.74	20.29	0.21	0.39	98.68
8	OM-4/846.6	69.97	50.55	0.73	3.13	11.31	0.25	14.78	19.06	0.25	0.01	100.08
9	OM-4/846.6	55.43	49.53	0.79	3.02	15.98	0.45	11.15	19.12	0.24	0.004	100.30
10	OM-4/846.6	76.03	51.53	0.56	3.11	8.99	0.20	16.00	19.68	0.20	0.05	100.34
11	OM-4/846.6	72.84	50.93	0.66	3.13	10.18	0.23	15.31	19.43	0.22	0.01	100.12
12	OM-4/846.6	79.50	51.18	0.51	3.38	7.49	0.17	16.29	20.24	0.22	0.21	99.71
13	OM-4/846.6	70.48	50.57	0.71	3.28	11.13	0.26	14.90	19.20	0.26	0.01	100.34
14	OM-4/846.6	73.37	51.07	0.52	2.80	9.63	0.26	14.88	20.31	0.22	0.02	99.72
15	OM-4/847.8	73.17	52.63	0.24	1.48	9.28	0.38	14.19	21.93	0.48	0.00	100.63
16	OM-4/847.8	70.38	52.90	0.29	1.48	10.39	0.44	13.85	21.38	0.53	0.006	101.28
17	OM-4/847.8	76.80	51.30	0.56	3.14	8.63	0.21	16.02	19.94	0.20	0.04	100.06
18	OM-4/847.8	77.29	52.38	0.35	2.03	8.18	0.25	15.62	21.28	0.18	0.04	100.32
19	OM-4/847.8	70.12	53.13	0.21	1.55	10.53	0.46	13.87	20.97	0.56	0.00	101.29
20	OM-4/847.8	73.27	53.28	0.15	1.53	9.40	0.42	14.44	21.39	0.54	0.005	101.17
21	OM-4/850.1	74.53	50.69	0.60	3.15	9.32	0.21	15.29	19.95	0.23	0.02	99.48
22	OM-4/850.1	76.72	51.42	0.53	3.07	8.49	0.20	15.69	20.11	0.22	0.20	99.95
23	OM-4/850.1	76.53	51.44	0.52	2.88	8.72	0.20	15.94	19.94	0.20	0.03	99.89
24	OM-4/850.1	72.92	50.92	0.63	3.06	10.09	0.23	15.24	19.61	0.25	0.01	100.06
25	OM-4/850.1	52.68	50.09	0.68	2.23	18.66	0.50	11.65	16.79	0.21	0.007	100.83
26	OM-4/850.7	67.67	51.13	0.78	3.11	12.41	0.28	14.57	18.55	0.28	0.008	101.13
27	OM-4/850.7	71.04	50.69	0.68	3.07	10.91	0.26	15.01	19.17	0.25	0.006	100.07
28	OM-4/850.7	73.18	51.07	0.63	3.13	10.21	0.23	15.62	19.20	0.20	0.01	100.32
29	OM-4/850.7	62.75	51.84	0.35	1.94	13.68	0.37	12.93	19.84	0.18	0.02	101.16
30	OM-4/850.7	79.47	51.78	0.52	3.48	7.57	0.18	16.43	20.10	0.20	0.21	100.48
31	OM-4/850.7	78.99	51.91	0.52	3.23	7.77	0.18	16.38	20.42	0.19	0.13	100.74
32	OM-4/850.7	66.16	52.01	0.07	1.38	11.61	0.34	12.73	22.57	0.03	0.02	100.78
33	OM-4/851.2	77.10	51.67	0.54	3.12	8.56	0.20	16.16	19.77	0.20	0.04	100.28
34	OM-4/851.2	79.49	51.64	0.47	2.88	7.53	0.18	16.36	20.32	0.18	0.12	99.70
35	OM-4/851.2	75.57	51.58	0.58	3.07	9.11	0.20	15.81	19.89	0.19	0.02	100.47
36	OM-4/851.2	66.78	51.32	0.60	2.90	12.39	0.32	13.97	19.15	0.18	0.01	100.86
37	OM-4/851.2	74.29	51.71	0.62	3.12	9.65	0.22	15.64	19.52	0.22	0.01	100.73
38	OM-4/851.2	79.22	51.94	0.50	3.18	7.61	0.17	16.27	20.48	0.18	0.13	100.49
39	OM-4/854.8	40.57	50.40	0.14	1.40	21.55	0.35	8.25	19.01	0.09	0.00	101.20
40	OM-4/854.8	76.61	52.35	0.51	2.78	8.80	0.21	16.17	19.90	0.20	0.02	100.96
41	OM-4/854.8	75.14	53.03	0.20	1.72	8.91	0.24	15.10	21.83	0.05	0.01	101.12
42	OM-4/854.8	59.56	49.98	0.70	3.14	14.86	0.39	12.27	19.07	0.24	0.01	100.69
43	OM-4/854.8	67.94	50.96	0.75	3.04	11.50	0.29	13.67	20.43	0.24	0.02	100.92
44	OM-4/856.1	79.23	52.51	0.51	3.09	7.75	0.18	16.59	20.52	0.17	0.11	101.46
45	OM-4/856.1	78.88	52.26	0.51	3.14	7.87	0.18	16.48	20.43	0.20	0.11	101.19
46	OM-4/856.1	66.51	51.88	0.56	2.68	12.16	0.34	13.55	20.47	0.23	0.008	101.90
47	OM-4/856.1	65.07	49.74	0.83	3.33	12.32	0.34	12.87	20.19	0.26	0.22	100.11
48	OM-4/856.1	79.09	53.27	0.55	2.44	7.59	0.18	16.11	21.57	0.21	0.23	102.18
49	OM-4/856.1	68.90	51.72	0.67	3.37	11.14	0.29	13.84	20.81	0.22	0.006	102.07
50	OM-4/856.1	53.97	48.37	0.61	3.09	16.61	0.46	10.92	18.61	0.21	0.01	98.91
51	OM-4/856.1	77.56	50.83	0.54	3.11	8.31	0.20	16.11	19.86	0.21	0.05	99.24
52	OM-4/856.1	65.35	49.96	0.75	3.56	12.72	0.30	13.46	19.31	0.26	0.01	100.35
53	OM-4/856.1	71.81	50.96	0.68	3.04	10.43	0.25	14.90	19.91	0.24	0.01	100.44

(continued)

Table 10.3 (continued)

No	No sample	MgO#	SiO_2	TiO_2	Al_2O_3	FeO	MnO	MgO	CaO	Na_2O	Cr_2O_3	Total
54	OM-4/856.1	52.43	51.74	0.20	1.40	17.41	0.40	10.77	19.47	0.19	0.01	101.60
55	OM-4/856.1	63.14	50.82	0.78	2.80	13.18	0.38	12.66	20.10	0.25	0.01	101.00
56	OM-4/858.3	79.27	52.48	0.46	2.95	7.69	0.18	16.50	20.31	0.21	0.10	100.90
57	OM-4/858.3	65.03	52.31	0.23	1.67	13.05	0.33	13.61	20.11	0.15	0.00	101.48
58	OM-4/858.3	70.75	51.48	0.60	2.99	11.01	0.26	14.94	19.45	0.22	0.01	100.98
59	OM-4/858.3	75.80	52.84	0.22	1.99	8.36	0.23	14.69	22.50	0.06	0.01	100.92
60	OM-4/858.3	79.88	51.43	0.47	3.17	7.42	0.17	16.52	20.31	0.18	0.16	99.85
61	OM-4/858.3	67.46	51.79	0.26	1.90	11.83	0.30	13.75	20.41	0.17	0.01	100.44
62	OM-4/859	79.71	51.43	0.47	3.12	7.39	0.17	16.29	20.48	0.18	0.16	99.71
63	OM-4/859	42.43	50.85	0.11	1.38	20.37	0.34	8.42	19.81	0.13	0.00	101.43
64	OM-4/859	79.44	52.16	0.48	3.12	7.53	0.17	16.31	20.41	0.19	0.15	100.54
65	OM-4/859	80.03	51.41	0.49	3.40	7.31	0.17	16.43	20.13	0.20	0.26	99.83
66	OM-4/859	72.98	50.75	0.61	3.05	9.90	0.27	15.00	19.90	0.21	0.01	99.72
67	OM-4/859	74.70	51.03	0.42	2.42	9.03	0.23	14.96	21.27	0.12	0.01	99.51
68	OM-4/861.1	67.74	50.78	0.70	3.07	11.22	0.38	13.21	20.73	0.24	0.01	100.35
69	OM-4/861.1	73.65	51.26	0.59	2.94	9.83	0.22	15.41	19.68	0.23	0.01	100.20
70	OM-4/861.1	81.81	52.15	0.39	3.02	6.68	0.16	16.84	20.42	0.19	0.49	100.35
71	OM-4/861.1	56.51	50.60	0.63	2.26	15.28	0.55	11.14	19.98	0.26	0.004	100.71
72	OM-4/861.1	70.22	50.99	0.72	3.09	10.68	0.37	14.12	20.35	0.25	0.01	100.59
73	OM-4/861.1	61.44	50.44	0.70	2.69	13.79	0.47	12.33	19.80	0.24	0.22	100.70
74	OM-4/876	80.61	51.43	0.45	3.32	6.94	0.16	16.19	20.30	0.20	0.35	99.36
75	OM-4/876	79.98	50.78	0.45	3.11	7.18	0.17	16.09	20.38	0.19	0.21	98.58
76	OM-4/876	80.87	50.65	0.44	3.33	6.83	0.15	16.18	20.45	0.20	0.34	98.60
77	OM-4/876	80.63	51.33	0.43	3.15	6.97	0.16	16.28	20.42	0.18	0.28	99.22
78	OM-4/876	66.18	50.44	0.77	2.72	12.05	0.34	13.23	20.04	0.26	0.11	99.97
79	OM-4/880.9	66.39	49.41	0.74	3.04	11.66	0.31	12.92	20.25	0.29	0.01	98.64
80	OM-4/880.9	75.87	50.72	0.53	3.07	8.74	0.19	15.42	19.73	0.22	0.03	98.67
81	OM-4/880.9	79.36	51.46	0.47	3.16	7.37	0.16	15.89	20.37	0.21	0.21	99.32
82	OM-4/880.9	79.52	50.85	0.50	3.71	7.29	0.18	15.87	19.84	0.19	0.52	98.97
83	OM-4/881.7	59.89	51.53	0.63	2.44	14.63	0.44	12.25	19.72	0.23	0.02	101.90
84	OM-4/881.7	64.61	51.48	0.66	2.55	13.24	0.36	13.55	19.43	0.19	0.02	101.50
85	OM-4/881.7	58.71	51.61	0.63	2.30	15.08	0.44	12.02	19.67	0.22	0.01	101.99
86	OM-4/881.7	56.51	51.20	0.56	2.02	16.94	0.46	12.35	17.86	0.17	0.003	101.58
87	OM-4/881.7	48.25	50.79	0.05	1.30	17.82	0.28	9.32	20.27	0.26	0.03	100.15
88	OM-4/888.2	77.07	51.40	0.47	3.16	8.39	0.19	15.82	20.10	0.20	0.22	99.97
89	OM-4/888.2	80.89	52.42	0.44	3.11	7.01	0.16	16.64	20.57	0.20	0.22	100.79
90	OM-4/888.2	66.04	52.52	0.23	1.36	12.18	0.29	13.28	20.22	0.20	0.02	100.31
91	OM-4/888.2	71.46	51.57	0.56	2.80	10.65	0.26	14.96	19.46	0.23	0.06	100.57
92	OM-4/888.2	79.91	52.84	0.43	3.01	7.29	0.23	16.25	20.67	0.17	0.21	101.12
93	OM-4/888.2	78.18	51.93	0.43	2.80	7.99	0.19	16.06	20.23	0.19	0.16	100.00
94	OM-4/901.9	69.53	50.48	0.90	3.23	11.72	0.29	15.00	18.42	0.30	0.02	100.38
95	OM-4/901.9	68.74	50.65	0.81	2.48	12.36	0.34	15.24	17.78	0.23	0.01	99.92
96	OM-4/901.9	77.10	50.41	0.77	3.95	8.35	0.20	15.77	19.76	0.24	0.33	99.80
97	OM-4/901.9	74.11	51.77	0.60	2.51	9.98	0.25	16.03	18.81	0.24	0.03	100.24
98	OM-4/903.5	77.39	51.19	0.69	3.28	8.46	0.21	16.23	19.50	0.25	0.11	99.94
99	OM-4/903.5	73.83	49.79	1.07	2.83	9.92	0.27	15.69	18.23	0.29	0.18	98.30
100	OM-4/903.5	75.08	50.85	0.81	2.75	9.59	0.25	16.20	18.57	0.26	0.05	99.35
101	OM-4/903.5	75.30	50.97	0.89	3.05	9.20	0.23	15.74	19.39	0.30	0.06	99.85
102	OM-4/903.5	78.86	51.88	0.63	3.14	7.96	0.22	16.65	19.23	0.23	0.10	100.06
103	OM-4/903.5	75.08	51.14	0.94	3.03	9.32	0.23	15.75	19.31	0.30	0.05	100.09
104	OM-4/903.5	66.96	52.17	0.25	1.46	11.79	0.25	13.40	20.55	0.39	0.01	100.30
105	OM-4/903.5	75.46	51.25	0.93	3.08	9.25	0.23	15.96	19.52	0.28	0.06	100.58

(continued)

Table 10.3 (continued)

No	No обр.	MgO#	SiO_2	TiO_2	Al_2O_3	FeO	MnO	MgO	CaO	Na_2O	Cr_2O_3	Total
106	OM-4/912.9	81.26	50.90	0.62	4.02	6.81	0.17	16.56	19.78	0.26	0.79	99.94
107	OM-4/912.9	81.14	51.07	0.62	3.94	6.84	0.17	16.52	19.70	0.24	0.80	99.92
108	OM-4/918.2	78.28	51.39	0.75	3.88	7.94	0.17	16.06	20.30	0.23	0.46	101.21
109	OM-4/918.2	76.48	52.05	0.67	3.27	8.76	0.20	15.98	20.20	0.23	0.23	101.60
110	OM-4/918.2	77.83	51.71	0.74	3.57	8.23	0.20	16.20	20.07	0.23	0.29	101.26
111	OM-4/918.2	77.50	51.13	0.75	3.73	8.36	0.20	16.15	19.75	0.24	0.30	100.63
112	OM-4/918.2	76.57	51.07	0.70	3.09	8.66	0.21	15.88	19.66	0.23	0.14	99.66
113	OM-4/918.2	76.46	50.81	0.70	3.02	8.77	0.22	15.98	19.48	0.25	0.11	99.36
114	OM-4/918.2	77.75	51.38	0.53	2.55	8.40	0.21	16.45	19.53	0.21	0.07	99.35
115	OM-4/920.6	79.90	51.59	0.63	3.63	7.49	0.18	16.70	20.02	0.22	0.46	100.94
116	OM-4/920.6	70.44	49.75	1.05	2.99	11.65	0.32	15.58	17.37	0.30	0.01	99.04
117	OM-4/920.6	74.50	50.02	0.79	3.16	9.62	0.24	15.77	18.97	0.30	0.02	98.91
118	OM-4/920.6	79.07	49.80	0.60	3.60	7.58	0.19	16.05	19.43	0.27	0.72	98.26
119	OM-4/920.6	79.34	50.91	0.62	3.57	7.47	0.18	16.08	20.24	0.26	0.38	99.73
120	OM-4/920.6	73.30	50.34	0.91	3.19	10.00	0.25	15.40	18.98	0.30	0.03	99.42
121	OM-4/920.6	72.49	50.59	1.00	2.95	10.37	0.26	15.33	19.00	0.32	0.02	99.85
122	OM-4/920.6	77.39	51.22	0.67	3.06	8.28	0.20	15.88	20.08	0.23	0.10	99.74
123	OM-4/920.6	78.70	51.47	0.63	3.13	7.80	0.18	16.15	20.27	0.24	0.18	100.07
124	OM-4/920.6	79.22	50.33	0.62	3.69	7.51	0.18	16.07	20.05	0.25	0.40	99.12
125	OM-4/926.2	76.94	51.74	0.98	4.43	8.31	0.20	15.54	19.40	0.48	0.70	101.80
126	OM-4/926.2	82.04	52.95	0.43	3.45	6.47	0.16	16.56	20.76	0.22	0.77	101.78
127	OM-4/926.2	82.93	52.95	0.41	3.55	6.10	0.14	16.63	20.81	0.20	0.85	101.66
128	OM-4/926.2	78.96	53.29	0.41	2.92	7.70	0.19	16.21	20.55	0.19	0.28	101.76
129	OM-4/926.2	83.43	53.66	0.37	2.88	6.12	0.15	17.29	20.57	0.21	0.53	101.80
130	OM-4/932.8	82.11	50.56	0.48	3.85	6.26	0.16	16.11	20.53	0.22	0.87	99.06
131	OM-4/932.8	82.32	51.62	0.46	3.84	6.16	0.14	16.09	20.75	0.21	0.98	100.27
132	OM-4/932.8	82.28	51.91	0.45	3.66	6.30	0.16	16.41	20.50	0.22	0.86	100.48
133	OM-4/932.8	82.61	52.31	0.39	3.25	6.14	0.14	16.36	20.84	0.22	0.78	100.44
134	OM-4/932.8	81.96	52.00	0.46	3.62	6.34	0.14	16.16	20.67	0.23	0.84	100.47
135	OM-4/935.9	83.48	52.03	0.39	3.02	6.01	0.15	17.03	20.85	0.18	0.67	100.36
136	OM-4/935.9	84.31	52.00	0.40	3.59	5.56	0.12	16.75	21.15	0.20	1.01	100.80
137	OM-4/935.9	83.53	51.93	0.39	3.61	5.84	0.14	16.61	21.06	0.21	0.98	100.78
138	OM-4/945	81.63	50.99	0.51	3.94	6.94	0.16	17.30	19.06	0.20	0.97	100.08
139	OM-4/945	79.54	52.33	0.51	3.53	7.75	0.16	16.90	19.28	0.23	0.95	101.66
140	OM-4/945	82.89	52.95	0.53	3.48	6.27	0.15	17.03	20.48	0.23	0.95	102.08
141	OM-4/945	82.21	52.18	0.58	3.68	6.55	0.16	16.98	19.80	0.21	0.94	101.09
142	OM-4/966.9	82.61	52.09	0.52	3.40	6.20	0.15	16.51	20.61	0.23	0.81	100.54
143	OM-4/966.9	82.39	51.71	0.55	3.55	6.22	0.15	16.33	20.44	0.24	0.90	100.10
144	OM-4/966.9	78.65	51.19	1.09	3.00	8.09	0.20	16.71	19.31	0.24	0.13	99.98
145	OM-4/966.9	82.70	51.49	0.52	3.71	6.20	0.16	16.63	20.27	0.23	0.90	100.12
146	OM-4/966.9	83.16	51.66	0.47	3.50	5.94	0.14	16.46	20.76	0.21	0.81	99.98
147	OM-4/966.9	83.01	51.19	0.51	3.68	6.01	0.14	16.47	20.55	0.22	0.79	99.59
148	OM-4/970	83.40	53.73	0.47	2.82	6.18	0.15	17.42	20.42	0.22	0.63	102.06
149	OM-4/970	83.68	53.68	0.46	2.80	6.12	0.15	17.60	20.40	0.22	0.65	102.10
150	OM-4/970	82.85	53.08	0.58	3.74	6.26	0.15	16.95	20.30	0.24	1.03	102.35
151	OM-4/974	83.95	51.29	0.56	4.15	5.77	0.13	16.94	20.17	0.24	1.18	100.45
152	OM-4/974	84.79	53.25	0.46	3.22	5.51	0.13	17.23	20.87	0.22	0.91	101.81
153	OM-4/974	84.23	52.78	0.55	3.65	5.64	0.13	16.89	20.75	0.23	1.03	101.68
154	OM-4/974	85.25	52.44	0.42	2.95	5.32	0.13	17.25	20.90	0.21	0.81	100.45
155	OM-4/976.4	83.22	52.32	0.71	3.30	6.01	0.14	16.73	20.32	0.29	0.93	100.78
156	OM-4/976.4	83.17	51.98	0.70	3.02	6.00	0.15	16.64	20.28	0.29	0.87	99.95
157	OM-4/981.2	84.04	53.24	0.68	2.95	5.84	0.15	17.23	20.27	0.29	0.80	101.46
158	OM-4/981.2	83.56	52.81	0.83	3.28	6.01	0.16	17.13	20.12	0.31	0.90	101.56

(continued)

Table 10.3 (continued)

No	No sample	MgO#	SiO_2	TiO_2	Al_2O_3	FeO	MnO	MgO	CaO	Na_2O	Cr_2O_3	Total
159	OM-4/981.2	84.33	53.31	0.58	3.04	5.72	0.14	17.26	20.50	0.27	0.83	101.66
160	OM-4/981.2	84.00	52.67	0.70	3.39	5.79	0.14	17.06	20.44	0.27	0.92	101.40
161	OM-4/981.2	83.61	52.31	0.74	3.53	5.98	0.15	17.10	19.98	0.29	0.95	101.05
162	OM-4/984.9	84.49	52.58	0.54	3.24	5.50	0.13	16.82	20.63	0.24	0.88	100.58
163	OM-4/984.9	83.97	51.26	0.63	3.69	5.64	0.13	16.56	20.51	0.27	1.01	99.71
164	OM-4/990.7	83.80	51.30	0.56	4.00	5.65	0.13	16.39	20.73	0.23	1.07	100.08
165	OM-4/990.7	84.68	52.08	0.64	3.04	5.24	0.14	16.26	21.54	0.37	0.96	100.29
166	OM-4/997.1	83.31	52.24	1.08	2.86	6.25	0.17	17.49	19.72	0.35	0.71	100.88
167	OM-4/998.3	82.82	50.11	0.67	3.72	5.97	0.15	16.15	20.16	0.36	0.98	98.29
168	OM-4/999.1	85.51	53.75	0.58	2.21	5.07	0.16	16.78	21.64	0.51	0.69	101.41
169	OM-24/506.2	62.66	49.99	1.01	2.47	13.74	0.46	12.94	18.25	0.46	0.00	99.34
170	OM-24/506.2	63.85	51.56	0.79	1.84	13.66	0.45	13.53	18.04	0.40	0.01	100.29
171	OM-24/506.2	64.26	51.25	0.86	1.98	13.62	0.46	13.73	17.99	0.40	0.00	100.30
172	OM-24/510.3	42.37	50.68	0.57	0.89	21.97	0.64	9.06	16.53	0.28	0.00	100.63
173	OM-24/510.3	34.95	49.98	0.49	0.79	23.91	0.66	7.21	16.99	0.30	0.00	100.35
174	OM-24/510.3	48.62	51.50	0.61	1.04	20.81	0.63	11.05	15.23	0.32	0.00	101.19
175	OM-24/510.3	51.42	51.72	0.65	1.15	18.74	0.55	11.13	16.87	0.29	0.00	101.10
176	OM-24/512	53.80	50.79	0.63	1.13	18.85	0.59	12.31	15.06	0.28	0.009	99.65
177	OM-24/513.2	54.36	50.37	0.61	1.18	18.11	0.51	12.10	16.62	0.16	0.00	99.68
178	OM-24/513.2	54.02	50.24	0.63	1.32	18.65	0.54	12.29	16.07	0.22	0.01	99.99
179	OM-24/513.2	54.40	49.19	0.69	1.45	18.17	0.52	12.16	16.44	0.23	0.00	98.87
180	OM-24/523	49.97	50.86	0.65	1.11	18.90	0.56	10.59	17.18	0.32	0.001	100.17
181	OM-24/523	51.82	50.78	0.67	1.20	18.63	0.55	11.24	16.79	0.28	0.00	100.16
182	OM-24/523	52.23	51.31	0.65	1.08	17.96	0.51	11.02	17.81	0.30	0.00	100.65
183	OM-24/523	54.10	52.23	0.68	1.20	16.77	0.59	11.08	18.59	0.32	0.001	101.48
184	OM-24/528.3	21.30	48.62	0.92	1.18	27.53	0.63	4.18	17.94	0.26	0.003	101.28
185	OM-24/528.3	30.95	50.04	0.58	0.82	25.38	0.64	6.38	17.54	0.21	0.002	101.61
186	OM-24/528.3	35.87	50.41	0.63	0.98	24.35	0.65	7.64	16.85	0.22	0.00	101.75
187	OM-24/528.3	41.96	50.27	0.71	1.29	21.28	0.54	8.63	18.28	0.22	0.00	101.24
188	OM-24/529.5	48.07	50.86	0.66	1.06	20.08	0.58	10.43	16.56	0.24	0.00	100.49
189	OM-24/529.5	51.96	50.92	0.75	1.27	18.85	0.57	11.44	16.37	0.25	0.00	100.42
190	OM-24/530.5	47.84	50.37	0.58	1.13	18.64	0.74	9.59	18.37	0.27	0.00	99.70
191	OM-24/530.5	50.29	50.90	0.64	1.19	19.35	0.52	10.98	17.12	0.24	0.00	100.96
192	OM-24/544.8	48.70	51.21	0.74	1.25	19.93	0.54	10.61	17.05	0.24	0.006	101.58
193	OM-24/544.8	45.20	50.75	0.51	0.95	19.24	0.71	8.90	19.25	0.20	0.00	100.53
194	OM-24/544.8	45.53	50.44	0.52	1.17	18.90	0.67	8.86	19.21	0.21	0.008	100.01
195	OM-24/568.3	60.36	51.34	0.71	1.49	15.13	0.41	12.92	18.24	0.28	0.003	100.55
196	OM-24/568.3	61.85	51.75	0.70	1.48	14.66	0.40	13.33	18.47	0.30	0.004	101.12
197	OM-24/568.3	58.40	51.49	0.65	1.13	16.64	0.47	13.10	16.82	0.27	0.00	100.58
198	OM-24/568.3	61.41	51.79	0.71	1.50	14.80	0.42	13.21	18.35	0.27	0.00	101.06
199	OM-24/568.3	61.34	51.64	0.71	1.44	14.72	0.41	13.10	18.41	0.24	0.00	100.69
200	OM-24/568.3	59.67	51.61	0.74	1.46	15.57	0.45	12.92	17.82	0.23	0.01	100.81
201	OM-24/570	62.63	52.21	0.70	1.43	14.37	0.40	13.51	18.61	0.26	0.001	101.50
202	OM-24/570	61.74	52.14	0.68	1.39	15.08	0.43	13.65	17.80	0.29	0.003	101.48
203	OM-24/570	62.13	52.10	0.69	1.44	14.77	0.42	13.59	18.19	0.26	0.006	101.48
204	OM-24/570	59.39	51.58	0.75	1.55	16.07	0.46	13.18	17.48	0.28	0.01	101.39
205	OM-24/570	62.37	52.16	0.72	1.44	14.51	0.40	13.49	18.53	0.25	0.005	101.52
206	OM-24/570	56.89	50.46	0.77	1.46	16.52	0.46	12.23	17.55	0.27	0.006	99.74
207	OM-24/589.2	65.83	50.99	0.68	1.62	13.25	0.42	14.32	17.01	0.30	0.001	98.61
208	OM-24/589.2	67.55	51.86	0.69	1.69	12.43	0.36	14.52	18.20	0.35	0.003	100.12
209	OM-24/589.2	67.02	52.44	0.62	1.24	12.79	0.39	14.58	18.27	0.35	0.00	100.70
210	OM-24/589.2	65.95	52.22	0.66	1.60	13.44	0.40	14.61	17.52	0.36	0.003	100.82
211	OM-24/589.2	66.89	51.77	0.68	1.64	12.76	0.38	14.46	17.95	0.33	0.007	99.98

(continued)

Table 10.3 (continued)

No	No sample	MgO#	SiO_2	TiO_2	Al2O_3	FeO	MnO	MgO	CaO	Na_2O	Cr_2O_3	Total
212	OM-24/607	69.49	51.11	0.67	2.11	11.56	0.29	14.77	18.94	0.29	0.007	99.77
213	OM-24/607	70.01	51.89	0.62	1.87	11.89	0.31	15.57	18.23	0.30	0.002	100.70
214	OM-24/607	68.73	51.75	0.68	2.02	12.20	0.33	15.04	18.64	0.31	0.001	100.98
215	OM-24/607	70.21	51.87	0.69	2.13	11.50	0.29	15.20	18.95	0.27	0.006	100.92
216	OM-24/607	61.41	51.38	0.73	1.94	15.82	0.48	14.12	16.33	0.31	0.004	101.13
217	OM-24/607	69.63	51.30	0.68	2.12	11.78	0.32	15.15	18.76	0.30	0.00	100.42
218	OM-24/607	69.11	50.48	0.70	2.22	11.58	0.30	14.53	18.76	0.34	0.002	98.92
219	OM-24/607	69.37	50.52	0.71	2.22	11.48	0.30	14.58	19.01	0.34	0.006	99.17
220	OM-24/607	69.33	50.41	0.70	2.23	11.46	0.29	14.53	18.96	0.33	0.00	98.92
221	OM-24/607	69.31	50.49	0.70	2.29	11.48	0.29	14.54	18.87	0.30	0.00	98.97
222	OM-24/607	69.64	50.66	0.70	2.23	11.38	0.29	14.64	19.06	0.31	0.00	99.28
223	OM-24/608	69.53	51.51	0.71	2.14	11.72	0.30	15.00	18.66	0.32	0.006	100.38
224	OM-24/608	68.85	51.75	0.66	1.95	11.90	0.32	14.75	18.72	0.30	0.002	100.37
225	OM-24/608	70.13	51.43	0.73	2.26	11.30	0.29	14.88	19.09	0.32	0.008	100.32
226	OM-24/608	69.36	51.76	0.66	2.06	11.80	0.30	14.98	18.60	0.33	0.01	100.52
227	OM-24/608	69.60	51.53	0.72	2.26	11.59	0.30	14.88	18.83	0.34	0.005	100.46
228	OM-24/608	70.06	51.62	0.71	2.23	11.33	0.28	14.87	19.16	0.29	0.003	100.50
229	OM-24/608	69.78	51.75	0.69	2.13	11.53	0.31	14.93	19.00	0.30	0.002	100.65
230	OM-24/608	69.74	51.88	0.72	2.23	11.52	0.29	14.89	19.01	0.32	0.001	100.88
231	OM-24/608	69.53	50.94	0.71	2.34	11.37	0.28	14.55	19.03	0.34	0.004	99.58
232	OM-24/608	68.41	51.38	0.76	2.24	11.89	0.31	14.44	19.13	0.33	0.00	100.49
233	OM-24/612	71.17	51.50	0.78	2.32	10.68	0.25	14.79	19.60	0.36	0.002	100.29
234	OM-24/612	68.81	51.09	0.72	2.29	11.79	0.29	14.59	18.62	0.39	0.002	99.79
235	OM-24/612	71.05	51.66	0.74	2.32	10.70	0.25	14.74	19.52	0.34	0.001	100.28
236	OM-24/612	69.96	51.19	0.79	2.21	11.46	0.35	14.97	18.51	0.65	0.004	100.14
237	OM-24/613.5	69.68	51.24	0.70	2.06	11.38	0.31	14.67	18.82	0.36	0.00	99.54
238	OM-24/613.5	68.48	51.29	0.72	2.28	11.65	0.32	14.20	18.65	0.42	0.001	99.53
239	OM-24/613.5	69.77	51.83	0.68	1.91	11.61	0.33	15.03	18.29	0.36	0.006	100.07
240	OM-24/613.5	69.50	51.36	0.69	1.99	11.47	0.32	14.66	18.65	0.37	0.01	99.53
241	OM-24/613.5	68.63	51.11	0.75	2.11	11.91	0.35	14.62	18.15	0.36	0.008	99.39
242	OM-24/616.5	70.14	51.26	0.78	2.29	11.03	0.31	14.54	19.37	0.38	0.011	99.97
243	OM-24/616.5	69.92	51.56	0.74	2.18	11.16	0.30	14.55	19.25	0.34	0.008	100.09
244	OM-24/616.5	69.81	51.62	0.71	2.07	11.44	0.30	14.84	18.68	0.33	0.000	100.00
245	OM-24/616.5	68.87	51.63	0.77	2.12	11.81	0.36	14.66	18.45	0.48	0.001	100.29
246	OM-24/620.5	76.57	51.84	0.58	2.37	8.67	0.22	15.89	20.06	0.25	0.02	99.91
247	OM-24/620.5	78.22	52.16	0.49	2.26	8.10	0.19	16.32	20.41	0.27	0.03	100.25
248	OM-24/630.2	71.21	51.56	0.76	2.50	10.62	0.25	14.73	19.72	0.37	0.01	100.54
249	OM-24/630.2	72.02	51.80	0.74	2.47	10.33	0.22	14.91	19.95	0.32	0.00	100.75
250	OM-24/630.2	71.91	51.69	0.75	2.43	10.30	0.24	14.79	19.73	0.36	0.007	100.30
251	OM-24/630.2	71.98	51.05	0.72	2.56	10.18	0.22	14.67	19.85	0.36	0.00	99.63
252	OM-24/630.2	68.20	50.80	0.83	2.69	12.54	0.32	15.09	17.28	0.39	0.008	99.96
253	OM-24/630.2	72.00	51.70	0.72	2.49	10.33	0.23	14.89	19.84	0.39	0.004	100.62
254	OM-24/630.2	69.95	52.10	0.60	1.68	11.86	0.33	15.48	17.95	0.33	0.002	100.35
255	OM-24/634	71.90	51.50	0.66	2.72	10.38	0.22	14.90	19.93	0.30	0.00	100.63
236	OM-24/655.2	76.59	52.53	0.55	2.06	8.64	0.22	15.85	20.17	0.29	0.01	100.34
237	OM-24/657	73.21	51.33	0.56	2.14	10.33	0.26	15.83	18.72	0.28	0.007	99.47
238	OM-24/657	74.40	51.65	0.58	2.37	9.54	0.22	15.55	19.94	0.29	0.01	100.16
239	OM-24/657	74.36	51.71	0.59	2.39	9.57	0.22	15.57	19.91	0.29	0.02	100.29
240	OM-24/657	75.20	51.50	0.60	2.43	9.19	0.21	15.63	20.04	0.28	0.006	99.91
241	OM-24/657	75.48	51.47	0.60	2.39	9.11	0.21	15.73	19.99	0.27	0.01	99.79
242	OM-24/657	74.97	51.73	0.58	2.28	9.48	0.23	15.93	19.59	0.25	0.02	100.11
243	OM-24/657	75.29	52.01	0.56	2.25	9.42	0.23	16.10	19.59	0.27	0.02	100.46
244	OM-24/657	75.34	51.72	0.58	2.36	9.23	0.22	15.82	19.83	0.27	0.01	100.05

(continued)

Table 10.3 (continued)

No	No sample	MgO#	SiO_2	TiO_2	Al_2O_3	FeO	MnO	MgO	CaO	Na_2O	Cr_2O_3	Total
245	OM-24/666.6	74.63	51.10	0.53	2.28	9.32	0.25	15.38	19.94	0.29	0.02	99.12
246	OM-24/666.6	74.48	50.31	0.59	2.43	9.45	0.22	15.47	19.42	0.26	0.01	98.17
247	OM-24/666.6	75.45	50.54	0.58	2.45	8.90	0.21	15.34	19.97	0.27	0.01	98.30
248	OM-24/666.6	75.46	50.42	0.55	2.39	8.91	0.22	15.37	19.94	0.28	0.02	98.12
249	OM-24/666.6	76.14	50.69	0.53	2.33	8.68	0.20	15.54	19.94	0.26	0.02	98.21
250	OM-24/666.6	76.13	51.18	0.55	2.22	8.82	0.21	15.78	19.93	0.31	0.01	99.02
251	OM-24/666.6	75.52	50.99	0.57	2.41	9.08	0.21	15.71	19.98	0.28	0.02	99.26
252	OM-24/666.6	76.36	51.17	0.54	2.32	8.72	0.20	15.80	20.06	0.27	0.03	99.12
253	OM-24/666.6	76.98	51.56	0.50	2.22	8.51	0.21	15.96	20.13	0.27	0.01	99.39
254	OM-24/666.6	75.63	51.29	0.58	2.38	9.00	0.21	15.67	19.99	0.29	0.03	99.46
255	OM-24/675.5	62.20	50.81	0.49	1.22	14.41	0.43	13.30	18.29	0.23	0.002	99.20
256	OM-24/675.5	62.57	50.71	0.67	1.49	14.62	0.43	13.71	17.53	0.27	0.005	99.44
257	OM-24/675.5	61.15	50.19	0.65	1.42	16.23	0.47	14.33	15.65	0.22	0.00	99.17
258	OM-24/675.5	44.62	43.72	1.81	4.90	23.90	0.39	10.80	14.04	0.88	0.006	100.45
259	OM-24/675.5	60.89	50.87	0.67	1.43	15.48	0.45	13.52	17.21	0.25	0.00	99.89
260	OM-24/685.4	78.33	51.48	0.49	2.26	7.92	0.19	16.06	20.42	0.24	0.05	99.13
261	OM-24/685.4	77.82	51.37	0.53	2.31	8.09	0.19	15.92	20.25	0.26	0.04	98.98
262	OM-24/685.4	78.14	51.33	0.52	2.31	8.02	0.20	16.08	20.26	0.24	0.06	99.03
263	OM-24/685.4	77.91	51.50	0.52	2.34	8.06	0.18	15.94	20.28	0.22	0.05	99.10
264	OM-24/685.4	76.94	51.18	0.54	2.35	8.41	0.19	15.74	20.12	0.21	0.04	98.81
265	OM-24/685.4	78.15	52.01	0.51	2.33	8.07	0.19	16.19	20.28	0.23	0.06	99.89
266	OM-24/685.4	75.28	51.78	0.54	2.20	9.22	0.22	15.75	19.80	0.29	0.01	99.83
267	OM-24/685.4	76.97	51.72	0.46	2.24	8.52	0.20	15.97	20.15	0.22	0.06	99.56
268	OM-24/709.2	73.15	51.16	0.61	2.19	10.21	0.26	15.60	19.05	0.28	0.01	99.38
269	OM-24/709.2	79.32	51.20	0.51	2.57	7.60	0.18	16.35	20.20	0.23	0.13	98.99
270	OM-24/709.2	79.81	51.06	0.44	2.21	7.38	0.18	16.36	20.38	0.23	0.13	98.39
271	OM-24/709.2	79.88	51.21	0.46	2.40	7.35	0.18	16.37	20.44	0.23	0.15	98.80
272	OM-24/711.4	78.97	51.30	0.56	3.29	7.71	0.19	16.24	20.01	0.22	0.23	99.76
273	OM-24/711.4	79.84	51.93	0.45	2.82	7.33	0.18	16.28	20.61	0.22	0.17	100.01
274	OM-24/711.4	79.69	51.57	0.48	2.77	7.35	0.17	16.18	20.70	0.21	0.15	99.60
275	OM-24/711.4	79.90	51.92	0.44	2.74	7.30	0.18	16.28	20.55	0.23	0.17	99.83
276	OM-24/711.4	79.72	51.58	0.45	2.73	7.34	0.17	16.18	20.58	0.23	0.15	99.43
277	OM-24/711.4	79.73	51.77	0.48	2.81	7.35	0.17	16.22	20.55	0.24	0.16	99.77
278	OM-24/711.4	79.63	51.49	0.49	2.81	7.38	0.18	16.18	20.61	0.23	0.15	99.54
279	OM-24/711.4	73.66	51.13	0.64	2.65	9.75	0.23	15.29	19.50	0.27	0.04	99.52
280	OM-24/711.4	79.97	52.08	0.42	2.47	7.42	0.18	16.62	20.15	0.22	0.13	99.71
281	OM-24/711.4	73.53	49.64	1.10	5.18	9.27	0.19	14.44	19.60	0.33	0.12	99.89
282	OM-24/715.4	79.70	52.02	0.40	2.12	7.62	0.19	16.78	19.45	0.22	0.12	98.94
283	OM-24/715.4	79.18	50.33	0.57	3.00	7.50	0.18	16.00	20.11	0.23	0.23	98.17
284	OM-24/715.4	79.44	50.47	0.50	2.66	7.50	0.18	16.25	20.04	0.22	0.17	98.00
285	OM-24/715.4	79.67	50.83	0.48	2.47	7.37	0.17	16.20	20.34	0.24	0.15	98.27
286	OM-24/731.1	79.88	52.14	0.56	2.87	7.45	0.17	16.59	20.30	0.22	0.27	100.60
287	OM-24/731.1	79.59	52.52	0.48	2.25	7.60	0.18	16.62	20.50	0.21	0.12	100.49
288	OM-24/731.1	80.76	52.42	0.43	2.35	7.10	0.17	16.72	20.65	0.22	0.27	100.35
289	OM-24/731.1	81.11	52.44	0.43	2.33	6.98	0.17	16.81	20.74	0.22	0.30	100.43
290	OM-24/731.4	81.23	51.79	0.42	2.22	6.88	0.16	16.70	20.44	0.23	0.30	99.15
291	OM-24/731.4	80.23	51.08	0.45	2.60	7.08	0.17	16.12	20.35	0.25	0.28	98.39
292	OM-24/731.4	81.59	51.63	0.37	2.23	6.72	0.17	16.70	19.67	0.26	0.33	98.10
293	OM-24/731.4	67.76	49.95	0.82	2.17	12.15	0.31	14.32	18.28	0.32	0.009	98.34
294	OM-24/733	80.79	51.59	0.44	2.53	7.02	0.16	16.56	20.50	0.24	0.33	99.39
295	OM-24/733	79.77	51.80	0.51	2.77	7.52	0.18	16.63	19.90	0.24	0.26	99.83
296	OM-24/733	69.58	51.13	0.71	2.14	11.51	0.28	14.77	18.77	0.28	0.01	99.62
297	OM-24/733	80.75	51.71	0.43	2.45	7.00	0.17	16.47	20.59	0.25	0.27	99.35

(continued)

Table 10.3 (continued)

No	No обр.	MgO#	SiO_2	TiO_2	Al_2O_3	FeO	MnO	MgO	CaO	Na_2O	Cr_2O_3	Total
298	OM-24/758.6	81.07	51.51	0.53	3.32	6.84	0.16	16.43	20.47	0.26	0.58	100.11
299	OM-24/758.6	81.80	52.18	0.42	2.35	6.70	0.17	16.89	20.29	0.24	0.46	99.72
300	OM-24/758.6	82.21	52.59	0.42	2.73	6.45	0.15	16.72	20.87	0.23	0.61	100.79
301	OM-24/758.6	82.09	52.81	0.41	2.23	6.59	0.16	16.94	20.49	0.23	0.42	100.30
302	OM-24/758.6	81.41	52.29	0.42	2.58	6.88	0.16	16.90	20.03	0.23	0.49	100.00
303	OM-24/758.6	82.14	52.69	0.36	2.02	6.65	0.17	17.16	20.13	0.23	0.39	99.82
304	OM-24/758.6	82.05	52.27	0.40	2.33	6.47	0.15	16.59	20.66	0.23	0.46	99.59
305	OM-24/758.6	81.41	52.06	0.46	2.51	6.69	0.15	16.43	20.68	0.25	0.35	99.60
306	OM-24/758.6	80.38	52.54	0.46	2.37	7.18	0.17	16.50	20.64	0.21	0.17	100.27
307	OM-24/758.6	80.04	52.88	0.48	2.34	7.38	0.17	16.60	20.45	0.24	0.14	100.70
308	OM-24/758.6	80.01	52.48	0.50	2.51	7.41	0.18	16.64	20.27	0.25	0.17	100.42
309	OM-24/758.6	79.51	52.40	0.52	2.66	7.59	0.19	16.52	20.24	0.23	0.15	100.51
310	OM-24/758.6	80.08	52.34	0.46	2.30	7.35	0.17	16.57	20.42	0.23	0.14	100.00
311	OM-24/758.6	80.49	52.52	0.46	2.35	7.14	0.17	16.52	20.67	0.23	0.18	100.26
312	OM-24/762.9	82.78	51.71	0.38	2.72	6.17	0.15	16.64	20.64	0.22	0.81	99.46
313	OM-24/762.9	82.13	51.47	0.48	3.39	6.29	0.14	16.21	20.63	0.24	1.01	99.89
314	OM-24/762.9	76.85	51.42	0.58	2.10	8.52	0.21	15.86	20.00	0.25	0.04	99.00
315	OM-24/762.9	78.08	51.50	0.44	2.88	7.92	0.19	15.82	20.27	0.27	0.73	100.03
316	OM-24/762.9	81.92	51.68	0.45	2.86	6.46	0.15	16.42	20.69	0.23	0.65	99.61
317	OM-24/762.9	81.18	51.83	0.41	2.95	6.64	0.15	16.06	20.69	0.26	0.88	99.88
318	OM-24/762.9	82.61	51.44	0.43	3.01	6.18	0.15	16.47	20.76	0.24	0.89	99.58
319	OM-24/762.9	82.16	51.11	0.47	3.31	6.30	0.15	16.27	20.69	0.23	0.90	99.45
320	OM-24/762.9	82.84	51.48	0.40	2.92	6.11	0.14	16.54	20.60	0.27	0.87	99.35
321	OM-24/762.9	82.10	51.41	0.47	3.46	6.37	0.15	16.39	20.45	0.25	1.02	99.99
322	OM-24/762.9	82.65	51.67	0.42	2.98	6.14	0.14	16.41	20.70	0.25	0.88	99.61
323	OM-24/762.9	82.59	51.33	0.45	3.17	6.13	0.15	16.31	20.91	0.27	0.91	99.65
324	OM-24/762.9	79.67	52.15	0.53	2.40	7.47	0.17	16.42	20.45	0.23	0.11	99.95
325	OM-24/762.9	76.81	51.76	0.53	2.36	8.49	0.21	15.77	20.16	0.25	0.17	99.71
326	OM-24/762.9	82.73	51.62	0.40	2.77	6.16	0.14	16.55	20.80	0.24	0.72	99.42
327	OM-24/762.9	82.47	50.51	0.46	3.34	6.09	0.14	16.07	20.85	0.22	1.02	98.71
328	OM-24/762.9	67.14	49.74	0.63	2.79	12.86	0.33	14.74	16.83	0.25	0.009	98.19
329	OM-24/762.9	82.66	51.56	0.42	2.82	6.18	0.14	16.52	20.78	0.21	0.76	99.41
330	OM-24/762.9	81.31	51.58	0.47	2.65	6.76	0.16	16.50	20.65	0.24	0.39	99.42
331	OM-24/769.8	81.00	51.28	0.55	3.34	6.78	0.16	16.21	20.26	0.22	0.72	99.55
332	OM-24/769.8	82.04	52.09	0.41	2.02	6.51	0.16	16.68	20.77	0.24	0.33	99.22
333	OM-24/769.8	79.71	51.85	0.52	2.94	7.59	0.19	16.72	19.44	0.26	0.37	99.90
334	OM-24/769.8	66.31	51.21	0.99	1.95	14.06	0.38	15.52	15.85	0.27	0.00	100.24
335	OM-24/774.7	80.93	51.38	0.49	2.71	6.93	0.16	16.50	20.73	0.28	0.44	99.64
336	OM-24/774.7	72.59	50.65	0.75	2.37	10.86	0.28	16.13	17.90	0.28	0.03	99.27
337	OM-24/774.7	76.77	51.36	0.68	2.64	8.60	0.21	15.94	19.78	0.29	0.09	99.60
338	OM-24/774.7	82.10	51.01	0.46	3.04	6.32	0.16	16.26	20.69	0.27	0.90	99.13
339	OM-24/774.7	82.65	51.86	0.40	2.74	6.15	0.14	16.43	20.80	0.26	0.80	99.61
340	OM-24/780.7	65.47	50.22	1.25	2.66	12.88	0.31	13.70	18.60	0.31	0.01	99.96
341	OM-24/780.7	69.76	50.55	1.00	2.61	11.26	0.28	14.57	18.92	0.31	0.02	99.53
342	OM-24/780.7	80.74	52.52	0.42	1.99	7.21	0.18	16.95	19.84	0.23	0.41	99.78
343	OM-24/780.7	79.84	51.81	0.57	2.87	7.30	0.18	16.22	20.21	0.25	0.59	100.02
344	OM-24/793.3	75.96	50.80	0.74	2.29	8.97	0.22	15.90	19.39	0.27	0.07	98.67
345	OM-24/793.3	77.88	50.98	0.64	2.36	8.15	0.19	16.09	19.88	0.27	0.13	98.70
346	OM-24/793.3	80.34	51.33	0.54	2.63	7.07	0.16	16.21	20.36	0.26	0.43	99.02
347	OM-24/793.3	81.80	51.33	0.55	3.34	6.46	0.15	16.29	20.58	0.24	0.65	99.61
348	OM-24/793.3	82.80	52.49	0.38	1.88	6.41	0.16	17.31	20.04	0.23	0.40	99.32
349	OM-24/793.3	81.77	51.01	0.56	3.38	6.45	0.14	16.23	20.55	0.26	0.68	99.29
350	OM-24/793.3	80.96	51.79	0.46	2.20	6.96	0.17	16.60	20.25	0.24	0.39	99.08

(continued)

Table 10.3 (continued)

No	No обр.	MgO#	SiO_2	TiO_2	Al_2O_3	FeO	MnO	MgO	CaO	Na_2O	Cr_2O_3	Total
351	OM-24/793.5	77.76	52.44	0.56	2.42	8.24	0.19	16.16	20.37	0.24	0.03	100.67
352	OM-24/793.5	77.66	52.10	0.53	2.28	8.22	0.20	16.03	20.31	0.23	0.04	99.96
353	OM-24/793.5	75.48	52.21	0.57	2.19	9.12	0.22	15.75	19.97	0.28	0.00	100.32
354	OM-24/798.3	77.49	51.39	0.62	2.62	8.56	0.23	16.53	18.41	0.27	0.27	98.91
355	OM-24/798.3	78.58	51.17	0.61	2.64	7.78	0.19	16.01	20.03	0.24	0.36	99.05
356	OM-24/798.3	73.84	51.31	0.83	2.35	10.39	0.27	16.45	17.91	0.27	0.05	99.84
357	OM-24/798.3	74.54	51.21	0.84	2.56	9.63	0.24	15.81	19.09	0.31	0.07	99.77
358	OM-24/798.3	74.53	51.52	0.83	2.53	9.67	0.24	15.87	19.07	0.31	0.07	100.12
359	OM-24/803.3	82.55	51.63	0.44	2.36	6.29	0.16	16.69	20.55	0.21	0.38	98.73
360	OM-24/803.3	81.78	50.80	0.54	3.12	6.59	0.16	16.59	19.93	0.25	0.60	98.59
361	OM-24/803.3	81.20	50.49	0.56	3.57	6.86	0.16	16.62	19.26	0.25	0.82	98.61
362	OM-24/803.3	81.24	51.11	0.53	2.86	6.89	0.17	16.73	19.91	0.25	0.56	99.03
363	OM-24/803.3	82.58	51.34	0.40	2.00	6.43	0.16	17.10	19.97	0.25	0.44	98.11
364	OM-24/803.3	77.05	50.51	0.77	2.84	8.52	0.20	16.04	19.32	0.28	0.20	98.69
365	OM-24/803.3	79.41	50.84	0.58	2.80	7.73	0.18	16.72	18.99	0.26	0.43	98.54

Table 10.4 Composition of rocks from Nadezhdinsky (nd) and Morongovsky (mr) Formations

No пп	1	2	3	4	5	6	7	8	9	10
No обр.	g-5	g-5_1	g-5_2	g-7	491/1	492/1	492/2	4,031/3	4,031/6	4,032
Свита	nd_3	nd_3	nd_3	nd_3	nd_1	nd_1	nd_1	mr	mr	mr
SiO_2	52.07	52.40	53.61	53.23	52.64	52.72	53.13	48.45	49.34	50.31
TiO_2	0.97	1.02	0.97	1.05	0.93	0.76	1.01	1.29	1.18	1.12
Al_2O_3	16.06	15.85	15.06	15.10	16.20	15.57	15.42	14.67	15.73	15.64
FeO	9.54	9.75	9.72	10.08	9.25	8.95	9.28	11.65	11.07	11.03
MnO	0.22	0.19	0.15	0.16	0.13	0.18	0.15	0.18	0.15	0.19
MgO	7.53	6.63	6.50	5.89	6.25	6.37	6.47	7.64	7.06	7.10
CaO	8.94	10.80	10.70	8.91	9.83	11.92	9.87	11.22	11.80	11.61
Na_2O	2.31	1.95	1.98	2.77	2.44	1.97	2.88	2.10	2.04	2.10
K_2O	1.68	0.94	0.91	1.94	1.65	0.96	1.49	0.39	0.17	0.34
P_2O_5	0.12	0.11	0.11	0.11	0.15	0.10	0.12	0.15	0.13	0.14
Сумма	99.44	99.64	99.71	99.24	99.49	99.51	99.82	97.73	98.67	99.56
Rb	50.86	18.21	27.85	72.31	44.66	33.06	45.52	5.72	2.90	3.78
Ba	639	392	350	495	602	274	371	126	99	120
Th	3.60	3.30	3.44	3.61	3.13	3.03	3.59	1.16	1.26	1.14
U	1.01	0.94	0.93	0.92	0.71	0.80	1.02	0.43	0.53	0.45
Nb	8.31	8.15	8.11	8.87	7.69	6.99	8.36	4.23	4.89	4.41
Ta	0.53	0.50	0.55	0.55	0.46	0.44	0.51	0.26	0.30	0.27
La	19.0	17.4	18.3	19.2	18.7	15.8	18.2	8.1	8.2	7.8
Ce	39.4	36.4	37.2	39.4	37.7	32.1	37.5	18.2	18.6	17.5
Pb	7.59	12.46	7.65	7.23	6.78	5.66	5.61	1.44	2.49	1.62
Pr	4.66	4.41	4.48	4.75	4.64	3.96	4.52	2.51	2.52	2.39
Nd	19.0	18.0	18.5	19.3	19.0	16.1	18.2	11.8	11.6	11.3
Sr	323	241	274	394	332	261	395	181	210	188
Sm	4.24	3.91	4.01	4.22	3.98	3.49	3.87	3.25	3.18	3.04
Zr	136	123	128	142	131	106	117	88	92	89
Hf	3.54	3.16	3.32	3.61	3.27	2.76	2.96	2.34	2.40	2.31
Eu	1.17	1.09	1.12	1.13	1.13	1.05	1.07	1.11	1.11	1.05
Ti	6,034	5,970	5,878	6,527	5,790	5,174	5,963	6,507	7,140	7,073

(continued)

Table 10.4 (continued)

No пп	1	2	3	4	5	6	7	8	9	10
No обр.	g-5	g-5_1	g-5_2	g-7	491/1	492/1	492/2	4,031/3	4,031/6	4,032
Свита	nd_3	nd_3	nd_3	nd_3	nd_1	nd_1	nd_1	mr	mr	mr
Gd	4.37	4.10	4.25	4.47	4.17	3.72	4.02	4.02	3.92	3.85
Tb	0.70	0.65	0.68	0.72	0.66	0.61	0.65	0.69	0.67	0.64
Dy	4.52	4.23	4.39	4.51	4.18	3.83	4.15	4.65	4.40	4.35
Ho	0.92	0.85	0.91	0.90	0.86	0.78	0.83	0.97	0.93	0.91
Y	25.1	23.2	23.9	24.8	23.0	21.1	22.8	26.0	24.7	25.2
Er	2.71	2.45	2.50	2.56	2.47	2.27	2.42	2.78	2.68	2.61
Tm	0.38	0.35	0.36	0.37	0.35	0.33	0.35	0.40	0.39	0.39
Yb	2.57	2.34	2.43	2.56	2.31	2.18	2.29	2.73	2.68	2.60
Lu	0.38	0.35	0.38	0.38	0.35	0.33	0.33	0.41	0.39	0.39
Ni	49	52	47	40	39	19	19	120	128	139
Cu	286	420	200	319	36	22	15	82	133	139
Zn	124	153	122	129	84	86	97	88	95	89
Mn	1,434	1,433	1,148	1,286	1,267	1,216	1,065	1,390	1,477	1,552
Sc	35	35	34	34	36	33	33	42	40	43
Co	136	123	128	142	40	35	39	47	51	54

Analyses were carried out in Max-Planck Institute of Chemistry (Mainz, Germany). Major elements were done by EPMA in glasses and rare elements by LA-ICP-MS. Analysts B. Stoll, D. Kuzmin and N. Krivolutskaya

MIX
Papier aus verantwortungsvollen Quellen
Paper from responsible sources
FSC® C105338

If you have any concerns about our products,
you can contact us on
ProductSafety@springernature.com

In case Publisher is established outside the EU,
the EU authorized representative is:
Springer Nature Customer Service Center GmbH
Europaplatz 3, 69115 Heidelberg, Germany

Printed by Libri Plureos GmbH
in Hamburg, Germany